Elsevier Studies in
Applied Electromagnetics in Materials, 5

SIMULATION AND DESIGN OF APPLIED ELECTROMAGNETIC SYSTEMS

Elsevier Studies in Applied Electromagnetics in Materials

Vol. 1. Electromagnetic Forces and Applications (J. Tani and T. Takagi, eds.)
Vol. 2. Electromagnetic Phenomena and Computational Techniques
(M. Enokizono and J. P. Nowacki, eds.)
Vol. 3. Nonlinear Phenomena in Electromagnetic Fields
(T. Furuhashi and Y. Uchikawa, eds.)
Vol. 4. Magnetoelastic Effects and Applications
(L. Lanotte, ed.)
Vol. 5. Simulation and Design of Applied Electromagnetic Systems
(T. Honma. ed.)

Elsevier Studies in
Applied Electromagnetics in Materials, 5

SIMULATION AND DESIGN OF APPLIED ELECTROMAGNETIC SYSTEMS

Proceedings of the Internaltional ISEM Symposium on
Simulation and Design of Applied Electromagnetic Systems
Sapporo, Japan, 26–30 January, 1993

Supplement of the
International Journal of
Applied Electromagnetics in Materials

Editor:
Toshihisa HONMA
Department of Electrical Engineering
Faculty of Engineering
Hokkaido University
Kita-ku, Sapporo, Japan

ELSEVIER
AMSTERDAM • LONDON • NEW YORK • TOKYO 1994

ELSEVIER SCIENCE B.V.
Sara Burgerhartstraat 25
P.O. Box 211,1000 AE Amsterdam, The Netherlands

ISBN: 0 444 81747 6

This book is printed on acid-free paper.

Printed in The Netherlands

PREFACE

The International Symposium on Simulation and Design of Applied Electromagnetic Systems (ISEM-Sapporo) was held from January 26 to 30, 1993 at Hokkaido Koseinenkin Kaikan, Sapporo, Japan. The conference was organized by Hokkaido University under the chairmanship of Prof. Toshihisa Honma with Drs. Masanori Tsuchimoto and Hajime Igarashi, as the General Secretaries.

This was the fifth ISEM meeting. The first was held in Tokyo, Japan in 1988, chaired by Prof. K. Miya (the University of Tokyo), the second in Kanazawa, Japan in 1990, chaired by Prof. K. Bessho (Kanazawa University), the third in Sendai, Japan in 1991, chaired by Prof. J. Tani (Tohoku University) and the fourth in Nagoya in 1992, chaired by Prof. Y. Uchikawa (Nagoya University). The next symposium will be held in June 1994 in Seoul, Korea, and will be chaired by Prof. S-Y. Hahn (Seoul National University).

The main theme of this symposium was the simulation and design of applied electromagnetic systems having interdisciplinary characters, thus providing an international forum for discussion on their present and future roles. The symposium was attended by 345 distinguished scientists including 72 from abroad (20 countries). The total number of papers was 212, with 63 submitted from 19 different countries for presentation at either oral or poster sessions. The plenary lecture "Towards microelectronics in quantum regime - key issues and roles of simulation-" was presented by Prof. H. Hasegawa, Director of the Research Center for Interface Quantum Electronics, Hokkaido University. The scope of the symposium covered a wide variety of topics in applied electromagnetic fields : advanced computer simulation, intelligent computer aided design, inverse problems, actuators, micromachines, superconductors, electromagnetic materials, surface acoustic and electromagnetic waves, biological systems, magnetic fluids, liquid metals, chaotic motions, magnetic levitations, linear drives, controls and magnetic bearings, electromagnetic mass accelerators, application of electromagnetic forces, electromagnetic propulsion and MHD generators.

The primary aim of the symposium is to provide an international crossroads for the above fields so that they can be bridged together to produce interdisciplinary fields, and hence also create a new field. It should also be noted that of all the cities in Japan, Sapporo is geographically the closest to North America and Europe in terms of air routes, and as such is attracting attention as an international crossroads for the academic networks of the world. This means that a new technology and science may be created through the international crossroads here. For our university, given its location, it was indeed a great honor to host the symposium on applied electromagnetic fields.

The proceedings of the fifth ISEM, published as a supplement to volume 4 of the Internal Journal of Applied Electromagnetics in Materials, contains 177 papers which were rigorously reviewed by the program committee. Some selected papers, as revised and extended, will be published in the upcoming special issue of this Journal : Vol. 4, No.4.

The growing number of participants during past ISEM meetings shows the worldwide interest that this symposium has created and the necessity for its continuation. The next ISEM meeting will be held in Seoul, Korea from June 22 to 24, 1994 and will focus on advanced computational and design technics in applied electromagnetic systems.

We would like to express our sincere gratitude to every participant of the symposium as well as the various cooperating societies, namely the Japan Society of Applied Electromagnetics, Institute of Electrical Engineers of Japan, Institute of Electronics, Information and Communication Engineers, Information Processing Society of Japan, Japan Society of Mechanical Engineers, Japan Society for Simulation Technology, Japan Society of Applied Physics, Magnetics Society of Japan, Japan Society of Magnetic Fluid Research and Cryogenic Association of Japan, and the principal sponsors, namely the Ministry of Education, Science and Culture in Japan, Hokkaido Government, Sapporo Municipal Government, Sapporo International Communication Plaza Foundation and Hokkaido Organization of Scientific and Technological Computation, without whose generous cooperation the symposium could not have been realized. Finally, on behalf of the organizing and steering committees, special thanks go to every contributor for their generosity in the financial support of this symposium. I am sure that ISEM-Sapporo was successful and that we can expect continued success with the sixth ISEM.

Sapporo, November 1993 *Toshihisa Honma*

CONTENTS

Plenary Lecture

Advanced Computer Simulation

Inverse Problem

Actuators

Micromachines

Superconductors

Electromagnetic Materials

Surface Acoustic and Electromagnetic Waves

Biological Systems

Magnetic Fluids

Liquid Metals

Linear Drive

Control and Magnetic Bearing

Electromagnetic Mass Accelerators

Application of Electromagnetic Forces

Electromagnetic Propulsion

MHD Generators

INTERNATIONAL STEERING COMMITTEE

T. Honma (Chairman)	Hokkaido University, Japan
K. Miya (Co-Chairman)	University of Tokyo, Japan
P. E. Allaire	University of Virginia, USA
O. Benda	Slovak Technical University, CSFR
K. Bessho	Kanazawa University, Japan
A. Bossavit	EDF, France
L. E. Cross	Pennsylvania State University, USA
R. Dang	Hosei University, Japan
M. Date	Osaka University, Japan
M. Enokizono	Oita University, Japan
T. Enoto	Hokkaido University, Japan
T. Fukuda	Nagoya University, Japan
V. V. Gogosov	Moscow University, Russia
S. Y. Hahn	Seoul National University, Korea
Z. Haznadar	University of Zagreb, Yugoslavia
K. A. Hempel	Technical Univ. Aachen, Germany
T. Higuchi	University of Tokyo, Japan
S. R. H. Hoole	Harvey Mudd College, USA
R. K. T. Hsieh	Royal Institute of Technology, Sweden
N. Ida	University of Akron, USA
D. H. Im	Hanyang University, Korea
Y. Kagawa	Okayama University, Japan
S. Kamiyama	Tohoku University, Japan
N. Kayukawa	Hokkaido University, Japan
A. Konrad	Toronto University, Canada
M. Koshiba	Hokkaido University, Japan
K. Kuriki	Institute of Space and Astronautical Science, Japan
S. Kuriki	Hokkaido University, Japan
M. A. Kurtz	Universität Stuttgart, Germany
F. Matsumura	Kanazawa University, Japan
G. A. Maugin	Universite Pierre - Marie Curie, France
D. H. Michael	University College London, UK
S. Minagawa	Electro - Commu. Univ., Japan
G. Molinari	University of Genova, Italy
F. C. Moon	Cornell University, USA
T. Morisue	Nagoya University, Japan
K. Mukasa	Hokkaido University, Japan
T. Nakata	Okayama University, Japan
J. P. Nowacki	Polish Academy of Sciences, Poland
T. Onuki	Waseda University, Japan
A. Prechtl	Tech. Univ. Wien, Austria
R. E. Rosensweig	Exon, Annandale, USA
G. Rubinacci	Univ. di Cassino, Italy
Y. Saito	Hosei University, Japan
I. Sebestyen	Technical University of Budapest, Hungary
J. N. Sheng	Xian Jiaotong University, PRC
J. Tani	Tohoku University, Japan
W. S. Trimmer	AT&T Bell Laboratory, USA

C. W. Trowbridge	Vector Fields Ltd., UK
Y. Uchikawa	Nagoya University, Japan
K. Uchino	Sophia University, Japan
S. Ueno	Kyushu University, Japan
A. A. F. van de Ven	Eindhoven Univ. of Technology, The Netherlands
K. Yamanouchi	Tohoku University, Japan
K. Yoshida	Kyushu University, Japan
M. Tsuchimoto (Secretary)	Hokkaido University, Japan
H. Igarashi (Secretary)	Hokkaido University, Japan

LOCAL STEERING COMMITTEE

T. Honma (Chairman)	Hokkaido University
K. Miya (Co-Chairman)	University of Tokyo
K. Bessho	Kanazawa University
J. Tani	Tohoku University
Y. Uchikawa	Nagoya University
K. Aisaka	Electric Power Development Co., Ltd.
K. Aoki	Nuclear Fuel Industries, Ltd.
T. Fujita	The Japan Atomic Power Co.
S. Fujiwara	Railway Technical Research Institute
K. Furukawa	Sumitomo Electric Industries
J. Hokkyo	Sony Corporation
A. Ishiyama	Waseda University
H. Ishizu	SPC Electronics Corporation
M. Karita	Shinko Electric Co., Ltd.
M. Kawase	Nippon Telegraph and Telephone Corporation
T. Kimura	Okuma Corporation
M. Kitano	Toyota Motor Corporation
I. Kobayashi	Chubu Electric Power Co., Inc.
M. Kobayashi	ALPS Electric Co., Ltd.
T. Koga	Toyo Denki Seizo K. K.
Y. Kohno	Ishikawajima-Harima Heavy Industries Co., Ltd.
M. Maekawa	Mayekawa Mfg. Co., Ltd.
F. Matsumura	Kanazawa University
E. Mikami	Mitsubishi Electric Corporation
K. Miki	Hitachi, Ltd.
K. Miura	TECNOVA Corporation
T. Morita	Fuji Electric Corporate Research and Development Ltd.
K. Nitadori	Oki Electric Industry Co., Ltd.
T. Obara	Hokkaido Railway Company
K. Ohnishi	The Japan Steel Works, Ltd.
T. Okabe	Nippondenso Co., Ltd.
Y. Okada	Ibaraki University
K. Oshitani	Mitsubishi Heavy Industries, Ltd.
N. Ozeki	Daido Steel Co., Ltd.
M. Rai	NSK Ltd.
K. Sangyoji	Brother Industries, Ltd.
M. Sato	The Hokkaido Electric Power Co., Inc.

T. Sawada	Canon Inc.
Y. Sawada	Toshiba Corporation
I. Shinohara	Tokyo Electric Power Company
S. Soejima	NGK Insulators, Ltd.
T. Takuma	Central Research Institute of Electric Power Industry
Y. Tanaka	NTT Data Communications Systems Corporation
T. Toda	Meidensha Corporation
T. Kataoka	Nippon Steel Corporation
T. Umino	Hitachi Seiko, Ltd.
K. Yada	OMRON Corporation
M. Yamada	Sanyo Electric Co., Ltd.
M. Yoshimoto	Oji Paper Co., Ltd.

LOCAL ORGANIZING COMMITTEE

T. Honma (Chairman)	Hokkaido University
S. Chonan	Tohoku University
M. Enokizono	Oita University
T. Furuhashi	Nagoya University
M. Hashimoto	University of Industrial Technology
A. Ishiyama	Waseda University
N. Kayukawa	Hokkaido University
M. Koshiba	Hokkaido University
S. Kubo	Osaka University
K. Kuriki	Institute of Space and Astronautical Science
F. Matsumura	Kanazawa University
M. Nagai	Tokyo University of Agriculture and Technology
K. Nagaya	Gunma University
T. Saito	Hosei University
Y. Saito	Hosei University
Y. Shindo	Tohoku University
T. Takagi	Tohoku University
T. Tanahashi	Keio University
S. Ueno	Kyushu University
M. Uesaka	University of Tokyo
R. Yamane	Tokyo Institute of Technology
K. Yoshida	Kyushu University
M. Yoshizawa	Keio University
R. Dang	Hosei University
Y. Ishihara	Doshisha University
E. Matsumoto	Kyoto University
S. Minagawa	Electro-Commu. University
I. Muta	Saga University
H. Tsuboi	Fukuyama University
S. Yamada	Kanazawa University
T. Enoto	Hokkaido University
T. Fukui	Hokkaido University
H. Igarashi	Hokkaido University

H. Kawaguchi	Hokkaido University
K. Mukasa	Hokkaido University
S. Oikawa	Hokkaido University
S. Tomioka	Hokkaido University
M. Tsuchimoto	Hokkaido University

SESSION ORGANIZERS

Advanced Computer Simulation	T. Takagi (Tohoku University) and M. Hashimoto (University of Industrial Technology)
Intelligent Computer Aided Design	T. Furuhashi (Nagoya University)
Inverse Problem	S. Kubo (Osaka University)
Actuators	S. Chonan (Tohoku University)
Micromachines	M. Uesaka (University of Tokyo)
Superconductors	A. Ishiyama (Waseda University)
Electromagnetic Materials	M. Enokizono (Oita University)
Surface Acoustic and Electromagnetic Waves	M. Koshiba (Hokkaido University)
Biological Systems	S. Ueno (Kyushu University) and Y. Saito (Hosei University)
Magnetic Fluids	T. Tanahashi (Keio University)
Liquid Metals	R. Yamane (Tokyo Institute of Technology)
Chaotic Motion	M. Yoshizawa (Keio University)
Magnetic Levitation	K. Yoshida (Kyushu University)
Linear Drive	M. Nagai (Tokyo Univ. of Agriculture and Technology)
Control and Magnetic Bearing	F. Matsumura (Kanazawa University) and D. Ebihara (Musashi Institute of Technology)
Electromagnetic Mass Accelerators	S. Usuba (National Chemical Laboratory for Industry)
Application of Electromagnetic Forces	K. Nagaya (Gunma University)
Electromagnetic Propulsion	K. Kuriki (Institute of Space and Astronautical Science)
MHD Generators	N. Kayukawa (Hokkaido University)

CONTRIBUTORS

ALPS Electric Co., Ltd.
Brother Industries, Ltd.
B. U. G. Inc.
Canon Inc.
Central Research Institute of Electric Power Industry
Chubu Electric Power Co., Inc.
Clarion Co., Ltd.
Daido Steel Co., Ltd.
Electric Power Development Co., Ltd.
Fora Company Ltd.
Fuji Electric Co., Ltd.
Fuji Electric Corporate Research and Development Ltd.
Fujitsu Denso Ltd.
Fujitsu Hokkaido Systems Engineering Ltd.
Fujitsu Tohoku Digital Technology Ltd.
Hitachi, Ltd.

Hitachi Building Systems Engineering and Service Co., Ltd.
Hitachi Cable Ltd.
Hitachi Chemical Co., Ltd.
Hitachi Information & Control Systems Inc.
Hitachi Seiko, Ltd.
Hokkaido Gas Corporation
Hokkaido Oki Electric Systems Co., Ltd.
Hokkaido Railway Company
Hokkaido Tecsys Corporation
HOXAN Corporation
IBM Japan Company Ltd.
Ishikawajima-Harima Heavy Industries Co., Ltd.
Isuzu Motors Ltd.
Japan Airlines
Japan Tobacco Inc.
JGC Corporation
Kirin Brewery Company Ltd.
Kokusai Electric Co., Ltd.
Matsushita Communication Industrial Co., Ltd.
Matsushita Electronics Industrial Co., Ltd.
Matsushita Systems Engineering Co., Ltd.
Mayekawa Mfg. Co., Ltd.
Meidensha Corporation
Mitsubishi Atomic Power Industries, Inc.
Mitsubishi Cable Industries, Ltd.
Mitsubishi Electric Corporation
Mitsubishi Heavy Industries, Ltd.
Mitsubishi Precision Co., Ltd.
Mitsumi Electric, Co. Ltd.
Morinaga Engineering Co., Ltd.
Morinaga Milk Industry Co., Ltd.
NEC Corporation
NEC Software Hokkaido Ltd.
NGK Insulators, Ltd.
Nikon Corporation
Nippondenso Co., Ltd.
Nippon Light Metal Company Ltd.
Nippon Mining and Metals Co., Ltd.
Nippon Sheet Glass Co., Ltd.
Nippon Steel Corporation
Nippon Steel Information & Communication Systems Inc.
Nippon Telegraph and Telephone Corporation
Nissan Motor Co., Ltd.
Nissin Electric Co., Ltd.
NSK Ltd.
NTT Data Communications Systems Corporation
Nuclear Fuel Industries, Ltd.
Oji Paper Co., Ltd.
Oki Electric Industry Co., Ltd.

Okuma Corporation
OMRON Corporation
PALOMA Industries Ltd.
Railway Technical Research Institute
Sanritz Automation Co., Ltd.
Sanyo Electric Co., Ltd.
SECOM Co., Ltd.
Seiko Instruments Inc.
Sharp Corporation
Shimadzu Corporation
Shimizu Corporation
Shin-Kobe Electric Machinery Co., Ltd.
Shinko Electric Co., Ltd.
Showa Electric Wire & Cable Co., Ltd.
Snow Brand Milk Products Co., Ltd.
Sony Corporation
SPC Electronics Corporation
Sumitomo Electric Industries
TECNOVA Corporation
The Furukawa Electric Co., Ltd.
The Hokkaido Electric Power Co., Inc.
The Japan Atomic Power Co.
The Japan Steel Works, Ltd.
Tohoku Electric Power Company
Tokyo Electric Power Company
Toshiba Corporation
TOSTEM Corporation
Toyoda Automatic Loom Works, Ltd.
Toyo Denki Seizo K. K.
Toyo Seikan Kaisha, Ltd.
Toyota Motor Corporation
Yokogawa Electric Corporation
Wavelabo K. K.
ZEROP Co., Ltd.

EXHIBITORS

CRC Research Institute, Inc.
Digital Equipment Corporation Japan
Kyokuto Boeki Kaisha Ltd.
MSC Japan Ltd.
Sumitomo Heavy Industries, Ltd.
Tokyo Denki Tokki Co., Ltd.
Yokogawa Hewlett - Packard, Ltd.

Plenary Lecture

Simulation and Design of Applied Electromagnetic Systems
T. Honma (Editor)

Towards Microelectronics in Quantum Regime
-Key Issues and Roles of Simulation-

Hideki Hasegawa

Research Center for Interface Quantum Electronics and Department of Electrical Engineering, Hokkaido University, Kita-13, Nishi-8, Sapporo 060, Japan

Abstract

The current status and future trends of research toward construction of novel microelectronics in the quantum regime are reviewed. Attempts are made to identify the key research issues and future roles of simulation for success of such research efforts.

1. INTRODUCTION

LSI technology has made a remarkable progress in these 40-50 years as roughly sketched in Fig.1 in terms of the number of devices on a chip. This progress has been brought about mainly by scale-down of device feature size as also shown in Fig.1. Thus, structures with the feature size of about 100 nm will be realized in production level around year 2000. Then, all the phenomena in devices will start to go into the quantum regime, since the de Broglie wavelength of electrons in solids falls generally in the range of 1-30 nm. This may impose fundamental limitations on the operation of existing devices based on "classical particle" picture of electrons and holes. However, viewed from a different angle, such a technological capability of miniaturization in turn may open up a possibility of constructing a novel "quantum microelectronics" based on quantum mechanical behavior of electrons.

The purpose of this paper is to briefly review the current status and future trends of such research efforts toward quantum LSIs, and to try to identify the key research issues and possible roles of simulation.

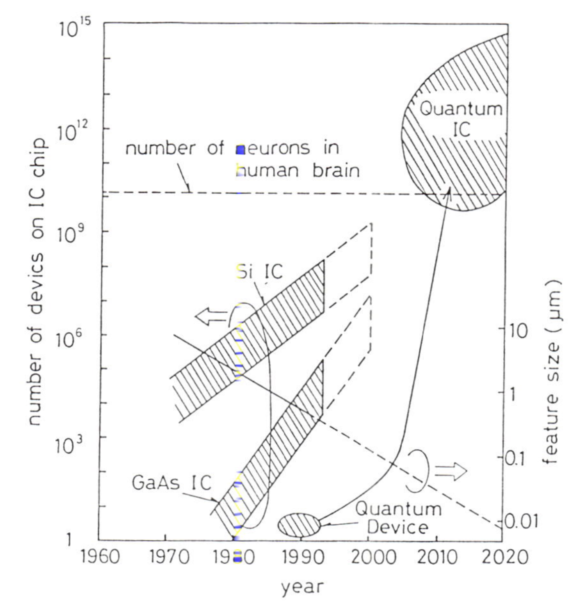

Fig.1 Progress of microelectronics.

2. LIMITATIONS OF PRESENT-DAY LSI ARCHITECTURE

The system architecture of present-day microelectronics is predominantly based on the Boolean logic. The device architecture to implement a such system architecture consists of an array of switches with resistors and capacitors where individual active transistors predominantly operate only as simple, small and robust switches of lumped nature.

However, this simple and forcible architecture will, upon further miniaturization, encounter grave and essential difficulties. Briefly, main material problems in semiconductors are threshold voltage fluctuation caused by random nature of doping and carrier injection caused by avalanche due to high field. In insulators, tunneling leakage of carriers thorough thin insulators and the interface instability issues at insulator-semiconductor interfaces are main problems. Extremely narrow wirings with high current densities cause reliability issues in metals due to electromigration and enhanced interfacial reactions. It is well known that the device power limits the achievable scale of integration, as is the case already today. Wiring capability is also a grave issue, since number of interconnects increase with the increase of number of devices on chip, resulting in increase of the wiring area on chip and the number of wiring levels.

Even if above limitations are removed by some ingenious device design, there remains a more fundamental limitation coming from the intrinsic nature of quantum transport as critically pointed out by Landauer[1]. As the number of the device dimension and the number of electronics are reduced, the system goes into the so-called "mesoscopic" regime where fluctuation becomes an essential and universal feature[2], making it extremely difficult to obtain the same electrical performance out of the devices having the same size.

In addition to the above material-device limitation, another recently emerging issue is the limited system-architecture capability of the present switch-array device architecture. It has been found that the present device architecture which has been so successful with von-Neumann architecture with Boolean logic is not suitable or extremely inefficient for implementing system applications such as neural networks, artificial intelligence, knowledge machines and video image processing where non-von-Neumann types of system architecture with massively parallel processing are involved.

3. QUANTUM STRUCTURES AND DEVICES

As the MOS and bipolar transistors are based on motions of electrons and holes in three dimensional device structures described by Newton' s equation for particle motion of electrons with effective mass, the devices in the quantum microelectronics will be based on "wave" motions of electrons in "quantum structures" described by Schrödinger's wave equation.

The formation principle and some typical examples of "quantum structures" are shown in Fig.2. By forming interfaces between different materials, potential energy difference for electrons and holes can be created at the interface. Then, by surrounding a small structure with a material having a higher potential energy, various quantum confinement structures shown in Fig.2 can be constructed. Similarly, various barrier structures can be produced. A periodic array of planar barriers and quantum wells is called superlattice.

In the quantum structures, new physical phenomena resulting from the wave nature of electrons appear, providing strong analogies between electronics and optics. The main effects include (1)formation of new confined quantum states with new energy levels, (2)single mode transport of electron waves, (3)field control of wavelength and phase of electron waves, (4)diffraction and Bragg reflection, (5) tunneling and resonant tunneling and (6)Coulomb blockade. Using these phenomena, many interesting devices have already been proposed and demonstrated, including quantum wire transistors, ballistic devices, quantum interference devices, resonant tunneling devices and single electron devices[3].

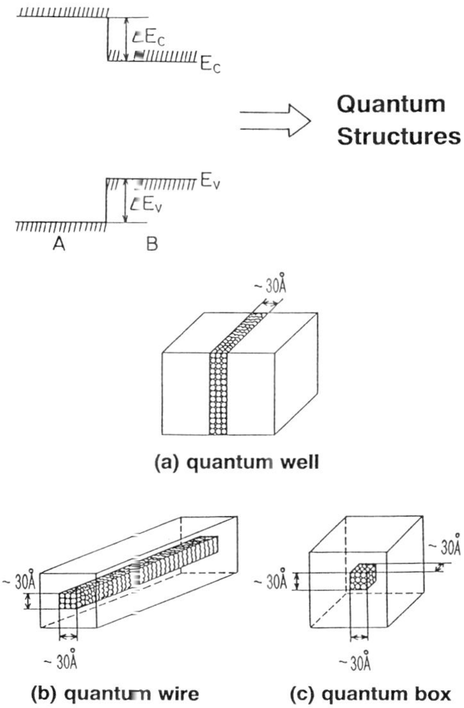

Fig.2 Basic quantum structures.

4. FABRICATION TECHNOLOGY AND KEY ISSUES

To realize quantum devices, technology should be mature enough to manufacture quantum structures with desired perfection and quality. Since an atomic-scale control of device dimension is necessary for fabrication of quantum structures, processing and characterization technologies will inevitably take a rather different style from those for present-day microelectronics. Use of a completely ultra-high vacuum (UHV) based fabrication/characterization system appears to be inevitable to avoid exposure to air during processing.

The major fabrication issue for quantum structures lies in the control of surface and interfaces of quantum structures[4]. Semiconductor interfaces such as p-n junction interface, heterojunction interface, metal-oxide-semiconductor interface and Schottky interface are already extremely important constituent elements of any of the present-day semiconductor devices, where the "particle" density and velocity of the carriers are controlled by these interfaces. On the other hand, the device interfaces in the quantum regime should now control the wave properties of electrons, such as amplitude, phase, reflection, transmission, interference and coherence. Due to reduced dimensions and different roles of the interfaces, interfaces play far more important roles in the quantum devices.

5. POSSIBLE LSI ARCHITECTURE AND ROLES OF SIMULATION

Obviously, possible LSI architectures include; (1)conventional Boolean logic based von Neumann type architecture implemented with conventional switch-array device architecture utilizing quantum transistors with improved switching performance, (2)conventional Boolean logic system architecture with new device architecture utilizing of quantum devices with increased functionality, and (3)new non-von-Neumann architecture with new device architecture utilizing of quantum devices with increased functionality.

Of these, the first alternative with the use of various wire transistors may be most practical in short terms. However, much of improvement over the conventional LSI technology cannot be expected since speed performance of an individual transistor does not account so much in a large integrated system being dominated usually by wiring. On the other hand, second and third alternatives seem to promise more improvements over the first alternative, since functional devices with reduced sizes imply much more increased functionality per area due to both of higher functionality per device combined with increased physical density of device. To realize the second and third alternatives, use of resonant tunneling based devices and single electron based devices seems to be most promising.

Simulation techniques will play extremely important roles in every aspect of the technologies used in the design and fabrication of quantum LSIs with new architectures. Since the whole system is extremely complicated, development of what one might call "a quantum compiler" is necessary, as schematically illustrated in Fig.3. For this, future close co-operation between system oriented workers and material/device workers is strongly needed. The hierarchy structure of the present LSI design/fabrication sequence has created a large gap between system oriented workers and material/device workers. If this gap is brought into the development of quantum LSIs, it will be impossible to arrive at new device and system architecture due to a very frustrating situation where system engineers know what systems should be realized, but do not know how to realize them with new, more functional and efficient devices. Also, device engineers know how electrons behave, but do not know how to utilize them for constructing new devices suited to new system architectures.

6. REFERENCES

1. R.Landauer.Physics Today, October, pp.119-121 (1989)
2. P.A.Lee, A.D. Stone and H. Fukuyama. Phys. Rev. B35, pp.1039-1070 (1987).
3. K.K.Likharev. IBM J. Res. Develop. 32, pp.144-158 (1988).
4. H.Hasegawa. Proc. of Royal Society Discussion Meeting on Semiconductor Growth, Surfaces and Interfaces, London, UK, March 3-4,1993. to appear in Proc. R. Soc.Lond.A, Sept. 1993 issue (1993).

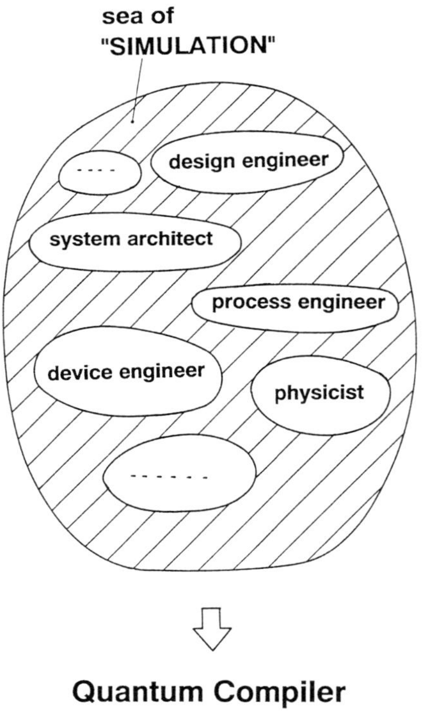

Fig.3 Roles of simulation.

Advanced Computer Simulation

Simulation and Design of Applied Electromagnetic Systems
T. Honma (Editor)

Concurrent multi-level simulation in computational prototyping

Meng H. Lean

Xerox Corporation, Webster Research Center, Mechanical Engineering Sciences Laboratory, North Tarrytown, New York, 10591 USA

Abstract

This paper has two important objectives. First, it advocates the role of computational prototyping as a parallel path for hardware product development in industry. It describes some current and future needs in the three essential components for computational prototyping: distributed computing environment; infrastructure of tools and information; and design methodologies. Second, a structured model for concurrent multi-level simulation to support the decision making process, is described. Its application to the system simulation of a xerographic print engine is discussed. Interactive steering of dynamical solutions and browsing of visual data are emphasized. PVM (Parallel Virtual Machine), from Oak Ridge National Labs is used for concurrent processing over networks.

1. INTRODUCTION

A major area of innovation in the next decade is the integration of the diverse computing technologies to construct practical and powerful multimedia environments to support the industrial product development process. Work productivity can be enhanced with desktop access to an electronic means to generate, process, archive, transmit, share, and reuse all forms of computer data. This *systems* approach will enable the sharing and interplay of both software and hardware embodiments and coordinate their interaction in producing rapid and meaningful solutions. A technology specific application in the use of this systems approach is *computational prototyping*, which is the role of computer simulation to augment the hardware prototyping process. Computational prototyping can be viewed as a complementary and parallel path where qualitative insights and trends gained from software experimentation of simulation models can be used to guide the hardware development process.

This paper will focus on two objectives: discuss the overall architecture of components to support a sharable environment for computational prototyping; and describe a concurrent multi-level simulation approach to support the decision making process in product design. Concurrency is used to denote two distinct meanings. *Concurrency in computing* is discussed in the parallel implementation of numerical algorithms for modeling xerographic subsystems on clusters of Unix workstations. *Concurrency in information and tools* is described in the context of multi-level simulation as a mechanism to support decision making. This model provides recourse to adequate information, tools, and resources, to enable the user to make the most reliable decision at any given instant in time. In this manner, a scalable path can be identified for any application that originates from a desktop.

2. COMPUTATIONAL PROTOTYPING

Three crucial components in support of computational prototyping are: the distributed computing environment; the tool and information infrastructure; and the design methodology that exercises the infrastructure within the environment.

The computing environment includes: distributed computing elements (DCE) and visualization elements on heterogeneous platforms interconnected with distributed information, computing, and visualization servers via wide bandwidth networks and interoperating through visual programming interfaces. Client-server implementations are suitable and allow the interchange and sharing of tools and information between users collaborating across networks spanning distant geographic locations. It also enables more efficient allocation of finite corporate compute resources by parceling and distributing them on servers for client access. This model reflects the trend towards distributed desktop computing where clusters of Unix workstations interoperate to achieve supercomputer-level performance at a small fraction of the cost.

This infrastructure includes simulation tools and corporate information accessed through an interactive query-and-search database front-end. Distributed file systems serve as repositories for the simulation design tools, image information, field data, design aids, and other corporate data. Other information on design rules, empirical data, knowledge base, costs tables, parts lists, etc, form the complementary set to represent corporate memory. Both kinds of information, leading to different data types, are stored in a hierachical evergreen structure so that over time, the volume of information grows with usage to capture corporate knowledge.

The dynamics of interactive design sessions is supported by providing interoperable functions, including: workstation video-conferencing, remote module execution, virtual collocation, shareable tool access, design process management, and automated multi-level simulation. The latter provides the most complete set of information to ensure reliability in the decision making process.

3. CONCURRENT MULTI-LEVEL SIMULATION

The objective of concurrent multi-level simulation is to provide a mechanism to exercise the multi-level structure during a computational prototyping session. Tools at each simulation level are represented as modular icons. Fig. 1 shows a particular application for the xerographic system. The schematic on the left represents a single pass in the xerographic loop where each clockwise rotation of the drum moves the process through the charging, imaging, development, transfer, and cleaning subsystems, respectively. This schematic is mapped into the network representation on the right where xerographic modules encapsulate the respective functions. The *Motion* module advances the drum in time to the *Control* module which increments system time. This network can be exercised interactively to predict and visualize system performance in the form of dynamic drum voltage and developed mass/area. More details on the system models are available in the literature.[1] Migration between the levels are via *detailing* and *abstraction* steps. *Detailing* incurs a downward move to regenerate new and more precise data. *Abstraction* is an upward move to capture and update functional models.

Fig. 2 shows the multi-level simulation layout. Suppose that an input data selection, say drum velocity, lies outside the range of validity for the prestored curves. A query of the imaging module allows the user to move down to the

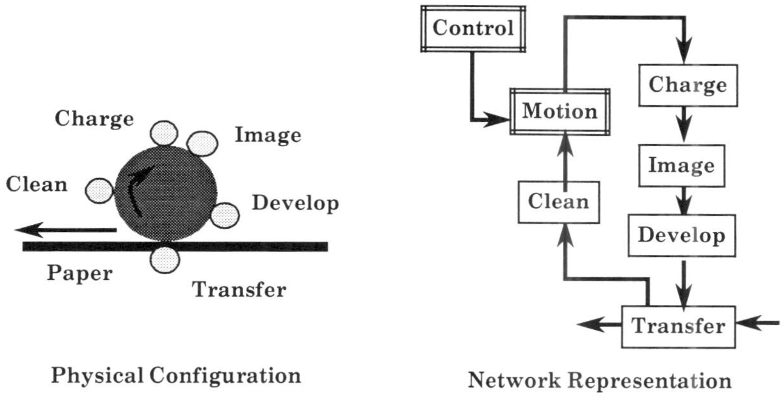

Fig. 1: Schematic to network representation.

device level via a detailing step. At this level, the embedded network can be exercised by varying run-time parameters to generate a new set of design curves. If necessary, geometries, material properties, and operating points, can all be varied to generate parametric information. The accumulated data from these experimental runs can then be encapsulated as new design curves and abstracted up to the process level for use.

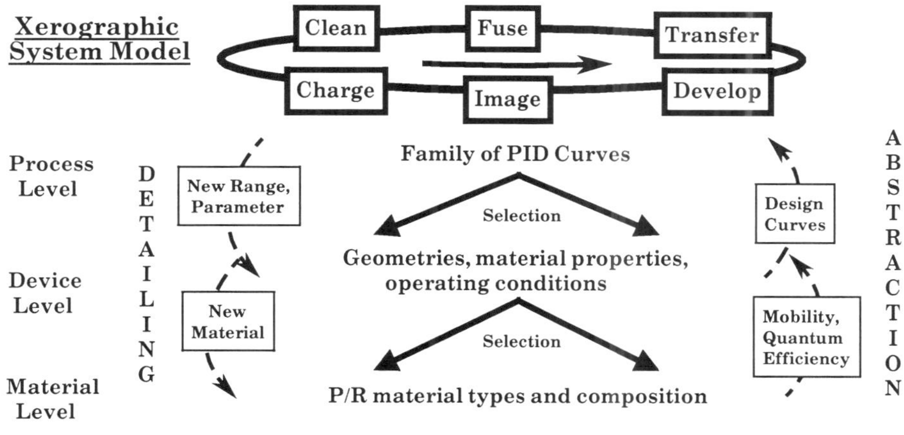

Fig. 2: Multi-level simulation of imaging subsystem.

4. IMPLEMENTATION

Details of simulation agorithms for charging,[2] imaging,[3] and development,[4] are discussed elsewhere. Here, we consider parallelization of these algorithms on a distributed computing environment using the Parallel Virtual Machine (PVM) toolkit. PVM is a public domain software available from Oak Ridge National

Laboratories that enables the utilization of loosely coupled heterogeneous networks of serial and parallel computers as a single computational resource. A PVM application consists of several subprograms which may be executed in parallel. Ideally, these subprograms should have a moderately large level of granularity. In PVM, the application program is split between the controlling host program and its node programs. The host program initiates PVM, sends copies of the node program to the available processors, communicates with the node processors by sending and receiving data, and performs disk I/O. The node programs contain the compute intensive portions of the algorithm which are executed in parallel in the node processors. Example applications are documented elsewhere.[1]

The test-bed implementation for a heterogeneous, distributed, network parallel, computing environment uses Unix as the operating system and X-windows as the terminal interface. The front-end user interface is based on AVS (Advanced Visual Systems), a visualization package with visual programming capabilities. We particularly emphasize the network connectivity of AVS which we enhance with additional simulation module libraries and tailor for the user to graphically assemble and/or merge and reconfigure network representations of computing and visualization tasks.

5. CONCLUDING REMARKS

This paper described the importance of computational prototyping as a parallel path for hardware product development where the implementation of a systems approach to engineering technology can integrate and sustain the dynamics of concurrent engineering. A test-bed implementation of an early distributed computing environment was described. A model for concurrent multi-level simulation (*process-device-material*) to support the design process, was introduced. Concurrency in computing was discussed in the parallel implementation of the simulation algorithms using PVM on a cluster of Unix workstations. Concurrency in information was described in the context of multi-level simulation to support the design process. It was shown that concurrent information ensures overall decision integrity by providing access to the most complete set of data and tools to support the decision making process. This model is scaleable and extensible to include all aspects of product development. This model also provided the structure and connectivity between levels of simulation detail to allow the sharing of high performance computing and communications capabilities with the practical computing requirements of industry.

6. REFERENCES

1. M.H. Lean,"Concurrent simulation of xerographic subsystems", Proc. IS&T/SPIE International Symposium on Electronic Imaging: Science and Technology, San Jose, Jan 31 - Feb 4, 1993.
2. M.H. Lean and P.S. Ramesh, "Simulation of electrohydrodynamic flows in charge generating devices", Proc. ASME Winter Meeting, Multidisciplinary Applications of CFD, FED., vol.129, Ed. O. Baysal,1991.
3. M.H. Lean and P.S. Ramesh, "Hybrid boundary integral equation method for charge injection and transport in electroreceptors", Boundary Element XIII, Eds. C.A. Brebbia and G.S. Gipson, pp. 395-396, Computational Mechanics Publications, UK, 1991.
4. M.H. Lean "Particle simulations of the development process", Proc. 8th Int. Congress on Advances in Non-Impact Printing Technology, IS&T, Williamsburg, VA, October 25-30, pp. 42-46, 1992.

Simulation and Design of Applied Electromagnetic Systems
T. Honma (Editor)

Inversion of eddy current data to reconstruct flaws by an optimisation process

J. R. Bowler

Department of Physics, The University of Surrey, Guildford, Surrey GU2 5XH, United Kingdom

Abstract

Eddy current probes for nondestructive testing have the potential to measure the shape of defects in metals. Here impedance changes due to eddy current probe interactions with simulated cracks in conductors have been used to reconstruct crack shapes using an optimisation technique. In the calculations, impedance predictions are compared with the measurements and a computed estimate of the crack shape varied until the predictions and observations are in tolerable agreement. The computed crack estimate then approximates the real crack. Results are presented comparing reconstructed defect shapes with a measured profile.

1. INTRODUCTION

In eddy current non destructive evaluation, the aim of inversion is to reconstruct an unknown flaw from a finite set of probe measurements. The probe signals, such as induced voltages or impedance changes, can be measured as a function of probe position and frequency or time. In an optimisation approach to inversion, predictions of these signals are made with a tentative estimate of the flaw and then compared with the measurements. If the agreement is inadequate, the estimated flaw is modified and new predictions are made. The process is repeated iteratively until the predictions match the observation to a reasonable approximation. Then the final flaw estimate should be an accurate representation of the real defect.

For the scheme to work, it is necessary to have a means of predicting the observations for an arbitrary flaw and a method of updating the flaw in such a way that the predictions are improved with each iteration. An improvement is achieved if a global error, quantifying the overall mismatch between predictions and observations, is reduced. Because of noise and inaccuracies in the measurement, it may not be possible to reduce the error to zero, but a minimum can be sought using one of the standard descent algorithms such as steepest descents or conjugate gradients. The error gradient depends firstly on the difference between the predictions and the measurements and secondly on the gradient of the predicted signals with respect to a variation of the flaw. A knowledge of this gradient is used to update the flaw in a "direction" that reduces the global error.

Mathematically a flaw such as a crack may be approximated using a representation that requires a finite number of parameters, in which case the search for an error minimum takes place in a finite dimensional parameter space. Alternatively the flaw may be defined by a flaw function and the search performed in a function space. An example of a flaw function is the equation of the bounding surface of a cavity in a conductor or the equation of the line that follows the edge of a crack. The functional gradient of the signal with respect to a flaw variation has been analyzed for flaw functions representing inhomogeneous defects in three dimensions[1], the surface of a homogeneous cavity or inclusion[1] and the line along the edge of an ideal crack[2]. In the present study the location of the crack edge is sought assuming that the plane of the crack is known.

Here we report the results of numerical calculations that have been carried out to reconstruct the profile of a crack from single frequency probe impedance data. The measurements were made on simulated cracks in the form of spark machined slots whose dimensions were known accurately. Details of the optimisation process and a brief account of the numerical scheme for predicting probe impedances are given in Section 2. In Section 3, comparisons of reconstructed flaw shapes and the measured crack profiles are reported followed by the conclusions, Section 4.

2. OPTIMISATION

The predicted probe impedance, Z, is considered to be a continuous function of probe position (x, y) and the excitation frequency ω. Together, these coordinates define a position in the observation space denoted by the vector $\mathbf{m} \equiv \{x, y, \omega\}$. The impedance is also dependent on the flaw function ξ. The difference between observations and predictions is gauged by a single real number, the global least squares "error" $\mathcal{E}[\xi]$, defined by

$$\mathcal{E}[\xi] = \sum_i |Z[\xi, \mathbf{m_i}] - Z_{obs}(\mathbf{m_i})|^2, \tag{1}$$

where the sum is over the observations and $Z_{obs}(\mathbf{m}_i)$ is the observed impedance at $\mathbf{m}_i$. The square brackets denotes a functional dependence on the flaw function ξ.

The equation of the line representing the crack boundary is $\xi(\mathbf{s}) = 0$, where $\mathbf{s}$ is a two-component vector coordinate in the crack surface. We seek the flaw function that minimises the global error and therefore consider the gradient of the error with respect to a variation in ξ. The error gradient, found by differentiating (1), is given by

$$\nabla_\xi \mathcal{E}(\mathbf{s}) = 2\mathrm{Re} \sum_i (Z[\xi, \mathbf{m}_i] - Z_{obs}(\mathbf{m}_i))^* \, \nabla_\xi Z(\mathbf{s}, \mathbf{m}_i), \tag{2}$$

where $\nabla_\xi Z(\mathbf{s}, \mathbf{m}_i)$ is the gradient of the impedance with respect to a variation of the flaw. In a calculation to find the function ξ that minimises the "error," the error gradient is evaluated and the boundary of the flaw is updated according to the chosen decent algorithm so that the new improved flaw function is $\xi + \delta\xi$. For steepest descents, the update of the flaw function is given by

$$\delta\xi(\mathbf{s}) = -\alpha \nabla_\xi \mathcal{E}(\mathbf{s}), \tag{3}$$

where the parameter α, governing the step size, is chosen to minimise $\mathcal{E}$ in the direction of the gradient. The flaw is updated until the condition $\mathcal{E}[\xi] < \epsilon$, has been satisfied, ϵ

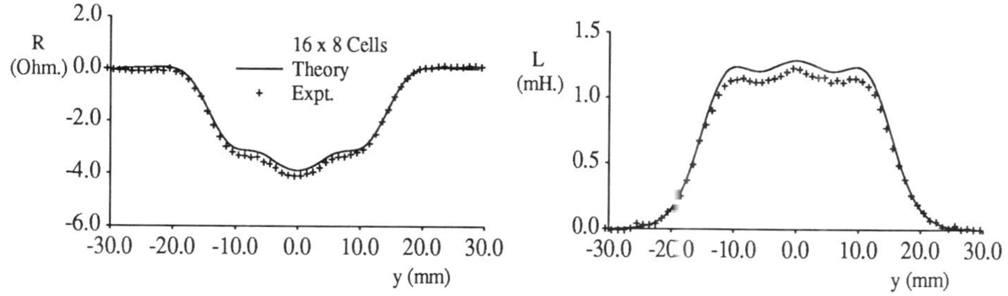

Fig. 1. Measured variation of resistance and self-inductance with probe position compared with predictions of a boundary element calculation.

being a suitably chosen real positive constant representing an acceptable residual error.

In order to evaluate the error gradient, it is necessary to calculate the impedance $Z[\xi, \mathbf{m_i}]$ and the impedance gradient, $\nabla_\xi Z(s, \mathbf{m}_i)$. The impedance gradient itself can be determined from the solution of multiple forward problems by using a finite difference scheme. However, the use of finite differences has a high computational cost and therefore should be avoided. A much more efficient approach, requiring only one forward problem for each observation point, makes use of the predicted field at the edge of a crack to determine the impedance gradient[2]. The theory considers an ideal crack defined as one with a negligible thickness yet acting as a perfect barrier to eddy current. An approximate solution of the electromagnetic field problem for such defects has been calculated using the boundary element technique[3] and this algorithm is incorporated into the inversion scheme described here.

3. INVERSION

Inversions have been performed on simulated cracks using a coil of rectangular cross section as a probe (4000 turns, I/D 5.02 mm, O/D 14.78 mm, length 4.99 mm, liftoff 0.30 mm). For the measurements, the coil was placed with its axis normal to the surface of the conductor (conductivity 38.7% IACS) in the plane of the crack. Its impedance recorded at a number of equally space positions using an impedance analyzer. The simulated crack was roughly semielliptical (22.10 mm long, 8.56 mm deep and 0.33 mm wide). The experimental results at a fixed frequency of 417 Hz, are shown in Figure 1 where they are compared with boundary element predictions computed on a 16x8 grid of rectangular elements.

The inversions are performed with a limited set of data, 11 points at 4 *mm* intervals, to keep the calculation time reasonably short. The calculation begins with an initial, possibly inaccurate, crack estimate, Figure 2. For each probe position the field on the surface of the crack is evaluated using the boundary integral code[3]. The probe impedance is calculated from the crack field and the solution near the edge of the crack is used to determine the impedance gradient. After the predictions have been computed they are compared with the observations, the error gradient evaluated and the flaw profile updated in accordance with equation (3). In this case symmetry about the crack centre has been

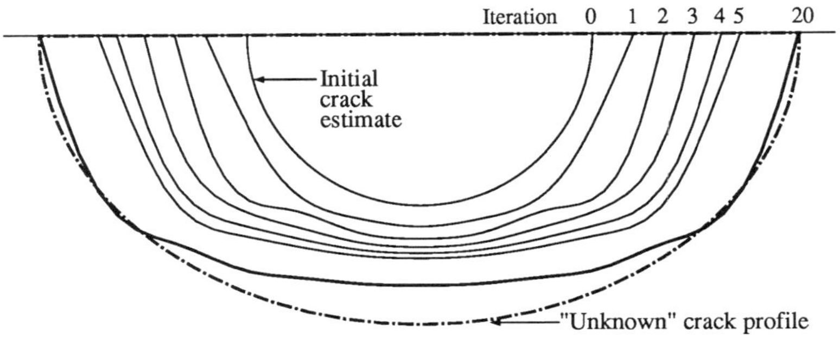

Fig. 2. Evolution of the crack profile through the iterative process of inversion.

enforced. Figure 2 shows the evolution of the estimated flaw profile.

4. CONCLUSION

Inversion often carries a high computational cost, therefore it is important to take advantage of analytical or computational methods that improve efficiency. By using a boundary integral technique to solve the forward problem for a ideal crack and making use of the computed field at the crack edge, it has been found that an iterative search for the crack profile can lead to a good approximation of the shape. The inversion gave a very close estimate of the crack length, 22.3 mm, which compares with an actual length of 22.1 mm. The crack depth however, was underestimated by 14%. By performing an inversion using computer generated impedances this error disappears, therefore the underestimate of depth it is due to imperfect predictions or experimental errors. Hence improving the accuracy of the experiments and predictions should reduce inversion errors.

Acknowledgement: This work was supported by Defence Research Agency, Farnborough and The Procurement Executive, MoD, UK. The author would like to thank Dr D. J. Harrison (DRA) for supplying experimental data.

1. S. J. Norton and J. R. Bowler, J. Appl. Phys. 73 (2) 501 1993.
2. J. R. Bowler and S. J. Norton, To be published.
3. J. R. Bowler, in Review of Progress in Quantitative Nondestructive Evaluation, Vol. 5A, D. Chimenti and D. Thompson Eds. (Plenum, New York, 1990) 149.

Simulation and Design of Applied Electromagnetic Systems
T. Honma (Editor)

The use of 3D finite elements in the simulation and design of electrical machines

D. Rodger, P.J. Leonard and H.C. Lai

School of Electrical Engineering, Bath University, BA2 7AY, U.K.

Abstract

Some techniques for modelling electrical machines using a 2 and 3D finite element package (MEGA) are outlined. These problems involve some interesting complexities which are best treated using a variety of different formulations. Some of these are described and some examples are given.

1. INTRODUCTION

Two dimensional finite element packages have been used in design work in the electrical machine area for over twenty years. More recently, the use of 3D finite element techniques has also become viable due to improvements in formulations, equation solving methods and computer hardware. At present many different approaches to formulating and solving general 3D electromagnetic field problems are being investigated worldwide. This contribution is limited to some of the formulations used in the MEGA finite element package. Some applications are described by way of illustration.

2. FIELD EQUATIONS

The volume in which a field solution is required is partitioned into regions which contain eddy currents and regions which do not.

Regions containing eddy currents are modelled using a three component vector, the magnetic vector potential A, either with or without an electric scalar potential V. Regions without eddy currents can be modelled also in terms of A, but it is more economic to use magnetic scalar potentials ψ as only one scalar variable is required.

Magnetostatic or electrostatic problems can be modelled as a subset of the above scheme, by using a scalar variable everywhere.

2.1. Field Equations For the Non-Conducting Regions

Using div B = 0 and H = - grad ψ, we can obtain: div μ grad ψ = 0 (1)

Regions which contain source coils may be modelled using the reduced scalar potential ϕ, defined as $H = -\text{grad}\,\phi + H_S$, where H_S is defined as curl $H_S = J_S$ in which J_S is the source current. This scheme can be modified to model voltage forced coils as well as current forced coils [1]. Voltage forced coils have been modelled earlier using a different scheme involving the use of the magnetic vector potential A [2] [3], but the advantage of the new technique is that the coils need not be modelled using a mesh which is the same shape as the coil, they just have to be immersed in an arbitrary mesh in which the reduced scalar potential is used. When closed magnetic circuits such as transformers are modelled in this way a cut is required which makes the two scalar regions simply connected.

2.2 Field equations for volume conducting regions

Low frequency eddy currents in 3D may be described using A (curl A = B) and V, the electric scalar potential.

The E field can be derived from A and V: $E = -\frac{\partial A}{\partial t} - \text{grad } V$ (2)

A region which is invariant in the cross section at right angles to a steady velocity u may be treated using the Minkowski transformation [4], in which a u $\times$ B term is added to (2):

$$E = -\frac{\partial A}{\partial t} - \text{grad } V + u \times \text{curl } A \qquad (3)$$

From curl H = J and div J = 0, we have, respectively:

$$\text{curl } \frac{1}{\mu} \text{ curl } A = -\frac{\partial A}{\partial t} - \sigma \text{ grad } V + \sigma\,(u \times \text{curl } A) \qquad (4)$$

$$\text{and} \quad \text{div } \sigma \left(-\frac{\partial A}{\partial t} - \text{grad } V + u \times \text{curl } A \right) = 0$$

The above equs (4) are solved simultaneously with equ (1) in non conducting regions. In order to ensure a unique solution, the divergence of A may be fixed using a penalty technique and A.n can be set to zero on the inside of conductors [4].

For stationary conductors, the u terms in equs (4) of course do not appear, usually it is more economic to dispense with the electric scalar term as well, this is straight forward if jumps in conductivity do not have to be modelled.

The electric scalar V can be eliminated from equs (4) for the moving case also if V is defined as V = A.u. Using u × curl A = grad (A.u) - (u.grad) A we can obtain, from equs (4):

$$\text{curl}\ \frac{1}{\mu}\ \text{curl}\ A\ =\ -\sigma\frac{\partial A}{\partial t}\ -\sigma\ (u\ .\ \text{grad})\ A \qquad (5)$$

This offers a very economic solution to problems like MAGLEV and some railgun electromagnetic launchers. Fig. 1 shows results from a small C core test rig [5] which illustrates that the two foregoing methods are equivalent.

2.3 Field equations for irregular shaped moving regions

For devices like squirrel cage induction machines the Minkowski transform method fails because there is no smoothness in cross section of any of the moving parts. In this case the moving problem must be solved as a full time transient problem. At each time step the rotor must be moved to a new position calculated from a mechanical equation and the electromagnetic forces. This can be accomplished using a recent sliding interface technique [6]. Here the meshes for rotor and stator are independently meshed but have an interface in common. Continuity of fields at the interface is ensured using Lagrange multipliers. This allows the rotor to slide relative to the stator (fig. 2).

Figure 3 shows a switched reluctance motor solved at one position using this technique. Any rotor position can be solved using the same mesh.

3. CONCLUSIONS

The electrical machines industry poses many interesting problems for the modeller. Computers are now becoming available which could usefully solve some of the complex coupled problems (mechanical, temperature, electromagnetics) which arise.

4. REFERENCES

1 P.J. Leonard and D. Rodger, Modelling Voltage Forced Coils Using the Reduced Scalar Potential Method, IEEE Trans Mag Vol. 28, No. 2, March 1992 pp. 1615-1618.

2 P.J. Leonard and D. Rodger, Voltage Forced Coils for 3D Finite Element Models, IEEE Trans Mag Vol. 24, No. 6, Nov. 1988 pp. 2579-2581.

3 K. Fujiwara, T. Nakata, N. Takahashi and A. Ahagon, 3D Finite Element Method for Analysing Magnetic Fields in Electrical Machines Excited from Voltage Sources, IEEE Trans Mag Vol. 24, No. 6, Nov. 1988 pp. 2582-2584.

4 D. Rodger, T. Karagular and P.J. Leonard, A Formulation for 3D Moving Conductor Eddy Current Problems, IEEE Trans Mag Vol. 25, No. 5, Sept 1989 pp 4147-4149.

5 D. Rodger, P.J. Leonard and T. Karagular, An Optimal Formulation for 3D Moving Conductor Eddy Current Problems with Smooth Rotors, IEEE Trans Mag Vol. 26, No. 5, Sept 1990 pp. 2359-2363.
6 D. Rodger, H.C. Lai and P.J. Leonard, Coupled Elements for Problems Involving Movement, IEEE Trans Mag, Vol. 26, No. 2, March 1990 pp. 548-550.

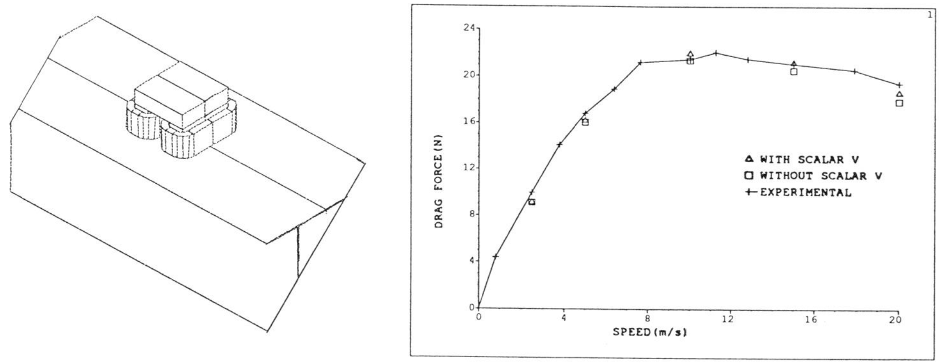

Fig. 1. Showing drag force versus speed for a small C core

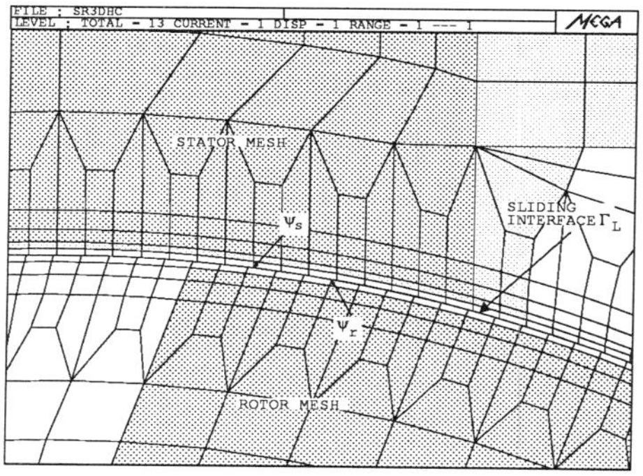

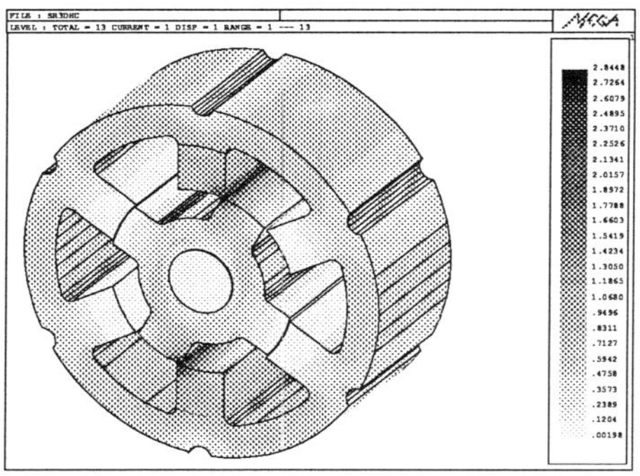

Fig. 2. Showing a sliding interface.

Fig. 3 Contour plot of magnitude of B for one position of a switched reluctance motor.

Simulation and Design of Applied Electromagnetic Systems
T. Honma (Editor)

New magnetic search coil system for two-dimensional magnetic measurement

M. Enokizono[a], K. Kawamura[a] and J. D. Sievert[b]

[a]Faculty of Engineering, Oita University, 700 Dannoharu, Oita 870-11, Japan
[b]Physikalisch Technische Bundesanstalt, 3300 Braunschweig, Germany

Abstract

The distribution of both the rotational magnetic flux density **B** and the rotational field strength **H** on three phase transformer core is investigated with a special sensor which is described in this paper. At first this special sensor for two-dimensional magnetic measurement and measurement system are shown. The components of the vectors of the magnetic flux density **B** and the magnetic field strength **H** are measured with four sensors during one period.

1. INTRODUCTION

In the past, the usual means of measuring the magnetic properties of magnetic materials was to detect the output corresponding to the magnetic flux density or other variables in the same direction as the applied magnetic field. In other words, magnetic properties have been measured using relations typified by the magnetic field strength **H** and the magnetic flux density **B** in a specific direction. That is, in the conventional magnetic measurements, the materials are excited and then measured in a single direction, so that the resulting magnetic properties are for the magnetic field **H** and magnetization **M** in the same direction (we call these one-dimensional measurements). The characteristic values obtained by such measurements represent only the properties in the given direction, but cannot reflect the overall properties or behavior of the material. But if the magnetic flux changes direction or is rotating, the applied magnetic field **H** and induction **B** (or magnetization) will not be parallel [1]-[3]. The inadequacy of such measurement methods is recognized from the fact that when magnetic materials are incorporated in actual electrical equipment as cores, as the T-joint of three-phase transformers or as the cores of electric motors for instance [4][5], they are subject to a rotating magnetic flux. In order to perform measurements of magnetic properties which include such relations, we have proposed that new magnetic search coils system for two-dimensional measurement is made.

With this study, we investigated what kind of action these rotating vectors **H** and **B** shows on the core of an actual electrical machinery and apparatus. The model core of a three-phase transformer was produced due to the measurement of the distribution. Also, a special sensor was developed, in order to measure these two-dimensional magnetic properties. This can be called a vector magnetic measurement. We call this sensor with a sensor for the measurement of a vector magnetic properties. Until now the measurement of rotational flux distribution only is reported. However about both of rotating magnetic field and rotational flux, these measurement results are nonexistent at all. If this both distributions become obvious, these results will give the important information to designers of an electrical machinery and apparatus.

2. SEARCH COILS AND MODEL CORE.

Our apparatus is based on the fieldmetric method using two H coils; it employs Hx and Hy coils for the measurement of field strength vector H and Bx and By coils to measure the magnetic flux density vector B. The construction of the measuring equipment apparatus in Fig. 1 and Photo 1. The H coils consist of 0.4 mm copper wire wound onto a bakelite sheet 1x18x18 mm^3, with the Hy coil positioned orthogonally over the Hx coil, and with effective area-turns of 104.826 cm^2(outer) and 95.0498 cm^2 respectively. The angle between Hx coil and Hy coil is 90°37'±12'. Magnetic flux density **B** is measured from the needles of 2 units. We call it double tip method. The disadvantages connected with the tip method for the measurement of **B** are the low voltage to be measured and the fact that a relatively high contribution from air flux may be picked up. To avoid these we led a copper band, closest to the specimen's surface, from one tip to the other and twisted the two wires from there on, thus avoiding a loop. The tips were made of brass; they are pressed against the specimen"s surface by springs. The distance between two tips for one direction is 28.0 mm. The holder for the tips was precisely machined so that the angle between the axes of the two pairs of tips were accuracy at 90°. The system of the two pairs of tips is rotatable as a whole. Low noise preamplifiers were installed close to the core sheet. Magnetic field strength is measured by a cross typed H-coil, that are set on the surface of the core sheet. Each value is obtained from the following equation[6].

$$B_i=\frac{2}{S_B}\int e_{Bi}dt \quad \ldots\ldots\ldots\ldots\ldots(1) \qquad H_i=\frac{1}{S_H N_H}\int e_{Hi}dt \quad \ldots\ldots\ldots\ldots\ldots(2)$$

where e_B and e_H are the induced voltage of B coil and H coil, respectively. S_B is cross area of the sheet . S_HN_H is effective area turns of H coil.i means x or y which shows direction..

Rotating magnetic field and magnetic flux are obtained from Lissajous's figure of each value. The Lissajous's figure means the locus of each vector. Figure 2 shows the basic circuit of the measurement apparatus. The measurements system is composed of microcomputer. 12 bit A/D,converter and amplifier. The outputs of H- and B-coils are fed into a computer through the A/D converter system.

By oriented silicon steel sheet (NSC HIB, 0.3 mm) model cores of three phase transformer were built up which exist of three limbs (equipped with magnetization coils of N windings), as shown in Fig. 3.

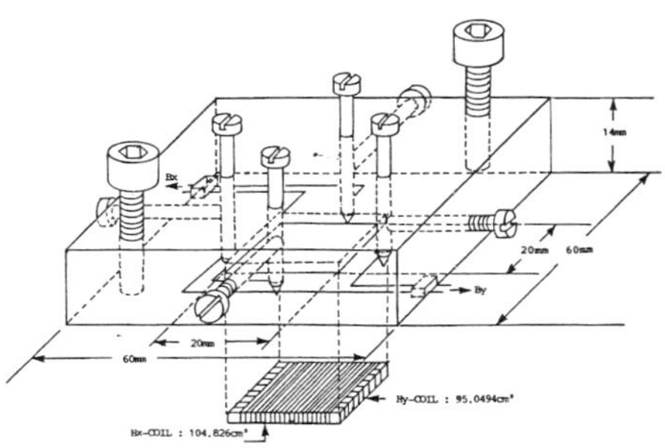

Fig. 1 Two-dimensional magnetic sensor.

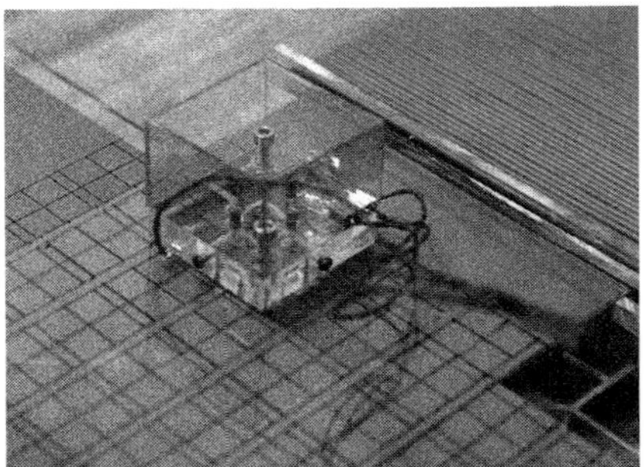

Photo 1 Sensor on model core.

3. MEASUREMENT RESULTS AND DISCUSSION.

Figures 4(a) and (b) show the distribution of the magnetic flux density **B** and magnetic field strength **H** in the neighborhood of the T-joint in the model core of a three-phase transformer. It is well known that a rotating magnetic flux is produced around the T-joint of the three-limbed core.. The manner of space variation of the shape of the elliptic locus is greatly affected by the joint angle. The axis ratio B_{min}/B_{max} of the magnetic flux density is very small except in the neighborhood of the joint lines in the laminated core of oriented silicon steel. Further, the major axis is nearly in the longitudinal direction of the core. On the other side, the axis ratio of the elliptic locus of magnetic field strength is larger than that. From these results, the shape of rotating magnetic field different from rotational flux entirely. This shows that the direction of magnetic field strength does not agree with one of the magnetic flux density. Also, because the core material has orientation in rolling direction, the locus shape of rotating magnetic field strength forms a flat ellipse toward a rolling. Rotational flux distribution is not symmetric right and left. This is the influence of rotational hysteresis. This hysteresis occurs toward rotating magnetic field that three phase current makes in the legs of transformer core. These result do not only agree with the result of the conventional magnetic field analysis but also cannot obtain from conventional formulation.. This is because the phase angle between H vector and B Vector is not considered. That is not also considered rotational hysteresis. In order to analyze the problem we need use the full permeability tensor.[7]

4. CONCLUSIONS

The distributions of the rotational flux and rotating magnetic field about a three-phase transformer were first become clear by the vector magnetic measurement method.

(1) The sensor system was developed for measuring the two-dimensional magnetic properties.

(2) Using the special sensor not only the rotational magnetic flux density but also the rotational magnetic field strength could be measured. And the measurement results and became clear the relation between the rotational magnetic flux density and rotational magnetic field strength.

5. REFERENCES

[1] M. Enokizono, T. Suzuki, J. Sievert and J. Xu: *IEEE Trans. Magn.*, **26**, 5, 2562 (1990)

[2] M. Enokizono, T. Suzuki and J. Sievert: *IEEE Trans. Magn.*, **26**,5, 2067 (1990)

[3] M. Enokizono, G. Shirakawa, T. Suzuki and J. Sievert: *J. Applied Magn. in Japan.*, **15**, 2, 261 (1991)

[4] K. Narita, T. Yamaguchi and I. Chida: *IEE Japan.*, **95-B**,**9**, 431 (1975)

[5] M. Enokizono, K. Kuwahara, M. Okabe and T. Yano: *IEE Japan*, **103-B**,2,844 (1983)

[6] W. Brix, K.A. Hempel and F.J. Schulte: *IEEE Trans. Magn.*, **20**, 5, 1708 (1984)

[7] M. Enokizono, T. Todaka and K. Yuki: *Elsevier Studies in Applied Electromagnetics in Materials* (ISEM-NAGOYA), **3**,
389 (1992)

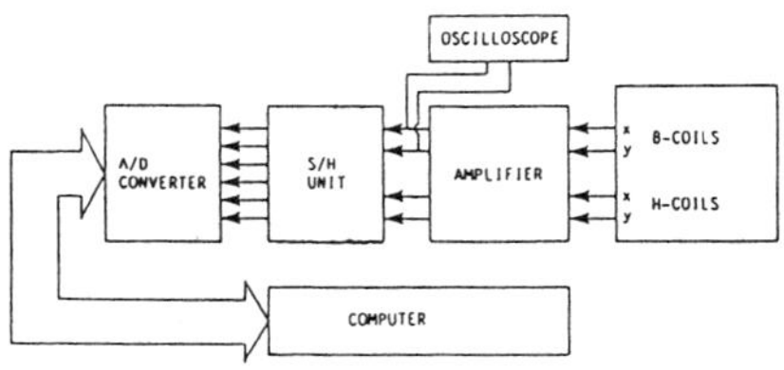

Fig. 2 Magnetic measurement.system

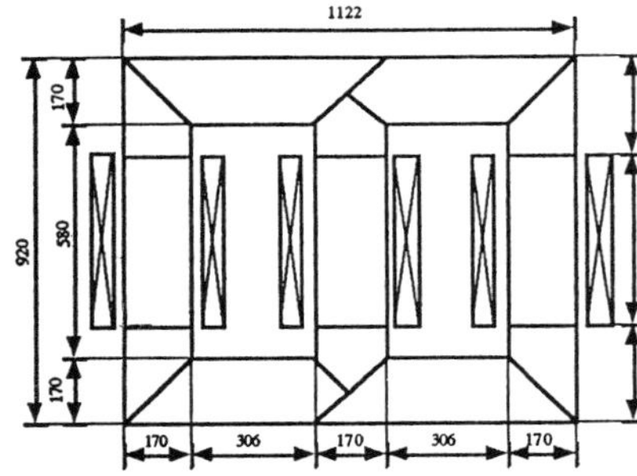

Fig. 3 Model core of three-phase transformer.

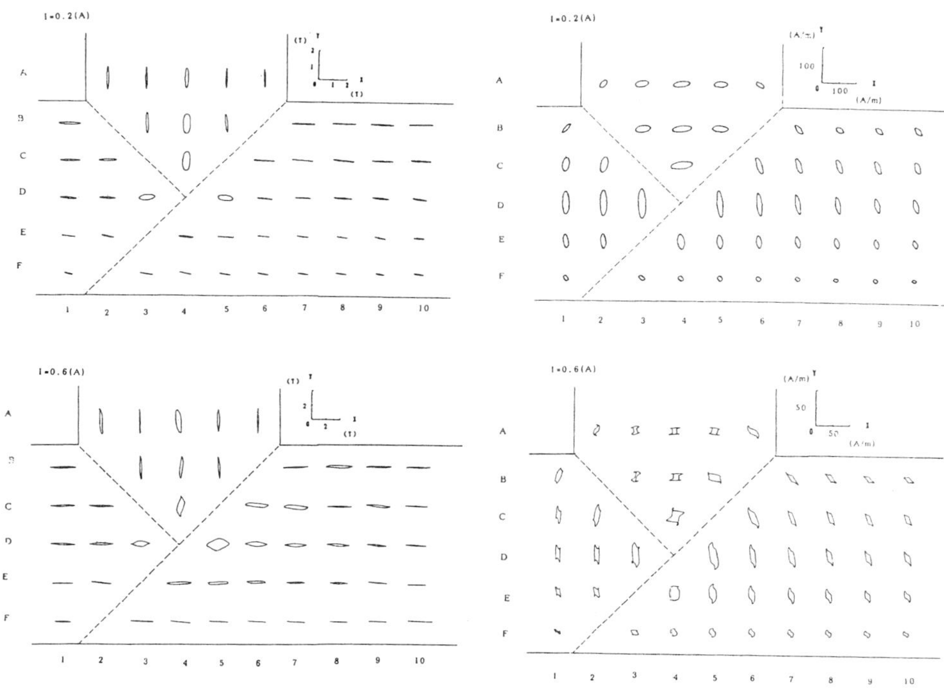

(a) Rotational magnetic flux density (b) Rotational magnetic field strength

Fig. 4 Distribution of rotational magnetic flux density and rotational magnetic field strength.

Simulation and Design of Applied Electromagnetic Systems
T. Honma (Editor)

Magnetic sensor using rotational magnetic flux for tubular material

S. Nagata[a] and M. Enokizono[b]

[a]Department of Electrical and Electronic Engineering, Faculty of Engineering, Miyazaki University, 1-1 Gakuenkibanadai-nishi, Miyazaki, 889-21 JAPAN

[b]Department of Electrical and Electronic Engineering, Faculty of Engineering, Oita University, 700 Dannoharu, Oita, 870-11 JAPAN

Abstract

This paper describes the new non destructive testing of tubular materials for defects. We have already reported about the experimental results about the magnetic sensor using rotational magnetic flux[1], and the range of defect detection was found to be broadened compared with conventional ac magnetic probes.

In this paper, this magnetic sensor is developed to detect a defect on a tubular conductor. In order to know the distribution of generated rotational magnetic flux, the boundary element method is employed. Moreover, this analysis reveals the relation between the output signal and the location of the defect.

1. INTRODUCTION

In the past, detection of defects has relied on visual inspections, but in recent years a need has arisen for more reliable and quantitative detection methods for use in ensuring the safety of the structural members of atomic reactors and for estimation of lifetimes. In response, a number of different methods for nondestructive testing (NDT) have been proposed. Analytical studies of methods employing electrical and magnetic fields have been conducted, and research is also underway on the reverse problem, estimation of defect shapes from the information obtained in experiments.

Eddy current methods are expected to find uses in numerous areas. However, such methods may not provide information on the defect depending on the direction of the detection, the range of detection is comparatively narrow, and they are not suited to detection of a wide range of defects or to estimation of the shape of the defect. These problems are regarded as arising from the fact that the direction of the applied magnetic field, which is either dc or ac, is unidimensional, as well as to the fact that the direction depends on the sensor scanning direction. When the object for NDT is a two-dimensional object such as a metal sheet, the defect will itself have a two-dimensional shape, position and direction. If a one-dimensional field is used to scan a two-dimensional defect, loss of information is inevitable.

We have proposed a method of defect detection which employs a rotating magnetic field. In this method, a rotating magnetic field is applied to the interior of a metal material, and defects are scanned and detected using the resulting magnetic flux or induced eddy currents.

In this paper, this rotational magnetic flux sensor is applied to a tubular conductor material, then we discuss the numerical result by means of two-dimensional boundary element method taking account of eddy currents.

2. MAGNETIC SENSOR USING ROTATIONAL MAGNETIC FLUX

Figure 1 shows the constructions of the magnetic sensor.

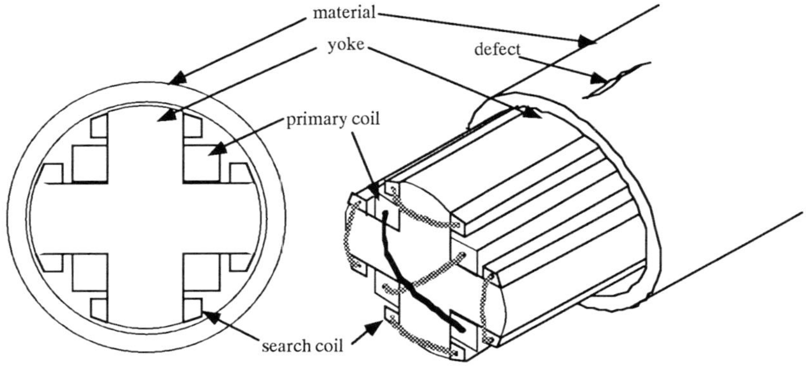

Fig. 1 Constructions of the magnetic sensor

Two-phase ac voltages are applied to the primary windings so that rotating magnetic flux is applied to the tubular conductor. The signal of the defect is picked up by the search coils. To extract a defect signal, two search coils are operated differentially to cancel out the exciting magnetic field.

3. RESULTS AND DISCUSSIONS

Figure 2 shows the analytical result. It shows the locus of generated rotational magnetic flux density when there is no defect on the tubular conductor.
Near the yoke, the direction of generated flux is unidimensional . It can be said that the defect near the yoke may not perturb the path of either flux or eddy current.

Figure 3 shows the positions of the defects on the tubular conductor, which are used in the numerical analysis.

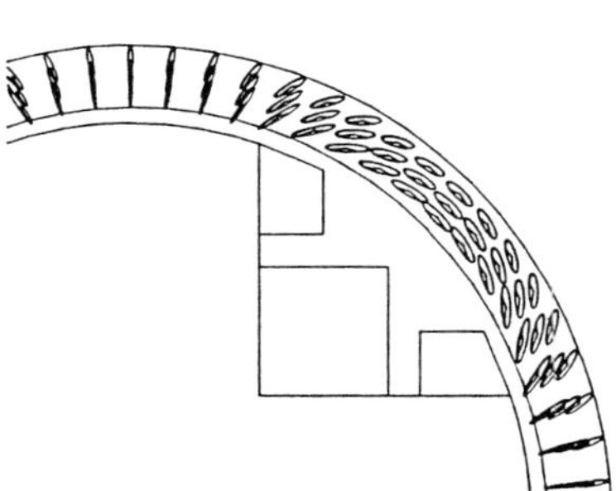

Fig. 2 The distribution of the rotational flux density.

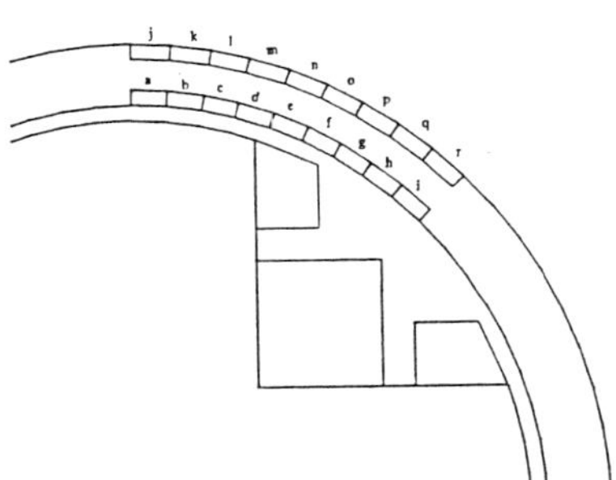

Fig.3 The conductor model include the defect.

Figure 4 indicates the Lissajou's figures which are obtained from the two differential voltages of the search coils. It is clear that very small signals are obtained when the defect is positioned near the yoke. On the other hand, signals of the defects are picked up at broadened range, whether the defects are positioned on inner or outer part of the tubular conductor.

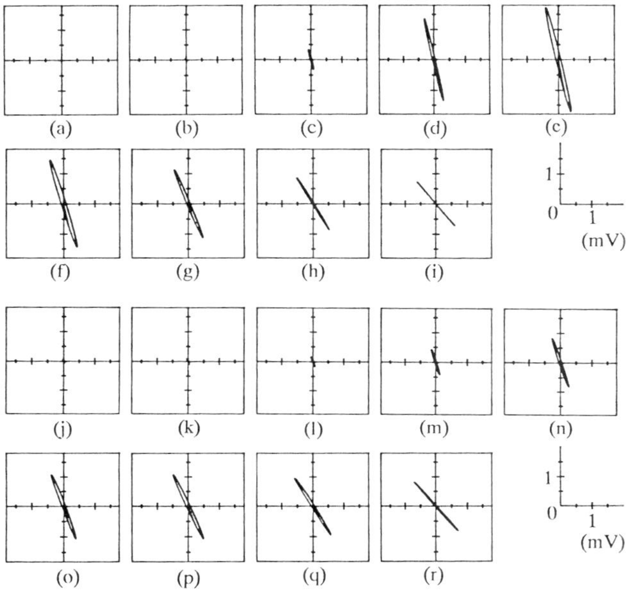

Fig. 4 Lissajou's figure obtained from the differential outputs.

From the Fig. 4, it is clear that the positions of the defects and the incline angles of the Lissajou's figure are correlated. Figure 5 shows the relationship between them. It can be said that the position of the defect can be estimated from the incline angle of the Lissajou's figure.

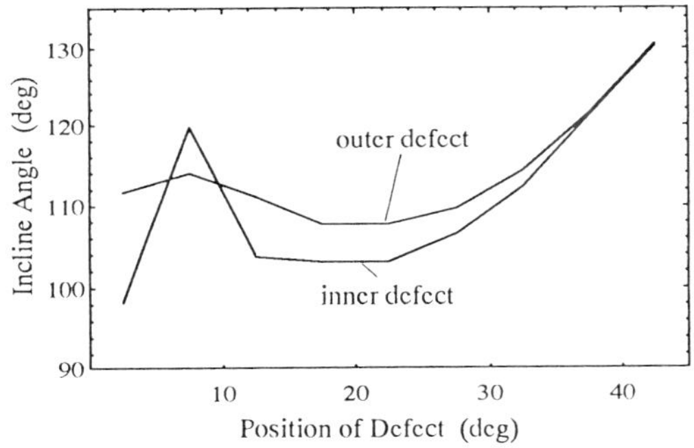

Fig. 5 The relationship between the defect position and the incline angle.

Finally, we investigate the locus of the system eigenvalues obtained from the output signals as shown in Fig. 6. To evaluate the phenomena which occurs inside the material, we employ the system eigenvalues obtained from the applied two-phase current and the output differential voltages. Figure 6 clearly shows the difference between the inner defect and outer defect.

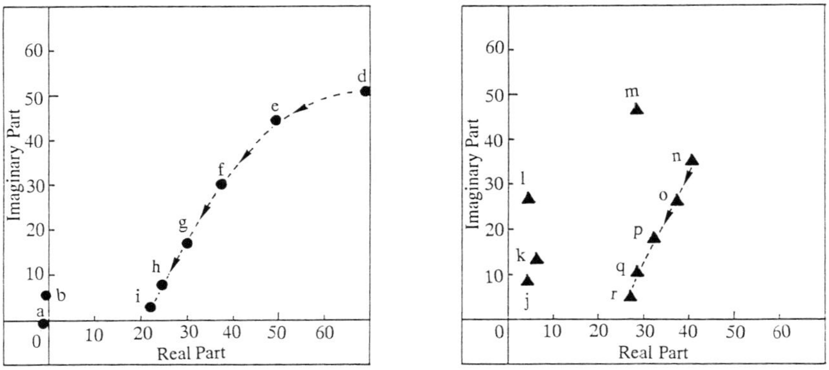

(a) inner defect (b) outer defect

Fig. 6 The locus of the system eigenvalues

4. CONCLUSIONS

We have proposed the magnetic sensor using rotational magnetic flux for the tubular material. We have clarified that this sensor is available to detect the defect on broadened range of the tubular material. The Lissajou's figure method is effective to estimate the position of the defect. Finally, we have proposed the system eigenvalues method to describe the condition of the material.

5. REFERENCES

1 M. Enokizono, S.Nagata : Non-Destructive Testing with Magnetic Sensor Using Rotational Magnetic Flux, Journal of the Magnetic Society of Japan, Vol.15, No. 2, pp455-460, 1991.

2 M. Enokozono, S. Nagata : Simulation Analysis of Magnetic Sensor for Non-destructive Testing by Boundary Element Method, IEEE Transactions on Magnetics, Vol. 26, No.2, 1990.

Simulation and Design of Applied Electromagnetic Systems
T. Honma (Editor)

Optimum shape design of electron lens using evolution strategy

Chang-Seop Koh [a], Ki-Ran Hong [a], Song-yop Hahn [a] and Hyun-Kyo Jung [b]

[a] Department of Electrical Engineering, Seoul National University, Seoul 151-742, KOREA

[b] Department of Electrical Engineering, Kangwon National University, Chuncheon 200-701, KOREA

Abstract

An electromagnetic lens for the Cathode Ray Tube of projection TV is designed so that minimum beam spot size on the screen is obtained. (1+1) evolution strategy algorithm is utilized as an optimization method. Through the numerical example, it is found that the spherical aberration should be considered in design and the usefulness of the proposed algorithm is verified.

1. INTRODUCTION

The projection TV uses three small Cathode Ray Tubes and optical magnification units to form the color picture on the screen. Each CRT is composed of electrostatic pre-focus lens(Gun), electromagnetic lens, convergence yoke and deflection yoke[1].

Conventionally the deterministic methods are used to find an optimum shape of electromagnetic lens that gives minimum shperical aberration. Hence the obtained shapes are dependent on the initial ones and many different initial shapes should be tried.

In this paper, (1+1) Evolution strategy is introduced to find the global optimum lens shape that gives minimum beam spot size on the screen in the presence of spherical aberration.

2. QUALITY FUNCTION

The electron beams generated at the cathode are accelerated and pre-focused in the electrostatic lens and enter the main electromagnetic lens. When the beams enter the electromagnetic lens, it is assumed that they have the Gaussian distribution in its density and radius of 3[mm] in its spot size which is defined in Fig.1.

In order to estimate the performance of the electromagnetic lens, the spot size of the electron beams on the screen is defined as the quality function[2].

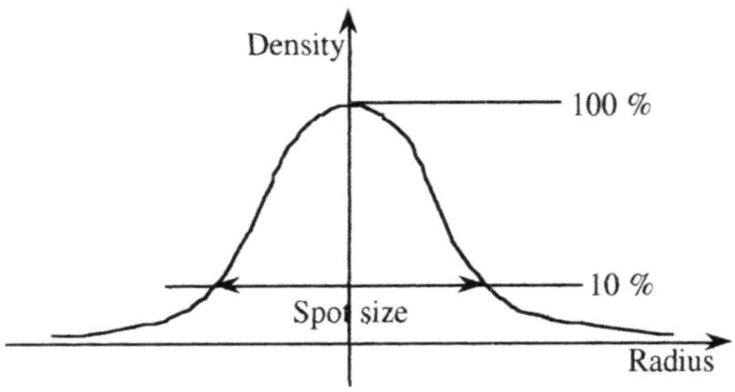

Fig.1 Beam spot size.

3. ELECTRON BEAM TRAJECTORY

In azimutally uniform system, the components of acceleration are given from Lorentz force equation as follows:

$$m(\ddot{r} - r\dot{\theta}^2) = q(E_r + r\dot{\theta}B_z) \qquad (1\text{-a})$$

$$m\ddot{z} = q(E_z - r\dot{\theta}B_r) \qquad (1\text{-b})$$

where dot(.) indicates the differential with respect to time t and the mass of an electron m assumed to be constant. The $\dot{\theta}$ in (1) can be computed by using Busch's theorem[3] i.e.,

$$\dot{\theta} = -\frac{q}{m} \cdot \frac{1}{2\pi r^2}(\Psi - \Psi_c) \qquad (2)$$

where ψ and ψ_c are the magnetic flux and reference magnetic flux at the point of $\dot{\theta} = 0$, respectively[3]. The 4th order Runge-Kutta algorithm is used to solve (1) and electric field intensity and magnetic flux density are computed by using standard finite element method.

In order to consider the interactions between the electron beams, the iteration method is used[3].

4. OPTIMIZATION ALGORITHM

In the view point of numerical optimization, the main feature of the electromagnetic lens design is that the quality function has no explicit expression with respect to the design variables and the electromagnetic fields. Hence the (1+1) evolution strategy is introduced, where optimization is achieved by repeating the regeneration, mutation and annealing processes.

New design variables are generated by the rule:

$$x_i^{new} = x_i^{old} + \alpha_i \cdot R_i \qquad (3)$$

where x_i^{new} and x_i^{old} are i-th, respectively, new and old design variables, and α_i is step width and R_i is random number with uniform distribution. The mutation is numerically simulated by adopting or rejecting the newly generated design variables according to the corresponding quality function values. The step width is also annealed by 0.85 rule[4].

5. NUMERICAL EXAMPLE

The proposed algorithm is applied to the optimum design of the electromagnetic lens, shown in Fig.2, used in 7" CRT of the projection TV. In order to obtain a minimum spot size, the lengths of the pure steel and the two permanent magnets and the two magnetic cores and the position of the electromagnetic lens are taken as the design variables.

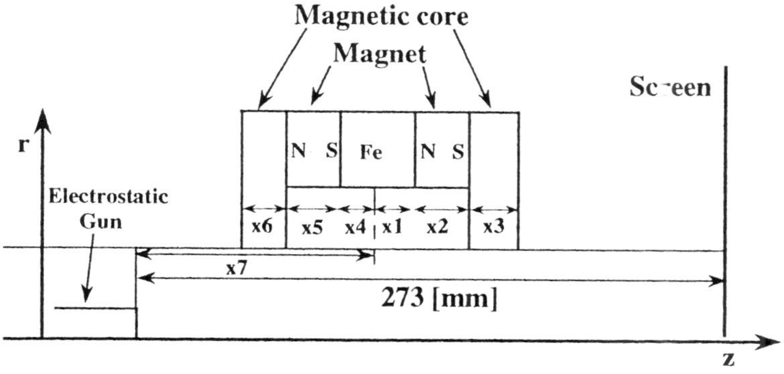

Fig.2 Electromagnetic lens configuration and the design variables.

The optimum design variables are found after 23 generations when the spherical aberration is ignored and after 44 generations when the spherical aberration is considered. Table 1 compares the optimum sizes of the electromagnetic lens with the initial size when the beam current is 3[mA]. It is found that when the spherical aberration is considered the main beams (started in the normal direction) are slightly over focused. Fig.3 shows the electron beam trajectories and the beam spot size with the optimized electromagnetic lens when the beam current is 3[mA].

Table 1
Comparison of optimized sizes with initial one [mm]

variable	initial	lower limit	upper limit	without spherical aberration	with spherical aberration
x_1	10.0	2.0	20.0	10.1	15.6
x_2	10.0	2.0	20.0	2.1	3.7
x_3	5.0	2.0	15.0	2.3	9.4
x_4	10.0	2.0	20.0	2.8	3.1
x_5	12.0	2.0	20.0	2.5	3.1
x_6	5.0	2.0	15.0	3.3	10.3
x_7	43.0	30.0	50.0	45.6	45.5

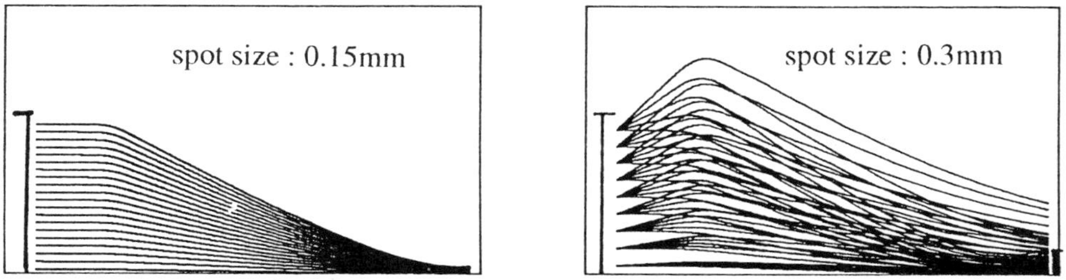

a) without spherical aberration b) with spherical aberration
Fig. 3 Electron beam trajectories.

6. CONCLUSION

An algorithm for the optimum size design of the electromagnetic lens is proposed. The algorithm combined the (1+1) evolution strategy with the computation of the beam trajectories. Through the numerical example, it is found that the spherical aberration should be considered in design and the usefulness of the algorithm is verified. The algorithm will be more useful and powerful if the electrostatic lens is included in the design.

7. REFERENCES

1 Ohsawa et al.,Journal of the SID, Vol.1 (1993) 23
2 D.J.Bechis et al., SID88DIGEST (1988) 122
3 X.Jiye, Aberration Theory in Electron and Ion Optics, Academy Press, Inc., 1986
4 A. Gottvald et al., IEEE trans. on Mag., Vol.28, No.2, MARCH 1992 1537

Simulation and Design of Applied Electromagnetic Systems
T. Honma (Editor)

Projective sampled pattern matching method for the defect recognition in conductive materials

T.Doi, H.Saotome and Y.Saito

College of Engineering, Hosei University
Kajino Koganei, Tokyo 184, Japan

Abstract

Previously, we have proposed the sampled pattern matching method in order to search for the current distributions in the biological objects from the local magnetic fields. Further, this method has been successfully applied to the defect identification problems from the local electric potential measurements.

In this present paper, we propose the simple implemented techniques of our sampled pattern matching method to improve the accuracy of the defect identifications.

1. INTRODUCTION

In order to protect the accidents of the aircraft, iron bridge, and nuclear reactor, crack or defect recognition is of paramount importance technology. Therefore, various nondestructive testing methodologies, e.g. the eddy current testing, X-ray computed tomography, ultrasonic imaging and electrical potential method, have been exploited and utilized. Among these methods, the electrical potential method requires a relatively simple device and measurement [1,2].

Previously, the current distributions in human heart as well as brain have been successfully estimated from the local magnetic fields by the sampled pattern matching (SPM) method [3,4]. The SPM method has been exploited for solving the inverse source problems also applicable to the defect recognition problems, i.e. medium parameter identification problems.

Denoting C as a system matrix, it is possible to write $C\mathbf{X}=\mathbf{Ys}$ as a discretized electric current flowing system equation of a conducive material having defect or crack. This system equation can be modified into $C_0\mathbf{X}=(C_0-C)\mathbf{X}+\mathbf{Ys}$, where C_0 is a system matrix without defect or crack. This means that the vector $\mathbf{X}(=C^{-1}\mathbf{Ys})$ is composed of the two input vectors $\mathbf{Y}$ and $\mathbf{Ys}$. The vectors $\mathbf{Y}$ and $\mathbf{Ys}$ are the equivalent field source vector $\mathbf{Y}=(C_0-C)\mathbf{X}$ caused by the defect or crack; and externally impressed field source vector $\mathbf{Ys}$, respectively. Thus, the defect or crack recognition problems can be reduced into the equivalent source vector $\mathbf{Y}$ searching problems. Obviously, this equivalent source vector $\mathbf{Y}$ depends on the vector $\mathbf{X}$. Further, this vector $\mathbf{X}$ is a function of the externally impressed field source vector $\mathbf{Ys}$. Thereby, the equivalent source vector $\mathbf{Y}$ can be expressed as a function of the impressed field source vector $\mathbf{Ys}$, i.e. $\mathbf{Y}=f(\mathbf{Ys})$. Introducing this functional relationship into our SPM method leads to the following advantages: this method requires a quite low CPU resource compared with those of the original SPM method because the SPM process has to be carried out only to the vectors satisfying the relationship $\mathbf{Y}=f(\mathbf{Ys})$; relatively highly accurate solutions can be expected, because one of the characteristics of solution vector $\mathbf{Y}$ is known; and changing the direction of vector $\mathbf{Ys}$ reaches the correct solution vector $\mathbf{Y}$ similar to the computed tomography. This new method is called the projective sampled pattern matching (PSPM) method because the functional relation between **Y** and **Ys** is that the direction of vector **Y** depends on the known vector **Ys**. In the present paper, we examine the characteristics of PSPM method. As a result, we propose here the simple implementation techniques to improve the accuracy of PSPM method.

2. THE PROJECTIVE SAMPLED PATTERN MATCHING METHOD

2.1. Basic equations

As described above, the original system equation is written down in a following form:

$$C_0 \mathbf{X} = \mathbf{Y}+\mathbf{Ys}, \tag{1}$$

where Co is m by m square matrix; **X**, **Y** and **Ys** are the m-th order column vectors. Taking the inverse matrix of C_0 and multiplying it to (1) yields

$$\begin{aligned}\mathbf{X} &= C_0^{-1}\mathbf{Y} + C_0^{-1}\mathbf{Ys} \\ &= C_0^{-1}\mathbf{Y} + \mathbf{X}_0,\end{aligned} \tag{2}$$

where $\mathbf{X}_0\,(=C_0^{-1}\mathbf{Ys})$ is a vector without defect or crack. Subtracting $\mathbf{X}_0$ from (2), we can get

$$\begin{aligned}\mathbf{X}_d &= \mathbf{X} - \mathbf{X}_0 \\ &= C_0^{-1}\mathbf{Y}.\end{aligned} \tag{3}$$

Generally, it is difficult to measure the entire difference vector $\mathbf{X}_d$ in (3) so that we can only obtain a part vector $\mathbf{X}_p$ of $\mathbf{X}_d$. This yields a system equation of the defect or crack recognition problem. Namely, denoting n as a number of measured points, we have

$$\begin{aligned}\mathbf{X}_p &= D\mathbf{Y}, \\ &= \sum_{i=1}^{m} y_i\,\mathbf{d}_i,\end{aligned} \tag{4}$$

where D and $\mathbf{X}_p$ are the n by m partial matrix in C_0^{-1} and measured difference vector with order n. The y_i and $\mathbf{d}_i$ in (4) are the i-th element of **Y** and i-th column vector in D, respectively.

2.2. Sampled pattern matching (SPM) method

If a point h in the target region takes the maximum of:

$$\gamma_i = \mathbf{X}_p^T\cdot\mathbf{d}_i/[\,|\mathbf{X}_p|\,|\mathbf{d}_i|\,],\quad i=1,2,\cdot,m, \tag{5a}$$

then h is called the first pilot point. If a point g takes the maximum of:

$$\gamma_{hj} = \mathbf{X}_p^T\cdot(\mathbf{d}_h+\mathbf{d}_j)/[\,|\mathbf{X}_p|\,|\mathbf{d}_h+\mathbf{d}_j|\,],\quad j=1,2,\cdot,m;\ j\neq h, \tag{5b}$$

then g is called the second pilot point. Similar process to (5a) or (5b) is continued up to the n-th pilot points. Thus, the element y_i in the vector **Y** has been assumed to

$$y_i = \begin{cases} 1, & \text{if i is a pilot point,} \\ 0, & \text{if i is not a pilot point,}\end{cases} \tag{6}$$
$$i=1,2,\cdots,m.$$

Pilot point solutions of (6) mean that the magnitude of solution y_i, $i=1\sim m$, in (4) is represented by the spatial concentrating rate of unit input.

2.3. Projective Sampled Pattern Matching (PSPM) Method

The SPM method is applicable to the inverse problems that the solution vector **Y** is not a function of the known function **Ys**, i.e.

$$\mathbf{Y} \neq f(\mathbf{Ys}). \tag{7}$$

However, if a relationship

$$\mathbf{Y} = f(\mathbf{Ys}), \tag{8}$$

is established, then the SPM processes are carried out only to the vectors $\mathbf{d}_i$ (i=1,2,···,m) satisfying the relationship (8). This method is called *the projective sampled pattern matching (PSPM) method.*

The PSPM method has the following features:

1) low cpu resource,
2) relatively highly accurate solutions,
3) changing the projective angles with $\mathbf{Ys}$ reaches the correct solution vector,
4) the pilot point solutions are available if the known vector $\mathbf{X}_p$ encloses a target area.

A curve obtained by plotting the pattern matching rate γ in (5a) or (5b) versus the pilot points reveals

1) if the curve monotonously increases in accordance with the addition of pilot points then the target area has single defect or indistinguishable plural defects,
2) if the curve has a deflection point then the object has the plural defects. In this case, each of the plural defects can be distinguished by shrinking the target area or removing the pilot points up to the first peak value of γ. This is because the pilot points up to the first peak value of γ provide the globally good pattern matching vectors common to the entire defects.

2.4. Examples

We chose the two dimensional electric potential method as the examples. The derivation of the system equation is described in detail in the Ref. [1].

Figure 1(a) shows one of the γ versus pilot points curves. Obviously, this curve monotonously increases in accordance with the addition of pilot points, so that the target area has single defect. In fact, the result of PSPM method shows the single defect as shown in Fig. 1(b).

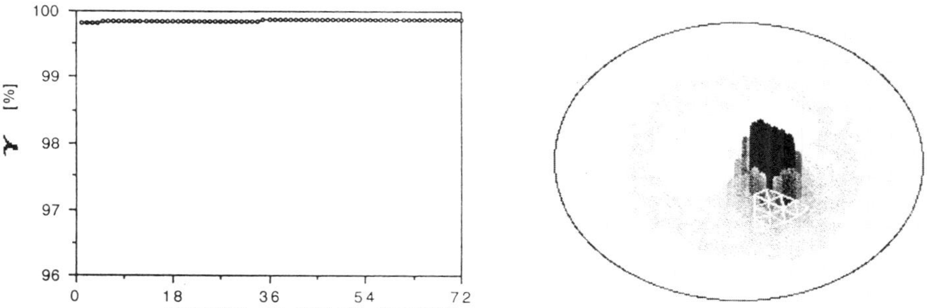

(a) (b)

Fig.1 Single defect recognition by measuring the electric potential around a target area. (a) γ vs. pilot points, and (b) recognized defect. The number of subdivision m=1801, and the number of measured points n=72. The direction of the externally impressed current input vector $\mathbf{Ys}$ was changed 72 times with 5 degree subdivision.

Figure 2(a) shows one of the γ versus pilot points curves. Obviously, this curve has a deflection point, so that the target area has the plural defects. In fact, the result of PSPM method shows the plural defects as shown in Fig. 2(b).

Figure 3(a) shows an improved plural defect recognition by shrinking the target area, and Fig. 3(b) shows the other improved plural defect recognition by removing the pilot points up to the first peak of γ.

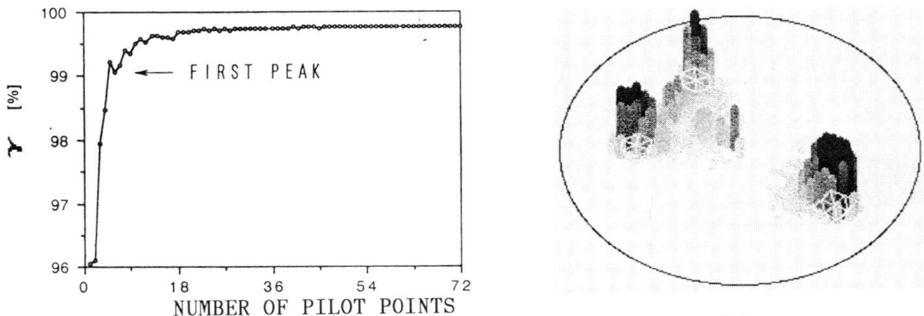

(a) (b)

Fig.2 Plural defects recognition by measuring the electric potential around a target area. (a) γ vs. pilot points, and (b) recognized defect. The number of subdivision m=1801, and the number of measured points n=72. The direction of the externally impressed current input vector **Ys** was changed 72 times with 5 degree subdivision.

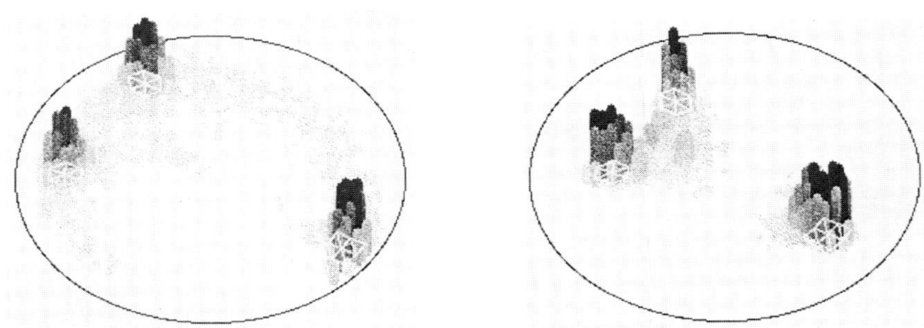

(a) (b)

Fig.3 Improved plural defect recognition. (a) Improvement by shrinking the target area, (b) improvement by removing the pilot points up to the first peak.

3. CONCLUSION

As shown above, we have examined the PSPM method and proposed the methods of improvement for the plural defect recognition problems. Examples have demonstrated that the plural defect recognition is dramatically improved by our method.

4. REFERENCES

1 T.Doi et al, Proceedings of the 2nd Japan-Hungary Seminar, Ed. T.Honma et al, Hokkaido Univ. Press, Sep.(1992)pp.61-66.

2 Y.Iwamura et al, IEEE Trans. Magn., Vol.26, No.2, Mar.(1990)pp.81-90.

3 H.Saotome,et al, Trans. IEEJ, Vol.112-A, No.4, April(1992).

4 H.Saotome,et al, Trans. IEEJ, Vol. C, to appear in March (1993).

Simulation and Design of Applied Electromagnetic Systems
T. Honma (Editor)

Leakage magnetic field source searching by the correlative analysis

K.Kitsuta[a], Y.Saito[a], S.Hayashi[b], K.Hatakeyama[b] and H.Tohya[b]

[a] Coll. Engng., Hosei University, Koganei, Tokyo 184, Japan
[b] Environment and Resource Lab. NEC Corp.,Miyamaeku Kawasaki, Kanagawa 213, Japan

Abstract

This paper describes a methodology of searching for the leakage magnetic field sources from the electronic devices by a correlative analysis. Previously, we have succeeded in searching for the magnetic field source from the measurement of local magnetic fields in biological systems. In the present paper, we now apply our method called the sampled pattern matching (SPM) method to the leakage magnetic field source searching problems. It is shown that a correlative analysis between the real and imaginary parts of magnetic field source yields an excellent result.

1. INTRODUCTION

With the development of modern digital computers, high frequency operation of the computers is essentially required in order to speed up the CPU and to realize the smaller size power supplies. This causes a problem so called the electromagnetic compatibility (EMC), i.e. the environmental effects caused by the electromagnetic fields leaking from the electronic devices. To remove this problem, at first, we have to find the leakage electromagnetic field source from the devices.

Previously, we have succeeded in searching for the magnetic field sources from the measurement of local magnetic fields in biological systems [1,2]. In the present paper, we now apply our method called the sampled pattern matching method to the EMC problems.

In most case, the electromagnetic wave can be represented in terms of the complex notation. This means that we have to independently searching for the real and imaginary parts of the leakage field source. Key idea of our approach is that both of the real and imaginary parts have to take the same path, thereby, a correlative analysis between the real and imaginary parts may yield the highly accurate result. Further, this removes a large amount of ambiguity associated with the sampled pattern matching solutions.

Thus, we have succeeded in searching for the leakage magnetic field source with high accuracy.

2. CORRELATIVE ANALYSIS OF THE LEAKAGE MAGNETIC FIELD SOURCE SEARCHING

2.1. Assumption

This paper assumes that the operating frequency of the target electronic devices is not so high. Moreover, it is assumed that the normal component of leakage magnetic fields is measured at a surface near to the target electronic devices. Thereby, the phase difference due to the space propagation time between the target and field measured points is assumed to be neglected. This assumption means that our problem is to searching for the low frequency alternating current density **J** distributions by measuring the local magnetic fields **H**.

2.2. Sampled Pattern Matching Method

The magnetic field **H** is related to the current density **J** by

$$\mathbf{H} = \nabla \times \int [\mathbf{J}/(4\pi|r|)]dv, \qquad (1)$$

where r is a distance between the field **H** and source **J** points. In (1), the volume V containing the current density **J** is subdivided into a large number of subdivisions V_i, i=1~m. In this small volume V_i, i=1~m, each of current densities $\mathbf{J}_i$, i=1~n, takes a constant value. Also, the number of field measured points is denoted by n, then (1) reduces into

$$\mathbf{U} = \sum_{i=1}^{m} \alpha_i \mathbf{d}_i, \qquad (2)$$

where

$$\mathbf{U} = [\mathbf{H}_1, \mathbf{H}_2, \cdots, \mathbf{H}_n]^T, \qquad (3a)$$
$$\mathbf{d}_i = (1/4\pi)[\mathbf{n}\times\mathbf{a}_{1i}/r_{1i}^2, \mathbf{n}\times\mathbf{a}_{2i}/r_{2i}^2, \cdots, \mathbf{n}\times\mathbf{a}_{ni}/r_{ni}^2]^T, \qquad (3b)$$
$$\alpha_i = J_i V_i, \quad i=1\sim m, \qquad (3c)$$
$$m >> n. \qquad (3d)$$

As shown in Fig.1, **n** is a unit vector in the direction of $\mathbf{J}_i$; $\mathbf{a}_{1i}, \mathbf{a}_{2i}, \cdot, \mathbf{a}_{ni}$ are the unit vectors from the source point i to the field points 1,2,·,n; $r_{1i}, r_{2i}, \cdot, r_{ni}$ are the distances from the source point i to the field points 1,2,···,n, respectively. Further, α_i, i=1~m, in (3c) is a magnitude of the current dipole $\mathbf{P}_i$ [1,3]. The condition m>>n in (3d) is always satisfied because the fields $\mathbf{H}_i$, i=1~n, can be measured on the limited surface but the field sources α_i, i=1~m, exist in the volume enclosed by the field measuring surface.

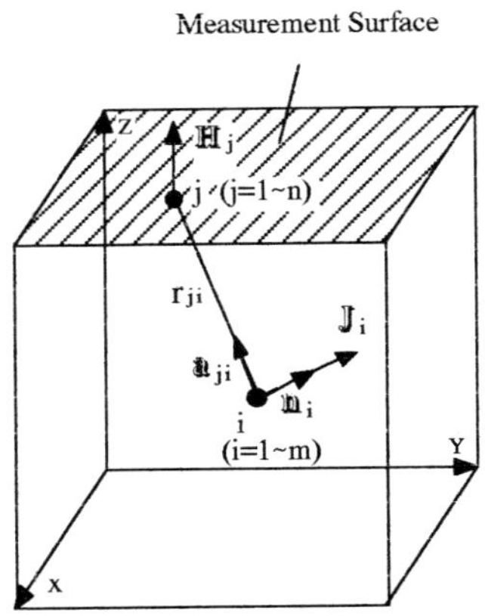

Figure 1. Relationship between the magnetic field and its source points.

Equation (2) means that the vector **U** is a linear combination of the vectors $\mathbf{d}_i$ (i=1~m). Thereby, (2) can be modified into

$$\mathbf{U} = \sum_{i=1}^{m} \{\beta_i \mathbf{d}_i + \sum_{j\neq i}^{m} \{\beta_{ij}(\mathbf{d}_i + \mathbf{d}_j) + \sum_{k\neq i, k\neq j}^{m} \{\beta_{ijk}(\mathbf{d}_i + \mathbf{d}_j + \mathbf{d}_k) + \cdots \}\}\}. \qquad (4)$$

Each of the terms in (4) has its distinct field pattern. Namely, the 1st, 2nd and 3rd groups on the right give the one, two and three pairs of the north (N) and south (S) magnetic poles, respectively. This means that the pattern matching analysis should be applied only to the 1st solution group but also the other remaining solution groups. After carrying out this sampled pattern matching analysis to all of the groups in (4), summation of their results will yield a unique field source pattern. Thus, the obtained field source pattern corresponds to the normalized current distributions [1].

The 1st solution group in (4) is

$$[\mathbf{U}/|\mathbf{U}|]^T[\mathbf{d}_1/|\mathbf{d}_1|],\ [\mathbf{U}/|\mathbf{U}|]^T[\mathbf{d}_2/|\mathbf{d}_2|],\ \cdots,\ [\mathbf{U}/|\mathbf{U}|]^T[\mathbf{d}_h/|\mathbf{d}_h|],\ \cdots,[\mathbf{U}/|\mathbf{U}|]^T[\mathbf{d}_m/|\mathbf{d}_m|]. \quad (5a)$$

If $[\mathbf{U}/|\mathbf{U}|]^T[\mathbf{d}_h/|\mathbf{d}_h|]$ takes the maximum in (5a), then 2nd solution group is

$$[\mathbf{U}/|\mathbf{U}|]^T[(\mathbf{d}_h+\mathbf{d}_1)/|\mathbf{d}_h+\mathbf{d}_1|],\ [\mathbf{U}/|\mathbf{U}|]^T[(\mathbf{d}_h+\mathbf{d}_2)/|\mathbf{d}_h+\mathbf{d}_2|],\ \cdots,\ 1,\ \cdots,[\mathbf{U}/|\mathbf{U}|]^T[(\mathbf{d}_h+\mathbf{d}_m)/|\mathbf{d}_h+\mathbf{d}_m|]. \quad (5b)$$

Similar procedure is continued until a peak value of inner product can be obtained. Thereby, the normalized solutions of (2) are

$$\alpha_1[|\mathbf{d}_1|/|\mathbf{U}|] \simeq [\mathbf{U}/|\mathbf{U}|]^T\{[\mathbf{d}_1/|\mathbf{d}_1|]+[(\mathbf{d}_h+\mathbf{d}_1)/|\mathbf{d}_h+\mathbf{d}_1|]+\cdots\}, \quad (6a)$$
$$\alpha_2[|\mathbf{d}_2|/|\mathbf{U}|] \simeq [\mathbf{U}/|\mathbf{U}|]^T\{[\mathbf{d}_2/|\mathbf{d}_2|]+[(\mathbf{d}_h+\mathbf{d}_2)/|\mathbf{d}_h+\mathbf{d}_2|]+\cdots\}, \quad (6b)$$
$$\cdots\cdots\cdots\cdots$$
$$\alpha_h[|\mathbf{d}_h|/|\mathbf{U}|] \simeq \{[\mathbf{U}/|\mathbf{U}|]^T[\mathbf{d}_h/|\mathbf{d}_h|]+1+1+\cdots\} \quad (6c)$$
$$\cdots\cdots\cdots\cdots$$
$$\alpha_m[|\mathbf{d}_m|/|\mathbf{U}|] \simeq [\mathbf{U}/|\mathbf{U}|]^T\{[\mathbf{d}_m/|\mathbf{d}_m|]+[(\mathbf{d}_h+\mathbf{d}_m)/|\mathbf{d}_h+\mathbf{d}_m|]+\cdots\}. \quad (6d)$$

Thus, it is possible to obtain the current dipole vectors from the magnetic field measurements. After obtaining the real and imaginary parts of the current dipole vectors independently, a correlation between them can be obtained by taking the inner product between the real and imaginary parts.

2.3. Example

Figure 2 shows a schematic diagram of a practically tested example [4]. The normal magnetic fields were measured at each of the grid points on the surface located 5cm above the target surface. The target device used in this experiment has a cross shape slit with 5mm width and includes a loop antenna located at the center of metallic box. The exciting frequency of the loop antenna is 1MHz. Therefore, the wave length is so long that the phase difference due to the space propagation time is negligibly small value.

In general, it may be considered that the position of target including the magnetic field source should be unknown. This means that we can measure the leakage magnetic fields but we do not know where they come from. Thereby, we assumed that the target having $40\times40\times12\text{cm}^3$ volume was located 5cm below the measurement surface. The target was subdivided into 153600 small rectangular prisms. Also, the angle in x-y surface is divided into 72 directions so that total number of unknowns m is 11059200 whereas the number of measured magnetic field n is 1681.

Figures 3(a) and 3(b) show the real and imaginary parts of the leakage magnetic field sources estimated by the sampled pattern matching method, respectively. They include a considerable ambiguity but a correlative analysis between them yields an excellent result as shown in Fig. 3(c).

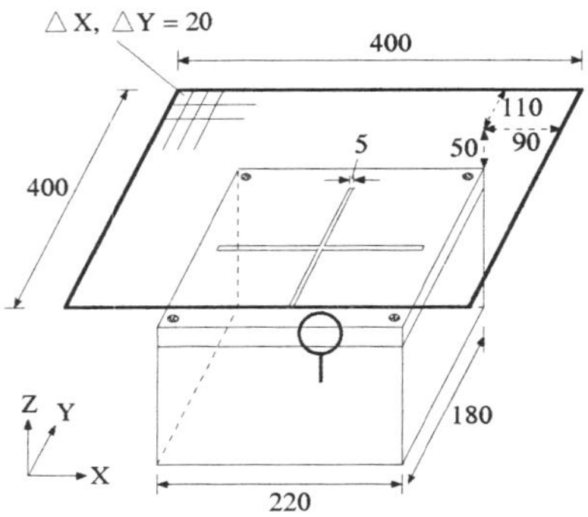

Figure 2. Schematic diagram of a practically tested example [unit mm].

(a) (b) (c)

Figure 3. Estimated leakage magnetic field source distributions. (a) real part, (b) imaginary part and (c) correlation.

3.CONCLUSION

As shown above, both of the real and imaginary parts include a considerable ambiguity but their correlation provides the reliable leakage magnetic field source distribution coincided with the results of previous work [4]. Thus, we have elucidated that our sampled pattern matching method can be available to solving for the EMC problems. Key idea to get the reliable results is similar to those of the space power distribution method [5].

4. REFERENCES

1. H.Saotome et al,"An Estimation of Current Distributions in Biological System by the Sampled Pattern Matching Method," Trans. IEEJ, Vol.113- C, No.1, January(1993) pp.101-108.
2. H.Saotome et al, "Inverse Problems in Biomagnetic Fields," Trans.IEEJ, Vol.112-A, No.4, April(1992)pp.279-286.
3. Y. Uchikawa, et al, Journal of Japan Applied Magnetics, Vol.13, No.3, (1989)pp.508-512.
4. S.Hayashi et al., Paper on Technical Meeting of IEICJ, EMCJ92-8, A-P92-25, May (1992).
5. K.Kitsuta et al., Proceedings of the 2nd Japan-Hungary Seminar, Ed. T.Honma et al, Hokkaido Univ. Press, Sep.(1992).

Simulation and Design of Applied Electromagnetic Systems
T. Honma (Editor)

Vector boundary element/finite element method for electromagnetic scattering

N. Ida[a] and J.S. Wang[b]

[a]Department of Electrical Engineering, The University of Akron, Akron, OH. 44325-3904

[b]Department of Electrical Engineering, University of Illinois, Urbana, Il. 61801.

Abstract

The use of boundary elements for computation of scattering parameters such as equivalent current sources and scattering cross section are important parameters in nondestructive testing with composite materials. Combination of boundary elements on the surface and finite elements in the interior allows accurate modeling of testing parameters. These are particularly useful when inclusions or defects in an otherwise uniform material are to be evaluated. The use of boundary element methods allows efficient modeling, in particular for small scatterers and for anisotropic materials. Results for scatterers in dielectrics are given as examples for solutions based on the coupled FEM/BEM method, in which the surface and volume vector finite elements are coupled. This method is particularly useful for modeling of inclusions and defects in composite materials, including lossy dielectrics.

1. THE FINITE ELEMENT MODEL

If materials are inhomogeneous, the following form is used for Helmholtz's equation

$$\nabla\times\left(\frac{1}{j\omega\mu}\nabla\times\mathbf{E}\right)+j\omega\varepsilon\mathbf{E}=0 \tag{1}$$

$$\nabla\times\left(\frac{1}{j\omega\varepsilon}\nabla\times\mathbf{H}\right)+j\omega\mu\mathbf{H}=0 \tag{2}$$

where we assume that ε and μ are anisotropic.

To formulate these equations in terms of finite elements in the domain, we use Galerkin's method. A set of approximating functions is used for **E** or **H** and a set of vector weighting functions $\mathbf{w}_m$, equal to the shape functions [1,2]. Integrating over the volume Ω of the interior domain gives

$$\int_\Omega \nabla\times\left(\frac{1}{j\omega\mu}\nabla\times\mathbf{E}\right)\cdot\mathbf{w}_m dv + \int_\Omega j\omega\varepsilon\mathbf{E}\mathbf{w}_m dv=0 \tag{3}$$

where $\mathbf{w}_m$ are the edge shape functions defined over the finite element. Using the following relations

$$\int_{\Omega} (\nabla\times\mathbf{a})\cdot(\nabla\times\mathbf{b}) - \mathbf{a}.\nabla\times\nabla\times\mathbf{b}dv = \int_S (\mathbf{a}\times(\nabla\times\mathbf{E})).\hat{\mathbf{n}}\, ds \tag{4}$$

$$\mathbf{a} = \mathbf{w}_m \qquad \nabla\times\mathbf{b} = \frac{1}{j\omega\mu}\nabla\times\mathbf{E} \tag{5}$$

gives the weak form of the equation as:

$$\int_{\Omega} \left(\frac{1}{j\omega\mu}\nabla\times\mathbf{E}\right)\cdot(\nabla\times\mathbf{w}_m)dv + \int_{\Omega} j\omega\varepsilon\mathbf{E}\cdot\mathbf{w}_m dv = \int_S (\hat{\mathbf{n}}\times\mathbf{H})\cdot\mathbf{w}_m ds \tag{6}$$

where S is the surface enclosing the volume. This form is advantageous because it only contains first order derivatives of the function to be evaluated. Similar steps lead to an equivalent formulation for **H** but will not be given here. The important point here is that the tangential term $\hat{\mathbf{n}}\times\mathbf{H}$ (or $\mathbf{E}\times\hat{\mathbf{n}}$ in the **H** formulation) are defined only on the surface and they represent surface current densities (or surface magnetization)

2. THE BOUNDARY INTEGRAL MODEL

The boundary integral equation is obtained from the basic equations in the exterior domain:

$$\nabla\times\mathbf{E}_t = -\mathbf{M} - j\omega\mu_0\mathbf{H}_t \tag{7}$$

$$\nabla\times\mathbf{H}_t = \mathbf{J} + j\omega\varepsilon_0\mathbf{E}_t \tag{8}$$

where t indicates the total field in the exterior domain an $(E_t=E_{inc}+E_{scat})$. Using this equation, Huygens equivalence principle, and the vector Green's theorem, the surface contribution is written as

$$\mathbf{E} = \mathbf{E}_{inc} - \int_S \left(j\omega\mu_0(\hat{\mathbf{n}}\times\mathbf{H})G_0 + (\hat{\mathbf{n}}.\mathbf{E})\nabla' G_0 - (\mathbf{E}\times\hat{\mathbf{n}})\times\nabla' G_0\right)ds' \tag{9}$$

where $\mathbf{E}_{inc}$ is the incident electric field intensity and is assumed to be known. The latter equations are assumed to apply in the exterior domain Ω_0 (and on the surface between the exterior and interior domains) and Eq (6) applies in the interior domain Ω. Eq (9) is used on the surface while the finite element equation in (6) is used in Ω. In addition, the equivalent surface sources are (the surface charge density n¥E is deleted through use of continuity relations):

$$\mathbf{E}\times\hat{\mathbf{n}} = \mathbf{M} \qquad \hat{\mathbf{n}}\times\mathbf{H} = \mathbf{J} \tag{10}$$

These are the surface current densities and surface magnetization. The surface current density is used to couple the finite element and boundary integral equations for the electric field while the surface magnetization is used for the magnetic field equations.

To do so, volume elements are used in the interior of the problem (tetrahedral or hexahedral elements) and equivalent surface elements (triangular elements or quadrilateral elements) [3]. The approximations in the volume and on the surface are:

$$\mathbf{E} = \sum_{j=1}^{m} E_j \mathbf{w}_j \qquad \hat{\mathbf{n}} \times \mathbf{H} = \sum_{j=1}^{n} J_j \mathbf{v}_j \tag{11}$$

where $\mathbf{w}_j$ are the edge shape functions for the volume element and $\mathbf{v}_j$ the edge shape functions for the boundary element. These are now substituted in the finite element equation (Eq. (6)) and the boundary element equation (Eq. (9)). After evaluation this leads to two systems of equations:
In the interior:

$$[S]_e\{E\}_e = [Q]_e\{J\}_e \tag{12}$$

On the surface:

$$[B]_e\{E\}_e - \{E\}_e^{inc} = [C]_e\{J\}_e \tag{13}$$

Combining the two equations we get

$$\begin{bmatrix} s_{11} & s_{12} \\ s_{21} & s_{22} - QC^{-1}B \end{bmatrix}_e \begin{Bmatrix} E \\ E_0 \end{Bmatrix}_e = \begin{Bmatrix} 0 \\ QC^{-1}E_0^{inc} \end{Bmatrix}_e \tag{14}$$

This system is unique in structure. s_{11} is $m \times m$ in size, symmetric and sparse while s_{22} is $n \times n$ and is nonsymmetric and dense. s_{12} and s_{21} are $n \times m$ and $m \times n$ and are symmetric (transpose of each other). The index *(1)* refers to interior edges while index *(2)* refers to surface edges. Since the incident electric field on the surface can be evaluated from independent calculations (analytical), the system above can be easily solved for the electric field in the interior and on the surface.

3. RESULTS

The general configuration used here is shown in Figure 1. The material sample is of arbitrary shape but the exterior domain is free space. The sample can be either isotropic or anisotropic while the flaw is almost always anisotropic. A simple geometry in *NDE* is shown in Figure 2. The figure shows the scattering cross section for a spherical dielectric of relative permittivity *4* and for the same dielectric with an embedded cubic lossy dielectric ($\varepsilon_r=80-j80$). The magnitude of the scattering cross section (as well as the shift in the dip can be used for testing purposes.

In addition, the equivalent surface current densities and equivalent surface magnetization may be calculated and are shown in figures 3 and 4.

4. CONCLUSIONS

The method for coupling of the finite element and boundary element methods offer a simple and effective way of modeling scattering applications in material. The application to nondestructive testing of materials, especially at higher frequencies was emphasized in this work although the method is general.

5. REFERENCES

1 J.S. Wang and N. Ida, "A 3-D FEM-BEM model for electromagnetic waves interaction with materials", in Review of Progress in Quantitative Nondestructive Evaluation, D. O. Thompson and D. E. Chimenti, Eds., Plenum Press, New York, 1990, Vol. 9A, pp. 311-318.

2 J.S. Wang, On Edge Elements and Method of Moments Solutions of Electromagnetic Scattering and Coupling, Ph.D Dissertation, The University of Akron, May 1992.

3 J.S. Wang and N. Ida, "Curvilinear and Higher Order Edge Finite Elements in Electromagnetic Field Computation", accepted for publication in the IEEE Transactions on Magnetics.

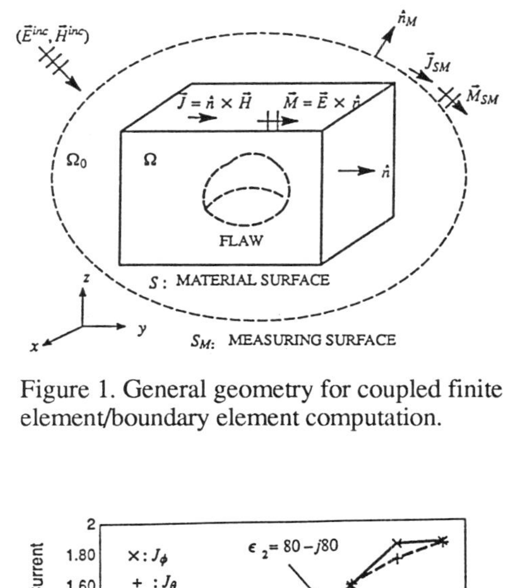

Figure 1. General geometry for coupled finite element/boundary element computation.

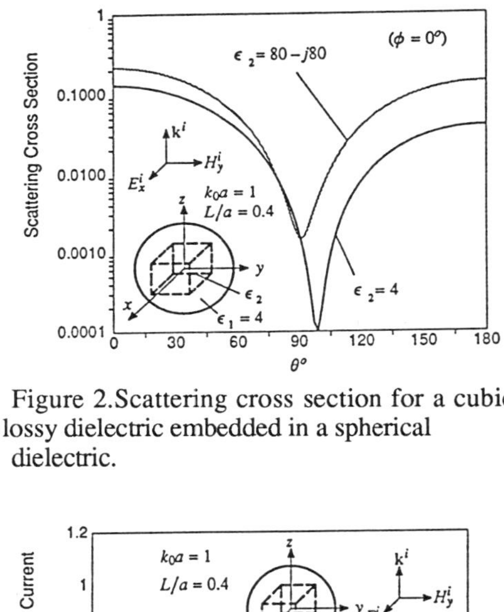

Figure 2.Scattering cross section for a cubic, lossy dielectric embedded in a spherical dielectric.

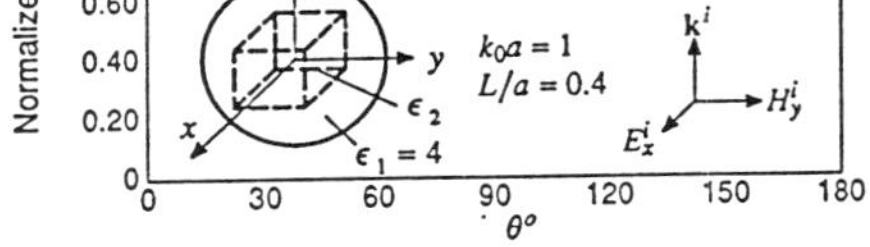

Figure 3. Normalized surface electric current density on the *yz* and *xz* planes.

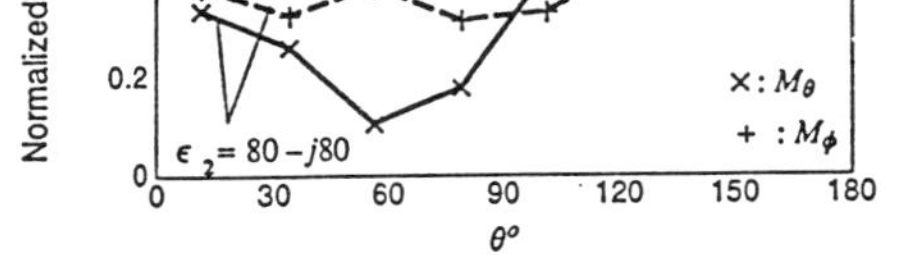

Figure 4. Normalized surface magnetic current density on the *yz* and *xz* planes.

Simulation and Design of Applied Electromagnetic Systems
T. Honma (Editor)

Application of the integrodifferential finite element - Green's function method for the study of eddy current shielding

Qiushi Chen, A. Konrad, J.D. Lavers and P.P. Biringer

Department of Electrical and Computer Engineering, University of Toronto, Toronto, Ontario M5S 1A4, CANADA

Abstract

This paper presents a study of eddy current shielding of steel beams using the integrodifferential finite element - Green's function method. The study involves the determination of the influences of structure dimensions on the effectiveness of eddy current shielding. Design guidelines for the shield and steel beams are established.

1. INTRODUCTION

Eddy current losses have long been problems facing designers of electric power apparatus. These losses occur in steel and metallic structures due to nearby time-varying magnetic fields. As the ratings of power devices increase, the associated eddy current losses become more significant. They have adverse effect on process efficiency, thermal capabilities and production costs. These problems are especially significant in electroheat engineering since power consumption is in the megawatts range.

The need for the understanding of eddy current effects in metallic structures in electroheat engineering prompted both experimental study [1] and theoretical investigations [2-4]. These investigations, while shedding light on the role and need of shielding, are valid for only certain applications. This paper focuses on the study of eddy current shielding of steel beams close to a current carrying conductor using the modified integrodifferential finite element - Green's function method introduced in [5]. Shielding is effective when the total loss in the steel and shield is less than that without shield. This paper also investigates the influences of structure dimensions on the effectiveness of eddy current shielding.

2. DESCRIPTION OF THE METHOD

In this section, the integrodifferential finite element - Green's function method is described. Since we are primarily concerned with the power loss in steel, the source current is assumed to be filamentary. The essence of this method is i) the use of the modified integrodifferential finite element method in the interior region which includes the steel beam, shield and air and ii) the use of the Green's function method in the exterior region which consists of air and source currents.

It is assumed that i) the magnetic field is linear, quasistatic and time harmonic with an angular frequency ω; ii) all currents are in the z direction and the field is two-dimensional.

In the interior region, the modified integrodifferential equation derived from the diffusion equation is given by

$$\frac{1}{\mu}\nabla^2 A - j\omega\sigma A + j\omega\sigma \frac{\int \sigma A \, dS}{\int \sigma \, dS} = 0 \tag{1}$$

The total magnetic vector potential (MVP) A_t is expressed as

$$A_t = A + A_0 \qquad (interior \;\; region) \tag{2}$$

where A is the z-component of the reduced MVP and A_0 is a constant [5].

In the exterior region, the field is decomposed into the source field and the reaction *Laplacian* field. For the exterior *Laplacian* field, the MVP is represented as follows:

$$A_e = \sum_{i=1}^{M} C_i G(\mathbf{r},\mathbf{r}_i) + C_0 \tag{3}$$

where G is the appropriate 2D Green's function given by (4) while C_i and r_i are the strength and location of the ith fictitious magnetic charge, respectively. C_0 is a constant and C_i satisfies the regularity constraint equation (5).

$$G(\mathbf{r},\mathbf{r}_i) = -\frac{1}{2\pi} \, ln \, |\mathbf{r}-\mathbf{r}_i| \tag{4}$$

$$\sum_{i=1}^{M} C_i = 0 \tag{5}$$

The source field is represented by A_s which is derived analytically [5]. The total MVP is

$$A_t = A_e + A_s \qquad (exterior \;\; region) \tag{6}$$

Equation (1) is transformed into the finite element matrix equation using the Galerkin criterion. The coupling of the interior and exterior regions is accomplished through the usual interface conditions. The resulting system of equations is expressed as

$$\begin{bmatrix} S & V \\ Q & G \end{bmatrix} \begin{Bmatrix} A \\ C \end{Bmatrix} = \begin{Bmatrix} F \\ 0 \end{Bmatrix} \tag{7}$$

The matrices in (7) are defined in [5].

3. SHIELDING EFFECTIVENESS

The steel beam being studied is placed close to a current carrying conductor as shown in Fig. 1. The conductivity of the shield is $5.8\times10^7 S/m$, the relative permeability of the steel is 100, its conductivity is $0.16\times10^7 S/m$. The thickness of the shield is 1.64 times its skin depth and envelopes the beam. The source current I is located $1m$ away from the centre of the steel with a frequency of $1300Hz$. The skin depth of the steel is δ.

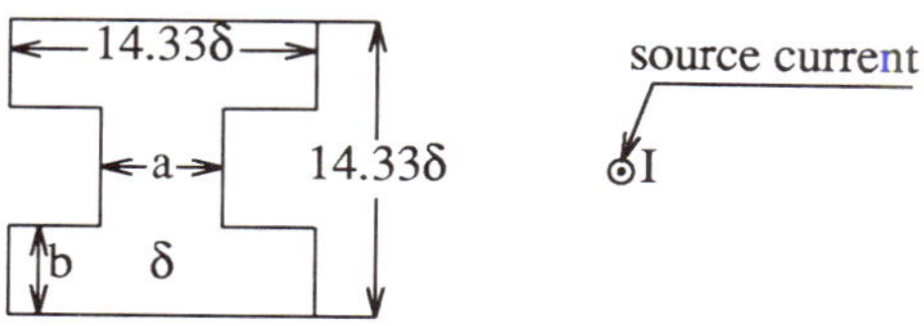

Figure 1. A steel beam in the vincinity of a current-carrying conductor.

The power ratio of the total loss in the beam and shield to the loss P_0 of the unshielded beam is obtained and tabulated in Table 1. Two kinds of shield are studied. One is I-shaped and the other is rectangular and both are in contact with the steel beam. P_I and P_R denote total power loss for the I-shaped and rectangular shields plus beams, respectively. A per unit base power loss P_U for the steel is defined as follows

$$P_U = \frac{I^2}{\sigma\mu_r} \tag{8}$$

Table 1
Power losses and shielding efficiency. (P_U=62.50W)

a/δ	2.87	2.87	2.87	2.87	5.73	5.73	5.73	5.73
b/δ	0.72	2.15	3.58	4.30	0.72	2.15	3.58	4.30
P_0/P_U	5.60	8.05	7.85	7.68	5.61	7.55	7.45	7.34
P_I/P_U	8.64	8.23	7.86	7.68	8.54	8.20	7.86	7.68
P_R/P_U	7.43	7.43	7.43	7.43	7.43	7.43	7.43	7.43
P_I/P_0	1.54	1.02	1.00	1.00	1.52	1.09	1.06	1.05
P_R/P_0	1.33	0.92	0.95	0.97	1.32	0.98	1.00	1.01

a/δ	8.60	8.60	8.60	8.60	11.46	11.46	11.46	11.46
b/δ	0.72	2.15	3.58	4.30	0.72	2.15	3.58	4.30
P_0/P_U	5.61	6.97	6.96	6.92	5.60	6.28	6.33	6.32
P_I/P_U	8.32	8.09	7.85	7.69	7.91	7.80	7.71	7.64
P_R/P_U	7.43	7.43	7.42	7.42	7.42	7.42	7.42	7.42
P_I/P_0	1.48	1.16	1.13	1.11	1.41	1.24	1.22	1.21
P_R/P_0	1.32	1.07	1.07	1.07	1.33	1.18	1.17	1.17

Table 1 shows the shielding efficiency of the beam with different dimensions. It is observed that rectangular shields are more effective than the I-shaped ones under the same conditions. This can be attributed to the fact that I-shaped shields have more surface area than the rectangular ones and the larger surface area results in more power loss. Therefore, it is desirable to use rectangular shields for the shielding of steel beams.

When the beam is unshielded, its power loss is dependent on its dimensions. A larger a will result in less loss. When b increases from 0.72δ to 2.15δ, loss increases sharply but when b increases from 2.15δ, loss decreases gradually. In addition, if the shield is I-shaped, the total power loss will be greater than the unshielded cases. If the shield is rectangular, the total power loss will be almost constant since the loss in the steel is very small. Power ratio is also dependent on the dimensions. It increases with increasing a. From the design point of view, it is desirable to have a larger a since this will enhance the mechanical strength of the structure while reducing power loss. The necessity of shielding also depends on material properties as well as excitation frequencies. The general rule of thumb is that shielding is more likely to be effective with a larger dimension, a smaller shield skin depth and a larger steel skin depth.

4. CONCLUSIONS

This paper presents a study of the eddy current shielding of steel beams using the integro-differential finite element - Green's function method. The study shows that rectangular shields are more effective than I-shaped shields. The larger the a, the smaller the power loss without shields. Because shielding is not always effective, it is necessary to determine the shielding feasibility for a particular case. When designing a steel beam in electroheat environment, it is preferable to chose a larger a since this will reduce the power loss and at the same time increase its mechanical strength.

5. REFERENCES

1 J.D. Lavers, P.E. Burke and P.P. Biringer, *Estimation of losses induced in roof girders of Q.I.T. furnaces SF-3 and SF-9 using scale model tests,* Report, Elec. Eng. Consociates Ltd., P.O. Box 940, Station F, Toronto, M4Y 2N9, Canada.

2 K.V. Namjoshi and P.P. Biringer, "Efficiency of eddy current shielding of structural steel surrounding large currents: A circuit approach," *IEEE Trans. Magn.,* vol. 27, No. 6, pp. 5417-5419, 1991.

3 K.V. Namjoshi and P.P. Biringer, "Power efficiency of eddy current shielding of structural steel in transverse magnetic field," *IEEE Trans. Magn.,* vol. 28, No. 5, pp. 2829-2831, 1992.

4 Qiushi Chen, A. Konrad and P.P. Biringer, "Eddy current shielding for the reduction of induced stray power losses in structural steel," in *Proc. Canadian Conf. on Electrical and Computer Engineering,* Toronto, Canada, Sep. 1992, pp. TM10.11.1-TM10.11.4.

5 Qiushi Chen, A. Konrad and P.P. Biringer, "An Integrodifferential Finite Element - Green's Function Method for the Solution of Unbounded Eddy Current Problems," to appear in *IEEE Trans. Magn.,* Vol. 29, No. 2, 1993.

Simulation and Design of Applied Electromagnetic Systems
T. Honma (Editor)

Nonlinear eddy current analysis by the integral equation method

Kazuhisa Ishibashi

Department of Mechanical Engineering, Tokai University
2-28-4 Tomigaya, Shibuya-ku, Tokyo, 151 Japan

Abstract

An integral equation method for a nonlinear eddy current analysis is proposed. The nonlinear eddy currents are formulated by surface and volume integral equations which are derived from Maxwell's equations by using Green's theorem. As the periodic electromagnetic quantities in a ferromagnetic conductor vary nonsinusoidally with time, even if the magnetic field on the conductor surface varies with time sinusoidally, the quantities are expanded by Fourier series and their fundamental and harmonic components are determined by solving the corresponding integral equations with the help of the Newton-Raphson method.

1. INTRODUCTION

The boundary element method (BEM) has been widely used for solving open boundary eddy current problems. In the formulation of the problems, two sets of surface integral equations are derived from Maxwell's equations by making use of Green's theorem and utilizing the boundary conditions on the surface of the conductor [1]. The integral equations contain two unknowns (the surface electric and magnetic currents) and can be solved effectively by an iterative method [2].

In the anlysis of eddy currents induced in a ferromagnetic conductor, as the permeability of the conductor varies with magnetic field intensity, a nonlinear eddy current analysis is required. As BEM is based on the principle of superposition, it is difficult to solve nonlinear problems by BEM. However, it is well known that even BEM can be applied to the analysis of nonlinear problems by considering nonlinear terms to be sources [3]. With the help of this concept, a nonlinear eddy current analysis is proposed. The eddy currents are analyzed by solving the surface integral equations, which are similar to those used in the linear analysis [2], and a volume integral equation for obtaining the magnetic field inside the conductor. In this paper, a method for determining the magnetic field inside the conductor is examined when the surface magnetic field is given.

2. INTEGRAL REPRESENTATIONS OF ELECTROMAGNETIC FIELDS

The electromagnetic fields inside the conductor satisfy Maxwell's equation.

$$\nabla\times\mathbb{E}+\partial\mathbb{B}/\partial t=0 \quad (1), \qquad \nabla\times\mathbb{H}-\sigma\mathbb{E}=0 \quad (2)$$

$$\nabla\cdot\mathbb{B}=0 \quad (3), \qquad \nabla\cdot\mathbb{E}=0 \quad (4)$$

where $\mathbb{E}$, $\mathbb{H}$ and $\mathbb{B}$ are the electric field, magnetic field and magnetic flux density and σ is conductivity. If $\mathbb{B}$ varies nonlinearly with $\mathbb{H}$, the electromagnetic fields inside the conductor vary nonsinusoidally with time even if the magnetic field on the conductor surface varies with time sinusoidally.

When the electromagnetic fields are in the steady state, namely, the transient fields have become negligible, the periodic electromagnetic quantities $\mathbb{E}$, $\mathbb{H}$ and $\mathbb{B}$ can be expressed by the Fourier series as

$$\mathbb{E}=\sum_{k=-\infty}^{\infty} \mathbb{E}_k \exp(jk\omega t), \quad \mathbb{H}=\sum_{k=-\infty}^{\infty} \mathbb{H}_k \exp(jk\omega t), \quad \mathbb{B}=\sum_{k=-\infty}^{\infty} \mathbb{B}_k \exp(jk\omega t) \tag{5}$$

where $j=\sqrt{-1}$, ω is the fundamental angular frequency defined as $\omega=2\pi/T$ with the period T and the k-th harmonic components of the electromagnetic quantities $\mathbb{E}_k$, $\mathbb{H}_k$ and $\mathbb{B}_k$ are expressed as

$$\mathbb{E}_k=\frac{1}{T}\int_0^T \mathbb{E}\exp(-jk\omega t)dt, \quad \mathbb{H}_k=\frac{1}{T}\int_0^T \mathbb{H}\exp(-jk\omega t)dt, \quad \mathbb{B}_k=\frac{1}{T}\int_0^T \mathbb{B}\exp(-jk\omega t)dt \tag{6}$$

In the nonlinear analysis, Green's theorem can not usually be applied to obtain integral representations. Therefore, we introduce a temporary permeability μ_t and rewrite (1) and (3) as

$$\nabla \times \mathbb{E}+j\omega\mu_t\mathbb{H}=-j\omega(\mathbb{B}-\mu_t\mathbb{H}) \quad (7), \qquad \nabla\cdot\mu_t\mathbb{H}=-\nabla\cdot(\mathbb{B}-\mu_t\mathbb{H}) \quad (8)$$

The equations (7) and (8) are for the fundamental components and the subscript k is omitted. Equations for the k-th harmonic components are given by replacing ω with $k\omega$. We can regard the right sides of (7) and (8) as the magnetic current and charge sources defined as $\mathbb{K}_i=j\omega(\mathbb{B}-\mu_t\mathbb{H})$ and $\rho_m=-\nabla\cdot\mathbb{K}_i/(j\omega)$. We call them the imaginary magnetic current and imaginary magnetic charge, respectively. Then, we can get the integral representations of the fields at an observation point P_o as follows [1].

$$\frac{\mathbb{E}_p}{T}=\mathbb{E}_e-\int_S[-j\omega\mu_t\mathbb{J}_s G_i-\mathbb{K}_s\times\nabla G_i-\frac{\nabla_s\cdot\mathbb{J}_s}{\sigma}\nabla G_i]dS \tag{9}$$

$$\frac{\mathbb{H}_p}{T}=\mathbb{H}_e-\int_S[-\sigma\mathbb{K}_s G_i+\mathbb{J}_s\times\nabla G_i-\frac{\nabla_s\cdot\mathbb{K}_s}{j\omega\mu_t}\nabla G_i]dS \tag{10}$$

where the subscript p denotes the value at P_o, S is the surface area of the conductor, $\mathbb{J}_s$ and $\mathbb{K}_s$ are the surface electric and magnetic currents defined as $\mathbb{J}_s=\mathbb{n}\times\mathbb{H}_s$ and $\mathbb{K}_s=-\mathbb{n}\times\mathbb{E}_s$ with the magnetic field $\mathbb{H}_s$ and electric field $\mathbb{E}_s$ on the surface of the conductor and the unit normal n pointed to the outside, G_i is Green's function defined as $G_i=\exp(-k_i r)/(4\pi r)$ with the distance r from a point within the integral region to P_o and the wave number given as $k_i=\sqrt{j\omega\mu_t\sigma}$, T is given as T=1 inside the surface and T=2 on the smooth surface and $\mathbb{E}_e$ and $\mathbb{H}_e$ are the electromagnetic fields produced by $\mathbb{K}_i$ and ρ_m:

$$\mathbb{E}_e=\int_{V_i}-\mathbb{K}_i\times\nabla G_i dV_i \quad (11), \qquad \mathbb{H}_e=\int_{V_i}[-\sigma\mathbb{K}_i G_i+\frac{1}{\mu_t}\rho_m\nabla G_i]dV_i \quad (12)$$

where V_i is the volume of the conductor.

3. FORMULATION OF ELECTROMAGNETIC FIELDS BY THE INTEGRAL EQUATION METHOD

From the representation of the fields given in (9) and (10), integral equations for determining $\mathbb{K}_s$ and $\mathbb{H}$ are derived as follows [1].

Positioning P_o on the surface of the conductor,

$$\frac{\mathbb{K}_{sp}}{2} - \mathbb{n} \times \int_S -\mathbb{K}_s \times \nabla G_i \, dS - \mathbb{n} \times \mathbb{E}_e = \mathbb{n} \times \int_S [-j\omega \mu_t \mathbb{J}_s G_i - \frac{\nabla_s \cdot \mathbb{J}_s}{\sigma} \nabla G_i] dS \tag{13}$$

$$\mathbb{n} \times \int_S [-\sigma \mathbb{K}_s G_i - \frac{\nabla_s \cdot \mathbb{K}_s}{j\omega \mu_t} \nabla G_i] dS - \mathbb{n} \times \mathbb{H}_e = -\frac{\mathbb{J}_{sp}}{2} - \mathbb{n} \times \int_S \mathbb{J}_s \times \nabla G_i \, dS \tag{14}$$

Positioning P_o inside the conductor,

$$\frac{\nabla \times \mathbb{H}_p}{\sigma} - \int_{V_i} -\mathbb{K}_i \times \nabla G_i \, dV_i + \int_S -\mathbb{K}_s \times \nabla G_i \, dS = -\int_S [-j\omega \mu_t \mathbb{J}_s G_i - \frac{\nabla_s \cdot \mathbb{J}_s}{\sigma} \nabla G_i] dS \tag{15}$$

$$\mathbb{H}_p - \int_{V_i} [-\sigma \mathbb{K}_i G_i + \frac{1}{\mu_t} \rho_m \nabla G_i] dV_i + \int_S [-\sigma \mathbb{K}_s G_i - \frac{\nabla_s \cdot \mathbb{K}_s}{j\omega \mu_t} \nabla G_i] dS = -\int_S \mathbb{J}_s \times \nabla G_i \, dS \tag{16}$$

4. NONLINEAR EDDY CURRENT ANALYSIS OF A STEEL PLATE

The validity of the proposed method is examined by a comparison with the result of the finite difference method [4]. A numerical example is quoted from [4], where a uniform magnetic field of $H_s = 6809 \sin(2\pi \times 200t)$ [A/m] is applied in parallel to the surface of an infinitely wide steel plate of ANSI-1026. Conductivity σ is equal to 6.485×10^6 [S/m] and the B-H curve is represented by Frohlich's equation as $B = 100000 \mu_o H/(19.4 + 0.0734 |H|)$ [Wb/m^2].

Any nonlinear eddy current problems can be analyzed by solving at least two equations: one equation is chosen from (13) and/or (14) and the other from (15) and/or (16).

Considering a closed surface of the plate as shown in Fig. 1, dividing the thickness of the plate into 30 layers, applying (14) to the center of the upper and lower surfaces, applying (16) to the center of the divided layers and using the linear element, we get a set of nonlinear simultaneous equations. The simultaneous equations are solved by making use of Newton-Raphson method [5]. In this examination, the fundamental and third harmonic components of the fields are considered.

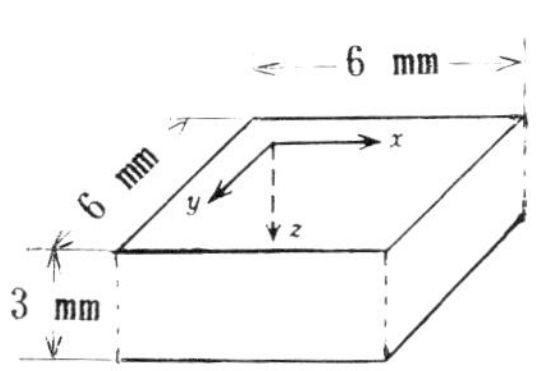

Fig. 1. A closed surface of the analysis region

Table 1. Number of times of iteration required for convergence

Temprary permeability μ_t [$\mu_0 \times 10^2$ H/m]	Surface magnetic field [$\times 10^2$ A/m]								
	1	2	4	6	10	20	40	60	100
1	5	∞	∞	∞	∞	∞	∞	∞	∞
2	5	5	7	∞	∞	∞	∞	∞	∞
4	4	5	6	6	7	9	9	8	8
6	4	5	5	6	5	6	8	7	6
10	4	4	4	5	6	5	6	6	6
20	4	4	4	4	5	5	6	6	7
40	3	3	4	5	5	6	7	7	8
60	4	4	4	5	4	6	7	7	8
100	4	4	5	5	6	7	8	8	9

∞ represents divergence.

In applying the Newton-Raphson method, the value of μ_t has to be chosen carefully to ensure quick convergence. The convergence is evaluated in terms of the number of iterations required for the relative error $\varepsilon = |K_1 - K_2| / |K_2|$ to become smaller than 0.001, where K_1 and K_2 represent the surface magnetic currents K_s calculated at n-th and (n+1)th iterations, respectively. The number of iterations required for deriving the converged value is given in Table 1. From the results in Table 1, we learn that the converged solution can usually be obtained by using the value of the initial permeability μ_i given at H=0 as that of μ_t.

Setting the value of μ_t equal to that of μ_i, calculated results are shown in Fig.2 where z denotes the distance from the surface of the conductor and The dotted and solid lines are those obtained by the proposed method (P.M.) and the finite difference method (F.D.M.) [4], respectively. This shows that the proposed method is sufficiently accurate from practical viewpoints.

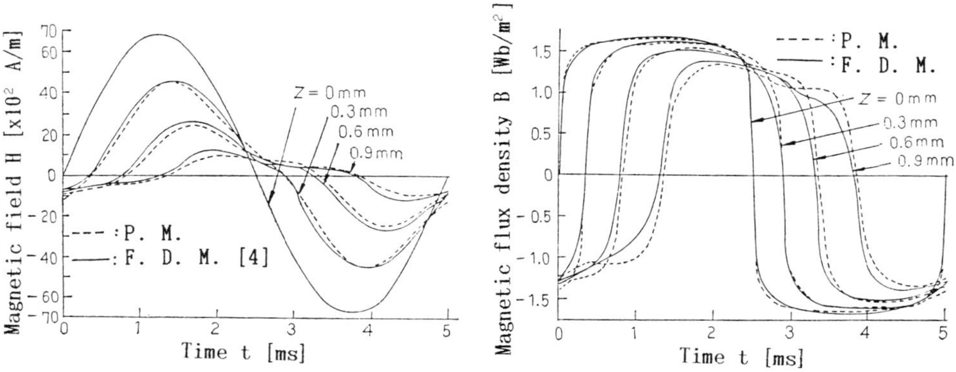

Fig. 2. Magnetic field and Magnetic flux density inside a steel plate

5. CONCLUSIONS

A method for analyzing eddy currents induced in a ferromagnetic conductor whose permeability varies with magnetic field magnitude are examined. In this method, introducing a temporary permeability μ_t, fixed constant value, and expanding the electromagnetic quantities by Fourier series, the fields are formulated by surface and volume integral equations whose unknowns are the surface magnetic and imaginary magnetic currents.

The magnetic field and the magnetic flux density inside a typical steel plate are analyzed with the help of the Newton-Raphson method. It is found that accurate solutions can usually be obtained by considering fundamental and third harmonic components and that converged solutions can usually be obtained using the value of the initial permeability μ_i as that of μ_t in the Newton-Rahpson method. If the value of μ_t is set equal to that of μ_i, this method is especially effective in solving eddy currents induced in a conductor which is thick compared with its skin depth because, at a sufficiently deep position where the permeability is almost equal to μ_t, the unknowns of the imaginary magnetic currents can be regarded as zero.

REFERENCES

1 R. Mittra (ed.), Computer Techniques for Electromagnetics, Pergamon Press, Oxford, 159 (1973)
2 K. Ishibashi, A Least Residual Approach for 3-D Eddy Current Analysis by BEM, IEEE Trans. Magn., Vol.29, No.2, March 1993 (to be published)
3 C. A. Brebbia and S. Walker, Boundary Element Techniques in Engineering, Butterworth & Co., London, (1980)
4 E. J. Woods, IEEE Trans. Power Apparatus Syst. PAS-100, 5, 2241 (1981)
5 T. Nakata and N. Takahashi, Finite Element Method in Electrical Engineering, Morikita Publishing, 201 (1982)

Simulation and Design of Applied Electromagnetic Systems
T. Honma (Editor)

Calculation of eddy currents in nonlinear media by a minimum order BEM formulation

A. Kost

Institute for Electrical Machines, Technical University of Berlin, Einsteinufer 11, D–W 1000 Berlin 10

Abstract

Whereas earlier methods to calculate nonlinear eddy currents by the boundary element method (BEM) need 6 unknowns per node on the nonlinear region's surface, in this paper a formulation is developed, which needs only 3 unknowns per node on the surface: 2 components of a virtual surface current density and 1 component of a virtual magnetic surface charge. As fully or widely populated matrices occur in BEM, the reduction of the number of unknowns is very important to handle complex 3–D problems.

1. INTRODUCTION

A method to calculate 3D nonlinear eddy currents by the boundary element method has been developed by the author in 1989 [1, 2]. It is based on a formulation for the vectors of the electric (E) and magnetic (H) field strength whereas the nonlinearity is focused in a part of the magnetic induction (B, nonl). The formulation is characterized by using 6 unknowns per node on the nonlinear region's surface and by belonging to the class of direct BEM formulations.

For linear eddy currents Mayergoyz [4] presented an indirect method, called "minimum order formulation", which needs only 4 unknowns per surface node, and Kost et al. [5] could improve the method by using only 3 unknowns per node.

The following chapters will show how the methods in [1], [2] and [5] can be combined to reduce the number of unknowns per node also for nonlinear eddy currents.

2. NONLINEAR FORMULATION WITH 6 UNKNOWNS PER NODE

The general problem is shown by Fig. 1. Examples for applications are ferromagnetic materials in electrical machines or in electromagnetic compatibility for screening.

In a paper of Kost [1] the following BEM–equation for 3–D nonlinear eddy currents, which is applied in Ω and on Γ, was formulated for H:

$$\frac{\Omega_i}{4\pi} \underline{\vec{H}}^{(K)}_{2n+1_i} = -\int_\Gamma [\kappa (\vec{n} \times \underline{\vec{E}}^{(K)}_{2n+1}) \Phi_{2n+1} + (\vec{n} \times \underline{\vec{H}}^{(K)}_{2n+1}) \times \text{grad}\, \Phi_{2n+1}$$

$$+ (\vec{n} \cdot \underline{\vec{H}}^{(K)}_{2n+1}) \,\text{grad}\, \Phi_{2n+1}] \, d\Gamma - \int_\Omega j(2n+1)\, \omega \kappa \underline{\vec{B}}^{(K-1)}_{\substack{nonl \\ 2n+1}} \Phi_{2n+1} \, d\Omega , \tag{1}$$

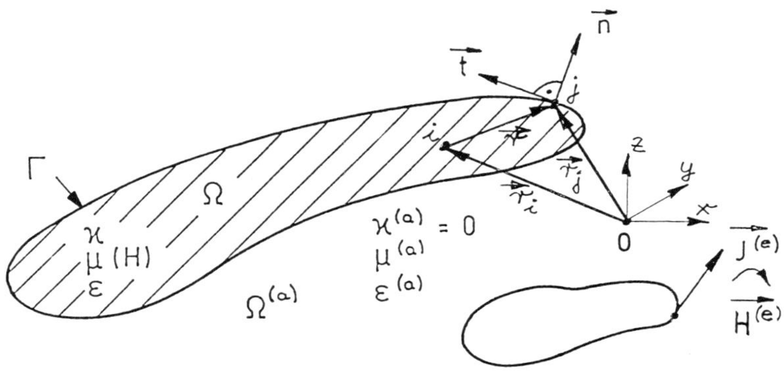

Fig. 1. Ferromagnetic material excited by a time periodic current density $\vec{J}^{(e)}$

with Ω_i: solid angle subtended by Γ at computation point i, weighting function $\Phi_{2n+1} = (4\ \pi\ r)^{-1} \exp\left(-\mathrm{j}\ \beta_{2n+1}\ r\right)$, $\beta^2_{2n+1} = -\,\mathrm{j}\ (2n+1)\ \omega\ p\ \kappa$, r: distance between source point and computation point, $\omega = 2\ \pi\ f$: angular frequency, κ: conductivity, p: permeability under a linear condition.

The electric and magnetic field strength $\vec{E}$ and $\vec{H}$, later needed virtual quantities $\vec{J}$ and σ, the scalar potential φ and the nonlinear part of the magnetic flux density $\vec{B}_{nonl}$ are described by

$$\vec{F}^{(K)} = \sum_{n=0}^{N} \mathrm{Re}\,\{\underline{\vec{F}}^{(K)}_{2n+1}\ (x, y, z)\ e^{j(2n+1)\omega t}\} \quad \text{with F replacing E, H, J, } \sigma \text{ or } \varphi, \tag{2}$$

$$\vec{B}^{(K-1)}_{nonl} = \sum_{n=0}^{N} \mathrm{Re}\,\{\underline{\vec{B}}^{(K-1)}_{nonl\,2n+1}\ (x, y, z)\ e^{j(2n+1)\omega t}\}\,, \tag{3}$$

in the step (K) of an iterative procedure resulting from the nonlinearity of the B(H)–curve. By this procedure which is explained in detail in [2] and reported in very abbreviated form in Fig. 2, first the 6 unknowns $\underline{\vec{E}}^{(K)}_{2n+1}$, $\underline{\vec{H}}^{(K)}_{2n+1}$ on Γ and then in Ω

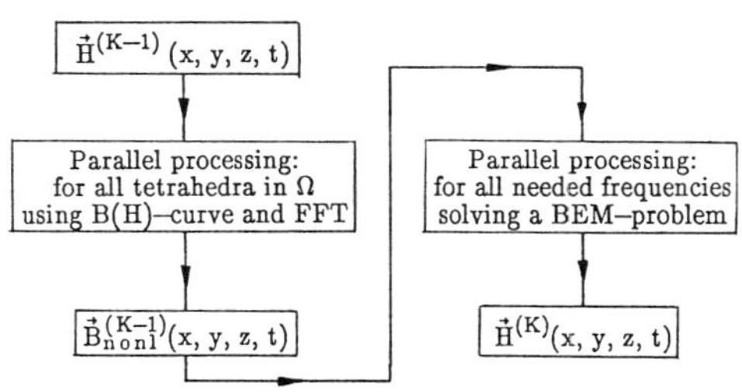

Fig. 2. Parallel computing of step (K)

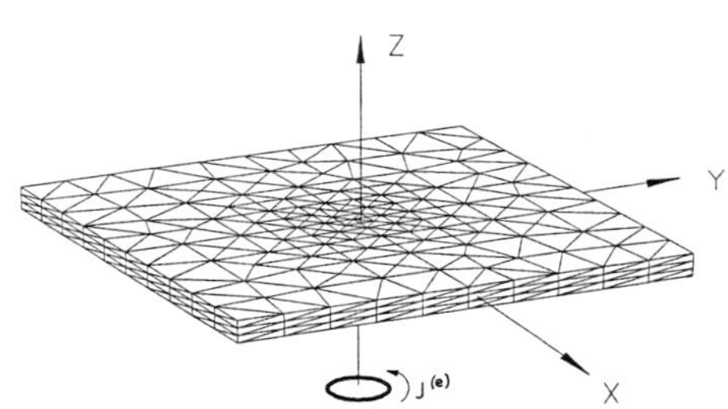

Fig. 3. Surface discretization of ferromagnetic screening plate

can be found. To compute them, the surface Γ is divided into triangular boundary elements (see Fig. 3) whereas the region Ω is divided into tetrahedra for evaluating the volume integral in eg. (1). The B(H)–curve is represented by

$$B = B_{lin} + B_{nonl} = pH + \alpha\,[B(H) - pH],\ 0 < \alpha \le 1\ , \tag{4}$$

and the influence and usefulness of the introduced parameters p, α is discussed and practically applied in [6].

3. NONLINEAR MINIMUM ORDER FORMULATION

The field describing differential equation in Ω together with eq. (4) is

$$\text{curl curl}\ \vec{H} = -\kappa\frac{\partial\vec{B}}{\partial t} = -\kappa\, p\frac{\partial\vec{H}}{\partial t} - \kappa\frac{\partial\vec{B}_{nonl}}{\partial t}\ . \tag{5}$$

With eqs. (2), (3) and the principle of the iterative solution process [1], namely handling the nonlinear parts as known (from the iterative step before) forcing functions, leads to

$$\text{curl curl}\ \underline{\vec{H}}^{(K)}_{2n+1} + j\,(2n+1)\,\omega\, p\,\kappa\,\underline{\vec{H}}^{(K)}_{2n+1} = -j\,(2n+1)\,\omega\,\kappa\,\underline{\vec{B}}^{(K-1)}_{nonl\,2n+1}\ . \tag{6}$$

It has been proved [3] for the linear case, that the solution $\vec{H}$ of the Helmholtz equation in Ω can be expressed by a virtual surface current sheet $\vec{J}$ on Γ, which consequently has to be valid also for the homogeneous solution of eq. (6):

$$\underline{\vec{H}}^{(K)}_{2n+1hom} = \int_\Gamma \text{curl}\,(\underline{\vec{J}}^{(K)}_{2n+1}\,\Phi_{2n+1})\,d\Gamma = -\int_\Gamma (\underline{\vec{J}}^{(K)}_{2n+1} \times \text{grad}\,\Phi_{2n+1})\,d\Gamma \tag{7}$$

This homogeneous solution now has to be identical with the surface integral of eq. (1), which occurs alone, when no nonlinear, inhomogeneous part exists. So eq. (1) with replacing the surface integral by eq. (7) gives the new formulation:

$$\underline{\vec{H}}^{(K)}_{2n+1} = -\int_\Gamma (\underline{\vec{J}}^{(K)}_{2n+1} \times \text{grad}\,\Phi_{2n+1})\,d\Gamma - \int_\Omega j(2n+1)\,\omega\,\kappa\,\underline{\vec{B}}^{(K-1)}_{nonl\,2n+1}\,\Phi_{2n+1}\,d\Omega\ , \tag{8}$$

$$\underline{\varphi}^{(K)}_{2n+1} = \frac{1}{4\pi}\int_\Gamma \frac{\underline{\sigma}^{(K)}_{2n+1}}{r}\,d\Gamma,\ \text{satisfying}\ \Delta\varphi = 0\ \text{outside}\ \Omega. \tag{9}$$

From the boundary conditions analogous to [4] and [5]

$$\vec{n} \times (\vec{H} + \text{grad}\,\varphi) = \vec{n} \times \vec{H}^{(e)},\ \vec{n}\cdot(\mu\vec{H} + \mu^{(a)}\,\text{grad}\,\varphi) = \mu^{(a)}\,\vec{n}\cdot\vec{H}^{(e)}\ , \tag{10}$$

the boundary integral equations in terms of the virtual quantities $\vec{J}$ and σ can be obtained:

$$\frac{1}{2}\vec{\underline{J}}^{(K)}_{2n+1}(\xi) + \int_{\Gamma} \vec{n}_{\xi} \times \left[\vec{\underline{J}}^{(K)}_{2n+1}(\eta) \times \text{grad}\, \Phi_{2n+1} \right] d\Gamma_{\eta}$$

$$- \int_{\Gamma} \underline{\sigma}_{2n+1}(\eta) \left[\vec{n}_{\xi} \times \text{grad}\, \Phi^{(a)} \right] d\Gamma_{\eta} + \int_{\Omega} \vec{n}_{\xi} \times j(2n+1)\, \omega\, \kappa\, \vec{B}^{(K-1)}_{\substack{\text{nonl}\\2n+1}}(\eta)\, \Phi_{2n+1}\, d\Omega_{\eta}$$

$$= - n_{\xi} \times \vec{\underline{H}}^{(e)}_{2n+1}(\xi)\,, \tag{11}$$

$$\frac{1}{2}\underline{\sigma}^{(K)}_{2n+1}(\xi) + \int_{\Gamma} \underline{\sigma}^{(K)}_{2n+1}(\eta) \left[\vec{n}_{\xi} \cdot \text{grad}\, \Phi^{(a)} \right] d\Gamma_{\eta}$$

$$+ \int_{\Gamma} M\, \vec{n}_{\xi} \cdot \left[\vec{\underline{J}}^{(K)}_{2n+1}(\eta) \times \text{grad}\, \Phi_{2n+1} \right] d\Gamma_{\eta} + \int_{\Omega} M\, \vec{n}_{\xi} \cdot j(2n+1)\, \omega\, \kappa\, \vec{B}^{(K-1)}_{\substack{\text{nonl}\\2n+1}}(\eta)\, \Phi_{2n+1}\, d\Omega_{\eta}$$

$$= - \vec{n}_{\xi} \cdot \vec{\underline{H}}^{(e)}_{2n+1}(\xi) \quad \text{with } M = \mu^{(K-1)}(\xi)/\mu^{(a)}. \tag{12}$$

In eqs. (11), (12) η and ξ denote the source point and computation point respectively. Besides the exciting field also the volume integrals are known from the step (K–1) before. So, by discretizing eqs. (11) and (12) into boundary element equations and following the method of [5] only 3 unknowns per node remain: 1 component σ per definitionem and 2 tangential components of J.

Having calculated H by eq. (8), afterwards in this step K the eddy current density $J = \kappa E$ can be found by solving an integral equation for E in [1], similar to eq. (1), which also contains 3 unknowns per node, namely the vector components of E.

5. REFERENCES

1 A. Kost: "Calculation of 3D Nonlinear Eddy Currents in an Electrical Machine by the Boundary Element Method", Proc. IV. Internat. Conf. on Boundary Element Technology, 1989, pp. 227–238

2 A. Kost, J. Shen: "Parallel computation of 3–D nonlinear eddy currents by the boundary element method (BEM) and the fast Fourier transform (FFT)", COMPEL, Vol. 9, Suppl. A, 1990, pp. 181–184

3 S. Kalaichelvan: "A boundary integral equation method for 3–D eddy current problems", Ph. D. Thesis, Univ. of Toronto, 1987

4 I. D. Mayergoyz: "3D eddy current problems and the boundary integral equation method", IMACS – Computational Electromagnetics, 1986, pp. 163–171

5 A. Kost, J. Yuan: "Berechnung dreidimensionaler Wirbelströme mit der Boundary Element Methode auf der Basis einer H, φ–Formulierung", Internat. Wiss. Koll. TH Ilmenau, 1992

6 A. Kost, J. Shen: "Influence of magnetization curve representation on parallel BEM computation of nonlinear eddy currents", IEEE Transactions on Magnetics, Vol. 27, No 5, 1991, pp. 3898–3901

Simulation and Design of Applied Electromagnetic Systems
T. Honma (Editor)

Improvement of the boundary element solution on the corner problem in compound materials

T. Takeda[a] and T. Kuwahara[b]

[a]Department of Electrical Engineering, Faculty of Engineering, Hokkaido University, Kitaku Kita-13 Nish-8, Sapporo, 060, Japan.

[b]Department of Electrical Engineering, Hakodate Technical Collage, Tokuracho 14-1, Hakodate, 042, Japan.

Abstract
Electrostatic potential problem which has the corner of dielectric materials is analyzed by boundary element method. The initial regular boundary mesh was modified to suit with the potential distribution near the corner. A method which generates a new mesh is proposed and the good result on potential was obtained using the modified mesh.

1. INTRODUCTION

This paper reports an investigation on the technique that improves the accuracy of BEM solutions without increasing new nodes or new elements from an initial mesh. The new boundary mesh which is generated by moving the position of nodes has a possibility to produce more accurate solutions than those by the the initial meshes. A simple method to move the mesh points is tested for the corner problem of electrostatic potential which has constituents of different medium. The error evaluation of BEM solution is building up in the present state[1-4]. Namely, some error estimators are proposed and used for introducing a position indicator to subdivide the mesh. However, the procedure comes to increase not a little number of boundary elements in the automatic mesh subdividing. Then, the analysis by the unchanged number of variables is preferable to save the computing cost if it is effective to get a more accurate solution.

2. MESH MODIFYING METHOD BY MESH POINT MOVING

The following method for mesh point (node) moving is based on the postulate that the direction of the movement and the length are roughly known. The movement is constrained on the boundary of the problem region. When a mesh point is moved, the key role of the method is not to make such irregular meshes having much distorted shape and never to make the mesh overlapped with each other.

2.1. Sub-region for point movement

The movement of the mesh point under consideration is limited in some (convex part of) sub-region of the boundary. The sub-region is naturally constituted by boundary elements which has the mesh point P in common (Figure 1). Then, no irregular mesh is made as far as the mesh point moves in the region. A new mesh point can be designated along movement vector M which is the projection of some attraction vector T onto the sub-region. A position of the new point is determined from the movable range multiplied by some weighting value. If the shape of sub-region is concave, the convex part of the region has to be the moving area as shown in Figure 2 because a new point out of the convex part should make elements overlapped with each other. Then, a judgment whether the point is inner or outer the convex part will be necessary in this situation. This process is applied to all node until the objective mesh is obtained.

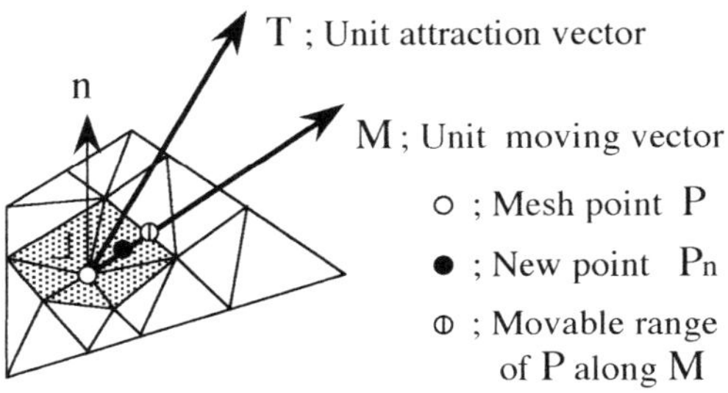

Figure 1. Mesh point moving in a sub-region of mesh

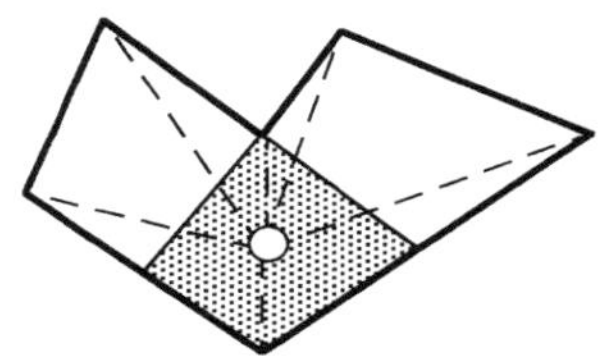

Figure 2. *Convex* part of *concave* sub-region

2.2. Attraction vector and moving extent

Strategy to modify the original mesh pattern into new one is necessary. Some error estimator on a trial solution will produce the indicator for mesh point moving or boundary element dividing in adaptive mesh generations. However, the procedure where the user in mesh generation for a problem can easily and arbitrarily designate the pattern of mesh is also important. The attraction vector of movement and moving extent of points here should be determined by the combination of some simple patterns as shown in Figure 3.

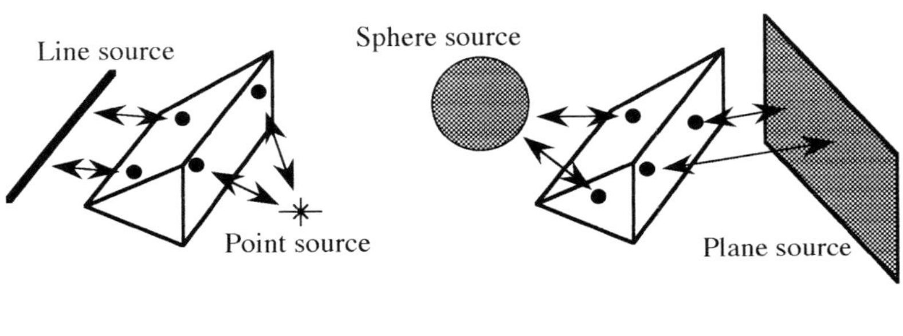

Figure 3. Attractive sources for mesh point moving

The position vector P_n to the new mesh point from the old one P, can be calculated as follows;

$$P_n = \{ (M_i \bullet T_i)\, f_i \, \mathrm{sign}|a_i|\, A \} M_i$$

$$\text{where } i = j, \text{ such that } \max_j [(M_j \bullet T_j) f_j]$$

$$f(|R_i|, a_i) = \frac{|a_i|}{|R_i| + |a_i|} \ldots \text{for example.}$$

$$0 \leq f_i \leq 1 \quad \text{for} \quad \infty > |R_i| \geq 0$$

where R_j, f_j and a_j denote the position vector of point P from j-th attractive or repulsive source, j-th weighting function which shows the territory of the source and the parameter for the extent, respectively. The position of a moving point should belong to the territory of sources. The moving direction is determined by the source which has the maximum of the weighting function multiplied by a factor. The factor denotes the directional cosine of attraction vector onto the surface at that point. The relaxation factor A and a_j can arbitrarily be specified by the user. All the mesh points of the boundary are repeatedly moved, except the vertexes which represent the geometrical feature of the problem.

3. FIELD CALCULATION ON THE CORNER OF DIELECTRICS BY A MODIFIED MESH

Figure 4 shows the geometrical configuration of an electrostatic problem where three dielectric materials come into contact on an edge. To solve the field near the edge is known as a difficult problem. Figure 5 and 6 represent the equally spaced (be called as regular, here after) meshes for the initial analysis and the modified mesh which is made by putting a line attractive source on the edge, respectively. The total meshes have 1174 mesh points and 270 boundary elements of third order interpolation.

The relative potential error of the solution by the modified mesh was 0.201% in average and it was less than one third of that of the regular mesh (0.706%). Then, the modified mesh is very effective in the potential analysis. However, the relative error of flux value (the normal component of potential gradient on the boundary) became slightly larger (5.98%) than that of the regular one (5.25%).

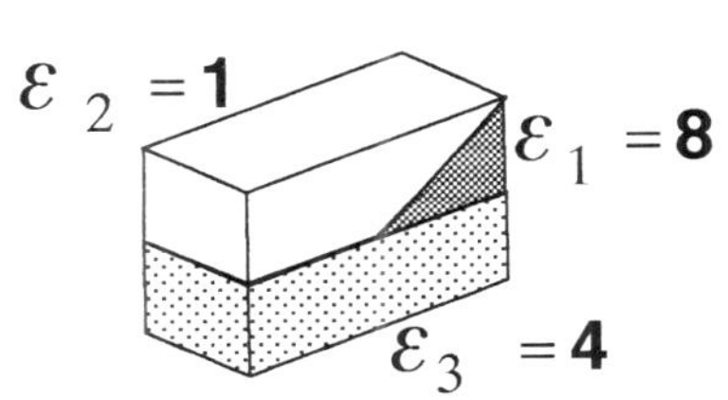

Figure 4. Sample problem; Electrostatic potential in dielectric materials

The flux on the upper side of one of the dielectric material (e_1= 8) is shown in in Figure 7 ; (left; by the regular mesh), (middle; by the modified one) and (right; the exact).

As seen in the figures, a non-physical 'waving' of flux can be found in the numerical solution. The 'waving' was not improved even by the modified mesh; besides, the height became larger as if the 'wave' had been compressed toward the edge of corner. The 'waving' phenomena which caused by the too small number of elements will be removed by adding some tiny new elements near the corner. However, the solution by the modified mesh here is the best one in the analysis where the number of mesh points is unchanged.

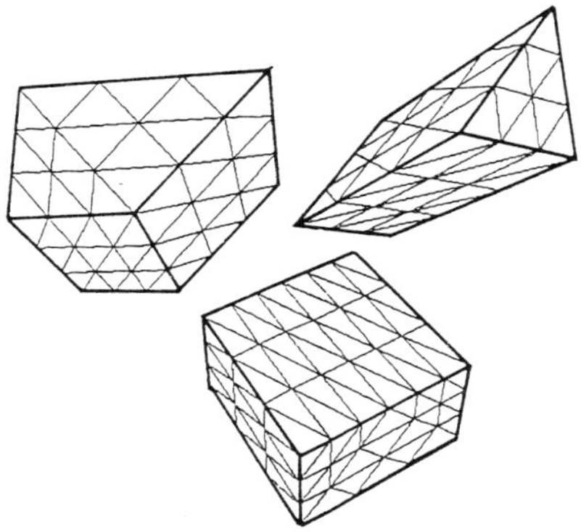

Figure 5. The initial and equally spaced mesh

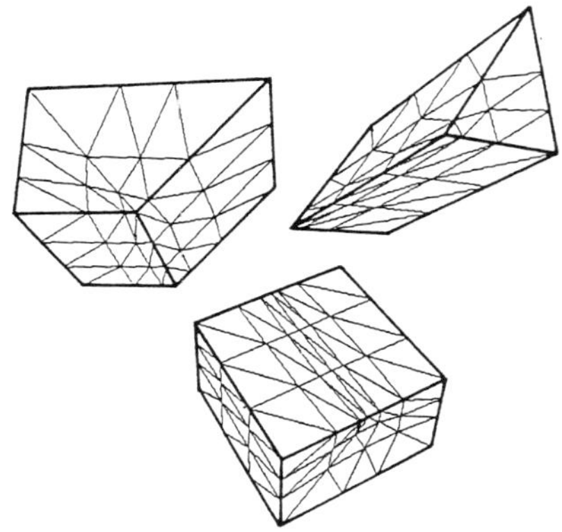

Figure 6. Modified mesh by moving points

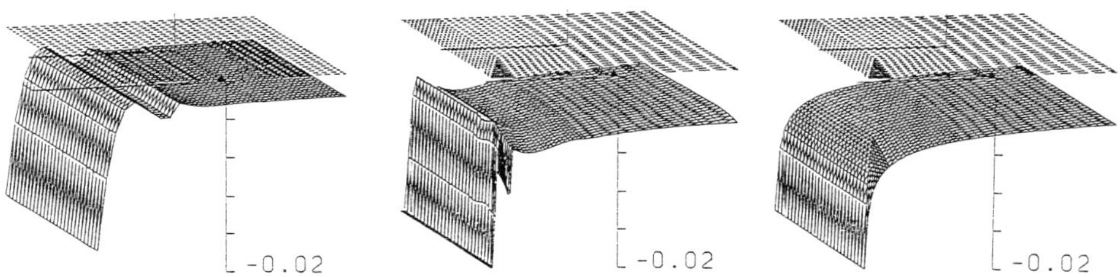

Figure 7. Flux distribution: (left) ; by regular mesh, (middle); by modified, (right); the exact

4. CONCLUSION

This paper described an analysis of the corner problem with dielectrics. In the numerical model, the mesh points are moved without increasing new elements by an adaptive manner to the problem. The example analysis on the potential evaluation showed that the modified mesh is more effective than that of equally spaced (regular) mesh.

5. REFERENCES

1 T. Takeda, T. Kuwahara, T. Miyata and M. Igarashi, IEEE tr. on Mag. Vol.26, No. 2(1990) 719-722.
2 K. Abe, Abstracts, Symp. of the Int. Asso. for Boundary Element Methods (IABEM-91), (1991)1-4.
3 M.S. Ingbar and A.K. Mitra, Eng. Analysis with Boundary Elements, No. 9(1992) 13-19.
4 K. Kawaguchi and N. Kamiya, ibid., 255-262.

Simulation and Design of Applied Electromagnetic Systems
T. Honma (Editor)

Attractive force analysis of flat type DC electromagnets by 3-D finite element method with edge elements

Y.Kawase[a], O.Miyatani[a] and S.Ito[b]

[a]Department of Electronics and Computer Engineering, Gifu University, Yanagido 1-1, Gifu, 501-11, Japan

[b]Tsuyama National College of Technology, Numa 624-1, Tsuyama, 708, Japan

Abstract

The Flat type dc electromagnets are widely used for saving space and high-speed operation, and were designed by empirical procedure using cut and try approach. In the optimum design of the electromagnets, however, it is necessary to obtain the attractive force accurately. In this paper, therefore, the newly developed 3-D finite element method with edge elements[1] is applied to the attractive force analysis of the electromagnets. The validity of the new method is confirmed by comparing with the measurements.

1. INTRODUCTION

It is difficult to analyze the attractive force characteristics of a flat type dc electromagnet accurately, because the structure is so flat and complex. Until recently, such electromagnets were designed by solving assumed magnetic circuits[2]. In this design method, it was impossible to find the optimum design, because the attractive force characteristics were obtained experimentally with several trial electromagnets. As the result, fabricating prototype models for testing and evaluation becomes very expensive, and the design requires long lead time.

In this paper, the developed 3-D finite element method with edge elements[1] has been applied to the analysis of the static attractive force characteristics of a flat type dc electromagnet. In addition, it is clarified that the residual stress caused by the cold working on iron cores in manufacturing has a large influence on the attractive force characteristics. And then, the usefulness of our method is confirmed by comparing the calculation and the experiment.

2. METHOD OF ANALYSIS

The fundamental equation of 3-D finite element analysis[3] is

$$rot(\nu\, rot\mathbf{A}) = \mathbf{J}_0 \tag{1}$$

where ν is the reluctivity, $\mathbf{A}$ is the magnetic vector potential and $\mathbf{J}_0$ is the current density.

The electromagnetic field is calculated by using edge elements in which the tangential components of the field across element boundaries are continuous. The 3-D region is divided into tetrahedral elements. In an edge element, the tangential projection of the vector potential the along each edge is interpolated. In case of tetrahedral edge elements, the vector potential $A^{(e)}$ of element e is denoted as follows:

$$\mathbf{A}^{(e)} = \sum_{ie=1}^{6} \mathbf{N}_{ie} A_{ie} \tag{2}$$

where $\mathbf{N}_{ie}$ is the vector interpolation function for the ie-th edge, and can be written as follows[4]:

$$\mathbf{N}_{ie} = V_{je}\, \mathit{grad} V_{ke} - V_{ke}\, \mathit{grad} V_{je} \tag{3}$$

where V_{je} is the barycentric coordinate of point (x,y,z) in the tetrahedron with respect to node je.

Galerkin's method was applied to obtain the system equations. The following equation is derived from equations (1),(2) and (3)

$$G_i = \iiint \mathit{rot}\mathbf{N}_i \cdot (\nu\, \mathit{rot}\mathbf{A})dv - \iiint \mathit{rot}\mathbf{N}_i \cdot (\mathbf{J}_0)dv = 0 \tag{4}$$

The attractive force is obtained by using the Maxwell stress tensor[5].

3. ANALYZED MODEL

Figure 1 shows the flat type dc electromagnet. Because of the non-symmetrical shape of the model, the whole region of the model including air is discretized using tetrahedral edge elements. Figure 2 shows the 3-D meshes of the cores and coil when the armature stroke G is 0.8mm. In order to obtain the steady state attractive force, the flux distributions are calculated taking into account the nonlinearity of the cores. Figure 3 shows the B-H curves used by the calculation. The curve α represents the B-H curve after annealing of a ring specimen, and the curve β is assumed to the actual B-H curve with residual stress in the cores. Furthermore, the flux distributions are obtained for the variable armature strokes. The discretization data and CPU time are shown in Table 1.

Table 1

Discretization data and CPU time (G=0.8mm, 100% of exciting current).

number of elements	91086
number of nodes	16800
number of edges	111003
number of unknown variables	101649
number of non-zero entries	857627
number of nonlinear iterations	6
CPU time (hours)	6.5

computer used : Sun Microsystems SPARC station 2

The nonlinear iteration is finished when the changes of the absolute values of flux densities in all elements become less than 50 G. The ICCG method is applied to solve the matrix. The convergence criterion for the ICCG method is 10^{-8}.

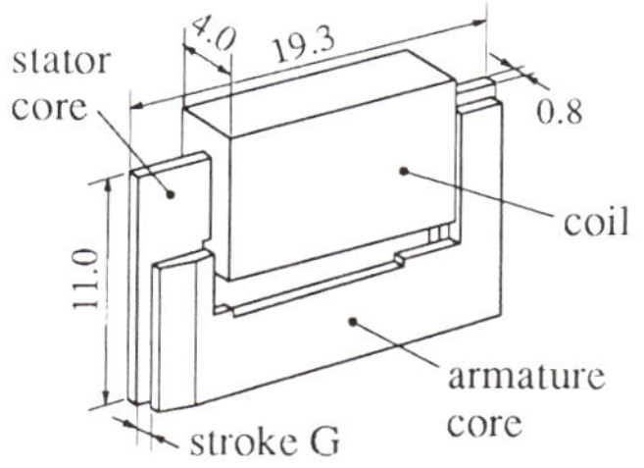

Figure 1. Flat type dc electromagnet.

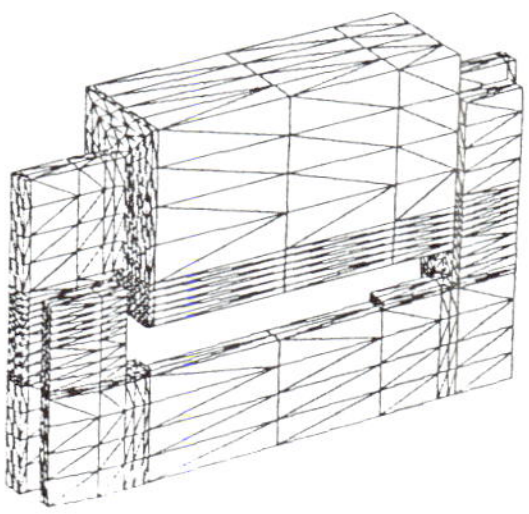

Figure 2. 3-D meshes of cores and coil. (G = 0.8mm)

4. CALCULATED AND MEASURED RESULTS

An example of the flux density vectors is shown in Fig.4. Figure 5 shows the relationships between the attractive force and stroke G for 50% and 100% of the exciting current. The measured attractive forces are also shown in the Figures.

When the exciting current is 50%, the linear analysis fairly agree well with the experiments. However, when the exciting current is 100%, the results of linear analysis are much higher than experiments due to the magnetic saturation of the cores. It shows that the nonlinearity of the core has a strong influence on the attractive force.

The attractive forces obtained by nonlinear analysis using B-H curve β agree well with the experiments at any exciting current. As shown in Fig.5, the attractive forces using B-H curve α show the some discrepancy between the calculation and the measurement, especially at the 100% of exciting current. Therefore, the B-H curve β represents the actual B-H curve of the cores.

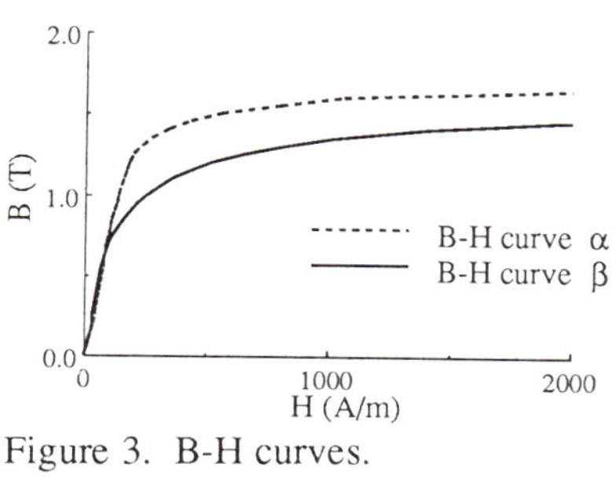

Figure 3. B-H curves.

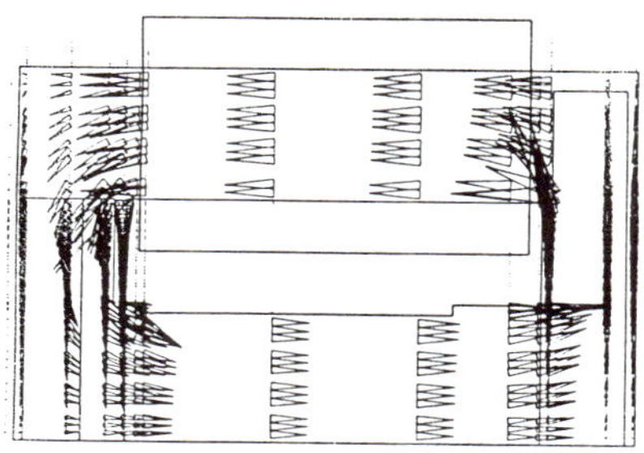

Figure 4. Distributions of flux density vectors. (G=0.8mm, 100%, B-H curve β)

The magnetization curve is reduced by cold working, that is, stamping or punching of cores while manufacturing. Therefore, the configuration of the magnetic circuit as well as the actual magnetization curve of the magnetic parts is very important for the design of flat type electromagnets with high performance.

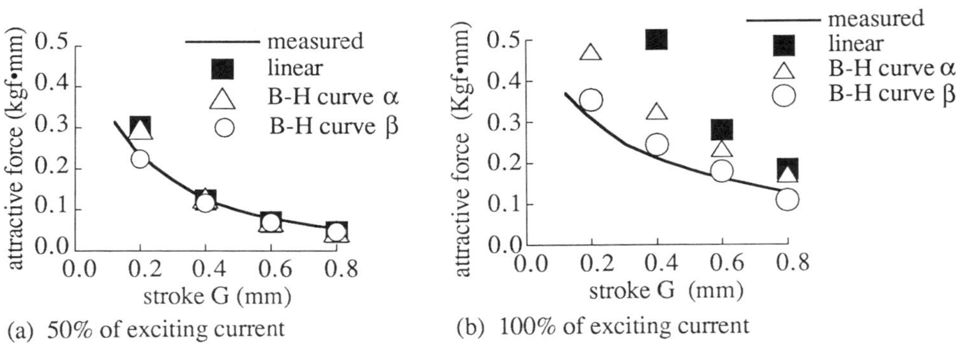

Figure 5. Attractive force characteristics.

5. CONCLUSIONS

It is concluded that the attractive force characteristics of the flat type dc electromagnet having complex shape can be obtained exactly by the finite element method with edge elements. Our method based on edge elements works well because only the tangential component of the vector potential at each element is continuous. The magnetization curve taking into account reduction occurred by the stamping, punching and so on should be considered for the calculation of the magnetic field analysis of the electromagnet, because the nonlinearity of the core has a strong influence on the attractive force characteristics.

Our new method can be applied by the small scaled CAD system composed of workstations. It will be used widely for designing any electromagnet having 3-D shape instead of the conventional trial and error approach. And, it reduces prototype developments and evaluation efforts for the design of electromagnet.

The authors would like to thank Mr. M.Ishikawa et al. at Fuji Electric Co, Ltd. for the measurements of the electromagnet.

6. REFERENCES

1 Y.Kawase, H.Kikuchi and Z.J.Cendes:"3-D finite element analysis of electromagnetic forces using edge elements",IEEE Conference on Electromagnetic Field Computation, WOB5 (1992).
2 H.C.Roters:"Electromagnetic devices",John Willey and Sons Inc., (1941).
3 T.Nakata, N.Takahashi and K.Fujiwara: "Investigation of a model to verify softwares for 3-D non-linear eddy current analysis",IEEE Trans. on Magnetics, Vol.26,No.2 (1990) 501.
4 A.Bossavit:"Whitney forms:a class of finite elements for three-dimensional computations in electromagnetism", IEE Proceedings, Vol.135, Pt.A, No.8 (1988) 493.
5 Y.Kawase, H.Kikuchi and S.Ito:"3-D non-linear analysis of dynamic behavior of the clapper type DC electromagnet",IEEE Trans. on Magnetics, Vol.27,No.5 (1991) 4238.

Simulation and Design of Applied Electromagnetic Systems
T. Honma (Editor)

Reduction of Mechanical Noise and Vibration for Permanent Magnet Electric Rotating Machines

K. J. Joo [a], H. S. Choi [a] and S. Y. Hahn [b]

[a] CAD/CAM Center, Samsung Electro-Mechanics Co., 314 Maetan-dong, Suwon, Korea

[b] Dept. of Electrical Engineering, Seoul National University, Shinrim-dong, Seoul, Korea

Abstract

Recently, a lot of efforts have been made to meet the noise and vibration performance requirements of the rotating electric precision machines by the use of various numerical and experimental techniques. This paper looks into the influence of the magnetization distribution of permanent magnet and the shape of airgap which play the role of primary variables of the electro-magnetic circuits, and finally, an experimental verification has been performed to illustrate the efficiency and the advantages of the proposed technique which lowers the level of mechanical noise and vibration via minimizing the cogging torque.

1. INTRODUCTION

The characteristics of the noise and vibration for the rotating electric machines are determined by the coupling effects of the electromagnetic and mechanical sources since the electromagnetic energy and the mechanical energy are complements in the system's energy. Hence, the problems of the mechanical noise and vibration require mechanical and electric treatments. The former deals with the mass unbalance, the misalignment, the rubs, the mechanical looseness, the cracks, the bearing and its preload, the mechanical properties, etc., while the latter does the materials of magnet, the shape and number of the slots, i.e. the shape of airgap, the magnetization distribution, the windings, etc.[1,2].

The cogging torque induced by the rate of change in the magnetic energy, which is a major portion of the electromagnetic energy, is the torque noise components which cannot be easily controlled by enternal units. Recently, efforts to solve the noise and vibration problems by reducing the magnetically induced cogging torque have been made along with researches on the magnetization, the shape optimization, etc.[3,4]. The cogging torque can be expressed as

$$T_c = \sum_{n=m}^{\infty} T_n (q_1, q_2, q_3, \dots) \cos(n\theta)$$

$$\text{where } m = m(q_1, q_2, q_3, \dots) \tag{1}$$

In the above expression, $q_1, q_2, q_3, \cdots$ indicate the parameters affecting the magnetic energy such as the magnetization distribution, the number of poles in a permanent magnet, the shape of airgap(e.g. no. of slots), etc. Minimizing the cogging torque implies decreasing T_n and increasing 'm', and the reduction of noise and vibration can be effectively achieved by an optimal determination of q_i's .

In this paper, significance of the magnetization distribution and the shape is assessed by investigating the cogging torque versus magnetization distribution and the radial force distribution between the stator and the rotor. The shape of the airgap is optimized with respect to the cogging torque by an application of the finite element sensitivity analysis. Finally, the proposed technique is applied to a small precision spindle motor with permanent magnet and the efficiency of the technique is illustrated by comparing the level of mechanical noise and vibration.

2. BASIC STATEMENTS

The stator and rotor are subjected to an travelling forces interacting between on the stator and the rotor surfaces. The structural behavior of the stator is solely governed by the interacting forces and can be expressed, in frequency domain, as

$$\mathbf{u}(\omega_k) = \mathbf{H}(\omega_k)\,\mathbf{f}(\omega_k) \qquad , k = 1,2, \ldots \tag{2}$$

where

$$\mathbf{H}(\omega_k) = 1/(\mathbf{K}_s + i\,\omega_k\,\mathbf{C}_s - \omega_k^2\,\mathbf{M}_s) \tag{3}$$

In the above expression, ω_k is the k-th frequency of the interacting forcing function, $\mathbf{H}(\omega_k)$ the frequency response function, $\mathbf{K}_s$ the stiffness matrix, $\mathbf{C}_s$ the damping matrix, $\mathbf{M}_s$ the mass matrix, $\mathbf{f}(\omega_k)$ the interacting force vector. The size of the matrix and vectors are determined by the degrees of freedom of the stator.

The torque induced by the electromagnetic sources can be expressed as

$$T(\theta) = -\frac{d}{d\theta}\left(\int_{vol}\left[\int_{0}^{B} \mathrm{H}\cdot d\mathrm{B}\right] dVol - E(\theta)\right) \tag{4}$$

where θ implies the rotor's relative rotation to the stator and $E(\theta)$.denotes the input electrical energy. The torque estimated by equation (4) characterizes the angular velocity of the rotor. Since the interacting forces and the torque are determined by the system's electromagnetic circuit, the frequency contents of those forcing functions are fully correlated. Therefore, the level of noise and vibration is directly influenced by the torque noise components especially on the frequency bands of the cogging torque function.

3. ASSESMENTS OF MAGNETIZATION DISTRIBUTION AND AIRGAP

As the megnetization magnitude(i.e.magnetization distribution) in the permanent magnets varies, the magnitude of the resulting cogging torque changes at a rate different from the total torque of the system. For a small permanent magnet motor as illustrated in Figure 1, assuming that the permanent magnet is magnetized in a trapezoidal form as shown in Figure 2, it is observed that the magnetization magnitude decreases as the value of α increases from 0 to 0.5. As shown in Figure 3, the cogging torque is reduced until α reaches 0.25 and then it grows up to have a peak around α = 0.3, while the torque costant, i.e. the total torque decreases monotonically. It can be understood that the wave form of the cogging torque gradually changes to take on a wave form different in phase of 180 degree around α = 0.25 with the minimum amplitude. The total torque was reduced by 23(%) over α = 0 ~ 0.4, while the cogging torque yield rapidly to be on the decrease of 10 times.

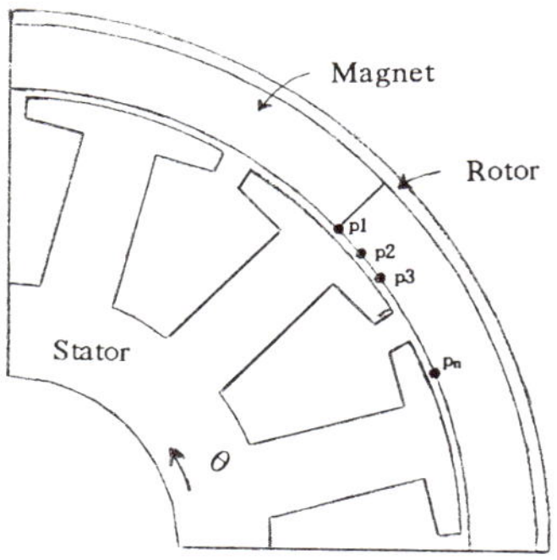

Figure 1. 1/4-part of cross section of a DC brushless motor

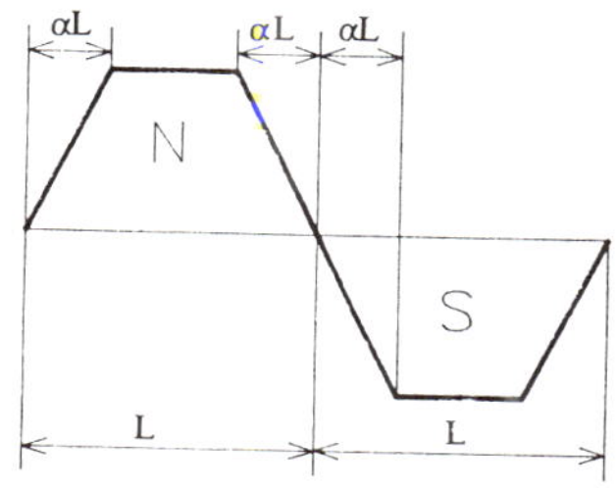

Figure 2. A distribution of magnetization

Radial traction forces acting on the surfaces of stator and rotor are considered as another crucial factor affecting the level of noise and vibration. The interacting forces excites the system's structural response at certain frequencies. When the rotor rotates, the force variation at a specific position of the rotor can be regarded as a travelling wave. The wave can be estimated by superposition of the harmonic force components at each integer multiple of frequencies which were resulted from the rotor's speed, magnetization distribution(or number of poles), shape of airgap(or number of slots) . The force distribution in radial direction acting over the points p_1, p_2, p_3, $\cdots$ in Figure 1, as the rotor rotates from 0 to 20 degree, was depicted in Figure 5. It is observed that the amplitude and the variation of the force distribution is the more reduced and the more uniform, respectively, for the optimized model of Figure 4.

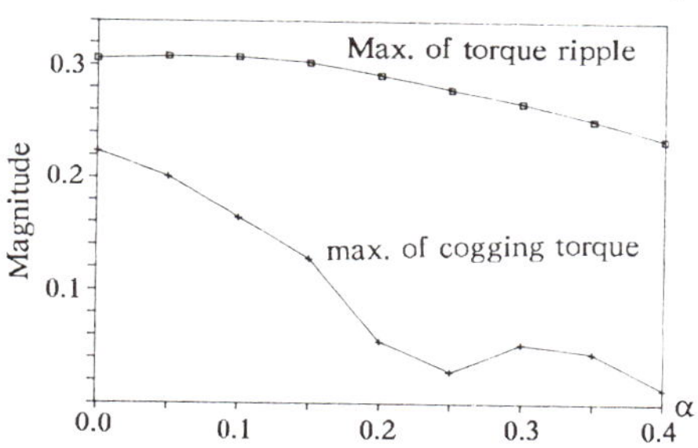

Figure 3. Torque constant and cogging torque vs. α value

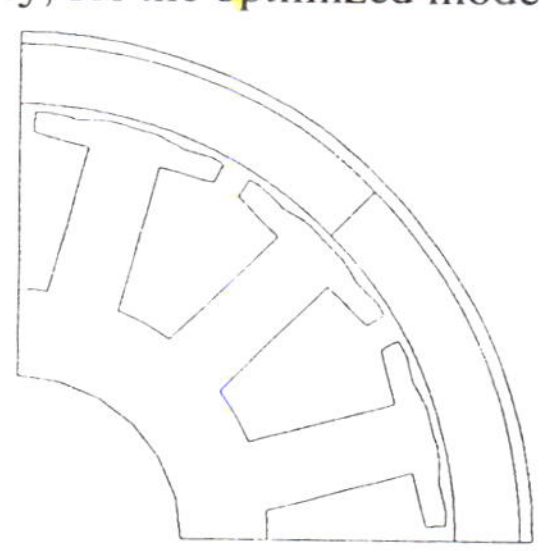

Figure 4 Optimized model for α=0.2 with respect to cogging torque

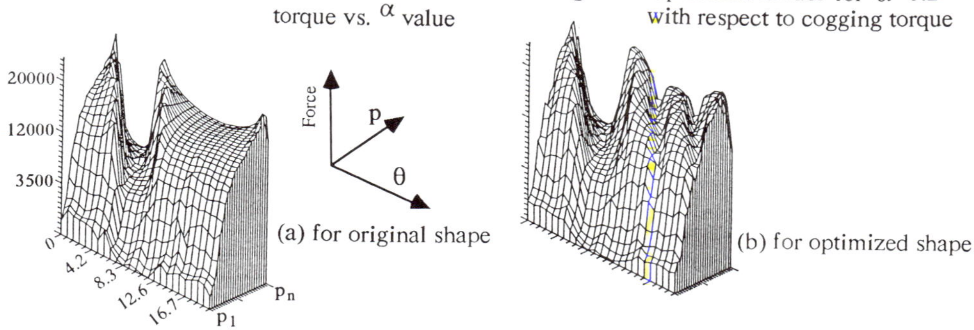

Figure 5. Distribution of interacting forces between rotor and stator for α=0.2

4. APPLICATION

As an illustration, a small precision spindle motor was optimized with respect to the cogging torque and the optimized sample was compared with the original.

The vibration was measured in circumferential direction at the outmost point of the motor unit running with 300 rpm and the noise was sensed at a distance of 30 cm away above the motor unit. The measured results are plotted in Figures 6 and 7. It can be observed that the levels of vibrations at cogging frequencies such as 360 Hz and 720 Hz are reduced by 6.7 times and 1.5 times , respectively, and the optimized model has the superior quality over the whole bands of frequencies. In the aspect of total noise the optimized model has a lower level than the original model by 6 dB .

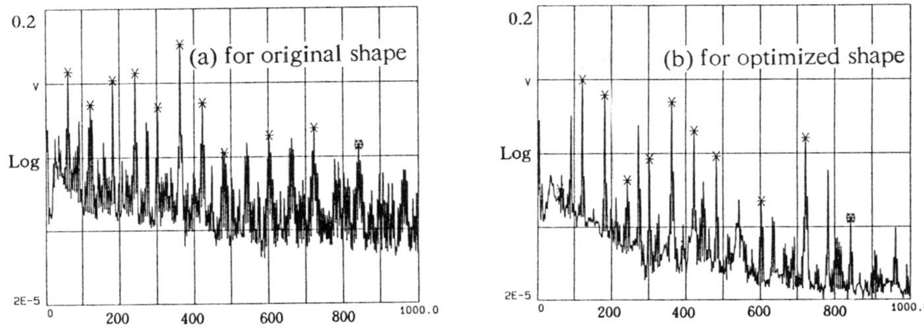

Figure 6. Vibration of motor unit with 300 rpm

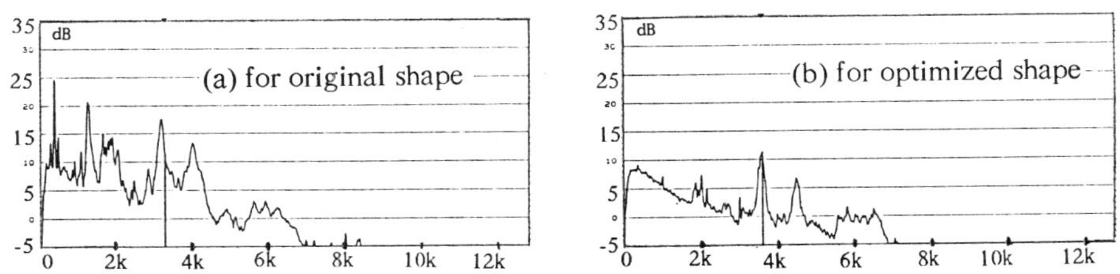

Figure 7. Noise of motor unit with 300 rpm

5. CONCLUSION

Assesments of the magnetization distribution and the shape of airgap were performed for the small precision motors with permanent magnet and it was experimentally verified that optimization with respect to the cogging torque is very effective for designing the low noise and vibration motor unit.

Since the trend of the precision rotating machines is toward the light weighted and small sized model featuring the low and high running speeds, the cogging torque reduction is becoming more critical with the advent of new high grade magnets, and the proposed reduction technique can be considered as an effective approach to achieve designing a silent small precision rotating machines.

6. REFERENCES

1 S. J. Yang, Low Noise Electrical Motors, Oxford Univ. Press, 1981
2 M. J. Goodwin, Dynamics of Rotor-bearing System, Unwin Hyman Ltd., 1989
3 T. Nakada, et.al., IEEE Trans. MAG, 22 (1986) 526
4 I. H. Park, et. al., J. Comp. Math. Electrical and Electric Eng., 9, Supp.A (1990) 111

Simulation and Design of Applied Electromagnetic Systems
T. Honma (Editor)

Analysis of iron losses in an induction motor at no load

T. Hanai, T. Tokumasu, E. Shimomura, and M. Kawakami

Toshiba Corporation, 2-4 Suehiro-cho, Tsurumi-ku, Yokohama 230, Japan

Abstract

Iron losses in an induction motor under operating conditions are generally not consistent with the loss characteristics of electromagnetic steel plate which were measured by the "Epstein Square" method. This stems from the losses being increased by such factors as flux waveform distortions, rotation of magnetizing direction and harmonic components of gap flux [1]. This report describes a calculation method for the iron losses including the loss increment. This method is based on a two-dimensional field analysis, and the harmonic components of the gap flux are dealt with separately from the fundamental component.

1. INTRODUCTION

It is important to evaluate the losses accurately in the design phase and in particular it is necessary to estimate local losses which were not dealt with by conventional methods. On the other hand, there has recently been rising a great demand for high efficiency of induction motors. From this point of view, it is also necessary to evaluate the motor losses accurately. In this paper, the iron losses among the motor losses are calculated from the results of an FEM analysis.

The iron losses are divided into two kinds; one is caused by the spatial fundamental component of the gap flux and the other is caused by the harmonic components of the gap flux generated by the effects of slots. There are several reports on the iron losses estimated from conventional FEM analysis [2]. However, little consideration has been paid to the losses due to the harmonic components of the gap flux. In this paper, the iron losses are calculated with such harmonic components, in addition to the fundamental flux.

2. CALCULATION OF IRON LOSSES

2.1. Iron losses due to fundamental flux

In the core of the motor, the locus of flux is rotational and its waveform is distorted. In addition, the magnetic characteristics of the electromagnetic steel plate vary due to cutting or tightening force. So, we calculated the iron losses using the flux density distributions obtained by field analysis, as follows:

(1) Iron losses in rotating elliptic field

The locus of flux is not a simple alternating field but a rotating elliptic one at the tooth roots and at the stator yoke. The iron losses in rotating fields are much larger than in alternating fields. We calculated them as the sum of the losses due to the alternating fields in the directions of the major and the minor axes of the ellipsis [3].

(2) Iron loss increment due to flux waveform distortions

Flux waveform distortions were generated by the nonlinearity of the electromagnetic steel plate, and they cause eddy current losses to increase. Here, the iron losses can be calculated as follows:

Iron losses = (Distortion factor)^2 × (Eddy current losses with sinusoidal waveform) + (Hysteresis losses for maximum of flux density)

where,

$$\text{Distortion factor} = \sqrt{\Sigma(n \times Bn)^2} / B_{fund} \tag{1}$$

n: order of harmonics

B_{fund}: spatial fundamental component (effective value) of flux density

Bn: spatial harmonic component (effective value) of flux density

(3) Iron loss increment due to building factors (cutting strains and tightening force)

This loss increment is caused in manufacturing process and they were recognized experimentally.

2.2. Iron losses due to harmonic flux

The surfacial and pulsating losses of the rotor are caused by the permeance harmonics of the stator. Thus, these losses were calculated on the following assumptions: (1) The stator was assumed to be a slotless core. (2) a harmonic magnetomotive force was arranged on the inner surface of the stator core as a current sheet. The losses were calculated by the same method as for fundamental wave.

3. ANALYSIS MODELS AND CONDITIONS

Studies were carried out using the following two models:

Model 1 (Fig. 1 (a)) is to evaluate iron losses including losses due to gap harmonic flux. It is a 4-pole, 2.2 kW squirrel-cage induction motor. For field

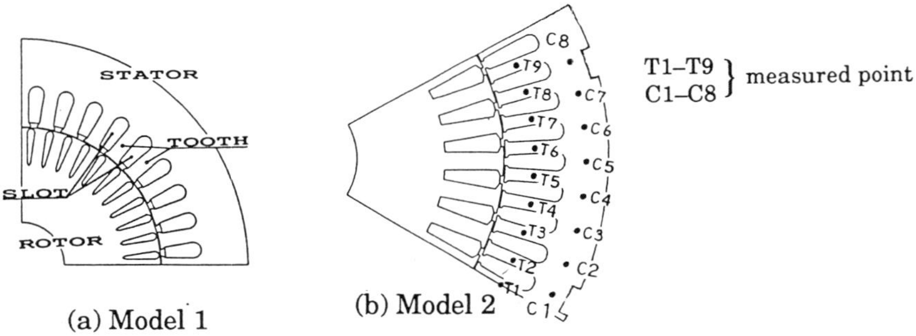

Figure 1. analysis model

analysis, a DC nonlinear field analysis was used and time variations of flux were obtained by replacing the spatial flux distribution with time variations. The coefficient of cutting strain and that of tightening force strain were assumed to be both 1.1.

Model 2 (Fig. 1 (b)) is for evaluating the calculations of iron losses due to fundamental flux. It is a 6-pole 2.2 kW wound-rotor type induction motor. The rotor was fixed and the rotor windings were excited by alternating current. For the field analysis, an AC nonlinear field analysis was used, and so the flux waveform distortions were taken into account. The building factors were the same as for model 1.

4. RESULTS OF ANALYSIS

With model 1, the flux contour, the iron losses due to fundamental flux, and the iron loss distribution generated in the rotor by the harmonics are shown in Figs. 2, 3, and 4 respectively. With model 2, a comparison of losses is shown in Table 1. The measured values in Table 1 were obtained with a loss evaluation method by temperature rise. These results prove the following,

- With model 1, the calculated value of iron losses was 104W and the measured value was 111W. There was a good agreement with a difference of about 7%.
- As for the results with model 2, there is a good agreement between analytical and measured values for iron losses due to the fundamental flux, despite neglecting magnetostriction. This is because the strain coefficient for the major axis is not so large. In case of model 1, the strain coefficient is around 1.0–1.1.
- Loss increment, which is caused by rotation of the flux locus, flux waveform distortion and building factors, is about 25%.
- The loss density at the rotor surface is very large. This might be because the flux is concentrated at these parts.

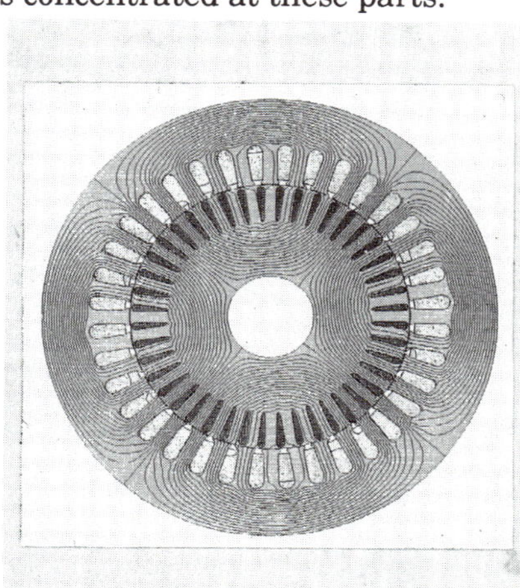

Figure 2. Flux contour (model 1)

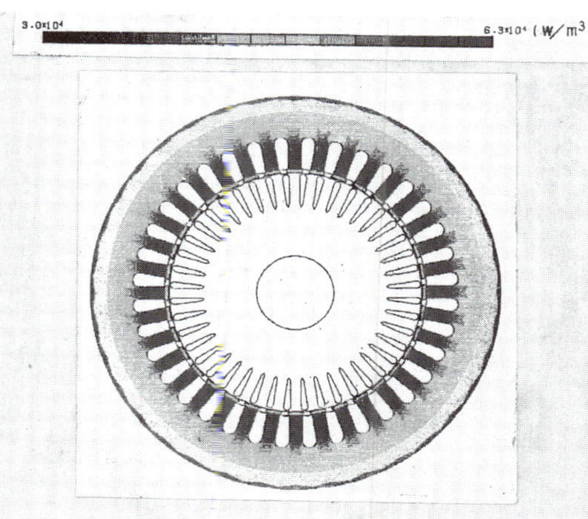

Figure 3. Loss distribution by fundamental flux (model 1)

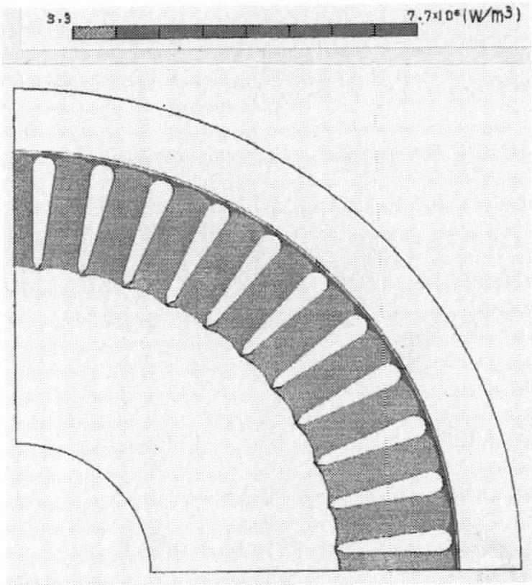

Figure 4. Loss distribution by harmonics of the flux (model 1)

Table 1
Loss comparison (model 2)

	No.	Calculated Value	Measured Value
Teeth	T1	4.1	3.9
	T2	2.9	2.9
	T3	3.1	3.2
	T4	4.5	4.5
	T5	3.3	3.6
	T6	3.4	3.3
	T7	4.5	error
	T8	2.9	3.3
	T9	2.9	3.2
Core Back	C1	2.3	2.5
	C2	2.1	2.0
	C3	2.3	1.9
	C4	2.2	1.9
	C5	2.0	1.7
	C6	2.2	1.8
	C7	2.2	2.4
	C8	2.3	2.4

5. CONCLUSION

In this work, we proposed the calculation method of iron losses included in harmonic components, and it was applied in induction motors. As the results, there was a good agreement between analytical and measured values, and it was shown that this method was useful.

6. REFERENCES

1 For example, Yano et al.: "Rotating flux and iron loss distributions at teeth of induction motor core," MAG-82-72, Technical meeting on Magnetics, I.E.E. Japan.

2 For example, Enokizono et al.: "Magnetic losses in induction motors," RM-82-22, Technical meeting on Rotational Machinery, I.E.E. Japan.

3 Yamaguchi et al.: "Rotating field iron losses in commercial silicon steel plates," Transaction of I.E.E. Japan 51-A44.

Simulation and Design of Applied Electromagnetic Systems
T. Honma (Editor)

Simulation of free-electron lasers systems using symplectic integrator

H. Takahashi, H. Kawaguchi and T. Honma

Department of Electrical Engineering, Faculty of Engineering,
Hokkaido University, Kita 13, Nishi 8, Kita-ku, Sapporo 060 Japan

Abstract

In this paper, a symplectic integrator for a simulation of the Free Electron Lasers (FEL) is presented. Firstly it is indicated that, for the FEL with a planer wiggler, one can implement the symplectic integrator very easily. And then, combining the integrator with radiation fields calculation, the FEL starting-up simulation is done.

1. INTRODUCTION

The relativistic electron motion in the Free Electron Lasers (FEL) system must be calculated precisely, because the velocity of an electron may exceed the light speed by numerical errors and the radiation fields emitted by the electron are quite sensitive to small changes of the electron energy. Generally, it is pointed out that popular integration schemes, including the Runge-Kutta class of algorithms, may produce numerical instability of particle energy [1]. This means that they can not simulate the FEL system precisely.

In this paper, a single particle model of FEL with planner wiggler (Fig.1) is analyzed using a simulation method based on the concept of the transfer matrix. At first, it is indicated that the equation of motion in the planar wiggler can be reduced into a matrix equation which means the canonical transformation. And then, one can find that this simulation method is the symplectic integrator. Moreover, combining with the calculation of radiation fields which are emitted from the electron, starting-up of the FEL is simulated.

2. TRANSFER MATRIX EQUATION

In a single particle picture of the FEL, the electron motion is described by the relativistic Hamiltonian,

$$H = [c^2(\mathbf{P} - e\mathbf{A})^2 + m^2c^4]^{1/2} = \gamma m c^2 \quad , \tag{1}$$

where m is the electron mass, c is the velocity of light, e is the elementary charge, $\gamma = (1 - v^2/c^2)^{-1/2}$, v is the velocity of an electron, $\mathbf{P} = \mathbf{p} + e\mathbf{A}$ is the canonical momentum and $\mathbf{p}$ is the ordinary momentum of the electron. Using the canonical equation of Hamilton, the time derivative of the momentum $\mathbf{p}$ is expressed as follows,

$$m\gamma \frac{dv_x}{dt} = eE_x - ev_z \frac{\partial A_x}{\partial z} \tag{2}$$

$$m\gamma \frac{dv_z}{dt} = \qquad ev_x \frac{\partial A_x}{\partial z} \tag{3}$$

where the y-component of eq.(2) is omitted since the vector potentials $\mathbf{A}$ of the magnetic fields and the laser electric fields in the planar wiggler (shown as Fig.1) have only an x-component ($\mathbf{A} = (A_x(z,t), 0, 0)$). Now, introducing the following complex variables ξ and ε [2],

$$\xi = z + \mathrm{i}x, \quad \varepsilon = \mathrm{i}E_x \quad , \tag{4}$$

one can rewrite the eqs.(2) and (3) as follows,

$$m\gamma \frac{d^2\xi}{dt^2} = -\mathrm{i}e \frac{\partial A_x}{\partial z} \frac{d\xi}{dt} + e\varepsilon \quad . \tag{5}$$

This equation can be solved as follows,

$$\mathbf{X}(t_2) = \mathrm{T}[t_2, t_1]\, \mathbf{X}(t_1) + \mathbf{S}(t_2, t_1) \quad . \tag{6}$$

where vectors $\mathbf{X}(t)$, $\mathbf{S}(t_1, t_2)$ and a matrix $\mathrm{T}[t_1, t_2]$ are defined by

$$\mathbf{X}(t) = \begin{vmatrix} \xi(t) \\ m\gamma \dfrac{d\xi(t)}{dt} \end{vmatrix} , \tag{7}$$

$$\mathrm{T}[t_2, t_1] = \begin{vmatrix} 1 & \dfrac{1}{m\gamma} \displaystyle\int_{t_1}^{t_2} exp\left(-\mathrm{i} \frac{e}{m\gamma v_{z0}} A_x \Big|_{t_1}^{\tau} \right) d\tau \\ 0 & exp\left(-\mathrm{i} \dfrac{e}{m\gamma v_{z0}} A_x \Big|_{t_1}^{t_2} \right) \end{vmatrix} , \tag{8}$$

$$\mathbf{S}(t_2, t_1) = \begin{vmatrix} \dfrac{e}{m\gamma} \displaystyle\int_{t_1}^{t_2} \int_{t_1}^{\tau} \varepsilon\, exp\left(\mathrm{i} \frac{e}{m\gamma v_{z0}} A_x \Big|_{\tau}^{u} \right) du\, d\tau \\ e \displaystyle\int_{t_1}^{t_2} \varepsilon\, exp\left(\mathrm{i} \frac{e}{m\gamma v_{z0}} A_x \Big|_{t_2}^{u} \right) du \end{vmatrix} , \tag{9}$$

where v_{z0} is the average velocity of z direction. For the integrals in eqs.(7) and (8), the paraxial approximation $A_x(z) = A_x(v_{z0}t)$ is used. The matrix equation (6) is just the canonical transformation from $\mathbf{X}(t_1)$ to $\mathbf{X}(t_2)$. One can obtain the variables $\mathbf{X}(t_2)$ at any time t_2 , calculating the transfer matrix $\mathrm{T}[t_1,t_2]$ and the vector $\mathbf{S}(t_1,t_2)$. The wiggler fields calculation is done by the following expression,

$$\mathbf{A}_w(\mathbf{r}) = \frac{\mu_0}{4\pi} \iiint \frac{\mathbf{r} \times \mathbf{M}}{|\mathbf{r}|^3}\, dv \tag{10}$$

This integral method is symplectic, because the determinant of the matrix $T[t_2,t_1]$ is always zero,

$$det(\mathrm{T}\,[\,t_2, t_1\,]) \equiv 0$$

Indeed, one can make sure numerically that the Liouville theorem is satisfied in this integrator. The figure 2 shows the volume changes in this integrator and the Runge-Kutta method using the Fehlberg formula. The notation $dV(t)$ is the infinitesimal phase space volume at the time t. And then $(dV(t)\text{-}dV(0))/dV(0)$ is the relative volume change from the initial volume $dV(0)$. The notation T is period of wiggler motion.

3. LIÉNARD-WIECHERT FIELDS

Radiation fields emitted from a relativistic electron are described by the Liénard-Wiechert potentials. In this case, the following Fourier form of the Liénard-Wiechert potentials [3] is convenient to analyze the fields,

$$\mathbf{A}_{rad}(\mathbf{r},t) = \frac{\omega e}{8\pi^2 \varepsilon_0 c^2} \sum_{n=-\infty}^{\infty} \int_0^{2\pi} \frac{exp\left[\,in\left(\omega t - \sigma - \frac{\omega}{c}|\mathbf{r}-\mathbf{r}'(\sigma)|\,\right)\right]}{|\mathbf{r}-\mathbf{r}'(\sigma)|} \frac{d\mathbf{r}'(\sigma)}{d\sigma}\, d\sigma \quad , \tag{11}$$

where ε_0 is the dielectric constant, ω is the angular frequency of the electron motion and $\mathbf{r}'(\sigma)$ is the electron orbit whose parameter is σ. The electromagnetic fields and poynting vectors, which are emitted from the electron, can be calculated by the vector potentials $\mathbf{A}_{rad}$ for each frequency. In the calculation, the laser electric fields E_x, which is used in the eq.(4), are assumed to have following form,

$$E_x = \sum_n E_{xn} \cos(\omega_{Ln} t - k_{Ln} z) \tag{12}$$

where the subscript n denotes n-th harmony.

4. NUMERICAL RESULTS

One can simulate the FEL starting-up using the integrator (6) and the radiation fields calculation (11) . The flow diagram of the starting-up simulation is shown in the Fig.3. And then, the Fig.4 shows the development of the radiation pattern for first iteration step. One can find the small enhancement of the radiation fields around the resonant frequency.

5. SUMMARY

In this paper, the symplectic integrator for the FEL simulation was presented. And then, combining the integrator with radiation fields calculation, the FEL starting-up simulation was done. In this simulation, the number of electron is one and the electron distribution effect is not considered. Therefore, the radiation fields saturate immediately in first or second iteration steps. The starting-up simulation considering the electron distribution effects is the future work.

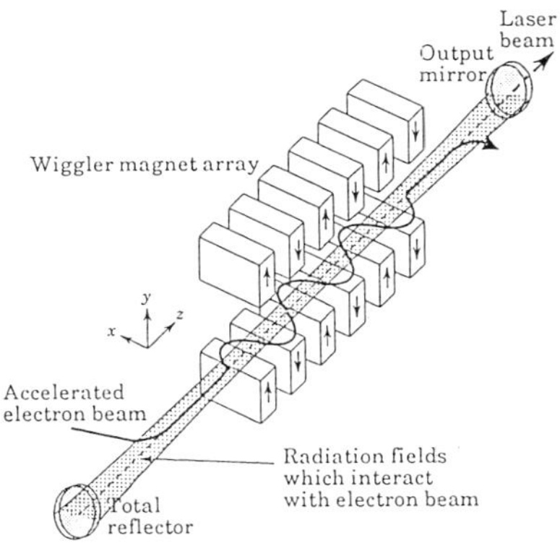

Fig.1 Schematic diagram of FEL

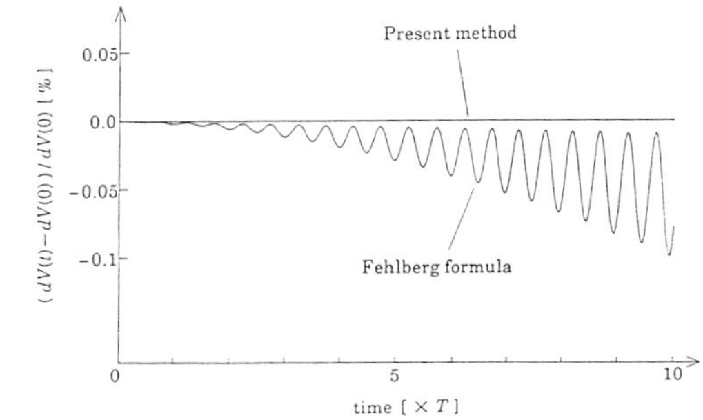

Fig.2 Phase space volume changes

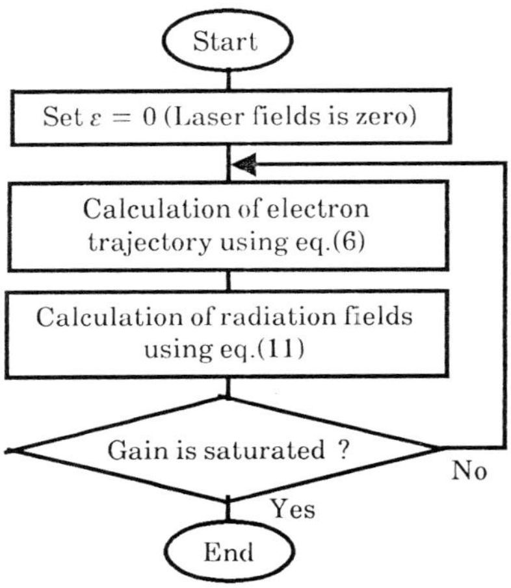

Fig.3 Flow diagram of starting-up simulation

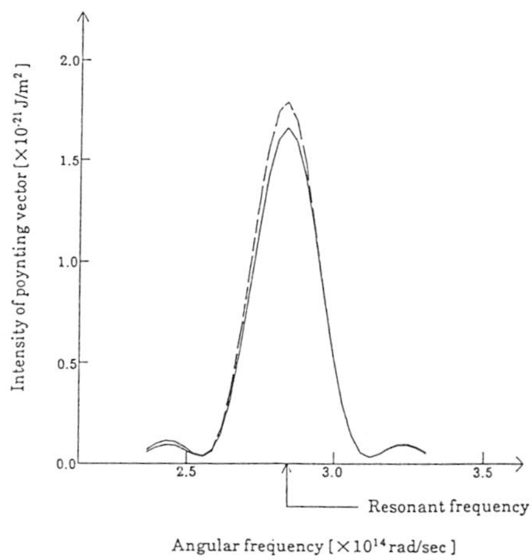

Fig.4 Development of radiation pattern

6. REFERENCES

1 J.Candy and W.Rozmus, J.Comp.Phys.,**92**, (1991) p.230

2 L.D.Landau and E.M.Lifshitz, Classical Theory of Fields, Pergamon Press, New York, 1971

3 H.Kawaguchi and S.Murata, J.Phys.Soc.Jpn, **58**[3] (1989) p.848

Simulation and Design of Applied Electromagnetic Systems
T. Honma (Editor)

Crosstalk simulation using 3D electromagnetic finite elements and SPICE

John R. Brauer[a], Eiji Nakamoto[b], and Pierre L. Allaud[a]

[a]The MacNeal–Schwendler Corp., 4300 W. Brown Deer Rd., Milwaukee, WI 53223 USA

[b]MSC Japan Ltd., 626, Shichikannon–Cho, Rokkakusagaru, Karasumadori, Nakagyou–Ku, Kyoto, JAPAN

Abstract

Electromagnetic finite element analysis is used to compute matrices and crosstalk of circuits. The matrices computed are C, G, L, and R. For the first time, frequency dependent L and R matrices are computed using 3D finite elements and then used in SPICE. The resulting voltages computed for three wires are shown to agree closely with measurements.

1. INTRODUCTION

To predict crosstalk, ringing, and pulse degradation in electric circuits such as high–speed computers, capacitance C and inductance L matrices can be used in circuit simulation programs. Very recent papers [1],[2],[3] have discussed how C and L matrices can be used in programs such as SPICE to simulate crosstalk. This paper discusses how resistance R and conductance G matrices are also important, and shows for the first time how to calculate *frequency–dependent* R and L matrices using 3D finite elements. The resulting crosstalk is then calculated using either SPICE or 0D finite elements.

The finite element software used here is MSC/EMAS™, which has 0D, 1D, 2D, and 3D electromagnetic finite elements [4],[5]. It is used to compute matrices C and G (both frequency–independent) and L and R (both frequency–dependent). Then the resulting behavior of an ac circuit is computed by SPICE and compared to measurements.

2. FINITE ELEMENT COMPUTATION OF STATIC C, G, L, AND R MATRICES

The finite element formulation used here [4] is based upon the potential vector {u} made up in general of the magnetic vector potential $\overline{A}$ and the time–integrated electric scalar potential ψ. It has in general three matrices making up its equation:

$$[M]\{\ddot{u}\} + [B]\{\dot{u}\} + [K]\{u\} = \{J\} \tag{1}$$

which can be solved for static, time domain, and frequency domain excitation vectors {J}. The [M] matrix is proportional to permittivity and accounts for electric field energy. The [B] matrix is proportional to conductivity and accounts for power loss. The [K] matrix is proportional to reluctivity (1/permeability) and accounts for magnetic field energy.

Static fields can be calculated using only one of the above three matrices at a time. Electrostatic fields and capacitance C can be computed [1],[3] using only the [M] matrix and the potential ψ. Steady currents and conductance G or resistance R can be computed using only the [B] matrix and ψ. G is nonzero if the conductivity of the PC board or other "insulative" material is greater than zero. Magnetostatic fields and L can be computed using only the [K] matrix and potential $\overline{A}$ if the current distribution {J} is known.

An entire circuit matrix can be computed with one finite element matrix decomposition. For example, an entire C matrix can be computed by constraining one trace (wire) at a time at 1 volt with all other traces grounded and noting the reaction charge Q computed at all the traces. The C matrix entries are each the respective values of Q/(V=1v) in farads.

3. FINITE ELEMENT COMPUTATION OF DYNAMIC L AND R MATRICES

The C and G matrices computed from the preceding static analyses are usually valid from DC to the frequency for which a dimension of the circuit analyzed becomes approximately 0.1 wavelength. However, due to skin effects, the L and R matrices computed from static analyses are usually correct only over a much lower frequency range.

To compute dynamic L and R matrices, we must solve (1) in the frequency domain neglecting displacement (capacitive) effects that are included in the C matrix. Thus (1) becomes at any frequency f:

$$[K]\{u\} + i2\pi f[B]\{u\} = \{J\} \tag{2}$$

where the excitation {J} can be input as an inhomogeneous Neumann *surface* excitation [5] at an end of each trace, and then the finite element software computes the ac current distribution in the traces and the resulting power losses and magnetic fields. One trace is energized at a time, and both entire L and R matrices can be obtained with only one dynamic matrix decomposition. With 1 ampere input for each trace, the power losses for each trace energized become the diagonals of the dynamic R matrix. Due to skin effects, the R matrix entries usually increase significantly with increasing frequency.

The frequency–dependent L matrix is computed using the definition for L:

$$L_{jk} = \lambda_k / I_j \tag{3}$$

where the flux linkage λ_k is that linking the kth trace. The flux linkage can be obtained using the definition:

$$\lambda = \int \overline{B} \cdot \overline{ds} \tag{4}$$

which is converted to a line integral by using Stokes' theorem:

$$\lambda = \int \overline{A} \cdot \overline{dl} \tag{5}$$

where with 1 ampere input for each trace, the flux linkages become the entries of the dynamic L matrix. Due to skin effects, the L matrix entries usually decrease with increasing frequency.

4. DYNAMIC L AND R MATRIX COMPUTATIONS FOR 3 WIRES PLUS GROUND

Figure 1 shows a cross section of a 3D finite element model of three–phase double–channel aluminum bus bars in air. This example is chosen because it has been the subject of previous investigations [6] and because its voltages have been measured [7]. It is analyzed here using frequency–dependent L and R matrices for the first time.

The dimensions of Figure 1 are given elsewhere [7], as are the applied unbalanced three–phase currents. The frequency is 60 Hz, which for the given dimensions is sufficiently high so that skin effects occur. Because the three currents are unbalanced and do not add up to zero, a ground must exist and is assumed at a radius of 2.54 m from the center. The model of Figure 1 has 3D finite elements within the aluminum that are thinner than one skin depth. The z extension of the 3D model of Figure 1 is 7.62 m. Figure 1 also shows the flux lines computed for the right bar (a) energized with 1 amp, with which the entire first row of the [L] matrix is obtained.

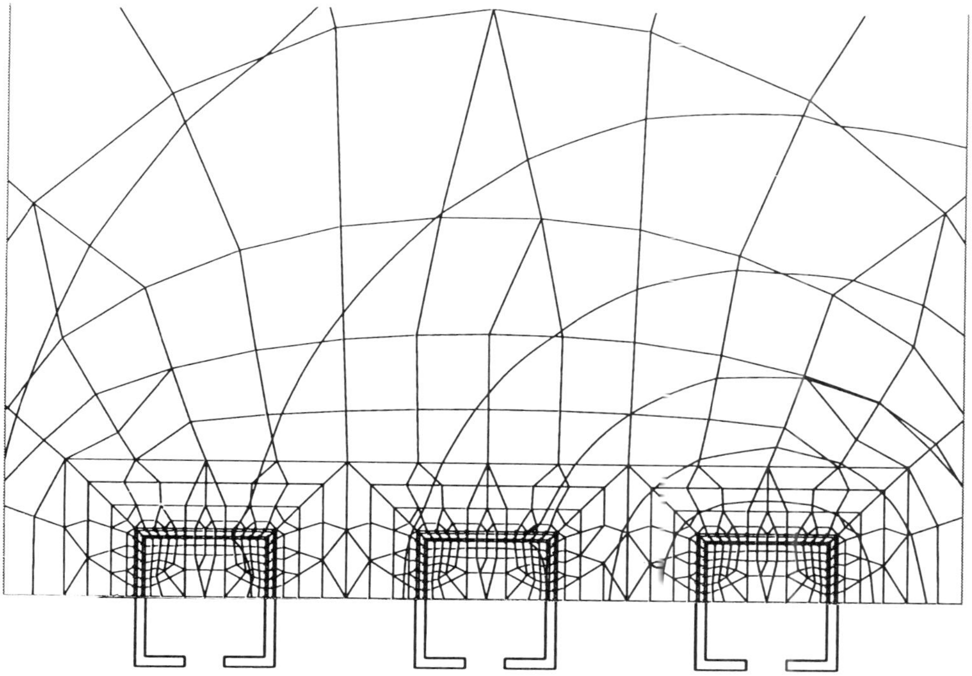

Figure 1. Detail of cross section of 3D finite element model of one−half of three double−channel bus bars, also showing the flux lines computed for the right bar (a) excited at 60 Hz, where bar bar (b) is the center bar and bar (c) is the left bar.

At 60 Hz, the L and R matrices computed by the techniques of (2) through (5) are:

$$[L] = 1.E-6H\begin{bmatrix} 10.0239 & 6.2933 & 4.3149 \\ 6.2933 & 10.0559 & 6.2917 \\ 4.3149 & 6.2917 & 10.0238 \end{bmatrix}, \quad [R] = 1.E-6\Omega\begin{bmatrix} 97.6615 & 0 & 0 \\ 0 & 97.5316 & 0 \\ 0 & 0 & 97.6646 \end{bmatrix} \quad (6)$$

which are different from those obtained at 0 Hz due to skin effects. For example, at 0 Hz the three diagonals of the [R] matrix are all 80.8775E−6 ohms, which are smaller and equal because at DC there are no skin and proximity effects.

5. VOLTAGES COMPUTED USING ABOVE [L] AND [R] MATRICES IN SPICE

The [L] matrix of (6) can be input into SPICE in several ways. The diagonals are input as self inductances called L in SPICE. To input the off−diagonal mutual inductances, one can either use coupling coefficients K or voltage−dependent−voltage sources E. Here we choose E because of the SPICE restriction that K must be positive, which in general is not true for 3D [L] matrices.

[R], [C], and [G] matrices are input more easily than [L]. Assuming [R] to be a diagonal matrix as in (6), it is input into SPICE as resistors R. The problem of Figure 1 has negligible [C] and [G] matrices, but in general [G] matrix entries are inverted in order to input them as SPICE resistors R, and [C] matrix entries (diagonal and off–diagonal) are directly input as SPICE capacitors C. The diagonals of C represent capacitance to ground.

The output terminals of the SPICE circuit of Figure 1 are all shorted together to correspond with test conditions [7]. The input terminals are given the three unbalanced currents of the specified amplitudes in amperes and phase angles in degrees:

$$I_a = 4395 \measuredangle 121^0, \quad I_b = 4370 \measuredangle 0, \quad I_c = 4420 \measuredangle 241^0 \tag{7}$$

The results of the SPICE analysis using the matrices of (6) are listed in Table 1.

Table 1
Line–to–line voltages for 3 double–channel bars of Figure 1.

Method	V_{ab}	V_{bc}	V_{ca}
MSC/EMAS and SPICE	5.76	5.49	8.22
Measurement	5.72	5.60	8.30
Percent difference	+0.7	−2.0	−1.0

6. CONCLUSIONS

C, G, L, and R matrices can be computed using 3D finite elements. The L and R matrices may be frequency dependent, and can be computed by inputting current using inhomogeneous Neumann surface excitations and by evaluating flux linkage using line integrals of magnetic vector potential. While C, G, and R are easily input into SPICE, the off–diagonal (mutual) inductances of L require either SPICE K or E inputs. After the matrices are all input, SPICE can be used to compute voltages and currents for various input and output sources and impedances. The example analyzed here of three wires (buses) carrying known ac currents gives computed voltages that agree within 2% with measured voltages.

7. REFERENCES

[1] J. R. Brauer, R. W. Nopper and D. C. Gates, *Digest of IEEE Mtg. on Electrical Performance of Electronic Packaging*, Tucson AZ, USA, April 22–24, 1992, 84–86.
[2] F. Broyde, E. Clavelier, and C. Hymowitz, *IEEE Circuits & Devices*, Sept. 1992, 8–16.
[3] J. R. Brauer and B. S. Brown, *IEEE Trans. Electromagnetic Compatibility*, May 1993, in press.
[4] J. R. Brauer, B. E. MacNeal, L. A. Larkin, and V. D. Overbye, *IEEE Trans. Magnetics*, Sept. 1991, 4085–4088.
[5] J. R. Brauer and F. Hirtenfelder, *IEEE Trans. Magnetics,* March 1992, 1659–1662.
[6] John R. Brauer, *IEEE Trans. Magnetics,* March 1982, 504–509.
[7] O .R. Schurig, discussion of paper by C. M. Siegel and T. J. Higgins, *AIEE Trans. P.A.S.*, Jan. 1952, 530–531.

Simulation and Design of Applied Electromagnetic Systems
T. Honma (Editor)

Effects of heat generation in MOSFET in non-equilibrium transport

H. Hayashi and R. Dang

College of Engineering, Hosei University 3-7-2 Kajino-cho, Koganei-shi, Tokyo 184, Japan

Abstract

Effects of non-equilibrium transport on MOSFET characteristics are reported using a hot-carrier simulator which also includes the effect of lattice temperature. Device characteristics based on this simulator are compared with those obtained by the conventional drift-diffusion model. We have confirmed that, the electron distribution extends deeper into the substrate at the drain side than the case of the drift-diffusion model. Velocity overshoot is also observed, using an energy-dependent mobility model.

1.Introduction

When the size of a semiconductor device reduces to the submicrometer dimension, it is necessary to review the conventional drift-diffusion model [1], since this model can not describe non-equilibrium transport phenomena, such as velocity overshoot and ballistic operation, which are apt to occur in miniaturized devices. To cope with this situation, the relaxation time approximation model [2] which also covers these non-equilibrium transport phenomena, has been recently developed.

Many hot-carrier analyses based on this model have been reported, where electron and hole temperatures are determined from the energy balance equations, but the lattice is fixed at room temperature, despite possible energy exchanges from hot-carriers to the lattice and vice-versa.

In this paper, the effects of hot-carrier induced lattice heating in MOSFET analysis are reported for non-equilibrium transport under non-isothermal conditions.

2.Model

The equations to be solved for a two dimensional steady state in a semiconducting environment, are the Poisson equation (1), the continuity equations for electrons (2) and holes (3), the energy balance equation for electrons (4), and the heat flow equation (5). They are listed, as follows, respectively.

$$div(\varepsilon \ grad \ \phi) = -q\ (p - n + N_D - N_A) \tag{1}$$

$$div(-\frac{J_n}{q}) = GR \tag{2}$$

$$div(\frac{J_p}{q}) = GR \tag{3}$$

$$div\ S_n = J_n \cdot E - n\ C_n + \xi_n\ GR \tag{4}$$

$$-div(k_L\ \ grad\ T_L) = Q \tag{5}$$

where ε, ϕ, n, p, N_D, N_A and GR represent the permittivity, electric potential, electron density, hole density, donor doping concentration, acceptor doping concentration, and generation-recombination of carriers, respectively. J_n, J_P are electron and hole current densities. S_n, C_n are energy flux and energy loss rate for electrons. ξ_n is the electron energy. k_L, Q and T_L stand for thermal conductivity, lattice heat generation rate and lattice temperature, respectively.

On the other hand, the carrier energy-dependent mobility is obtained from the following equations, after Cook [3].

$$\mu_n(E_D(T_n), E_G, N) = \frac{C_n(T_n, T_L)}{q\ E_D(T_n)^2} \tag{6}$$

$$E_D(T_n)^5 + \frac{G\ \upsilon_C}{\mu'} E_D(T_n)^4 - \frac{C_n^2}{q^2 \upsilon_S^2} E_D(T_n)^3 - \frac{C_n^2}{q^2 \mu'}\left(\frac{1}{\upsilon_C} + \frac{G\ \upsilon_C}{\upsilon_S^2}\right) E_D(T_n)^2 - \frac{C_n^2}{q^2 \mu'^2} E_D(T_n) - \frac{C_n^2 G\ \upsilon_C}{q^2 \mu'^3} = 0 \tag{7}$$

where E_G, E_D are perpendicular and parallel components of the electric field to the current density vector, respectively, and all other symbols have their conventional meaning. The carrier-energy dependent mobility is obtained by substituting this function $E_D(T_n)$ for E_D in the conventional field-dependent mobility [4]. Here, the energy loss rate C_n is given as follows.

$$C_n(T_n, T_L) = \frac{\xi_n - \xi_o}{\tau_{wn}} = \frac{3}{2} k_B \frac{T_n - T_L}{\tau_{wn}} \tag{8}$$

where τ_{wn} is energy relaxation time for electrons, and ξ_o is the lattice energy. T_n and k_B are the electron temperature and Boltzmann's constant.

As for the energy balance equation, we consider the electrons only. Note that all other physical models, such as generation-recombination, heat generation and so forth, are identical with those in Ref. [5].

3. Solution Strategy

The following solving strategy has been adopted. The total set of five unknowns, namely, ϕ, n, p, T_n, T_L are divided into two subsets, (ϕ, n, p) and (T_n, T_L), which are to be iteratively solved as the independent groups but entire loop is coupled, as illustrated in the flow chart on the right.

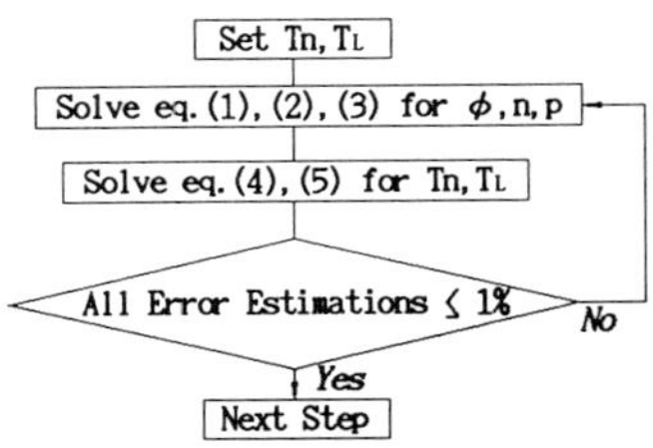

Flow chart

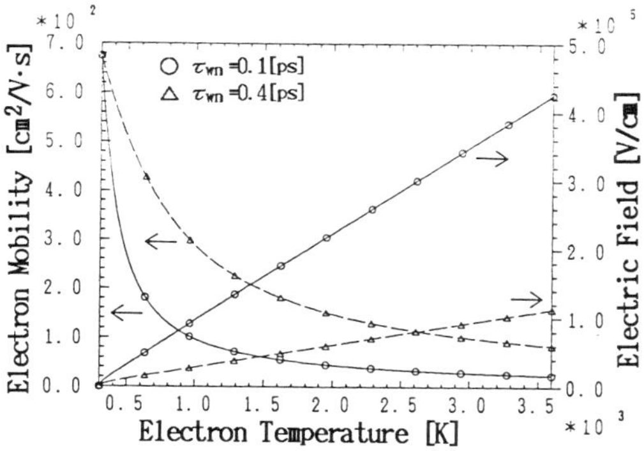

Figure 1. Energy-dependent mobility and parallel component of electric field, in terms of electron temperature.

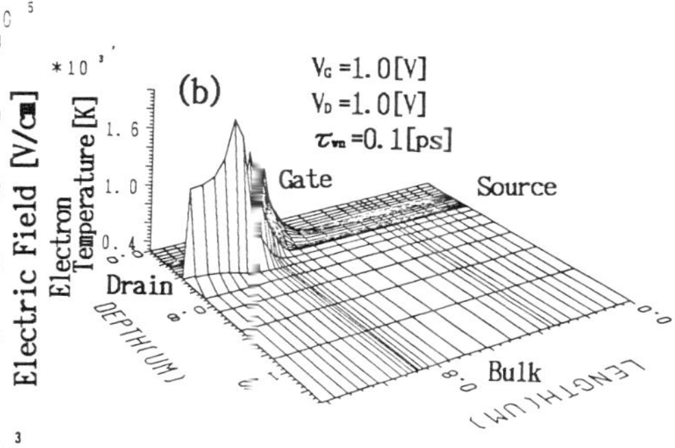

Figure 3. Electron temperature distribution.

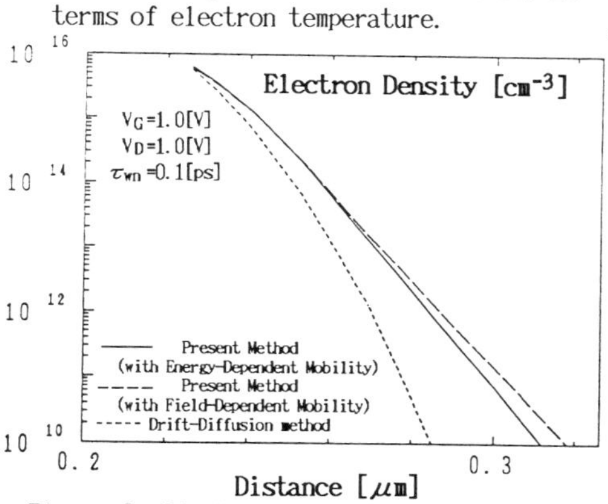

Figure 2. Electron density distributions in the indepth direction at drain edge of the channel.

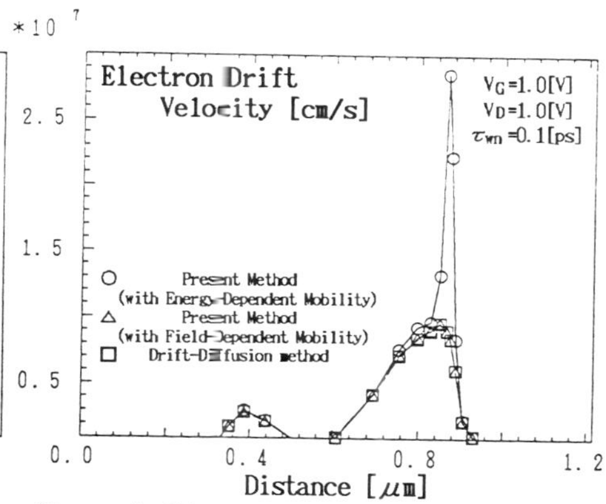

Figure 4. Electron drift velocity profile along Si-SiO2 interface.

4.Simulation Results

An n-channel MOSFET with effective channel length of 0.5 μm is chosen as test device. Major device parameters include an oxide thickness of 26nm, a source and drain junction depth of 0.2 μm and a uniform bulk impurity concentration of $10^{17}cm^{-3}$. The impurity concentration profile for the diffusion regions, namely, drain and source, is generated, using Gaussian distribution for the indepth direction and complementary error function for the transverse direction. On the other hand, two cases of constant relaxations time are considered, namely, τ_{wn} =0.1ps and 0.4ps.

In Fig.1, the simulated energy dependent-mobility, $\mu(E_D(T_n),E_G,N)$, and the parallel component of electric field, $E_D(T_n)$, in terms of the electron temperature, are shown together.

It is found that, as the electron temperature rises, $E_D(T_n)$ becomes smaller at τ_{wn}=0.4ps than at τ_{wn}=0.1ps. This results in a higher mobility, because a large relaxation time reduces carrier scattering effect. This agrees with results obtained by Monte Carlo simulations [6].

Fig.2 shows the electron density distributions in the indepth direction at the drain edge of the channel. It is found that, due to electron heating, the extension of the electron concentration into the substrate is deeper in the present case than in the case of the drift-diffusion model. In other words, the channel current spreads broader, due to the effect of temperature gradient. It is to note that the results obtained using the conventional field-dependent mobility show a slight difference with our results as shown in Fig.2.

Fig.3 shows the electron temperature distribution. Electrons have high energy in the vicinity of the drain, since the electrons are subjected to a high electric field. As a result, the electron temperature is about 1600K at peak value.

In Fig.4, the electron velocity profiles computed from hot-carrier simulation are compared to that obtained from the drift-diffusion model. One can see that, in the present model, velocity overshoot is clearly demonstrated and the maximum velocity is about 2.7 times, as compared with the conventional case. This is larger than previously reported values [2]. This is attributed to an overestimation of energy based on the energy-dependent mobility in use. Note that, in our prediction of the velocity overshoot effect, we have tried to kept the physical model as rigorous as possible and made use of more realistic values of various physical parameters, such as energy relaxation time, etc. These seem to have been overlooked previously. Also, note that the energy-dependent expression for the Yamaguchi mobility as shown above is novel, and a mere application of the same mobility model in its original form can not be expected to produce the same velocity overshoot effect as reported in the present paper.

On the other hand, although not shown, we have confirmed that, for a bias of V_G=1.0V and V_D=2.0V, the simulated heat generation is small. As a result, the lattice temperature is almost identical with the room temperature. Also, the terminal current values based on the present method are similar with the conventional cases.

5.Conclusions

A non-isothermal device simulator has been developed for non-equilibrium transport phenomena under non-isothermal conditions. Our results were compared with those simulated from the drift-diffusion model, and we have confirmed that the electron density distributions based on the present model spread broader into the substrate. Furthermore, velocity overshoot is also predicted for a silicon MOSFET transistor, using an energy-dependent mobility model.

6.References

1 See, for example, S. R. H. Hoole, *IEEE Trans. Magn.*, Vol.25, No.4, pp.2831-2833, 1989.

2 K. Cook and J. Frey, *IEEE Trans. Electron Devices*, Vol.ED-29, No.6, pp.970-977, 1982.

3 R. K. Cook, *IEEE Trans. Electron Devices*, Vol.ED-30, No.9, pp.1103-1110, 1983.

4 K. Yamaguchi, *IEEE Trans. Electron Devices*, Vol.ED-30, No.6, pp.658-663, 1983.

5 H. Hayashi and R. Dang, *IEEE Trans. Magn.*, (to appear).

6 C. Jacoboni, *Solid-State Electronics*, Vol.20, pp.77-89, 1977.

Simulation and Design of Applied Electromagnetic Systems
T. Honma (Editor)

Magnetohydrodynamic equilibrium configuration of a flux-core spheromak by grid generation method

T. Kanki, M. Nagata, T. Uyama, M. Miyauchi[a], and A. Kamitani[a]

Dept. of Electrical Engineering, Faculty of Engineering, Himeji Institute of Technology
Shosha 2167, Himeji, Hyogo 671-22, Japan

[a]Dept. of Electrical and Information Engineering, Faculty of Engineering, Yamagata University
Jonan 4-3-16, Yonezawa, Yamagata 992, Japan

Abstract

A method for automatic numerical generation of a boundary-fitted curvilinear coordinate system is applied to the solution of the Grad-Shafranov equation. By use of the method, the influence of the divertor coil current on magnetohydrodynamic (MHD) equilibrium configurations of a flux-core spheromak (FCS) is investigated. The results of computations show that no magnetic surfaces close in the flux conserver (FC) if $I_c/I_p>2$ is fulfilled. Here I_c and I_p denote the divertor coil current and the total plasma current, respectively.

1. INTRODUCTION

The spheromak is an axisymmetric low-aspect-ratio configuration which has both toroidal and poloidal magnetic fields in a plasma. It has many potential advantages as a fusion reactor: compactness and simplicity of design of external coils and metallic containers. In addition, the spheromak has a possibility of the translation of a toroidal plasma, which allows the formation chamber to be separated from the burn chamber and, therefore, enables a spheromak to be merged with a tokamak into one with a larger current. One of the methods for producing a spheromak is to use a magnetized coaxial plasma gun. In the gun, a toroidal plasma with poloidal and toroidal fields is formed. The poloidal fields are provided by a bias poloidal coil in the inner electrode, whereas the toroidal fields are produced by gun current. After formed in the gun, the toroidal plasma is ejected into the metallic container which is called a flux conserver and relaxes into a spheromak configuration.

Recently a FCS has attracted great attention because of its possibility of the sustainment by helicity injection. The FCS is a spheromak configuration in which the externally imposed open fields exist outside the separatrix. When poloidal currents are applied along the open fields by use of DC voltage, the toroidal flux is induced and the FCS is sustained by MHD relaxations involving a flux conversion process. The sustainment of the FCS has been successfully achieved by helicity injection in the FACT device at Himeji Institute of Technology [1].

The purpose of this paper is to determine equilibrium configurations of the FCS in the FACT device by use of the grid generation method [2,3] and to investigate the effects of the bias field on them. In the next section, the model of the FC and the DC helicity injection system is introduced, and a curvilinear coordinate system whose coordinate lines coincide with surfaces of the divertor bias coil and the FC is numerically generated. This coordinate system is called a boundary-fitted curvilinear coordinate system [2,3]. Conclusions are presented in the third section. The MKS units with $\mu_0=1$ are used throughout this paper, where μ_0 is the permeability of vacuum.

2. CALCULATION METHOD

A drum-type FC has been used in the FACT device. Although the plasma does not exist around the bias divertor coil in the actual experiments, we assume that it exists there. In addition, we assume that the FC and the cathode electrode are made of perfect conductors. In Figure 1, we show the model of the FC and the helicity injection system which is used throughout this paper.

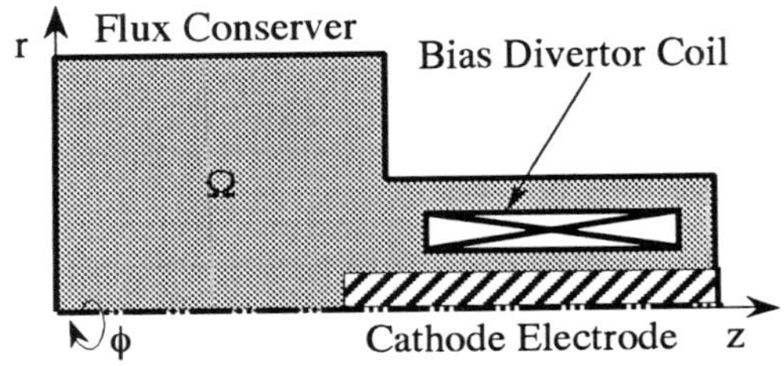

Figure 1. The model of the FC and the helicity injection system.

Let us use the cylindrical coordinate system (z,r,ϕ), and take the symmetry axis of the FC as z-axis. Since the equilibrium state of the spheromak plasma is axisymmetric, we can determine it by solving the Grad-Shafranov equation in the z-r plane. In the following, this plane is called a physical plane and the region, where the equilibrium configuration is determined, is denoted by Ω. Let us solve the Grad-Shafranov equation in Ω by use of the boundary-fitted curvilinear coordinate system.

First of all, we introduce a numerical plane (ξ,η) and map the region Ω onto a square region in the numerical plane (see Figures 2(a) and (b)). The inverse mapping functions, $z=z(\xi,\eta)$ and $r=r(\xi,\eta)$, are numerically generated by the solutions to the boundary value problem of the partial differential equations:

$$\alpha r_{\xi\xi} - 2\beta r_{\xi\eta} + \gamma r_{\eta\eta} = -J^2[P(\xi,\eta)r_\xi + Q(\xi,\eta)r_\eta], \tag{1}$$

$$\alpha z_{\xi\xi} - 2\beta z_{\xi\eta} + \gamma z_{\eta\eta} = -J^2[P(\xi,\eta)z_\xi + Q(\xi,\eta)z_\eta], \tag{2}$$

where $\alpha=z_\eta^2+r_\eta^2$, $\beta=z_\xi z_\eta+r_\xi r_\eta$, $\gamma=z_\xi^2+r_\xi^2$, and $J=\partial(z,r)/\partial(\xi,\eta)$. Further, inhomogeneous functions $P(\xi,\eta)$ and $Q(\xi,\eta)$ adjust the spacing of the ξ- and η-curves in the physical plane [3]. The boundary conditions to eqs.(1) and (2) are chosen such that the boundary of the square region in the $\xi-\eta$ plane is mapped onto that of the region in the z-r plane. By solving numerically eqs. (1) and (2) under these Dirichlet boundary conditions, we obtain the inverse mapping functions, $z=z(\xi,\eta)$ and $r=r(\xi,\eta)$. Figure 2(a) shows coordinate lines in the z-r plane.

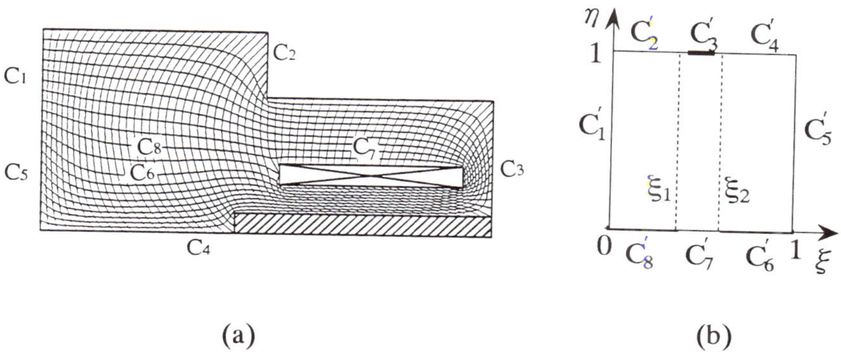

Figure 2. (a) Coordinate lines in the physical plane. (b) The numerical plane.

Next, we solve the Grad-Shafranov equation in the numerical plane. In terms of the boundary-fitted curvilinear coordinate system, the equation can be written as

$$\frac{1}{J^2}\bar{\Delta}\psi + \frac{1}{rJ}\frac{\partial(\psi,z)}{\partial(\xi,\eta)} - \frac{1}{J^3}\left[\frac{\partial(\psi,r)}{\partial(\xi,\eta)}\bar{\Delta}z - \frac{\partial(\psi,z)}{\partial(\xi,\eta)}\bar{\Delta}r\right] = -r^2\frac{dp}{d\psi} - I\frac{dI}{d\psi}, \tag{3}$$

where ψ is the poloidal flux function and $\bar{\Delta}$ is a differential operator defined by

$$\bar{\Delta} \equiv \alpha\frac{\partial^2}{\partial\xi^2} - 2\beta\frac{\partial^2}{\partial\xi\partial\eta} + \gamma\frac{\partial^2}{\partial\eta^2} \quad . \tag{4}$$

The boundary conditions for eq. (3) is as follows.

$$\psi=0 \text{ on } C'_1,\ C'_2,\ C'_4 \text{ and } C'_5, \tag{5}$$

$$\partial(\psi,r)/\partial(\xi,\eta)=0 \text{ on } C'_3, \tag{6}$$

$$I_C=\int_{C_7}\frac{1}{r}\frac{\partial\psi}{\partial n}dl \quad \text{on } C_7, \tag{7}$$

where $\partial/\partial n$ expresses the normal derivative. Further, ψ is considered to be an unknown constant on C_7. Since the curves C_6 and C_8 are identical in the physical plane, the reentrant conditions are imposed on C_6' and C_8'. We assume $dp/d\psi=0$ and $IdI/d\psi=\lambda\psi/R^2$ as $p(\psi)$ and $I(\psi)$. Here λ is constant and R denotes a radius of the FC. Substituting these assumptions into eq. (3), we obtain the eigenvalue equation. By solving the equation numerically, we get the solution $\psi(\xi,\eta)$ in the numerical plane. The solution $\psi(z,r)$ in the physical plane is obtained by use of the inverse mapping functions, $z=z(\xi,\eta)$ and $r=r(\xi,\eta)$. A typical example of the solution is shown in Figure 3. Figure 3 shows the poloidal flux contours in the physical plane.

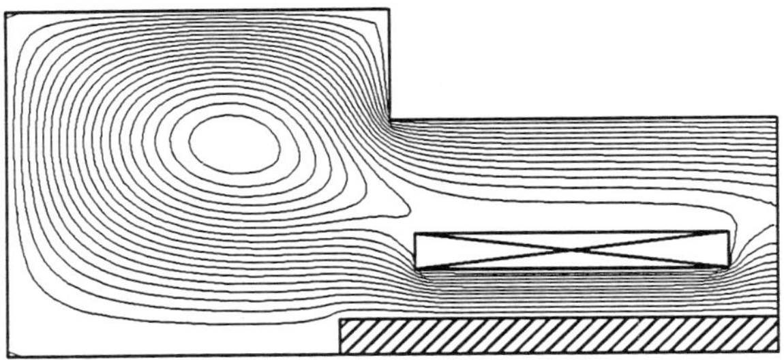

Figure 3. The equilibrium configuration in the physical plane with $I_c/I_p=1.0$.

3. CONCLUSIONS

By using the method explained in the previous section, we have investigated the equilibrium configurations of the FCS plasma sustained by the helicity injection system. Conclusions obtained in the present study are as follows.

1. No magnetic surfaces close in the FC if the current applied in the divertor coil satisfies $I_c/I_p>2$.
2. The ratio of the magnetic flux in the separatrix to the total one depends strongly on the coil current.

REFERENCES

1 M. Nagata, T. Masuda, S. Naito, and T. Uyama, J. Phys. Soc. Jpn., **60** (1991) 3203.
2 A. Kamitani, S. Kaneko, T. Uyama, and K. Watanabe, J. Phys. Soc. Jpn., 56 (1987) 2755.
3 J. F. Thompson, F. C. Thames, and C.W. Mastin, J. Comp. Phys., **24** (1977) 274.

Simulation and Design of Applied Electromagnetic Systems
T. Honma (Editor)

Automatic numerical generation of finite elements for two-dimensional region of arbitrary shape

A. Kamitani and M. Miyauchi

Department of Electrical and Information Engineering, Faculty of Engineering,
Yamagata University, Johnan 4-3-16, Yonezawa, Yamagata 992, Japan

Abstract

A method for automatic numerical generation of finite elements in the two-dimensional multiply connected region of an arbitrary shape is presented. The region of a complex shape is divided into some simply connected subregions and each subregion is mapped onto the square region in the transformed plane. The inverse functions of the mapping are determined by the numerical solutions to the elliptic partial differential equations with Dirichlet boundary conditions. After the mapped square regions are systematically divided into a set of triangular elements, each element is inversely mapped onto the physical plane by use of the functions. Finite element data for the global region are determined from those of the subregions.

1. INTRODUCTION

Finite Element Method (FEM) has been widely used in all fields concerned with the numerical solution to the partial differential equations. The reason for this is that the region of a complex shape is easily treated by means of FEM. However, the region must be divided into a set of elements before the FEM program is applied. Although this process has been usually performed by use of a man-machine interface program, it costs us much time and labor to generate finite elements.

The purpose of the present paper is to develop the method for dividing a two-dimensional region of an arbitrary shape into a set of triangular elements. In the next section, we explain the basic idea of the method for dividing the simply connected region into a set of triangular elements. In the third section, the method for generating FEM data for the multiply connected region is summarized. Conclusions are presented in the final section.

2. BASIC IDEA

In using the FEM code, we need to have the following data ready beforehand as input data.

(i) total number of nodes N_{node}

(ii) total number of elements N_{elem}

(iii) coordinate of each node $\mathbf{r}_k$ $(k=1,2,\cdots,N_{node})$

(iv) global node number of the g-th local node in the e-th element $\sigma(e,g)$ $(e=1,2,\cdots,N_{elem})$
In the following, we call the above data finite element data or FEM data. Let us explain the method for dividing the simply connected region into a set of linear triangular elements and generating FEM data. The basic idea of our method consists of three steps.

First, a two-dimensional region of a complex shape in the physical plane (x,y) is mapped onto a square region in the transformed plane (ξ,η). The inverse functions, $x(\xi,\eta)$ and $y(\xi,\eta)$, of the mapping are numerically generated by the solution to the following elliptic partial differential equations[1, 2]:

$$\alpha\, \mathbf{r}_{\xi\xi} - 2\beta\, \mathbf{r}_{\xi\eta} + \gamma\, \mathbf{r}_{\eta\eta} + J^2[\, P(\xi,\eta)\, \mathbf{r}_\xi + Q(\xi,\eta)\, \mathbf{r}_\eta\,] = 0\,, \tag{1}$$

where $\mathbf{r}(\xi,\eta) = [x(\xi,\eta),\, y(\xi,\eta)]^T$, $\alpha = \mathbf{r}_\eta \cdot \mathbf{r}_\eta$, $\beta = \mathbf{r}_\xi \cdot \mathbf{r}_\eta$, $\gamma = \mathbf{r}_\xi \cdot \mathbf{r}_\xi$ and $J = x_\xi y_\eta - x_\eta y_\xi$. The functions $P(\xi,\eta)$ and $Q(\xi,\eta)$ are defined by

$$\begin{aligned} P(\xi,\eta) = &-\sum_{i=1}^{n} a_i^{(P)} \,\mathrm{sgn}(\xi - \xi_i^{(P)}) \exp(-c_i^{(P)}|\xi - \xi_i^{(P)}|) \\ &-\sum_{j=1}^{m} b_j^{(P)} \,\mathrm{sgn}(\xi - \bar{\xi}_j^{(P)}) \exp\{-d_j^{(P)}[(\xi - \bar{\xi}_j^{(P)})^2 + (\eta - \bar{\eta}_j^{(P)})^2]^{1/2}\}, \end{aligned} \tag{2a}$$

$$\begin{aligned} Q(\xi,\eta) = &-\sum_{i=1}^{n} a_i^{(Q)} \,\mathrm{sgn}(\eta - \eta_i^{(Q)}) \exp(-c_i^{(Q)}|\eta - \eta_i^{(Q)}|) \\ &-\sum_{j=1}^{m} b_j^{(Q)} \,\mathrm{sgn}(\eta - \bar{\eta}_j^{(Q)}) \exp\{-d_j^{(Q)}[(\xi - \bar{\xi}_j^{(Q)})^2 + (\eta - \bar{\eta}_j^{(Q)})^2]^{1/2}\}, \end{aligned} \tag{2b}$$

where $a_i^{(P)}$, $a_i^{(Q)}$, $b_j^{(P)}$, $b_j^{(Q)}$, $c_i^{(P)}$, $c_i^{(Q)}$, $d_j^{(P)}$ and $d_j^{(Q)}$ are positive constants, and $\xi_i^{(P)}$, $\bar{\xi}_j^{(P)}$, $\bar{\eta}_j^{(P)}$, $\eta_i^{(Q)}$, $\bar{\eta}_j^{(Q)}$ and $\bar{\xi}_j^{(Q)}$ are constant value within interval [0,1]. These functions adjust the spacing of the η- and ξ-curves in the x-y plane. The boundary conditions to eq.(1) are chosen such that the boundary of the square region in the ξ–η plane is mapped onto that of the region in the x-y plane. Equation (1) is solved numerically under these Dirichlet conditions and the inverse mapping functions are obtained.

Next, triangular elements are generated in the square region. By dividing the region into $N_\xi \times N_\eta$ pieces of small rectangles and inserting a diagonal line to each rectangle, the square region is divided into a set of triangular elements. Each rectangle has two diagonal lines and which of them to be inserted depends on the shape of its inverse image. The diagonal line, the end of which is the vertex of a maximum interior angle, is adopted to divide the quadrilateral in the physical plane. The small rectangle in the transformed plane is separated into two triangular elements by means of the image of the diagonal line. By using the above method, *we can avoid, to some extent, the appearance of deformed triangular elements in the physical plane.* Let us determine FEM data of the square region divided by this method. The total number of nodes and elements are easily evaluated as $N_{node} = (N_\xi+1)(N_\eta+1)$ and $N_{elem} = 2 N_\xi N_\eta$. Further, the coordinate of the k-th node can be determined as $(\Delta\xi\, \mathrm{Mod}(k-1, N_\xi+1), \Delta\eta\, [(k-1)/(N_\xi+1)])$. Here, $\Delta\xi$ and $\Delta\eta$ denote height and width of each rectangle, respectively. Throughout this paper, the symbol "/" means the division operator of integer. After inserting diagonal lines into all rectangles, four types of elements appear as shown in Figure 1. The global node number $\sigma(e,g)$ of the g-th local node in the e-th element depends on the type of the element. For the element of type (1), function $\sigma(e,g)$ is determined by

$$\sigma(e, 1)=(N_\xi+1)\{[(e-1)/2]/N_\xi\} + \mathrm{Mod}((e-1)/2, N_\xi) + 1, \tag{3a}$$

$$\sigma(e, 2)=(N_\xi+1)\{[(e-1)/2]/N_\xi\} + \mathrm{Mod}((e-1)/2, N_\xi) + 2, \tag{3b}$$

$$\sigma(e, 3)=(N_\xi+1)\{[(e-1)/2]/N_\xi+1\} + \mathrm{Mod}((e-1)/2, N_\xi) + 1. \tag{3c}$$

For the elements of the other type, function $\sigma(e,g)$ can be determined in a similar way.

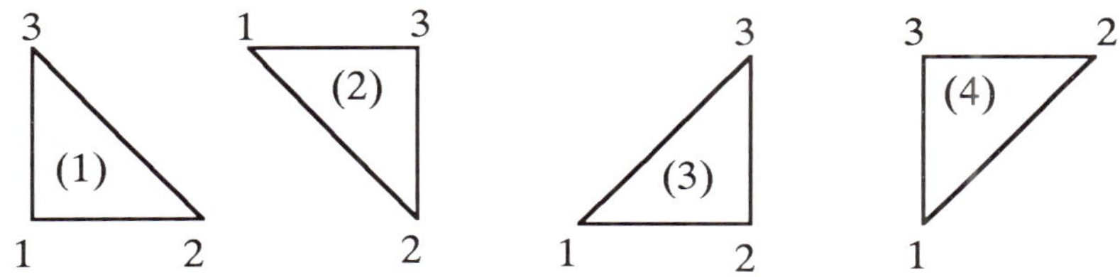

Figure 1. Four types of triangular elements. Numerals beside vertices indicate local node numbers.

Finally, we inversely map each element in the transformed plane onto the physical plane. In the inverse mapping, the total number of nodes and elements and function $\sigma(e,g)$ do not undergo a change. Therefore, we have only to determine the coordinates of nodes. By using the inverse mapping functions and the (ξ,η) coordinate of the k-th node, we can determine its (x,y) coordinate as

$$\mathbf{r}_k = \mathbf{r}\,(\Delta\xi\, \mathrm{Mod}(k-1, N_\xi+1),\ \Delta\eta\, [(k-1)/(N_\xi+1)]\,). \tag{4}$$

Thus, all FEM data in the simply connected region have been determined by use of the above three steps.

3. ELEMENT GENERATION IN MULTIPLY CONNECTED REGION

In general, a multiply connected region can be divided into some simply connected subregions which do not overlap with each other. By generating FEM data of each subregion and subsequently assembling all FEM data, we can constitute FEM data for the multiply connected region. The assembling method is written in detail in Ref.3.

By using the above method, a multiply connected region can be easily divided into a set of triangular elements. Typical examples of the element generation in multiply connected regions are shown in Figures 2 and 3. The method explained in the present section is also effective to generate FEM data for a simply connected region with an extremely complex shape.

Figure 2. A plate containing many holes

Figure 3. A gear wheel

4. CONCLUSIONS

A general method for generating triangular elements in the two-dimensional multi-connected region of an arbitrary shape has been developed. With its use, the element generation in the region with a complex shape and any number of holes is not inherently more difficult than the problem with simple geometry. Hence, we may conclude that our method is powerful to generate finite elements in a region with complex geometry.

Although present effort is confined to the generation of linear elements in the interest of space of this paper, *our method is easily extendable to that of higher order elements*.

REFERENCES

1 J. F. Thompson, F. C. Thames and C. W. Mastin, J. Comp. Phys. **15** (1974) 299.
2 J. F. Thompson, F. C. Thames and C. W. Mastin, J. Comp. Phys. **24** (1977) 274.
3 M. Miyauchi and A. Kamitani, to be published in Trans. Jpn. Soc. Ind. Appl. Math.

Simulation and Design of Applied Electromagnetic Systems
T. Honma (Editor)

Preprocessor for the 3D Boundary Element Method Based on the Constructive Solid Geometry

Masakuni Nishino and Tsuyoshi Takeda

Dept. of Electrical Eng., Faculty of Eng., Hokkaido University
Kita 13, Nishi 8, Kita-ku, Sapporo, 060, Japan

ABSTRACT

This paper describes a technique of the preprocessing for the 3D boundary element method. When we solve the problem having 3D territories by the boundary element method, it is difficult for us to input correctly a large quantity of data for boundary elements by the human power.

Then we need the preprocessor which automatically generates the geometrical mesh data by inputting the minimum information about the problem. We have developed the preprocessor for the 3D boundary element method based on the constructive solid geometry (CSG method). The method combines primitives to represent the territory of the problem under consideration using logical operations such as AND, OR and SUBTRACT between primitives . The primitives have meshes as the boundary elements on the surface. However, the method can't make the complex territory having many materials because all primitives is assumed to have the same material.

To deal with the complex material, we introduced the material constant as parameter into usual logical operations of the constructive solid geometry. The parameter is used for the information about OR operation. If the parameters of the two primitives under operation are different, a new boundary surface between the two primitives is made with meshes. Namely, the new method of the mesh generation proposed here can make meshes on the boundary surface between the complex territories.

1. PREPROCESSOR

We have developed the preprocessor for the 3D boundary element method based on the CSG method. The method combines six primitives (1. an half space, 2. a rectangular parallelepiped, 3. an ellipsoid, 4. a circular cone, 5. a cylinder, 6. a sweep) each other to represent the territory of the problem using logical operations such as AND, OR, and SUBTRACT between primitives.

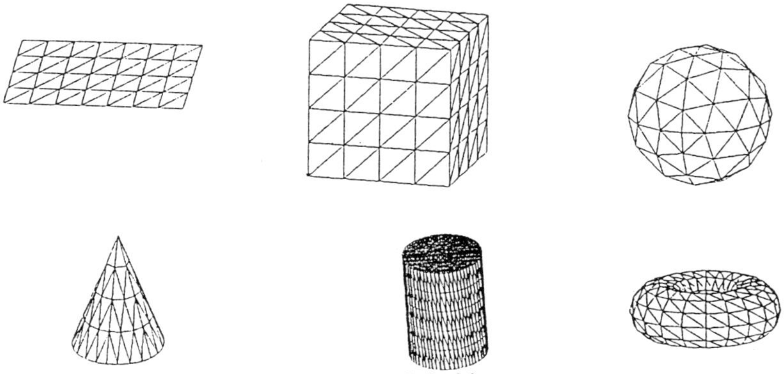

Figure 1. Primitives

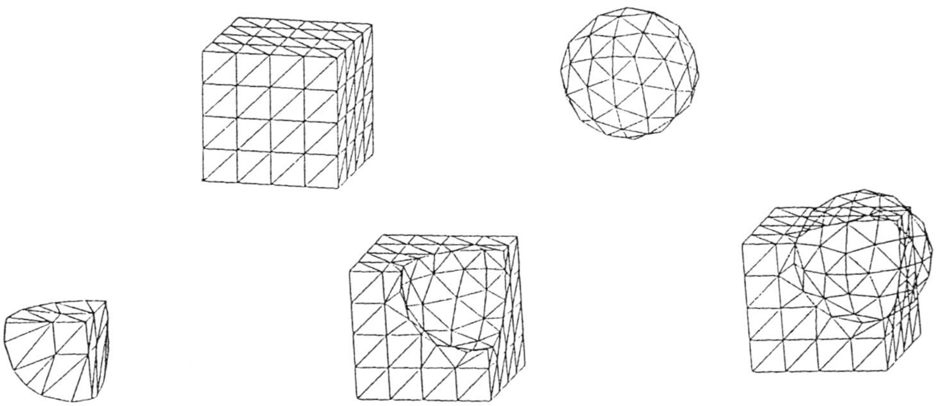

Figure 2. The result of the logical operations

2. THE MESH GENERATION ON THE BOUNDARY SURFACE BETWEEN THE COMPLEX TERRITORIES

In the CSG method , we represent the territory of the problem as the result of the logical operation. However it is unnecessary for us to have the concept of the complex materials about the logical operation such as AND and SUBTRACT. In opposition to these, the shape of the territory of the problem is different from either single material or complex materials under OR operation.

When we consider the problem about the complex materials, the order of the OR operation; either (A OR B) or (B OR A), is important. If we change the order of OR operation, a different primitive is made.

In this paper, we define that the late operated primitive has priority over the primitives operated previously.

We discuss the case of the complex materials as OR operation as follows.

3. THE PROCESS OF THE LOGICAL OPERATION

3.1 Let us perform OR operation between primitive A having material ***a*** and primitive B having material ***b***.
3.2 If both A and B are made by the single material, furthermore if ***a***=***b***, the preprocessor calculates In-Out function and carry out usual OR operation.
If both A and B, or either A or B made by the complex materials, or ***a*** isn't equal to ***b***, it calculates In-Out function corresponds to the complex materials. So it generates the geometrical mesh to the inner boundary surface.
In these operation , the In-Out function takes a role that it evaluates the co-ordinates of mesh point either the inside of the primitive, on the boundary surface or the outside of the primitive, element by element .
3.3 It matches the points near the cross line between primitive A and primitive B .
3.4It matches the adjacent elements of the cross line between primitive A and primitive B.

4. IN-OUT FUNCTION OF THE COMPLEX MATERIALS

We assume here to do the logical operation (OR); the primitive ***B*** to the primitive ***A*** , where ***A*** is made by material **a** and ***B*** is made by material **b** as shown in Fig.3.

The inside point such as **a**, **b** or **c** remains in the inside after the OR operation. The point on the boundary such as **d**, **e** or **f** is still on the boundary after the OR operation, and **f** is also the edge point of the inner boundary surface. The inside point of the primitive ***A*** and the boundary surface point of the primitive B such as **g** is the inner boundary surface after the OR operation. The point on the boundary of the primitive ***A*** and the inside point of the primitive ***B*** such as **h** is deleted because the primitive ***B*** has priority over the primitive ***A***, as shown in Fig.4 .

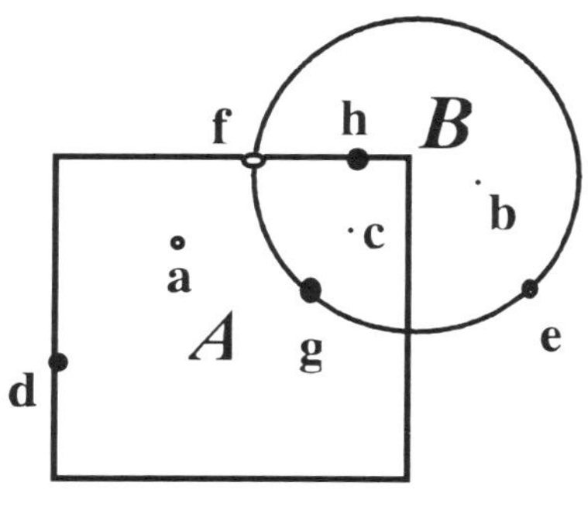

Figure 3.The point of the OR operation

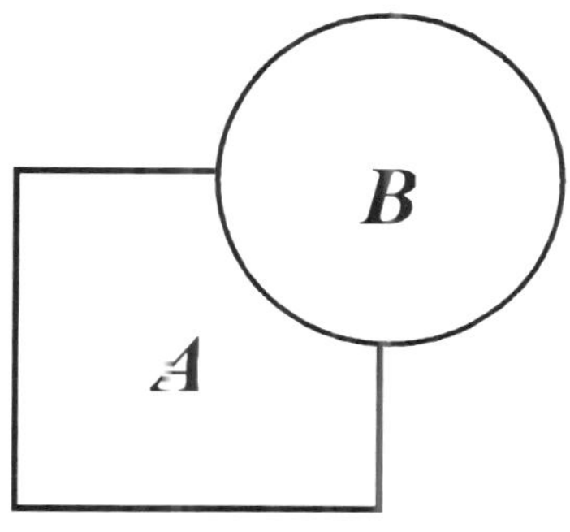

Figure 4.The result of the OR operation

Next, we consider OR operation; the primitive C having material 2 onto the primitive having complex materials. The primitive having complex materials has been made by the another OR between the primitive A having material 1 and the primitive B having material 2 as shown in Fig. 5. As the result, we obtain a new one as shown in Fig. 6. In these procedures, first ,we compare the material A with C. The material isn't equal, so the boundary surface between A and C is remained as the inner boundary surface. Next, we compare the material B with C, in this case, the material is equal, so the boundary surface between B and C is deleted. The result is shown in Fig.7. Figure.8 shows that of Fig. 7 in 3D. Figure.9 represents the separated constituents of Fig. 8 to see the inner boundary surface.

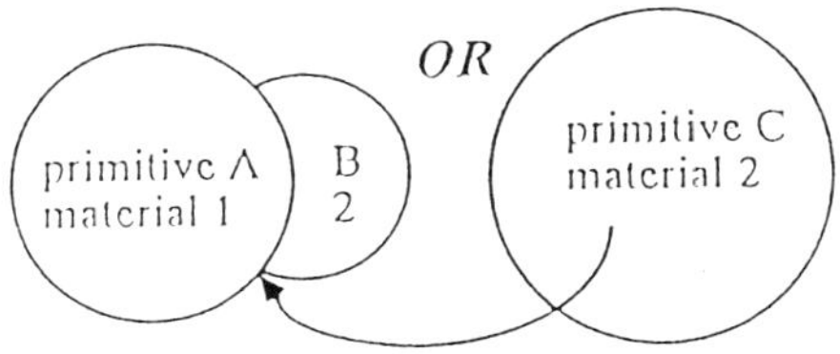

Figure 5.

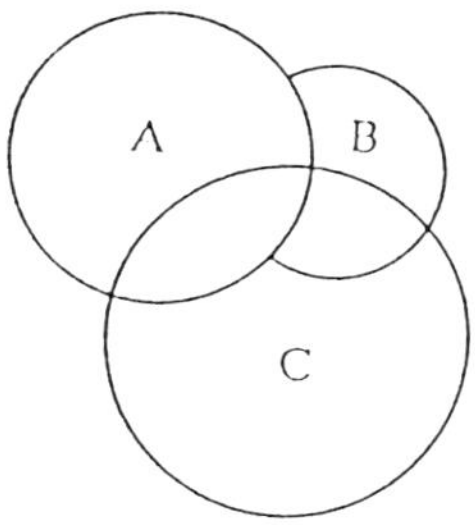

Figure 6.

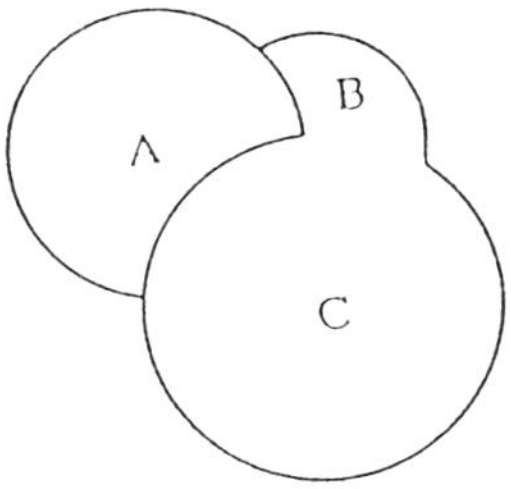

Figure 7.

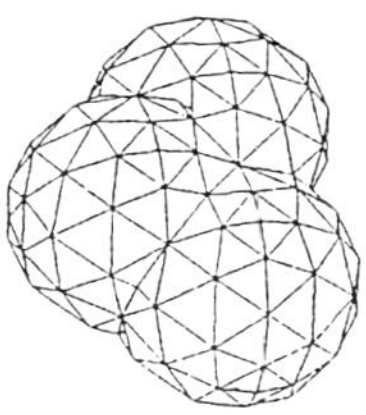

Figure 8.

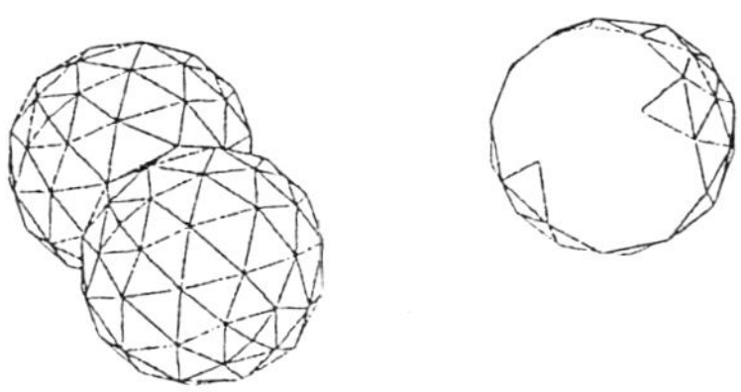

Figure 9.

5. CONCLUSION

We could generate the geometrical mesh on the boundary surface between the complex territories. The preprocessor is extended to treat the complex materials.

References

1. N. Okinno, "Methodology for Computer Aided Design", Yokendo Tokyo(1982).
2. H.Honma, T. Kuwahara and T. Takeda," Three Dimensional Automatic Mesh Generation for the Boundary Element Method", I.E.E. of Japan , Vol.110A, No.9(1990), pp.623-629,

Simulation and Design of Applied Electromagnetic Systems
T. Honma (Editor)

Non-planar lateral oscillation of the wire in a system of wire and plate electrodes

Y. Itoh[a] , M. Yoshizawa[a] , Q.Z. Du[a] , M. Sugawara[b]

[a] Department of Mechanical Engineering, Keio University, 3-14-1 Hiyoshi, Kouhoku-ku, Yokohama-shi, Kanagawa, Japan

[b] Diafoil-Hoechst Co., Ltd, 347 Inokuchi, Santo-cho, Sakata-gun, Shiga, Japan

Abstract

It is well-known that the self-excited lateral oscillation of a wire electrode occurs when high electric voltage is applied between a wire and infinite plate electrodes in an electrostatic field. In order to clarify the outbreak mechanism of the above self-excited oscillation, the non-planar lateral oscillation of the wire is examined, in the case where it is excited by the electrostatic force in the absence of corona discharge.

1 Theoretical Analysis

A system of wire and plate electrodes, as shown in Figure 1, consists of a plate electrode of length L and a wire electrode of length L, static tension T, diameter R, cross-sectional area A, mass per unit length ρA and Young's modulus E, placed horizontally above the plate electrode. The distance between the wire and the plate electrodes is H. u, v and w are the displacements measured from the static equilibrium position in the horizontal, vertical and longitudinal directions respectively, and functions of the coordinate z and time t. c is the coefficient of viscous damping. v_{st} is the static lateral deflection of the wire due to gravity. In the absence of corona discharge, if the applied voltage of the wire electrode is $\phi_0 \sin \nu t$, then the electrostatic force acting on the wire is given by [1]

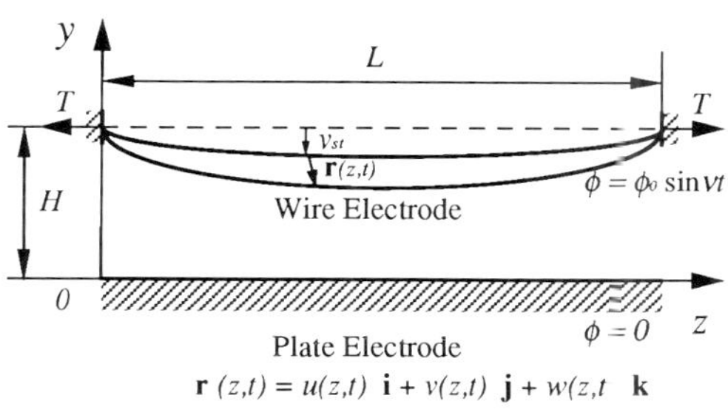

Figure 1 Analytical Model

$$F_e = -\frac{\pi \varepsilon_0 \phi_0^2}{H\{\ln(2H/R)\}^2} \sin^2 \nu t \ , \tag{1}$$

where ε_0 is the permittivity of free space (8.854×10^{-12} N·m^2/C^2). It is clear from equation (1) that the frequency of the electrostatic force is twice the frequency of the applied voltage of the wire electrode.

When the dimensionless variables which carry the asterisk

$$u = Hu^* \ , \ v = Hv^* \ , \ v_{st} = Hv_{st}^* \ , \ z = Lz^* \ , \ t = \frac{L}{\pi}\sqrt{\frac{\rho A}{T}} t^* \ , \tag{2}$$

are introduced, the dimensionless equations governing the motions of the wire are expressed as follows [2]:

$$\frac{\partial^2 u}{\partial t^2} + 2\mu \frac{\partial u}{\partial t} = \frac{1}{\pi^2}\frac{\partial^2 u}{\partial z^2} + \frac{1}{\pi^2} h^2 \gamma \frac{\partial^2 u}{\partial z^2} \int_0^1 \left[\frac{dv_{st}}{dz}\frac{\partial v}{\partial z} + \frac{1}{2}\left(\frac{\partial u}{\partial z}\right)^2 + \frac{1}{2}\left(\frac{\partial v}{\partial z}\right)^2 \right] dz \ , \tag{3}$$

$$\frac{\partial^2 v}{\partial t^2} + 2\mu \frac{\partial v}{\partial t} = \frac{1}{\pi^2}\frac{\partial^2 v}{\partial z^2} + \frac{1}{\pi^2} h^2 \gamma \left(\frac{d^2 v_{st}}{dz^2} + \frac{\partial^2 v}{\partial z^2} \right) \int_0^1 \left[\frac{dv_{st}}{dz}\frac{\partial v}{\partial z} + \frac{1}{2}\left(\frac{\partial u}{\partial z}\right)^2 + \frac{1}{2}\left(\frac{\partial v}{\partial z}\right)^2 \right] dz + \frac{f}{h}\sin^2 \Omega t \ , \tag{4}$$

$$v_{st} = -\frac{\Gamma}{2h}(z^2 - z) \ , \tag{5}$$

$$u(0,t) = u(1,t) = v(0,t) = v(1,t) = 0 \ , \tag{6}$$

where $\mu = \frac{1}{2\pi}\frac{cL}{\sqrt{\rho AT}}$, $h = \frac{H}{L}$, $\Gamma = \frac{\rho ALg}{T}$, $\gamma = \frac{EA}{T}$, $\Omega = \frac{L}{\pi}\sqrt{\frac{\rho A}{T}}\nu$, $f = \frac{1}{\pi^2}\frac{L}{T}F_e$. The longitudinal inertia is neglected as is done in the case of strings. The asterisks indicating the dimensionless variables are omitted in equations (3) through (6) and hence forward.

It is assumed that the wire oscillates with the first mode of the wire. Therefore the solutions of the linearized equations (3) and (4) subject to the boundary conditions (6) are assumed

$$u(z,t) = \xi(t) \sin \pi z \ , \ v(z,t) = \zeta(t) \sin \pi z \ . \tag{7}$$

Substituting equation (7) into equations (3) and (4),

$$\ddot{\xi} + 2\mu\dot{\xi} + \xi = -2q\xi\zeta - p\xi^3 - p\xi\zeta^2 \ , \tag{8}$$

$$\ddot{\zeta} + 2\mu\dot{\zeta} + \zeta = -\beta\zeta - q\xi^2 - 3q\zeta^2 - p\xi^2\zeta - p\zeta^3 + 2\delta \sin^2 \Omega t \tag{9}$$

are obtained, where $\beta = \frac{8}{\pi^4}\gamma\Gamma^2$, $p = \frac{1}{4}\pi^2 h^2 \gamma$, $q = \frac{1}{\pi} h\gamma\Gamma$, $\delta = \frac{2f}{\pi h}$. As a typical case, the prediction derived from equations (8) and (9) is as follows: if the excitation frequency of the electrostatic force is varied near fundamental natural frequency of the wire oscillation, i.e. $2\Omega = 1 + \varepsilon\sigma$, then the horizontal component of the wire oscillation except the vertical component is excited with the fundamental natural frequency of the wire oscillation due to the term $p\xi\zeta^2$ in equation (8). ε is a dimensionless small parameter, and σ is the detuning parameter. Therefore non-planar lateral oscillation of the wire will occur.

The method of the multiple scale is used to obtain the approximate solutions of equations (8) and (9). The coefficients of the equations (8) and (9) are assumed as follows: $\mu = \varepsilon\hat{\mu}$, $p = \hat{p}$, $q = \varepsilon^{1/2}\hat{q}$, $\beta = \varepsilon\hat{\beta}$, $\delta = \varepsilon^{3/2}\hat{\delta}$. Letting $\xi = \varepsilon^{1/2}a_1\cos(t+\varphi)$, $\zeta = \varepsilon^{1/2}a_2\cos(t+\psi)$, then the steady-state solutions are obtained. Theoretical result for the steady-state amplitudes of the wire oscillation is shown in Figure 2.

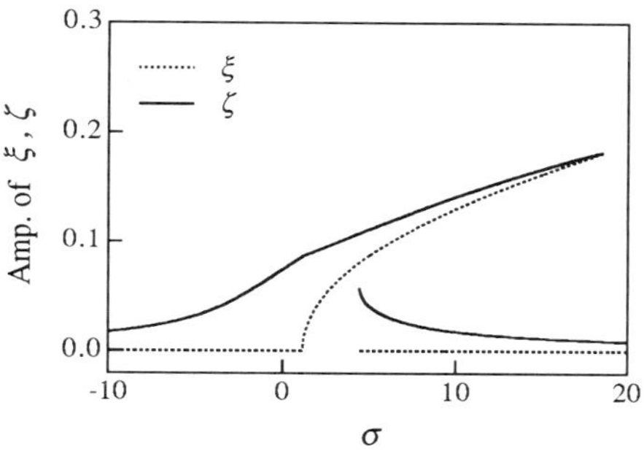

Figure 2 Theoretical Result: $\hat{\mu} = 0.419$, $\hat{\beta} = 0.0301$, $\hat{p}$=1.11, $\hat{q}$= 0.176, $\hat{\delta}$=11.1, $\varepsilon = 0.001$

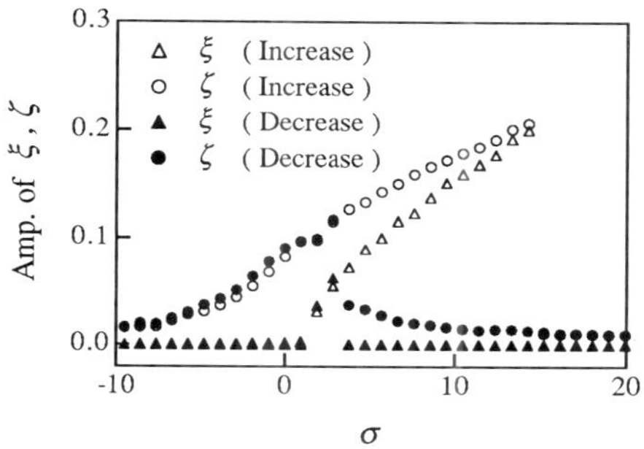

Figure 3 Experimental Result: $\hat{\mu} = 0.419$, $\hat{\beta} = 0.0301$, $\hat{p}$=1.11, $\hat{q}$= 0.176, $\hat{\delta}$=11.1, $\varepsilon = 0.001$

2 Experiment

Experiment on the non-planar lateral oscillation of the wire electrode is conducted to confirm the theoretical prediction in the previous section. It is conducted with the wire electrode of 85mm length, 0.15mm diameter, static tension 4.4N, mass per unit length 3.40×10^{-4}kg/m, Young's modulus 3.63×10^{11}N/m^2. The fundamental natural frequency of the wire oscillation is 67.0Hz. The distance between the wire and the plate electrodes is 15mm. The displacements of u, v are measured by the optical displacement meters.

Figure 3 shows the amplitudes of the steady-state lateral oscillation of the wire when the frequency of the electrostatic force is varied very slowly back and forth through resonance of the wire, i.e. the frequency of the applied voltage is varied half of natural frequency of the wire oscillation, while the amplitude of the applied voltage is held constant 3.1kV. Figure 4 shows the motions of the wire in $x - y$ plane in the case where σ increases and decreases.

When σ increases, non-planar lateral oscillation of the wire occurs at the value of $1.86 < \sigma < 14.25$, and the jump phenomenon occurs and the motion changes from non-planar back to planar. When σ decreases, the jump phenomenon occurs at the value of $\sigma = 3.80$, and the motion changes from planar to non-planar.

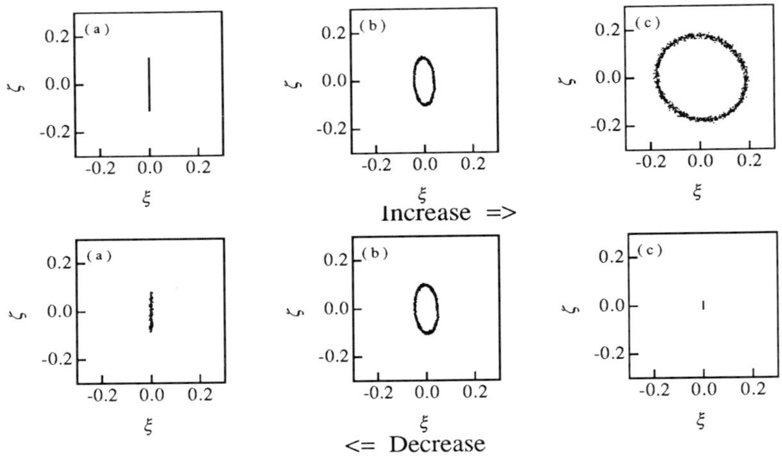

Figure 4 Motions of the wire electrode: (a) $\sigma = -3.80$ (b) $\sigma = 2.85$ (c) $\sigma = 12.35$: $\hat{\mu}$=0.419, $\hat{\beta}$=0.0301, $\hat{p}$=1.11, $\hat{q}$= 0.176, $\hat{\delta}$=11.1 , $\varepsilon = 0.001$

3 Discussion

It is predicted from the equations (8) and (9) that if the excitation frequency of the electrostatic force is varied near twice natural frequency of the wire oscillation, i.e. the excitation frequency of the applied voltage is varied near the natural frequency of the wire oscillation, then the vertical component of the wire oscillates with the twice fundamental frequency of the wire oscillation. The horizontal component of the wire oscillates with the fundamental natural frequency of the wire due to the term $2q\zeta\xi$ in equation (8). However, non-planar lateral oscillation of the wire could not be observed in the experiment.

4 Conclusion

Non-planar lateral oscillation of the wire electrode is studied both theoretically and experimentally, in the case where the wire is excited by the electrostatic force.

As a result, the horizontal component of the wire oscillates with the fundamental natural frequency of the wire oscillation, when the applied voltage is varied near half of the natural frequency of the wire oscillation. When the frequency of the electrostatic force reaches the critical value of the excitation frequency, the non-planar lateral oscillation of the wire was observed in the experiment.

Furthermore, the theoretical result agrees with the experimental one qualitatively.

5 Reference

1. YOSHIZAWA, M., Trans. Jpn. Soc. Mech. Eng.,(in Japanese), vol.58,548,C(1992-4)
2. RAO, G.V. and IYENGER, R.N., (1991), J. Sound Vib., 149(1), 24-41

Simulation and Design of Applied Electromagnetic Systems
T. Honma (Editor)

Evaluation of critical magnetic viscous damping ratios for plate vibrations with different boundary conditions

T. Takagi and J. Tani

The Institute of Fluid Science, Tohoku University
Katahira 2-1-1, Aoba-ku, Sendai 980, Japan

Abstract

This paper describes the numerical evaluation of magnetic viscous damping coefficients of thin plate vibration under magnetic field. The authors of this paper apply the MMD (Modal magnetic damping) method to the evaluation of critical magnetic viscous damping ratios of a plate with different boundary conditions. The relationship between the evaluated values and the boundary conditions from the view point of vibration modes, material constants, size and natural frequencies are discussed.

1. INTRODUCTION

The coupling effect between eddy current and deformation of structure plays an important role for the dynamic behavior in magnetic field. The coupling effect of structure under transient field was investigated by some researchers [1-4]. In the previous numerical studies concerning the magnetic viscous damping the eddy current was calculated from the governing equation considering deflection. Lorentz force and deflection have been analyzed using the calculated eddy current in each time step. Hence the calculations of eddy current and deformation have been fully coupled.

The authors of this paper have already proposed a new method, "MMD (Modal magnetic damping) method," for electromagnetomechanical coupling analysis [5]. In the proposed method eddy current and plate bending analyses can be performed separately but considering the coupling effect between them. They developed the method for a thin plate and applied it successfully to the TEAM problem #12 "FELIX copper cantilevered beam plate [1-3]." They also clarified that magnetic viscous damping coefficient is proportional to the square of magnetic induction. The proposed method can evaluate magnetic viscous damping coefficients for natural vibration modes of a structure in the field. In this paper the authors apply the method to the evaluation of critical magnetic viscous damping ratios of a plate with different boundary conditions. They discuss the relationship between the evaluated values and several boundary conditions from the view point of vibration modes, material constants (Young's modulus, mass density, conductivity), thickness and natural frequencies.

2. NUMERICAL METHOD

Since the authors of this paper have already presented the MMD method [5], it is briefly explained in this section. Deformation of a plate can be analyzed using FEM. Matrix form of the equation of motion becomes the following expression.

$$\mathbf{M}\ddot{\mathbf{a}} + \mathbf{C}\dot{\mathbf{a}} + \mathbf{K}\mathbf{a} = \mathbf{f} \tag{1}$$

Neglecting viscous damping and force terms in eq. (1), we can obtain eigenvalues (natural frequencies) and eigenvectors (natural modes). In the present method an equivalent magnetic viscous damping coefficient can be obtained from the condition that dissipated energy due to magnetic damping is equal to joule heating due to induced current. When a plate vibrates on a natural vibration mode, the work done by an equivalent magnetic damping force is expressed as,

$$W_1 = \int (\mathbf{C}\dot{\mathbf{a}}_j)^T d\mathbf{a}_j. \tag{2}$$

Then the work done by the force during a period can be expressed and the relationship between the work W_1 and magnetic viscous damping coefficient c_j is given as,

$$W_1 = \int_0^{\frac{2\pi}{\omega_j}} \dot{\mathbf{a}}_j^T \mathbf{C}\dot{\mathbf{a}}_j dt = \pi\omega_j c_j. \tag{3}$$

On the other hand energy lost as joule heating during a period is given as follows:

$$W_2 = \frac{1}{\sigma}\int dV \int_0^{\frac{2\pi}{\omega_j}} J^2 dt. \tag{4}$$

The relationship, $W_1 = W_2$, gives an equivalent magnetic viscous damping coefficient c_j and equivalent magnetic critical damping ratio ζ_j as,

$$c_j = \frac{W_2}{\pi\omega_j}, \quad \zeta_j = \frac{c_j}{2m_j\omega_j} = \frac{W_2}{2\omega_j^2\pi} \tag{5}$$

where $m_j = 1$ [5]. Since the induced current density is proportional to magnetic induction, W_2 ,c_j and ζ_j are proportional to the square of magnetic induction.

3. NUMERICAL RESULTS AND DISCUSSION

The method described above is applied to the numerical evaluation of critical magnetic viscous damping ratios of a thin plate. In finite element calculation 65 nodes and 96 triangular finite elements are used here. Length and width of a plate are 0.48 and 0.1 m. Static uniform field is applied along the longitudinal direction of the plate. Table 1 shows six cases of considered boundary conditions. Figure 1 shows that as the case number increases, ζ_j becomes smaller (cf. Table 1). This means that case 1 (a cantilever plate) has strong magnetic viscous damping effect. In the following the change of critical magnetic viscous damping ratios are discussed for the case 1.

Natural vibration modes obtained for a cantilever plate (case 1) are shown in figure 2. First, second and fourth modes show bending modes and third one is a torsional one. The ratio for the third mode is very small because external field acts on a plate in longitudinal direction. Figures 3 - 6 shows the change of critical magnetic viscous damping ratio, ζ_j, and natural frequency for $B = 0.2$T when Young's modulus, mass density, thickness and conductivity change. Natural frequency increases in proportion to $E^{\frac{1}{2}}$, and ζ_j is proportional to $E^{-\frac{1}{2}}$ (cf. Figure 3). Natural frequency decreases in proportion to $\rho^{-\frac{1}{2}}$, ζ_j is also proportional to $\rho^{-\frac{1}{2}}$ for lower mode (cf. Figure 4). Natural frequency increases in proportion to thickness, h. But ζ_j non-linearly decreases as thickness increases (cf. Figure 5). When conductivity, σ , increases, natural frequency is constant. ζ_j increases in proportion to conductivity in the first mode (cf. Figure 6). The relationship obtained above holds for the lower modes (lower frequencies) because the higher frequency makes the non-local term (eddy current due to current itself) larger and the relationship does not always hold.

4. CONCLUSIONS

Critical magnetic viscous damping ratios, ζ_j, were evaluated for a plate under uniform magnetic field with several boundary conditions. As the natural frequency increases, ζ_j becomes smaller. The relationship among ζ_j, E , ρ , h and σ were discussed. It was found that ζ_j is almost proportional to $E^{-\frac{1}{2}}\rho^{-\frac{1}{2}}\sigma$, and ζ_j non-linearly decreases as h increases.

This work is partly supported by Grant-in-Aid for Scientific Research (C) (No. 04650208), the Ministry of Education, Science and Culture in Japan.

Table 1. Considered boundary conditions

Case	Boundary Conditions
1	
2	
3	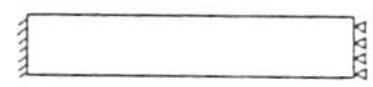
4	
5	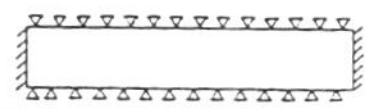
6	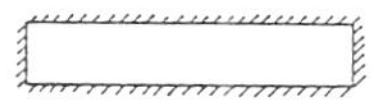

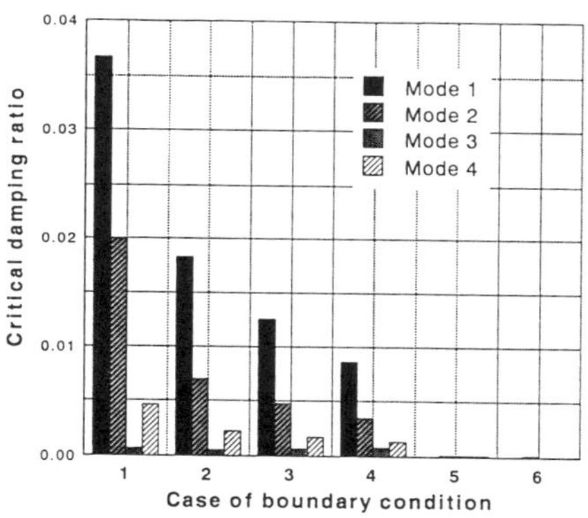

(a) Critical magnetic viscous damping effect

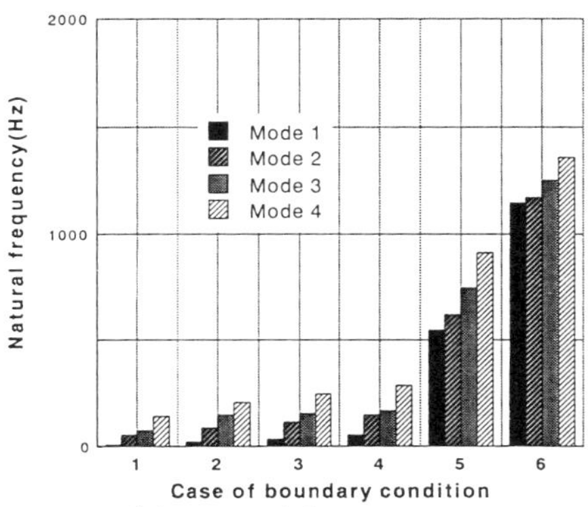

(b) Natural frequency

Figure 1. Boundary conditions and damping effect

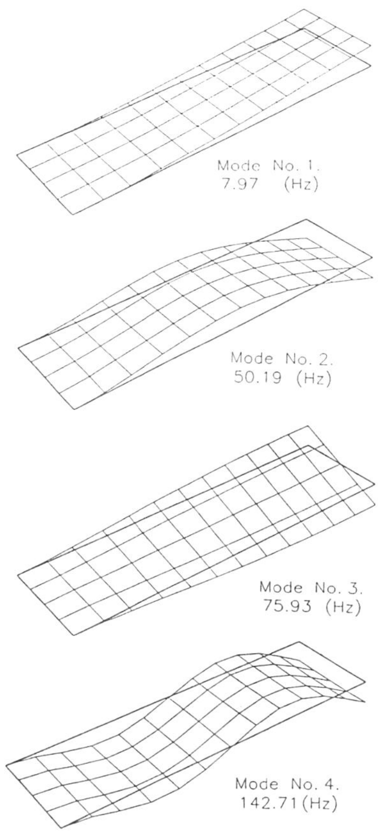

Figure 2. Natural frequency and mode

5. REFERENCES

1 T. Q. Hua, M. J. Knott, L. R. Turner, and R. B. Wehrle, Fusion Techn., 10 (1986) 1047.
2 T. Morisue, IEEE Trans. Magn., 26(1990) 540.
3 T. Takagi, J. Tani, S. Matsuda and S. Kawamura, IEEE Trans. Magn., 28 (1992)1259.
4 T. Takagi, J. Tani and S. Kawamura, Trans. Japan Soc. Mech. Eng., 58 (1992) 3537.
5 T. Takagi and J. Tani, Intern. J. Appl. Electromagn. in Mater. (in press).

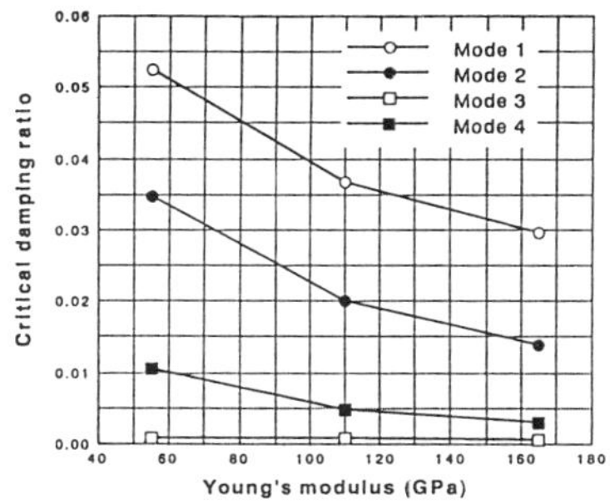

(a) Critical magnetic viscous damping ratio

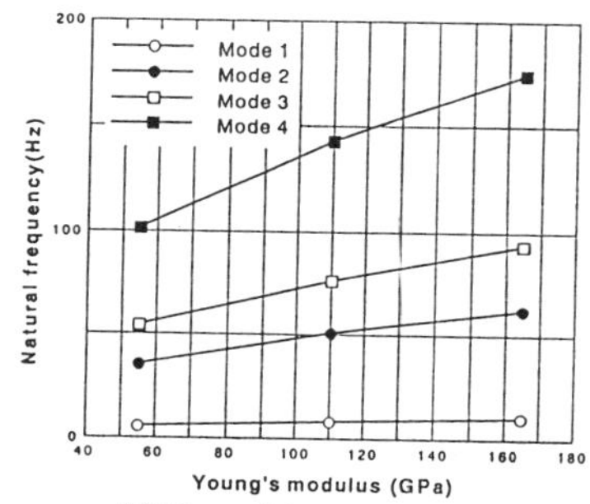

(b) Natural frequency

Figure 3. Young's modulus and damping effect

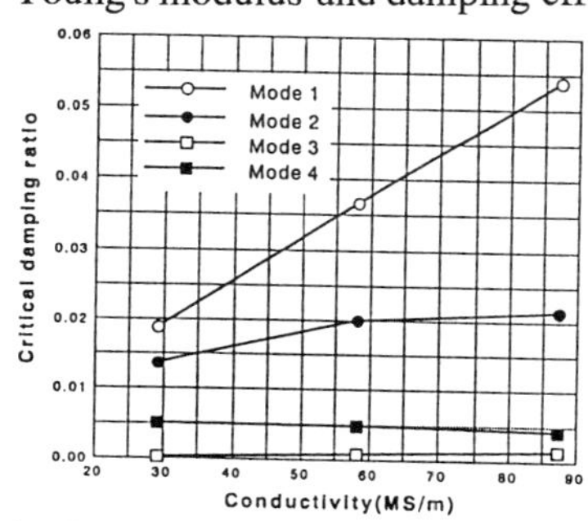

Figure 6. Conductivity and damping effect

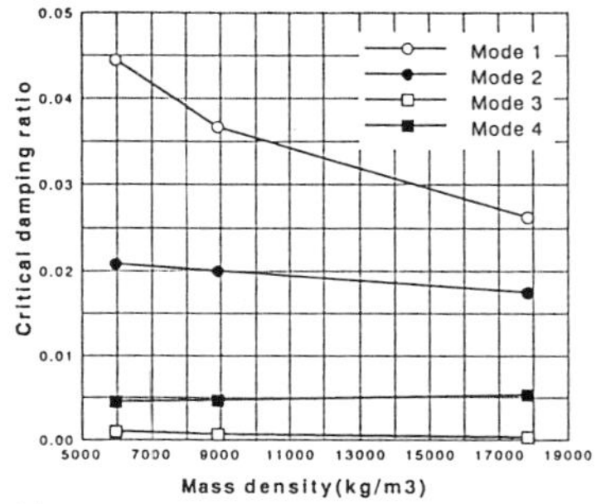

(a) Critical magnetic viscous damping ratio

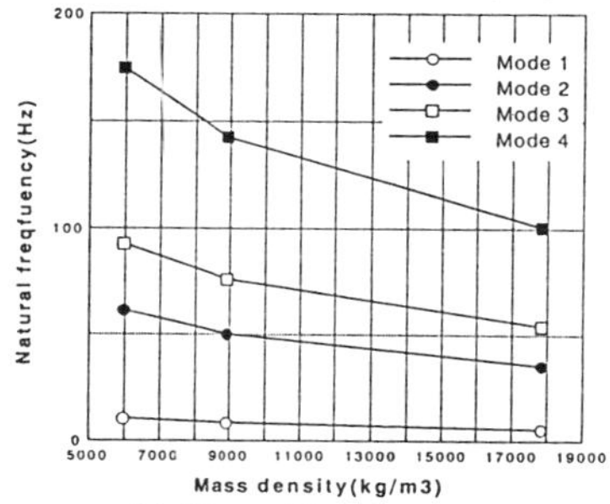

(b) Natural frequency

Figure 4. Mass density and damping effect

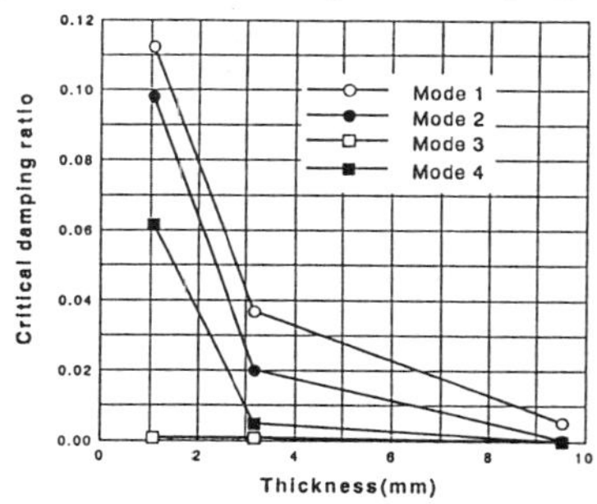

(a) Critical magnetic viscous damping ratio

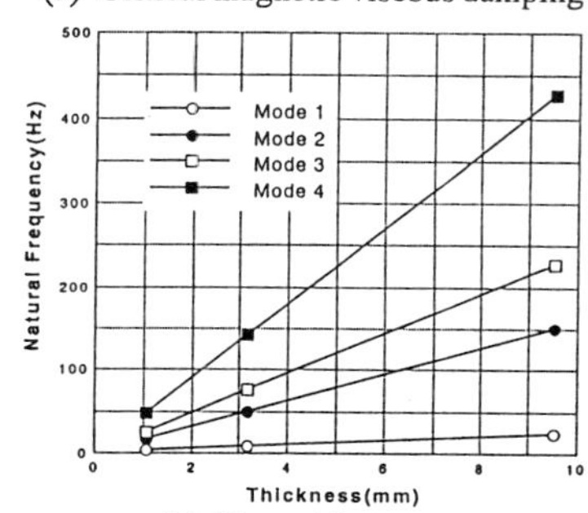

(b) Natural frequency

Figure 5. Thickness and damping effect

Simulation and Design of Applied Electromagnetic Systems
T. Honma (Editor)

A Large Deformation Theory of Inelasticity Subjected to Electromagnetic Forces

I. Nishiguchi and J. Yamamoto

Department of Mechanical Systems Engineering,
Kanagawa Institute of Technology,
1030 Shimoogino, Atsugi, Kanagawa, 243-02, Japan

Abstract

A large deformation theory of inelasticity is proposed in which interaction between electromagnetic fields with deformable solids is taken into account. The weak forms of Maxwell's equations together with the principle of virtual work (i.e. weak forms of the balance of momentum) in a reference configuration constitute governing equations for the FE formulation. By choosing a reference configuration that is instantaneously coincident with the current configuration, the updated Lagrangian formulation is obtained.

In the present formulation, the load correction matrix is generalized to incorporate the deformation dependent electromagnetic forces. In the community of the FEM computation, the load correction matrix has been employed mainly in the simulation of metal forming to take into account the surface forces such as a hydrostatic pressure loading or tool forces [1].

1. WEAK FORM OF MAXWELL'S EQUATIONS

Weak forms of the Ampere's law and the Faraday's law in a fixed reference configuration can be derived by the direct transformation from spatial weak forms. The results coincide with the weak forms that are obtained from the local expressions by Lax and Nelson[4] though we made a distinction between the covariant and contravariant vector explicitly. Note it is not necessary the reference configuration of these equations coincides with the undeformed configuration.

1.1 Ampere's Law

Local form of the Ampere's law is given in [2]

$$\nabla\times\boldsymbol{\eta}-\tilde{\mathbf{d}}=\boldsymbol{\iota} \tag{1}$$

with

$$\boldsymbol{\eta}=\mathbf{h}-\mathbf{v}\times\mathbf{d},\quad \tilde{\mathbf{d}}=\frac{\partial\mathbf{d}}{\partial t}+\nabla\times(\mathbf{d}\times\mathbf{v})+(\mathbf{v}\cdot\mathbf{d})\mathbf{v},\quad \boldsymbol{\iota}=\mathbf{i}-q_e\mathbf{v} \qquad (2)(3)(4)$$

where $\mathbf{h}$ is the magnetic field, $\mathbf{v}$ the velocity, $\mathbf{d}$ the electric displacement vector, $\mathbf{i}$ the total current density, q_e the charge density. $\nabla\cdot\mathbf{d}=q_e$ is assumed in the derivation of (1). We note $\tilde{\mathbf{d}}$ is the vector equivalent of the Truesdell stress rate[3] and the following relation holds:

$$\tilde{\mathbf{d}}=\frac{1}{J}L_{\mathrm{v}}(J\mathbf{d})=\frac{1}{J}F\frac{d}{dt}(JF^{-1}\mathbf{d}) \tag{5}$$

where F is the deformation gradient, $J=\det(F)$, L_{v} the Lie derivative.

Contracting (1) with a vector function $\mathbf{w}$ and integration by parts give

$$\int_{B(t)}\boldsymbol{\eta}\cdot(\nabla\times\mathbf{w})\cdot dv-\int_{B(t)}\tilde{\mathbf{d}}\cdot\mathbf{w}dv-\int_{B(t)}\boldsymbol{\iota}\cdot\mathbf{w}dv+\int_{S\eta(t)}\boldsymbol{\eta}_{\tan}\cdot\mathbf{v}ds=0 \tag{6}$$

where the boundary condition $\mathbf{n}\times\boldsymbol{\eta}=\boldsymbol{\eta}_{\tan}$ on S_η is appended and $\mathbf{w}$ is assumed to be zero on the surface $S-S_{\boldsymbol{\eta}}$.

Eq.(6) is the transposed weak form of the Ampere's law in the current configuration and it can be transformed to a fixed reference coordinate $(X^1,\ X^2,\ X^3)$ with $\mathbf{W}=W_K$ as a vector field:

$$\int_{BR}\mathbf{H}\cdot\mathrm{CURL}\mathbf{W}dV-\int_{BR}\frac{\partial\mathbf{D}}{\partial t}\cdot\mathbf{W}dV-\int_{BR}\mathbf{I}\cdot\mathbf{W}dV+\int_{SR}\mathbf{H}_{\tan}\cdot\mathbf{W}dS=0. \tag{7}$$

In this expression, the reference forms of $\boldsymbol{\eta}$, $\mathbf{d}$ and $\boldsymbol{\iota}$ are introduced:

$$\mathbf{H}=\mathrm{H}_K=\frac{\partial x^i}{\partial X^K}\eta_i=\phi^*(\boldsymbol{\eta}) \tag{8}$$

$$\mathbf{D}=D^K=J\frac{\partial X^K}{\partial x^i}d^i=J\phi^*(\mathbf{d}),\qquad \mathbf{I}=I^K=J\frac{\partial X^K}{\partial x^i}\iota^i=J\phi^*(\boldsymbol{\iota}) \tag{9)(10}$$

where $\phi^*(\cdot)$ is the pull back operation, $\mathbf{D}$ and $\mathbf{I}$ are the Piola transformation [4] of $\mathbf{d}$ and $\boldsymbol{\iota}$, respectively. The transformation from Eq.(6) to Eq.(7) can be done by 'bare hands', though the use of some differential geometry(e.g. [5]) may be helpful.

1.2 Faraday's Law

A local form of the Faraday's law of induction is [2]

$$\nabla\times\boldsymbol{\varepsilon}+\tilde{\mathbf{b}}=\mathbf{0} \tag{11}$$

where

$$\boldsymbol{\varepsilon}=\mathbf{e}+\mathbf{v}\times\mathbf{b} \tag{12}$$

with $\mathbf{e}$ the electric field, $\mathbf{b}$ magnetic field. Note $\nabla\cdot\mathbf{b}=0$ is used in the derivation of (11). Taking a similar procedure, we get the transposed weak form of the Faraday's law in both the current and a reference configuration.

$$\int_{B(t)}\boldsymbol{\varepsilon}\cdot(\nabla\times\mathbf{w})\cdot dv+\int_{B(t)}\tilde{\mathbf{b}}\cdot\mathbf{w}dv+\int_{S\eta(t)}\boldsymbol{\varepsilon}_{\tan}\cdot\mathbf{w}ds=0 \tag{13}$$

$$\int_{BR}\mathbf{E}\cdot\mathrm{CURL}\mathbf{W}dV+\int_{BR}\frac{\partial\mathbf{B}}{\partial t}\cdot\mathbf{W}dV+\int_{SR}\mathbf{E}_{\tan}\cdot\mathbf{W}dS=0 \tag{14}$$

$$\mathbf{E}=\mathrm{E}_K=\frac{\partial x^i}{\partial X^K}\varepsilon_i=\phi^*(\boldsymbol{\varepsilon})\quad \mathbf{B}=B^K=J\frac{\partial X^K}{\partial x^i}b^i=J\phi^*(\mathbf{b}) \tag{15)(16}$$

1.3 Poynting's Energy Balance Equation

An energy balance equation that is similar to Poynting's can be obtained from Eqs.(1) and (11) by taking the inner product of Eq.(1) with $\boldsymbol{\varepsilon}$ and Eq.(11) with $\boldsymbol{\eta}$. After the integration,

$$\int_{B(t)}\boldsymbol{\varepsilon}\cdot\boldsymbol{\iota}dv+\int_{B(t)}(\boldsymbol{\varepsilon}\cdot\tilde{\mathbf{d}}+\boldsymbol{\eta}\cdot\tilde{\mathbf{b}})dv=-\int_{S(t)}(\boldsymbol{\varepsilon}\times\boldsymbol{\eta})\cdot\mathbf{n}ds \tag{17}$$

where $\mathbf{n}$ is the normal vector in the current configuration.
Eq.(17) can be represented in a reference form as

$$\int_{BR}\mathbf{E}\cdot\mathbf{I}dV+\int_{BR}\left(\mathbf{E}\cdot\frac{\partial\mathbf{D}}{\partial t}+\mathbf{H}\cdot\frac{\partial\mathbf{B}}{\partial t}\right)dV=-\int_{SR}(\mathbf{E}\times\mathbf{H})\cdot\mathbf{N}dS \tag{18}$$

with $\mathbf{N}$ the normal vector in the reference configuration.

1.4 Potential

A scalar and vector potential ϕ - $\mathbf{a}$ defined by

$$\mathbf{b}=\nabla\times\mathbf{a} \tag{19}$$

$$\mathbf{e}=-\left(\nabla\phi+\frac{\partial \mathbf{a}}{\partial t}\right) \tag{20}$$

may also be transformed to a reference form.

Substituting Eqs.(19) and (20) into Eq.(12) gives,

$$\boldsymbol{\varepsilon}=\mathbf{e}+\mathbf{v}\times\mathbf{b}=-\nabla\phi-\frac{\partial \mathbf{a}}{\partial t}+\mathbf{v}\times(\nabla\times\mathbf{a})=-\frac{\partial\phi}{\partial x^i}-\dot{a}_i+v^j\frac{\partial a_j}{\partial x^i} \tag{21}$$

On the other hand, we define a scalar and vector potential in a reference form by

$$\mathbf{B}=\mathrm{CURL}\mathbf{A} \tag{22}$$

$$\mathbf{E}=-\left(\frac{\partial \mathbf{A}}{\partial t}+\mathrm{GRAD}\Phi\right) \tag{23}$$

Adopting the pull back of $\mathbf{a}$ for $\mathbf{A}$

$$\mathbf{A}=A_K=\frac{\partial x^i}{\partial X^K}a_i=\phi^*(\mathbf{a}) \tag{24}$$

with Eqs.(15) and (23) gives

$$\mathbf{e}=\phi_*(\mathbf{E})=\frac{\partial X^K}{\partial x^i}E_K=-\left(\dot{a}_i+\frac{\partial v^j}{\partial x^i}a_j+\frac{\partial\Phi}{\partial x^i}\right)=-(L_{\mathbf{v}}(\mathbf{a})+\nabla\Phi). \tag{25}$$

Eq.(21) and Eq.(25) coincide when

$$\Phi=\phi-\mathbf{v}\cdot\mathbf{a}. \tag{26}$$

2. WEAK FORM OF LINEAR MOMENTUM BALANCE

The momentum balance of a deformable body is

$$\nabla\cdot\boldsymbol{\sigma}+\rho\mathbf{f}_\rho+\mathbf{f}_v=\rho\dot{\mathbf{v}} \tag{27}$$

in local form and

$$\int_{B(t)}\boldsymbol{\sigma}:\nabla\mathbf{w}\,dv+\int_{B(t)}\left[\rho(\dot{\mathbf{v}}-\mathbf{f}_\rho)-\mathbf{f}_v\right]\cdot\mathbf{w}\,dv-\int_{S(t)}\bar{\mathbf{t}}\cdot\mathbf{w}\,ds=0 \tag{28}$$

in spatial form, where $\boldsymbol{\sigma}$ the Cauchy stress, ρ the density.

Representation of Eq.(28) in a reference form is

$$\int_{B_R}\mathbf{P}:\mathrm{GRAD}\mathbf{W}dV+\int_{B_R}\left[\rho_R(\dot{\mathbf{V}}-\mathbf{F}_\rho)-\mathbf{F}_v\right]\cdot\mathbf{W}dV-\int_{S_R}\bar{\mathbf{T}}\cdot\mathbf{W}dS=0 \tag{29}$$

where $\rho_R=J\rho$ and $\mathbf{P}$ is the first Piola-Kirchhoff stress[3]:

$$\mathbf{P}=P^{iJ}=J\sigma^{ik}\frac{\partial X^J}{\partial x^k}=J\boldsymbol{\sigma}\boldsymbol{F}^{-T}. \tag{30}$$

3. FINITE ELEMENT FORMULATION

For the deformable body subjected to the electromagnetic fields, weak form of the Ampere's law and/or the Faraday's law, when combined with the weak form of the balance of momentum, can serve as the governing equation of the FE analysis.

In the plasticity theory, incremental or rate form of the balance of momentum is inevitable by the rate nature of the inelastic constitutive relation, where no potential exits and the final stress state depends on the loading history. Even for elasticity, the rate form is important since in certain situations it corresponds to the linearized form of the balance of momentum and connection with the buckling theory or numerical methods such as the Newton-Raphson scheme is apparent. Analogy may hold for the electromagnetism.

A reference form can be differentiated straightforwardly. For example, rate form of the Ampere's law, Eq.(7), can be rewritten as

$$\int_{B(t)} L_v(\boldsymbol{\eta})\cdot(\nabla\times\mathbf{w})dv - \int_{B(t)} \tilde{\boldsymbol{\iota}}\cdot\mathbf{w}\,dv = 0 \tag{31}$$

where the electric displacement is neglected and the updated Lagrangian formulation (ULF) is employed. In the ULF, the reference configuration is selected to instantaneously coincide with the reference.

Eq.(31) may suggest simple constitutive equations such as

$$L_v(\boldsymbol{\eta}) = \mathbf{v}\cdot\tilde{\mathbf{b}} \qquad \tilde{\boldsymbol{\iota}} = \boldsymbol{\tau}\cdot L_v(\boldsymbol{\varepsilon})$$

where $\mathbf{v}$ and $\boldsymbol{\tau}$ are tensors that may depend on the mechanical and electromagnetic history in general.

As for the momentum balance, we need the dependency of stress rate on the deformation and constitutive relations of inelasticity (e.g. [6]) or hypoelasticity or rate forms of hyperelastic constitutive equations are available. It should be pointed out, however, the power of external forces in the momentum balance depends on the deformation and beyond the control of the experimenter in general. In regard to the surface traction, loading by a uniformly pressurized fluid, loading by contact with a tool and follower forces of elastic instability are among known examples and the stiffness matrix that comes from the deformation dependency is called the load correction matrix.

We have generalized the loading term in the rate form of the momentum balance to include the effect of the deformation dependent loading and the result for the ULF is the following.

$$\int_{S_R} \frac{\partial \overline{\mathbf{T}}}{\partial t}\cdot\mathbf{W}dS = \int_{S(t)} \mathbf{w}\cdot\dot{\boldsymbol{\sigma}}\cdot\mathbf{n}\,ds + \int_{S(t)} \mathbf{w}\cdot[(\nabla\cdot\mathbf{v})\boldsymbol{\sigma} - \boldsymbol{\sigma}\cdot\boldsymbol{l}^T]\cdot\mathbf{n}ds \tag{32}$$

$$\int_{B_R} \frac{\partial \mathbf{F}_v}{\partial t}\cdot\mathbf{W}dV = \int_{B(t)} \mathbf{w}\cdot[\dot{\mathbf{f}}_v + (\nabla\cdot\mathbf{v})\mathbf{f}_v]dv \tag{33}$$

The last term of the Eq.(32) is the source of the surface load correction matrix. When the body force is given as the divergence of a tensor field such as

$$\mathbf{f}_v = \nabla\cdot\boldsymbol{\sigma}_v = \frac{\partial \sigma_v^{\,ij}}{\partial x^j}, \tag{34}$$

Eq.(33) can be rewritten as

$$\int_{B_R} \frac{\partial \mathbf{F}_v}{\partial t}\cdot\mathbf{W}dV = -\int_{B(t)} \nabla\mathbf{w} : \dot{\boldsymbol{\sigma}}_v\,dv - \int_{B(t)} \nabla\mathbf{w}\cdot[(\nabla\cdot\mathbf{v})\mathbf{I} - \nabla\mathbf{v}]\cdot\boldsymbol{\sigma}_v\,dv \tag{35}$$

where $\mathbf{w}$ is assumed to be zero on the surface. The first term in Eq.(35) is the rate form of Eq.(1) in [7] and the second term is the deformation dependent term and a source of the load correction matrix to take into account the unknown deformation dependency. Under certain conditions this term is linear in the velocity and linearization is possible.

The support of the Grant-in-Aid for Scientific Research from the Japanese Ministry of Edication is acknowledged.

4. REFERENCES

1 Y. Yamada and T. Hirakawa, Application of Numerical Methods to Forming Processes, AMD-Vol.28, ASME (1978) 27.
2 G.A. Maugin, Continuum Mechanics of Electromagnetic Solids, North-Holland, Amsterdam, 1988.
3 J. E. Marsden and T.J.R. Hughes, Mathematical Foundations of Elasticity, Prentice-Hall, Englewood Cliffs, N.J., 1983.
4 M. Lax and D. F. Nelson,Physical Review B, Vol.13, No.4 (1976) 1777.
5 A. Bossavit, IEEE Trans. Mag. No.2 (1992) 1263.
6 I. Nishiguchi, T.-L. Sham and E. Krempl, ASME J. Applied Mech. Vol.57 (1990) 548.
7 A. Kameari, in Nonlinear Phenomena in Electromagnetic fields, Elsevier, Amsterdam (1992) 449.

Simulation and Design of Applied Electromagnetic Systems
T. Honma (Editor)

Electromagnetic and Mechanical Interaction Analysis of a Thin Shell Structure Vibrating in an Electromagnetic Field

T. Horie and T. Niho

Dept. of Mechanical Systems Engineering, Kyushu Institute of Technology, 680-4 Kawazu, Iizuka, Fukuoka 820, Japan

Abstract

This paper describes a finite element formulation of the eddy current and structure coupled problem for curved shell structures. Coupling terms produced by curved geometry as well as flat plate geometry were obtained. Both matrix equations for eddy current and structure were solved simultaneously using coupling sub-matrices. TEAM Workshop bench mark problem #16 was solved to verify the formulation and the computer code. Agreement with experimental results was very good for such plate problem. Influence of each coupling term was also examined.

1. INTRODUCTION

Development of a design method which introduces electromagnetic and mechanical coupling effect or electromagnetic damping effect is needed for the mechanical design of thin shell structures located in an electromagnetic field such as fusion reactor components. Some investigations have been performed[1,2] both analytically and experimentally for thin flat plates with the coupling effect of out-of-plane deflection and external magnetic field. Since the fusion reactor components are made up of curved thin shell structures without rigid supports to reduce the thermal stress, other coupling effects related to in-plane deformations and induced magnetic field as well as nonlinear effect by $\dot{\mathbf{u}} \times \mathbf{B}$ should be taken into account for the coupled analysis.

2. METHOD OF THE ANALYSIS

The coordinate system used for the thin shell structure is shown in Fig.1. At any point on the shell mid-surface, $\bar{x}$ and $\bar{y}$ are tangent axes, and $\bar{z}$ is the normal axis to the surface.

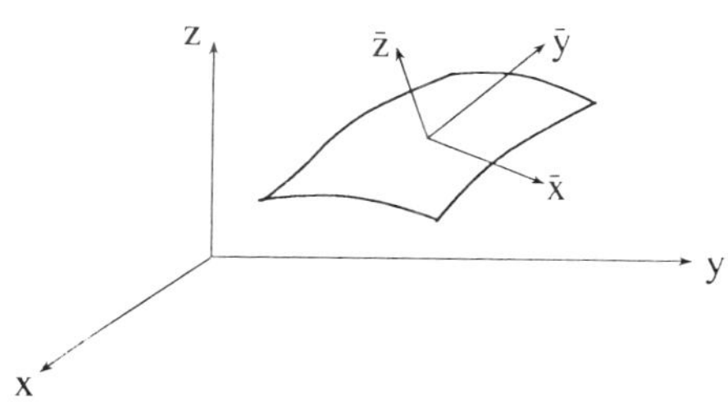

Fig.1 Coordinate system of the shell

2.1 Eddy current analysis

Considering a conductor moving in an electric field **E** and a magnetic field **B**, current density **j** is given by

$$\mathbf{j} = \kappa \left(\mathbf{E} + \dot{\mathbf{u}} \times \mathbf{B}\right) \tag{1}$$

where κ, **u** and the notation $(\cdot)$ are electric conductivity, displacement of the conductor, and the time derivative, respectively. By using Faraday's law and the current vector potential **T**, eq.(1) becomes

$$\nabla \times \nabla \times \mathbf{T} = -\kappa \dot{\mathbf{B}} + \kappa \nabla \times (\dot{\mathbf{u}} \times \mathbf{B}) \tag{2}$$

where the magnetic induction field **B** is the sum of the external magnetic field $\mathbf{B}^{ex}$ and the field $\mathbf{B}^{in}$ induced by the eddy current.

Since the eddy current is expressed by only the normal ($\bar{z}$) component of the current vector potential T for thin shell structures, eq.(2) is reduced to

$$\dot{B}_{\bar{z}}^{in} - \frac{1}{\kappa}\left(\frac{\partial^2 T}{\partial \bar{x}^2} + \frac{\partial^2 T}{\partial \bar{y}^2}\right) - \{\nabla \times (\dot{\mathbf{u}} \times \mathbf{B})\}_{\bar{z}} = -\dot{B}_{\bar{z}}^{ex} \tag{3}$$

$$\dot{B}_{\bar{z}}^{in} = \frac{\mu_0}{4\pi}\frac{\partial}{\partial \bar{x}}\int_{s'}\left\{\frac{\partial \dot{T}'}{\partial \bar{y}'}(\mathbf{e}_{\bar{x}'}\cdot\mathbf{e}_{\bar{y}}) - \frac{\partial \dot{T}'}{\partial \bar{x}'}(\mathbf{e}_{\bar{y}'}\cdot\mathbf{e}_{\bar{y}})\right\}\frac{h'}{r}ds' - \frac{\mu_0}{4\pi}\frac{\partial}{\partial \bar{y}}\int_{s'}\left\{\frac{\partial \dot{T}'}{\partial \bar{y}'}(\mathbf{e}_{\bar{x}'}\cdot\mathbf{e}_{\bar{x}}) - \frac{\partial \dot{T}'}{\partial \bar{x}'}(\mathbf{e}_{\bar{y}'}\cdot\mathbf{e}_{\bar{x}})\right\}\frac{h'}{r}ds' \tag{4}$$

$$\{\nabla \times (\dot{\mathbf{u}} \times \mathbf{B})\}_{\bar{z}} = \underbrace{B_{\bar{x}}\frac{\partial \dot{u}_{\bar{z}}}{\partial \bar{x}} + B_{\bar{y}}\frac{\partial \dot{u}_{\bar{z}}}{\partial \bar{y}}}_{(a)} + \underbrace{\dot{u}_{\bar{z}}\frac{\partial B_{\bar{x}}}{\partial \bar{x}} + \dot{u}_{\bar{z}}\frac{\partial B_{\bar{y}}}{\partial \bar{y}}}_{(b)} \underbrace{- B_{\bar{z}}\frac{\partial \dot{u}_{\bar{x}}}{\partial \bar{x}} - B_{\bar{z}}\frac{\partial \dot{u}_{\bar{y}}}{\partial \bar{y}}}_{(c)} \underbrace{- \dot{u}_{\bar{x}}\frac{\partial B_{\bar{z}}}{\partial \bar{x}} - \dot{u}_{\bar{y}}\frac{\partial B_{\bar{z}}}{\partial \bar{y}}}_{(d)} \tag{5}$$

where **e**, h and r mean the unit vector, shell thickness and the distance between the points (x,y,z) and (x',y',z'), respectively. All coupling terms together with their nonlinearity in eq.(5) are necessary for curved shell geometry although last four terms and the nonlinearity disappear for a flat plate. Finite element discritization with 8-node shell element leads to

$$[\mathbf{U}]\{\dot{T}\} + [\mathbf{R}]\{T\} + [\mathbf{C}_e]\{\dot{u}\} = \{\dot{B}^{ex}\} \tag{6}$$

Matrices $[\mathbf{U}]$, $[\mathbf{R}]$ and $[\mathbf{C}_e]$ are inductance matrix, resistance matrix and coupling matrix, respectively.

2.2 Structural analysis

The isoparametric 8-node shell elements [3] with the reduced integration technique is used for the structural analysis. The electromagnetic force $\mathbf{j} \times \mathbf{B}$ is the coupling term with the eddy current. Since **j** and **B** are functions of unknown variable T, $\mathbf{j} \times \mathbf{B}$ force is also nonlinear for curved geometry. However, the nonlinearity together with the in-plane force components vanishes if the geometry is flat. The finite element equation is expressed as

$$[\mathbf{M}]\{\ddot{u}\} + [\mathbf{K}]\{u\} + [\mathbf{C}_s]\{T\} = \{F^{ex}\} \tag{7}$$

where matrices $[\mathbf{M}]$, $[\mathbf{K}]$, $[\mathbf{C}_s]$ and vector $\{F^{ex}\}$ are the mass matrix, stiffness matrix, coupling matrix and the external mechanical force, respectively.

2.3 Method for coupled analysis

We can rewrite the eqs. (6) and (7) as

$$\begin{bmatrix}\mathbf{M} & \mathbf{0}\\ \mathbf{0} & \mathbf{0}\end{bmatrix}\begin{Bmatrix}\ddot{u}\\ \ddot{T}\end{Bmatrix} + \begin{bmatrix}\mathbf{0} & \mathbf{0}\\ \mathbf{C}_e & \mathbf{U}\end{bmatrix}\begin{Bmatrix}\dot{u}\\ \dot{T}\end{Bmatrix} + \begin{bmatrix}\mathbf{K} & \mathbf{C}_s\\ \mathbf{0} & \mathbf{R}\end{bmatrix}\begin{Bmatrix}u\\ T\end{Bmatrix} = \begin{Bmatrix}F^{ex}\\ \dot{B}^{ex}\end{Bmatrix} \tag{8}$$

for the eddy current and structure coupled system. Using step-by-step time integration technique based on both Newmark's β and Crank-Nicolson methods, eq.(8) results in the coupled equation system

$$\begin{bmatrix} \mathbf{K} + \dfrac{\mathbf{M}}{\beta(\Delta t)^2} & \mathbf{C}_s \\ \mathbf{C}_e & \mathbf{U} + \alpha\,\Delta t\,\mathbf{R} \end{bmatrix} \begin{Bmatrix} u \\ T \end{Bmatrix}_{n+1} = \begin{Bmatrix} F_1 \\ F_2 \end{Bmatrix} \tag{9}$$

where Δt, β and α are the time increment and parameters of Newmark's β and Crank-Nicolson methods.

3. RESULTS

To verify the finite element code based on the curved shell formulation, we analyze the TEAM Workshop benchmark problem #16 (Magnetic damping in torsional mode)[4] and consider the effect of each coupling term. A cantilever plate is placed in a crossed time-changing field and a steady uniform field as shown in Fig.2. The coil current profile which produces the time-changing field is shown in Fig.3.

Figure 4 shows the mesh subdivision of the plate. Number of the total degrees of freedom is 563. Figure 5 shows the time change of the plate deflection. Analytical results agree very well with the experimental results by Takagi et.al.[2] for both deflectional (B_0=0.0 T) and torsional (B_0=0.1 T) cases. Figure 6 shows the influence of each coupling term on the magnetic damping of the plate. The effect of coupling terms (a) of eq.(5) is large compared with that of coupling terms (b). Each value of the coupling terms is shown in Fig.7. Since the terms (a) are produced mainly by the steady magnetic field, these are dominant all the time. The terms (b) are small and decrease rapidly with the decrease of the transient magnetic field. The terms (c) and (d) are negligibly small because of the flat plate geometry of the problem. Figure 8 shows the **j** × **B** force components along the plate width. Although both out-of-plane torsional force and in-plane force exist, displacement by the in-plane force is very small for such plate problem. The in-plane force may be important for curved shell structures.

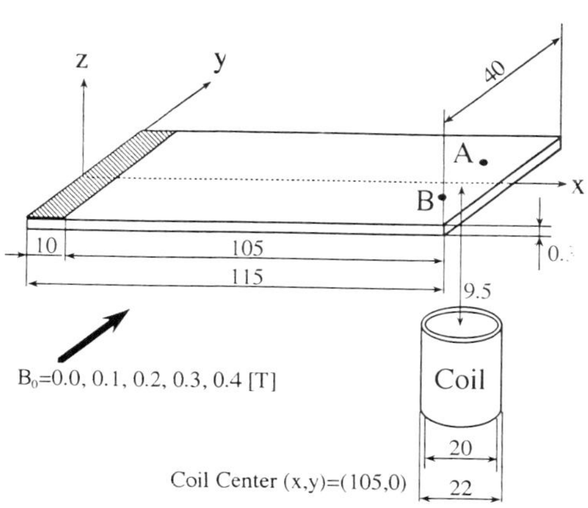

Fig.2 Cantilever plate in electromagnetic field

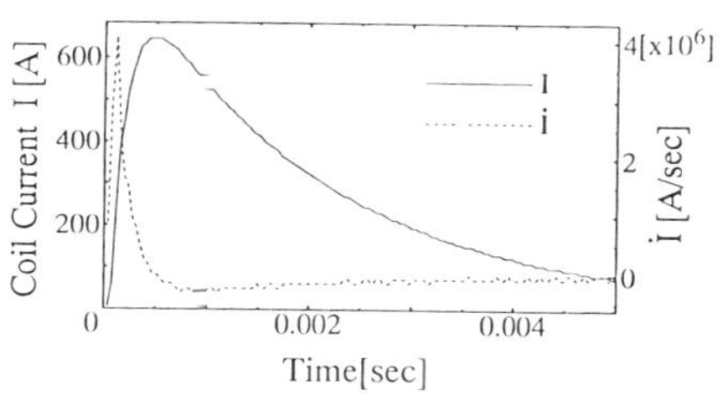

Fig.3 Current profile of the coil

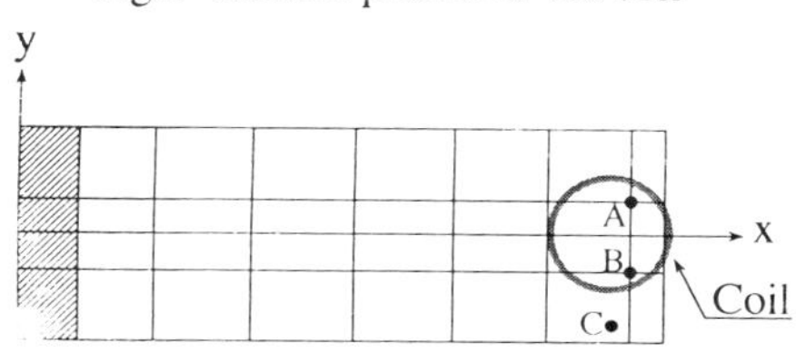

Fig.4 Finite elements for the plate

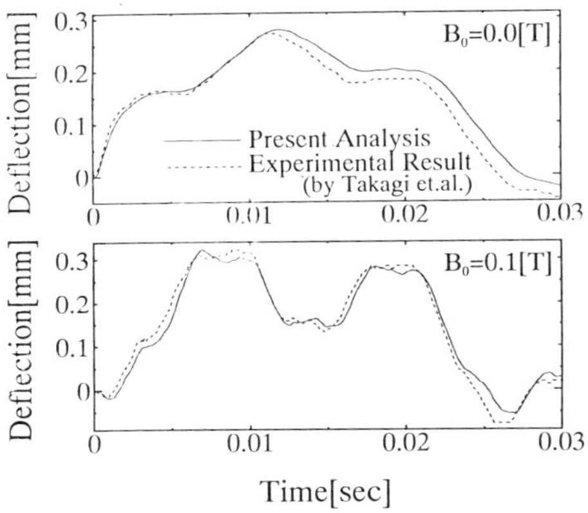

Fig.5 Deflection of the plate (Point A)

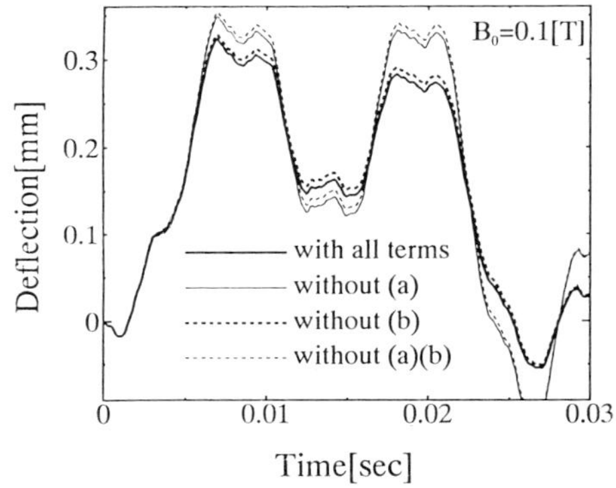

Fig.6 Effect of each coupling term to the deflection (Point A)

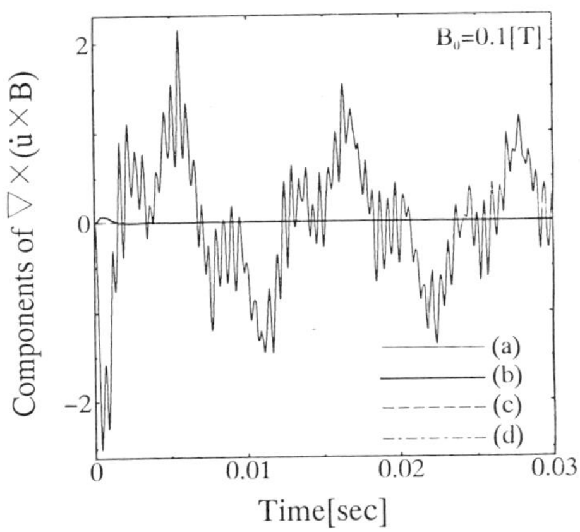

Fig.7 Value of each coupling term (Point C)

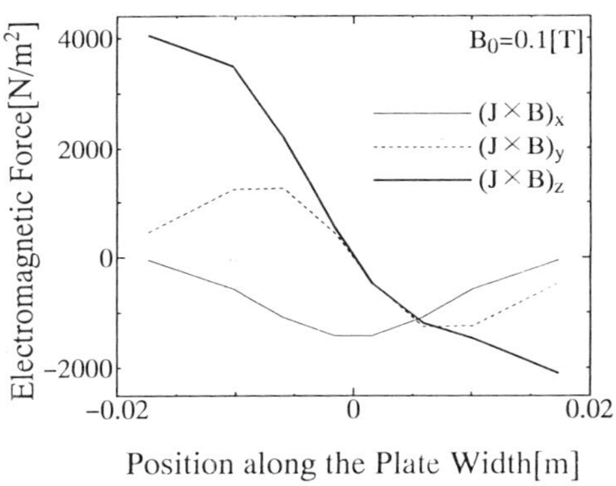

Fig.8 **j** × **B** force components (x = 105 mm, t = 0.1955 msec)

4. CONCLUSION

A finite element analysis method of the eddy current and structure coupled problem was developed for curved shell structures. The agreement between the analysis results and the experimental results is very good for the plate problem. Benchmark analyses for curved shell structures are still to be investigated.

5. REFERENCES

1 L. R. Turner and T. Q. Hua, COMPEL, 9 (1990) 205.
2 T. Takagi, J. Tani and S. Kawamura, Trans. of JSME, 58-C (1992) 3537.
3 O. C. Zienkiewicz, J. Too and R. L. Taylor, Int. J. Num. Meth. Eng., 3 (1971) 275.
4 T. Takagi, S. Kawamura and J. Tani, Proc. Third Intern. TEAM Workshop, EUR 14173 EN (1992) 275.

Simulation and Design of Applied Electromagnetic Systems
T. Honma (Editor)

Magnetic viscous damping effects in ITER blanket

K.Demachi[a], Y. Yoshida[a], M.Kurokawa[a], K.Miya[a] and S.Nishio[b]

[a]Nuclear Engineering Research Laboratory, The University of Tokyo
Tokai-mura, Naka-gun, Ibaraki 319-11, Japan

[b]Naka Fusion Research Establishment, Japan Atomic Energy Research Institute
Naka-machi, Naka-gun, Ibaraki 310-11, Japan

Abstract

The magnetic damping effect due to the speed electromotive force is an important effect from the viewpoint of magnetostructural design of the first wall and blanket of fusion reactors. A computer code based on A-ϕ method was developed to evaluate the dynamic stress induced at the thin shell structures of fusion reactors. Numerical analyses of the coupled vibration of the ITER first wall during a plasma disruption were made to investigate the magnetic damping effect on vibration of the first wall.

1. INTRODUCTION

For a tokamak fusion reactor, it is one of the most important subjects to establish the structural design in which the first wall and blanket structures can withstand against strong magnetic force induced by plasma current termination. For estimation of the magnetic force, it is essential to take into account the magnetic damping effect caused by the speed electromotive force $\boldsymbol{v} \times \boldsymbol{B}$, where $\boldsymbol{v}$ is the velocity and $\boldsymbol{B}$ is the magnetic flux density. Consideration of this effect gives us a new guide line for the design of the ITER blanket such as the adoption of flexible thin structures.

In literatures induced stresses are usually calculated without consideration of the magnetic damping[1]. In this paper we evaluated the magnetic damping effects on vibration of the first wall of the ITER (International Thermonuclear Experimental Reactor) by three-dimensional numerical calculations.

2. NUMERICAL METHODS

The A-ϕ method was applied for the calculation of eddy currents. The governing equations including $\boldsymbol{v}\times\boldsymbol{B}$ are written as follows [2]:

$$\nabla \times \frac{1}{\mu}\nabla \times \boldsymbol{A} = -\sigma(\dot{\boldsymbol{A}} + \nabla\phi - \boldsymbol{v} \times \boldsymbol{B}) \qquad (1)$$

$$\nabla \cdot \sigma(\dot{\boldsymbol{A}} + \nabla\phi) = \nabla \cdot \sigma(\boldsymbol{v} \times \boldsymbol{B}). \tag{2}$$

The FEM-BEM coupled method[3] was used to discretize the above equations. The magnetic force was calculated as the Lorentz force like

$$\boldsymbol{J} \times \boldsymbol{B} = -\sigma(\dot{\boldsymbol{A}} + \nabla\phi - \boldsymbol{v} \times \boldsymbol{B}) \times \boldsymbol{B} \tag{3}$$

and the vibration of the blanket was analyzed according to the equation of motion of the shell as

$$[M]\{\ddot{u}\} + [K]\{u\} = \{f\} \tag{4}$$

where $[M]$ and $[K]$ are the mass and stiffness matrices, respectively. $\{u\}$ and $\{f\}$ are the displacement and body force vectors corresponding to the Lorentz force, respectively.

The direct integration method was applied for the time integrations in both electromagnetic field and vibration analyses.

3. RESULTS AND DISCUSSION

Fig.1 shows a computational model of a side module of the outboard blanket of the ITER. This model is based on the ITER CDA (Conceptual Design Activity) design[4]. The plasma current was approximated as 36 loop currents whose total intensity is 22MA and disrupts to zero linearly during 20msec as shown in Fig.1. The magnetic field conditions are such that the toroidal magnetic field at the first wall is 4 Tesla while the poloidal field is 0.5 Tesla. A fixed condition was applied as mechanical constraint to a boundary denoted with "abcd" in Fig.1. Computations were carried out for three cases of different thicknesses of the first wall (7, 15 and 25mm).

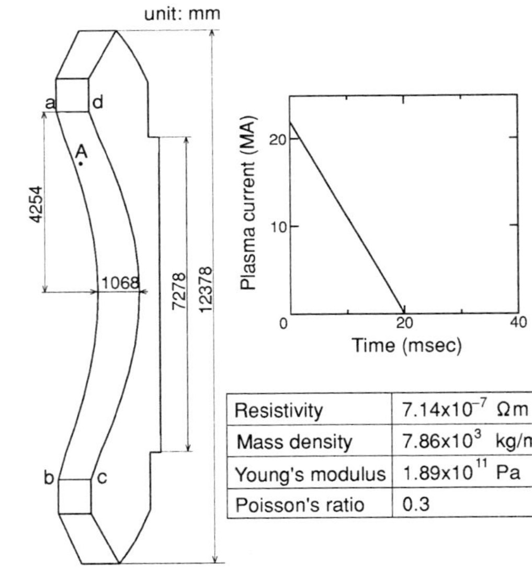

Resistivity	7.14×10^{-7} Ωm
Mass density	7.86×10^{3} kg/m^3
Young's modulus	1.89×10^{11} Pa
Poisson's ratio	0.3

Fig.1 Computational model

Fig.2 shows the numerical results of time-variation of eddy current density at "A". The coupled results of 7mm thick wall are very different from the uncoupled results while the results computed by both methods are almost the same for the thicker wall (15mm and 25mm).

Fig.3 shows the numerical results of dynamic response of deflection at location "A". Though the results by uncoupled calculation show vibration, the coupled results show that the vibration is suppressed completely and quickly by the strong magnetic damping effect.

The numerical results of time-variation of maximum principal shear stress on the ITER first wall are shown in Fig.4. In case of 7mm thick first wall, the uncoupled calculation 40% overestimate the coupled results. On the other hand, in case of 25mm thick wall the difference between the stresses calculated by two methods is small. Namely the magnetic damping effect becomes stronger when the first wall is thinner. This is because the speed of deformation of the thin wall is faster than that of the thick wall.

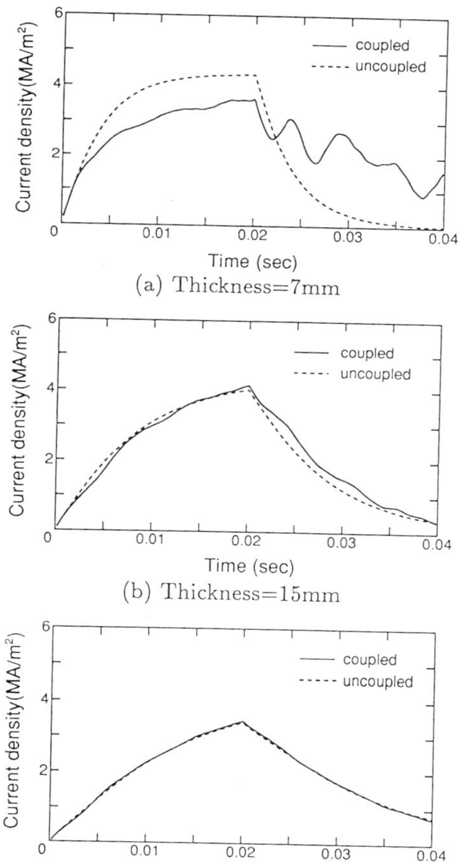

(a) Thickness=7mm

(b) Thickness=15mm

(c) Thickness=25mm

Fig.2 Time-variation of eddy current at "A"

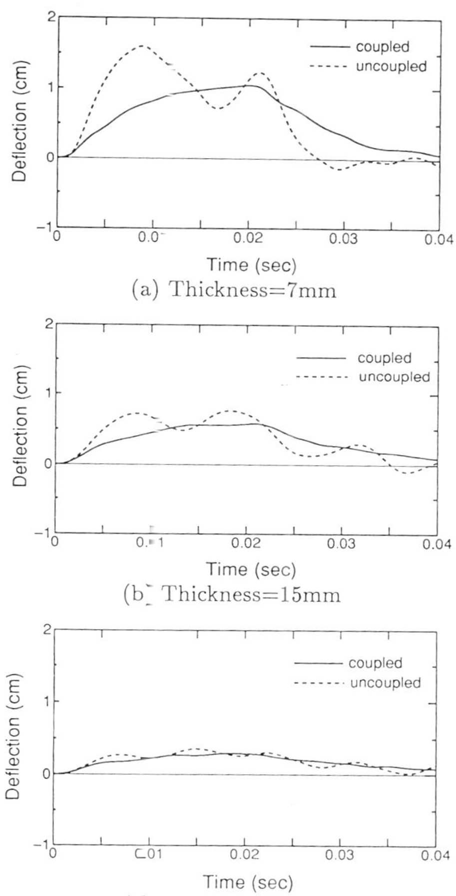

(a) Thickness=7mm

(b) Thickness=15mm

(c) Thickness=25mm

Fig.3 Time variation of deflection at "A"

Fig.5 shows the relation between the maximum value of principal shear stress and the thickness of the first wall. The difference between the coupled and uncoupled results becomes larger when the thickness of the wall is thinner. On the other hand, the thermal stress induced at the first wall is, as well-known, proportional to the thickness of the wall. This is why flexible thin structures are better candidates for the construction of the blanket components and the first wall. Since the stress in thin structure computed by the coupled method is smaller than that calculated by conventional uncoupled one, numerical results depicted in Fig.4 prove the applicability of flexible thin structures.

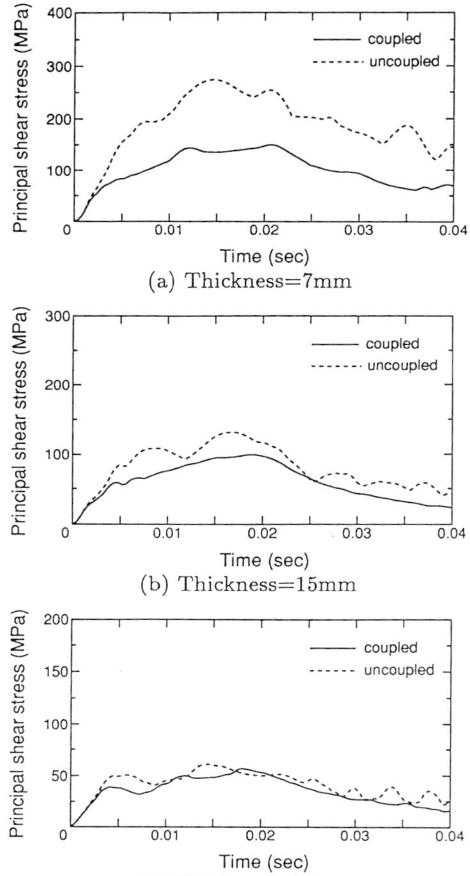

(a) Thickness=7mm

(b) Thickness=15mm

(c) Thickness=25mm

Fig.4 Time-variation of principal shear stress

4. CONCLUSIONS

A numerical method to solve the coupled problem between the eddy current and the motion of shell was applied to the analyses of the magnetomechanical coupled vibration of the ITER first wall. It is found that the magnetic damping is an important effect because the conventional uncoupled method overestimates the induced stress. A flexible thin structure can be favored for magnetostructural design of the fusion reactor first wall.

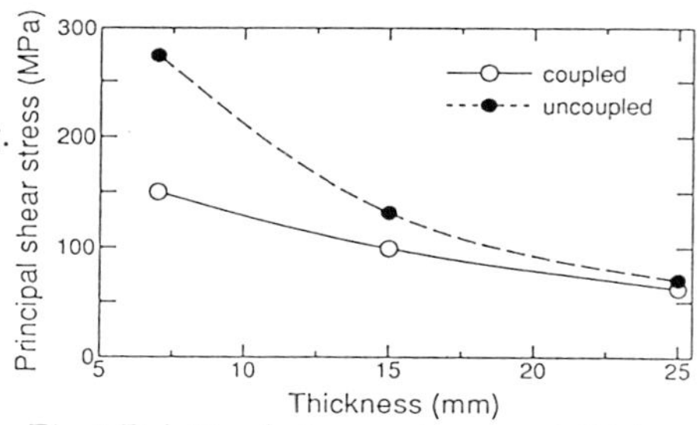

Fig.5 Relation between stress and thickness

5. REFERENCES

1. Y.R.Crutzen et al., Fusion Eng. Design, Vol.11, pp.293-303, 1989
2. Y.Yoshida et al., Int. J. Appl. Elec.Mag. in Materials, Vol.4, No.1, in press
3. F.Matsuoka, Proc. IUTAM on electromagnetomechanical interactions in deformable solids and structures, Elsevier Science, pp.169-174, 1987
4. G.Vieider et.al., Fusion Eng. Design, Vol.16, pp.23-34, 1991

Simulation and Design of Applied Electromagnetic Systems
T. Honma (Editor)

Postprocessing for Boundary Element Analysis in Electromagnetic Field Problems

H. Tsuboi[a], T. Sato[a], T. Horino[a], M. Tanaka[a], T. Misaki[a], K. Sano[b] and H. Watanabe[b]

[a] *Department of Information Engineering, Fukuyama University, Gakuencho, Fukuyama 729-02, Japan*

[b] *Kurabo Industries Ltd., Neyagawa, Osaka 572, Japan*

Abstract

This paper describes a postprocessing system for the boundary element methods. In the system, scalar variables and vector variables on the boundary surfaces or the arbitrary planes are displayed by contour lines and arrows respectively. Furthermore, an efficient display method of the contour lines on the arbitrary planes is proposed.

1. INTRODUCTION

Boundary element method is one of the practical numerical method for electromagnetic field analysis. The postprocessing for finite element method has been discussed in Ref. [1]. In this paper, the authors propose a postprocessing system for the boundary element methods.

Using no color scale, we can classify the post-processes for the boundary element method into four types shown in Fig. 1. The scalar variables are displayed by contour lines. The vector variables are displayed by arrows, which are represented by a cone and a rod so that the magnitude and direction can be indicated. Furthermore, the contour lines and arrows are displayed on the boundary surfaces or the arbitrary planes.

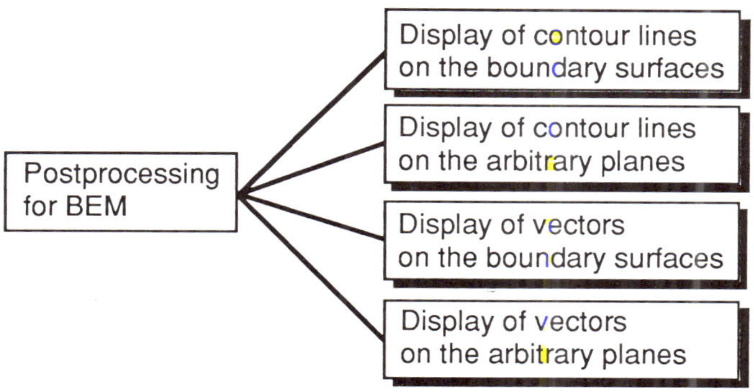

Fig. 1 Postprocessing for boundary element methods.

In the case of the boundary element methods, long computation time is required to calculate accurate contour lines on the arbitrary planes because there is no node on which variables are defined. The authors propose an efficient method to obtain the accurate contour lines on the arbitrary planes. On the other hand, the arrows are drawn by the hidden-surface techniques based on the z-buffer algorithm.

2. SCALAR VARIABLES

The scalar variables are displayed by using contour lines. The contour lines on the boundary surfaces can be obtained by the linear interpolation of the values of the variable on the nodes. However, it is not easy to obtain the accurate contour lines on the arbitrary planes because there is no node. We propose an efficient method to calculate the contour lines on the arbitrary planes.

Generally, we use two kinds of methods to obtain the contour lines on the arbitrary planes in the boundary element method. One is the method using fine triangular mesh. The values of the scalar variable on the all mesh points are calculated and the contour lines are obtained by the linear interpolation on the mesh. In this case, long calculation time is required to calculate the scalar values on the mesh points. The other is the successive approximation method[2]. The successive method requires long calculation time for the iterative calculations to determine the contour paths.

Accordingly, we propose a method to trace the contour lines using a virtual triangular mesh. The flow chart of the proposed method is shown in Fig.2. We consider a virtual triangular mesh which is derived form a rectangular mesh. We can trace a contour line from the starting point P_0 by making triangular element T_1 on the virtual triangular mesh as shown in Fig. 3. The values of the scalar variable on the nodes of the triangular elements are calculated and the segment of the contour line P_0-P_1 is determined by the linear interpolation using the values of the nodes N_1, N_2, N_3. In next step, P_1 becomes starting point. This procedure will be done iteratively. In the proposed method, the computation time can be reduced because only the required values of the scalar variables on the nodes are calculated.

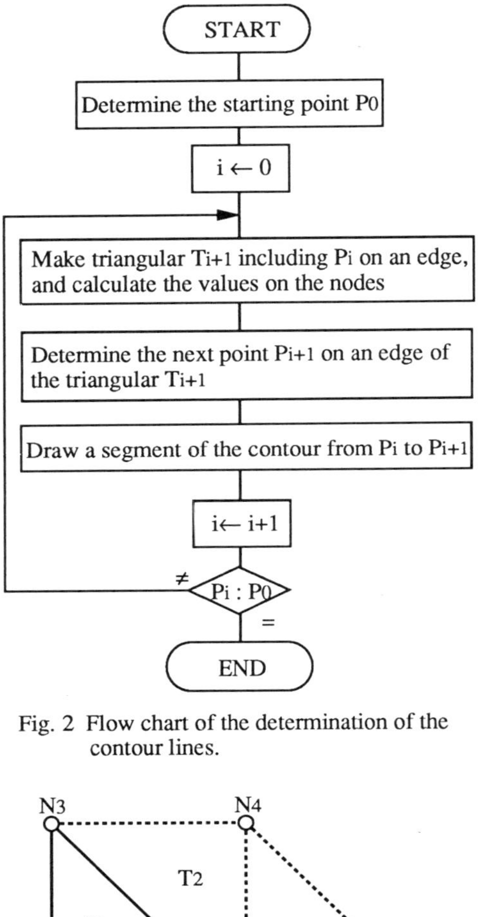

Fig. 2 Flow chart of the determination of the contour lines.

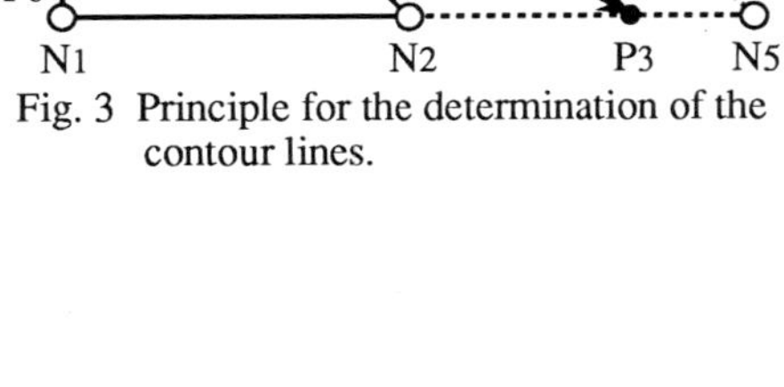

Fig. 3 Principle for the determination of the contour lines.

Figure 4 shows an example of the display of contour lines. The contour lines are the equipotential lines of the three-phase infinite conductor model. The interval of the equipotential lines is 10 (V) and the number of pixels is 1024x1024. In this case, iterative calculation is not required and the computation time is reduced to 11 (%) compared with the case in which the potentials on the all nodes are calculated. The method is effective for the display of the accurate contour lines.

Fig. 4 Example of the contour lines: equipotential lines.

3. VECTOR VARIABLES

Vector variables are displayed by using arrows. An arrow is constructed of a cone and a rod. The diameter of the cone indicates the magnitude of the vector and the view of the arrow indicates the three-dimensional direction of the vector as shown in Fig.5.

The arrows are drawn by the hidden-surface techniques based on the z-buffer algorithm. In other words, after sorting by z-coordinates of the center of gravity, we display the arrows and the boundary elements in the sorted order. Furthermore, the boundary element to be displayed are selected by the direction of the normal vectors: the inner product of the normal vector of the element and the unit vector along the z-axis is positive.

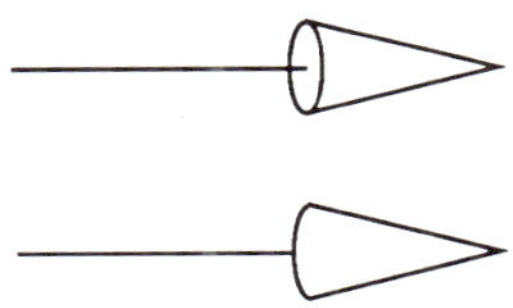

Fig. 5 Arrows representing vector variables.

4. EXAMPLES

Figure 6 shows an examples of the postprocessing of three-dimensional electric field problem which is calculated by the surface charge simulation method. Figure 6(b) shows the electric field vectors on the boundary surfaces. Figure 6(c) shows the equipotential lines on the x-y plane. The results are printed by a *PostScript* printer.

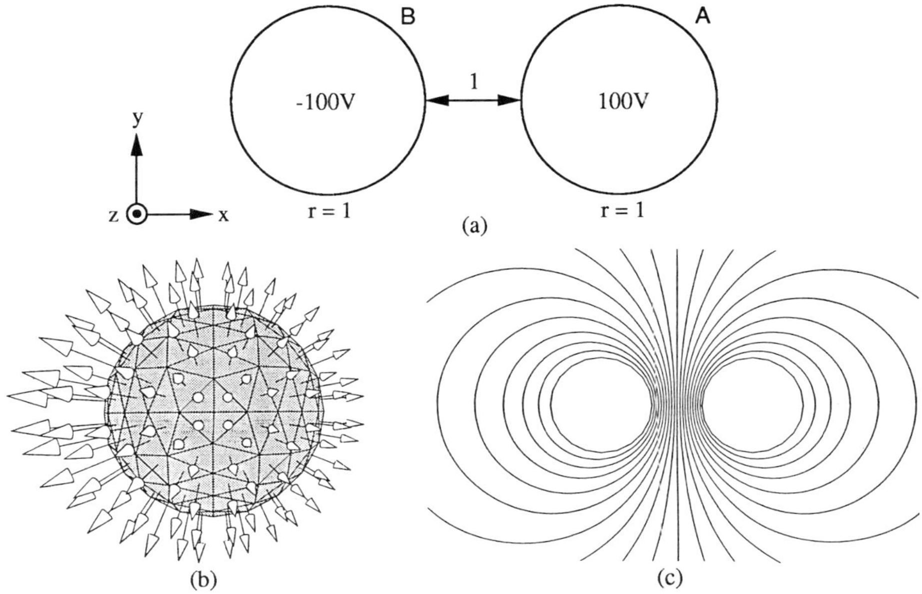

Fig. 6 Two conducting sphere model, (a)computation model, (b)electric field vectors on the sphere surface A, (c)equippotential lines on the x-y plane.

5. CONCLUSIONS

The postprocessing system for the boundary element method is described. The scalar variables are displayed by the contour lines on the boundary surfaces or the arbitrary planes. The efficient method to display the contour lines on the arbitrary planes was proposed. The method is using a virtual triangular mesh to reduce computation time and it is effective for the display of the accurate contour lines. On the other hand, the vector variables are displayed by the arrows on the boundary surfaces or the arbitrary planes so that the magnitude and the direction of the vectors can be indicated.

The proposed method can be extend to color display and used effectively in three-dimensional boundary element analysis.

6. REFERENCES

1 E. Nakamae, "Display of Distributed Scalars and Vectors," IEEE Trans. on Magnetics, Vol.26, No.2, pp.755-760, 1990.

2 L. Krahenbuhl and A. Nicolas, "Efficient Techniques for Boundary Integral Equation Methods," IEEE Trans. on Magnetics, Vol.MAG-19, No.6, pp.2667-2669, 1983.

Simulation and Design of Applied Electromagnetic Systems
T. Honma (Editor)

A BEM analysis of space charge field in a corotron

T. Morinaga, H. Igarashi and T. Honma

Department of Electrical Eng., Faculty of Eng., Hokkaido University
Kita 13, Nishi 8, Kita-ku, Sapporo 060, Japan

Abstract

This paper describes an analysis of space charge fields in the corotron which is used to electrify a photoconductor in electrophotographical processes. The space charge fields in the device are analyzed by the combination of the boundary element method (BEM) with the method of characteristics. In particular, the non-uniform voltage applied on the wires is shown to give nearly uniform current profiles on a metallic plate in the device.

1. INTRODUCTION

The corotron electrifying the photoconductor plays an important role in electrophotographical processes in copy machines and laser printers. Figure 1 illustrates a corotron under consideration, which has three wires and a metallic shield above a metallic plate. In this device, a strong, non-uniform electric field in the vicinity of the thin wires causes corona discharge, which produces a space charge field between the metallic shield and plate. Since the electric field and space charge distribution in this device strongly affect each other, they must be self-consistently determined.

The space charge fields have been so far calculated by simultaneously solving the Poisson equation and the equation of current continuity. There are a number of papers on the numerical analysis of those equations for various systems such as high voltage DC transmission lines [1, 2], electric precipitators [3-5], corona devices [6] and printers [7]. Moreover, the combination of boundary element method (BEM) with the method of characteristics (MOC) has been shown to effectively give the current-voltage characteristics of corona devices, which are in good agreement with experimental results [8]. In Ref. 8, it was also pointed out that the center wire did not efficiently produce the corona current compared with the side wires when the same voltage is applied on the wires.

In this paper, the space charge field in the corona device is analyzed by the combination of BEM with MOC to investigate the dependence of the field distribution and current profile on the voltage applied on the wires.

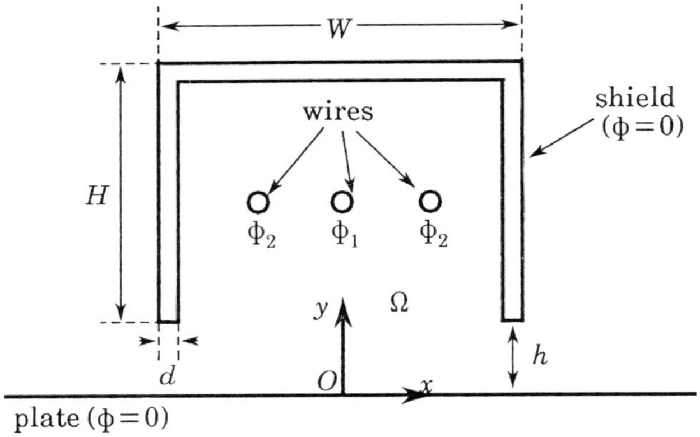

Figure 1. A numerical model of a corotron.

2. FORMULATION AND SOLUTION PROCEDURE

The unipolar space charge fields are governed by

$$\nabla^2\phi(\mathbf{r}) = -\frac{\rho(\mathbf{r})}{\varepsilon_0}\,, \tag{1}$$

$$\nabla \cdot \mathbf{J}(\mathbf{r}) = 0\,, \tag{2}$$

$$\mathbf{J}(\mathbf{r}) = \mu\,\rho(\mathbf{r})\,\mathbf{E}(\mathbf{r})\,, \tag{3}$$

where ϕ is the electric potential, $\mathbf{E}$ the intensity of electric fields, ρ the space charge density, ε_0 the permittivity of free space, $\mathbf{J}$ the current density, and μ the mobility of ions. The Poisson equation (1) is solved by BEM to obtain the potential ϕ in the open solution region Ω provided that the charge density ρ is known. On the other hand, the substitution of (1) and (3) into (2) yields

$$\mathbf{v}(\mathbf{r}) \cdot \nabla\rho(\mathbf{r}) + \mu\,\frac{\rho(\mathbf{r})^2}{\varepsilon_0} = 0\,, \tag{4}$$

where $\mathbf{v}$ is the velocity of ions given by $\mu\mathbf{E}$. On characteristic lines (lines of electric force), a solution to (4) can be expressed in the form

$$\rho(\mathbf{r}) = (1\,/\,\rho_W + \mu\,t(\mathbf{r})\,/\,\varepsilon_0)^{-1}\,, \tag{5}$$

where ρ_W is the charge density on the discharge wire and t denotes the time required to move along a characteristic line from a field point $\mathbf{r}$ to the surface of the wire. The space charge density ρ in the solution region Ω can be obtained from (5) providing that ρ_W and the electric field $\mathbf{E}$ are known. The solution procedure to obtain the self-consistent relation among ϕ, ρ and ρ_W is as follows. At the initial stage, for $\rho=0$, the electric fields in Ω are computed by solving (1).

A virtual ion is then moved from a nodal point to the wire surface at a velocity of $-\mu E$ and the moving time t is obtained from

$$t(\mathbf{r}) = \sum_i t_i = \frac{1}{\mu}\sum_i \frac{l_i}{|\mathbf{E}_i|}, \tag{6}$$

where l_i represents the length of a segment of the characteristic line. The substitution of the time (6) into (5) yields the charge density ρ at the nodal point for a tentative value ρ_W. The space charge density ρ is determined by this procedure at all nodal points. The electric fields are calculated again by solving (1) with the new ρ obtained above. If the reputation of these procedures gives a convergent solution, then the charge on the wire surface ρ_W is modified so that the electric field $|\mathbf{E}|$ on the wire coincides with Peek's corona onset field E_C given by

$$E_C = 3.1\times10^6\delta\left(1.0 + \frac{3.08\times10^{-2}}{(\delta a)^{1/2}}\right)[V/m], \tag{7}$$

where δ is the relative air density and a is the radius of the wire. The charge on the wire surface ρ_W is determined by numerically solving the nonlinear equation $|\mathbf{E}(\rho_W)| = E_C$.

3. NUMERICAL RESULTS AND CONCLUSION

The dimensions of the corotron under test are as follows : the width W, height H and thickness d of the shield are taken to be 40, 18 and 1 [mm], respectively. The height h is 4 [mm] and the three wires, 0.08 [mm] in diameter, are placed at (-10,12), (0, 12) and (10, 12) [mm], respectively. In addition, the potential ϕ_2 on the side two wires is held at 5.5 [kV] and ϕ_1 is varied from 5.5 to 7.5 [kV].

Figure 2 shows the resultant current profiles on the plate. We see that the current profile strongly depends on ϕ_1 and it becomes nearly uniform when ϕ_1 is equal to 6.0 [kV]. Figure 3 shows the dependence of the electric field distribution on ϕ_1. When the same voltage is applied on the three wires (see (a)), the greater part of ions, which fly along the characteristic lines represented by the solid lines, are emitted from the side wires. In contrast to this, the increase in ϕ_1 spreads the characteristic lines coming from the center wire and presses those from the side wires outward, as shown in (b). From these results, we conclude here that non-uniformity of the applied voltage may improve the efficiency of the corotron.

4. REFERENCES

1 T. Takuma, T. Ikeda and T. Kawamoto, IEEE Trans. Power. App. Sys., PAS-100 (1981) 4802.
2 J. L. Davis and J. F. Hoburg, J. Electrostatics, 13 (1986) 1.
3 P. L. Levin and J. F. Hoburg, IEEE Trans. Ind. Applications, 26 (1990) 662.

4 B. E. Springett, F. M. Tesche, A. R. Davies and J. A. L. Thompson, Photo. Sci. Eng., 22 (1978) 200-212.
5 S. Cristina, G. Dinelli and M. Feliziani, IEEE Trans. Industry Applications, 27 (1991) 147.
6 C. J. Budd and A. A. Wheeler, J. Comp. Phys., 97 (1991) 1.
7 M. H. Lean and G. A. Domoto, J. Appl. Phys., 61 (1987), 3931.
8 H. Igarashi, S. Sakai, T. Nakamura, T. Morinaga and T. Honma, IEEE Trans. Mag., 29 (1993) 1508.

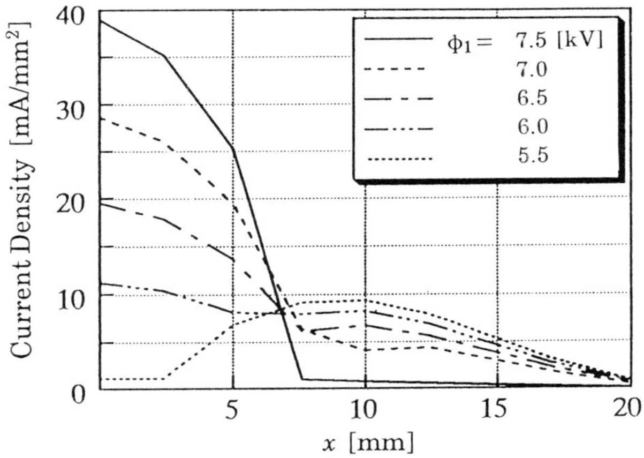

Figure 2. Dependence of the current profile on ϕ_1.

(a) $\phi_1 = 5.5$ [kV] (b) $\phi_1 = 7.5$ [kV]

Figure 3. Electric field distributions.

Simulation and Design of Applied Electromagnetic Systems
T. Honma (Editor)

Defect searching in ferromagnetic materials

Y. Oohira[a], Y. Hirama[a], S. Hayano[b] and Y. Saito[b]

[a]R & D Hitachi Building Systems Engineering and Service, 16-23 Nakagawa, Adachi-ku, Tokyo 120, Japan
[b]Coll. of Engng., Hosei Univ., Koganei, Tokyo 184, Japan

Abstract
In the present paper, theoretical and experimental work concerning defect searching in ferromagnetic materials is carried out. Key idea of this searching method is as follows. When the DC magnetic field is applied to the ferromagnetic materials, their magnetization vectors are aligned in a direction of the applied DC field. However, if there is a defect in the material, then alignment of magnetization vectors is distorted at the defect position. This distorted magnetization vectors are detected by the moving search coils. Thus, it is possible to detect the defect in ferromagnetic materials in a quite simple manner.

1. INTRODUCTION

Defect searching in the ferromagnetic materials is of paramount importance because most of the structural materials in various artificial objects, e.g. car, train, escalator and elevator, are composed of iron or its compounds, i.e. ferromagnetic materials. Sometimes, the defect of structural materials leads to the serious accidents. In order to protect such tragic accidents, it is essentially required to exploit a reliable defect inspection device for the ferromagnetic materials. Thereby, various nondestructive testing methods, e.g. eddy current testing, electric potential method, ultrasonic wave imaging, and X-ray tomography, are currently used [1-3]. Even if any of these existing methods are applied to the defect searching in ferromagnetic materials, it is difficult to expect the reliable results because of their own magnetization vectors. In the present paper, the theoretical and experimental works concerning with the defect searching in ferromagnetic materials are carried out. Key idea of this searching method is as follows. When the DC magnetic field is applied to the ferromagnetic materials, their magnetization vectors are aligned in a direction of the applied DC field. However, if there is a defect in the material, then alignment of magnetization vectors is distorted at the defect position. This distorted magnetization vectors are detected by the moving search coils. Thus, it is possible to detect the defect in ferromagnetic materials in a quite simple manner [4]. Based on the principle described above, a defect inspection device for the wire ropes suspending the elevator cages has been successfully realized and practically used for the daily maintenance work. This paper describes the optimum design and usefulness of our new device.

2. THEORETICAL BACKGROUND

As is well known, ferromagnetic materials are composed of magnetic domains. This leads to a following constitutive relationship:

$$B = \mu_0 H + M, \tag{1}$$

where B, H, M and μ_0 are the magnetic flux density, magnetic field intensity, magnetization vector and permeability of air, respectively. Because of existence of magnetic domains, anomalous eddy currents flow between the domain walls when alternating magnetic field is applied to the ferromagnetic materials. Thereby, it is difficult to apply conventional eddy current testing (ECT) method to the ferromagnetic materials, because the input impedance of the ECT device is dominated by the domain structure of the materials. In order to overcome this difficulty, two representative methods may be considered. One is the DC magnetic field superposition in addition to the alternating magnetic fields in order to saturate the material, i.e. reducing into nonmagnetic material property [1]. The other is the DC magnetic field application to the materials in order to align the magnetization vector in a direction of the applied DC field, if there is a defect in the material, then the magnetization vector could not align at the defect position. This magnetic field distortion can be picked up by moving the search coils in a parallel direction to the applied DC magnetic field. When we introduce the permeabilty μ into (1), (1) is modified to

$$B = \mu_0\{1+[M/(\mu_0 H)]\}H = \mu H, \tag{2}$$

where the permeability μ is defined by

$$\mu = \mu_0\{1+[M/(\mu_0 H)]\}. \tag{3}$$

Mathematically, former method is based on the following fact:

$$\lim_{H\to\infty} \langle\mu\rangle = \lim_{H\to\infty} \langle\mu_0\{1+[M/(\mu_0 H)]\}\rangle = \mu_0. \tag{4}$$

Equation (4) means that the strong magnetic field application makes the ferromagnetic material property possible to reduce nonmagnetic property. Thus, the ECT sensibility can be enhanced. On the other side, latter one has to measure the permeability discontinuity due to the defect. As shown in Fig.1, when a search coil moves from left to right sides, obviously a voltage is induced by the magnetic field distortion at the defect position. This magnetic field distortion may be regarded by the permeability difference at the defect position.

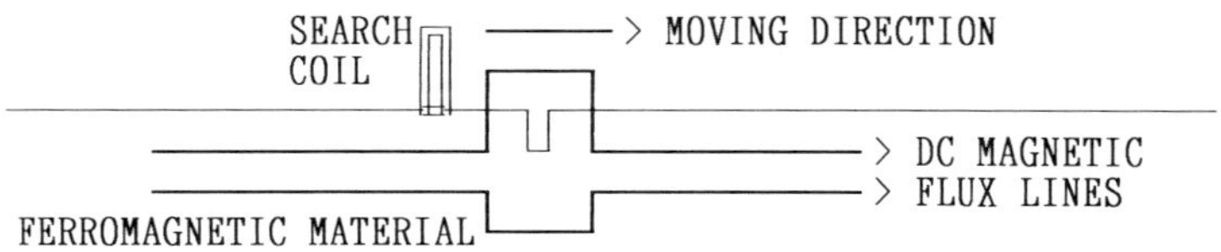

Figure 1. Principle of the DC magnetic field method.

Thus, this latter method is based on the quite simple principle even though the target is limited to the objects having flat surface. Further, taking the difference of induced voltages in the two aligned search coils along the target surface enhances the sensibility of this DC magnetic field method. In order to satisfy the following condition,

$$\nabla\cdot B = 0, \tag{5}$$

we introduce a vector potential A defined by

$$\nabla\times A = B. \tag{6}$$

Combining (2),(6) with Maxwell's equation:

$$\nabla\times H = J, \tag{7}$$

it is possible to obtain a following governing equation:

$$\nabla\times(1/\mu)\nabla\times A = J, \tag{8}$$

where J denotes a current density. The details of magnetic field distortion at the defect in Fig. 1 can be obtained by solving (8).

3. DEFECT INSPECTION DEVICE FOR WIRE ROPES

3.1. Wire rope tester

We have exploited a defect inspection device for the wire ropes suspending the elevator cage, because we have an urgent need of wire rope testing particularly used for the modern high speed elevators. Figure 2 shows a schematic diagram of the device. The left and right electromagnets makes the magnetization vectors in the wire rope align along the rope. The two heads are located at the center to picking up the difference between the induced voltages, and the distance between these heads coincides with the strand pitch of wire.

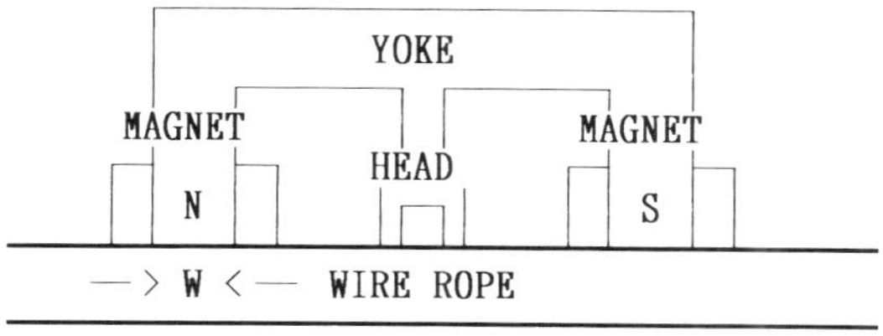

Figure 2. Schematic diagram of a wire rope tester. (W denotes the width of magnetic pole)

3.2. Experimental

We have carried out the experiments changing the width of magnetic poles. One has W=4cm and the other has W=8cm width of magnetic poles. The diameter of wire rope used in the both experiments is 1.2cm. The moving velocity of the wire was set to 30m/s. Moreover, the wire rope has six defects located at the different positions. The experimental results are summarized in Table 1.

Table 1 : Experimental results of the exploited wire rope testers.

	OUTPUT PEAK VOLTAGE [mV]					
WIDTH	DEFECT 1	DEFECT 2	DEFECT 3	DEFECT 4	DEFECT 5	DEFECT 6
W=4cm	15.5	6.5	12.0	8.0	6.5	0.0
W=8cm	19.0	9.0	13.0	8.5	6.0	5.0

Figure 3 shows an experimental recorder output signals, where upper and lower signals are the direct induced voltage difference and filtered outputs, respectively. Consideration of the results in Table 1 and Fig. 3 obviously shows that the tester equipping the wider magnetic pole has high sensibility. This is because the wider magnetic pole provides the uniform magnetic field distribution in the wire rope cross section. This fact was confirmed by obtaining the FEM solutions of (8). Figure 4 shows an example of magnetic field distribution in the wire rope.

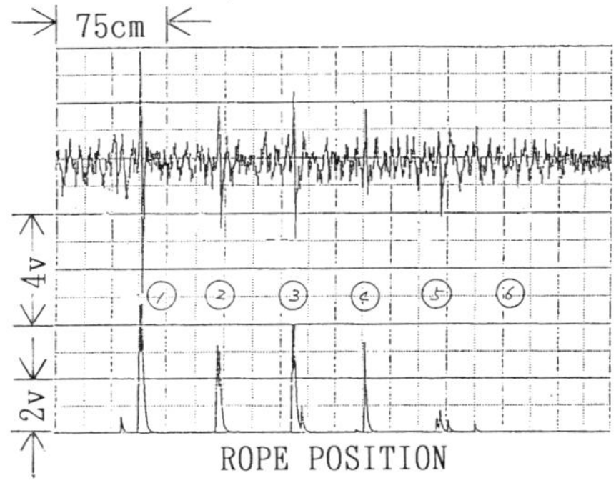

Figure 3. An experimental recorder output signals. The upper and lower signal are the direct induced voltage difference and filtered outputs, respectively.

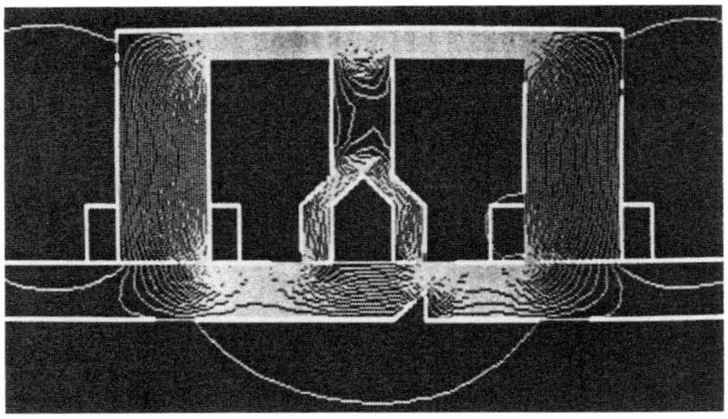

Figure 4. Computed magnetic field distribution in wire rope. Two dimensional model and relative permeability of wire rope was assumed to 100.

4. CONCLUSION

As shown above, we have succeeded in realizing one of the defect searching devices for the ferromagnetic materials. The exploited wire rope tester is practically used for the daily maintenance work of the elevator. Furthermore, it has been revealed that the optimum width of magnetic pole exists depending on the diameter of wire rope. According to our research, at least, the magnetic pole having the width W = 6～7×(wire rope diameter) is required to get the reliable results.

5. REFERENCES

1 Y.Hirama et al, The method for Eddy Current Sensor added to Magnetic Bias, Paper on Technical Meeting of Magnetics in IEEJ, MAG-92-73, Feb.(1992) pp.143-148.

2 T.Doi et al, Inverse Analysis of Defect Identification based on the Electric Potential Method, Proc. of 2nd Japan-Hungary Joint Seminar, Published by Hokkaido University, (1992).

3 H.Saotome et al, Crack Identification in Metallic Materials, IEEE Trans. Magn. to be published.

4 W.Lord et al, The generated reaction field method of detecting defects in steel bars, J. of Mater Evaluation, 29(2),(1971) pp.21-27.

Simulation and Design of Applied Electromagnetic Systems
T. Honma (Editor)

An application of DC magnetic fields to eddy current sensor

Y.Hirama[a], Y.Oohira[a], N.Sahoda[a], S.Hayano[b] and Y.Saito[b]
[a]R & D Hitachi Building Systems Engineering and Service, 16-23 Nakagawa, Adachi-ku, Tokyo 120, Japan
[b]Coll. of Engng., Hosei Univ., Koganei, Tokyo 184, Japan

Abstract

In the present paper, a pure experimental work concerns with the eddy current sensor utilizing the DC magnetic fields. The operation principle of the eddy current testing (ECT) is based on the input impedance deflection caused by the defect or crack. Therefore it is essential to taking into account the magnetization characteristics of the objects. In order to overcome this difficulty, we have applied the DC magnetic fields superposing to the alternating magnetic fields. The DC magnetic field makes the magnetic materials possible to saturate. This means that the magnetic materials may be regarded as the nonmagnetic materials. Thus, we have succeeded in increasing the sensibility of ECT up to the three times.

1. INTRODUCTION

Crack or defect searching is of paramount importance for the safety check of the elevator and escalator in modern buildings, because the defect and crack of these objects possibly cause the serious accidents. Thereby, various nondestructive testing methods, e.g. eddy current testing, electric potential method, ultrasonic wave imaging, and X-ray tomography, are currently utilized [1-3]. Among these methods, the eddy current testing method does not require the direct contact to object and the complex electronic circuits. Further, the defect or crack in the objects whose major frame parts are composed of the conductive metallic materials can be selectively inspected by the eddy current sensor.

In the present paper, a pure experimental work concerns with the eddy current sensor utilizing the DC magnetic fields [4]. Most of the main frame parts of the elevator and escalator consist of iron or its compounds so that the objects are in essence the magnetic materials. The operation principle of the eddy current testing is based on the input impedance deflection caused by the defect or crack. Therefore it is essential to taking into account the magnetization characteristics of the objects. In order to remove this problem, we have applied the DC magnetic fields superposing to the alternating magnetic fields. The reason why is applied the DC magnetic fields is quite simple. The applied DC magnetic field makes the magnetic materials possible to saturate. This means that the magnetic materials may be regarded as the nonmagnetic materials. Our experimental work suggests that the sensibility of our eddy current sensor can be enhanced up to the three times higher than the conventional eddy current sensor. The sensibility of this DC biased ECT is not lineally related to the biasing DC magnetic fields but nonlinear saturation curve similar to those of B-H curve of the magnetic materials.

Thus, we have succeeded in increasing the sensibility of ECT up to the three times.

2. THE DC BIASED EDDY CURRENT SENSOR

2.1. Basic Equations

As is well known that the magnetic materials are composed of the magnetic domains. This leads to a following constitutive relationship:

$$B = \mu_0 H + M, \tag{1}$$

where B, H, M and μ_0 are the magnetic flux density, magnetic field intensity, magnetization vector and permeability of air, respectively. Because of the existence of magnetic domains, the anomalous eddy currents flow between the domain walls when the alternating magnetic field is applied to the ferromagnetic materials. Thereby, it is difficult to apply the conventional eddy current testing (ECT) method to the ferromagnetic materials, because the input impedance of the ECT device is dominated by the domain structure of the materials.

2.2. Operation Principle

In order to carry out the nondestructive testing for the ferromagnetic materials, two representative methods may be considered. One is the DC magnetic field superposition in addition to the alternating magnetic fields in order to saturate the material, i.e. reducing the magnetic materials into nonmagnetic materials [1,4]. The other is the DC magnetic field application to the materials in order to align the magnetization vector in a direction of the applied DC field. If there is a defect in the material, then the magnetization vector could not align at the defect position. This magnetic field distortion can be picked up by moving the search coils in a parallel direction to the applied DC magnetic field [5].

When we introduce the permeabilty μ, (1) is modified to

$$B = \mu_0\{1+[M/(\mu_0 H)]\}H = \mu H, \tag{2}$$

where the permeability μ is defined by

$$\mu = \mu_0\{1+[M/(\mu_0 H)]\}. \tag{3}$$

Mathematically, former method is based on the following fact:

$$\lim_{H\to\infty} \langle\mu\rangle = \lim_{H\to\infty} \langle\mu_0\{1+[M/(\mu_0 H)]\}\rangle = \mu_0. \tag{4}$$

Equation (4) means that the application of strong magnetic field H makes the magnetic materials possible to saturate. This means that the magnetic materials may be regarded as one of the nonmagnetic materials. Thus, the ECT sensibility can be enhanced. On the other side, latter one has to measure the permeability discontinuity due to the defect. Even though latter method is based on the quite simple principle and requires the relatively simple electronic circuits but the target is limited to the objects having flat surface. Also, its determinative weak point is that the error operation is caused by the residual magnetic flux in the magnetic materials.

2.3. Exploited DC biased ECT Device

We have exploited a DC biased eddy current testing device for the defect or crack inspection in the thin iron plate. This is because we have an urgent need of the iron plate inspection used for the modern escalator steps. If the iron plate is broken, then the escalator becomes one of the slow speed guillotines.

Figure 1 show a schematic diagram of the device. The left and right electromagnets make the magnetic materials having plate shape to saturate. The head of eddy current sensor is located at the center, and alternating current is fed to the coils enclosing the head. The difference of input impedance of this coils caused by the defect or crack indicates the existence of defect or crack in the iron plate.

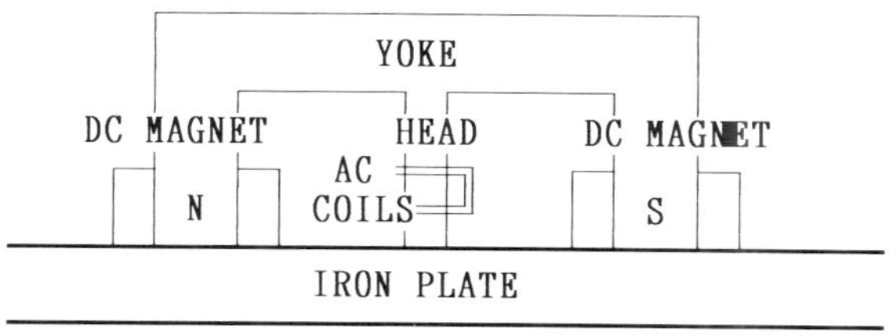

Figure 1. Schematic diagram of a DC biased ECT.

2.4. Experimental

We have carried out the experiments changing the size of defect. Figure 2 shows a sample iron plate used in the experiment. The length, thickness and width of this iron plate are 30, 2.5 and 10cm, respectively. Also, the details of defects in this sample are shown in Fig. 2.

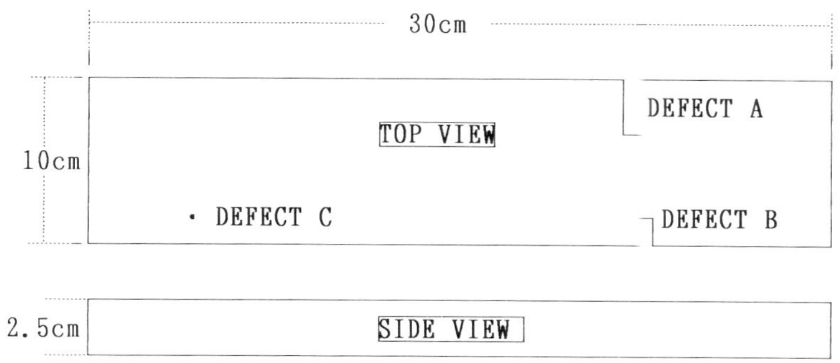

Figure 2. A sample iron plate used in the experiment.
DEFECT A:length=30mm, width=0.5mm, thickness=25mm.
DEFECT B:length=6mm, width=2mm, thickness=25mm.
DEFECT C:hole shape crack, diameter=1.5mm, depth=25mm.

Using the 50kHz exciting frequency, we have checked up the sensibility of our DC biased eddy current sensor. As shown in Fig.3, the sensibility for the relatively large defects A and B were not so enhanced. However, the sensibility for the small defect C was increased up to the three times. The results of our experiments reveal that the large defects, e.g. defect A and B in Fig.2, dominate the input impedance of the exciting coils but the effect of small defect C in Fig. 2 may be lost by the ferromagnetic property of the sample plate. The sensibility for the small defect such as C in Fig.2 is not lineally related to the biassing DC magnetic fields but nonlinear saturation curve similar to those of B-H curve of the magnetic materials. This fact verifies our theoretical consideration.

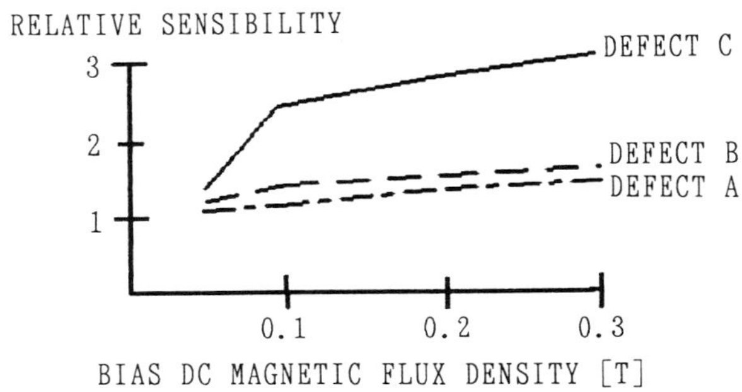

Figure 3. Sensibility vs. DC bias flux density.

3. CONCLUSION

As shown above, a pure experimental work has been carried out to the eddy current sensor utilizing the DC magnetic fields. In order to enhance the ECT sensibility for the ferromagnetic materials, we have applied the DC magnetic fields superposing to the alternating magnetic fields. The DC magnetic field successfully has made the magnetic materials to saturate. Thus, we have succeeded in increasing the sensibility of ECT up to the three times.

4. REFERENCES

1 Y.Hirama et al, The method for Eddy Current Sensor added to Magnetic Bias, Paper on Technical Meeting of Magnetics in IEEJ, MAG-92-73, Feb.(1992)pp.143-148.

2 T.Doi et al, Inverse Analysis of Defect Identification based on the Electric Potential Method, Proc. of 2nd Japan-Hungary Joint Seminar, Published by Hokkaido University, (1992).

3 H.Saotome et al, Crack Identification in Metallic Materials, IEEE Trans. Magn. to be published.

4 G.Cecco et al., Eddy Current Detection of Corrosion Damage in Heat Exchanger Tubes, At Energy Can Ltd., No.AECL-6965, pp.33, (1980).

5 Y.Oohira et al, Defect Searching in the Ferromagnetic Materials, Companion paper of the ISEM-Sapporo, Jan. (1993).

Simulation and Design of Applied Electromagnetic Systems
T. Honma (Editor)

2-D sensor of magnetic material using magnetic field visualization technique

M. Hashimoto

Department of Electrical Engineering, University of Industrial Technology, 4-1-1 Hashimotodai, Sagamihara 229, Japan

Abstract

The author developed a magnetic field visualization system that can measure all the three components of an arbitrary magnetic field using a Hall-device-array-probe and demonstrated visual images of the magnetic field on a computer display. This paper describes a detection method of shape and position of magnetic material helping with this system.

1. INTRODUCTION

In the magnetic flux measurement method, an apparatus equipped with a sensor such as a Hall device or a pick-up coil is used and measurement can only be made at one point and in one direction. However, a magnetic field is a vector quantity having three components. Therefor, it is difficult to estimate magnetic field if it is measured by one sensor, and the measurement of three components is required [1]. This is especially true when the magnetic field shows nonlinear behavior, as in ferromagnetic materials. Therefore, it is necessary not only to measure the distribution of a magnetic field but also to develop models for the intuitive understanding of magnetic behavior. In light of the above situation, we have developed the magnetic field visualization system [2,3]. In this paper I make an attempt to develop a 2-D magnetic material sensor that can measure the shape and size of the magnetic material.

2. MEASURING SYSTEM

The magnetic field visualization system is shown schematically in Figure 1. The system consists of many components such as the 1-D probe, the control and amplifier equipment, the probe driving device, the personal computer and the display.

The developed probe to measure the magnetic material consists of magnetic flux sensors, a permanent magnet, multiplexers and pre-amplifier. As a magnetic sensor GaAs type Hall devices are selected because of their good linearity in sensitivity with respect to magnetic flux and because of their good temperature characteristic. The probe is constructed of ninety six Hall devices in total, thirty two of which are arranged in three columns corresponding to the three directions, i.e. to the three components of the arbitrary magnetic field to be measured. The permanent magnet is arranged on the Hall-device-array and supplies magnetic field to detecting material. The multiplexers which can switch each output of the devices, and the pre-amplifier connected after the multiplexers are also set in the probe. As a result handling of the probe become easy, because the output line from the probe is only one cable which contains the line of the switching control signals, the power supply and the output of pre-amplifier. An arrangement of the Hall device array and the magnet is shown in Figure 2.

Offset revision process is needed for deduction the magnetic field made by the magnet in the probe and for revision of offset of each Hall device because the offset values of each Hall device are always different from itself. In this equipment, the offset revision are fixed using a feed-back circuit with a D/A converter by the CPU process. These offset data are saved in the memory of the equipment, the offset processing are handled automatically on measuring since then. Therefore this probe can detect only distortion of the magnetic field.

Graphic programs for visualization of magnetic fields were developed for easy understanding of the magnetic field distribution with the help of computer graphic technique.

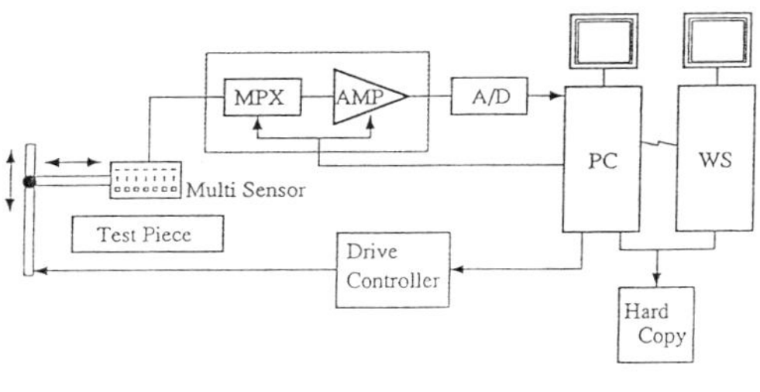

Figure 1. Block diagram of magnetic flux visualization system.

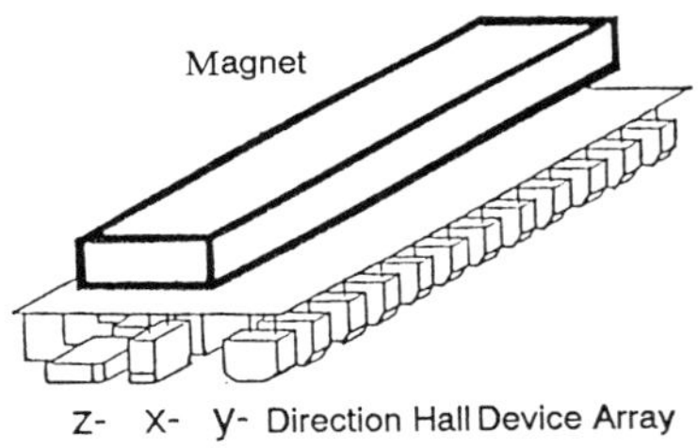

Figure 2. Magnetic material sensor.

3. MEASUREMENT

The examples of 2-D measurement of magnetic material are introduced as follows. In Figure 3 the detected model consisted of a plate and two cubes made of ferromagnetic material is shown. In Figure 4 the full color graphics of the three components (Bx,By,Bz) of perturbed field by detected material is shown, where it is very easy to identify the location and source intensity of the magnetic material. In this figure the line measurement of magnetic flux distribution along the x- and y-directions chosen by a mouse is calculated and also displayed. Figure 5 shows a bird's-eye view of wire-frame graphics of the amplitude of the perturbed magnetic flux density of the z-component above the full color graphic of the magnetic field.

In Figure 6 an outline (+ marks) on the Bz graphic is presumed by a method to plot the points of maximum and minimum of Bx and By along x-line or y-line, respectively. The outline expresses properly the shape of the magnetic material.

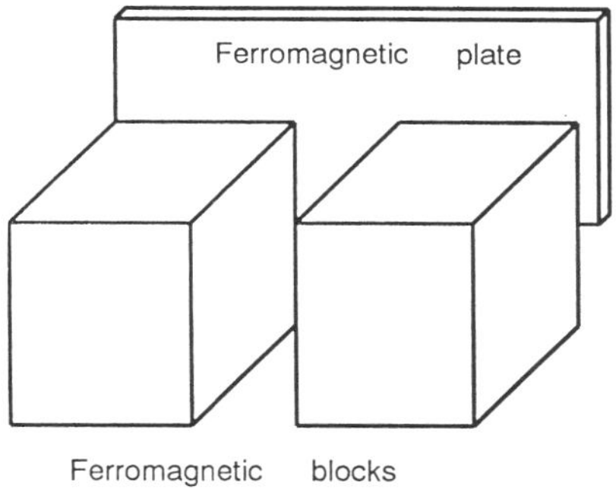

Figure 3. Arrangement of ferromagnetic.

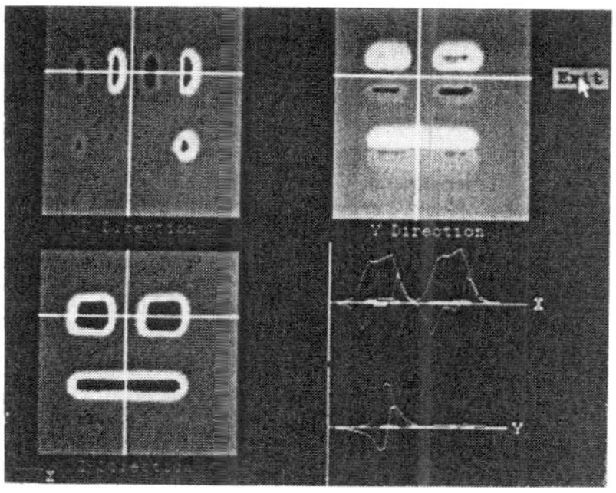

Figure 4. Visualization graphics (Bx,By,Bz) and line measurements.

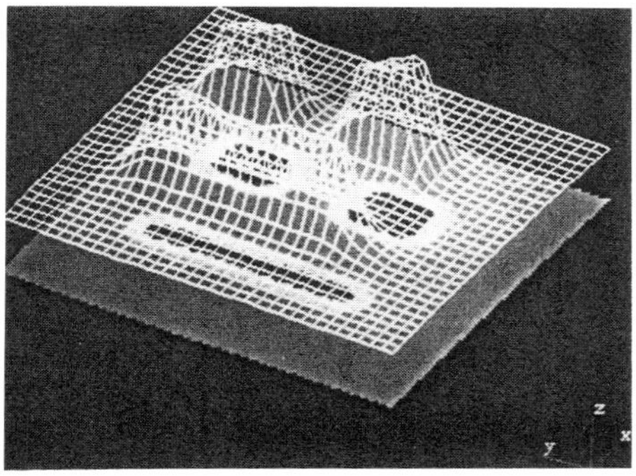

Figure 5. Wire-frame graphics of Bz.

Figure 6. Outline presumption of magnetic material.

Then two experiments are performed in order to confirm the property of the developed sensor. The detecting magnetic material is a plate (80mm x 35mm x 0.2mm thickness). When we detect magnetic material, we usually use a method to measure magnetic flux over the material in fixed magnetic flux. Figure 7 shows the result by this method, and indicates perturbed magnetic field by the detecting material using a compensation method that is carried out deduction operations to null data of each position previously measured at magnetic field without detecting material. On the other hand, Figure 8 also shows detected result for the same plate using by the developed sensor. Comparing these with it, the result shown in the Figure 8 using the developed probe expresses more exactly than the method using fixed magnetic field shown in the Figure 7. By the results, we can understand that the developed sensor can detect in proportion to permeability of material i.e. shape of it, whereas the method using fixed magnetic flux detects the poles of magnetized material that is not the shape of it i.e. the method is influenced the shape effect.

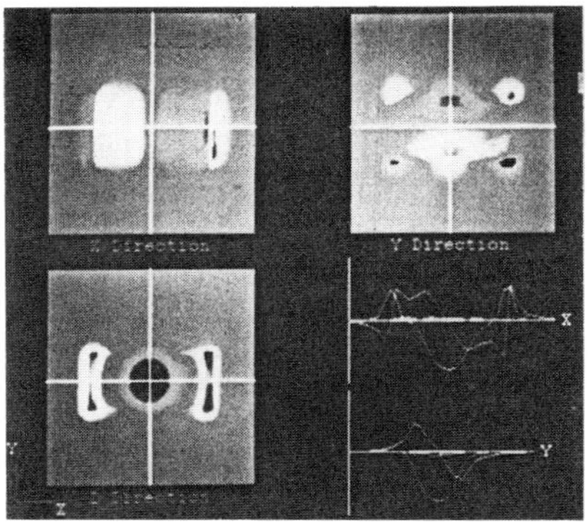

Figure 7. Distortion magnetic field over a sheet.

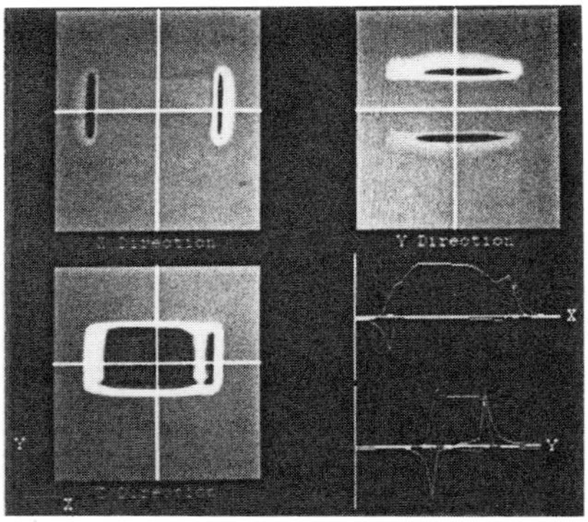

Figure 8. Measurement result of a sheet using the developed sensor.

4. CONCLUSIONS

The following conclusions are obtained by the present study.

(1) I developed the magnetic material sensor.

(2) The shape and position of magnetic material was clearly expressed on the visualization graphics using this system.

REFERENCES

1 K.Maenaka, T.Ohgusu and T.Nakamura, IEICE, J74-C-II No.5 (1991) 325.

2 M.Hashimoto, Y.Yamada, T.Takagi and K.Miya, QNDE 9 (1990) 585.

3 M.Hashimoto, M.Uesaka and K.Miya, The 11th Sensor Symposium (1992) 67.

Intelligent Computer Aided Design

Simulation and Design of Applied Electromagnetic Systems
T. Honma (Editor)

Solving inverse problems in non-linear PDEs by recurrent neural networks

Tadasu Uchiyama and Noboru Sonehara

NTT Human Interface Laboratories, Nippon Telegraph and Telephone Corp., 1-2356 Take, Yokosuka-shi, Kanagawa, 238-03 JAPAN

Abstract

A neural network approach for determining coefficients in non-linear partial differential equations on the basis of known boundary values is proposed and a computer simulation based on this approach is described.

1. INTRODUCTION

Problems of determining variable coefficients of a certain partial differential equation (PDE) on the basis of known values of the equation at fixed points for all times are a kind of inverse problems. Various classes of the *linear* PDEs have been comprehensively investigated [1]. There are, however, very few studies concerning non-linear inverse problems.

A certain class of PDE can be approximated by a differential difference equation (DDE) describing the dynamics of a recurrent neural network. The network is trained so that its output and the known boundary values of the equation are consistent for all times.

In the following, the DDE approximating a certain class of PDEs is derived and a recurrent neural network is designed based on the DDE. This is followed by a computer simulation of identifying variable coefficients of a PDE.

2. DERIVATION OF DIFFERENTIAL DIFFERENCE EQUATION

Consider the one-dimensional problem to find solution $u(x, t)$ $(0 \le x \le 1, t \ge 0)$ which satisfies the following equations:

$$u_t(x, t) = p(x)\, z_{xx}(x, t) + q(x)\, z_x(x, t) + r(x)\, z(x, t) + s(x)$$

$$z(x, t) = f(u(x, t), x), \tag{1}$$

$$\eta_0\, z(0, t) + \zeta_0\, z_x(0, t) = \varphi_0(t)$$

$$\eta_1\, z(1, t) + \zeta_1\, z_x(1, t) = \varphi_1(t), \tag{2}$$

$$z(x, 0) = \phi(x), \tag{3}$$

where $p(x)$, $q(x)$, $r(x)$, $s(x)$, $\varphi_0(t)$, $\varphi_1(t)$ and $\phi(x)$ are real functions and η_0, η_1, ζ_0 and ζ_1 are real constants. Equations (2) and (3) give the boundary conditions and the initial condition, respectively. The function f represents the relationship between $u(x, t)$ and $z(x, t)$ at x. Some practical problems described by the same class of equation as (1) can be seen in [2].

The *inverse* problem arising from the above mentioned direct problem is to determine appropriate $p(x)$, $q(x)$, $r(x)$ and $s(x)$ so that $z(0, t)$ and $z(1, t)$ approximate the boundary values at $x = 0$ and $x = 1$ observed when $p(x)$, $q(x)$, $r(x)$ and $s(x)$ are set to correct values.

Define $u_n(t) = u(x_n, t)$ and $z_n(t) = z(x_n, t)$, where x_n denotes $x_n = n/N = nh$. Denoting the function values at x_n as $p_n = p(x_n)$, etc., and replacing the spatial differentials in (1) with the central differences, the DDE in terms of $u_n(t)$,

$$\frac{d}{dt} u_n(t) = (2p_n + q_n)/(2h)\, z_{n+1}(t) + (2p_n - q_n)/(2h)\, z_{n-1}(t) + (-2p_n + hr_n)/h\, z_n(t) + s_n$$

$$z_n(t) = f_n(u_n(t)) = f(u_n(t), x_n) \qquad n = 1, \ldots, N-1, \tag{4}$$

is derived. The boundary conditions (2) are transformed to

$$z_0(t) = -\zeta_0/(h\eta_0 - \zeta_0)\, z_1(t) + h/(h\eta_0 - \zeta_0)\, \varphi_0(t)$$

$$z_N(t) = \zeta_1/(h\eta_1 + \zeta_1)\, z_{N-1}(t) + h/(h\eta_1 + \zeta_1)\, \varphi_1(t) \tag{5}$$

by replacing the differentials at $x = x_0$ and $x = x_N$ with the forward and backward differences, respectively. The initial condition (3) is transformed to

$$z_n(0) = \phi_n = \phi(x_n). \tag{6}$$

3. UNIT MODEL AND NETWORK ARCHITECTURE

Equations (4) and (5) have the forms of

$$\frac{d}{dt} u_n(t) = \sum_m w_{nm}\, z_m(t) + v_n$$

$$z_n(t) = f_n(u_n(t)) \qquad n = 1, \ldots, N-1 \tag{7}$$

and

$$u_n(t) = \sum_m w_{nm}\, z_m(t) + v_n$$

$$z_n(t) = f_n(u_n(t)) = u_n(t) \qquad n = 0, N, \tag{8}$$

respectively, where w_{nm} and v_n are constant values. Interpreting $u_n(t)$ and $z_n(t)$ as an internal state and an output value of the unit composing a neural network, (7) and (8) provide the dynamics of a recurrent network where the output function of each unit is given by f_n. In this case, w_{nm} and v_n represent connection weights and threshold values, respectively.

The recurrent network composed of these units is shown in Figure 1. The initial output of each unit is set according to (6).

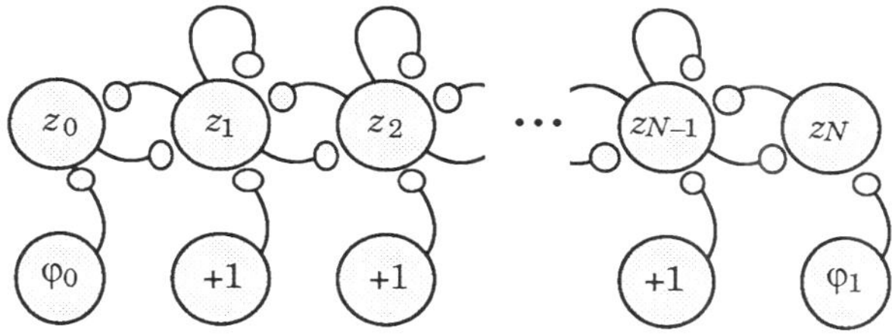

Figure 1. The network architecture. Small circles represent connection weights and threshold values, each of which equals the corresponding coefficient in (4) and (5).

Suppose the physical system described by (1)-(3). In this system, the identification problem is to find appropriate $p(x)$, $q(x)$, $r(x)$ and $s(x)$ so that $z(0, t)$ and $z(1, t)$ approximate $\Phi_0(t)$ and $\Phi_1(t)$, respectively, each of which is the boundary value at $x = 0$ and $x = 1$ observed when $p(x)$, $q(x)$, $r(x)$ and $s(x)$ are set to correct values. In the discrete system, this means to find appropriate p_n, q_n, r_n and s_n so that $z_0(t)$ and $z_N(t)$ approximate $\Phi_0(t)$ and $\Phi_1(t)$, respectively.

To solve this problem, the learning algorithm proposed by Williams and Zipser [3] is applied, where the objective function is defined as

$$E = \int_{t_1}^{t_2} dt\, (z_0(t) - \Phi_0(t))^2 + \int_{t_1}^{t_2} dt\, (z_N(t) - \Phi_1(t))^2, \tag{9}$$

where t_1 and t_2 determine the time interval for calculating the objective function. p_n, q_n, r_n and s_n are derived through the gradient descent.

4. SIMULATION

A one-dimensional identification problem is simulated. The following non-linear PDE is assumed for the system to be identified:

$$u_t(x, t) = -\, z_x(x, t) + r(x)\, z(x, t) + s(x)$$

$$z(x, t) = \tanh u(x, t). \tag{10}$$

The boundary conditions and the initial condition are also assumed as $z(0, t) = \varphi_0(t)$, $z_x(1, t) = 0$ and $z(x, 0) = 0$, respectively.

The problem is to reconstruct these coefficients on the basis of the known boundary values at $x = 0$ and $x = 1$, each of which is defined by $\Phi_0(t)$ and $\Phi_1(t)$, respectively, where $\Phi_0(t)$ is given by $\varphi_0(t)$ and $\Phi_1(t)$ is numerically calculated according to (10) by setting $r(x)$ and $s(x)$ to original values given by the continuous curves in Fig. 2.

A network composed of 32 units is designed by dividing the section $0 < x < 1$ into 31 subsections. The initial values of r_n and s_n, which are to be reconstructed, are set to zero. 512 pairs of the boundary values are used for the learning. After 2500 iterations, E becomes close to zero andr_n and s_n are derived as shown in Fig. 2. If $E = 0$ holds for arbitrary function $\varphi_0(t)$, the reconstructed values are guaranteed to agree with the original ones in this kind of problems in non-linear gradient flow [4]. Thus the network succeeded in reconstructing the approximate values of the original coefficients.

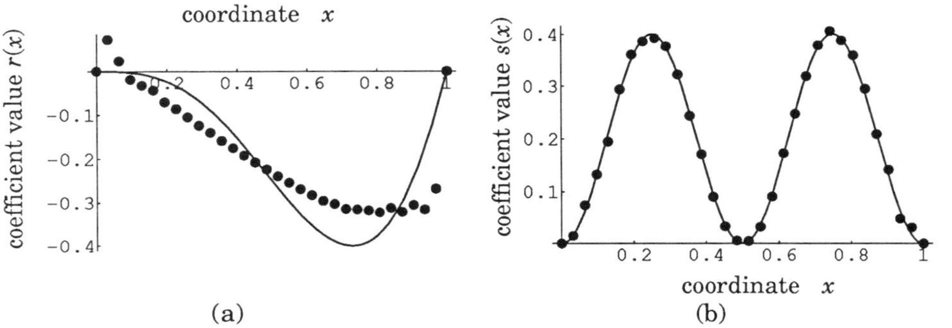

Figure 2. Original and reconstructed values of (a) $r(x)$ and (b) $s(x)$. Each value is represented by a continuous curve and filled circles, respectively.

5. CONCLUSION

A neural network approach for determining coefficients in non-linear PDEs on the basis of known boundary values is proposed. The network is designed based on the DDE approximating the PDE and trained so that its output and the known boundary values of the equation are consistent. The computer simulation shows that the adjustable connections converge to the approximate values of the original coefficients.

6. REFERENCES

[1] Tijhuis, A.G., *Electromagnetic Inverse Profiling*, VNU Science Press 1987

[2] Ames, W.F., *Nonlinear Partial Differential Equations in Engineering*, Academic Press 1965

[3] Williams, R.J. and Zipser, D., *Neural Computation* **1**, 270-280 (1989)

[4] Uchiyama, T. and Sonehara, N., in preparation.

Simulation and Design of Applied Electromagnetic Systems
T. Honma (Editor)

A neuro-fuzzy approach to optimization and inverse problems

H. Ichihashi

Department of Industrial Engineering, College of Engineering, University of Osaka Prefecture, 1-1 Gakuen-cho, Sakai, Osaka 593, JAPAN

Abstract

The usual cost function (objective function) in supervised learning is an error function. A more general function such as a functional of integral form can be used in the Neuro-Fuzzy approach. The functional is generally expressed as a time integral of a performance index and can be minimized by the method of variational calculus.

By the simple formula of the integration of Gaussian functions, the integration of the function representing Neuro-Fuzzy model can be written simply and is applied to computerized tomography.

1. INTRODUCTION

Supervised learning in artificial neural networks provides a convenient method to find an approximate function which is composed of some kind of basis functions (e.g., sigmoid function). In this paper a Neuro-Fuzzy model is obtained in the form of Gaussian Radial Basis Functions(RBF) which can be regarded as a three-layered neural network[4] and fuzzy reasoning rules too[2,3].

To solve the differential equation in the method of variational calculus, the Neuro-Fuzzy model is used for approximating the solution. We apply it to a traditional problem of finding a line of swiftest descent.

The integration of Neuro-Fuzzy model is applied to the computerized tomography. A straight-line ray optics model is assumed for the propagation mechanism. A numerical example of computed tomography by the direct method of solution using the steepest descent method is presented.

2. NEURO-FUZZY OPTIMIZATION BY THE METHOD OF VARIATIONAL CALCULAS

Let $A_{i,k}$ denotes the membership function of the k-th fuzzy rule in the domain of the i-th input variable x_i. The k-th rule is written as

"If x_1 is $A_{1,k}$ and x_2 is $A_{2,k}$ ... and x_n is $A_{n,k}$, then y is w_k."

The conclusion part of the fuzzy reasoning rule which infers the output y is simplified as real number w_k. The final output y is written as

$$y = \sum_{k=1}^{K} \mu_k \cdot w_k = \sum_{k=1}^{K} \{ \prod_{i=1}^{N} A_{i,k}(x_i) \} \cdot w_k. \tag{1}$$

where μ_k is the compatibility degree of the premise part of the k-th fuzzy rule, which is computed with the algebraic product operator. This model is a simplified fuzzy model. In the case of Gaussian membership function, $A_{i,k}$ is defined by $A_{i,k}(x_i) = \exp\{-(x_i-a_{i,k})^2 / b_{i,k}\}$, where the parameters $a_{i,k}$ and $b_{i,k}$ $(i=1,\ldots,N)$ are given for each k and are changed in the training procedure. μ_k have the form of Gaussian functions $F(\|\mathbf{x}-\mathbf{a}\|)$ and y in equation (1) is its linear combination, i.e. Gaussian Radial Basis Functions (GRBF). In this way the GRBF is reinterpreted as a set of fuzzy rules. The normalization by the sum of μ_k (i.e., the centroid defuzzification method of fuzzy reasoning) is omitted for simplicity in the supervised learning.

For applying the methods of variational calculus, firstly we describe how to solve differential equations. Since a simplified fuzzy model consists of Gaussian functions, it is an infinitely differentiable function. Hence, any nonlinear ordinary differential equations and boundary conditions can be written using y in equation (1) as :

$\sum_{r=1}^{R} B_r(x)\frac{\partial^r}{\partial x^r} y=f(x)$ and $\sum_{r=1}^{R} C_r(x)\frac{\partial^r}{\partial x^r} y=g(x)$, respectively. B_r and C_r are the coefficients in a differential equation and a boundary condition respectively. The problem is to find the values of parameters $a_{i,k}$, $b_{i,k}$ and w_k in y. Let errors between both sides of the equations at given collocations be $R_0(x)=F_0(x)-f(x)$ and $R_1(x)=G_0(x)-g(x)$ respectively, where F_0 and G_0 are the left sides. Let a cost function be $J(x)=R_0(x)^2+\gamma \cdot R_1(x)^2$ where γ is a positive constant. Let δ, δ_a and δ_b be

$$\delta = R_0(x) \cdot \frac{\partial}{\partial w_k} F_0(x) + \gamma \cdot R_1(x) \cdot \frac{\partial}{\partial w_k} G_0(x)$$

$$\delta_a = R_0(x) \cdot \frac{\partial}{\partial a_{1,k}} F_0(x) + \gamma \cdot R_1(x) \cdot \frac{\partial}{\partial a_{1,k}} G_0(x)$$

$$\delta_b = R_0(x) \cdot \frac{\partial}{\partial b_{1,k}} F_0(x) + \gamma \cdot R_1(x) \cdot \frac{\partial}{\partial b_{1,k}} G_0(x)$$

respectively. By applying the steepest descent method, we have a learning rule such as $w_k^{NEW}=w_k^{OLD}-\tau \cdot \delta$, $a_{1,k}^{NEW}=a_{1,k}^{OLD}-\tau \cdot \delta_a$ and $b_{1,k}^{NEW}=b_{1,k}^{OLD}-\tau \cdot \delta_b$, where τ denotes positive learning constant.

Let us take an example of ordinary differential equation of fourth order:

$$\frac{d^4y}{dx^4} + y = 0 \ (x \in [0, \sqrt{2\pi}\,]), \quad y(0) = y'(0) = y(\sqrt{2\pi}) = 0, \ y'(\sqrt{2\pi}) = -1$$

21 collocations (x) were uniformly spaced in the domain interval $[0,\sqrt{2\pi}\,]$. Figure 1 shows the simulation results by the conjugate gradient method of Fletcher-Reeves. The circles ○ indicate theoretical values of y at several data points. ①-④ are Gaussian functions after learning. y' and y'''' are the first order and the fourth order derivatives of the approximate solution y respectively.

The problem to find the line of swiftest descent was presented by Bernoulli in 1696 and is an origin of calculus of variations. In Figure 2 a mass point at 0 falls down to A by its gravity along any path. This problem is to find a path y(x) which minimize the elapsed time J from 0 to A :

$$J(y(x)) = \int_0^a \frac{1+y'^2}{y} dx$$

Hence, we have a solution by solving Eulerian equation $y(1+y'^2)=C$ where C is an integration constant (i.e. an unknown constant). The theoretical solution of y is a cycloid. Let a cost function be

$$E=\sum_{z\in I}\{y(z)(1+y'(z)^2)-y(z+1)(1+y'(z+1)^2)\}^2+\gamma\sum_{z\in I'}\{(y(z)-y_z)^2+(x(z)-x_z)^2\}$$

where, $I=\{0,1,\cdots,9\}$, $I'=\{0,10,11,12,13,14\}$, $(x_0,y_0)=(0,0)$, $(x_{10},y_{10})=(1,1)$, $(x_{11},y_{11})=(1.2,0.9)$, $(x_{12},y_{12})=(1.28,0.75)$, $(x_{13},y_{13})=(1.2,0.6)$, $(x_{14},y_{14})=(1.0,0.5)$. z is introduced for parametric representation of y. In Figure 2, points B-E are given to find an approximate function y which smoothly interpolates those points. Since this function can not be represented by a single variable x, the parameter z is introduced. y and x are represented as $y(z)=\sum_{k=1}^{K}\mu_{1k}(z)\cdot w_k$ and $x(z)=\sum_{k=1}^{K}\mu_{2k}(z)\cdot w_k$ respectively.

In Figure 2 the simulation results after 40,000 iterations of the learning (the steepest descent method) are shown where $\gamma=100$. The approximate solution is similar to the theoretical solution (○ in Figure 2) between 0 to A and smoothly interpolates points B to E.

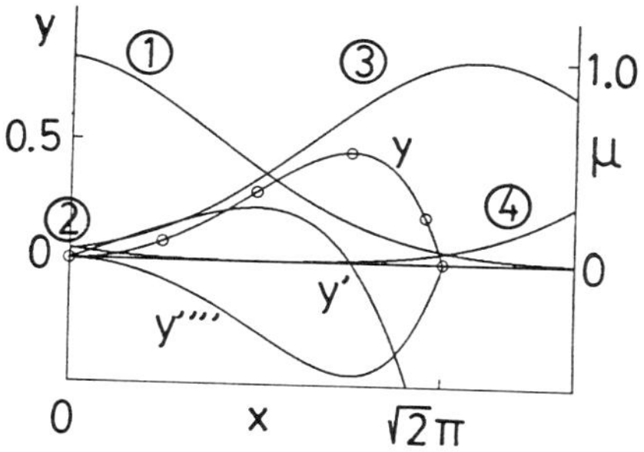

Figure 1. Approximate solution of an ordinary differential equation.
$\gamma=50$

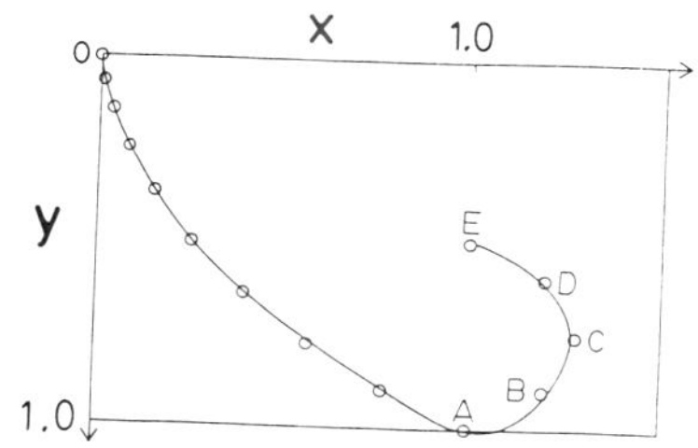

Figure 2. Approximate solution representing the line of swiftest descent from 0 to A, which smoothly interpolates points B to E.

3. NEURO-FUZZY COMPUTED TOMOGRAPHY

We here consider an application such as the geophysical tomography. In Figure 3, a typical situation, the data collection system scans the rectangular region (CDEF) with a signal source located in one borehole (CD) and a receiver in the other (EF). A multitude of source and receiver locations are selected to sample the region with relatively small number of transmission paths. The receiver signals carry time-of-flight information for velocity measurements and amplitude information for attenuation measurements. In this section we describe a Neuro-Fuzzy approach to computed tomography and test its performance on computer generated data. A region of probing (cross section) is shown in Figure 3, where a typical ray path is indicated (AB) for signal propagation. In the region of probing the distribution of a material parameter, such as velocity or attenuation rate, is to be calculated. We assume the straight line ray-optic model of the propagation mechanism.

Let $y(x_1,x_2)$ be the output of Neuro-Fuzzy model in equation (1). The line from A $(0,x_2^R)$ to B $(1,x_2^L)$ can be written as $x_2=\alpha x_1+x_2^L$ where $\alpha=x_2^R-x_2^L$. Let (x_1,x_2) be coordinate of a point on the line AB. Then $x_1=z/\sqrt{1+\alpha^2}$, $x_2=\alpha z/\sqrt{1+\alpha^2}+x_2^L$, where z is a parameter denoting the length from A to the point (x_1,x_2). Let a Gaussian function be $\mu(x_1,x_2)=\exp\{-((x_1-a_{1K})^2+(x_2-a_{2K})^2)/b_{1K}\}$, where it is assumed that b_{1K} equals b_{2K} for simplicity. The circles in Figure 3 represent contour curves of a Gaussian function with two variables. $\mu(z)=\zeta\exp\{-(z-a_3)^2/b\}$, where $\zeta=\exp\{-(\alpha a_1+x_2^L-a_2)^2/((1+\alpha_2)b)\}$ and $a_3=(\alpha(x_2^L-a_2)-a_1)/\sqrt{1+\alpha^2}$. And, the line integral of $y(x_1,x_2)$ along the path AB can be written as $\int_{-\infty}^{\infty} y(z)dz=\sum_{K=1}^{K}\zeta_K\sqrt{\pi b_K}\cdot w_K$.

The line segment AB is a propagation path L_P. We express the distribution of some material parameter such as attenuation rate by the Neuro-Fuzzy model $y(x_1,x_2)$. Attenuation rate is a measure of how rapidly plane wave energy is dissipated with distance in the medium. If the refractive-index changes are very small we can assume the straight line ray-optic model. And, the projection along a path L_P can be written as $I_P=\int_{L_P}y(z)dz$, $(p=1,\cdots,P)$. This equation holds regardless of whether we are dealing with attenuation or velocity. A transmitter is located on CD $(x_1=1)$ and A receiver is on EF $(x_1=1)$ as shown in Figure 4. Synthetic data corresponding to Q equally spaced transmitter locations and R receiver locations can be generated. We can have I_P^* for each of P $(=Q\times R)$ propagation paths.

Let a cost function be $J=\sum_{P=1}^{P}(I_P^*-I_P)^2$ and formulate a nonlinear mathematical programming problem to find the values of parameters $a_{i,K}$, $b_{i,K}$ and w_K in the Neuro-Fuzzy model, which minimize the cost function.

36 propagation paths are chosen as in Figure 4 from 6 equally spaced transmitter and receiver locations. In Figure 5, (a) the simulated test distribution of a material parameter and (b) the output of Neuro-Fuzzy model $y(x_1,x_2)$ after learning are shown. Both 3-D graphic and contour curves (an image of the cross section) are shown.

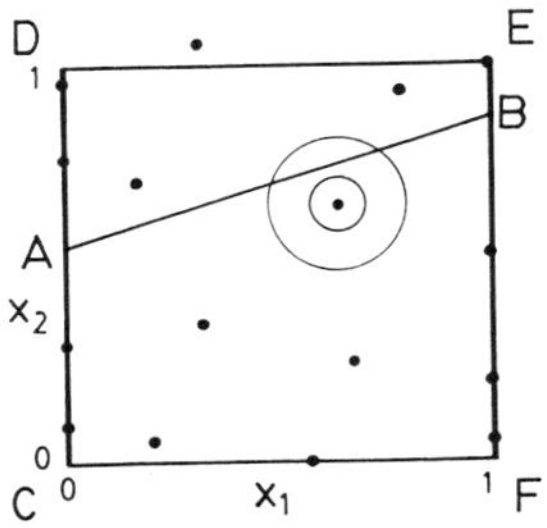

Figure 3. Region of probing

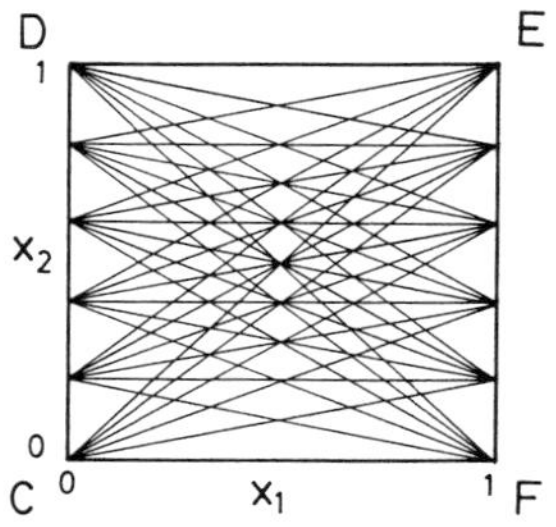

Figure 4. Propagation paths

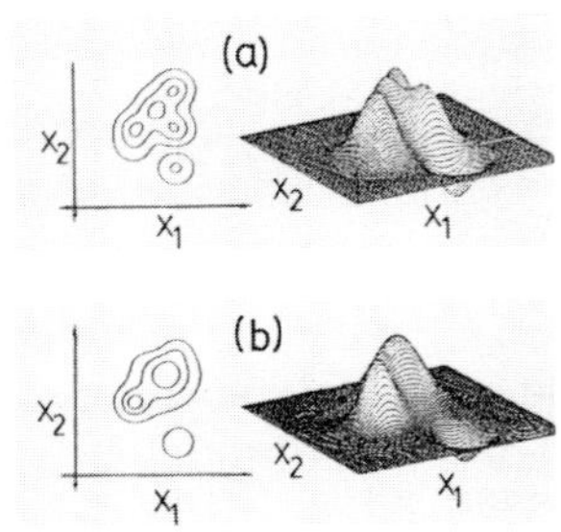

Figure 5. Simulation results

REFERENCES

1 K.A.Dines and R.J.Lytle, Proc. of IEEE, 67 (1979) 1065.
2 H.Ichihashi, Proc. of IFSA'91, vol.Eng (1991) 49.
3 H.Ichihashi, Proc. of Japan U.S.A. FA Sympo, San Francisco (1992) 877
4 T.Poggio and F.Girosi, Sciences, 247 (1990) 978.

Simulation and Design of Applied Electromagnetic Systems
T. Honma (Editor)

Shape Optimization of Rotor Slot in Cage Induction Motors

Il-han Park, Hyang-beom Lee, In-gu Kwak and Song-yop Hahn

Department of Electrical Engineering, Seoul National University, Shilim-dong, Kwanak-gu, Seoul 151-742, KOREA

Abstract

In eddy current problem, the design sensitivity evaluation algorithm based on finite elememt discretization is presented. It is applied to the rotor slot shape design of cage induction motors. The design objective is to obtain the optimal rotor slot shape which generates the desired speed-torque characteristics curve.

1. INTRODUCTION

Electrical machines have until now been designed using magnetic circuit method, a kind of lumped parameter approximation. However that method does not provides accurate field distribution and it is almost impossible to analyze various local properties; force density, loss distribution, etc. As accurate field analysis tool, finite element method has been much developed and recently a new design method of coupling finite element with optimization algorithm is being studied as the precision design tool of electrical machines with improved performance.

In this paper we present a design sensitivity evaluation procedure based on finite element discretization[1-2], which can be applied to rotor slot shape design of cage induction motors. The numerical example presented is simple one which validates the sensitivity analysis of eddy current problems, but this algorithm can be applied to various and complex design models of induction motors.

2. SENSITIVITY ANALYSIS IN COMPLEX VARIABLE SYSTEM

The objective function representing performance of a design is a scalar-valued function of state variable. Then since the state variable in eddy current problem is complex value the sensitivity derivation requires careful manipulation. Such an objective function can be written as

$$F = F(A_r, A_i) \tag{1}$$

where A_r and A_i are the real and imaginary parts of magnetic vector potential $A = A_r + jA_i$ and j is unit imaginary number. The sensitivity is represented as

$$\frac{dF}{dp} = \frac{\partial F}{\partial A_r}\frac{\partial A_r}{\partial p} + \frac{\partial F}{\partial A_i}\frac{\partial A_i}{\partial p} \tag{2}$$

In (2), in order to obtain the derivative of state variable to design parameter we differentiate the system equation (3) of state variable to provide (4)

$$[S][A] = [f] \tag{3}$$

$$\frac{\partial A}{\partial p} = [S]^{-1} \frac{\partial}{\partial p} \{[f] - [S][\tilde{A}]\} \tag{4}$$

where $[\tilde{A}]$ is a solution at the nominal state. And the left-hand side of (4) can also be written as

$$\frac{\partial A}{\partial p} = \frac{\partial A_r}{\partial p} + j\frac{\partial A_i}{\partial p} \tag{5}$$

And we define a notation expressing the derivative of scalar variable to complex variable as

$$\frac{\partial F}{\partial A} \equiv \frac{\partial F}{\partial A_r} - j\frac{\partial F}{\partial A_i} \tag{6}$$

With only the complex variables of (5) and (6), the sensitivity (2) can be written as

$$\frac{dF}{dp} = Re\ [\frac{\partial F}{\partial A} \frac{\partial A}{\partial p}] \tag{7}$$

And inserting (4) into (7) gives

$$\frac{dF}{dp} = Re\ [\frac{\partial F}{\partial A}[S]^{-1} \frac{\partial}{\partial p} \{[f] - [S][\tilde{A}]\}\ ] \tag{8}$$

Here we introduce the adjoint equation of adjoint variable as

$$[S]\ [\lambda] = [\frac{\partial F}{\partial A}] \tag{9}$$

Using (9), finally the sensitivity (8) is obtained as

$$\frac{dF}{dp} = Re\ [[\lambda]^T \frac{\partial}{\partial p} \{[f] - [S][\tilde{A}]\}\ ] \tag{10}$$

The above overall procedure for sensitivity evaluation is summarized as follows (I) Solve the system equation (3) for the state variable of magnetic vector potential. (II) Solve the adjoint equation (9) for the adjoint variable. (III) Calculate the sensitivity coefficients (10) with the obtained state and adjoint variables.

3. SENSITIVITY FOR ROTOR SLOT SHAPE DESIGN

The rotor slot shape of induction motor determines its speed-torque characteristics curve. In this paper we search a slot shape to improve the speed-torque curve. Since the torque at a given speed is linealy proportional to input power from external source and input power is easier to deal with than Maxwell stress tensor, we define the objective function with input power. The stator part supplyng input power is modeled as the surface current J of ideal travelling wave[3], so the input power at a given speed (slip=s) is written as

$$P_s = Re\{ \int_\Gamma j\omega s\ A\ J^*\ dl\} \tag{11}$$

where Γ is the outer boundary of surface current and ω is the angular velocity of surface current and s is slip and $*$ means the conjugate of complex variable.

The objective function is defined as

$$Min : F = \sum_{s}(P_s - P_{os})^2 \tag{12}$$

where P_{os} is a target input power at slip=s. Its sensitivity is written as

$$\frac{dF}{dp} = 2 \sum_{s}(P_s - P_{os})\frac{dP_s}{dp} \tag{13}$$

where the derivative of input power P_s with respest to p is calcualted using (10). And the forcing term of adjoint equation (9) for the k-th node is expressed as

$$\frac{\partial P_s}{\partial A_k} = \int_{\Gamma} j\omega s\, N_k J^* \, dl \tag{14}$$

where N_k is the shape function of the k-th node.

4. NUMERICAL EXAMPLE

As in Fig.1, the anlysis model is one-fourth section of a 4-pole cage induction motor with 24 rotor slots and semi-periodic boundary condition is applied to analyze it. The analysis domain is discretized with triangular elements and the second order basis function is used. And the total number of unknowns is about 3600. In this example we search the slot shape which increases only the starting torque. So the objective function is defined in the form of (12). As shown in Fig.2 the 10 design parameters are taken on the interface between slots and iron core to generate any possible deformed shape and each design parameter has many corresponding movable nodes since all slots are identical and its shape is symmetrical to the center line. After about 20 iterations the shape of slot is changed as Fig.3 and we can see that it resembles the double cage rotor. At this shape the speed-torque curve is obtained using Maxwell stress tensor as Fig.4 where we can see much increase 67[%] of starting torque and a little decrease of the torque at slip=0.04.

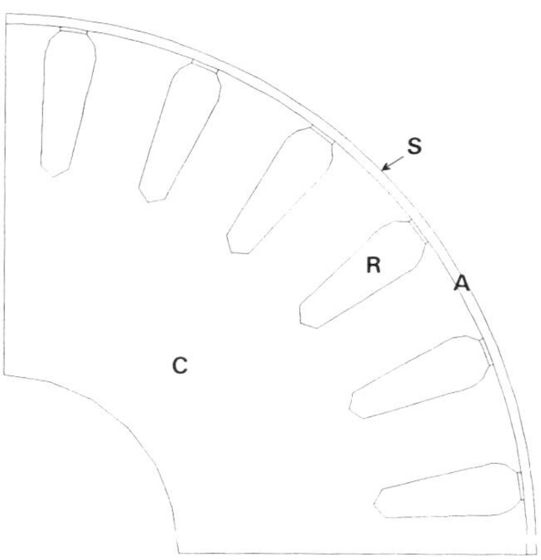

R: aluminuum rotor bar of resistivity $3.33*10^{-8}$ [ohm-meter]

C: iron core of relative permeability $1.0*10^{3}$

A: air gap from Carter coefficient

S: surface current $J = J_0\, e^{j(\omega t - p\theta)}$

Fig. 1 Analysis model

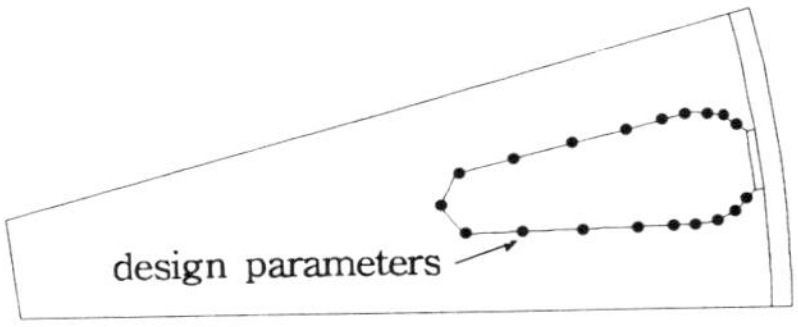

Fig. 2 Inital shape of slot

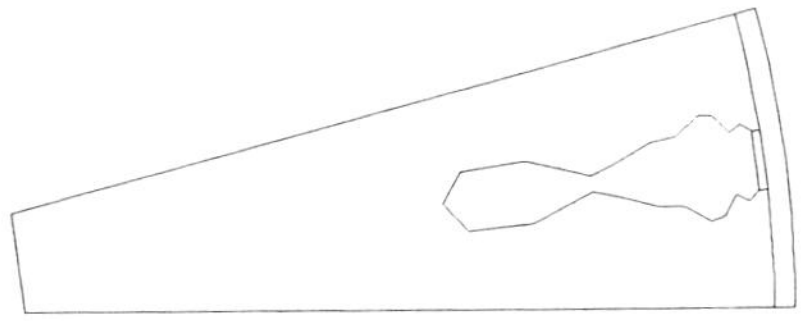

Fig. 3 Final shape of slot

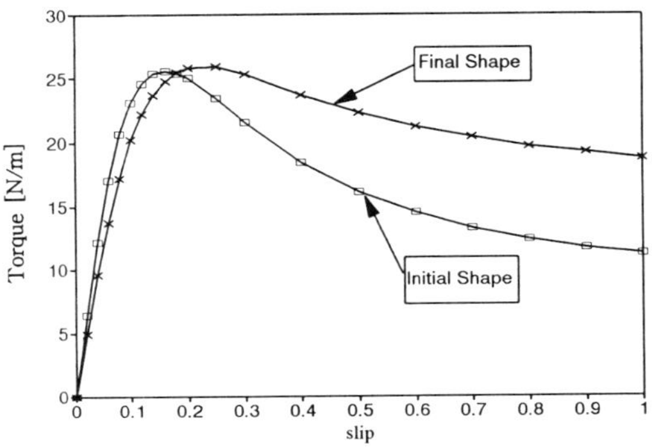

Fig.4 Speed-torque characteristics curves

5. CONCLUSION

In eddy current problem the sensitivity analysis based on finite element method was verified and it is expected to make much progress which has not been achieved by the lumped parameter approximation of magnetic circuit method. The numerical example presented is a simple test problem but this sensitivity algorothm can be utilized as a precision design tool for improvement of various performances in induction motors.

6. REFERENCES

1. Il-han Park, Beom-taek Lee and Song-yop Hahn, " Pole Shape Optimization for Reduction of Cogging Torque by Sensitivity Analysis," COMPEL,Vol.9,No.5, pp.111-114, 1990
2. Il-han Park, Beom-taek Lee and Song-yop Hahn, " Design Sensitivity Analysis for Nonlinear Magnetostatic Problems Using Finite Element Method," IEEE Trans. on Magnetics Vol.28,No.2,pp.1533-1536, 1992
3. S. Williamson and M.J.Robinson, " Calculation of Cage Induction Motor Equivalent Circuit Parameters Using Finite Elements," IEE Proceedings-B, Vol.138,No.5, pp.246-274, 1991

Simulation and Design of Applied Electromagnetic Systems
T. Honma (Editor)

A Mesh Refinement For Boundary Element Method Using An Error Estimating Index

Masashi Nakahira and Tsuyoshi Takeda

Faculty of Eng., Hokkaido University, Kita 13, Nishi 8, kita-ku Sapporo 060, Japan

ABSTRACT

This paper describes a mesh refinement for boundary element solution using an error estimating index. The numerical calculation based on the boundary element method can analyze the whole region by using only the boundary value in the model of the problem. Then, the method doesn't need so much calculating cost and input data, when it is compared with the finite element method which have to divide the whole region into elements. Numerical simulations have necessarily the approximation error by an interpolation of the solution. Thus it is important how the boundary elements should be constructed. The accuracy of the analysis depends on the layout of boundary elements. And some effort has been made on it[1][2][3].

1. INTRODUCTION

To reduce the error of analysis, the boundary elements should be refined by using some methods. Then the error estimating index[4][5] has been proposed. We try to change the mesh pattern by subdividing the mesh which the error estimating index designates.

By showing the relation between the error estimating index and the potential error, we can make use of the error estimating index in practice.

As the another layout of boundary elements, "Homogeneous meshing" and "Moving mesh near edges" are shown to make comparison with the above. In their layout, the error estimating index isn't used, all of the boundary elements are refined into smaller elements.

The result will be shown through the numerical solution of a three-dimensional sample problem.

2. ERROR ESTIMATING INDEX

We can obtain two potential values by applying BEM. Then, for the potential problem, an error estimating index is proposed as follows:

$$\eta = \frac{\left(U^I - U^A \right)}{\left| U^I \right|_{\text{average}}}$$

where

U^I :the potential value which is interpolated by the nodal potential of the model.

U^A:the limit value of an inner potential onto the boundary which is calculated by the boundary integration

3. MESH SUBDIVISION

The methods which change the mesh pattern are followings:

1. The error estimating index designates the element to be subdivided as shown in Figure 1. A designated element is subdivided into two (type A), three (type B), or four elements (type C).
2. Homogeneous meshing as shown in Figure 2.
3. Moving mesh near edges as shown in Figure 3.

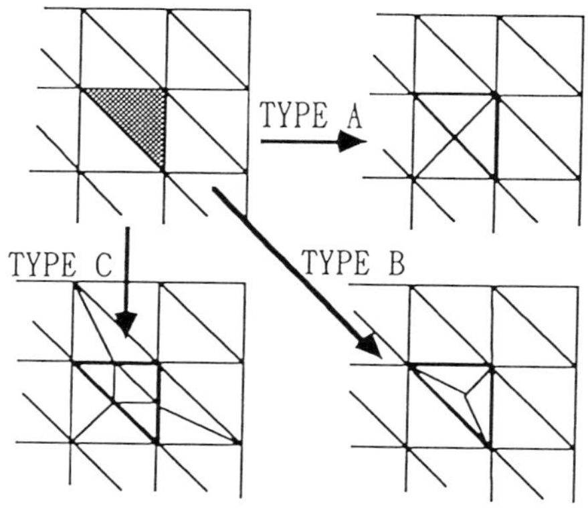

Figure 1. Subdivision of designated elements, type A, type B, type C

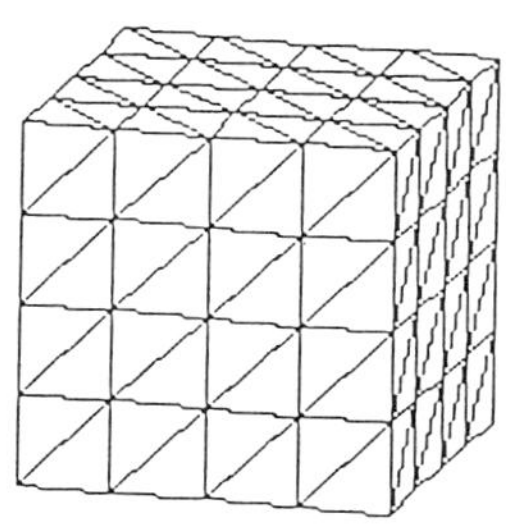

Figure 2. Homogeneous meshing

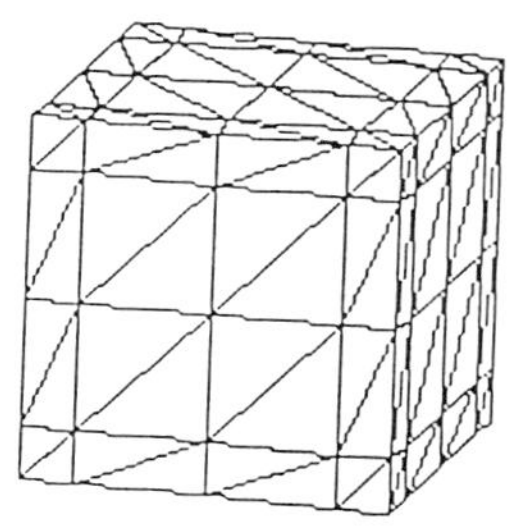

Figure 3. Moving mesh near edges

4. NUMERICAL CALCULATION

We analyzed a three-dimensional sample problem as shown in Figure 4, and used first order boundary elements, where Dirichet boundary condition is given to "S3" and "S6", Neumann boundary condition is given to the other surface, respectively. The exact solution of the present problem is a cubic function with respect to coordinates.

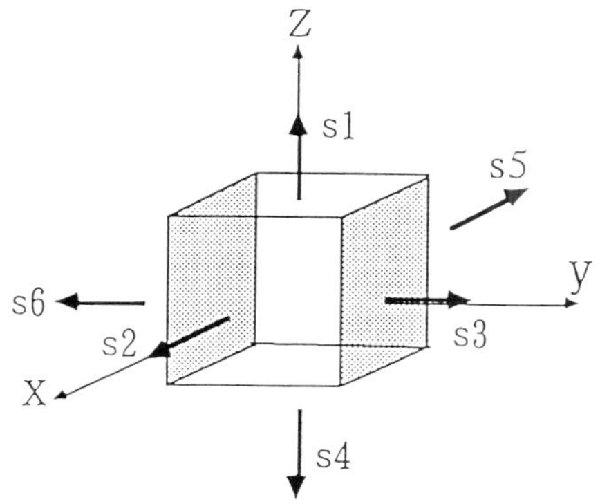

Figure 4. Boundary condition of a three-dimensional sample problem

5. COMPARISON BETWEEN " η " AND "POTENTIAL ERROR"

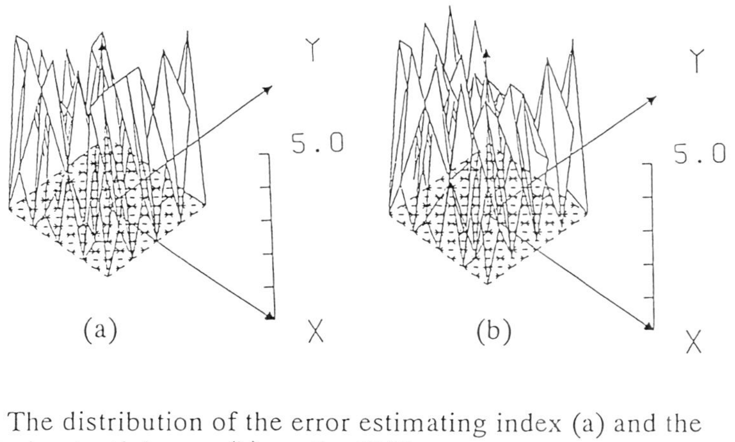

Figure 5. The distribution of the error estimating index (a) and the interpolated potential error (b) on the "S5".

6. CONCLUSION

In Figure 5, the distribution of the error estimating index is similar to the interpolated potential error. This means that the index can almost show the potential error.

As sown in Fig.6, the accuracy of the analysis can be improved by these proposed methods, but the effect is different among them. "Moving mesh near edges" has not been more effective than "Homogeneous meshing". Type A, B and C can divide the local part intensively, where the potential error exist. And we guess the more intensive division will be effective. But contrary to the prediction, our experiment has shown that the less intensive division (type A) has been better than type B and C. And the effectiveness of homogeneous meshing and "type A" is nearly same. These show that "type A" is not an intensive division type but a whole boundary division type.

This numerical solution has shown that the most effective method is the whole homogeneous boundary divisions type in the sample problem.

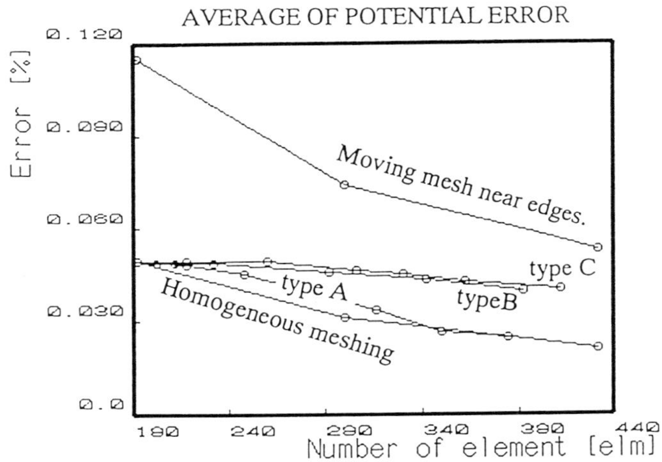

Figure 6. The error of type A, B, C, Homogeneous meshing and Moving mesh near edges.

REFERENCES

1 E.Kita, N. Kamiya:"An Adaptive Boundary Mesh",Proceedings of the Seventh Japan National Symposium on Boundary Element Methods(1990)29-34

2 M.S.Ingber, A.K.Mitra:"Grid Redistribution Based on Measuradle Error Indicators for the Direct Boundary Element Method",Enginering Analysis with Boundary Elements 9(1992)13-19

3 G.Yagawa:"Fuzzy and Neuro VS. Computational Mechanics",12th Symp. on Computing Electrical & Electronic Eng. (1991)1-21

4 T.Takeda,T.Kuwahara:"Estimation of The Accuracy on BEM Solution",Proceedings of the Fourth Japan National Symposium on Boundary Element Methods(1987)11-15

5 T.Takeda,T.Kuwahara:"On An Estimating Method of Boundary Element Method",Theory and Applications of Boundary Element Methods,June(1987)347-353

Simulation and Design of Applied Electromagnetic Systems
T. Honma (Editor)

Optimal design of capacitor-driven coil gun by genetic algorithm

Seog Whan Kim[a], Song-Yop Hahn[a], Hyun Kyo Jung[b], Tae Kyung Chung[c]

[a] Dept. of Electrical Eng., Seoul National Univ., Seoul 151-742, KOREA

[b] Dept. of Electrical Eng., Kangwon Univ., Chuncheon 200-701, KOREA

[c] Dept. of Electrical Eng., Chung Ang Univ., Seoul 156-756, KOREA

Abstract

This paper presents the analysis of a capacitor driven inductive coil gun and the optimal design of the coil gun. An equivalent circuit is used for modeling the coil gun. The circuit equations of the equivalent circuit and the equation of motion for the projectile are solved by using Runge-Kutta method. For the optimal design, the genetic algorithm and a launching simulation are used. The specifications of the coil gun designed by the proposed algorithm are obtained and compared with those designed by approximate equations.

1. INTRODUCTION

Electromagnetic launching systems have advantages compared to existing chemical launching systems; they can be easily installed and repeatedly used and do not damage the launching devices, and the force exerted on the projectile is distributed uniformly[1][2][3]. Among the systems, the coil gun is suitable for launching heavy projectile. Furthermore, for rapid acceleration and the reduction of friction, a capacitor-driven inductive coil gun is recommendable[4].

This paper describes a launching simulation and an optimal design of the capacitor-driven inductive coil gun.

2. STRUCTURE AND MECHANISM OF COIL GUN

The system of a capacitor-driven coil gun is shown in Fig. 1. The coil gun consists of some driving coils fixed on barrel and sleeve. The capacitor bank is connected to the coils as energy source. According to the sequential switching from switch S_1 to S_n, a traveling field is established inside the barrel and eddy current is induced on the surface of the conductive sleeve. So the projectile can be accelerated by the electromagnetic force from the interaction between the traveling magnetic field and the eddy current on the sleeve. In this paper, the coil gun has 6 coils connected in three phase.

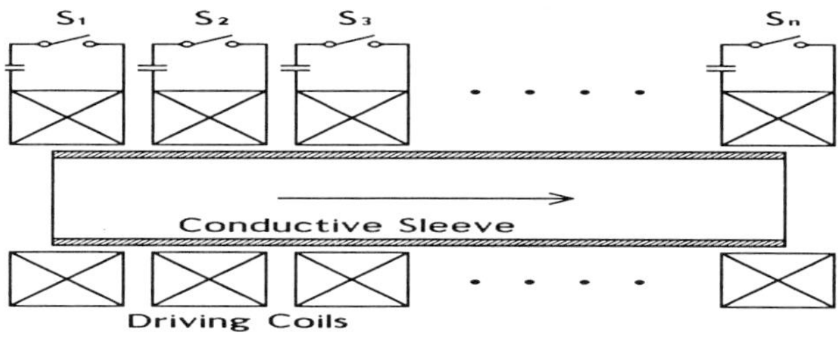

Fig. 1. Structure of a capacitor-driven coil gun.

3. LAUNCHING SIMULATION

Since the distribution of eddy current on the sleeve is not uniform, the sleeve is assumed to be composed of shorted coils. The inductances of the equivalent circuit are calculated by using flux integration method. The circuit equations and equation of motion for the projectile are represented as follows;

$$\{[L]+[M]\}\frac{d}{dt}[I]=[V]-[R][I]-v_p[G][I], \tag{1}$$

$$[C]\frac{d}{dt}[V_c]= -[I_d], \tag{2}$$

$$M_p\frac{dv_p}{dt}=\frac{1}{2}[I]^T[G][I], \tag{3}$$

$$\frac{dx}{dt}=v_p, \tag{4}$$

where $[G]=\frac{d}{dx}[M]$, and M_p and v_p are the mass and velocity of projectile, respectively.

By solving (1),(2),(3) and (4) by using the Runge-Kutta method, all parameters describing the characteristics of the coil gun are obtained.

4. OPTIMAL DESIGN

Barrel length, gap size between two coils, coil turns, firing interval and initial position of projectile are taken as design variables. They are the same ones as the input variables of the simulation program. The five design variables are shown in Fig. 2. In this figure, l_b is barrel length, l_i gap size between two coils, n coil turns and S_c cross sectional area of coil. Length of one coil l_c and thickness of coil t_c can be calculated from the following equations.

$$l_c = \frac{l_b - 5 \times l_i}{6} \tag{5}$$

$$t_c = \frac{S_c \times n}{l_c} \tag{6}$$

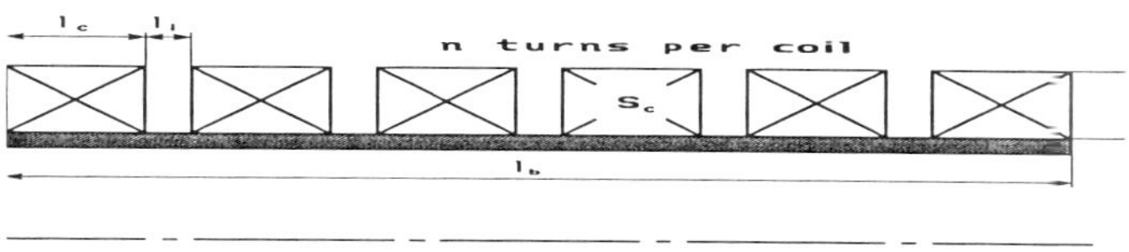

Fig. 2. Design variables for the coil gun.

Since the design variables are influenced one another in the course of design, it is very difficult to obtain the optimal results by existing methods. And also in the optimal design problem for the coil gun, the global optimal point can not be obtained easily with the deterministic methods, because the problem has many local optima. To get around the above difficulties, numerical simulation and genentic algorithm for optimization are used. The genetic algorithm is based on the evolution through generations.

The evolution consists of three steps: reproduction, crossover and mutation. In this paper, "biased roulette wheel spinning" method for reproduction is used and a method similiar to uniform crossovecr method for Crossover. The chromosome of the coil gun consists of 48 bits, eight bits for coil turns and ten bits for the rest four variables: length of one coil, thickness of one coil gap between each coil and initial position of the projectile. And the population of one generation is 100.

Table 1 shows the optimal design result of the coil gun. For given charging voltage, capacitance and dimension of projectile, the barrel length, gap size between two coils, coil turns, firing interval and initial position of projectile are obtained for the maximum muzzle velocity.

The coil gun in ref. [4] has been designed by approximate equations, and the muzzle velocity of the projectile is 3.5 [m/sec]. When the coil gun is designed with same charging voltage and capacitance by the proposed algorithm, the muzzle velocity becomes 4.7 [m/sec]. From these results, it is shown that the performance of the coil gun designed by genetic algorithm becomes better.

Table 1
Result of optimal design for 1 section coil gun

aluminium projectile	length [mm]	100
	thickness [mm]	1.6
	outer diameter [mm]	58.4
	mass [g]	76
	initial position [mm]	14.66
driving coils	turns	36
	thickness [mm]	3.48
	length [mm]	20.715
	gap size [mm]	2.572
	inner diameter [mm]	61.4
capacitance per phase [μF]		760
charging voltage [V]		600
firing interval [μs]		363.64
muzzle velocity [m/s]		18.57

5. CONCLUSION

This paper describes a launching simulation and an optimal design method for the capacitor-driven inductive coil gun. For the simulation an equivalent circuit is employed. And for the optimal design of the capacitor-driven coil gun, the optimization is carried out by genetic algorithm. From the comparison between the results obtained by the proposed algorithm and those by other algorithms, it is concluded that the optimal design method employing genetic algorithm is more powerful than other existing methods.

6. REFERENCES

1 H. Kolm, and P. Mongeau, IEEE Trans. Magnetics, MAG-20(2), 227-230(1984).

2 Harry D. Fair, Phil Coose, C. P. Meinel, D. A. Tidmin, IEEE Trans. Magnetics, 25(1), 9-16(1989).

3 Hans Mark, and Chancellor, IEEE. Magnetics, 25(1), 17-19(1989).

4 Jianling He, IEEE Trans. Plasma Science, 17(3), 429-438(1989).

Simulation and Design of Applied Electromagnetic Systems
T. Honma (Editor)

Meta-optimization in Electromagnetics: An Application in Eddy-Current Testing

Aleš Gottvald*, Kenzo Miya

Nuclear Engineering Research Laboratory, Fac. of Engineering, University of Tokyo, Tokai-mura, Ibaraki, 319-11, Japan

* *On leave from:* Inst. of Scientific Instruments, Czech Academy of Sciences, Královopolská 147, CS-612 64 Brno, Czech Republic

Abstract

Meta-optimization means optimizing optimization itself. Varying auxiliary parameters governing optimization trajectories, a new meta-optimization criterion (quality of local minimizers, CPU-time, ...) can be extremized. Generation - Mutation - Annealing strategies can be used on both levels of optimization. Self-adaptive algorithms arise. The principle of meta-optimization is of very general applicability: a problem of inverting reduced magnetic field data in Eddy-Current Testing shows an example of meta-optimization in action.

1. INTRODUCTION: OPTIMIZING OPTIMIZATION ITSELF

Global optimization methods are a fundamental tool for solving *inverse problems* of field theory [1 - 4]. Nevertheless, a level of success in this area is often critically dependent on overall tuning of both optimization and analysis procedure. A mechanical copying of an optimization method X recommended elsewhere is no guarantee of success in other applications.

The above difficulty can be ascribed to the fact that the individual applications require specific *meta-optimization* scheme (optimizing conditions - auxiliary parameters). The concept of meta-optimization means *computerized optimization of optimization itself*. The difference between a *meta-optimized* and some *ad hoc* optimizing system can be dramatic. In applications as ECT, NMR or MRI, it can often increase performance of the optimizing system in several orders, in terms of speed or accuracy. Some qualitative characteristics of primary importance are also enforced, as reliability or robustness.

2. THE PRINCIPLE OF META-OPTIMIZATION

The principle of meta-optimization is based on a parametric dependence of optimization trajectories [1]. Consider a (global) optimization problem. Let $\mathbf{x} := (x_1, \ldots, x_n)$ be *degrees of freedom* (DOF), and $\mathbf{a} := (a_1, \ldots, a_q)$ be any *auxiliary parameters* governing optimization trajectories. In standard techniques, these auxiliary parameters $\mathbf{a}$ are treated as *static* and their values are *fixed*. In the context of meta-optimization, on the contrary, these auxiliary parameters are *dynamic*, and their values can be *varied* (up-dated) after every local optimization cycle. Consequently, a *meta-optimization problem* arises:

$$\text{minimize } F(\mathbf{a}) \tag{1}$$

where the function $F: R^q \rightarrow R$ defines a *meta-optimization criterion*. In other words, the parameters $\mathbf{a}$ are adopted as new

DOF. These DOF **a** act on a *meta-optimization* level, however, while the standard DOF **x** are varied on an *optimization* level as usually. If only quality of local minimizers is of concern, this function F is a mapping $(a_1, \dots, a_q) \rightarrow y^*$, where $y^* := f(\mathbf{x}^*)$ is a local minimizer value due to a (standard) optimization criterion $f: R^n \rightarrow R$. In this case, meta-optimization actually reduces to global optimization.

The function F is recursive and typically too complicated to be optimized via higher-order deterministic methods. However, robust and reliable zeroth-order probabilistic methods can be applied for this task very effectively. *Generation - Mutation - Annealing (GMA) Strategies*, (e.g., Evolution Strategy [5, 4], Simulated Annealing [2]) are studied for meta-optimization in our work. In a context of GMA-strategies, the above auxiliary parameters can be an initial configuration, a number and type of DOF, parameters of mutation and annealing, etc.

The rationale under the concept of meta-optimization consists also in implementation and methodological aspects: (a) the process of up-dating the auxiliary parameters can be localized in one module of the algorithm whilst most of the original optimization software is left intact; (b) unified theory of many global optimization methods arises.

3. EXAMPLE: META-OPTIMIZATION IN ECT

Let us outline typical auxiliary parameters convenient for meta-optimization in one particular application, and namely in *Eddy-Current Testing* (ECT) (Figure 1) [6, 7, 8].

Analysis: Let the magnetic field variation $\delta\mathbf{B}$ due to the conductivity-flaw $\delta\sigma$ in an eddy-current domain V be sought in the form of time-harmonic complex Green's solution:

$$\delta\mathbf{B} = -i\omega\mu_0 \int_V G\cdot(\delta\sigma\mathbf{A}_0 + \sigma\delta\mathbf{A})\, dV \qquad (2)$$

(i := imaginary unit, ω := angular frequency, $\mu_0 = 4\pi\cdot10^{-7}\ \mathrm{Hm}^{-1}$, G := Green's function, δ := symbol of variation, σ := conductivity, $\mathbf{A}_0$:= flaw-free vector potential (effective)). In (2), the (dominant) first term in the integrand is localized on the flaw domain. The (correcting) second term, which is non-zero only in the conductive domain V, can be approximated on V with

$$\delta\mathbf{A} = -i\omega\mu_0 \int_V G_0\cdot \delta\sigma\mathbf{A}_0\, dV \qquad (3)$$

An adequate choice of Green's functions G and G_0, as well as other physical considerations, will be discussed elsewhere.

Optimization: Discretizing the integral (2) in terms of DOF **x**, we arive at the global optimization problem of the type:

$$\underset{\mathbf{h}_a<\mathbf{x}<\mathbf{h}_b}{\text{minimize}}\ \|\delta\mathbf{B}(\mathbf{x}) - \delta\mathbf{B}_i\| := f(\mathbf{x}) \qquad (4)$$

which defines an optimal fit between the model field variation, $\delta\mathbf{B}(\mathbf{x})$, and the input field variation, $\delta\mathbf{B}_i$, using a norm $\|\cdot\|$ (not critical, an Euclidean will do). A standard linear way discretizes the unknown $\delta\sigma$ in terms of equidistant pixels. However, a dramatic reduction of the number of DOF, as well as the number of input field data, is possible only via a *non-linear way of discretization* (geometry of pixels varied). Adopting this way, the optimization problem (4) is no more linear. Consequently, *GMA-Strategies* with *Meta-optimization* have been applied to solve (4).

Generation of feasible configurations (trials) has been specified using the n-dimensional Gaussian probability density function ρ of the type:

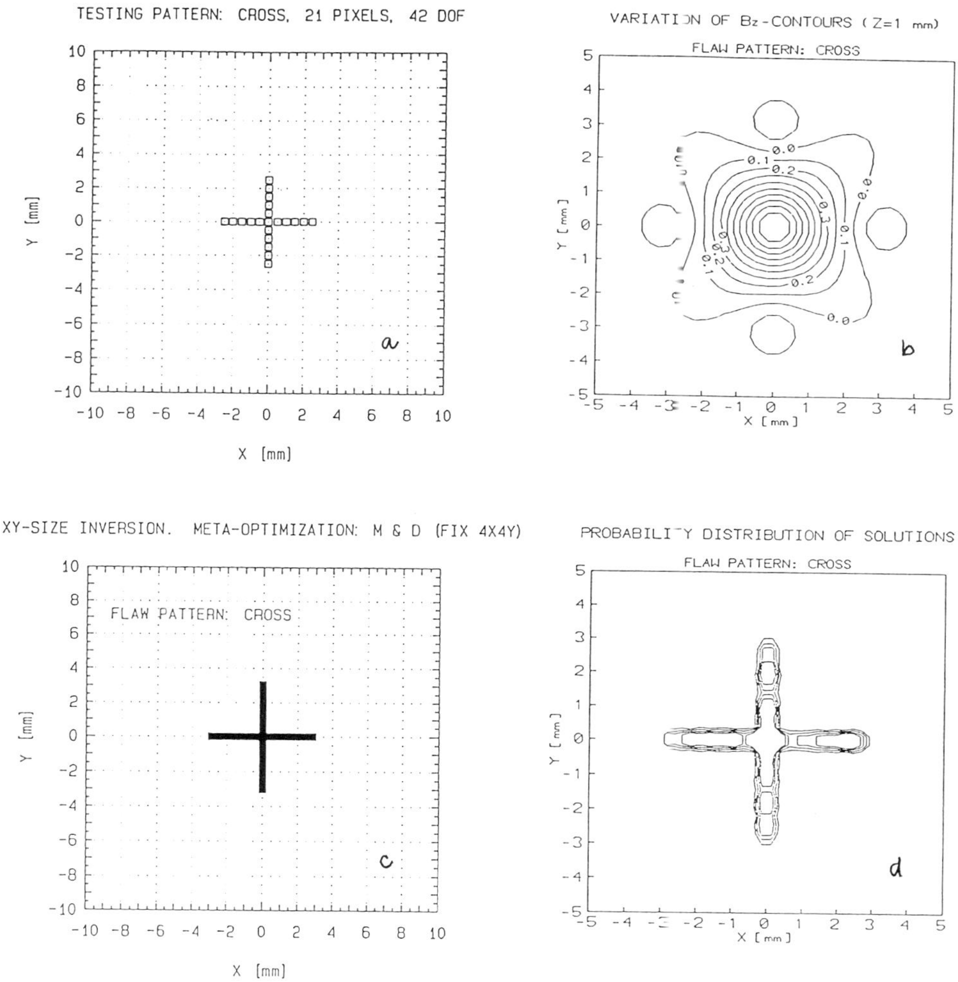

Figure 1. Inverting a conductivity flaw from merely 16 input field data using GMA-strategies with meta-optimization: an infinitesimally thin conductive plate at z=0, flaw-free eddy-currents being z-axisymmetric & radially constant, δB_z at z=1 mm plane being inverted: *(a)* testing pattern; *(b)* normalized normal component of magnetic field variation due to the testing pattern - imaginary part; *(c)* result of XY-size inversion for 8 x 20 pixels generated; *(d)* f-weighted probability distribution of pixels due to local minimizers. (σ_o = 10^6 S/m, ω ~ 50 kHz).

$$\rho(\mathbf{x}, \mathbf{m}, \mathbf{d}) \sim \exp\{-\sum_i [(x_i - m_i)/d_i]^2\} \qquad (5)$$

where the components of the vectors **m** and **d** are the mean values and dispersions of the individual DOF, respectively.

Mutation has been specified using a (1+1)-Evolution Strategy scheme associated with the optimization criterion (4):

$$\mathbf{m}^{(k+1)} := \begin{cases} \mathbf{x}^{(k)} & \text{if } f_x < f_m \\ \mathbf{m}^{(k)} & \text{if } f_x >= f_m \end{cases} \qquad (6)$$

where k := index of iteration, $f_x := f(\mathbf{x}^{(k)})$, $f_m = f(\mathbf{m}^{(k)})$.

Annealing has been specified using a 2-parametric scheme:

$$\mathbf{d}^{(k+1)} := \begin{cases} \mathbf{d}^{(k)} * Q & \text{if probability}\{f_x < f_m\} < P \\ \mathbf{d}^{(k)}/Q & \text{if probability}\{f_x < f_m\} >= P \end{cases} \qquad (7)$$

where P ~ 0.2 <= 1, Q ~ 0.9 <= 1 is a standard choice [5].

Meta-optimization: The following auxiliary parameters **a** are now good candidates of meta-optimization, as they can be updated (varied) after every local optimization cycle:

(a) Initial mean value vector $\mathbf{m}^{(0)}$ in (6).
(b) Initial dispersion vector $\mathbf{d}^{(0)}$ in (7).
(c) Parameters of annealing P and Q in (7).
(d) Borders $\mathbf{h}_a$, $\mathbf{h}_b$ in constraint inequalities for DOF in (4).
(e) Parametrically dependent input field $\delta \mathbf{B}_i$ in (4) due to multi-view or multi-frequency measurements.
(f) Parametrically dependent exciting field $\mathbf{A}_0$ in (2).
(g) Number n of DOF (model size) in (5).
(h) Small correcting terms as (3), if considered at all.

A typical result of data inversion is shown in Figure 1. A meta-optimization scheme for $\mathbf{m}^{(0)}$ and $\mathbf{d}^{(0)}$ has been applied. Inverting any 4 ÷ 16 geometric flaw-parameters requires CPU-time ~ 60 ÷ 600 seconds on a PC of about 1 MFLOPS. Only 16 input field data (computed) is used for the inversion.

4. CONCLUSIONS

Computerized optimization of optimization itself is a step to a large area of many vistas. As shown in an example from ECT, many sensitive auxiliary parameters naturally arise in inverse problems. Consequently, meta-optimization can be very effective for enforcing global convergence, reducing computing demands, or achieving specific inversion goals. Such qualities of optimizing systems are of primary importance in many advanced applications, as ECT, NMR, MRI, etc.

5. REFERENCES

1 A. Gottvald, Inverse Problems, Non-linear Phenomena and Global Optimization Methods, In: Nonlinear Phenomena in Electromagnetic Fields (ed. T. Furuhashi), Elsevier, 1992
2 J. A. Scales et al., J. Comput. Phys. 103 (1992) 258
3 A. Gottvald, IEEE Trans. Magn., 1 (1988) 411
4 A. Gottvald et al., IEEE Trans. Magn., 2 (1992) 1538
5 H. P. Schwefel, Numerical Optimization of Computer Models, Wiley, 1981
6 H. Sabbagh - D. Sabbagh, IEEE Trans. Magn., 4 (1986) 282
7 J. R. Bowler - S. A. Jenkins, J. Appl. Phys., 3 (1991) 1107
8 T. Morisue, IEEE Trans. Magn., 2 (1992) 1324
9 A. Gottvald, IEEE Trans. Magn., 2 (1990) 399

Simulation and Design of Applied Electromagnetic Systems
T. Honma (Editor)

A method of estimation of current distribution using genetic algorithms with variable-length chromosomes

Y.Tanaka, A.Ishiguro and Y.Uchikawa
Dept. of Electronic-Mechanical Eng., School of Eng., Nagoya University
Furo-cho, Chikusa-ku, Nagoya, 464-01 Japan
Phone: +81-52-781-5111 ext.6460
Fax: +81-52-781-9263

Abstract
Most of the inverse problems are suffering from ill-posed problems. As a result, they are difficult to solve analytically and uniquely. But the necessity of analytical method for inverse problems has become much greater because of its wide application fields. This paper presents the use of Genetic Algorithms for estimating current distribution. We also propose an efficient searching strategy by devising the structure of chromosomes. Finally, the feasibility of the proposed method is confirmed by simulations.

1. Introduction

Recently, the necessity of analytical method for inverse problems has become much greater because of their usefulness in various application fields[1]-[3]. Most of the inverse problems are, however, suffering from ill-posed problems. As a result, they are difficult to solve analytically and uniquely.
We have been investigating an estimation of static two-dimensional current distribution in materials, using the observed external magnetic fields as a practical example[4]. By discretizing the above problem, it can be regarded as a combinatorial optimization problem, viewed from a mathematical standpoint. But if the number of the discretization becomes large, the computation time to obtain an optimal current distribution will be explosively increased. To overcome this obstacle, we proposed the estimation method by incorporating Genetic Algorithms[5][6].
The obtained current distribution by using the proposed method, however, was expressed by scattering unit currents due to the coding method[4]. Thus, it was difficult to comprehend the current distribution quantitatively. Moreover, using the previous coding method, we could not incorporate physical constraints such as continuity of current distribution and so on.
To solve these problems, this paper presents the use of variable-length chromosomes with devised coding method for solving inverse problems in electromagnetics. And the feasibility of the proposed method is confirmed by simulations.

2. How Do We Transform Inverse Problems into Optimization Problems?

In this paper, as a rudimentary stage of investigation, we treat a very simple example as shown in Fig.1. And we assume that a uniform electric current flows from behind toward the front of this figure, in an infinitely long straight conductor with a rectangular heavily-shaded cross section. Magnetic sensors of k in number, at the sensing points denoted by the solid disc, can only remotely detect the normal component of the excited magnetic field with respect to the sensing surface. Our inverse problem is aimed at estimating the current distribution(*i.e.* reconstructing the conductor configuration) from the detected data of the magnetic field at a limited number of sensing points. For simplicity, we assume that the material's boundary is depicted by a rectangle, and we divide it into sub-squares of equal sizes as shown in Fig.1. To each sub-square we assign the state in the following way:

If a sub-square is a conductor, then the state is 1.

If it is an insulator, then the state is 0.

For further simplicity, we assume that each sub-square conductor carries a unit current. Through these procedures, we can represent arbitrary current distribution by locating sub-square conductors. Thus this problem can be regarded as one of optimization problems.

For optimization, we need a measure with which we can evaluate the suitability of solutions, namely *fitness*. In this paper, we simply utilize the following similarity as the fitness:

Let $\boldsymbol{H_O}$ be the whole set of the detected data at every sensing points (see Fig.1). Next, we randomly distribute unit currents to sub-squares, and we assume as a candidate of the solutions as shown in Fig.2. Let $\boldsymbol{H_A}$ be the whole set of the normal components of the calculated magnetic field generated by the current distribution in the candidate of solutions (see Fig.2). The fitness between $\boldsymbol{H_O}$ and $\boldsymbol{H_A}$ are written as:

$$\text{fitness} = \frac{\boldsymbol{H_0}^{\mathrm{T}} \cdot \boldsymbol{H_A}}{\|\boldsymbol{H_0}\| \, \|\boldsymbol{H_A}\|}$$

From the above arrangements, the current problem is transformed into the problem of how to locate unit currents with maximization of the fitness. But there is an inevitable problem that candidates of solutions explosively increase as the number of sub-squares increases. As a result, it will be difficult to find an optimal solution by deterministic approaches. To overcome these obstacles, we incorporate GA into this problem.

3. Incorporating Method of GA

3-1. Proposed Coding

When we apply GA into the problem, we must consider how to code the candidate of solutions into the chromosomes. In our previous proposed method, the obtained current distribution resulted in the scattering of unit currents due to the coding method. As a result, it was difficult to comprehend the obtained distribution quantitatively. Thus, in this paper, we devise the coding by incorporating the size and the position of the conductor into genes, as shown in Fig.3. In this research, for simplicity, the region of current distribution is represented by a rectangle with variable length Rect-x and Rect-y. And we allow the length of each chromosome to be variable since the number of regions of current distribution is different in each searching iteration and chromosome.

Next, we must consider how to transform Genotype into Phenotype. In this paper, this process is simply carried out by mapping the information in each gene of chromosomes onto the material surface as shown in Fig.4. Note that each state of the sub-squares just inside the mapped conductor should be turned to 1.

With regard to the details of GA, we recommend seeing the listed reference[5].

3-2. Proposed Algorithm

In this research, we use the following algorithm:

Firstly, we prepare 50 chromosomes with randomly decided length between 10 and 20 as initial candidates of solution. Then we randomly choose three pairs of chromosomes, and operate *preselection*[5] to them by comparing each fitness value in each generation. By using this operation, we can avoid *premature convergence*[5] in the evolutional process. On the other hand, we randomly choose five chromosomes and operate the following mutation to maintain the diversity of population, changing the size and the position of the conductor written in the randomly chosen gene.

4. Proposed Local Search Technique

Whereas GA has a strong global search ability, some reports have been made on the weakness of its local search ability. To overcome this problem, we propose to divide the searching process into the following two phases:

(i) The 1st Phase

In this phase, the search for current distribution proceeds until the prespecified generation. After the above process, we extract the global information on current distribution such as the number of estimated regions and their rough position.

(ii) The 2nd Phase

We carry out the search with GA again after deciding the structure of the chromosomes based on the obtained global information in the 1st phase. In this phase, efficient search can be realized since the space being searched is diminished by the previously obtained information. Fig.5 shows the procedure in the case of two regions of current distribution.

5. Simulation Results

Fig.6 shows the simulation results of our proposed method. In the same figure (a), the black portions denote the regions where currents exist, and the vertical lines, with respect to the material's boundary, represent the observed magnetic fields caused by the current distributions in the materials. Figures (b) and (c) are the results obtained after the 1st and 2nd phases. From these figures, it is understood that the search in the 2nd phase can proceed efficiently using the global information obtained in the 1st phase. The parameters of this simulation are listed in Table 1.

6. Conclusions and Further Works

In this paper, we have proposed the use of Genetic Algorithms with two phases of searching for estimating current distribution in materials. And we have shown the validity of this method by simulations. Many issues remain open for research. Among the most important is the development of the coding method which realizes the quantitative estimation of current value. This topic is currently under investigation.

Acknowledgement

The authors are grateful to The NAITO FOUNDATION where part of this work was supported by an initiative grant.

References

1 K.Kishimoto, et al., JSME International Journal, Series I, 32-2 pp.256-262(1989).
2 G.M.L.Gladwell, "Inverse Problems in Vibrations, Dordrecht", MartinusNijhoff(1986).
3 Y.Saito, et al., "A Formulation of the Inverse Problems in Magnetostatic Fields and Its Application to a source Position Searching of Human Eye Fields", Journal of Applied Physics, Vol.67, No.9, pp.5830-5832(1990).
4 A.Ishiguro, et al., "An Estimation of Current Distribution Using Genetic Algorithms", ISEM, Paper AP-2-10(1992).
5 D.E.Goldberg, "Genetic Algorithms in Search Optimization and Machine Learning", Addison-Wesley(1989).
6 R.K.Belew, et al., Proc. of the fourth International Conference on Genetic Algorithms, Morgan Kaufmann(1991).

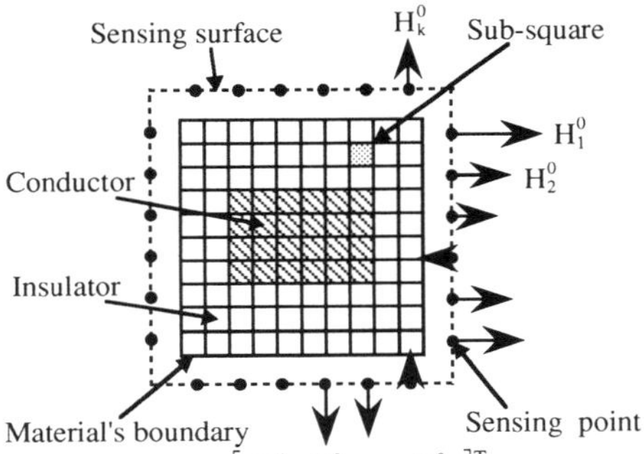

$$H_0 = [\, H_1^0, H_2^0, \cdots, H_k^0 \,]^T$$

Fig.1:A practical example of static two-dimensional current distribution.

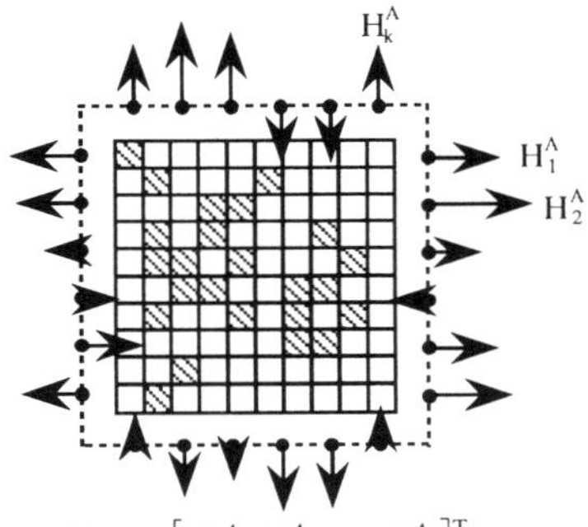

$$H_A = [\, H_1^A, H_2^A, \cdots, H_k^A \,]^T$$

Fig.2:An example of randomly generated candidate of solution.

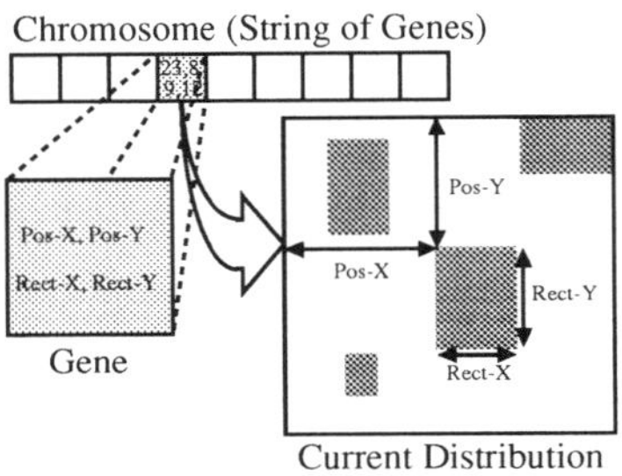

Fig.3:Coding and mapping to phenotype.

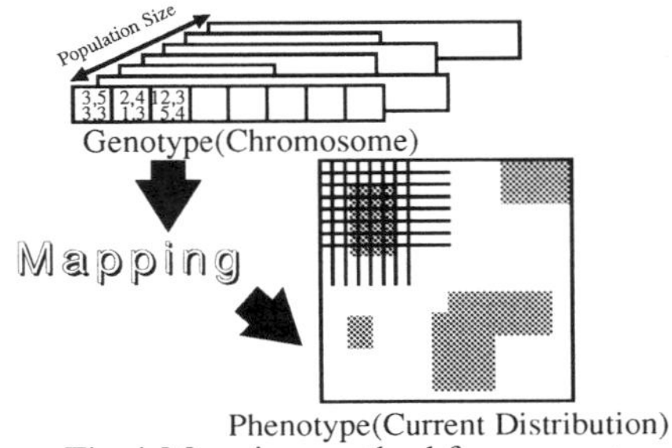

Fig.4:Mapping method from genotype to phenotype.

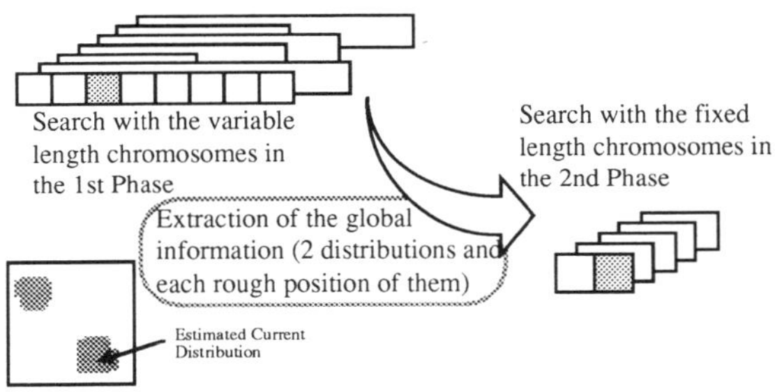

Fig.5:Procedure of the 1st and the 2nd phases.

Table 1:Parameters of simulations.

Material	40 wide × 40 long
Sensing surface	42 wide × 42 long
	42 (Total)
Sensing points	11 (Horizontal)
	11 (Perpendicular)
Sub-squares	2601 (51 × 51)
Population	50

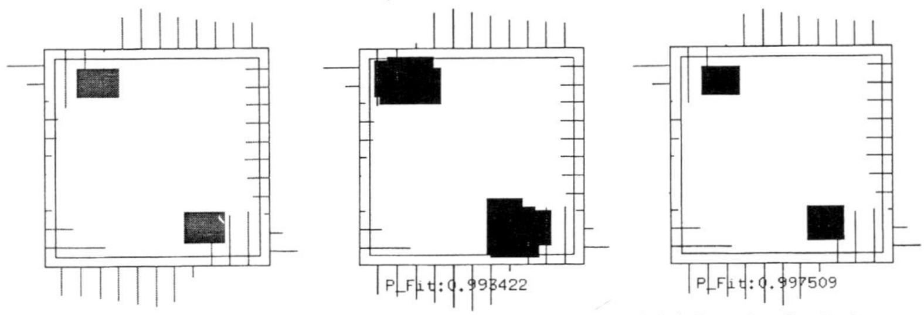

(a)Real current distribution. (b)After the 1st phase. (c)After the 2nd phase.

Fig.6:Simulation results.

Simulation and Design of Applied Electromagnetic Systems
T. Honma (Editor)

Massively parallel neural networks for the inverse source problems

K.Kitsuta and Y.Saito

Coll. of Engng., Hosei University, Kajino, Koganei, Tokyo 184, Japan

Abstract

We have previously proposed an approach to searching for the current distribution patterns in human heart. In the present paper, a theoretical background of our method is given by means of the neural networks. Our method yields the field source patterns having somewhat ambiguity so that it is required to pick up only the major patterns. This means that considerable experience is required to decide the major parts. In order to remove this difficulty, we propose here a new method that the magnetic and electric field source searchings are carried out in parallel by the massively parallel neural networks. After obtaining their independent results, a correlative analysis between them provides a highly accurate electromagnetic field source pattern.

1.INTRODUCTION

Most of the modern electronic medical systems are based on the applications of the inverse problems, for examples, X-ray computed tomography and ultrasonic imaging. Successful results of these first stage applications of the inverse problems are now leading to the second stage applications of the inverse problems, i.e. searching for the electric current paths in human heart and clarifying the neural behaviors in human brain. Definitely difference between the first and second stages is that the first stage is mainly concerned with the medium identification problems and the second is the electrical signal source searching problems , i.e. inverse source problems. Mathematically it has been proofed that the three dimensional current distributions in human body is not analytically evaluated by measuring the magnetic fields around the body. To overcome this difficulty, we have previously proposed an approach to searching for the current distribution patterns in human heart [1-3].

In the present paper, we show that a theoretical background of our method called "*Sampled Pattern Matching*" method can be interpreted as one of the neural networks. One of the weak points of our sampled pattern matching method is that the method yields the field source patterns having somewhat ambiguity. Thereby, it is required to pick up only the major patterns. This means that considerable experience is essentially required to decide the major parts. In order to remove this difficulty, we propose here a new method that the magnetic and electric field source searchings are carried out in parallel by the massively neural networks. After obtaining their independent results, a correlative analysis between them provides a highly accurate electromagnetic field source pattern. Thus, we have succeeded in realizing a new system for human heart diagnosis.

2.THE INVERSE SOURCE PROBLEM

2.1 System equations

Most of the electromagnetic field problems are reduced to solving the following form equation:

$$\lambda\nabla^2\psi = -\sigma, \tag{1}$$

where λ,ψ and σ are the medium parameter, potential and source density, respectively. Using the Green's function $G=1/(4\pi|r|)$, the integral form of (1) is

$$\psi = \int [\sigma/(4\pi|r|\lambda)]dv. \tag{2}$$

In the current fields, the parameters λ,ψ and σ in (1) correspond to the conductivity κ[s/m], electrical potential ϕ[V] and time derivative of electrical charge density $-\partial\rho/\partial t[C/(m^3 s)]$, respectively. By means of (2), the electric potential ϕ is

$$\phi = -\int [(\partial\rho/\partial t)/(4\pi|r|\kappa)]dv. \tag{3}$$

Discretizing (3) into the small subdivisions, we have

$$\mathbf{U} = \sum_{i=1}^{m} \alpha_i \mathbf{d}_i, \tag{4}$$

where

$$\left.\begin{array}{l} \mathbf{U} = [\phi_1,\phi_2,\cdots,\phi_n]^T, \\ \alpha_i = -[(\partial\rho_i/\partial t)/\kappa_i]\Delta v_i = P_{Ui},\ i=1\sim m, \\ \mathbf{d}_i = [1/(4\pi)][(\mathbf{n}_i\cdot\mathbf{a}_{1i}/r_{1i}),(\mathbf{n}_i\cdot\mathbf{a}_{2i}/r_{2i}),\cdots,(\mathbf{n}_i\cdot\mathbf{a}_{ni}/r_{ni})] \\ \text{n:number of measured points}, \\ \text{m:number of subdivisions: } v=\Delta v_i+\Delta v_2+\cdots+\Delta v_m. \end{array}\right] \tag{5}$$

In (5), α_i (=P_{Vi} [Vm]),$\mathbf{n}_i$,$\mathbf{a}_{ji}$ and r_{ji} ($i=1\sim m, j=1\sim n$) are the voltage dipole, unit vector in the direction of voltage dipole, unit directional vector from the source point i to measured point j and distance from the source point i to measured point j, respectively. Further, the electrical field vector **E** on the conductor surface can be obtained by taking the gradient of scalar potential ϕ, i.e. $\mathbf{E}=-\nabla\phi$.

By means of the Biot-Savart's law, the system equation of magnetic fields is written in much the same way as (4). In this case, the vectors and parameters are

$$\left.\begin{array}{l} \mathbf{U} = [H_1,H_2,\cdots,H_n]^T, \\ \alpha_i = |\mathbf{J}_i|\Delta v_i = P_{Ii},\ i=1\sim m, \\ \mathbf{d}_i = [1/(4\pi)][(\mathbf{n}_i\times\mathbf{a}_{1i}/r_{1i}^2),(\mathbf{n}_i\times\mathbf{a}_{2i}/r_{2i}^2),\cdots,(\mathbf{n}_i\times\mathbf{a}_{ni}/r_{ni}^2)]^T. \end{array}\right] \tag{6}$$

In (6), $|\mathbf{J}_i|\Delta v_i$ (= P_{Ii} [Am], $i=1\sim m$) is the current dipole where $\mathbf{J}_i$ is a current density in a volume Δv_i; $H_1,H_2,\cdots,H_n$ are the measured magnetic fields; and the others are similarly defined as those of (5). Generally, the field or potential measured region is limited in a some surface and the source region is limited in the volume enclosed by the measuring surface so that a following condition is commonly held on both systems.

$$m >> n. \tag{7}$$

2.2 Neural nets solutions

Key idea of the neural networks is that the parameters called weight functions of the networks having the multiple input and single output terminals are determined by the various input pattern vectors and their corresponding supervisor outputs. After finishing this learning process, any input pattern can be recognized and identified in accordance with the previous learning.

The leaning process of our problem is carried out by

Learning

		INPUT		SUPERVISOR
				1 2 m
$\mathbf{d}_1'$	:	Γ_1	=	$[1,0,..,0]^T$,
$\mathbf{d}_2'$	:	Γ_2	=	$[0,1,..,0]^T$,
....				
$\mathbf{d}_m'$	:	Γ_m	=	$[0,0,..,1]^T$,

where $\mathbf{d}_i'=\mathbf{d}_i/\|\mathbf{d}_i\|$ and Γ_i $(i=1,2,,\cdots,m)$ are the normalized input and supervisor vectors, respectively. The first synapse combination is analytically carried out by

$$W=\sum_{i=1}^{m}\Gamma_i(\mathbf{d}_i')^T=\begin{vmatrix}\mathbf{d}_1'^T\\ \mathbf{d}_2'^T\\ \cdot\\ \mathbf{d}_m'^T\end{vmatrix} \tag{8}$$

The input of the normalized vector $\mathbf{U}'=\mathbf{U}/\|\mathbf{U}\|$ to this perceptron W yields the first most dominant point h taking the maximum of output. Namely, threshold value of this perceptron is the maximum value in output $W\mathbf{U}'$. This corresponds to the first pilot point of the sampled pattern matching method [1,2]. Learning of the second perceptron unit is carried out in much the same way as those of the first perceptron unit:

Learning

		INPUT		SUPERVISOR
				1 2 $m-1$
$(\mathbf{d}_h+\mathbf{d}_1)'$	:	Γ_1'	=	$[1,0,..,0]^T$,
$(\mathbf{d}_h+\mathbf{d}_2)'$	:	Γ_2'	=	$[0,1,..,0]^T$,
....				
$(\mathbf{d}_h+\mathbf{d}_m)'$	:	Γ_m'	=	$[0,0,..,1]^T$,

where h refers to the first perceptron output point;and $(\mathbf{d}_h+\mathbf{d}_j)'=(\mathbf{d}_h+\mathbf{d}_j)/\|\mathbf{d}_h+\mathbf{d}_j\|$ $(j=1,2,,\cdots,m-1;j\neq h)$ is a normalized input vector. The synapse combination is given by

$$W'=\sum_{j\neq h}^{m}\Gamma_j'(\mathbf{d}_h+\mathbf{d}_j)'^T=\begin{vmatrix}(\mathbf{d}_h+\mathbf{d}_1)'^T\\ (\mathbf{d}_h+\mathbf{d}_2)'^T\\ \cdots\cdots\\ (\mathbf{d}_h+\mathbf{d}_m)'^T\end{vmatrix}. \tag{9}$$

The input of $\mathbf{U}'=\mathbf{U}/\|\mathbf{U}\|$ to this second perceptron W' yields the second most dominant point g taking the maximum in the output $W'\mathbf{U}'$. This corresponds to the second pilot point of the sampled pattern matching method [1,2]. Thus, our sampled pattern matching method is one of the neural networks with following modifications: first, learning process with supervisor is analytically carried out. Second, threshold value is the maximum value not a constant. Third, same input vector for any unit.

Summation of the entire output $(W+W'+\cdots)\mathbf{U}'$ yields the normalized solution of (4) as

$$\left.\begin{aligned}
&\alpha_1[|\mathbf{d}_1|/|\mathbf{U}|]\simeq[\mathbf{U}/|\mathbf{U}|]^T\{[\mathbf{d}_1/|\mathbf{d}_1|]+[(\mathbf{d}_h+\mathbf{d}_1)/|\mathbf{d}_h+\mathbf{d}_1|]+\cdots\},\\
&\alpha_2[|\mathbf{d}_2|/|\mathbf{U}|]\simeq[\mathbf{U}/|\mathbf{U}|]^T\{[\mathbf{d}_2/|\mathbf{d}_2|]+[(\mathbf{d}_h+\mathbf{d}_2)/|\mathbf{d}_h+\mathbf{d}_2|]+\cdots\},\\
&\cdots\cdots\cdots\cdots\\
&\alpha_h[|\mathbf{d}_h|/|\mathbf{U}|]\simeq\{[\mathbf{U}/|\mathbf{U}|]^T[\mathbf{d}_h/|\mathbf{d}_h|]+\overline{1+1}+\cdots\}\\
&\cdots\cdots\cdots\cdots\\
&\alpha_m[|\mathbf{d}_m|/|\mathbf{U}|]\simeq[\mathbf{U}/|\mathbf{U}|]^T\{[\mathbf{d}_m/|\mathbf{d}_m|]+[(\mathbf{d}_h+\mathbf{d}_m)/|\mathbf{d}_h+\mathbf{d}_m|]+\cdots\}.
\end{aligned}\right\} \tag{10}$$

2.3 Massively parallel neural nets solutions

In physical system containing the field sources, it is possible to measure the different quantities caused by the same field sources. For example, in electromagnetic fields, we can measure the electric and magnetic fields distributions, independently. This leads to the massively parallel neural networks method. Namely, after evaluating the independent solutions for the electric and magnetic systems by their own neural networks, correlation between them may yield the electromagnetic field source distribution without a large amount of ambiguity.

Figures 1 and 2 show the voltage, current dipoles and correlation factor distributions in the human hearts [4]. Obviously, the correlation factor distributions in these figures demonstrate a definitely difference between the normal and abnormal heart conditions.

3. CONCLUSION

As shown above, we have clarified that our sampled pattern matching method is one of the neural networks. In order to evaluate the electromagnetic field source without ambiguity, we have proposed the massively parallel neural networks method. As a result, it has been suggested that the human heart diagnosis may be carried out with higher accuracy by combining the magnetocardiogram and electrocardiogram.

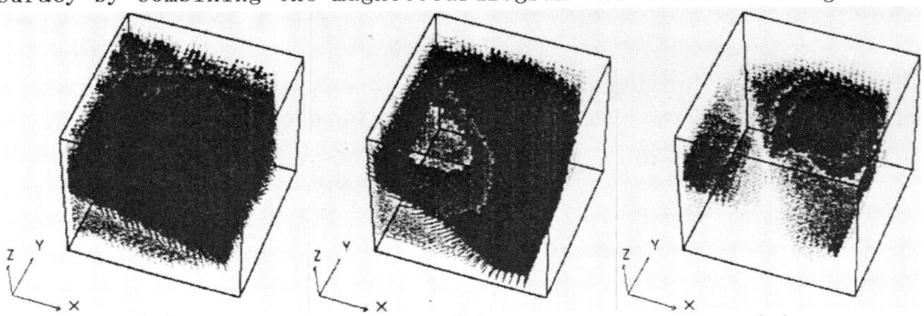

Fig.1 Normal human heart. (a)Estimated by the voltage dipole method using 6×6=36 points electric potentials measured at the top surface; (b) estimated by the current dipole method using 6×6=36 normal magnetic fields measured over 5mm above the top surface; and (c) correlation factor distribution obtained by taking the inner product between the current and voltage dipole vectors.

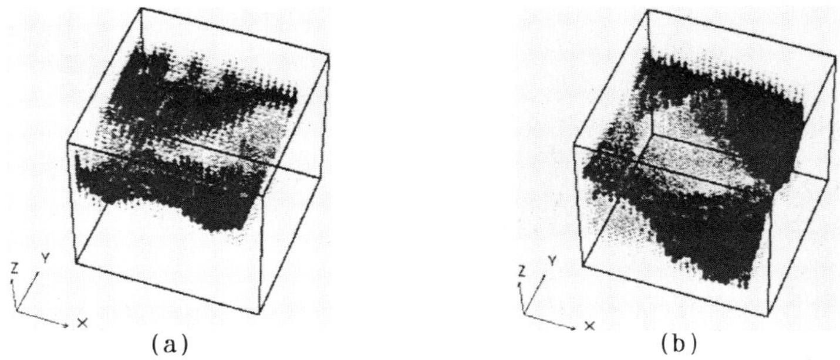

Fig.2 Correlation factor distributions in the abnormal human hearts. (a) Pulmonary hypertension and (b) Pulmonary stenosis.

4. REFERENCES

1 H.Saotome,et al, Trans. IEEJ, Vol.113-C, No.1, January(1993).
2 H.Saotome,et al, Trans. IEEJ, Vol.112-A, No.4, April(1992).
3 K.Kitsuta,et al., Proceedings of the 2nd Japan-Hungary Seminar, Ed. T.Honma et al, Hokkaido Univ. Press, Sep.(1992).
4 Y.Nakaya et al., Journal of Electrocardiology, Vol.21, No. 2 (1988).

Simulation and Design of Applied Electromagnetic Systems
T. Honma (Editor)

An Adaptively Trained Neural Network for the Optimal Design of Electromagmetic Devices

Osama A. Mohammed, Dong C. Park, Riaz Merchant and Fuat G. Üler

Department of Electrical and Computer Engineering
Florida International University, Miami, Florida 33199, U. S. A.

Abstract

An adaptively trained Artificial Neural Network (ANN) for design optimization of electromagnetic devices is presented. The adaptation increases the performance of the ANN specially to data which were not originally present in the training set of the ANN. Data can be input by an experienced designer in addition to a case generator from FE solutions. The proposed adaptive ANN forms the heart of an ANN based design optimization environment. Optimal designs are obtained quickly once the ANN is trained with a variety of topologies. Results of implemented examples are provided to show the effectiveness of the proposed adaptive algorithm for design optimization of electromagnetic devices.

1. INTRODUCTION

Optimization of geometric boundaries in electromagnetic devices is receiving widespread attention and interest by many researchers. Recently, several design optimization techniques have appeared in the literature ranging from evolution strategies to simulated annealing, and search techniques for global minima [1-4]. During the past few years, ANNs have been successfully applied to many engineering problems involving associative operations, search operations, classification and regression, as well as pre-processing [5].

The optimal design problem seeks an optimal output (minimum size, best performance, etc.) for a given set of constraints (inputs). In order to apply the ANNs to the design optimization of electromagnetic devices, consider the optimal design problem as an inverse problem: given an output specification ($\boldsymbol{O}$), find the input geometries ($\boldsymbol{I}$) which satisfy given specification in an optimal fashion. The steps involved can be described as follows:

1. Create N data sets by applying sets of inputs to the set of governing equations F; (I_i, O_i), $i=1,\dots,N$.
2. Train the ANN with O_i as input and I_i as desired output of the ANN.
3. Give the desired performance for which optimal geometry is sought to the ANN. The ANN responds with a practical solution which may or may not be the optimal one.
4. The ANN output is passed on to the optimization module. This step involves optimization and testing till the optimal geometry is found.
5. The optimal geometry is appended to the training set of step 2 and the ANN adapted with this new data.

Although the first two steps are computationally extensive, these are done off line. However, once trained, the ANN provides a response within a few milliseconds.

This paper explores the suitability of applying adaptively trained ANNs to the optimal design problem in electromagnetic devices. It will be shown that with adaptation, the results obtained are more accurate than conventional back-propagation algorithm.

2. DESCRIPTION OF AN ADAPTIVE ANN ENVIRONMENT

The proposed adaptive design environment consists of four modules as shown in Figure 1. The first module deals with the preparation of the ANN training data based on a particular topology. The training set is prepared by using an experts input or by equations governing the topology. This procedure is done only at the beginning of the training procedure and could be deemed as a system initialization. The second module is the trained ANN. Test examples are given to the ANN and the response is fed into the third module which is the optimization module. The optimization module responds with an optimal solution. This optimal geometry is later incorporated into the ANN training set by the adaptation algorithm. The input module consists of an example generator and off-line training of the ANN. The examples are generated by using equations governing the geometry of the device through FE solutions and inputs from experts. Sufficient examples are generated to adequately train the network for a good response. The trained ANN is put on-line after the training is complete. The output from the ANN is a practical one. This is a safe assumption because the training set consists of practical examples. The ANN serves to reduce the search time encountered in other methods [1-4].

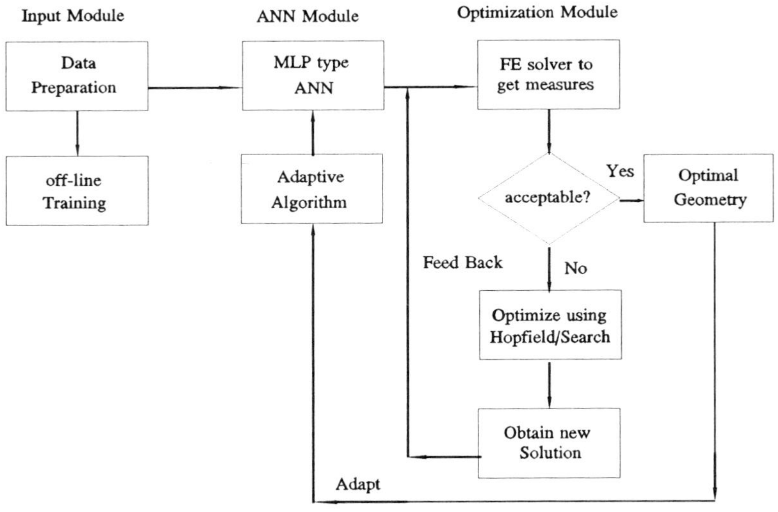

Fig. 1. System flow diagram of the design environment

The optimization procedure can now be applied to the output from the ANN.

Optimization can be done by using any search algorithm. In this paper Hopfield Network/Simulated Annealing is utilized as a combined search mechanism. The main aim of optimization is to reduce the cost function which reflects minimization of core losses, geometry etc. in addition to size minimization.

4. THE ADAPTIVE MECHANISM

The final stage is the adaptation of the optimal geometry. This is done by appending the optimal geometry, obtained after the optimization routine, into the training set of the ANN. The ANN is then trained for a few more iterations with the new training set, and the weights allowed to change, to learn the new data. The adaptation is important for obtaining a quicker response from the system whenever a similar performance measure or optimal design is required.

3. RESULTS AND OBSERVATIONS

The above procedure is put into application for the optimal design of a 2-pole dc motor, Figure 2. The geometrical features w1-w8 are optimized. The objective is to obtain the minimum size that could minimize the mmf per pole while maintaining uniform flux density. The ANN is trained with variables defining the shape of the device as inputs and the performance criteria required as the output. The input variables of the ANN are derived from finite element solutions representing the performance of the magnetic circuit for the various shapes included in the training set. The output variables of the ANN represent coordinate points defining the geometries. The shape could be allowed to change and the design could be made within tolerances in the optimization module. More details on this will be presented in future work.

Fig. 2. 2-pole dc motor

Fig. 3. Optimization results

The results of the ANN module, optimization module and the adaptation module are given in Tables I and II and the shapes are illustrated in Figure 3. The iron area in Table II is calculated by adding element areas of associated finite element grid. These results are shown in comparison with the results obtained from using a conventional search technique [4]. It can be seen from these results that adaptation improves the results obtained from ANN and minimized the shape while satisfying the objective performance.

TABLE I
COMPARISON OF OPTIMAL DESIGN RESULTS FOR THE 2-POLE DC MOTOR

Optimization Method	Design Parameters in cm							
	W1	W2	W3	W4	W5	W6	W7	W8
ANN Module	0.462	0.54	0.149	1.007	3.333	2.588	1.429	2.951
Optimization Module	0.4503	1.5510	0.1501	1.0091	3.3499	2.5625	1.423	3.0471
Adaptive ANN Module	0.4623	1.5526	0.1473	1.0169	3.3344	2.5931	1.4311	2.8714
Conventional Search Method	0.4781	1.5633	0.1461	1.0172	3.3278	2.6021	1.4312	2.8842

TABLE II
COMPARISON OF CORE AREAS* OBTAINED WITH DIFFERENT METHODS(cm^2)

ANN Module	Optimization Module	Adaptive Module	Search Method
136.361	135.621	135.033	134.792

* *Iron area is calculated by adding element areas of associated FE grid.*

5. CONCLUSION

A new method of design optimization using an adaptively trained ANN is presented. A major advantage of the method is its quite short response time in obtaining the optimal geometry. The adaptation makes it possible to respond even more quickly to new data not present in original training set and provides a dynamic feature to the design optimization process.

6. REFERENCES

1 K. Preis and A. Ziegler, "Optimal design of electromagnetic devices with evolution strategies," *Proc. of International Symposium on 3-D Electromagnetic Fields*, Okayama, Japan, pp. 119-122, 1990.

2 J. Simkin and C. W. Trowbridge, "Optimization problems in electromagnetics," *Digests of Computational Electromagnetic Fields Conference*, Toronto, Canada, 1990.

3 Gottwald, "Comparative analysis of optimization methods for magnetic," *IEEE Trans. Mag.*, vol. 24, no. 1, pp. 411-414, 1988.

4 O. A. Mohammed and W. K. Jones, "A dynamic programming -finite element procedure for the design of nonlinear magnetic devices," *IEEE Trans. Mag.*, vol. 26, no. 2, pp. 666-669, March 1990.

5 D. E. Rummelhart, G.E. Hinton, and R.J Williams, "Learning internal representations by error propagation", *Parallel Distributed Processing*, vol. 2, pp. 318-362, MIT Press, Cambridge, 1987.

Simulation and Design of Applied Electromagnetic Systems
T. Honma (Editor)

Autotuning method for different nonlinear functions of a neural network controller

Takayuki Yamada and Tetsuro Yabuta
NTT Telecommunication Field Systems R&D Center
Tokai, Ibaraki, 319-11, Japan

Abstract

A method for autotuning the uniform nonlinear function shapes of a neural network has been proposed. This paper proposes new autotuning methods for different nonlinear function shapes in order to improve nonlinear mapping capability.

1. Introduction

Many neural network controllers have been proposed that provide both flexibility and learning capability in servo controllers.[1][2] However, many parameters, such as the optimum shape of the sigmoid function (nonlinear saturated function), are determined by trial and error. Thus, our previous study proposed an autotuning method for uniform nonlinear function shapes to improve the neural network nonlinear mapping capability.[3] However, if we can use different nonlinear function shapes, better mapping capability can be obtained. Therefore, this paper proposes two autotuning methods for different nonlinear function shapes of a neural network. With type 1, the nonlinear function shapes within one layer are the same and nonlinear function shape of each layer is tuned. With type 2, the nonlinear function shape of each neuron is different and is tuned. These methods have been developed by extending our previous work. The proposed new autotuning algorithm is based on the steepest descent method. The proposed autotuning methods are simulated through the use of a learning type direct controller. Simulation results confirm that the neural network mapping capability using the proposed methods is better than that using the previous method.

2. Autotuning method for different nonlinear functions

Our servo controller neural network has three layers with no inner feedback loop and no direct connection from the input layer to the output layer. Since our object plant is a SISO (Single input single output) plant, the output layer has one neuron. Both the input and hidden layer neurons have a nonlinear function as shown in the following equation.

$$f(x)=2\{1-\exp(-xYg(P))\}/Yg(P)\{1+\exp(-xYg(P))\} \tag{1}$$

where x is the input of the nonlinear function, Yg(P) is the parameter determining its shape and P is the trial number in the learning type.

Figure 1 shows this nonlinear function shape. As shown here, we can improve the nonlinear mapping capability of the neural network controller through the use of this nonlinear function because parameter Yg can change the nonlinear function shape .

This paper proposes the following two autotuning methods for the nonlinear function shape shown in eq.(1).

[Type 1] The nonlinear function shapes within one layer are the same and the sigmoid function shape of the output layer is different from that of the hidden layer. The nonlinear function shape of each layer is tuned based on the steepest descent method as shown in the following equation.

$$Yg_i(P+1)=Yg_i(P)-\eta_{2i}(\partial J(P)/\partial Yg_i(P)) \quad [i=1,2] \tag{2}$$

where η_{2i} is a parameter determining the convergence speed of the sigmoid function tuning, subscript i=1 indicates the hidden layer, subscript i=2 indicates the output layer and J is the cost function. The purpose of the servo controller is to minimize J.

[Type 2] The nonlinear function shape of each neuron is different and is tuned. Each nonlinear function shape is tuned as shown in the following equation.

$$Yg_{1j}(P+1)=Yg_{1j}(P)-\eta_{21j}(\partial J(P)/\partial Yg_{1j}(P)) \quad [j=1,2,...] \tag{3}$$

$$Yg_2(P+1)=Yg_2(P)-\eta_{22}(\partial J(P)/\partial Yg_2(P)) \tag{4}$$

where eqs.(3) and (4) respectively express the autotuning laws of the hidden and output layers. Subscript j indicates the number of hidden layer neuron.

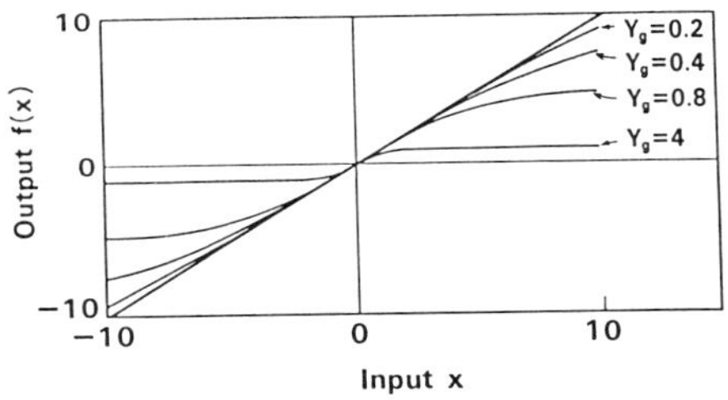

Fig.1 Sigmoid function.

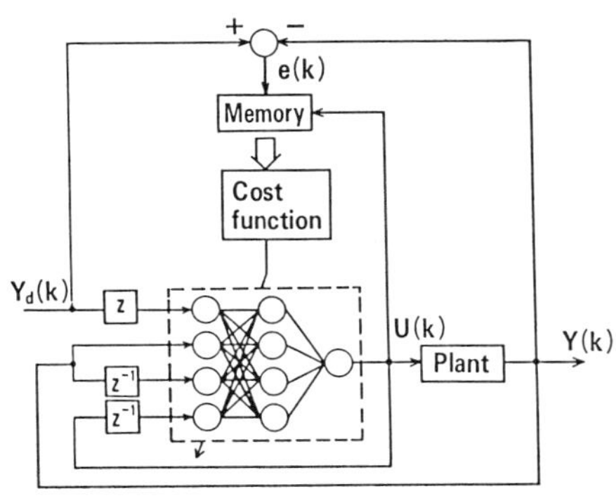

Fig.2 Learning type direct controller.

3. Learning type direct controller

In order to verify the usefulness of the proposed autotuning methods, we apply them to a learning type direct controller.[3] When we select a predominantly linear SISO second order discrete time system as an object plant, the neural network input vector I and the plant input are defined as follows;

$$I^T(k)=[Yd(k+1), Y(k), Y(k-1), U(k-2)] \quad (5)$$

$$U(k)=f\{\omega^T(P)f(W(P)I(k))\} \quad (6)$$

where Yd is the desired control value, Y is the plant output, k is the sampling number, ω is the weight vector whose elements are the weights from the hidden layer to the output layer and W is the weight matrix whose elements are the weights from the input vector to the hidden layer. Both the weight vector and matrix are tuned so as to minimize the cost function as follows;

$$W(P+1)=W(P)-\eta_1 (\partial J(P)/\partial W(P)) \quad (7)$$

$$\omega(P+1)=\omega(P)-\eta_1 (\partial J(P)/\partial \omega(P)) \quad (8)$$

$$J(P)=\frac{1}{2}\sum_{k=1}^{\rho} \varepsilon^2(k)=\frac{1}{2}\sum_{k=1}^{\rho} \{Yd(k)-Y(k)\}^2 \quad (9)$$

Where ρ is the sampling number within one trial and η_1 is the parameter determining the convergence speed of the weight tuning. Figure 2 shows a block diagram of the learning type direct controller.

4. Simulation

In order to confirm the characteristics of the proposed autotuning methods, they are applied to a second order plant described by the following equation.

$$Y(k)=-a_1Y(k-1)-a_2Y(k-2)+U(k-1)+bU(k-2)+C_{non}Y^2(k-1) \quad (10)$$

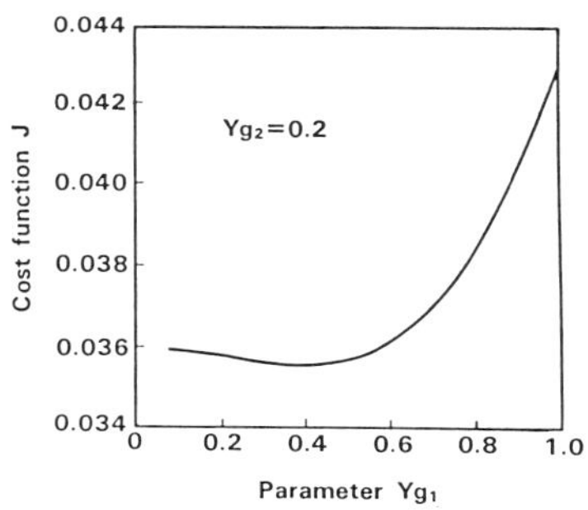

Fig.3 Energy curve with respect to Yg_1.

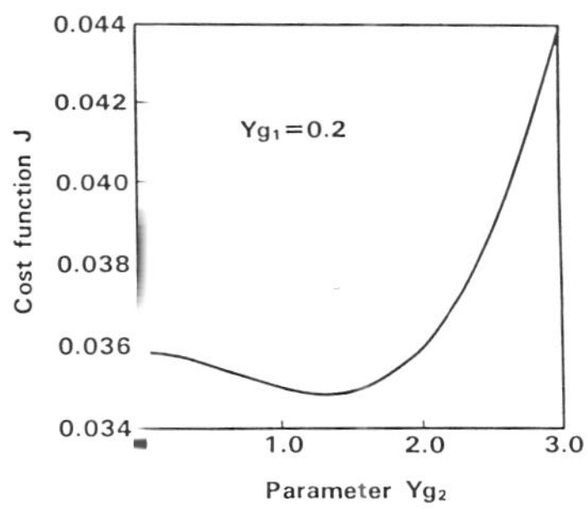

Fig.4 Energy curve with respect to Yg_2.

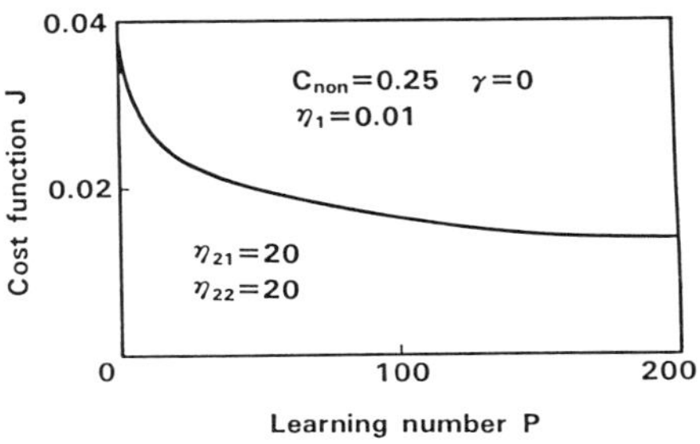

Fig.5 Autotuning process with type 1.

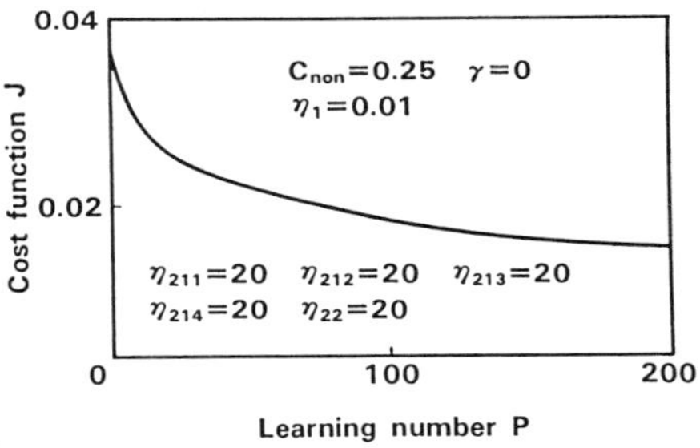

Fig.6 Autotuning process with type 2.

where C_{non} is the coefficient of the nonlinear term. In this simulation, values of a_1=-1.3, a_2=-0.3, b=0.7, a_3=-0.02 and C_{non}=0.25 are selected. Figures 3 and 4 respectively show the energy curves with respect to parameters Yg_1 and Yg_2. As shown here, the energy curves are parabolic. This means that the proposed autotuning methods based on steepest descent method can be applied to this control system. Figures 5 and 6 respectively show the autotuning processes with types 1 and 2. As seen here, the cost functions decrease as tuning progresses. The respective cost functions of types 1 and 2 after 200 trials are about 87% and about 79% of that with the previous method. This means that our proposed autotuning methods are useful for improving nonlinear mapping capability.

5. Conclusion

This paper proposed autotuning methods for different nonlinear function shapes in order to improve the nonlinear mapping capability of a neural network servo controller. The proposed methods were applied to a learning type direct controller and simulated for a second order discrete time plant. The simulation results confirmed that the nonlinear mapping capability of the proposed methods is better than that of the previous autotuning method.

Acknowledgment

The authors express their thanks to Dr.K.Ishihara and Dr.M.Kawase NTT Telecommunication Field Systems R&D Center, for their continued encouragement. They also thank Mr.Terakado, NTT Advanced Technology Corporation for his assistance in this research.

REFERENCES

[1] Kumpati S. Narendra and Kannan Paratharathy, "Identification and Control of Dynamical Systems Using Neural Networks", IEEE Transactions on Neural Network, Vol.1, No.1, pp4-27(1990)

[2] M.Kawato, Y.Uno, M.Isobe and R.Suzuki, "A Hierarchical Model for Voluntary Movement and its Application to Robotics", Proceedings of 1987 IEEE International Conference on Network, Vol.IV, pp573-582 (1987)

[3] Takayuki Yamada and Tetsuro Yabuta, "Neural Network Controller Using Autotuning Method for Nonlinear Functions", IEEE Transactions on Neural Networks, Vol.3, No.4, pp595-601(1992)

Simulation and Design of Applied Electromagnetic Systems
T. Honma (Editor)

A hierarchical fuzzy modeling method with comprehensible fuzzy rules — An approach to inverse problems —

Shoichi Nakayama[a], Takeshi Furuhashi[b] and Yoshiki Uchikawa[b]

[a]Fujitsu VLSI Limited, Kozoji-cho, Kasugai, Aichi 487, Japan

[b]Department of Electronic Mechanical Engineering, Nagoya University, Furo-cho, Chikusa-ku, Nagoya, 464-01, Japan

Abstract

This paper presents a hierarchical fuzzy modeling method using fuzzy neural networks. The method identifies nonlinear plant with multi-inputs precisely. The acquired fuzzy rules with the new modeling method are easy to comprehend.

1. INTRODUCTION

Inverse problems have been widely studied by many researchers and engineers in various fields. Identification of complex nonlinear plant model is one of the inverse problems and has strong needs in the industry. In the case where the plant has close correlationships in its inputs and/or outputs, the identification of the plant model is a difficult work.

Fuzzy modeling is a method to describe the characteristics of complex systems using fuzzy inference[1]. The method has a distinguishing feature in that it can express complex nonlinear systems linguistically. Rules in linguistic expressions are easy for human beings to understand. The fuzzy inference is an effective tool for human - computer interactions. However, when the system has multi-inputs, the number of the acquired fuzzy rules becomes large and the rules are hard to understand. And when it is difficult to obtain sufficient data from the system with multi-inputs, the fuzzy model cannot be made precise.

This paper presents a hierarchical fuzzy modeling method. The proposed fuzzy modeling method uses fuzzy neural networks(FNNs)[2]. The identified fuzzy rules with the new method are easy to understand. The method enables to obtain a precise fuzzy model even with limited number of input-output data.

2. HIERARCHICAL FUZZY MODEL USING FNNS

Figure 1 shows a basic configuration of the proposed hierarchical fuzzy model. The figure shows the case where a system with three inputs x_1, x_2, x_3 and one output y is modeled as a two stage fuzzy system. In this example, the inputs x_1, x_2 mainly determine the input-output characteristics of the system, and the input x_3 complements the error of the first stage model. The hierarchical method requires less number of fuzzy rules than the conventional single stage fuzzy model does.

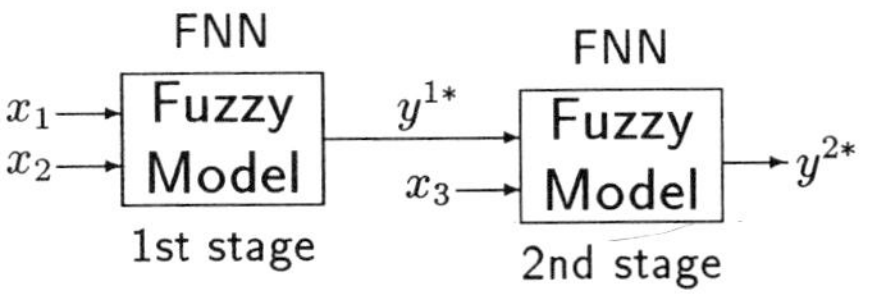

Figure 1. Hierarchical Fuzzy Model

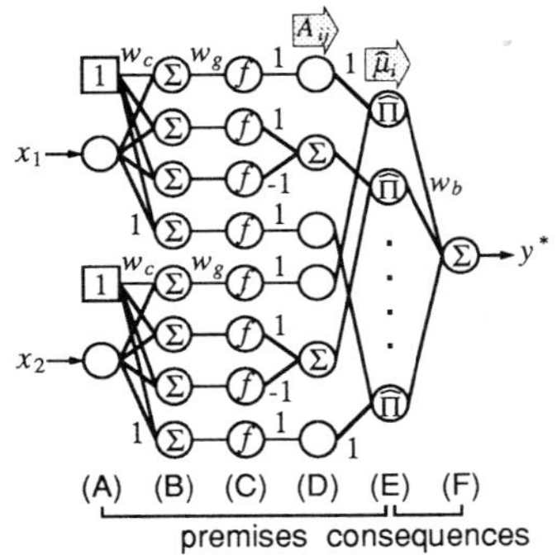

Figure 2. Fuzzy Neural Network (Type I)

This paper uses "Type I" of the FNNs[2] as shown in Figure 2. The figure shows the case where the FNN has two inputs x_1, x_2 and one output y^* and three membership functions in each premise. The symbols ○, □ in the figure mean units of the neural network and the denotation w_c, w_g, w_b and $1, -1$ are the connection weights. The FNN realizes a simplified fuzzy inference written as follows :

$$R^i\text{: If } x_1 \text{ is } A_{i1} \text{ and } x_2 \text{ is } A_{i2} \text{ then } y = b_i \qquad (i = 1, 2, \cdots, n) \tag{1}$$

$$y^* = \sum_i \mu_i b_i \Big/ \sum_i \mu_i = \sum_i \widehat{\mu}_i b_i \qquad \widehat{\mu}_i = \mu_i \Big/ \sum_k \mu_k \qquad \mu_i = A_{i1}(x_1)A_{i2}(x_2) \tag{2}$$

where R^i is the i-th fuzzy rule. A_{i1}, A_{i2} are fuzzy variables. b_i is a constant. n is the number of fuzzy rules. μ_i is the truth value of R^i. $\widehat{\mu}_i$ is the normalized truth value so that the sum of $\widehat{\mu}_i$ is unity. y^* is the inferred value.

The connection weights w_b correspond to the singletons in the consequences b_i. The FNN tunes the membership functions in the premises and identifies the fuzzy rules by adjusting the connection weights w_c, w_g and w_b, through the Back-Propagation(BP) algorithm. The learning algorithm for the FNN is well described in [2].

3. HIERARCHICAL FUZZY MODELING

This paper uses the fuzzy modeling method in [2]. The input-output data of modeling object are divided into two groups(group A and group B) of which the statistical characteristics are the same. The structure in the premise part, i.e. the number of the input variables and that of the membership functions, is identified first using the parameter increasing method in [2]. The identified models using the data of group A, B are called model A, B, respectively. The precision and the generality of the model are evaluated with the criterion C[2] :

$$C = \sqrt{\sum_{i=1}^{n_A}(y_i^A - y_i^{AA})^2 + \sum_{i=1}^{n_B}(y_i^B - y_i^{BB})^2} + \sqrt{\sum_{i=1}^{n_A}(y_i^{AB} - y_i^{AA})^2 + \sum_{i=1}^{n_B}(y_i^{BA} - y_i^{BB})^2} \tag{3}$$

where n_A and n_B are the number of the data of group A, B, respectively. y_i^A, y_i^B are the outputs of group A, B, respectively. y_i^{AA}, y_i^{BA} are the inferred values corresponding to y_i^A, y_i^B with the model A, and y_i^{AB}, y_i^{BB} are the inferred values to y_i^A, y_i^B with the model B, respectively. The first term on the right hand side in (3) means the precision of the models and the second term evaluates the generality of the acquired models.

This parameter increasing method takes influential input variables one by one into the model. The larger the number of input variables, the more the number of fuzzy divisions. The large number of fuzzy divisions makes many input sub-spaces without any data, and makes the generality worse. The modeling is, then, stopped at the minimum criterion C.

The premise portions and the constants in the consequences of the selected fuzzy model with the best structure is finely tuned with the BP learning algorithm. The fuzzy model in the first stage is obtained.

Then, fuzzy modeling for the second stage is done using the output of the fuzzy model in the first stage and the input variables not used in the first stage. Through the above process, hierarchical fuzzy model with multiple stages will be acquired. The model has a small number of fuzzy divisions of input space and a high generality.

4. SIMULATION

This paper uses the nonlinear system given by

$$y = (1 + x_1^{0.5} + x_2^{-1} + x_3^{-1.5} + x_4^{-2.5} + x_5^{1.5})^2 \tag{4}$$

as the object of the proposed hierarchical fuzzy modeling. The system has five inputs $x_1 \sim x_5$ and one output y. The input-output data used for the modeling are listed on Table 1. A variable(x_6) which has no relationship with the output y is included. The group A has the data No.1~40 and the group B has the data No.41~80.

The learning rate for the connection weights w_c, w_g, w_b were set to be 3×10^{-5}, 3×10^{-5}, 0.05, respectively. The learning was done 1000 times at the most. In one time of learning, every pair of data are used once.

Table 2 shows the process of the structure identification in the premises. The parameters(number of input variables and membership functions) were increased step bye step. $x_j[k]$ means that the number of membership functions for the input variable x_j is k. The structures with the circles ○ were selected and that with the double circles ◎ was finally chosen. The fuzzy model in the first stage has the structure with $x_1[2]$ and $x_5[2]$.

The selected model was then finely tuned. The maximal learning time was 2000. The data of group A were used for this fine tuning. Those of group B were used for evaluating the generality of the final model.

In the same way, $x_2[2]$ and $x_4[2]$ were selected in the second stage model. $x_3[4]$ was chosen at the third stage. The variable x_6 was taken into the model at the fourth stage.

Table 1 Input and output data

No.	x_1	x_2	x_3	x_4	x_5	x_6	y	No.	x_1	x_2	x_3	x_4	x_5	x_6	y
1	1	3	1	5	0.9	1	17.68	41	1	1	5	1	0.1	1	16.98
2	1	5	2	2	0.3	1	8.38	42	1	3	4	4	0.7	1	9.46
3	1	1	3	1	0.5	5	20.67	43	1	5	3	5	0.5	5	7.64
4	1	3	4	3	0.7	5	9.66	44	1	1	2	3	0.3	5	12.83
5	1	5	5	5	0.1	1	5.47	45	1	3	1	1	0.9	1	26.91
6	5	1	4	4	0.9	1	27.52	46	5	5	2	2	0.1	1	15.98
⋮	⋮	⋮	⋮	⋮	⋮	⋮	⋮	⋮	⋮	⋮	⋮	⋮	⋮	⋮	⋮
40	5	3	2	3	0.1	5	16.15	80	5	1	4	3	0.9	5	27.87

Table 2 Identification of structures of fuzzy model (1st stage)

step	premise structure	C
1	○ $x_1[2]$	62.10
	$x_2[2]$	72.40
	$x_3[2]$	100.83
	$x_4[2]$	83.63
	$x_5[2]$	69.82
	$x_6[2]$	109.13
2	$x_1[3]$	62.11
	$x_1[2], x_2[2]$	56.58
	$x_1[2], x_3[2]$	87.65
	$x_1[2], x_4[2]$	62.18
	◎ $x_1[2], x_5[2]$	55.01
	$x_1[2], x_6[2]$	93.25
3	○ $x_1[3], x_5[2]$	55.01
	$x_1[2], x_2[2], x_5[2]$	59.24
	$x_1[2], x_3[2], x_5[2]$	79.92
	$x_1[2], x_4[2], x_5[2]$	56.85
	$x_1[2], x_5[3]$	56.39
	$x_1[2], x_5[2], x_6[2]$	87.46

Table 3 Identified hierarchical fuzzy rules

1st stage y^1		x_1 Small	x_1 Big
x_5	Small	10.5 [S]	17.3[MB]
	Big	15.3[MS]	27.3 [B]

↓ y^{1*}

2nd stage y^2				x_2 Small	x_2 Big
y^{1*}	S	x_4	S	18.4 [M]	12.4[MS]
			B	12.1[MS]	6.3 [S]
	B	x_4	S	38.6 [B]	29.4[MB]
			B	30.0[MB]	20.8 [M]

↓ y^{2*}

3rd stage y^3		x_3 Small	x_3 MS	x_3 MB	x_3 Big
y^{2*}	Small	9.3	4.2	3.3	3.8
	Big	43.8	39.0	36.2	32.4

The criterion C of the model was not better than the previous model. The variable x_6 was judged to be irrelevant to the system.

The precision of the model was judged by eq.(5)

$$J_A = \frac{1}{40}\sum_{i=1}^{40}\frac{|y_i - y_i^{3*}|}{y_i} \times 100 \quad (\%) \qquad J_B = \frac{1}{40}\sum_{i=41}^{80}\frac{|y_i - y_i^{3*}|}{y_i} \times 100 \quad (\%). \tag{5}$$

The hierarchical model has the following values : $J_A = 3.8$ (%) and $J_B = 6.4$ (%). When a single-stage fuzzy model with input $x_1 \sim x_5$ was made, the mean errors J_A, J_B became 2.4 (%), 20.3 (%), respectively. The single-stage model had many input sub-spaces without any data. This model had a poor generality.

Table 3 shows the identified hierarchical fuzzy rules. The labels on the table S, MS, MB, B are Small, Medium Small, Medium Big, Big, respectively. The number of rules was 20 less than 32 of the single stage model. From the rule table, the output y is monotonically increasing with x_1 and x_5 and monotonically decreasing with x_2, x_3, x_4. These rules well describe the eq.(4).

5. CONCLUSIONS

This paper presented a hierarchical fuzzy modeling method with comprehensible fuzzy rules. The model had a higher precision than the conventional single-stage model in the case where there was a small number of data from the multi-inputs system.

6. REFERENCES

1 T. Hasegawa, et. al, A Fuzzy Modeling of BOF Using Fuzzy Neural Network, Elsevier Studies in Applied Electromagnetics in Materials 3 (1992) 61–64.

2 S. Horikawa, T. Furuhashi, Y. Uchikawa and T. Tagawa, A Study on Fuzzy Modeling Using Fuzzy Neural Networks, Proceedings of the IFES'91 (1991) 562–573.

Simulation and Design of Applied Electromagnetic Systems
T. Honma (Editor)

Knowledge based techniques in electromagnetic analysis systems

D.A.Lowther

Electrical Engineering Department, McGill University, 3480 University Street, Montreal, Quebec, H3A 2A7, Canada.

Abstract
The use of various knowledge based techniques, such as neural networks and blackboards, in the design process of electromagnetic devices is discussed. The gains in computation time in various phases of device analysis are demonstrated.

1. INTRODUCTION

Since the earliest days of electromagnetics design, two paradigms have been followed in creating devices. The first is an experience-based model, i.e." we have built one similar to this before and this is the required structure"; the second is an analysis-based model which attempts to explicitly or implicitly determine relationships between design quantities and specification parameters. Over the past few decades, the analysis paradigm has received great attention because gains in computational power provided the possibility for refining and improving modelling capabilities based solely on the laws of physics. Thus it is now possible to simulate electromagnetic systems with reasonable verisimilitude on a computer and this can avoid costly mistakes which may occur in constructing physical devices. However, many of the analysis systems have become complex enough that they require considerable expertise to run them effectively and this requires that an engineer becomes highly trained in these areas.

To combat some of this requirement, systems have been developed which attempt to refine their own answers based on error analyses. Two current examples of this are the use of iterative schemes to solve non-linear problems and adaptive processes to refine the input data description for analysis [1]. In the former example, an iterative scheme can be shown to be effective because the shape of the function space is well understood in the neighbourhood of the solution. In the latter case, the iteration is based on some physical quantity which should be single valued and have a distinct minimum.

Developments in knowledge based techniques have resulted in several representation schemes which may be used to record some of the experience gained about solution behaviour from iterative processes. This knowledge can be combined with conventional numerical techniques to improve computation times and memory requirements. One example of this has been the use of neural network technology in mesh generation [2].

2. WHAT IS DESIGN?

Design has traditionally consisted of two components. The first of these is a synthesis component which is generally performed by an engineer who takes the specifications, which are frequently in terms of the terminal quantities, and attempts to determine the appropriate values for all the possible parameters in the device. The second component is a 'build and test' phase where the device is constructed and tested to see if it meets the specifications. Over time, the second phase has been largely replaced by computer based analysis tools which simulate the performance of the device at ever increasing levels of complexity and reduce the need to construct intermediate devices for laboratory testing. The combination of synthesis and analysis phases in an iterative loop forms the fundamental design process. Both the synthesis and analysis processes are examples of searching systems, although their goals are very different. In a synthesis phase, information about a device is expanded, in the analysis phase it is contracted. The fundamental goal in both phases is to reduce the time taken in searching for an appropriate solution.

3. SEARCH SPACES

The search spaces used in designing an electromagnetic device are extremely large and complex - a simple transformer may need hundreds of parameters to be specified before it can be constructed. Many of the parameters are not independent but the relationships between them may be qualitative as much as quantitative. Thus information about the structure of the design space is extremely difficult to obtain and varies from device to device.

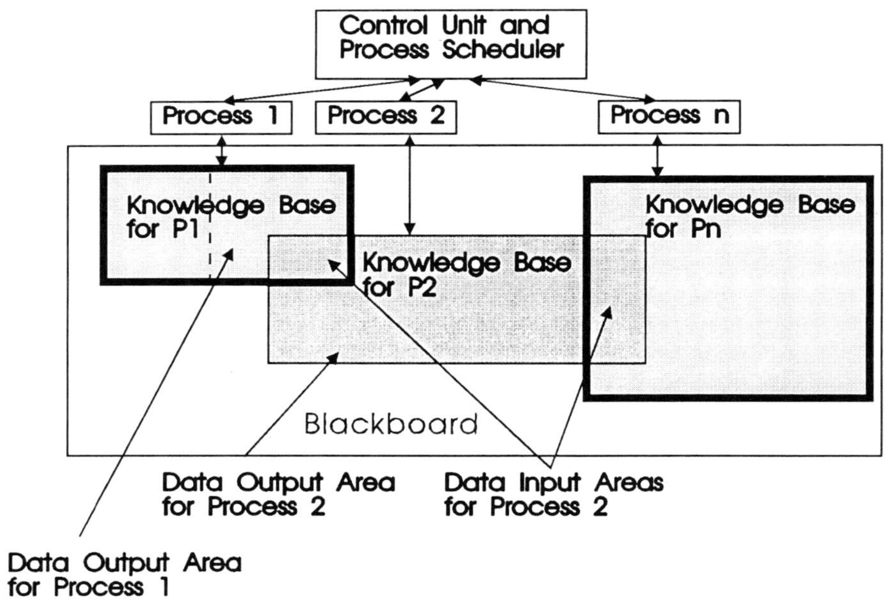

Figure 1. Blackboard architecture.

However, when a device is analyzed, the search spaces for its performance are well understood because of the basic physics underlying its operation. In this phase much can be done to optimize the searching strategies and equation solvers are good examples of some of the techniques available for searching spaces with well defined structures. The goal in any searching process is to reduce the space of solutions as quickly as possible.

4. SEARCH SPACES IN SYNTHESIS

Synthesis is the process of creating a complete device from a set of specifications. Controls are placed on the search space by several means. The most obvious of these is the underlying physics. Equally important to the complete design are controls resulting from economic, legal, sociological and aesthetic considerations. These are much harder to quantify and are generally those which require the expertise of the engineer. Searching a space with dependencies created by these requires mechanisms other than the traditional mathematical techniques.

Each of the components of the synthesis process requires a different approach and an architecture is needed which will allow these techniques to work together. One candidate for this is the blackboard architecture, Figure 1, [3]. In effect, it provides a software operating system which allows many different problem solving approaches to cooperate in completing a design.

5. SEARCH SPACES IN ANALYSIS

There are many ways to determine the solution to a particular problem in electromagnetics such as the distribution of magnetic flux in an iron core. These may be broadly classified as quantitative and qualitative. In the former approach, algebraic equations are constructed which constrain the solution to obey the underlying physics; in the latter, the approximate solution is created based on a set of rules such as "if there is a corner in a permeable structure, then the flux density will be high". These rules allow a qualitative solution to the field problem to be determined very quickly. Such rules can provide a short cut to finding an accurate solution.

An analysis system based on finite elements requires that there is some knowledge of the field solution prior to creating the elements. Thus the mesh must be designed based on the potential field solution to achieve optimum results. This mesh design may be performed in one of two ways. Conventionally, an experienced engineer creates the structure and the finite element analysis, in a sense, merely refines the solution based on the expectations of the user. In doing this, prior knowledge is used heavily. An alternate approach is based on an adaptive strategy - the system learns about the solution from an examination of the errors, [1]. In this case, the errors can be determined because of an understanding of the constraints imposed by the underlying physics. Where these constraints are violated indicates regions of high error.

A more appropriate approach can try to use prior knowledge in the way that an engineer does in designing the initial mesh. The problem is in how to classify and structure the knowledge. A rule based system can be used and some success has been achieved with this approach [4]. An alternate method is based on a learning approach using neural networks coupled with adaption. As an example, Figures 2 and 3 show the mesh generated by a neural network for a recording head structure and the corresponding flux plot. To obtain an accurate solution using a neural network coupled with adaption took less than half the CPU time that pure adaption required.

There are several areas within the domain of analysis tools where similar requirements arise. For example, consider the problem of post-processing results from analysis. Whereas mesh generation expands the initial data for solution, post-processing contracts it. For example, an entire field solution may be expressed by a single number, the inductance. However, inductance is not uniquely defined hence a decision has to be made as to which definition is appropriate. Here, again, a knowledge based approach can be used to help in making the appropriate choice.

Frequently, analysis programs deal with one aspect of a device, e.g. its magnetic, thermal or structural performance. There is a requirement to couple these programs to more closely simulate interactions between the different aspects. This results in a loosely coupled system where a blackboard architecture may again be usefully employed.

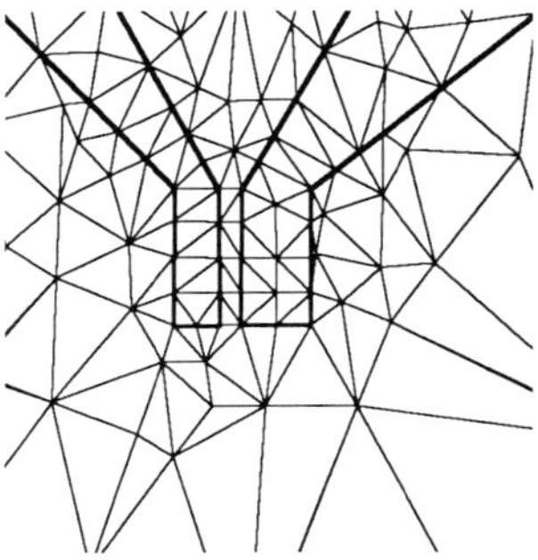

Figure 2. Mesh generated by neural network system.

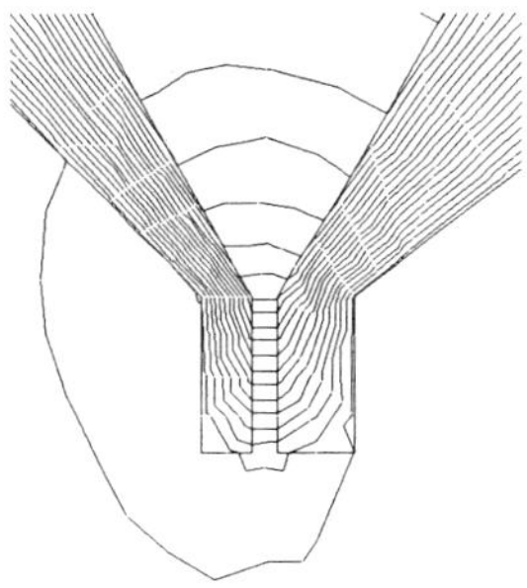

Figure 3. Flux plot on neural network mesh.

6. CONCLUSIONS

A complete design system requires both synthesis and analysis capabilities to be present. Both synthesis and analysis phases can benefit by using knowledge based techniques to provide relationships between parameters in a design which may result in overall performance improvements of conventional algorithms.

7. REFERENCES

1. P. Fernandes, P. Girdinio, P. Molfino and M. Repetto, IEEE Transactions on Magnetics, 5 (1990), pp. 2187-2189.
2. D.N.Dyck, D.A.Lowther and S.McFee, IEEE Transactions on Magnetics, 2 (1992), 1767-1770.
3. R. Engelmore and T. Morgan (eds.), Blackboard Systems, Addison-Wesley, Massachusetts, 1988.
4. K.Reichert, J.Skoczylas, T.Tarnhuvud, IEEE Transactions on Magnetics, 5 (1991), 4197-4200.

Inverse Problem

Simulation and Design of Applied Electromagnetic Systems
T. Honma (Editor)

Numerical simulation of probe-crack interaction of three-dimensional NDE models

Z Badics[a], H Komatsu[a], H Mototsuji[a], K Aoki[a], F Nakayasu[a] and K Miya[b]

[a] Nuclear Fuel Industries, LTD.
950, Ohaza-noda, Kumatori-cho, Sennan-gun, Osaka 590-04 Japan

[b] Nuclear Engineering Research Laboratory, The University of Tokyo
7-3-1 Hongo, Bunkyo-ku, Tokyo 113, Japan

Abstract

A finite element model has been developed to analyze the electromagnetic interaction between probe coils and crack-type defects. The crack-type irregularities are treated here as non-conducting thin sheets with negligible thickness. To verify the method a benchmark problem has been solved and the calculated and measured data have been compared.

1. INTRODUCTION

This paper focuses on crack-type irregularities and deals with the analysis of three dimensional eddy current distributions disturbed by SCC (Stress Corrosion Crack) or fatigue type cracks. These cracks have an extreme characteristic, i.e. one of their dimensions - width - is much smaller than the other extents - length and depth. (The width is usually less than 0.1% of the other two dimensions.)

Since the tangential electric field jumps on the opposite sides of the crack, the width of the crack has to be subdivided into several finite elements to describe the jump accurately. So the fine resolution in this direction limits the size of the elements in the other two directions, i.e. along the depth and the length. To avoid the discretization inside the crack a TSC (thin sheet crack) model has been developed. This model has been incorporated into a finite element potential formulation without disrupting the primary finite element scheme.

Fig. 1 shows the arrangement of a probe-crack benchmark problem. We have calculated the crack impedance ΔZ - the impedance change of the probe due to the presence of the crack - at different probe locations. Furthermore, the calculated and the measured values of the crack impedances have been compared.

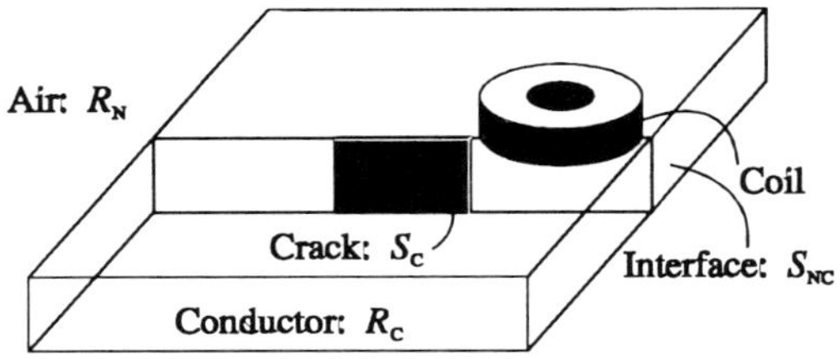

Figure 1. Typical probe-crack problem

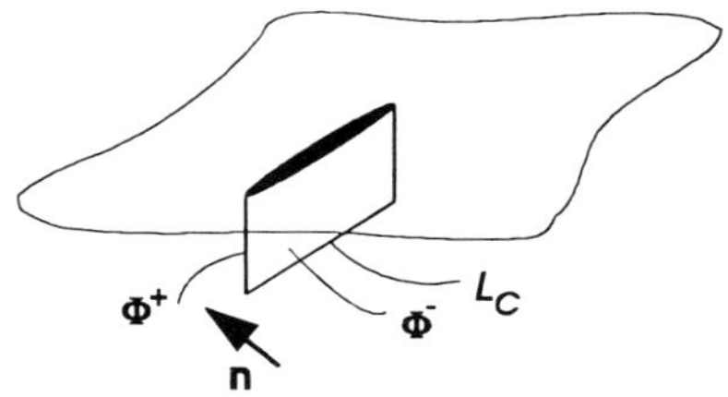

Figure 2. Thin sheet crack model

2. FINITE ELEMENT FORMULATION

In the conductor the electromagnetic field is described by the magnetic vector potential **A** and the electric scalar potential Φ, and in the non-conducting region the reduced magnetic scalar potential Ω is used. Considering the Maxwell's equations with displacement currents neglected, one can obtain the governing equations for the defined potentials. Following [1], the weak form of these governing equations can be derived as

$$\int_{R_C} \left(\tfrac{1}{\mu}\nabla\times\mathbf{A}\cdot\nabla\times\mathbf{N}+\tfrac{1}{\mu}\nabla\cdot\mathbf{A}\nabla\cdot\mathbf{N}+\sigma\dot{\mathbf{A}}\cdot\mathbf{N}+\sigma\mathbf{N}\cdot\nabla\Phi\right)dv-\int_{S_{NC}}\mathbf{H}_{NC}\times\mathbf{n}\cdot\mathbf{N}ds=0 \tag{1}$$

$$\int_{R_C}\left(\sigma\dot{\mathbf{A}}\cdot\nabla N+\sigma\nabla\Phi\cdot\nabla N\right)dv+\int_{S_{NC}+S_C}\mathbf{J}_{NC}\cdot\mathbf{n}N ds=0 \tag{2}$$

$$\int_{R_N}(\mu\nabla\Omega\cdot\nabla N)dv-\int_{S_{NC}}\mathbf{B}_{NC}\cdot\mathbf{n}N ds=0 \tag{3}$$

where **N** and N are vector and scalar weighting functions and suffix $_{NC}$ denotes values on the interface S_{NC}. (The formulation ensures the uniqueness of the potentials [1].)

3. CRACK AS A SECONDARY SOURCE

According to the requirements the crack is treated here as a non-conducting region of negligible thickness. Hence, the eddy current has no normal component on the crack surface:

$$\mathbf{J}\cdot\mathbf{n}=0 \tag{4}$$

Suppose that the conductor and the crack have the same permeability. Then, the tangential component of the magnetic field and the normal component of the magnetic flux density are continuous here, i.e.

$$\mathbf{H}^+\times\mathbf{n}-\mathbf{H}^-\times\mathbf{n}=\mathbf{0} \tag{5}$$

$$\mathbf{B}^+\cdot\mathbf{n}-\mathbf{B}^-\cdot\mathbf{n}=0 \tag{6}$$

The plus and minus signs denote limiting values on the opposite sides of the crack. (See in Fig. 2)

It can be proved that the conditions (4-6) are consistent and ensure the uniqueness of the field quantities. As a result, the tangential component of the electric field is not continuous, its jump can be related to a magnetic surface current density $\mathbf{K}^{(m)}$ as

$$\mathbf{n} \times \mathbf{E}^{+} - \mathbf{n} \times \mathbf{E}^{-} = -\mathbf{K}^{(m)} \tag{7}$$

4. THIN SHEET CRACK MODEL

The mathematical model that satisfies the conditions (4-6) is the following:

- the normal component of the magnetic vector potential is set to zero, and
- a doubled layer of electric scalar potential is introduced on the crack.

Since the magnetic vector potential is still continuous, the application of the weak form (1) results the weak continuity of the tangential component of the magnetic field along the crack surface too. The essential condition

$$\mathbf{A} \cdot \mathbf{n} = 0 \tag{8}$$

with omitting the surface term of the weak form (2) yields the vanishing normal component of the current density.

This implies that the normal component of the electric field is also zero and, therefore, the jump of the tangential electric field can be expressed only by the electric scalar potential as

$$\mathbf{E}^{+} - \mathbf{E}^{-} = -\nabla\left(\Phi^{+} - \Phi^{-}\right) = -\nabla_t \chi \tag{9}$$

The Eq. (9) means that the jump can be described by a surface scalar χ which is represented here by a doubled layer of electric scalar potential. (The ∇_t denotes the tangential gradient on the crack surface.) From this restriction the continuity of the normal component of the magnetic flux density follows [2].

The electromagnetic field components perpendicular to the edge of the crack tend to infinity as $r^{-1/2}$ when r tends to zero [3]. (r is the distance from the edge of the crack.) This implies that χ has to tend to zero for vanishing r following from Eq. (9). Therefore, the continuity of the scalar potential Φ has to be retained along the boundary curve L_c of the crack. (See in Fig.2.)

5. RESULTS

Fig. 3 shows the experimental arrangement that has been set up to test the new crack model. This is a metal plate with a thin crack (black area) excited by a cylindrical coil. δ denotes the skin depth. Fig. 4 depicts the variation of the absolute value of the ΔZ along the axis x while Fig. 5 shows the trajectory of the ΔZ on the complex plane. The calculated and the measured ΔZ values have been normalized by the absolute value of the calculated ΔZ at $x = 0$.

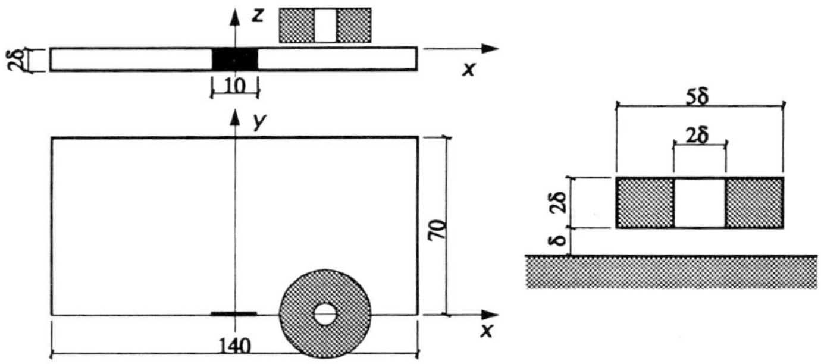

Figure 3. Arrangement of the benchmark problem

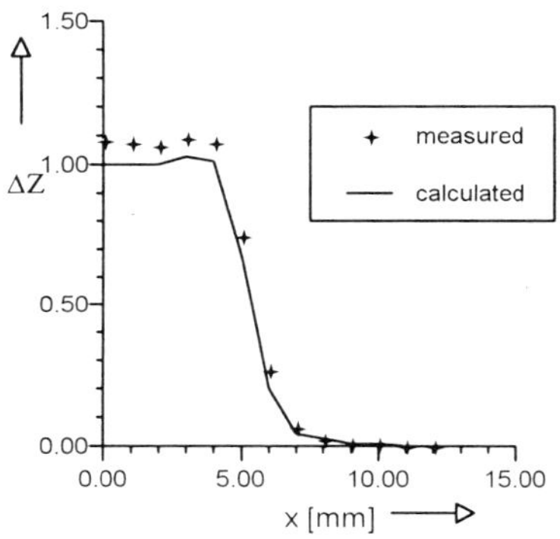

Figure 4. Absolute value of ΔZ vs. coil location

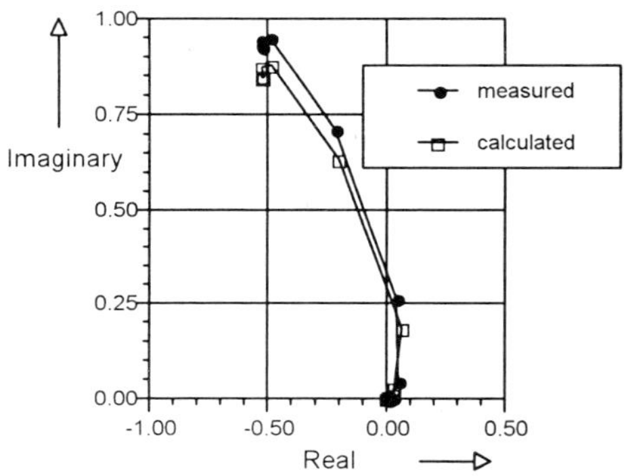

Figure 5. ΔZ trajectory

6. CONCLUSION

The paper has introduced a thin sheet crack model that can describe eddy current distribution disturbed by a crack with extreme thinness. Furthermore, the application of the model ensures the right behavior of the electromagnetic field near the edge of a crack as well. The TSC model can be incorporated into a finite element system efficiently. Although the ΔZ values are very small comparing to the total impedance of the coil, relatively good agreement has been reached between the calculated and the experimental values.

7. REFERENCES

1 O Biro and K Preis, IEEE Trans. on Magn., Vol.25, No.4 (1989) 3145
2 J R Bowler, Electrosoft, Vol.2, No.2/3 (1991) 142
3 A H Kahn, R Spal and A Feldman, J.Appl.Phys., Vol.48, No.11 (1977) 4454

Simulation and Design of Applied Electromagnetic Systems
T. Honma (Editor)

Defect Reconstruction From Eddy Current Data Using Fourier-Transform Analysis Method

J. Pavo and K. Miya

Nuclear Engineering Research Laboratory, Faculty of Engineering,
The University of Tokyo, Tokai, Ibaraki, 319-11, Japan

Abstract

A method is presented for the shape reconstruction of surface defects in conducting materials. The input data of the inversion is the normal component of the magnetic field measured in different locations between the exciting coil and the conducting surface. The actual two-dimensional defect shape is approximated by a trapezoid. Nonlinear optimization procedures are used for the inversion. First the so-called pre-inverted defect shape is calculated using Simulated Annealing optimizing procedure and a pre-calculated data base containing the magnetic field of different defect prototypes. The pre-inverted shape is improved by the use of the Downhill Simplex algorithm as local optimizer. Numerical examples are presented at the end of the paper.

1. INTRODUCTION

The determination of the characteristic data of a defect on a conducting surface is required in the solution of various practical engineering problems. Such kind of inverse problems have been under extensive investigation for some years. Usually eddy current is generated in the conducting material and the electromagnetic field is measured on/above the surface of the conductor. From this measured data the defect has to be characterized. The published reconstruction procedures use different calculation methods for the solution of the relating direct problem (e.g. FEM [1], method of volume integral equations [2], method of mapping [3,4], etc.). There are also different optimizing methods are used for the inversion but usually linear and first order optimization procedures are applied (e.g. [1,2]) It is difficult to compare the performance of the published methods because the reconstructed result depends on the quality of the used measurement data. Large numbers of reliable data can be the input of fast and accurate inversion methods, on the other hand, we cannot expect high accuracy if only a small number of noisy data is available.

In this paper a two-dimensional arrangement is investigated, although the presented method can be applied for three-dimensional configurations without the change of the applied theory. For the solution of the direct problem (i.e. the calculation of the magnetic field of the defect) the Fourier-transform method was used [5]. The inversion is made in two steps. First a fast and rough approximation of the defect shape is given using a Simulated Annealing optimizing procedure and a pre-calculated data base containing the magnetic field of properly chosen defect prototypes. The information obtained by this so-called pre-inversion can be used as initial points for individual local optimizers (Downhill Simplex algorithm). Using the results obtained by this local optimizers the reconstructed shape can be enhanced. The numerical examples presented at the end of the paper show the reliability of the method even if the input data is very noisy and measurements are given only in small numbers of points.

2. FIELD SIMULATION

The theory used for the calculation of the magnetic field generated by the interaction of the exciting coil and the defect was presented for three-dimension configurations in [5]. Here brief summary of this Fourier-transform method is given for two-dimensional configuration in order to help the explanation of the inversion procedure.

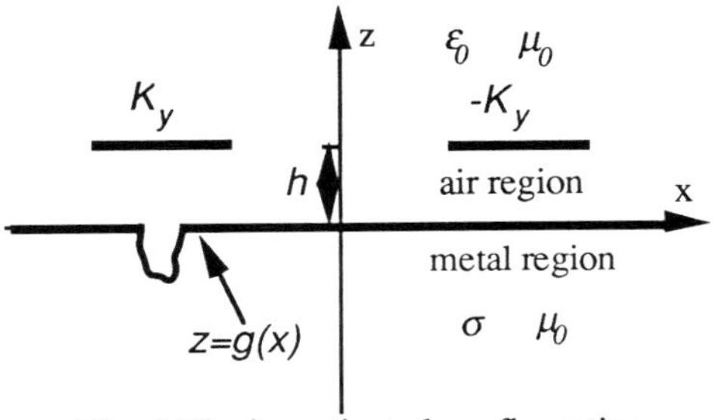

Fig. 1 The investigated configuration

The investigated configuration is shown in Fig. 1. The exciting coil represented by a current sheet in the $z=h$ plane with the surface current (K_y) flowing to the positive and negative y directions. For better coil modelling the number of current sheets can be increased without increasing the number of unknowns needed for the solution of the problem. The conductor-air interface is described by the $z=g(x)$ function.

Considering the Fourier-transform of the Helmholtz's and the wave equations in the conductor and air regions, respectively, and also taking into account the radiation conditions in the $z\to\infty$ and $z\to-\infty$ directions, the electromagnetic field can be described in the different regions of the configuration by the following four functions of the Fourier-transformed plane:

$$\mathcal{H}_o(\alpha) \text{ in } h<z; \quad \mathcal{H}_1^+(\alpha), \mathcal{H}_1^-(\alpha) \text{ in } g(x)<z<h; \quad \mathcal{H}_2(\alpha) \text{ in } z<g(x) \tag{1}$$

where α is the pair of the x co-ordinate variable in the Fourier-transform plane. These functions are determined by the interface conditions in the $z=h$ and $z=g(x)$ surfaces:

$$H_{o,x}(z=h)-H_{1,x}(z=h)=K_y; \quad E_{o,y}(z=h)-E_{1,y}(z=h)=0 \tag{2}$$

$$\overline{H}_1(z=g(x))\times\overline{n}-\overline{H}_2(z=g(x))\times\overline{n}=0; \quad E_{1,y}(z=g(x))-E_{2,y}(z=g(x))=0 \tag{3}$$

where *0, 1* and *2* indexes refer to the field components in the regions $h<z$, $g(x)<z<h$ and $z<g(x)$, respectively. The x and y subscripts mean the x and y components of the field vectors, respectively. $\overline{n}$ is the normal vector of the $z=g(x)$ surface.

Eqs. (2) and (3) are sufficient for the numerical evaluation of the functions listed in Eq. (1), consequently the electromagnetic filed can be evaluated everywhere. After taking the Fourier-transform of Eqs. (2), they can be solved analytically (that is the reason why the number of the current sheets can be increased without increasing the number of the unknowns). Eqs. (3) are integral equations that can be solved numerically using the point matching technique along the $z=g(x)$ line. By taking advantage of the sampling theory the proper distance between the sampling points used for the point matching method can be determined.

Once the functions listed in Eq. (1) are given by the solution of the integral equations (3), the magnetic field in each region can be calculated analytically in the form of Fourier-integral that can be evaluated numerically by FFT. For example, in the air region the z component of the magnetic field will be the following:

$$H_{1,z}(x,z)=\int_{-\infty}^{\infty}\frac{\alpha}{\gamma}\left[-\mathcal{H}_1^+(\alpha)\cdot e^{-j\gamma z}+\mathcal{H}_1^-(\alpha)\cdot e^{j\gamma z}\right]\cdot e^{-j\alpha x}\,d\alpha \tag{4}$$

where $\gamma=[\omega^2\mu_0\varepsilon_0-\alpha^2]^{1/2}$ if $\omega^2\mu_0\varepsilon_0>\alpha^2$ and $\gamma=-j[\omega^2\mu_0\varepsilon_0-\alpha^2]^{1/2}$ if $\omega^2\mu_0\varepsilon_0<\alpha^2$.

Functions (1) can be calculated analytically if there is no defect (i.e. if $g(x)=0$). Knowing this solution the integral equations to be solved in the presence of a defect can be modified to integral equations that contain the perturbation of the field (the difference of the electromagnetic field with and without defect) as unknowns. This is very important from the numerical point of view because the perturbation is concentrated to a small area in space.

In cases when narrow defects or cracks are considered difficulties arising with the presented calculations. The difficulties can be overcome by treating the defect as a proper boundary condition imposed on the plane of it. The above described theory can be applied for such configurations after small modifications.

3. INVERSION

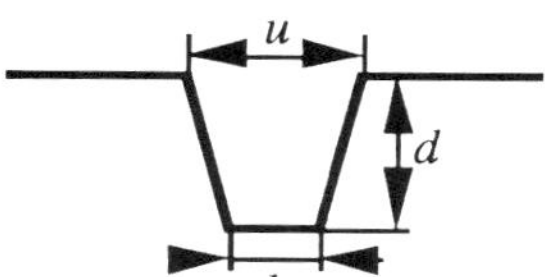

Fig. 2 The approximating defect shape

For the defect reconstruction the following assumptions are made:

•The arrangement shown in Fig. 1 is considered. The z component of the magnetic field perturbation (the difference of the magnetic field with an without defect) is measured above the defect in different points in the air region. The measurement is simulated by calculation.

•The location of the middle of the defect is known.

•The reconstructed defect is approximated by the shape shown in Fig. 2, i.e. it is described by three characteristic data (u, l, d).

The object function of the optimizing procedures used for the reconstruction is:

$$\sum_{f=0}^{N_f}\sum_{m=0}^{N_m}\frac{\left|H_z^{m,f(measured)}-H_z^{m,f(calculated)}\right|}{\left|H_z^{m,f(measured)}\right|} \tag{5}$$

where $H_z^{m,f(measured)}$ and $H_z^{m,f(calculated)}$ are the z components of the magnetic field in the mth measurement point taken on the fth measurement frequency that is measured and calculated by assuming the approximating defect shape, respectively. N_f and N_m are the number of measurement frequencies and the number of measurement points on one frequency, respectively.

The reconstruction is made in two successive steps. The first step is called pre-inversion. In this step a fast calculation is made to give a first approximation of the defect shape and to get initial points for further optimizations. The pre-inversion is in fact a search in a pre-calculated data base. The magnetic field of possible defect prototypes (covering the range of $1<u<10$, $0<l<10$, $0<d<2$ [mm]) are calculated and the values of the $\mathcal{H}_l(\alpha)$ functions in the sampling points are stored in the data base. Using these values the magnetic field of the given defect can be calculated analytically in the whole air region (see Eq. (4)). It is seen that only low amount of data is needed to be stored (e.g. in the presented examples only 64 complex values were used for the description of the magnetic field of one defect prototype). The search in the data base is made by Simulated Annealing optimizing procedure [6]. In most cases the result of the pre-inversion can be accepted as the final result of the reconstruction, although in the case of small numbers of noisy measurement data this pre-inverted result is not reliable.

In the second step of the reconstruction procedure individual local optimizers are started with the initial points selected by the pre-inversion step. Since the initial points are usually close to a local minima the typical number of function calls are between 30 and 50. Downhill Simplex algorithm [6] is used for optimization in this step. The results of this local optimizers can be processed statistically to improve the shape of the reconstructed defect. This second step of the reconstruction is specially useful when only small number and noisy data is available for the inversion.

4. NUMERICAL EXAMPLE

The results of defect shape reconstruction from 10 and 4 measurement data are shown in Figs. 3 and 4. The input data was the *z* component of the magnetic field perturbation due to the defect. 5, 15 and 25 percent random noise was added to the simulated measurement data. 10 local optimizers were started in each example. The results obtained by the individual optimizers were averaged. In this way it was possible to obtain as a final result not just trapezoidal but also other kind of symmetrical shapes. In Figs. 3 and 4 the original shape and the reconstructed shapes obtained from the 5, 15 and 25 percent noisy data were plotted.

It can be seen that the results are acceptable even in the case when 25% noisy data was given in 4 measurement points. Figs. 3 and 4 imply that the reconstruction method is robust and reliable.

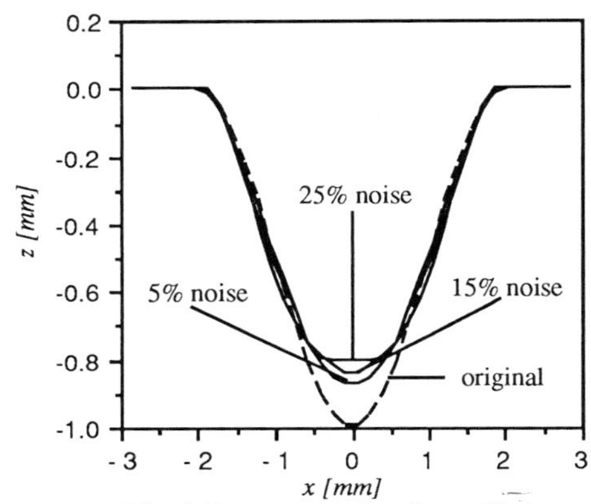

Fig. 3 Reconstruction from 10 data

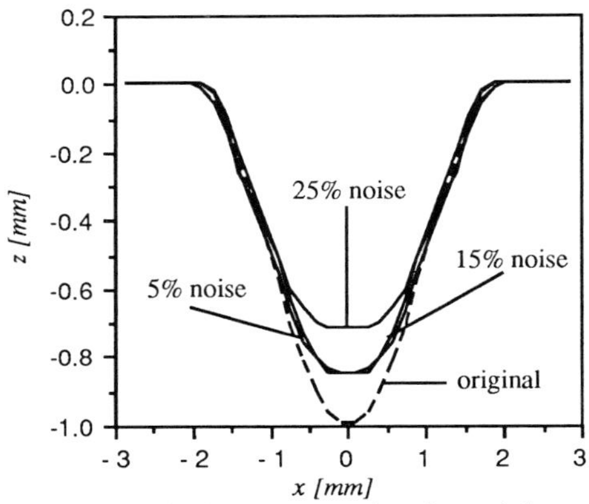

Fig. 4 Reconstruction from 4 data

5. CONCLUSIONS

A method was presented for two-dimensional defect reconstruction. The electromagnetic field of the defect was simulated by the Fourier-transform method. The reconstruction is made in two steps. In the first pre-inversion step a fast result of the approximating defect shape and initial points for further local optimizers are obtained. In the second step individual local optimizers are started using the initial points presented by the first step. The results of this optimizers can be used statistically to improve the reconstructed shape. The reliability of the inversion is shown by the examples.

6. REFERENCES

1 S.R.H. Hool et al., IEEE Trans. Magn., vol.MAG-27, no.3, (1991) 3433
2 L.D. Sabbagh and H.A. Sabbagh, J. Nondestructive Evaluation, vol.7, nos.1/2, (1988) 95
3 S.H.H. Sadeghi and D. Mirshekar-Syahkal, IEEE Trans. Magn.,vol.MAG-28, no.2 (1992) 1008
4 H. Takahashi and K. Miya, Electrosoft, vol.2. nos.2/3 (1991) 104
5 J. Pavo and K. Miya, J. Applied Electromagnetics in Materials (submitted)
6 W.H. Press et al., Numerical Recipes, Cambridge University Press, Cambridge, 1986

Simulation and Design of Applied Electromagnetic Systems
T. Honma (Editor)

An inverse analysis with the current vector potential T

Masaaki Kurokawa, Kanjiro Sugiyama and Kenzo Miya

Nuclear Engineering Research Laboratory, Faculty of Engineering,
University of Tokyo, 2-22 Shirakata-Shirane, Tokai-mura, Ibaraki, Japan

Abstract

A method of solving inverse problems of electromagnetic field is described. This method involves the current vector potential **T**. In this study, we solved direct problems to obtain magnetic field in the air region, which was used as input data for the inverse problem. From the results of the inversion, It is revealed that the position and length of through defects can be identified by the present method.

1. INTRODUCTION

Conventional eddy current testing (ECT) methods utilize the effects on coil impedance of variations in the conductance of the material adjacent to the probe coil. The presence of a defect in the material affects the conductance and, consequently, the coil impedance. The ECT generally use defect sizing algorithms based on data from calibration defects. Recently, an improvement of current ECT is one of the most important and interesting problems in electromagnetic nondestructive evaluation to obtain more informations of defects. Therefore, several computational models have been proposed so far[1,2,3]. In this study, we introduce an inverse method based in term of the current vector potential **T** and show some numerical results. Magnetic induction **B** in the air region used in the inverse problem is calculated by newly developed three dimensional eddy current code based on **A**-ϕ formulation.

2. FORMULATION OF DIRECT PROBLEM

The governing equations that describe the eddy current in conductors are written as follows,

$$\frac{1}{\mu_0}\nabla^2\mathbf{A} = \sigma\left(\frac{\partial\mathbf{A}}{\partial t} + \nabla\phi\right), \qquad \nabla\cdot\sigma\left(\frac{\partial\mathbf{A}}{\partial t} + \nabla\phi\right) = 0 \tag{1),(2}$$

For an external region where a current source exist, the governing equation is given by

$$\frac{1}{\mu_0}\nabla^2\mathbf{A} = -\mathbf{J}_f \tag{3}$$

where $\mathbf{A}$, ϕ and $\mathbf{J}_f$ are the magnetic vector potential, the electric scalar potential and the source current density. In this study we deal with an AC problem, so eqs.(1) and (2) are rewritten for sinusoidal field using imaginary unit, j and angular frequency, ω.

$$\frac{1}{\mu_0}\nabla^2\mathbf{A} = \sigma(j\omega\mathbf{A}+\nabla\phi), \quad \nabla\cdot\sigma(j\omega\mathbf{A}+\nabla\phi)=0 \qquad (4),(5)$$

Eqs.(3), (4) and (5) were discretized by BEM and FEM, respectively[4].

3. FORMULATION OF INVERSE PROBLEM

From Helmholtz theorem, we can obtain the following equation for the current vector potential $\mathbf{T}$.

$$C(p)\mathbf{T}(p) = \frac{1}{4\pi}\left[\int_v(\nabla\cdot\mathbf{T}+\nabla\times\mathbf{T}\times)\nabla\frac{1}{R}dv - \int_s(\mathbf{n}\cdot\mathbf{T}+\mathbf{n}\times\mathbf{T}\times)\nabla\frac{1}{R}ds\right] \tag{6}$$

where, coefficient C(p) takes

C(p)=1 :where point p is included in the region;
C(p)=0.5 :where point p is located on the boundary;
C(p)=0 :where point p is located outside the region under consideration.

Applying the Coulomb gauge($\nabla\cdot\mathbf{T}=0$) and boundary condition ($\mathbf{n}\times\mathbf{T}=0$, on surface), eq.(6) can be rewritten as follows.

$$\mathbf{B} = \mu_0 C(p)\mathbf{T}(p) + \frac{\mu_0}{4\pi}\int_s(\mathbf{n}\cdot\mathbf{T})\nabla\frac{1}{R}ds \tag{7}$$

When p is located outside the conductor region, C(p) becomes zero.

$$\mathbf{B} = \frac{\mu_0}{4\pi}\int_s T_n\nabla\frac{1}{R}ds \tag{8}$$

where T_n represents normal component of $\mathbf{T}$. By discretizing the above equation, we can obtain the following vector matrix equation.

$$\{B\} = [A]\{T_n\} \tag{9}$$

where $\{T_n\}=\{T_{n1}...T_{nn}\}$ is the unknown vector, $\{B\}=\{B_1..B_m\}$ is the data vector, [A] the {m x n} system matrix. In order to obtain least squares solutions of eq.(9), we used modified Gram-Schmidt method which belongs to the QR factorization, together with the thin plate approximation.

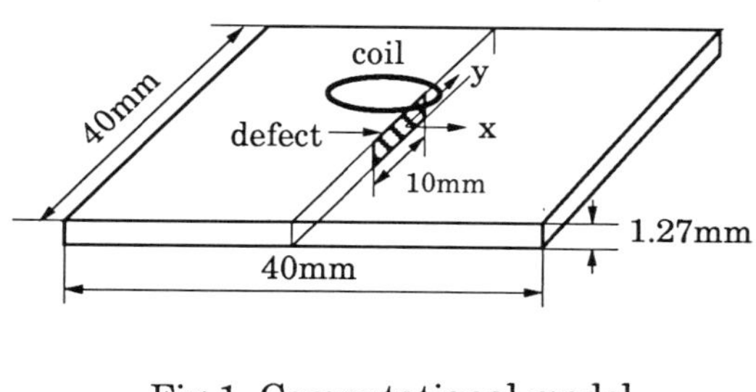

Fig.1 Computational model.

4. NUMERICAL RESULTS

Table 1. Parameters of calculations

The coil	
Radius	:1.5mm
Lift off:	:1.0mm
Frequency:	:100KHz
The test specimen	
Rectangular thin plate (40mm x 40mm x 1.27mm)	
Conductivity	: 1.0x10^6
The defect	
Straight, Cross-shape H-shape, T-shape	
Depth: through-thickness	
Width: 0.2mm	

The calculation model is shown in Fig.1. We considered a rectangular thin plate with defects. One turn coil with 100kHz frequency was located 1mm above the center of the plate surface. The conductivity and the thickness of the plate are similar to those of a tube used in a steam generator of nuclear power plants. In direct problems, for the symmetry, one half of the plate was considered. The total elements and nodes were about 500 and 800, respectively. The defect was treated as a low conductive region and included in the FEM region. Three components of magnetic induction B were calculated at 225 (15 x 15) points located 1mm above the plate surface. They were used as input data in the inverse problems. Calculations were carried out for 4 cases of defects listed in Table.1. Fig.2 shows arrangements and lengths of straight, T-shape, Cross-shape and H-shape defects. In the inversion, number of unknown Tn is 169 (13 x 13) in every case. Figs.3-4 show the reconstructed Tn distribution as a contour map on the plate for 4 cases of defects. We can recognize the crack shapes from reconstructed Tn shown in these figures.

5. CONCLUSIONS

1 Three dimensional eddy current code based on **A**-ϕ formulation was successfully developed and applied to obtain magnetic field as input data for the inverse problems.

2 It is revealed that the position and length of through defects can be identified by the present inversion method even for complex arrangements of defects.

6. REFERENCE

1 K. Miyata, K. Miya and T. Sugiura, 'Crack Shape Recognition Based on T-Method', Proc. 9th. Int. Conf. NDE in the Nuclear Industry, Tokyo, 1988, 25-28.

2 H.A Sabbagh and D. Sabbagh, 'Eddy-current model for Three-Dimensional Inversion', IEEE Trans., Mag 22, 1986, 282-291.

3 T. Takagi et al., 'An inverse approach to Eddy current Testing using current Vector Potential', Proc. 1st. Japan-Czechoslovak Joint Seminar, 1992, 29-32.

4 F. Matsuoka, 'Calculation of A Three Dimensional Eddy Current by The FEM-BEM Coupling Method.', IUTAM Proc, Tokyo, 1987, 169-174.

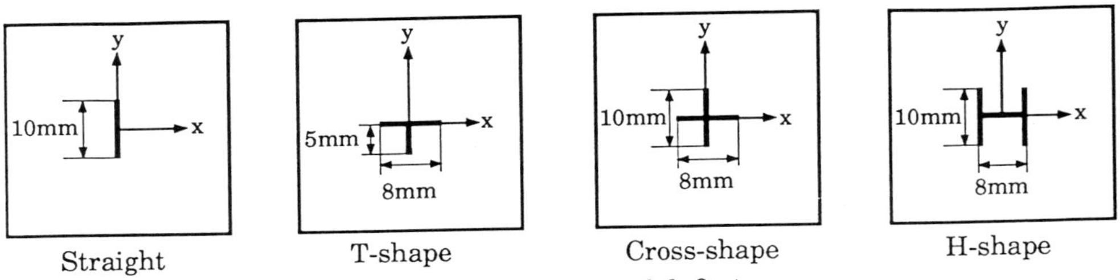

Fig.2 Configurations of defects.

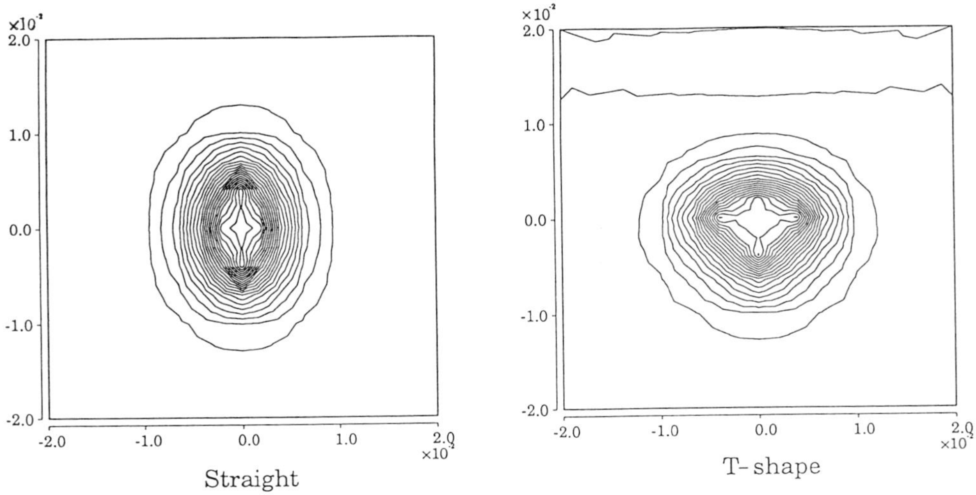

Fig.3 Reconstructed T_n for the straight and T-shape defects.

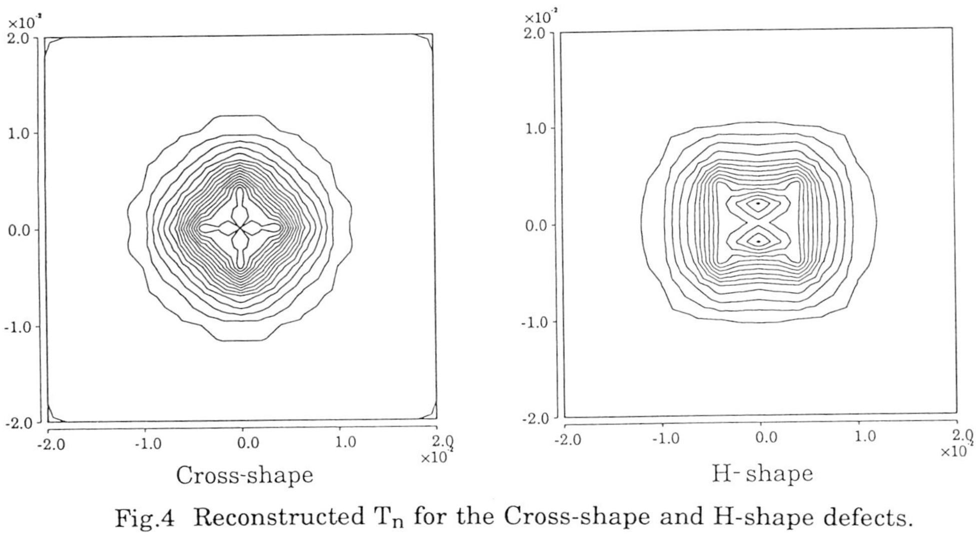

Fig.4 Reconstructed T_n for the Cross-shape and H-shape defects.

Simulation and Design of Applied Electromagnetic Systems
T. Honma (Editor)

The locally orthogonal coordinate systems for inverse problem analysis

H. Saotome and Y. Saito
College of Engineering, Hosei University, 3-7-2 Kajino, Koganei, Tokyo 184, Japan

Abstract

Solving the inverse problem of magnetic field requires a large quantity of computation because solutions of this problem, i.e. current element (or dipole) vectors, are not only a function of their positions but also their directions. This paper proposes locally orthogonal coordinate systems with respect to magnetic field patterns on a measurement surface, such as the cylindrical and spherical coordinate systems, which can remove a repetitive calculation for determining the spatial angles of current dipole vectors. As a result, we have succeeded in realizing an efficient methodology for the inverse problems.

1. INTRODUCTION

Analyzing inverse problems is one of the most crucial future works or business in engineering because it can provide solutions for identification, optimization, automatic designing by computers, non-destructive testing, medical diagnosis and so forth. We have previously proposed a promising method called the sampled pattern matching (SPM) method for solving inverse problems in electromagnetic fields [1-6]. In the magnetostatic field, the inverse problem of obtaining current distribution from a locally measured magnetic field is carried out by estimating current element (or dipole) positions as well as the direction of each current dipole vector. Evaluating a spatial angle of each current dipole vector needs a repetitive process depending on the required angle resolution. For example, if five-degree resolution on a plane is required, then the SPM process of 72 times is repetitiously necessary at each of positions in order to cover the entire angle 360 degrees. In the previous studies, we have inevitably carried out the repetitive process with the Cartesian coordinate system [1-6]. However, this algorithm requires considerable CPU time.

The purpose of this paper is to find out the optimal coordinate systems that can remove the repetitive process for the current dipole angle evaluation. As a result, the locally orthogonal coordinate systems with respect to magnetic field components on a measurement surface, such as the cylindrical and spherical coordinate systems, are proposed in order to reduce the CPU time due to the current dipole angular division. This is because the magnetic field pattern on the measurement surface of these coordinate systems can be decomposed into two orthogonal field pattern components whereas it cannot be done in the Cartesian coordinate system.

2. INVERSE PROBLEMS OF HOMOGENEOUS MAGNETOSTATIC FIELD

2.1. Formulation

In the homogeneous open boundary magnetostatic field, the relation between the magnetic field intensity **H** and its source, the current density **J**, is expressed by a volume integral:

$$\mathbf{H} = \nabla \times \int_{v} G\, \mathbf{J}\, dv, \tag{1}$$

where G is the Green function depending on the geometric relation between **H** and **J**.

In (1), the product $\mathbf{J}\,dv$ corresponds to the current dipole vector.

Employing the cylindrical coordinate system as shown in Figure 1, the magnetic field intensity H_z normal to the measurement surface can be decomposed into two components H_{zr} and $H_{z\theta}$:

$$H_z = H_{zr} + H_{z\theta}, \tag{2}$$

where H_{zr} and $H_{z\theta}$ are caused by r and θ components of the current dipole vector $\boldsymbol{\alpha}$ (= $\mathbf{J}$ dv) shown in Fig. 1, respectively. With unit vectors of the cylindrical coordinates, $\mathbf{e}_r$, $\mathbf{e}_\theta$ and $\mathbf{e}_z$, we have

$$H_{zr} = \frac{(\alpha_r\ \mathbf{e}_r)\times\mathbf{s}}{4\pi s^3}\cdot\mathbf{e}_z = \frac{\alpha_r\ \xi}{4\pi s^3}, \tag{3a}$$

$$H_{z\theta} = \frac{(\alpha_\theta\ \mathbf{e}_\theta)\times\mathbf{s}}{4\pi s^3}\cdot\mathbf{e}_z = \frac{\alpha_\theta\ \eta}{4\pi s^3}, \tag{3b}$$

$$\alpha_r = \boldsymbol{\alpha}\cdot\mathbf{e}_r, \tag{3c}$$

$$\alpha_\theta = \boldsymbol{\alpha}\cdot\mathbf{e}_\theta, \tag{3d}$$

$$s = |\mathbf{s}|, \tag{3e}$$

where the distance vector $\mathbf{s}$, ξ and η are shown in Fig. 1. Moreover, we obtain

$$\int_P H_{zr}\ H_{z\theta}\ dS = 0, \tag{4}$$

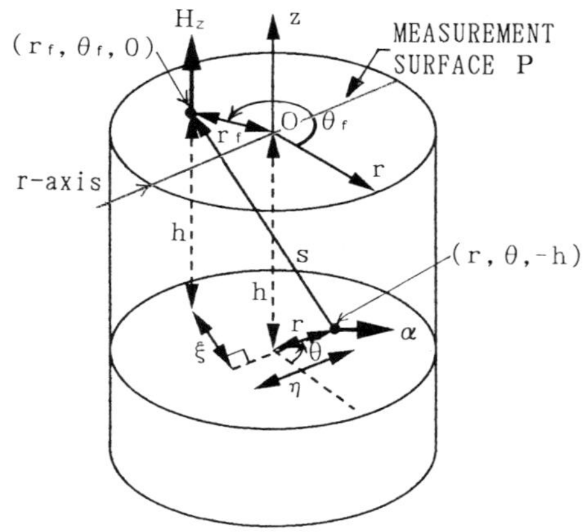

Figure 1. Cylindrical coordinate system.

where P refers to a circular measurement surface area shown in Figure 1. This is caused by the fact that the magnetic field patterns composed of H_{zr} and $H_{z\theta}$ in Figure 1 are odd and even symmetrical functions with respect to the r-axis, respectively.

Similarly, the following formulation can be carried out with the spherical coordinate system shown in Figure 2. The magnetic field intensity normal to the spherical measurement surface in Figure 2, H_r, can be represented by

$$H_r = H_{r\theta} + H_{r\phi}, \tag{5}$$

where $H_{r\theta}$ and $H_{r\phi}$ are caused by θ and ϕ components of the current dipole vector $\boldsymbol{\alpha}$ shown in Figure 2, respectively. Using the distance vector $\mathbf{s}$, unit vectors $\mathbf{e}_r$, $\mathbf{e}_\theta$ and $\mathbf{e}_\phi$ in the spherical coordinate system shown in Figure 2, we get

$$H_{r\theta} = \frac{(\alpha_\theta\ \mathbf{e}_\theta)\times\mathbf{s}}{4\pi s^3}\cdot\mathbf{e}_r, \tag{6a}$$

$$H_{r\phi} = \frac{(\alpha_\phi\ \mathbf{e}_\phi)\times\mathbf{s}}{4\pi s^3}\cdot\mathbf{e}_r, \tag{6b}$$

$$\alpha_\theta = \boldsymbol{\alpha}\cdot\mathbf{e}_\theta, \tag{6c}$$

$$\alpha_\phi = \boldsymbol{\alpha}\cdot\mathbf{e}_\phi, \tag{6d}$$

and

$$\int_P H_{r\theta}\ H_{r\phi}\ dS = 0, \tag{7}$$

where P refers to a spherical measurement surface area shown in Figure 2.

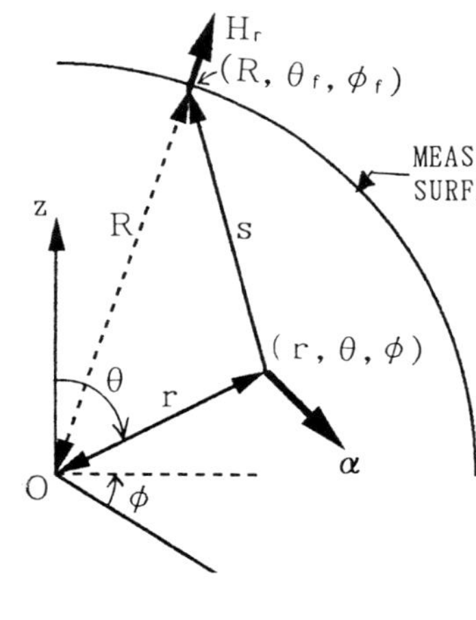

Figure 2. Spherical coordinate system.

Discretizing (1) into m volume subdivisions gives a system equation of the inverse problem:

$$\mathbf{u} = \sum_{j=1}^{m} (\alpha_{pj}\ \mathbf{d}_{pj} + \alpha_{qj}\ \mathbf{d}_{qj}); \tag{8}$$

p=r, q=θ for the cylindrical coordinate system;
p=θ, q=ϕ for the spherical coordinate system;

where the elements of $\mathbf{u}$ are measured magnetic field intensities normal to the measurement surface, α_{pj} and α_{qj} are p and q components of a current dipole vector at the point j. The vectors $\mathbf{d}_{pj}$ and $\mathbf{d}_{qj}$ show the estimated magnetic field patterns on the measurement surface due to α_{pj} and α_{qj}, respectively.

2.2. Faster SPM method

The most notable point of our new formulation is the orthogonality between two magnetic field component patterns expressed by (4) and (7). This orthogonality is not satisfied in the commonly used Cartesian coordinate system. Equations (4) and (7) give

$$\mathbf{d}_{pj}^{T} \cdot \mathbf{d}_{qj} = 0; \quad j=1,2,\cdots,m. \tag{9}$$

This relation makes it possible to independently evaluate the estimated magnetic field pattern matching rates of p and q component patterns with respect to the measured magnetic field pattern. Namely,

$$\gamma_{pj} = \mathbf{u}^{T} \cdot \mathbf{d}_{pj} \,/\, (\|\mathbf{u}\|\ \|\mathbf{d}_{pj}\|), \tag{10a}$$

$$\gamma_{qj} = \mathbf{u}^{T} \cdot \mathbf{d}_{qj} \,/\, (\|\mathbf{u}\|\ \|\mathbf{d}_{qj}\|), \tag{10b}$$

where j=1,2,$\cdots$,m. Therefore, current dipole searching is carried out by finding out the maximum of

$$\gamma_{j} = \sqrt{\gamma_{pj}^{2} + \gamma_{qj}^{2}}; \quad j=1,2,\cdots,m. \tag{10c}$$

After finding out the first current dipole vector, we continue with the procedures similar to (10a)-(10c) up to the peak of γ [1-6]. Making the most of (9), the SPM algorithm is carried out only in p and q directions instead of taking all the divided spatial angles of a current dipole into account.

3. Examples

Applying the proposed formulation, current dipole distribution is quickly

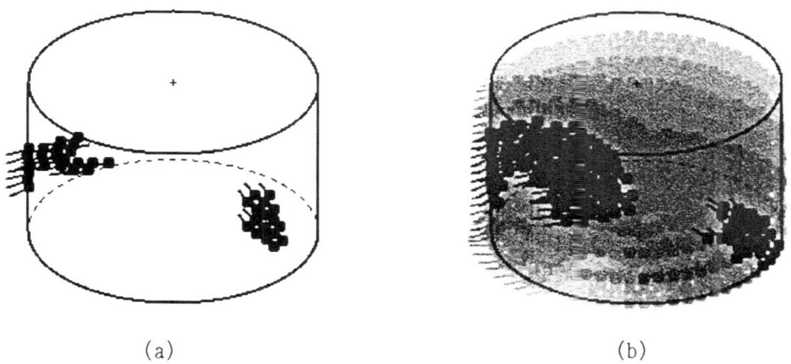

(a) (b)

Figure 3. Current dipole distributions in the cylindrical coordinate system. (a)Correct distribution; (b)estimated distribution (n=37, m=2046).

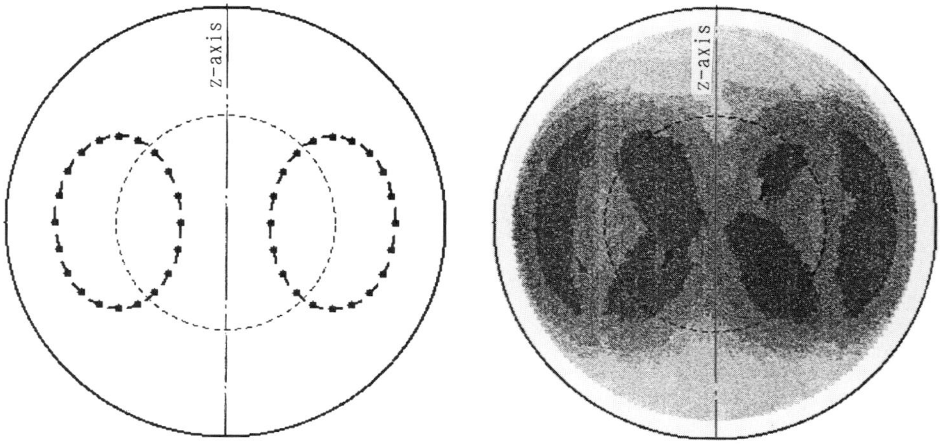

(a) (b)

Figure 4. Current dipole distributions in the spherical coordinate system. (a)Correct distribution; (b)estimated distribution (n=407, m=105245).

obtained without the CPU time required to determine each angle of current dipoles. Figure 3 shows correct and estimated current dipole distributions obtained by employing the new method with the cylindrical coordinate system. Figure 4 shows an example of correct and estimated current distributions in the spherical coordinate system. The field source searching region of this spherical example ranges in radial direction from 0.5 to 0.9 regarding a radius of the sphere as one. The number of measurement points, n, and the number of unknowns, m, are n=37, m=2046 in Fig. 3 and n=407, m=105245 in Fig. 4, respectively.

4. CONCLUSION

As shown above, we have formulated the inverse problem of magnetostatic field by means of the locally orthogonal coordinate systems. This enables us to obtain current dipole distribution faster than the conventional algorithm employing the Cartesian coordinate system. The novel algorithm proposed in this paper leads to fast medical diagnosis with a magnetocardiogram or a magnetoencephalogram.

5. REFERENCES

1 Y.Saito, et al., J. Appl. Phys., 67, No.9, (1990), pp.5830-5832.
2 H.Saotome, et al., Trans. IEE, Japan, Vol.112-A, No.4, (1992), pp.279-286.
3 H.Saotome, et al., Trans. IEE, Japan, Vol.113-C, No.1, (1993), pp.101-108.
4 H.Saotome, et al., A Neural Behavior Estimation by the Generalized Correlative Analysis, IEEE Trans. Magn., March, (1993), in Press.
5 H.Saotome, et al., Crack Identification in Metallic Materials, IEEE Trans. Magn., March, (1993), in Press.
6 H.Saotome, et al., Electromagnetic Field Source Searching from the Local Field Measurement, Int. J. Appl. Electromag. Matrls, (1993), in Press.

Simulation and Design of Applied Electromagnetic Systems
T. Honma (Editor)

A designing strategy based on the inverse analysis

S.Hayano, T.Katoh and Y.Saito

College of Engineering, Hosei University, Kajino, Koganei, Tokyo 184, Japan

Abstract
This paper proposes a new designing strategy based on the inverse analysis approach. As a concrete example, we try to decide the size and arrangement of the exciting coils in order to realize a locally concentrated magnetic field distribution along by the center axis for the local induction heating and hyperthermia use. As a result, it is found that the sampled pattern matching method previously proposed by us is a quite useful methodology for obtaining the desired magnetic field distributions.

1. INTRODUCTION

Conventional designing strategy of the magnetic devices, for example MRI magnet, magnetic shielding and electric machines, is that the desired magnetic field distribution is iteratively evaluated by solving a governing equation with electric current given condition.

On the contrary, in the present paper, the electric current distribution is estimated by the sampled pattern matching (SPM) method with the magnetic field distribution given condition [1-5]. This means that one of the designing strategies is proposed here by means of the inverse analysis approach. Conventional approach to the regular or forward problems yields a unique solution so that an iterative approach is essentially required to reaching a final goal of the designing. On the other side, our inverse approach based on the SPM method provides a unique solution pattern [2-5]. This means that our inverse approach allows some degree of the designer's creative freedom. As a concrete example, we try to decide the layout of the exciting coils in order to realize a locally concentrated magnetic field distribution along by the center axis for the hyperthermia and local induction heating use. As a result, it is found that our SPM method is a quite useful methodology for obtaining the desired magnetic field distributions.

2. THE INVERSE ANALYTICAL APPROACH FOR THE MAGNETIC DEVICES DESIGNING

2.1 Basic equations

Since most of the magnetic devices are composed of the loop windings, then let us consider a current filament in the form of a circle of radius R carrying the current I on the r-z cylindrical coordinate system. We are to find the magnetic field H at a general position of r and z. In the given case, the radially directed magnetic field H_r and the z-axis directed magnetic field H_z are respectively given by

$$H_r = \frac{I}{2\pi r} \frac{z}{\sqrt{(R+r)^2+z^2}} \left[\frac{R^2+r^2+z^2}{(R-r)^2+z^2} E(\kappa) - K(\kappa)\right] = If(r,z), \tag{1}$$

$$H_z = \frac{I}{2\pi} \frac{1}{\sqrt{(R+r)^2+z^2}} \left[\frac{R^2-r^2-z^2}{(R-r)^2+z^2} E(\kappa) + K(\kappa)\right] = Ig(r,z), \tag{2}$$

where $K(\kappa)$ and $E(\kappa)$ are the complete elliptic integrals of the first and second kinds, respectively [6]. Also κ in these integrals is defined by

$$\kappa = \sqrt{4rR/[(r+R)^2+z^2]}. \tag{3}$$

Let us consider a problem that the magnetic fields H_{r1}, H_{r2}, $\cdot$, H_{rn}, H_{z1}, H_{z2}, $\cdot$, H_{zn} caused by the m-th unknown loop currents I_1, I_2, $\cdot$, I_m are given quantities and we have to evaluate the m-th unknown loop currents with the condition $n \neq m$. Then by means of (2), it is possible to write a following system equation:

$$U = \sum_{i=1}^{m} I_i d_i, \tag{4}$$

where

$$U = [H_{r1}, H_{r2}, \cdot, H_{rn}, H_{z1}, H_{z2}, \cdot, H_{zn}]^T, \tag{5a}$$

$$d_i = [f(r_1,z_1), f(r_1,z_2), \cdot, f(r_n,z_n), g(r_1,z_1), g(r_1,z_2), \cdot, g(r_n,z_n)]^T, \tag{5b}$$

$i=1,2, \cdot, m$.

2.2 Sampled pattern matching method

<a>Algorism The formal algorism of our SPM method is as follows.

At first, we calculate a pattern matching figure γ_i using the column vector d_i of (5b) and given vector U of (5a), and then find the maximum pattern matching figure, i.e. if a point h takes the maximum of

$$\gamma_i = U^T \cdot d_i / [|U||d_i|], \quad i=1,2,\cdot,m, \tag{6a}$$

then we call h as a first pilot point and associated vector $\mathbf{d}_h$ is called the first pilot pattern vector.

The second step is carried out by combining the first pattern vector $\mathbf{d}_h$ and remaining pattern vectors in (5b), i.e. we search for the maximum of

$$\gamma_{hj} = U^T \cdot (d_h+d_j) / [|U||d_h+d_j|], \quad j=1,2,\cdot,m; \; j\neq h. \tag{6b}$$

If a point g takes the maximum in (6b) then g is the second pilot point and $\mathbf{d}_g$ is the second pilot pattern vector. Similar processes are continued up to the peak pattern matching figure γ.

<b>**Solutions** Generally, the number of equations n is not equivalent to the number of unknowns m so that it is difficult to obtain a unique solutions I_i $(i=1,2,\cdot,m)$ in (4). Obviously, Our SPM algorism of (6a) and (6b) has been based on the following assumptions.

At first, the solution I_i ($i=1,2,\cdot,m$) in (4) takes the value of 1 or 0. This means that the pilot points take the unit value 1 and the other remaining points take the value 0. Secondly, each of the magnitudes I_i ($i=1,2,\cdot,m$) in (4) could not be evaluated uniquely but is represented by the concentrating ratio of the unit value 1 in space. In the other words, a region comprising of a large number of unit currents represents a large current flowing region and a region comprising of a small number of unit currents represents a small current flowing region.

<c>**Theoretical background** Mathematically, the pattern matching figure γ_i in (6a) corresponds to the Cauchy-Schwarz relation so that the angle θ_i between the vectors U and d_i in a linear space is given by

$$\theta_i = \cos^{-1}[\gamma_i]$$
$$= \cos^{-1}\{U^T \cdot d_i / [|U||d_i|]\}, \quad i=1,2, \cdot, m. \tag{7}$$

Obviously, (7) shows that the maximum value of γ_i corresponds to the minimum angle θ_i. Thereby, the maximum value of γ_i indicates which vector d_i $(i=1,2, \cdot, m)$, i.e. input source position, coincides with the vector U regardless of the input source magnitude.

In statistically, the operation of pattern matching corresponds to the factor analysis. Conventional factor analysis is always carried out only to the first step of our sampled pattern matching method. Second and third steps of our sampled pattern matching process are obviously generalization of the conventional factor analysis. Therefore, our sampled pattern matching method may be regarded as one of the generalized factor analysis methods.

The other theoretical background is the incomplete Fourier series. Equation (4) means that the vector U is a linear combination of the vectors $d_i\,(i=1\sim m)$. Thereby, (4) can be modified into

$$U=\sum_{i=1}^{m}\{\beta_i d_i+\sum_{j\neq i}^{m}\{\beta_i(d_i+d_j)+\sum_{k\neq i,k\neq j}^{m}\{\beta_i(d_i+d_j+d_k)+\cdots\}\}\}. \tag{8}$$

Each of the terms in (8) has its distinct field pattern. Namely, the 1st, 2nd and 3rd groups on the right give the one, two and three pairs of the N(north)-S(south) magnetic pole pairs, respectively This means that the SPM method is one of the Fourier analysis methods if the vectors on the right of (8) are the orthogonal functions. However, they are not necessarily always the orthogonal function so that the SPM method is a kind of incomplete Fourier series [3].

2.3 An example

In order to demonstrate a concrete example of our designing strategy, let us consider a realization of a device giving a locally high magnetic field distribution along the center axis. Our problem is to realize the magnetic field distribution shown in Fig. 1(a) by the symmetrically arranged four winding. Figure 1(b) shows a schematic diagram of the device. The conditions of designing this device are as follows:

1) Target region is the 10cm line along with the z-axis in Fig. 1(b).
2) Each of the windings consists of the six loop coils.
3) The radius of each loop coil can be changed from 0.7 to 1.0cm.
4) The current of each loop coil has to take the same value.
5) The angle ϕ to the z-axis shown in Fig. 1(b) has to take the value within 30~80 degree.

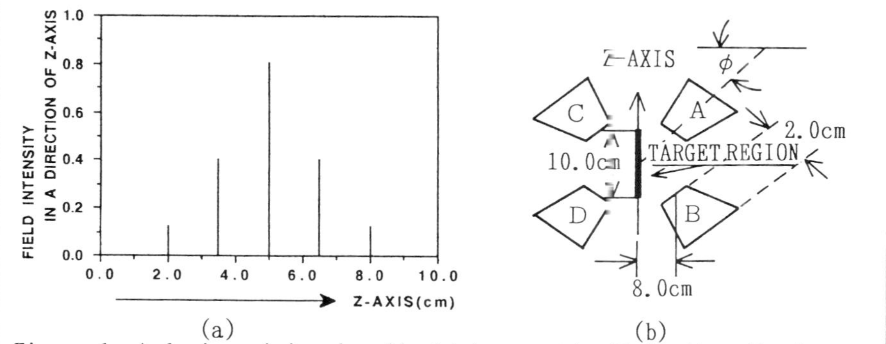

Figure 1. A device giving locally high magnetic field distribution. (a) Desired magnetic field distribution in the direction of z-axis. (b) Schematic diagram of the device.

The first step is to decide the most reasonable winding angle ϕ and the radius of each loop coil. The algorism of this step is as follows:

1) Set a winding angle ϕ_i within 30~80 degree.
2) Select one of the six loop coils of each winding.
3) Set a radius r_i of loop coil within 0.7~1.0cm.
4) Calculate the pattern matching figure γ_i by 6a) using the given field

vector U in Fig. 1(a) and pattern vector d_i in (5b).

5) Decide each of the six coil radii by selecting the maximum value of pattern matching figure.
6) Decide the most reasonable angle ϕ by selecting a loop coil taking the maximum value of pattern matching figure.

The second step is to decide which coil should be connected to the current source. This process was also carried out by the SPM procedure of (6a) and (6b) using the given field vector U in Fig. 1(a) as well as pattern vector d_i in (5b).

As a result, the magnetic field distribution shown in Fig. 2(a) was obtained. Figure 2(b) shows the schematic diagram of finally designed device. To summarizing our designing strategy, at the first step, the electric current distribution was estimated by the SPM method with the given magnetic field distribution. This lead to the position and shape determination of winding. At the second step, the coil connection was also determined by the SPM method in order to fit in excellently the given magnetic field distribution. Thus, our designing strategy proposed here needs not an iterative process. Obviously, a definitive difference between the conventional and our designing strategies is caused by the forward or inverse analytical approach.

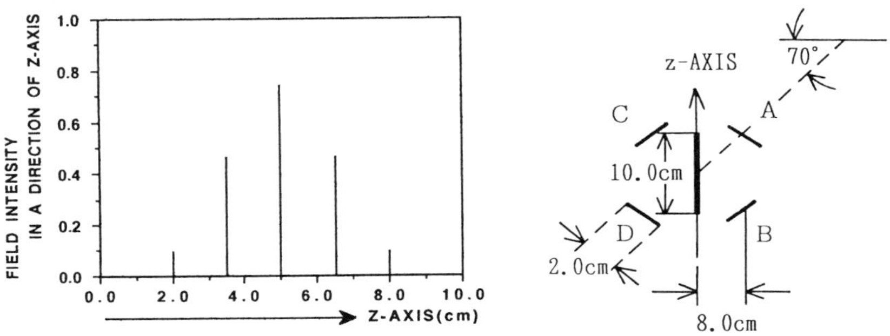

(a) (b)

Figure 2. Results of the designing.
(a) Obtained field distribution corresponding to Fig. 1(a).
(b) Schematic diagram of the designed device.

3. CONCLUSION

As shown above, we have clarified that a completely different designing strategy to the conventional one is possible by means of the inverse analytical approach. One of the best merits of this new approach is that a fairly good result can be expected even if no iteration is carried out.

Thus, it has been suggested that the inverse analytical approach makes it possible to design the electromagnetic devices in a highly efficient manner.

4. REFERENCES

1 Y.Saito, et al., J. Appl. Phys., 67, No.9 pp.5830-5832, (1990).
2 H.Saotome, et al., Trans. IEE, Japan, Vol.112-A, No.4, pp.279-286, (1992).
3 H.Saotome, et al., Trans. IEE, Japan, Vol.113-C, No.3, (1993), in Press.
4 H.Saotome, et al., IEEE Trans. Magn., March, (1993), in Press.
5 H.Saotome, et al., IEEE Trans. Magn., March, (1993), in Press.
6 M.Stafl, Electrodynamics of Electrical Machines, (Iliffe Books Ltd., London, 1967), pp.124-126.

Simulation and Design of Applied Electromagnetic Systems
T. Honma (Editor)

Application of Genetic Algorithm to Inverse Estimation of Current Distribution from Magnetic Fields

Maki Kishimoto, Kaoru Sakasai and Katuyuki Ara

Japan Atomic Energy Research Institute
Tokai-mura, Naka-gun, Ibaraki-ken,319-11,Japan

Abstract

Genetic algorithm is a search algorithm based on the mechanism of evolution of living organisms. This method was applied to inverse estimation of current distribution from induced magnetic fields with an intention to develop a non-distructive beam profile monitor for intense ion accelerators. Case studies with computer simulations showed its effectiveness.

1. Introduction

Genetic Algorithm (GA) is a new type algorithm on the basis of the mechanism of evolution of living organisms [1]. Normally, most organisms adjust themselves to the environment through the processes of selection, reproduction and mutation. GA is search and optimization algorithm that utilizes these mechanisms of organisms and has wider application than do the conventional ones. The feature of GA is easy to avoid local minimum problem because GA searches from a population of points simultaneously. In this paper, the authors present the method for inverse estimation of current distribution from induced magnetic fields by use of GA and inspect the feasibility of application of it to the beam profile measurement for intense ion accelerators.

2. Beam profile monitor for intense ion accelerators

In the Japan Atomic Energy Research Institute (JAERI), the development of the intense proton accelerator (1.5GeV, 10mA) for the OMEGA project (Options Making Extra Gains of Actinides and Fission Products has been continued. The conventional type of beam profile measuring methods adopt the measurement of the signals from the strip lines or of fluorescence from the fluorescent screen inside the beam vacuum pipe [2]. But, for intense proton beam these invasive measuring methods can not be adopted and new beam measuring instruments need to be developed.

The authors have been studying about the selective magnetic field detection with high-temperature superconductor [3]: that method can be applied to the new type of non-invasive beam profile monitor. The schematic diagram of the beam profile monitor is shown in Fig.1. This beam monitor consists of two high temperature superconductor disks, which are seperated with a gap of very short distance.

The magnetic sensors are settled in that gap so as to surround the beam vacuum pipe. By the magnetic shield effect of these disks due to perfect diamagnetism of superconductor, the magnetic field induced from ion beam in the vicinity of the superconductor disks can be detected selectively and in high sensitivity. Then, the current distribution of ion beam is estimated from the magnetic fields measured by this instrument by GA.

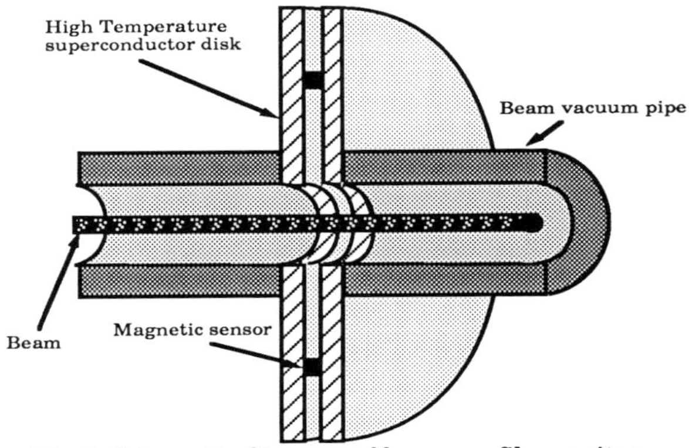

Fig.1. Schematic diagram of beam profile monitor

3. Application of genetic algorithm to inverse problem

We consider the coss-section of beam vacuum pipe and that the currents flows in the direction orthogonal to it. The cross-section is divided into small segments and then the magnetic field at a measuring point is represented by the sum of the magnetic fields induced from the current of each segment. The estimation problem of the current distribution as a source of magetic field can be regarded as the optimum allocation problem of the currents in the cross-section of beam vacuum pipe. In other word, it is to arrange the currents in the cross-section so that the magnetic field calculated from them can be equal at each measuring point to the measured.

3.1 Design of the gene

A research group of Prof. Y.Uchikawa of Nagoya university tried to apply GA to estimate an area where electric currents flow uniformly in a conductor [4]. In application of beam profile estimation, the authors have introduced an unique trial; introduction of " multi-value gene" in the definition of the gene.

We divide the current distribution area into M×N segments and define the coordinate system as shown in Fig.2. Here, the current Iij which flows through the segment (i,j) is represented as :

$$I_{ij} = n_{ij} J_u \, dxdy$$

where, Ju is unit current density and nij is an integer from 0 to 255, which represents the strength of the current flowing through the segment (i,j).

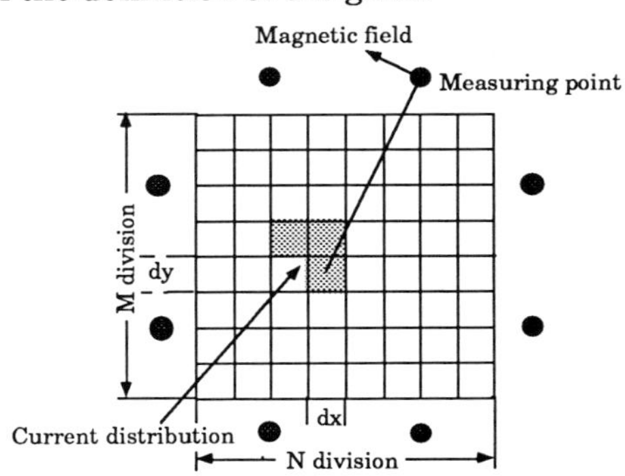

Fig.2. System of Measurement

Then, the state of the current distribution area is represented by the M×N state matrix which element is the current strength nij. Then, We let one state matrix correspond to one gene which has M×N multi-value element. In Fig.3, we show the example of current distribution, its state matrix and the population.

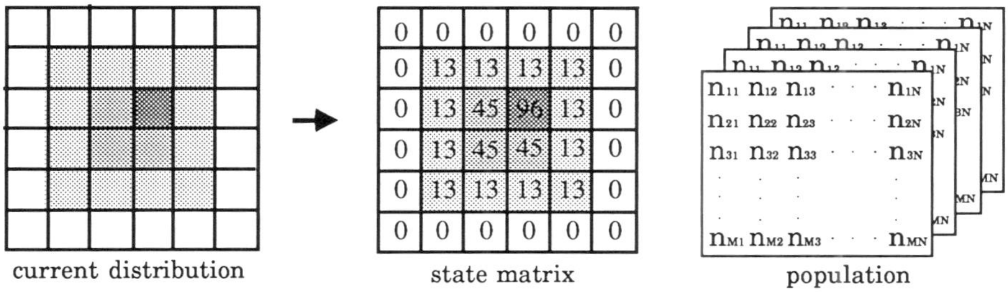

Fig.3. State Matrix

3. 2 Evaluation

As fitness, which is an index of evaluation of gene in genetic algorithm, we consider the error evaluation function. In Fig.2, the magnetic field that the currents induce at the measuring point α is given by

$$\mathbf{B}^{\alpha} = \sum_{i}^{M}\sum_{j}^{N} \frac{\mu_0 dxdy}{4\pi} \frac{\mathrm{J}ij \times (\mathbf{r}_{ij} - \mathbf{r}^{\alpha})}{|\,\mathbf{r}_{ij} - \mathbf{r}^{\alpha}\,|^3}$$

When the magnetic field measured at point α is $\mathbf{B}^{\alpha}{}_{m}$, we define the error evaluation function to evaluate the fitness of the gene as :

$$\frac{1}{2}\sum_{\alpha} |\,\mathbf{B}^{\alpha}{}_{m} - \mathbf{B}^{\alpha}\,|^2$$

we try to find out the current distribution of ion beam to make this error evaluation function equal to zero by means of GA.

4. Computer Simulation

To verify the effectiveness of the estimation method of current distribution by GA, computer simulations were carried out. The parameters and the results of the simulation are shown in Table.1 and Fig.4. Fig.4 (a) shows the real current distribution and Fig.4 (b),(c),(d) show the current distributions which estimated by using GA, which the number of the measuring points is 8, 16, 24 respectively. The hight of the figures represents the current strength. From the simulation results, we find that GA shows the very good convergence even in the case of very few measuring points.

Table 1. Parameters of the simulation

Segments	256 (16 × 16)
Measuring points	8,16,24 points
Current area	10 segments
Population	100

5. Conclusions

In this paper, we present the method of inverse estimation of current distribution from magnetic fields and the application of this method to the beam profile monitor for intense ion acceralators. Computer simulations showed that GA has the capablities to determine the current distribution even in case of very few measuring points.

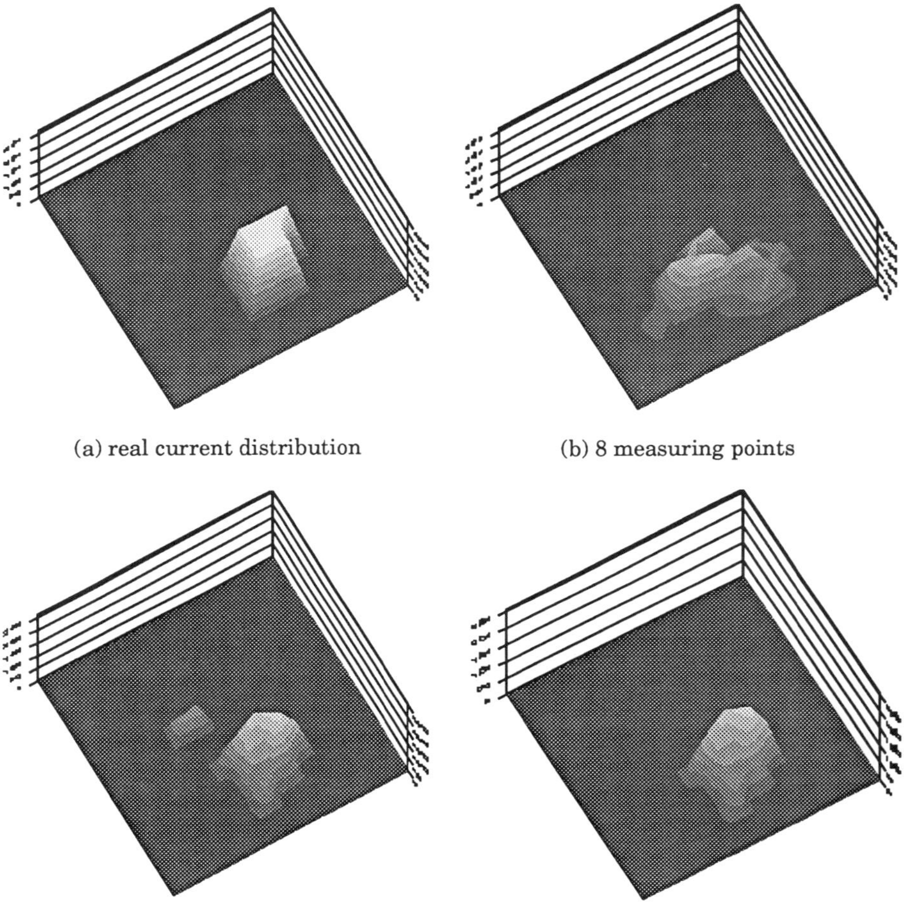

(a) real current distribution

(b) 8 measuring points

(c) 16 measuring points

(d) 24 measuring points

Fig.4. Simulation results for different numbers of measuring points

6. References

[1] John.H.Holland, Scientific American, July 1992,44-50

[2] Proceeding of the Workshop on Advanced Beam Instrumentation, KEK, Japan,1991

[3] K.Ara, K.Sakasai and M.Kishimoto, Nunlinear Phenimena in Electromanetic Fields, Elsevier Science Publishers B.V.,1992, 193-196

[4] A.Ishiguro et.al., AP-2-10, ISEM'92-Nagoya

Simulation and Design of Applied Electromagnetic Systems
T. Honma (Editor)

Study on neural network for automatic ECT data evaluation

H.Komatsu[a], Y.Matsumoto[a], Z.Badics[a], K.Aoki[a], F.Nakayasu[a]
M.Hashimoto[b] and K.Miya[c]

[a] Nuclear Fuel Industries, Ltd.
950, Ohaza-noda, Kumatori-cho, Sennan-gun Osaka, Japan

[b] University of Industrial Technology, 4-1-4 Hashimotodai, Sagamihara, Kanagawa, Japan

[c] Nuclear Engineering Research Laboratory, The Faculty of Engineering, University of Tokyo, Tokai-mura, Ibaraki, Japan

Abstract

At the in-service inspection of the steam generator (SG) tubings in Pressurized Water Reactor (PWR) plants, eddy current testing (ECT) has been widely used at each outage.

At present, ECT data evaluation is mainly performed by ECT data analyst, therefore it has the following problems.

① Only ECT signal configuration on the impedance trajectory is used in the evaluation.
② It is an enormous time consuming process.
③ The evaluation result is influenced by the ability and experience of the analyst.

In this work, we performed the preliminary study on the possibility of the application of neural network to ECT data evaluation.

1. INTRODUCTION

At present, ECT data evaluation is mainly performed by ECT data analyst with the support of automatic data screening system based on the ECT signal configuration on the impedance trajectory. However, this method makes it difficult to identify the true defect signal hidden in background signals such as lift-off noise and deposit signals. Therefore, the multi-frequency method is utilized to resolve this problem, but much time is needed for this treatment.

Recently, several basic studies have been performed to categorize ECT signals automatically by using various mathematical algorithms after the parameterization of raw ECT singals. The K-means algorithm, neural network and AI method are included in such mathematical treatment. These new ECT data evaluation methods seem attractive due to the fact that of defect compared with the usage of only signal amplitude and phase angle in the final output on impedance trajectory.

Therefore, the appropriate mathematical algorithm with good parameterization of raw ECT signals have the following benefits.

① Superior capability of ECT signal categorization.
② Prevention of human error.
③ Reduction of time for ECT data evaluation.

2. DATA – ACQUISITION

The conventional eddy current probe, bobbin type probe, was utilized in this work. This probe consists of two annular shaped coils which are excited by a sinusoidal alternative current and connected differentially in the electronic instrumentation.

We prepared various mock – up SG tube samples including machined defects (EDM notch) in both axial and circumferential direction on ID/OD surface, Cu and Fe_3O_4 deposits and dent samples.

3. RESULTS

Usually, ECT data analyst evaluates ECT signal by only amplitude and phase angle on the impedance trajectory. However, if we notice a raw ECT signal, we can find various kinds of parametric features, including time domain features, frequency domain features and lissajous features. Among nearly fifty candidates constituting characteristic feature of a raw ECT signal of each defect, we selected several optimum parameters.

In the first trial, we investigated the capability of neural network to classify various ECT signals into defect, deposit and dent. (see Fig.1)

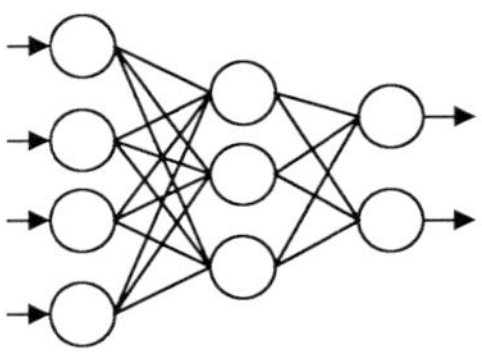

- Input data parameters
  - ① Time domain parameter No.1
  - ② Time domain parameter No.2
  - ③ Phase angle (400KHz)
  - ④ Phase angle (100KHz)
- Learning data sets
  - Defect : 12sets
  - Dent : 4sets
  - Cu deposit : 4sets
  - Fe_3O_4 deposit : 4sets
- Output vector
  - Defect : $\begin{pmatrix}0\\0\end{pmatrix}$
  - Dent : $\begin{pmatrix}1\\0\end{pmatrix}$
  - Cu deposit : $\begin{pmatrix}0\\1\end{pmatrix}$
  - Fe_3O_4 deposit : $\begin{pmatrix}1\\1\end{pmatrix}$

Fig.1 Neural network for classification

Table 1 Results of classification

Unknown input	Output Vector (*)	Result (**)	
Circumferential defect (ID 80% TW)	$\begin{pmatrix}0.005\\0.006\end{pmatrix}$	Defect	○
Circumferential defect (OD 80% TW)	$\begin{pmatrix}0.003\\0.001\end{pmatrix}$	Defect	○
Axial defect (ID 80% TW)	$\begin{pmatrix}0.005\\0.005\end{pmatrix}$	Defect	○
Axial defect (OD 80% TW)	$\begin{pmatrix}0.003\\0.001\end{pmatrix}$	Defect	○
Cu deposit	$\begin{pmatrix}0.005\\0.997\end{pmatrix}$	Cu deposit	○
Cu deposit	$\begin{pmatrix}0.004\\0.996\end{pmatrix}$	Cu deposit	○
Dent	$\begin{pmatrix}0.992\\0.003\end{pmatrix}$	Dent	○
Dent	$\begin{pmatrix}0.992\\0.004\end{pmatrix}$	Dent	○
Fe_3O_4 deposit	$\begin{pmatrix}0.996\\1.000\end{pmatrix}$	Fe_3O_4 deposit	○
Fe_3O_4 deposit	$\begin{pmatrix}0.995\\1.000\end{pmatrix}$	Fe_3O_4 deposit	○

(*) Definition of output vector at learning process.
(**) ○ : correct judge, × : misjudge

In this case, four parameters are utilized among optimum parameters to characterize feature of raw ECT signal and 24 ECT signals are used in the learning process. As shown in Table 1, this neural network could classify 10 unknown ECT signals perfectly.

Secondly, we tested the capability of neural network to identify the location (ID/OD), direction and depth of defect. (see Fig.2)

Five input parameters and twelve known ECT signals were utilized for feature extraction and learning process, respectively. Table 2 showed good results of the neural network for four unknown ECT signals.

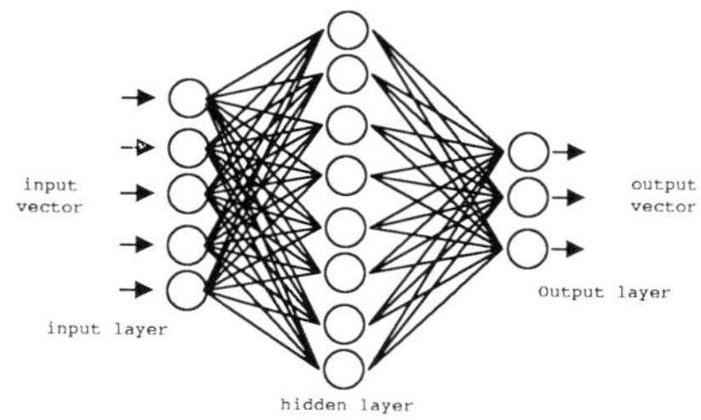

- Input data parameters
  ① Time domain parameter No.1
  ② Time domain parameter No.2
  ③ Frequency domain parameter No.1
  ④ Phase angle (400KHz)
  ⑤ Phase angle (100KHz)

- Learning data sets
  Defect : 12sets

- Output vector

·Circumferential defect
- ID
  - 40%TW $\begin{pmatrix}0\\0\\0\end{pmatrix}$
  - 80%TW $\begin{pmatrix}0\\0\\1\end{pmatrix}$
- OD
  - 40%TW $\begin{pmatrix}0\\1\\0\end{pmatrix}$
  - 80%TW $\begin{pmatrix}0\\1\\1\end{pmatrix}$

·Axial defect
- ID
  - 40%TW $\begin{pmatrix}1\\0\\0\end{pmatrix}$
  - 80%TW $\begin{pmatrix}1\\0\\1\end{pmatrix}$
- OD
  - 40%TW $\begin{pmatrix}1\\1\\0\end{pmatrix}$
  - 80%TW $\begin{pmatrix}1\\1\\1\end{pmatrix}$

Fig.2 Neural network for identification of location (ID/OD), direction and depth of defect

Table 2 Results of identification of the location (ID/OD), direction and depth of defect

Unknown input	Output vector(*)	Result(**)	
Circumferential defect (ID 80%TW)	$\begin{pmatrix}0.000\\0.000\\1.000\end{pmatrix}$	Circumferential defect (ID 80%TW)	○
Circumferential defect (OD 80%TW)	$\begin{pmatrix}0.010\\1.000\\0.990\end{pmatrix}$	Circumferential defect (OD 80%TW)	○
Axial defect (ID 80%TW)	$\begin{pmatrix}0.957\\0.000\\1.000\end{pmatrix}$	Axial defect (ID 80%TW)	○
Axial defect (OD 80%TW)	$\begin{pmatrix}1.000\\1.000\\0.989\end{pmatrix}$	Axial defect (OD 80%TW)	○

(*) Definition of output vector at learning process.
(**)○ :correct judge × :misjudge

4. CONCLUSION

There are several candidate parameters to characterize the nature of defect in raw ECT signals, as compared with the ECT signal configuration in the final output on impedance trajectory.

In our preliminary study, even if neural network of small size is trained with very limited number of learning data set, it showed good identification capability of defects with the usage of good input parameters.

In conclusion, neural network has high potential as a tool for automatic ECT data evaluation system.

5. REFERENCE

(1) Shankar,R.,Brown,C.L.,Mucciardi,A.N.,and Davis,T.J., "Feasibility of Using Adaptive Learning Networks for Eddy – Current Signal Analysis," NP – 723, Electric Power Research Institute, Palo Alto, Cal., 1978.

(2) Doctor,P.G.,Harrington,T.P.,Davis,T.J.,Morris,C.J.,and Fraley, D.W., "Pattern – Recognition Methods for Classifying and Sizing Flaws Using Eddy – Current Data," Eddy – Current Characterization of Materials and Structures, ASTM STP 722, George Birnbaum and George Free, Eds., American Society for Testing and Materials, 1981, pp.464 – 483.

(3) Flord,J.H., "Computer Intelligence Design for Automatic Inspection of Steam Generator Tubes", American Society of Mechanical Engineers for presentation at the Joint Conference of the Pressure Vessels and Piping, Materials, Nuclear Engineering and Solar Divisions, 1981, No.81 – NE – 2

(4) Brown,C.L.,Defibaugh,D.C.,Morgan,E.B.and Mucciardi,A.N., "Automatic Detection, Classification, and Sizing of Steam – Generator Tubing Defects by Digital Signal Processing," Eddy – Current Characterization of Materials and Structures, ASTM STP 722, George Birnbaum and George Free, Eds., American Society for Testing and Materials, 1981, pp.484 – 493.

(5) Simard,P.and Gaillard., "Automatic Classification and Recongnition of Defects in an Eddy Current Non Destructive Testing, " Proceeding of IEEE, ICASSP 86, Vol.1986, pp.1441 – 1444.

(6) Udpa,S.S. and Udpa,L., "Eddy Current Defect Characterization Using Neural Networks," Materials Evaluation, Vol.48, March 1990, pp.342 – 347.

Actuators

Simulation and Design of Applied Electromagnetic Systems
T. Honma (Editor)

THIN–LAYER PIEZOELECTRIC ACTUATORS: THEORY AND PRACTICE

H. S. Tzou

Department of Mechanical Engr. and Ctr. for Robotics & Manufacturing Systems
University of Kentucky, Lexington, Kentucky 40506–0046 U.S.A.

Abstract

Artificial distributed sensors and actuators integrated with an elastic continuum provide a life–like sensation and action/reaction capability for the continuum system – the integrated sensor/actuator/continuum system. In this study, thin–layer distributed piezoelectric actuators are laminated with a generic elastic thin shell continuum and a distributed actuation theory is proposed. Electromechanics of the distributed actuators are analyzed; resultant effects are integrated with the generic shell system equations. With four system parameters, the system equations can be directly simplified to a number of geometries. Distributed convolving ring actuators are designed and their performance studied.

1. INTRODUCTION

Recent studies on precision electromechanical systems and "intelligent" structural systems have shown the versatilities of piezoelectric materials in both sensor and actuator applications [1–3]. Theories of distributed sensing and control of shell continua using distributed piezoelectric layers were proposed [3]. Piezoelectric bimorph actuators were studied and applications discussed [3,4]. This paper is concerned with thin–layer distributed actuators laminated on shell continua. Generic actuation theory is proposed, and application to ring actuators is demonstrated.

2. DISTRIBUTED SHELL ACTUATORS

It is assumed that the distributed actuators are made of piezoelectric materials ant they are laminated on the bottom surface of an elastic shell continuum, Figure 1. (Note that the actuators can also be embedded in a shell laminate.) It is assumed that the piezoelectric actuator layer is unconstrained and is free from external in–plane normal forces. Thus, the stress effects can be neglected in the derivation. The induced strains due to imposed control voltages, the *converse piezoelectric effect*, are used to control of the shell motion. It is also assumed that the applied control voltage ϕ^a is much more significant than the self–generated voltage ϕ, from the *direct piezoelectric effect*, in the distributed actuator. Thus, the self–generated voltage is neglected in the distributed actuator. ϕ^a can be either a reference voltage, an open–loop control, or a distributed sensor signal ϕ^s, a closed–loop control. The sign of the feedback voltage should be carefully controlled so that the induced moments can control the motion and counteract the shell oscillation. System dynamic equations including all control effects are derived.

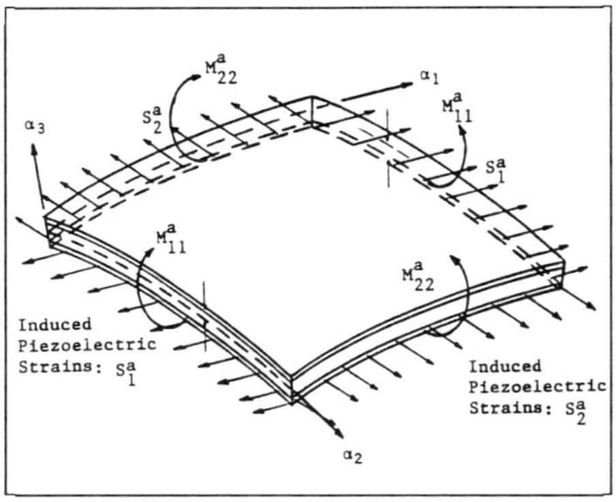

Fig.1 Distributed actuator on a shell.

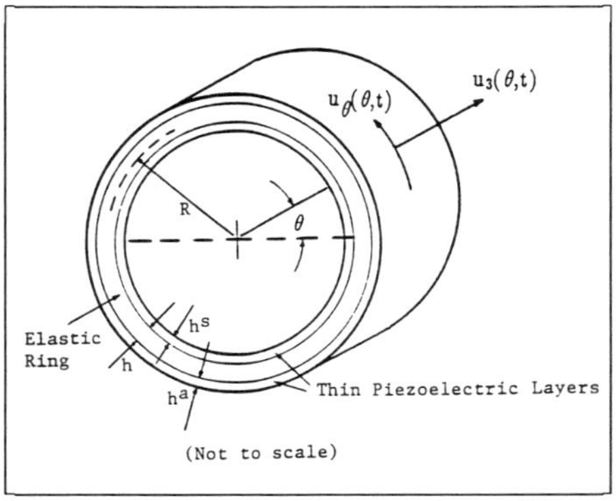

Fig.2 A circular ring.

It is assumed that the distributed piezoelectric actuator is made of a bi–axially polarized piezoelectric material. Thus, a control voltage ϕ^a applied to the distributed actuator layer introduces two in–plane normal strains S^a_{ii} (α_1 and α_2 directions) in the actuator layer due to the *converse piezoelectric effect.*

$$S^a_{ii} = (d_{3i}\ \phi^a)/\ h^a\ , \quad i=1,2\ , \tag{1}$$

where d_{3i} is the piezoelectric strain constant; h^a is the actuator thickness; the superscript "a" denotes the distributed piezoelectric actuator. (The induced strains and the resultant effect are illustrated in Figure 1.) Note that these strains are generated in the distributed actuator layer which is located a distance away from the shell neural surface. Thus, these strains introduce counteracting control moments for the shell structure. The in–plane effective forces N^a_{ii} and moments M^a_{ii} induced by the imposed control voltage can be expressed as

$$N^a_{ii} = d_{3i} Y_p \phi^a\ , \quad \text{and} \quad M^a_{ii} = r^a_i\ d_{3i}\ Y_p\ \phi^a\ , \quad i=1,2\ , \tag{2)\&(3}$$

where Y_p is Young's modulus of the piezoelectric actuator and r^a_i is the effective moment arm (distance measured from the neutral surface to the mid–plane of the piezoelectric actuator). The in–plane twisting (shear) effect is not considered. Note that the imposed control voltage ϕ^a is determined by control algorithms [3]. Considering these induced control normal forces and counteracting moments, one can derive the governing equations, in tensor expression, of the laminated shell:

$$-(\tilde{N}_{11}A_2)_{,1} - (\tilde{N}_{21}A_1)_{,2} - \tilde{N}_{12}(A_{1,2}) + \tilde{N}_{22}(A_{2,1}) - \frac{1}{R_1}[(\tilde{M}_{11}A_2)_{,1}$$
$$+ (\tilde{M}_{21}A_1)_{,2} + \tilde{M}_{12}(A_{1,2}) - \tilde{M}_{22}(A_{2,1})] + A_1A_2\,\rho\,h\,\ddot{u}_1 = A_1A_2\,F_1\ , \tag{4}$$

$$- (\tilde{N}_{12}A_2)_{,1} - (\tilde{N}_{22}A_1)_{,2} - \tilde{N}_{21}(A_{2,1}) + \tilde{N}_{11}(A_{1,2}) - \frac{1}{R_2}[(\tilde{M}_{12}A_2)_{,1}$$
$$+ (\tilde{M}_{22}A_1)_{,2} + \tilde{M}_{21}(A_{2,1}) - \tilde{M}_{11}(A_{1,2})] + A_1A_2\,\rho\,h\,\ddot{u}_2 = A_1A_2\,F_2\,, \tag{5}$$

$$- \{[(\tilde{M}_{11}A_2)_{,1} + (\tilde{M}_{21}A_1)_{,2} + \tilde{M}_{12}(A_{1,2}) - \tilde{M}_{22}(A_{2,1})]/A_1\}_{,1}$$
$$- \{[(\tilde{M}_{12}A_2)_{,1} + (\tilde{M}_{22}A_1)_{,2} + \tilde{M}_{21}(A_{2,1}) - \tilde{M}_{11}(A_{1,2})]/A_2\}_{,2}$$
$$+ A_1A_2\,[\tilde{N}_{11}/R_1 + \tilde{N}_{22}/R_2] + A_1A_2\,\rho\,h\,\ddot{u}_3 = A_1A_2\,F_3\,, \tag{6}$$

where A_1 and A_2 are Lame's parameters; R_1 and R_2 are radii of two in–plane α_1 and α_2 axes; ρ is the mass density; h is the shell thickness; $\ddot{u}$ is the acceleration; the ~ terms include the feedback control effects induced by the converse piezoelectric effect. These resultant forces and moments are modified to include the induced normal forces and counteracting moments, i.e., $\tilde{N}_{ij} = N_{ij} - d_{ij}h^aY_p\phi^a$ and $\tilde{M}_{ij} = M_{ij} - M^a_{ij}$. N_{ij} and M_{ij} are the mechanical membrane forces and bending moments [3]. Note that the in–plane twisting effect is assumed negligible, i.e., $\tilde{M}_{12} = M_{12}$ and $\tilde{N}_{12} = N_{12}$. In addition, material properties of the piezoelectric actuator layer is not considered. The feedback voltage ϕ^a is determined by the control algorithms, e.g., the direct proportional feedback, negative velocity feedback, and Lyapunov feedback [3]. Note that depending on four system parameters, two Lame's parameters and two radii, the system equations can be simplified to many common geometries, e.g., spherical shells, cylindrical shells, conical shells, plates, non–shell type continua, etc. [2,3].

3. DISTRIBUTED CONVOLVING PIEZOELECTRIC RING ACTUATORS

The Lame's parameters for a circular ring are $A_1 = 1$ and $A_2 = R$ where R is the ring radius; the radii $R_1 = \infty$ and $R_2 = R$, Figure 2. Based on the four parameters and modal orthogonality, a cosine shaped modal actuator can be designed such that it is only effective to the k–th mode [5]. Thus, the k–th modal equation of the ring with cosine shaped distributed actuators can be written as

$$\ddot{\eta}_k + \frac{c}{\rho h}\dot{\eta}_k + \omega_k^2\eta_k = \frac{e_{31}\phi_3^a}{\pi R\rho h}\left(1 + \frac{k^2(h+h^a)}{2R}\right), \tag{7}$$

where η_k is the modal participation factor; c is the viscous damping; ω_k is the natural frequency; h is the ring thickness; h^a is the actuator thickness. The first term is related to the in–plane membrane control effect; the second is the bending control effect. Note that for a free–floating ring, the first mode k = 0 is a breathing mode and the second mode k = 1, is a translational rigid–body mode [3]. For $k \geq 2$, sets of transverse and circumferential modes, with distinct natural frequencies, appear. Thus, distributed cosine shaped piezoelectric actuators are primarily designed for $k \geq 2$ and control effectiveness are evaluated for $k \geq 2$ modes. Spatially distributed cosine shaped actuators for k = 2 and 3 modes can be plotted based on the cosine functions. Note that the polarity changes need to be considered in the actuators. The feedback voltage ϕ_3^a can be setup as a velocity feedback; their first two modal damping control effects and contribution of the circumferential and bending control effects are studied, Figures 3a and 3b. It is observed that the circumferential control effect is much more significant than the bending control effect for the second and third modes. The overall control effect is a sum of these two effects.

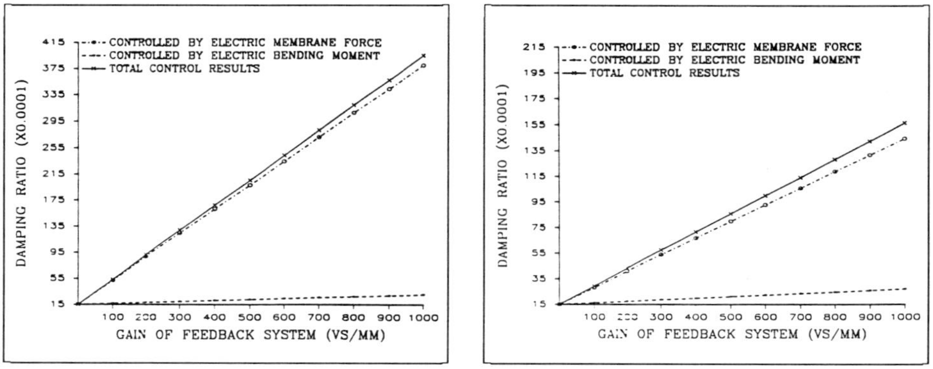

Fig.3 Damping control of the ring, a) k = 2 and b) k = 3 modes.

4. CONCLUSIONS

Distributed piezoelectric actuators for elastic shell continua were proposed; their resultant control effects and the system governing equations were derived. As an application of the generic theory, distributed convolving ring actuators were proposed and performance evaluated. It was noted that the distributed modal actuators are effective for modal control of the circular ring.

5. ACKNOWLEDGEMENT

This research was supported, in part, by a grant from the Army Research Office (DAAL03–91G–0065) (1991–1994), Technical Monitor: Dr. Gary L. Anderson. A grant from the National Science Foundation (No. RII–8610671) and the Commonwealth of Kentucky (1986–1991) is also gratefully acknowledged. A travel grant supported by the ISEM Organizing Committee is very much appreciated.

6. REFERENCES

1 H.S. Tzou and T. Fukuda, *Precision Sensors, Actuators, and Systems*, Kluwer Academic Publishers, Dordrecht/ Boston/ London, August (1992).

2 H.S. Tzou and G.L. Anderson, *Intelligent Structural Systems*, Kluwer Academic Publishers, Dordrecht/ Boston/ London, December (1992).

3 H.S. Tzou, *Piezoelectric Shells: Distributed Sensing and Control of Continua*, Kluwer Academic Publishers, Dordrecht/ Boston/ London, February (1993).

4 Z. W. Jiang, S. Chonan, H. Abe, and J. Tani, ASME *J. of Dynamic Systems, Measurements, and Control*, Vol.(113), pp.327–329, (1991).

5 H.S. Tzou, J.P. Zhong, J.P., and Natori, M.C., ASME *J. of Vibration and Acoustics*, Vol.115, No.1, pp.40–46, (1993). (Jp–ThnLyrAct'93.PzShl193)

Simulation and Design of Applied Electromagnetic Systems
T. Honma (Editor)

Ultrasonic linear actuator operation using a cylindrical piezoelectric ceramic vibrator

K. Toda and H. Tokumaru

Department of Electronic Engineering, National Defense Academy, Hashirimizu, Yokosuka 239, Japan

Abstract

A compact linear actuator is described by using a cylindrical piezoelectric ceramic vibrator with an asymmetrical electrode configuration. The unidirectional displacement induced on the side surface of the ceramic vibrator plays the key role for the operation as a stator. The directionality of linear actuator operation is changeable, depending on which of two electrodes on the top surface of the ceramic vibrator is under the ON condition. A practical linear actuator is demonstrated by introducing an oscillation circuit including the ceramic vibrator as a resonant element.

1. INTRODUCTION

Recently, interest in ultrasonic actuators has been increasing as a new type of torque source because of their high torque under low speed condition [1,2]. The actuator operation is essentially related with the unidirectional displacement on a surface of a vibration medium as a stator element. Various types of actuators have been reported, including a Langevin-type vibrator and an annular plate with non-antisymmetric modes [3,4].

They have some difficulties in constructing concise power feeding circuits. In this paper, we report a method of constructing a linear actuator using a cylindrical piezoelectric ceramic vibrator in a combination with a simple circuit for power supply, which is available for continuous as well as for stepping operations.

2. OPERATION PRINCIPLE OF LINEAR ACTUATOR

A cylindrical piezoelectric ceramic vibrator as the key element is schematically shown in Fig. 1 (a), where two half-moon-shaped electrodes are mounted on one of the two flat surfaces. The polar axis is normal to the electrode surfaces. An application of voltage between one of the two half-moon-shaped electrodes and the electrode on the other surface induces two vibration modes, asymmetrical longitudinal mode in the length direction and radial mode in the direction normal to the length, as shown in Fig. 1 (b). The mechanical coupling between the two modes produces a unidirectional mechanical force as a stator on the side surface of the ceramic vibrator.

The surface displacement distribution, examined by contacting a ball-bearing on the side surface, is also shown in Fig. 1 (a), in which the direction and the length of the arrow correspond to the direction and the strength of the mechanical displacement, respectively. In this case, the size of the cylindrical vibrator is 10 mm in length and 10 mm in diameter. The vibrator material is a 91A piezoelectric ceramic with the k_{33} value of 70%, produced by TDK Company. The measured resonant frequency of the ceramic vibrator is 139 kHz, which is essential for the unidirectional mechanical motion. It is noticeable that the directions of two unidirectional surface displacements, at the opposite side surfaces, are anti-parallel along the poling axis.

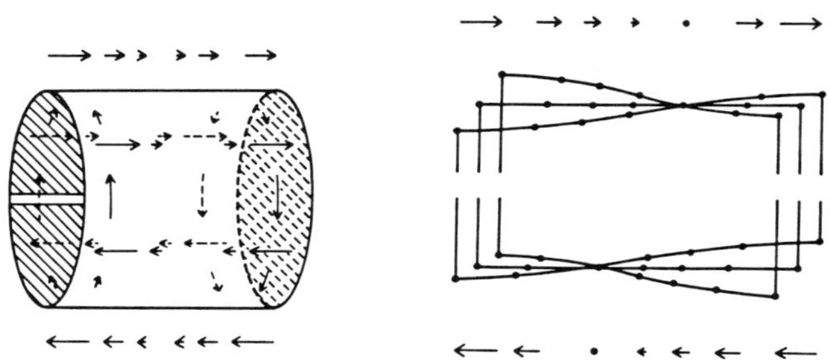

(a) (b)

Figure 1.(a) Cylindrical piezoelectric ceramic vibrator with two half-moon-shaped electrodes on one end surface and displacement at side surface. (b) Schematic of vibration modes at resonant frequency.

A typical linear actuator construction using the above-described unidirectional surface displacement of the ceramic vibrator is schematically shown in Fig. 2, where the ceramic vibrator plays the role of a stator. A metal linear slider (THK: LS1077) cemented with a ceramic friction material moves lineally along the side surface of the ceramic vibrator.

A block diagram for constructing an oscillation circuit including the piezoelectric ceramic vibrator as a resonance element is shown in Fig. 3. The current-detector section works for detecting the resonant frequency component of the current flowing through the ceramic vibrator. The detected signal is fed back to the ceramic vibrator via a voltage and power amplifiers. This feed-back loop is regarded as an oscillator operating at the resonant frequency.

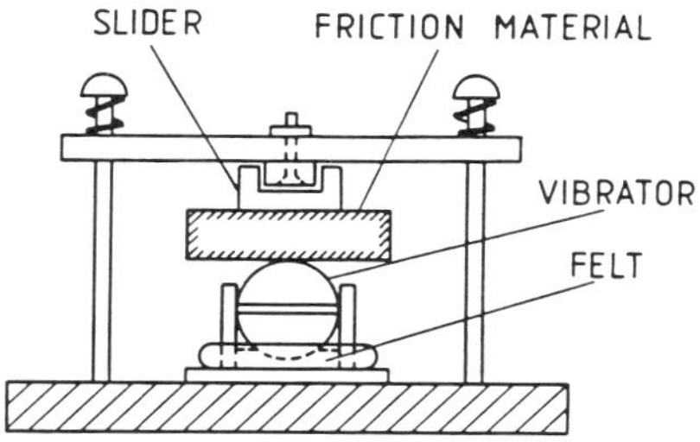

Figure 2. Schematic view of ultrasonic linear actuator

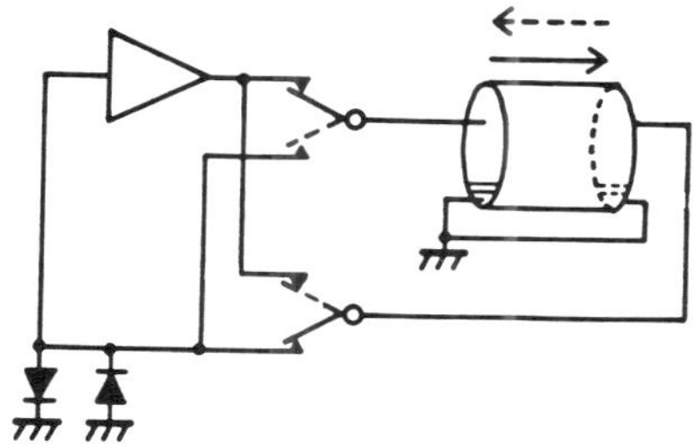

Figure 3. Block diagram of oscillation circuit

3. EXPERIMENTAL RESULTS AND DISCUSSION

Based on the design method described above, the linear actuator performance has been evaluated experimentally under a mechanical line contact between the ceramic stator and the slider. Figure 4 shows the relationships between the speed and the applied dc voltage for two moving directions, measured under no loading condition.

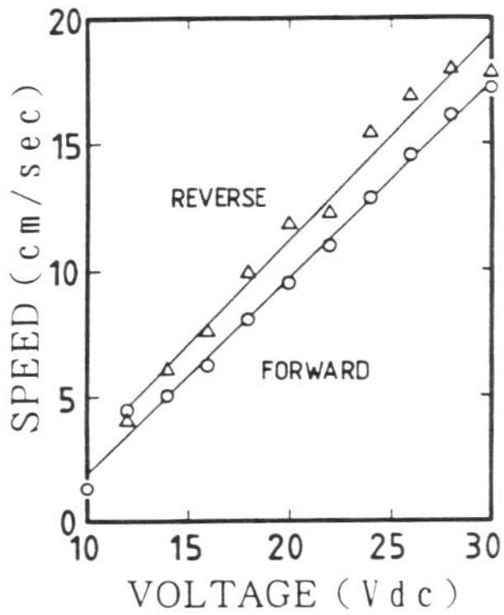

Figure 4. Measured applied dc voltage dependence of slider speed.

The moving direction of the linear actuator is controllable by incorporating a switching circuit as shown in Fig. 5. With a 139 kHz voltage applied to the piezoelectric vibrator, the slider speed increases with the applied voltage.

The slider speed measured as a function of load is shown in Fig. 6 for two applied voltages, which are similar to those of conventional dc motors except for low speed and high torque. The decrease of slider speed with loading results in the high torque operation.

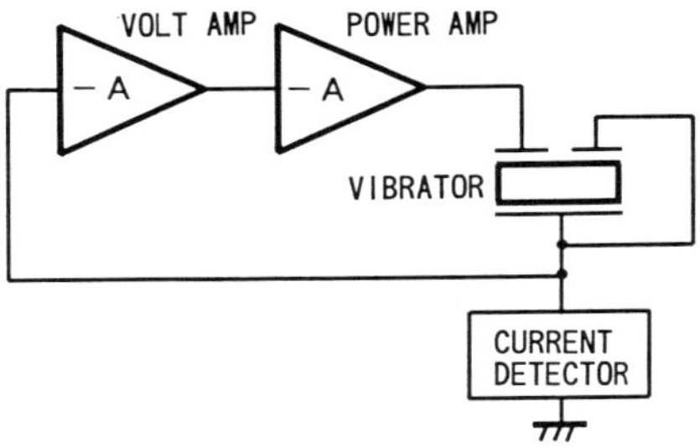

Figure 5. Schematic view of circuit for switching moving direction of actuator

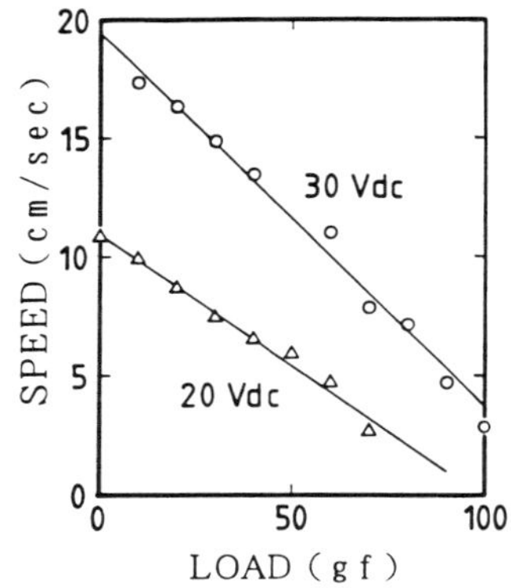

Figure 6. Measured relationship between slider speed and load.

4. CONCLUSIONS

The mechanical displacement excited on the side surface of a piezoelectric ceramic vibrator with two half-moon-shaped electrodes on the top surface was used for linear actuator operation. The construction of an oscillation circuit including the ceramic vibrator as a resonant element works successfully stable actuator operation. The present linear actuator is promising as a small, compact linear actuator.

5. REFERENCES

1 T. Sashida, Appl. Mecha. Eng., 5 (1983) 88.
2 M. Kuribayashi, S. Ueha and E. Mori, J. Acoust. Soc. Am., 77 (1985) 1431.
3 A. Kumada, Jpn. J. Appl. Phys., 24, Suppl. 24-2 (1985) 739.
4 T. Takano, Y. Tomikawa, T. Ogasawara, S. Sugawara and M. Konno, Jpn. J. Appl. Phys. 27, Suppl. 27-1 (1988) 192.

Simulation and Design of Applied Electromagnetic Systems
T. Honma (Editor)

Micro Mass Axis Alignment Device with Piezo Elements for Optical Fibers and Its Applications

S. Aoshima, N. Yoshizawa, and T. Yabuta

NTT Telecommunication Field Systems R&D Center
Tokai-Mura, Naka-Gun, Ibaraki-Ken, 319-11, Japan

Abstract

We propose a micro mass axis alignment device which can independently align individual optical fibers. Axis alignment experiments are performed with the fabricated device. This device can be applied to the axis alignment of fusion mass-splice devices for optical fiber and devices for coupling optical fibers to a semiconductor laser array.

1. Introduction

Fusion mass-splice devices for optical fibers have been developed for splicing fiber ribbons [1]. Conventional fusion mass-splice devices use only V-grooves to align the optical fiber axes. However, for these devices, axis misalignment errors can easily occur, because of the following reasons. (1) It is very difficult to fabricate a highly accurate V-groove. (2) Optical fibers have core eccentricity. These both cause an increase in splice loss. The splice loss is about 0.1 dB (average value for 1.55 μm DSF) for conventional fusion mass-splice devices with V grooves. The warst splice loss is about 0.4 dB (eccentricity 1 mm). This loss is acceptable for ordinary optical fibers. However, a lower splice loss is required for submarine optical fibers, because they must transmit signals over very long distances. So, an axis alignment device which can independently align individual optical fibers is required in order to reduce the splice loss. So, we propose a mass axis alignment device, which can independently align individual optical fibers.

2. Mass Axis Alignment Device

Figure 1 shows the proposed mass axis alignment device. This device is constructed with optical fibers, a V groove unit, piezo elements, micro arms and a base. The V grooves hold each optical fiber of a fiber ribbon. In this figure, the V groove unit has only four grooves, however, the number of grooves can be greatly increased. The V-groove unit is fixed to the base. One end of each piezo element is also fixed to the base and the other end is connected to an L-shaped arm. As shown in the top right of Fig.1, projections are attached to the bottom of each L-shaped arm. These micro arms are positioned side by side and fixed to the base. Piezo elements are located compactly in this design. The angles at the top of the micro arms are the same as the V groove angles. Each arm can be moved up and down with micron order precision, by adjusting the voltage applied to the two piezo elements. The core centers of individual optical fibers can be moved independently on a two dimensional surface by the up and down movement of the two micro arms. Figure 2 shows the method by which an optical fiber is moved. The optical fiber moves upwards or downwards when both piezo elements are extended or contract, as shown in Fig.2(a) and (b), respectively. It moves upwards to the right or left when only the right or left element is extended, as shown in Fig.2(c) and (d), respectively. Thus, the core center of an individual optical fiber can be moved on a two dimensional surface by controlling the degree to which the right and left piezo elements are extended.

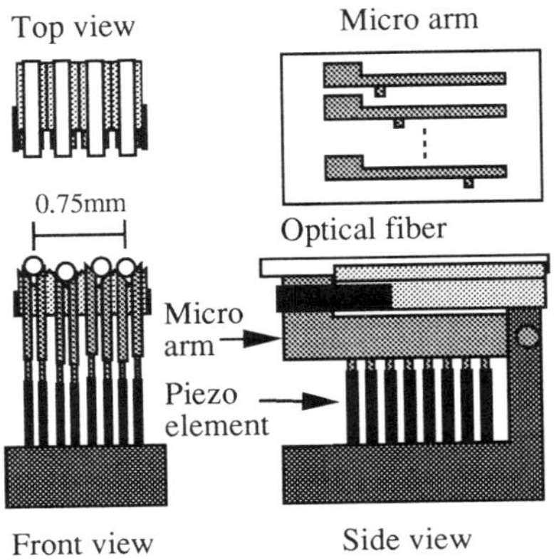

Fig. 1 Proposed mass axis alignment device.

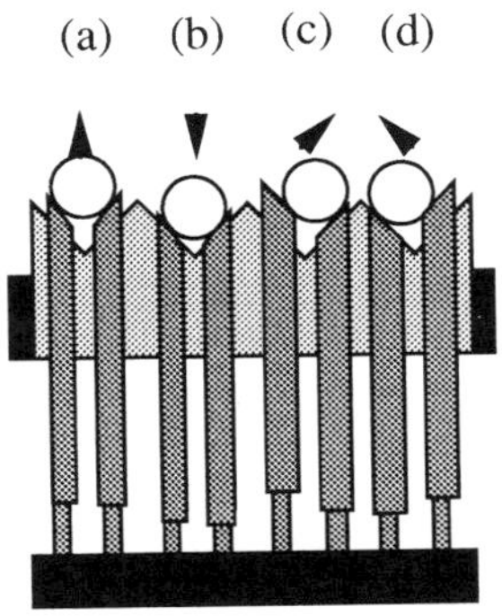

Fig. 2 Method by which an optical fiber is moved.

3. Experimental Results

A photograph of the fabricated axis alignment device is shown in Fig. 3. This device was designed for use with four-fiber tape. An axis alignment experiment was performed with the fabricated device to confirm its effectiveness. Microscope photograph of optical fiber cross section and two micro arms are shown in Fig. 4. Their movements were observed on a TV monitor. We investigated the relationship between the center of the fiber core and the voltage applied to the piezo element. The center of the core moved in the X-Y plane as shown in Fig. 5 when the voltage applied to the left piezo element was increased. In this experiment, the right arm was fixed. Voltages of 0 to 150 V were applied to the left piezo element. The position of the center of the core was measured every 10 V. It was

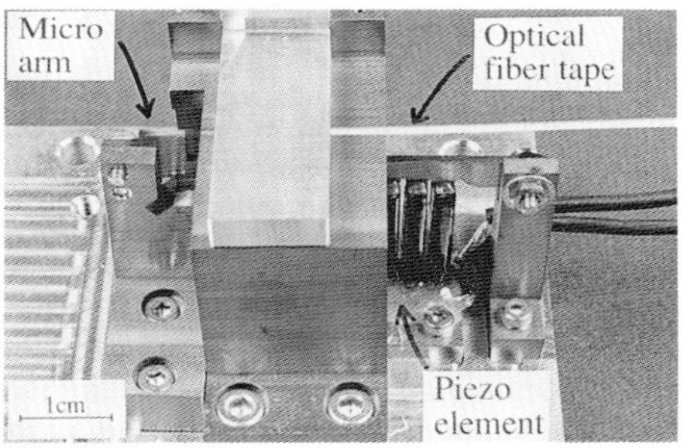

Fig. 3 Photograph of the fabricated axis alignment device.

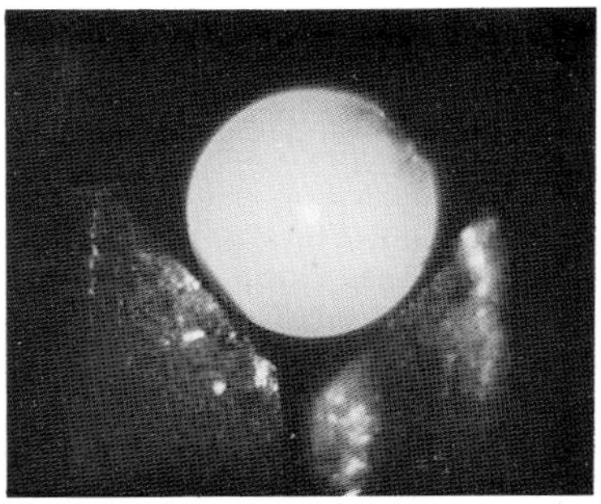

Fig. 4 Microscope photograph of optical fiber cross section and the two micro arms.

confirmed that the core center moves smoothly upward and to the right. The fiber and the micro arms were always in contact. This device realized axis alignment with submicron order precision by using open loop control. It will be possible to control the fiber core with 0.1μm order precision by using feedback control. Figure 6 shows experimental results on the axis alignment range. The following voltages were applied to the left and right piezo elements at the points shown in the figure.

Point (Left , Right): O(0 V, 20 V), A(100 V, 20 V), B(100V, 120 V), C(0 V, 120 V)
The device we fabricated can move the center of the core to any position within the area OABC (about 20x20 mm) as shown in figure.

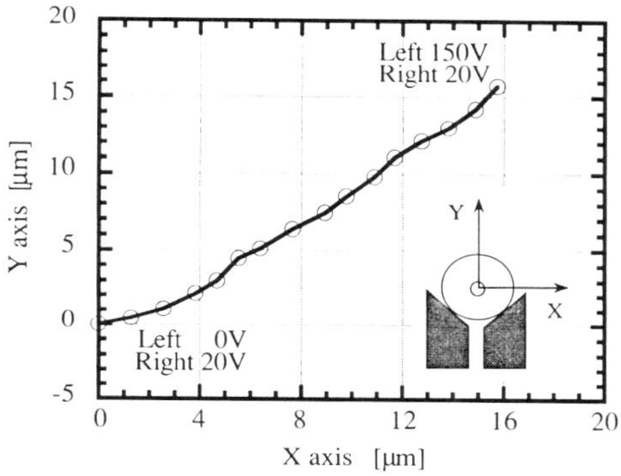

Fig. 5 Relationship between the center of he fiber core and the voltage applied to the left piezo element.

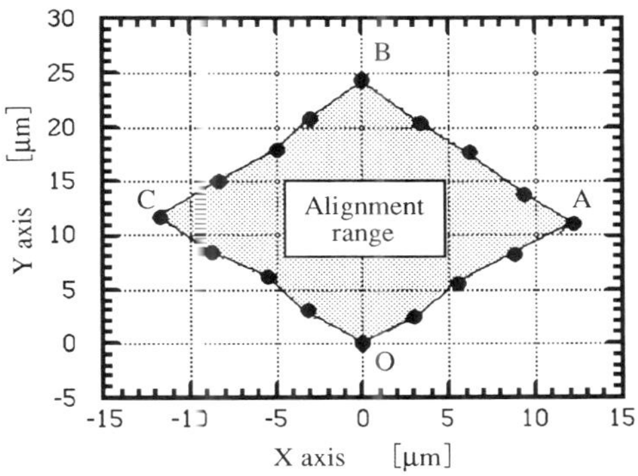

Fig. 6 Experimental results of the axis alignment range.

4. Application to optical coupling between a laser array and optical fibers

The conventional method[2] of optical coupling between a semiconductor laser array and optical fibers can perform axis alignment by using V grooves etched in silicon. However, axis misalignment errors can easily occur as a result of errors in the size of the V grooves and core eccentricity. So, we proposed a device for low loss optical coupling between a laser array and optical fibers, which is shown in Fig. 7. This device can reduce axis misalignment errors between the laser array and the optical fibers. The device is composed of a laser array attached to a silicon waferboard which is etched with V grooves and has a hole in it for axis alignment, optical fibers, a laser power supply, an optical power sensor, a piezo element controller and the proposed alignment device. The silicon waferboard is fixed on a base. The axis alignment device consists of piezo elements, micro arms, and an up-and-down micro stage. The micro stage is used for setting the initial position of the micro arms. The piezo elements are controlled by feedback control using information provided by an optical power sensor. Low loss coupling is achieved as follows.

Process(1): The silicon waferboard is positioned on the base.
Process(2): The ends of the micro arms is inserted into the hole in the silicon waferboard. The piezo elements are controlled to maximize the optical power of each fiber. Then, the centers of laser and fiber axes are aligned.
Process(3): Optical fibers are fixed on the silicon waferboard with adhesive when all the fiber axes have been aligned.

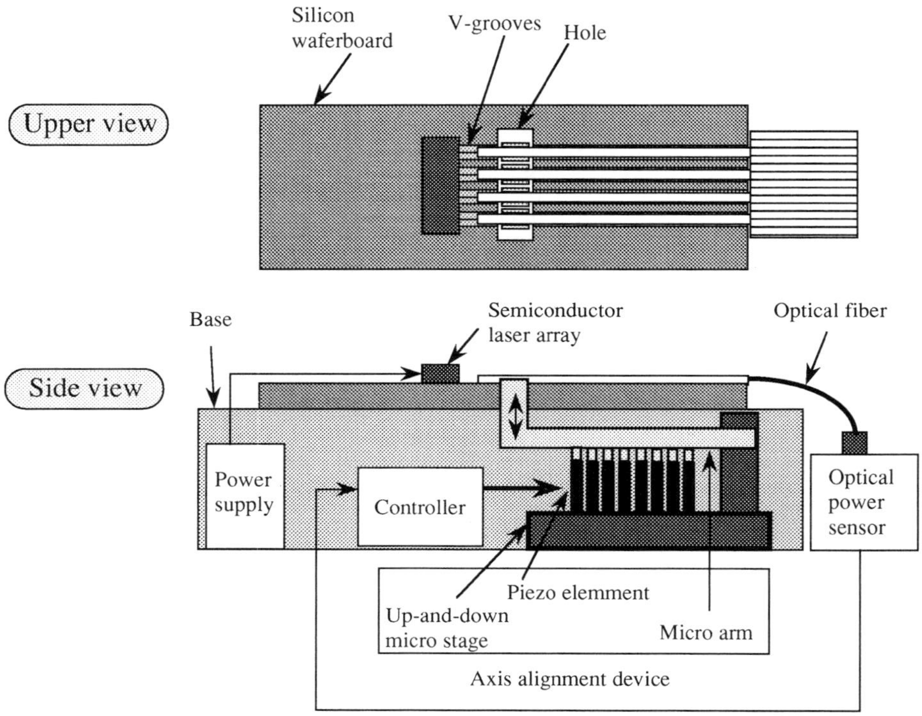

Fig. 7 Proposed device for low loss optical coupling between a laser array and optical fibers.

5. Conclusion

We proposed a micro mass axis alignment device with piezo elements which can independently align individual optical fibers. Experimental results indicate the following: (1) It will be possible to align a fiber core axis with 0.1μm order precision by using feedback control. (2) The axis alignment range is about 20x20 μm. These characteristics are sufficient for an axis alignment device for fusion mass-splice devices. Moreover, we proposed an application of this device for optical coupling between a laser array and optical fibers, and an additional axis alignment device.

6. References

1 M. Miyauchi, M. Matsumoto, T. Haibara, and N. Kawasaki, "Design and Performance of Mass Fusion Splice for Single-Mode Fiber Ribbon," The Transactions of The IEICE, Vol. E 72, No. 8, pp. 912-918, 1989.

2 C. A. Armiento, M. Tabasky et al., "Passive coupling of InGaAsP/InP laser array and singlemode fibers using silicon waferboard," Electronics Letters, Vol. 27, no. 12, pp.1109-1111, 1991.

Simulation and Design of Applied Electromagnetic Systems
T. Honma (Editor)

Shaking table tests of a piping system supported by a semiactive damper using piezoelectric actuators

H. Shimoda[a], K. Ohmata[a] and F. Okamoto[b]

[a]School of Science and Technology, Meiji University, 1-1-1 Higashimita, Tama-ku, Kawasaki 214, Japan

[b]JR Higashinihon, 1-6-5 Marunouchi, Chiyodaku, Tokyo 100, Japan

Abstract

The present report is concerned with the results of shaking table tests of an L-shaped piping supported by the semiactive damper which the authors have developed for piping systems. The damper is composed of a ball screw and a disk brake consisting of a brake disk, a brake shoe, a brake lever and piezoelectric actuators. The resistance force characteristics, the effect of vibration suppression of the damper applied to the L-shaped piping and the effective control law of the damper are discussed. Moreover, the seismic responses of the L-shaped piping, which have been already calculated by the authors, are verified by the shaking table tests.

1. INTRODUCTION

The mechanical snubber, which consists of mechanical elements, has been used in nuclear power plants to safeguard the piping systems. The mechanical snubber is required to restrain the deflection of the piping during an earthquake. It is not allowed, however, to restrict thermal expansion of the pipings being a gradual movement. In recent years, some active control type dampers for piping systems are proposed to provide better performances while simplicating the structures and reducing the costs[1-2].

In this paper, a prototype of the semiactive damper, which has been already proposed by the authors[3], is made in order to verify the performance of the damper by means of the shaking table. In the first place, braking ability and responsibility of the disk brake are discussed. Next, seismic responses of an L-shaped piping are measured when it is restrained at both ends and supported by the damper at its bent section. The responses are compared with calculated results. The effective control law for the damper is also discussed.

2. CONSTRUCTION OF THE SEMIACTIVE DAMPER

Figure 1 shows the construction of the semiactive damper which the authors have developed. If a relative linear motion is given between the base(1) and the load column(2) by an earthquake motion, the ball nut(3) attached to the linear motion ball bearing(4) moves linearly. The linear motion is transformed into a rotary motion of the screw shaft(5) and the brake disk(6). If a voltage is applied to two piezoelectric actuators(7) which are connected in series, they expand and push against the brake lever(8). Then, the brake shoe(9) is pushed against the brake disk and a brake torque is generated in the brake disk.

The main features of the damper are as follows:

1. Brake force is controlled correponding to the response of the piping or an input acceleration.
2. Resistance force caused by a thermal expansion is very small because the disk brake does not work during a gradual movement.
3. The construction of the damper is simpler than that of a convensional mechanical snubber.

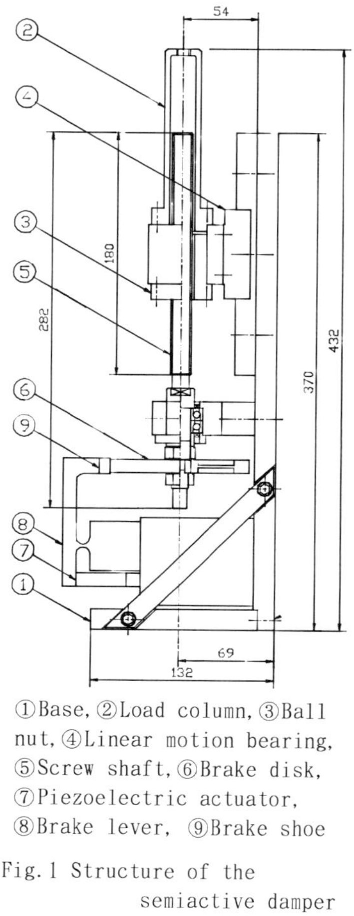

①Base, ②Load column, ③Ball nut, ④Linear motion bearing, ⑤Screw shaft, ⑥Brake disk, ⑦Piezoelectric actuator, ⑧Brake lever, ⑨Brake shoe

Fig. 1 Structure of the semiactive damper

3. PERFORMANCE OF THE DISK BRAKE AND THE DAMPER

The braking ability and responsibility of the disk brake are examined. When the disk brake has an air gap of 10 μm, the piezoelectric actuators generate the braking force of 17N/100V as shown in Fig. 2. The generated braking torque in the brake disk with the same gap are shown in Fig. 3. It is confirmed that the braking torque can be considered to be constant for the rotational speed and the braking ability of

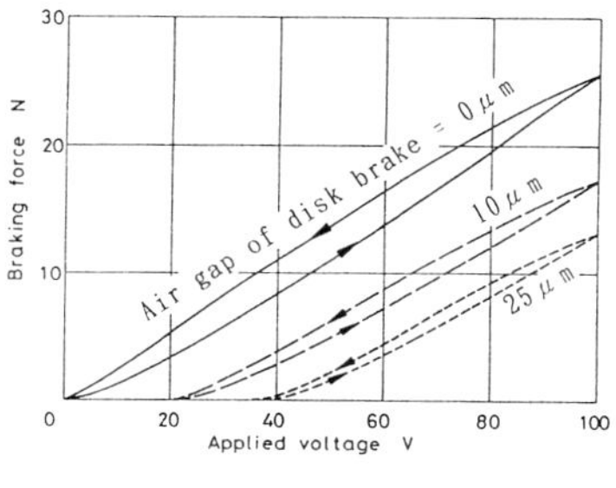

Fig. 2 Braking force of the disk brake

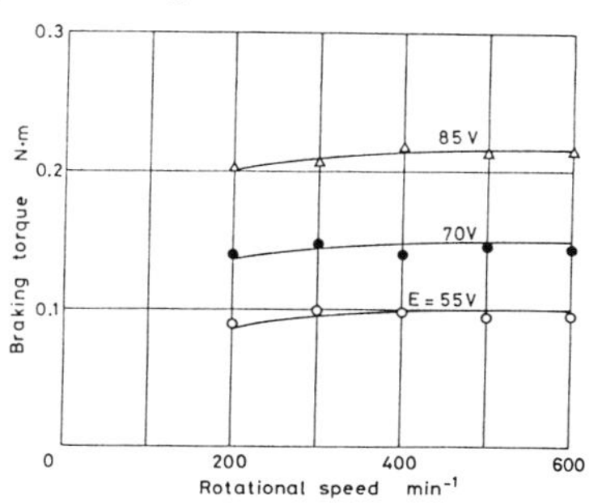

Fig. 3 Braking torque of the disk brake

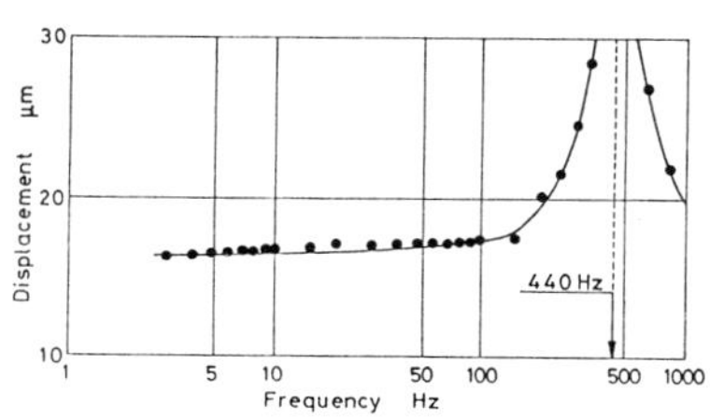

Fig. 4 Resonance curve of the disk brake

the disk brake meets our requirements.

Figure 4 shows a resonance curve of the brake shoe when a sinusoidal voltage is applied to the piezoelectric actuators. The responsibility of the disk brake is sufficient because the resonance frequency of the disk brake is significantly higher than the range of seismic frequencies which must be taken into account.

Relationship between the resistance force and displacement are measured when a sinusoidal displacement is given between the base and the load column of the damper. It can be seen from Fig.5 that the resiatance force of the damper is practically sufficient.

4. SHAKING TABLE TEST

The L-shaped piping model which is restrained at both ends on the shaking table and supported by the semiactive damper at its bent section is shown in Fig.6. The piping is length of 3m×3m and outerdiameter of 60.5mm. The specifications of the damper and the L-shaped piping are shown in Table 1. Input accelerations used here are El Centro(1940) NS and Akita (1983) NS normarized to be $2m/s^2$ at the maximum acceleration.

The damper is controlled by the following control laws:

(1)On-off control corresponding to the velocity $|\dot{u}_6|$

(2)On-off control corresponding to the input acceleration $|\ddot{y}_H|$

(3)Proportional control corresponding to the velocity $|\dot{u}_6|$

(4)Proportional control corresponding to the input acceleration $|\ddot{y}_H|$

where $|\dot{u}_6|$ is the relative velocity of the L-shaped piping at its bent section.

Figure 7 shows the time histories of the displacement response u_6 of the piping, the resistance force Fs of the damper and the brake force Fb.

It can be seen from Fig.7 that the seismic displacement response of the piping can be successfully reduced by the damper and the experimental wave forms agree well with the calculated wave forms. The maxima of the relative displacement u_6, acceleration $\ddot{y}_6$ and resistance force Fs for four control laws are summa-

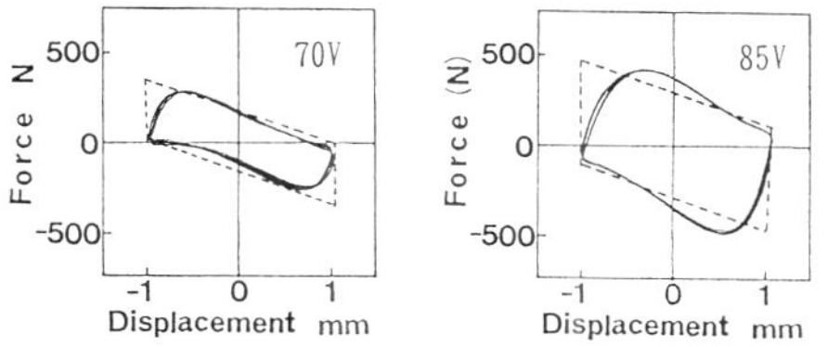

Experiment: ——— , Simulation: --------

Fig.5 Resistance force vs. displacement

Fig.6 L-shaped piping on the shaking table

Table 1 Specifications of the damper and the L-shaped piping

Damper	Lead of ball screw	10 mm
	Moment of inertia	548 kg•mm^2
	Radius of brake disk	50 mm
	Coefficient of friction	0.3
	Displacement of PZT	30μm/100V
	Force of PZT	3430N/100V
	Dimension of PZT	10×10×36mm
Piping	Material	STPG42
	Young's modulus	208 GPa
	Outer diameter	60.5 mm
	Wall thickness	5.5 mm
	Length	3m×3m
	Weight per length	7.46 kg/m

rized in Table 2. It is apparent from Table 2 that the control law 3 is the most effective among the four control laws, and the maximum values of Fs are approximately equal regardless of the control laws.

5. CONCLUSIONS

The results may be summarized as follows: (1)The deflection of the L-shaped piping can be reduced successfully by the semiactive damper. (2)The semiactive damper has the effect of vibration suppression even by simple control method such as on-off control or proportional control. (3)The experimental results of the response of the piping agree approximately with the calculated results. (4)The proportional control corresponding to the relative velocity of the piping is most effective for suppressing the deflection of the piping.

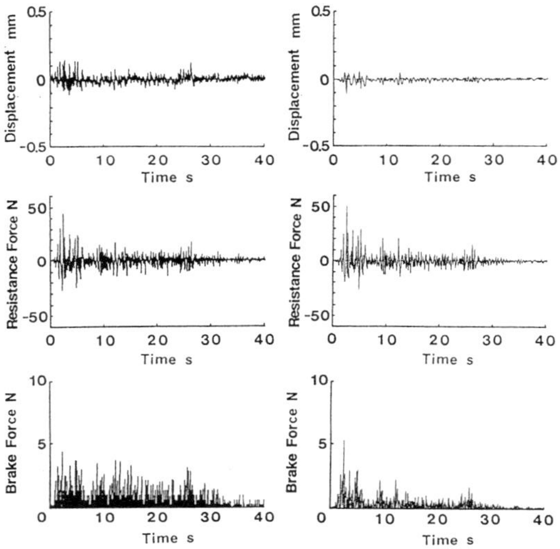

(Experiment) (Simulation)

Fig. 7 Time histories of u_6, Fs and Fb (Control condition 3)

6. REFERENCES

1 K. Ohmata and Y. Nakahara, Proc. of the 4th ISEM, (1992), 569.
2 Mechatro Damper, Sanwa Tekki Corp., Document No. RA-D1-075 (1991).
3 H. Shimoda, K. Ohmata and F. Okamoto, Proc. of the 10th WCEE, Vol.4(1992), 2173.

Table 2 Maxima of relative displacement, acceleration and resistance force

Control condition		El Centro (1940) NS			Akita (1983) NS		
		$\lvert u_6\rvert max \times 10^{-3}$(m)	$\lvert \ddot{y}_6\rvert max$ (m/s²)	$\lvert Fs\rvert max$ (N)	$\lvert u_6\rvert max \times 10^{-3}$(m)	$\lvert \ddot{y}_6\rvert max$ (m/s²)	$\lvert Fs\rvert max$ (N)
Without damper	Exp.	4.7	9.98	—	8.5	10.34	—
	Cal.	5.3	7.27	—	6.6	8.57	—
Brake force of an actuator(N)							
① Fb= { 10($\lvert\dot{u}_6\rvert \geq 0.0005$m/s)	Exp.	0.20	2.38	45.3	0.29	2.56	45.7
0 ($\lvert\dot{u}_6\rvert < 0.0005$m/s)	Cal.	0.14	2.20	52.3	0.21	2.89	52.3
② Fb= { 10($\lvert\ddot{y}_H\rvert \geq 0.1$m/s²)	Exp.	0.29	2.39	44.9	0.37	2.54	45.3
0 ($\lvert\ddot{y}_H\rvert < 0.1$m/s²)	Cal.	0.28	2.20	50.8	0.29	2.21	45.4
③ Fb= 4000× $\lvert\dot{u}_6\rvert$	Exp.	0.15	2.38	43.2	0.19	2.52	42.9
	Cal.	0.11	2.00	50.2	0.15	1.99	49.8
④ Fb= 0.25× $\lvert\ddot{y}_H\rvert$	Exp.	0.25	2.40	42.3	0.30	2.54	43.3
	Cal.	0.23	1.99	51.0	0.27	2.00	48.4

Simulation and Design of Applied Electromagnetic Systems
T. Honma (Editor)

Flutter control of cantilever thin plates with electromagnetic and piezoelectric actuators

Y.Z. Liu, J. Tani, and S. Chonan

The Institute of Fluid Science, Tohoku University
2-1-1 Katahira, Aoba-ku, Sendai 980, Japan

Abstract

This paper presents a method to control the flutter vibration of a cantilever thin plate exposed to a uniform smooth wind stream. A digital PID control algorithm is introduced to this active vibration control problem. As the actuator, two kinds of actuators in which one is an electromagnetic coil and the other is a piezoelectric device, were used and whose control effects were compared. It is found that a good effect of vibration suppression is achieved in both cases.

1. INTRODUCTION

Thin plates are extensively used in aeronautical and space structures, where the reduction of the weight per strength is of paramount importance. When the panels are subjected to aerodynamic loads larger than the critical value, it loses its stability and flutter occurs [(1)]. It is very dangerous if the flutter occurs, because a large amplitude of vibration might be brought to the structures, and sometimes there is possibility which the structures will be destroyed. Therefore, it is important and necessary to suppress the flutter.

This paper presents a method to control the flutter vibration of a cantilever thin plate exposed to a uniform smooth wind stream. A digital PID control algorithm is introduced to this active vibration control problem. As the actuator, two kinds of actuators in which one is an electromagnetic coil and the other is a piezoelectric device, were used and whose control effects were compared. It is found that a good effect of vibration suppression is achieved in both cases.

According to the control scheme designed, the vibration control was carried out experimentally. The plate was placed paraller to the flow in a low turbulent wind tunel. Flutter vibration of the plate was measured by using a photonic displacement vibrometer. The velocity distributions and fluctuations of the wake were measured by using a hot-wire anamometer for the plate with and without control. The frequency spectrum of fluctuating velocity was computed by using the Fast-Fourier-Transform methods. It is found that the velocity fluctuations of the wake increase with flutter.

2. EXPERIMENT

The equipment used in the experiment is depicted schematically in figure 1.

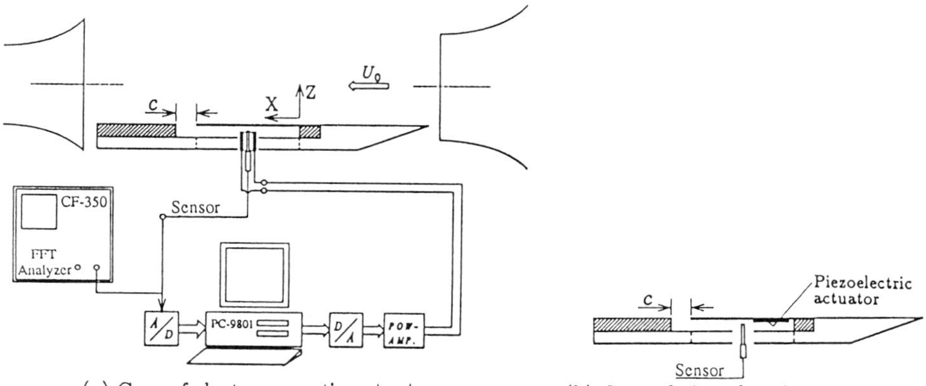

(a) Case of electromagnetic actuator (b) Case of piezoelectric actuator

Fig.1 Analytical model and coordinate system

The experiment was carried out by using a wind tunnel which is set up at the Air Current Measure Research Institution of the Institute of Fluid Science, Tohoku University. The test model is a cantilever thin plate which is exposed to a low turbulent wind tunnel parallel to the flow, and is made of stainless steel (Young's modulus is 205.8 GPa and density is $7.9 \times 10^3 \text{kg/m}^3$). The length, width and thickness of the plate are 150 mm, 60 mm and 0.3 mm respectively. In order to control the flutter induced by the aerodynamic loads on the plate, two kinds of actuators were used. One is the electromagnetic actuator which was established noncontacting under the plate, and another one is piezoelectric actuator which was put on the lower surface of the plate. The control input could be determined by using digital PID control theory.

Usually, the PID control algorithm in discrete time can be shown as [(2)]

$$u(k) = K_p\{y(k-1) - y(k)\} + K_i\{r(k) - y(k)\} + K_d\{2y(k-1) - y(k-2) - y(k)\} \tag{1}$$

where, u is control input, r is objective input, y is displacement of the plate, and K_p, K_i, K_d are P, I, D gains respectively and determined by test.

3. RESULTS AND DISCUSSION

In this study, only the fundamental vibration mode was taken into account, because this mode produced a large amplitude oscillation comparatively. When velocity of the wind was increased up to $U_0 = 9.0$ m/s (without piezoelectric actuator) or $U_0 = 12.3$ m/s (with piezoelectric actuator), the plate lost its stability and flutter occured, where the frequencies were 10.25 Hz without piezoelectric actuator and 11.75 Hz with piezoelectric actuator. The time histories of displacement of the plate in the both cases are shown in figure 2 (a) and (b) respectively. It is shown that the flutter oscillation of the plate is sufficientlly suppressed when the control is applied. In this figure, symbol ↑ indicates the starting time of the control.

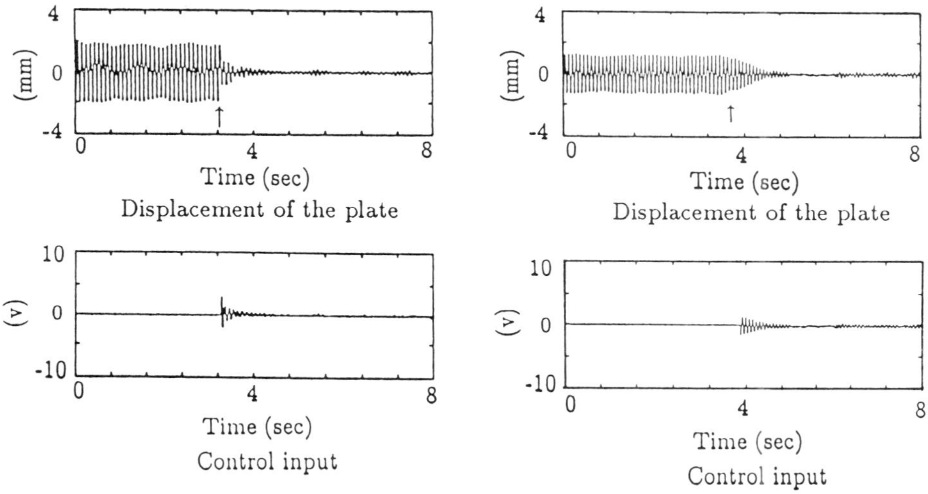

(a) With electromagnetic actuator (b) With piezoelectric actuator

Fig.2 Time histories of displacement of plate and control input

Where, the gains of the PID control are shown in table 1.

Table 1. Gains of the PID control

	Gains		
Kind of actuator	K_p	K_i	K_d
Electromagnetic actuator	0.20	0.25	0.22
Piezoelectric actuator	0.12	0.15	0.08

On the other hand, in order to elucidate how the air flow is affected by the structure vibration, we examined the velocity fluctuation of the wake for the case without piezoelectric actuator. The output signal of the fluctuating velocities measured with a hot-wire anemometer are shown in figure 3. The hot-wire probe was fixed at $x/L = 1.083$, and $y/L = 0.0$, then it was traversed in z-direction. Here, z is the vertical distance from the plate. Without control, the values of the velocity fluctuation became large when the probe approached to the plate, because there was a border layer of velocity fluctuation. Therefore, when the probe came to nearby $z/L = 0.0$, a oscillation with lower freqency appeared. It is clear that this frequency is the same as that of the plate and the velocity fluctuation was induced by the vibration of the plate. However, the vibration compoents with the low freqency had disappered when the control was applied.

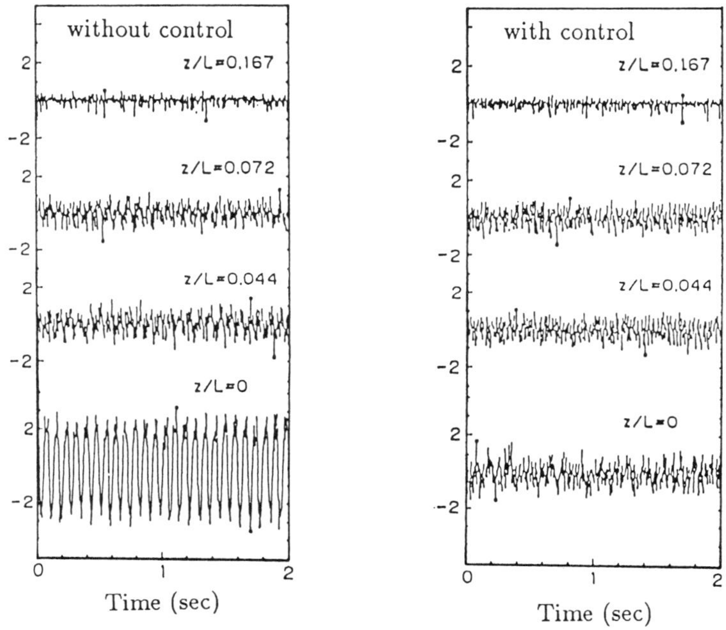

Fig.3 Velocity fluctuation of the wind stream

4. CONCLUSIONS

In this paper, the flutter control for cantilever plates in a uniform smooth wind stream was carried out by using the PID control theory. The results obtained are summarized as follows:

1. The flutter oscillation of the plate is sufficientily suppressed by using ether electromagnetic actuator or piezoelectric actuator.
2. The velocity fluctuation of the wake becomes very samll when the control is applied.

5. REFERENCES

(1) Weaver, D. S. and Unny, T. E., "The Hydroelastic Stability of a Flat Plate," Journal of Applied Mechanics, SEPT, 1970, pp. 823-827.

(2) Yasuhito, T., "Digital Control," Iwanamishoten, In Japanese, 1985.

Simulation and Design of Applied Electromagnetic Systems
T. Honma (Editor)

Force Control of a Piezoelectric Bimorph Cantilever Introduced as a Finger in the Miniature Gripper

Z.W.Jiang[a], S.Chonan[a] and S. Sakuma[b]
[a]Department of Engineering Science, Tohoku University, Sendai, Japan
[b]Sumitomo Metal Industries Ltd., Japan

Abstract

A theoretical and experimental study is presented for the force control of a piezoelectric bimorph cantilever introduced as a finger in the miniature gripper. Both the PID and the digital optimal control schemes are introduced to drive the finger so that the tip produces 0.03N and then maintains the force at the constant magnitude. Results obtained show that the experimental and theoretical results are in good agreement with each other, further the piezoelectric cell beam works well as a finger in the dexterous minute force control of the gripper.

1. Introduction

It was shown in the previous papers[1-3] that the piezoelectric bimorph cell works well as an actuator to control actively the position and the vibration of a flexible robotic miniature arm. The cells were bonded on both sides of the arm. This paper is a further study on the use of a single piezoelectric bimorph cell as a robotic finger in the mechanism of a miniature gripper holding the object with a constant minute force. One finger of the gripper is a piezoelectric bimorph cantilever with a force sensor at the tip while the mate is a finger with adequate stiffness which is modeled by a rigid beam in the theoretical analysis of the system.

2. Theoretical Analysis

Fig. 1 shows a cantilevered piezoelectric bimorph cell of length l with a force sensor of mass m at the tip. The cell plays a role of a finger in the miniature gripper designed. An applied voltage to the cell produces a constant continuous bending moment M_p in the cell and then deforms the cell itself in the xy-plane. Considering the equilibria of moments and forces acting on the small element of the cell, one has the equation of motion of the cell beam as

$$[\rho A + m\delta(x-l)]\frac{\partial^2 y}{\partial t^2} + \frac{\partial^2}{\partial x^2}[EI(1+c\frac{\partial}{\partial t})\frac{\partial^2 y}{\partial x^2}] = M_p[\delta'(x) - \delta'(x-l)] - F_s, \quad (1)$$

where ρ is the mass density, A the cross-sectional area, E Young's modulus, I moment of inertia, c coefficient of structural damping, $\delta(\cdot)$ Dirac delta function and M_p the bending control moment; F_s is the force signal from the tip sensor and it is given by

$$F_s = K_s y(l,t) = K_s y(x,t)\delta(x-l) \quad (2)$$

Here, K_s is the stiffness of the tip force sensor.

The solution of eq. (1) is assumed in the form

$$y(x,t) = \sum_{n=1}^{N} w_n(x) f_n(t), \tag{3}$$

where

$$\begin{aligned} w_n(x) =& A_n S(\xi_n x) + B_n T(\xi_n x) + C_n U(\xi_n x) \\ &+ D_n V(\xi_n x) + \alpha_n^4 \xi_n w_n(l) V[\xi_n (x-l)] H(x-l), \end{aligned} \tag{4}$$

$$(\xi_n)^4 = -(\rho A) p_n^2 / EI(1 + cp_n), \quad \alpha_n^4 = -(mp_n^2 + K_s)/EI(1 + cp_n). \tag{5}$$

Here, the eigenvalues ξ_n and the unknowns A_n - D_n are determined from the system boundary conditions.

Substituting eq. (3) into eq. (1) and then applying the Galerkin's method to the resulting equation, one has the equation of motion in the matrix form

$$\mathbf{M}\ddot{\mathbf{x}} + \mathbf{\Gamma}\dot{\mathbf{x}} + \mathbf{K}\mathbf{x} = \mathbf{U}M_p, \quad \mathbf{x} = [f_1, \dot{f}_1, ..., f_n, \dot{f}_n]^{\mathrm{T}}. \tag{6}$$

The state equation discretized in time is

$$\mathbf{x}(i+1) = \mathbf{A}\mathbf{x}(i) + \mathbf{B}M_p(i), \quad F_s(i) = \mathbf{C}\mathbf{x}(i) \tag{7}$$

In the following analysis, both the PID control and the digital optimal control scheme are introduced to drive the finger so that the tip produces a commanded force and then maintains the force at the constant magnitude. In the optimal control scheme, the control moment is produced as

$$M_p(i) = -\mathbf{H}\mathbf{x}(i) + \mathbf{K}z(i), \quad z(i+1) = z(i) + F_d - [F_s(i) + F_s(i-1)]/2. \tag{8}$$

Here, $F_d(i)$ is the commanded force and the gain matrix $\mathbf{H}$ and $\mathbf{K}$ are obtained by minimizing the quadratic performance function with a set of weight $(R, \mathbf{Q})$

$$J = \frac{1}{2} \sum_{i=0}^{\infty} [\mathbf{x_e}(i)^T \mathbf{Q} \mathbf{x_e}(i) + M_e(i+1) R M_e(i+1)], \tag{9}$$

where the space variable error $\mathbf{x_e}$ and the input error M_e are given by

$$\mathbf{x_e}(i) = \mathbf{x}(i) - \mathbf{x}(\infty), \quad M_e(i) = M_p(i) - M_p(\infty). \tag{10}$$

Here, $\mathbf{x}(\infty)$ and $M_p(\infty)$ are the final values of $x(i)$ and $M(i)$ when $F_s(i) \to F_d$.

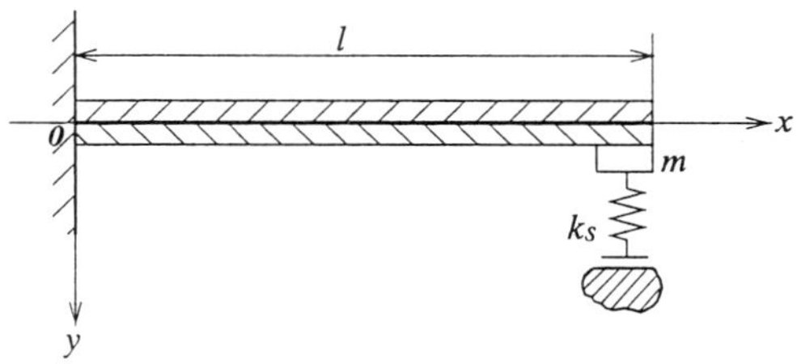

Fig. 1. Geometry of problem and coordinates.

Tab. 1. Physical parameters of bimorph finger.

Length(l)	36.97 mm
Width(b)	13.43 mm
Thickness(h)	0.93 mm
Density(ρ)	7.92×10^3 Kg/m^2
Young's modulus(E)	61.28 GPa
Damping coefficient(c)	1.89×10^{-5} Sec.
Mass of sensor(m)	5.05 g
Stiffness of sensor(K_s)	2110 N/m

For the case of the PID feedback control, the control moment is generated depending on the measured force error between the tip and the object, i.e.

$$M_p(i) = G_p[F_d - F_s(i)] + G_v[F_s(i) - F_s(i-1)]/T + \frac{1}{2}G_I \sum_{k=0}^{i}[2F_d - F_s(k) - F_s(k-1)]. \tag{11}$$

Here, G_p, G_v and G_I are the proportional, derivative and integral feedback gains.

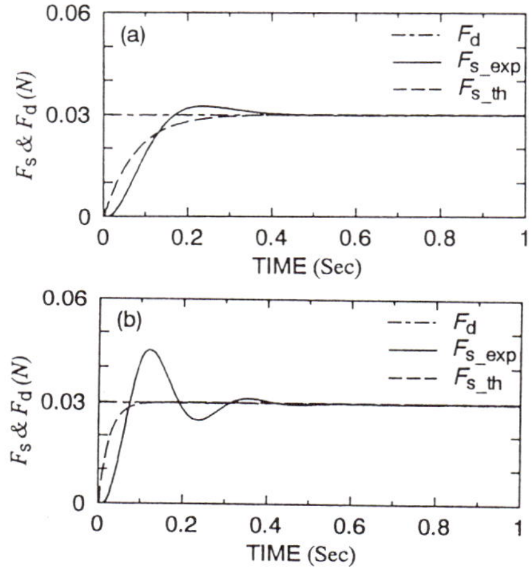

Fig. 2. Finger tip force obtained by optimal control scheme. The weights are (a) R=10^{-5} and Q=diag($10\ 10^3\ 10^2\ 10^6$) and (b) R=10^{-5} and Q=diag($10\ 10^3\ 10^2\ 10^5$).

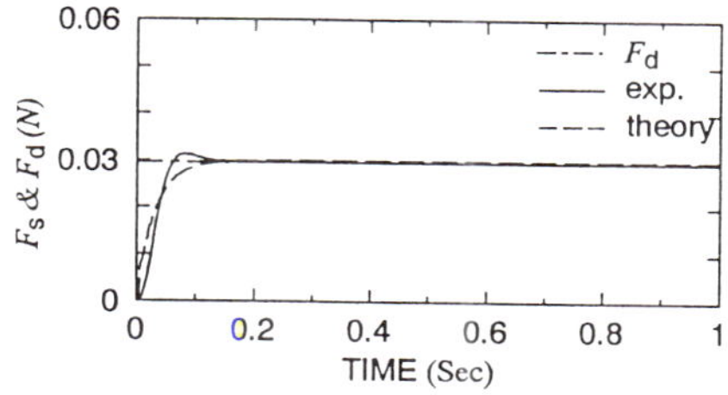

Fig. 3. Finger tip force obtained by PID control scheme. The feedback gains are G_p=0.07, G_v=0.006 and G_I=0.05.

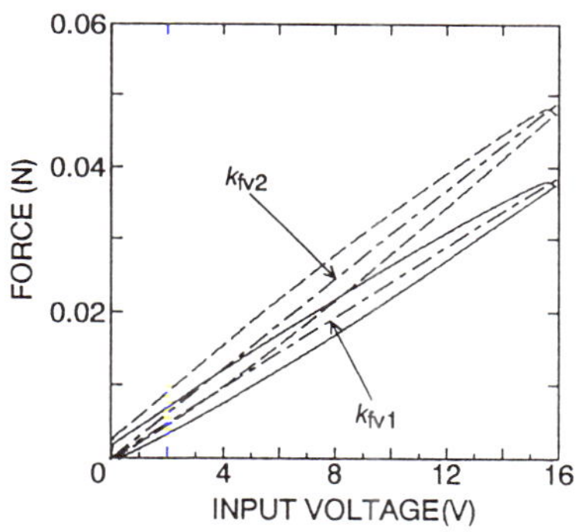

Fig. 4. Finger tip force for two different base mounting conditions; k_{fv1}=2.41× 10^{-3}(N/V), k_{fv2}=3.06× 10^{-3}(N/V).

3. Results and Discussions

In the following, one investigates the problem of controlling the finger so that the tip produces a small prescribed force and then maintains the force at the commanded constant magnitude. The physical parameters of the finger system are as given in Table 1.

Fig. 2 shows the results when the finger was controlled by the optimal control law. The one-dotted chained line shows the commanded constant force 0.03N. The dashed line is the simulated output from the force sensor, while the solid line is the corresponding result obtained experimentally. It is found from Fig.2a that the experimental result is in good correspondence with the theoretical one when the finger was driven so that the tip force produced comes to the desired value slowly. However, if the control is fast and the selected weight $(R, \mathbf{Q})$ is large, the control effect is degenerated in the experiment and the experimental result obtained shows no good correspondence with the theoretical one as evident from Fig.2b.

Fig. 3 is the case when the PID control law was introduced. Both the simulation and experimental results compare favorably. It is also noted that the PID control scheme works better than the optimal control law for the tip force control of the finger system under consideration.

Fig. 4 shows the variation of the force produced at the finger tip when the piezoelectric finger was loaded and then unloaded under two different clamped conditions. Since the bimorph finger is small, the mounting is rather difficult and the same mounting condition is not reproduced. The dashed lines are the linearized ratios obtained by applying the least square approximation to the hysteretic curves. The ratios are respectively $k_{fv1} = 2.41 \times 10^{-3}$(N/V) and $k_{fv2} = 3.06 \times 10^{-3}$(N/V), which are quite different from each other.

The robustness of the system under different mounting conditions was examined by using both the optimal control law and the PID control scheme. Fig.5 is the time variation of the finger tip force obtained for the two mounting conditions given in Fig.4. Fig.5a is the results when the finger was driven by the optimal control law with a constant set of weight(R, $\mathbf{Q}$). It is found that the controller works quite differently for the two cases. However, when the system was driven by the PID control law the results obtained are nearly the same even if the same feedback gains were used(Fig.5b). Thus, it is said that the PID control scheme is more suitable than the optimal control law for driving the piezoelectric bimorph finger that performs the task of the tip force control.

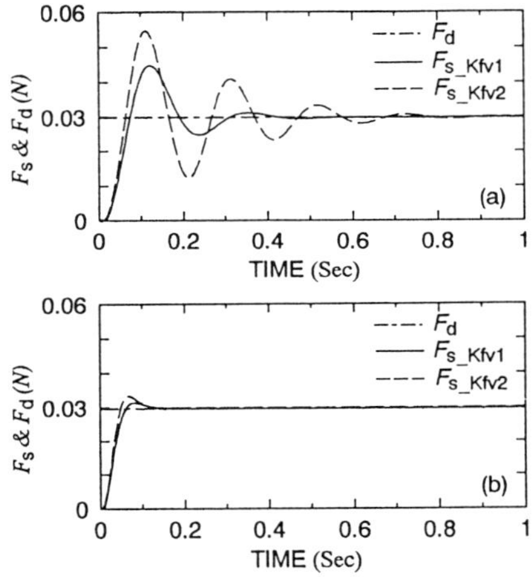

Fig. 5. Robustness of the finger under different mounting conditions. (a), optimal control, $R=10^{-5}$, $Q=diag(10\ 10^3\ 10^2\ 10^5)$; (b), PID control, $G_p=0.07$, $G_v=0.006$ and $G_I=0.05$.

4. Conclusions

1) The piezoelectric bimorph cell works well itself as a miniature finger holding a small object with a prescribed minute force.
2) Both the optimal and the PID control schemes have been tested to control the tip force. It has been found that the PID control is more effective for the force control of the bimorph finger.
3) Theoretical results are in good correspondence with the experimental ones, particularly for the case of the PID control, which means that the miniature gripper using the piezoelectric bimorph cell as a finger can be designed systematically with the aid of the computer simulation.

5. REFERENCES

[1] Z.W.Jiang, S.Chonan and J.Tani, "Position Control of a Flexible Arm Using Piezoelectric Bimorph Cells", ASME, J. Dyn. Sys., Meas. , and Cont., 113(2), 327-329, 1991.

[2] Z.W.Jiang, S.Chonan and J.Tani, "Tracking Control of a Miniature Flexible Arm Using Piezoelectric Bimorph Cells", Inter. J. Robot. Res., 11, 260-267, 1992.

[3] Z.W.Jiang, S.Sakuma, S.Chonan and J.Tani, "Control Ability of a Flexible Robot Arm Driven by Piezoelectric Bimorph Cells", Electromagnetic Forces and Applications, Elsevier Science Publishers, 23-26, 1992.

Micromachines

Simulation and Design of Applied Electromagnetic Systems
T. Honma (Editor)

First Functional Planar Rotational Magnetic Micromotors

H. Guckel, T.R. Christenson, K.J. Skrobis, J. Klein

Wisconsin Center for Applied Microelectronics
Department of Electrical and Computer Engineering
University of Wisconsin
Madison, WI 53706

ABSTRACT

Planar magnetic micromotors hold great promise for novel sensor systems. They can be fabricated by thick photoresist technologies such as LIGA to which surface micromachining and assembly is added in order to provide high quality magnetic materials and submicron bearing tolerances. Processing of this type has been used to fabricate reluctance machines with fully integrated current windings. Maximum achieved rotational speeds of 34,000 rpm with output torques of roughly 20 x 10^{-9} N-m have been achieved for preliminary structures which can be further improved with magnetic bearings and hysteresis motor designs. The reported structures contain photodiode shaft encoders which facilitate motor dynamic data extraction and offer the possibility of closed loop operation.

1. INTRODUCTION

Micromotors are electromechanical devices which are physically small. They therefore produce small torque outputs which implies that they can be influenced by small externally applied loading forces. This attribute suggests that micromotors will make excellent sensors and, in particular, will produce a family of sensors with autoranging capabilities if the transduction sensitivity is rotational speed sensitive. This statement is supported by research from the late 1950's in which magnetic levitation and rotation are used to produce rotational speeds of up to 1.6 million rps in accelerometers which were limited in sensitivity by Brownian motion [1].

Micromotors are components for micro electro mechanical systems or MEMS. They can be implemented via electrostatic or magnetostatic energy conversion schemes. Since silicon technologies do not normally involve magnetic materials the appearance of electrostatic silicon motors is not surprising. These machines have several drawbacks when compared to magnetic actuators. They cannot participate in the richness of possible motor types which magnetics allows and are in essence restricted to magnetic reluctance motor equivalents which produces maximum speed restrictions due to torque ripple. Magnetic bearing issues present another problem area and, as far as MEMS are concerned, are joined by microelectronic difficulties with high speed, high voltage control circuitry for high impedance loads. The observation that a magnetic micromotor implies an electric microgenerator tilts the scale towards a serious effort in micromagnetics.

2. TOOL ISSUES FOR PLANAR MAGENTIC MICROMOTORS

Construction techniques for micromagnetic motors for MEMS applications require first of all a material base which provides precision ferromagnetic parts. There is a need for both soft and hard magnetic material with reproducible and predictable B-H behavior. If one envisions planar construction techniques the structural height of the materials should be as large as possible. The shape of these prismatic structures must not deviate with height if precision bearing surfaces are to be produced which is required for controlled frictional behavior. The anticipated need for coils for current to flux conversion can be accommodated in planar processing if multiple mask levels can be used to produce these non-prismatic structures. An additional requirement originates from intended MEMS rather than micromotor applications: the entire process must be compatible with silicon microelectronics technology to provide for the needed sensors for closed rather than open loop control.

The majority of the above requirements have been met in the processing sequence which is now functional at Wisconsin. A major component of this construction technique involves a modified form of LIGA, a processing procedure which originated in Germany [2]. This process, LIGA, falls under the general heading of thick photoresist technology: up to 500 μm in thickness. Photoresist thicknesses of this magnitude require high energy photons for exposure which in the present case are provided by highly collimated x-ray fluxes from a synchrotron source. Since exposures of this type use atomic absorption processes standing wave problems are absent and prismatic definitions with run-outs of less than 0.1 μm per 100 μm of photoresist thickness are possible after developing of the exposed photoresist pattern. The end result is a prismatic photoresist shape with large structural height and nearly perfect edge acuity. If this polymer structure is attached to a substrate which is planar and is covered with a non-patterned plating base, say 150 Å of Ti followed by 150 Å of Ni, the photoresist recesses can be filled with electro plated metal which shares the geometric properties of the photopolymer.

The electroplated metal which this process uses can of course be ferromagnetic. Nickel, in particular, and iron have been studied in some detail and are augmented by Ni-Fe alloy deposits. The magnetic behavior of these materials have been studied by fabricating miniature cores with an airgap, winding them with several turns of wire and measuring the gap field with a gauss meter as a function of coil current. The procedure produces B-H data with a reasonable accurate H-axis and an uncalibrated B-axis which may be calibrated by a known material sample which has been measured on a suitable magnetometer. Based on these measurements it is concluded that electro plated nickel as plated makes a rather poor magnetic material: high coercivity and low permeability. Annealing in nitrogen or hydrogen improves the magnetic behavior significantly [3]. This is in part due to strain reduction and in part due to hydrogen removal and wetting agent removal during the anneal cycle. Coercivity reductions to 2 oersted and increases of saturation flux densities to 6000 - 8000 gauss occur and make the material acceptable for any magnetic applications. The anneal behavior for pure iron is similar to nickel. Alloy plating is not mature enough to yield significant conclusions.

The need to anneal and the need to use a silicon substrate are in conflict because of thermal expansion and metallurgy difficulties. This conflict can be resolved if metal parts which are to be annealed can be separated from the substrate, are then annealed and are assembled on the substrate to form the actuator. This approach requires a modified version of LIGA which permits fully attached metal parts, partially attached structures and fully freed metal shapes. This approach appears to be quite complex and indeed it is. However, it also contains an important benefit: the achievement of submicron bearing tolerances by the subtraction of two optically defined mask dimensions. The procedure is best illustrated by example. Thus, a shaft with a 100 μm diameter can be fabricated with a tolerance of 100 ± 0.05 μm without much difficulty. A hole through a gear which is to be mounted on this shaft can be produced with a dimension of 100.5 ± 0.05 μm. The assembled part which takes advantage of the high edge acuity of the process ends up with a nominal bearing tolerance of

± 0.25 μm which is adequate for reproducible friction in the assembly. Figure 1 illustrates the result.

The Wisconsin processing sequence is summarized as follows:

(1) Silicon substrate with pre-fabricated electronics and lower coil sections
(2) Optically defined polyimide sacrificial layer
(3) Unpatterned plating base application
(4) Basic LIGA process
(5) Partial sacrificial layer removal
(6) Assembly and packaging

3. MICROMOTOR RESULTS

The processing tool as outlined in the previous section has been applied to three types of magnetic micromotors.

The first motor, a reluctance machine without current windings was constructed to measure the frictional behavior of nickel - nickel interfaces. For this purpose a two-phase or four pole magnetic stator circuit was designed in which the rotor is located. The rotor interior was designed with an easy-axis / hard-axis interior. The rotor circumference was decorated with gear teeth to allow for outside gear loading. Magnetic drive was supplied by locating a rotating permanent magnet or electromagnet on the bottom side of the substrate and observing rotor rotation at minimum required magnetic field. The measured field strength to overcome friction was found to be roughly 6 gauss. This would suggest that frictional losses for the device consume roughly 0.1% of the saturation flux density or that the nickel - nickel system

Figure 1. SEM photograph of assembled rotor with a rotor-shaft tolerance of 0.5μm.

Figure 2. SEM photograph of completed magnetic micromotor.

can support three decades of load torque. This is adequate for practical applications and is far superior to silicon micromotor performance where 50% of the available torque is typically consumed by friction.

The goals for the second planar magnetic micromotor were very different from that of the first machine. Again annealed nickel is used for the magnetic material. This time a three phase design was implemented for the stator and a four pole rotor was used leading to a 6 : 4 machine. All six stator segments were furnished with 9 turn windings for current to flux conversion. These windings are implemented as a U-shape during processing and are completed by wire bonding with 30 µm diameter aluminum alloy wire. The substrate was pre-processed with pn-junction photodetectors which allow for the measurement of the rotor dynamics and for all practical purposes form a shaft encoder. The stator thickness was chosen at 200 µm, the rotor thickness as 100 µm. The shaft was fluted to produce an air bearing effect. Figure 2 illustrates the machine.

The machine was tested with 3 phase non-overlapping, bi-polar square waves. Operation of the reluctance motor confirmed calculations which pointed towards stepping motor performance with significant over and under shoot. Frictional effects were found to be very small as evidenced by similar performance in room ambient and hard vacuum. This observation is in part related to the fact that the rotor levitated by some 45 µm as a consequence of the vertical reluctance variation. Maximum rotational speed scaled with the moment of inertia of the rotor and non-skipping operation to 34,000 rpm was achieved for a 280 µm rotor design. The upper speed in this open loop test is limited by severe torque ripple which is typical for this type of machine. An estimated output torque of 20 x 10^{-9} N-m was produced and exceeds silicon micromotor performance by at least three decades. Additional torque production will occur if the stator reluctance losses are decreased which will permit gap reductions for better performance. Further speed increases are possible if the rotor moment of inertia is reduced still further. However, it is also clear that very high speed operation for a low inertia reluctance or electrostatic motor will always be impeded by large torque ripple for open loop operation. A more constant torque versus rotation angle characteristic will avoid this difficulty.

The above consideration led to a preliminary investigation of a classic hysterisis motor which uses a disc shaped rotor with four small holes for photodiode interrogation. A device diameter of 280 µm with a 70 µm nickel thickness produced stable operation to 8000 rpm. Higher speeds produced vertical instabilities which are related to eddy currents in this non-optimized structure. The results indicate that the magnetic material is good enough to pursue this type of machine which holds the promise of high speed operation.

4. CONCLUSIONS

The very preliminary results which have been reported here indicate that planar magnetic micromotors are feasible and will eventually reach high rotational speeds. This goal will be achieved with better motor designs which will use improved magnetic materials, and certainly will involve permanent magnets and novel magnetic bearings. Their application to new sensor concepts which cannot be duplicated by other techniques will assure their economic viability.

5. REFERENCES

1 J.W. Beams, Physics Today, v 12, n 7, Jul. 1959, pp. 20-27.
2 E.W. Becker, W. Ehrfeld, P. Hagmann, A. Maner, D. Münchmeyer, *Microelectronic Engineering*, 4, 1986, pp. 35-56.
3 R.M. Bozorth, *Ferromagnetism*, 1951, Van Nostrand.

Simulation and Design of Applied Electromagnetic Systems
T. Honma (Editor)

Design of An Optical Servo System
(Applications to An Optical Microgripper and An Optical Walking Robot Using Bimorph PLZT Actuators)

T. Fukuda[a], S. Hattori[b], F. Arai[a], H. Matsuura[a] and S.Nagamori[c]

[a]Department of Mechano-Informatics and Systems, Nagoya University, Furo-cho, Chikusa-ku, Nagoya 464-01 Japan

[b]Central R&D Laboratories, Nippon Densan Corporation, Tsutsumisotomachi 10, Nishikyogoku, Ukyo-ku,Kyoto 615, Japan

[c]Nippon Densan Corporation, Karasuma-dori, Oike, Nakagyo-ku, Kyoto Japan

Abstract
We focus on an optical servo system using bimorph type optical piezoelectric actuator (PLZT) which has relatively big displacement. We made two application examples of the optical servo system. One is the optical microgripper, and the other is the walking robot. We propose conceptual design of these applications, and show an experimental results of the optical servo system applied to these examples.

1. INTRODUCTION

Research works of a micro machine are performed prosperously in recent years. Many kinds of actuators have been considered as an actuator of the micro machine [1-4]. Almost of these actuators are driven by electric energy.

Recently in the field of micro electro mechanical systems, a few studies have been made at an optical actuator which can be operated by optical energy source. Utilization of optical energy as driving energy source of the actuator has many advantages compared with utilization of electric energy. When we consider speedy changing of informational communication technology, it is thought that these merit especially works more advantageously.

As the research works of the optical piezoelectric element [5,6], there have been reported about relation between material, composition and response characteristics, manufacturing method, application to the optical switch, application to the hydroric control system as the micro valve, and so on. We proposed a photo response model of board like rectangular parallelogram [7].

2. CONCEPTUAL DESIGN OF OPTICAL SERVO SYSTEM

Figure 1 shows conceptual design of the optical servo system. In this system, feedback controller is employed, and position control system is designed. Displacement of the bimorph optical piezoelectric actuator is measured by the non-contact sensor, and U.V. beam irradiation intensity is adjusted by controlling diaphragm of the optical system. For precise feedback control, the other sensing information such as temperature at surface, intensity of illumination, and so on can be used.

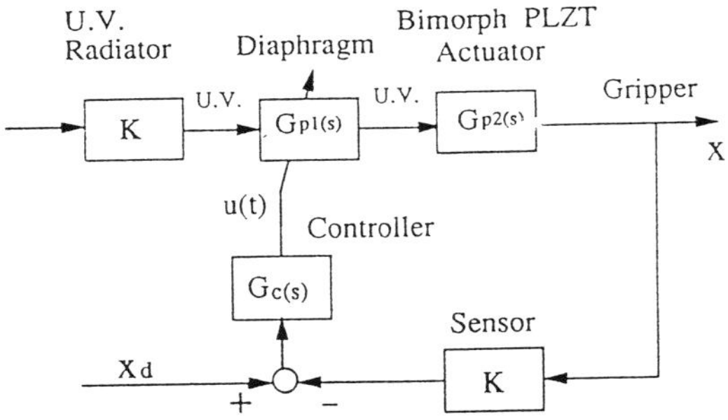

Figure 1. Block diagram of optical servo system.

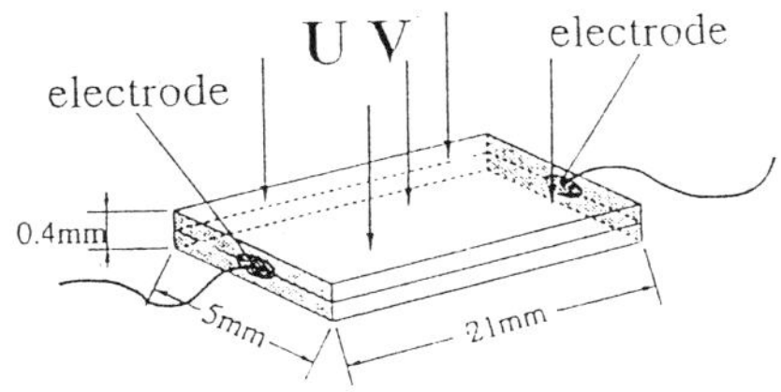

Figure 2. Bimorph-PLZT actuator.

3. BIMORPH OPTICAL PIEZOELECTRIC ACTUATOR

In this optical servo system, we used the bimorph-PLZT actuator. An PLZT is piezoelectric element of $(Pb,La)(Zr,Ti)O_3$ ceramics. This PLZT element is a ceramic polycrystalline material made of $Pb(Zr,Ti)O_3$, which is solid solution of $PbTiO_3$ and $PbZrO_3$, and La_2O_3. In the experiments, the test piece we have used is PLZT(3/52/48) which is made of 3 % of La in total weight and whose mol ratio of $PbZrO_3$ and $PbTiO_3$ is 52/48. La is thought having the highest photostrictive effect.

Figure 2 shows configuration of the bimorph-PLZT actuator. Two pieces of PLZT after polarization whose configuration is 21mm x 5mm x 0.2mm is glued so that each polarization direction is opposite to the other's. Polarization is finished at 50 °C by applying 25kV. Electrodes are set at both end of the test piece by vaporization of silver. Each PLZT thin plates are conducted [8].

Figure 3 shows time history of displacement at tip of the bimorph-PLZT actuator by U.V.Beam Irradiation. Total measurement time is 200 seconds, and U.V.beam is irradiated constantly for first half 100 seconds and it put out for last half 100 seconds. The experiments are conducted for three cases of illumination intensity. From experimental results, the maximum displacement of the bimorph-PLZT actuator can be over 50 μm within several seconds in response time in all cases.

4. APPLICATIONS

We made two application examples of the optical servo system using bimorph-PLZT actuator.

4.1 Optical microgripper

Figure 4 shows configuration of the optical microgripper. This optical microgripper has two bimorph actuators each of which has a needle at the tip, and two mirrors which reflect U.V. beam at the surface of the bimorph-PLZT from the optical fiber are set at both side of the actuators. The size of this microgripper is 45mm (or 80mm is including the length of a needle) in length, 32mm in width, and 10mm in height. U.V. beam is irradiated by the optical fibers to each actuators of the optical microgripper through the mirror independently. The displacement of the microgripper at tip is 100 μm.

Figure 5 shows an experimental result of the optical microgripper controlled by the proposed optical servo controller. Tip position of the PLZT actuator is controlled with this control system. This gripper has strong anti-noise characteristics and is prosperous for the medical application such as the micro surgery.

4.2 Optical walking robot

Figure 6 shows the optical walking robot using the bimorph-PLZT actuator. It has two bimorph actuators, and U.V. beam is irradiated by the optical fibers to the actuators. There is no power supply cable required for actuation of this mobile robot. This robot moves by controlling the U.V. beam irradiation pattern to the actuator, and maximum speed is 1.6 μm/s.

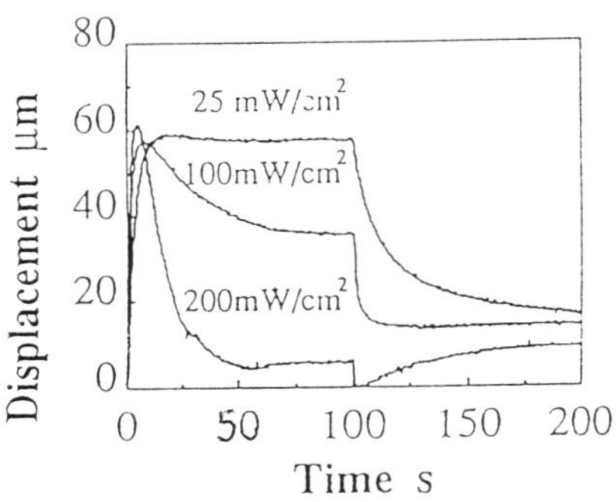

Figure 3. Time history of displacement of bimorph-PLZT actuator.

Figure 4. Prototype of optical microgripper.

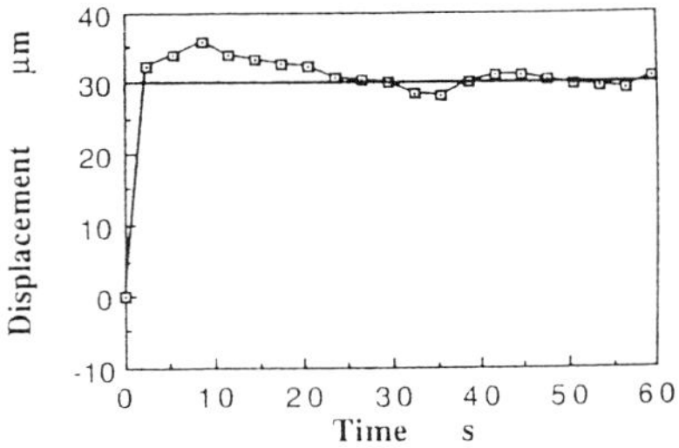

Figure 5. A result of servo control experiments.

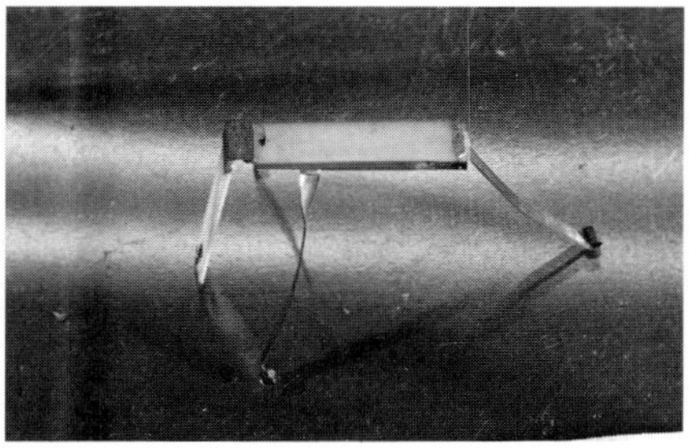

Figure 6. Prototype of optical walking robot.

5. CONCLUSIONS

We made the bimorph optical piezoelectric actuator which has relatively big displacement and measured its response characteristics for U.V. beam irradiation. Response time of this actuator is improved in the experiments compared with the previous type whose configuration is board like rectangular parallelogram. And we proposed the conceptual design of the optical actuator - servo system, and showed the application examples for the optical microgripper and the optical walking robot. The position control experiments based on the proposed control system is carried out. These results convince us applicability of the optical actuator in our future.

6. REFERENCES

1 T. Fukuda, T. Tanaka, " Micro Electro Static Actuator with Three Degree of Freedom", Proc. of IEEE Int. Workshop on Micro Electro Mechanical Systems (MEMS), pp. 153-158,1990.

2 S.Egawa, T. Higuchi, " Multi-Layered Electro-static Film Actuator", Proc. of IEEE Micro Electro Mechanical Systems Workshop, pp.17-24,1989.

3 T. Fukuda, H. Hosokai," Gaiant Magnetostrictive Alloy (GMA), Applications to Micro Mobile Robot as a Micro Actuator without Power Supply Cables", Proc. of IEEE int. Workshop on Micro Electro Mechanical Systems (MEMS), pp.210-215,1991.

4 T. Fukuda, M. Fujiyoshi, K. Kosuge and F. Arai,"Electrostatic Micro/Nano Manipulator with 6 D.O.F.",ASME,DSC-Vol.32, Micromechanical Sensors, Actuators, and Systems, pp.197-202,1991.

5 K. Uchino and M. Aizawa," Photostrictive actuator using PLZT ceramics', Jpn. j. Appl. Phys.,24, Suppl. 42-3, pp.139-142,1985.

6 M. Inoue, T. Sada and K. Uchino, " Photo driven relay using PLZT ceramics", ISAF'86, Proc. of the Sixth IEEE Int. Symp. Appl. Ferroelectrics,pp.16-19,1986.

7 S. Hattori, T. Fukuda, S. Nagamori,"A Study on Optical Piezo-electric Actuator (Response Experiments by U. V. Beam and Photo Response Model)",Journal of Robotics and Mechatironics, Vol.4, No.4, pp.321-329,1992.

8 T.Fukuda, S.Hattori, F.Arai, H.Matsuura, T. Hiramatsu, Y.Ikeda and A.Maekawa, " Optical Servo System Using Bimorph Optical Piezo-electric Actuator", Proc. of Third Int. Symp. Micro Machine and Human Science p.45-50, 1992.

Simulation and Design of Applied Electromagnetic Systems
T. Honma (Editor)

Electro-Thermomechanical Modeling and Simulation of Dynamic Behavior of a Shape Memory Alloy Actuator

S.Biwa, K.Yamada, E.Matsumoto and T.Shibata

Department of Mechanical Engineering, Kyoto University,
Kyoto 606-01, Japan

Abstract

This paper describes an attempt of continuum-mechanical modeling of the TiNi shape memory alloy wire and its application to the simulation of the dynamic response of an electrically driven actuator. The one-dimensional rate-type constitutive equation and the energy equation are reduced to forms appropriate for the dead-weight bias actuator and the material constants are experimentally determined. The response of the actuator to stepwise voltage change is calculated and an example of feedback control is simulated.

1.INTRODUCTION

Shape memory alloy (SMA) wires can be used as elements for the actuator converting the electric power into the mechanical one due to the electrically produced Joule heat through the thermoelastic martensitic transformation. Many elements have been designed so far, and their characteristics have been examined theoretically as well as experimentally. In these studies, however, theoretical analyses are mainly based upon simple approximation in which the actuator is, for example, regarded as a transfer element with first-order lag [1].

The attempts to analyze the behavior of the electrically driven SMA actuators based on the continuum-thermomechanical models have been done by several authors [2,3]. In this paper we use the thermomechanical constitutive equation and the energy equation [4] to derive the equations governing the variation of the temperature and the strain. This enables us to simulate the transient behavior of an SMA actuator, by solving the derived equations for the electric voltage given as a function of time.

2.THERMOMECHANICAL CONSTITUTIVE EQUATION AND ENERGY EQUATION

The energy equation for the one-dimensional material can be written as

$$\dot{U} = \frac{1}{\rho_R}\sigma\dot{\varepsilon} - \frac{1}{\rho_R}q' + r, \tag{1}$$

where U denotes the internal energy density, σ — stress, ε — strain, q — heat flux, r — heat supply and ρ — mass density. The suffix R denotes a quantity in the reference state, and the prime and the dot denote differentiation with respect to space and time, respectively. By introducing the Helmholtz free energy density, $\psi(\varepsilon, \theta, \xi)$, where θ and ξ are the temperature and the martensite fraction, and using the Clausius-Duhem inequality, we obtain the following constitutive law:

$$\sigma = \rho_R \frac{\partial \psi}{\partial \varepsilon}, \tag{2}$$

or, in a rate type,

$$\dot{\sigma} = D\dot{\varepsilon} + \Theta\dot{\theta} + \Omega\dot{\xi}, \quad D = \rho_R \frac{\partial^2 \psi}{\partial \varepsilon^2}, \quad \Theta = \rho_R \frac{\partial^2 \psi}{\partial \varepsilon \partial \theta}, \quad \Omega = \rho_R \frac{\partial^2 \psi}{\partial \varepsilon \partial \xi}. \tag{3}$$

According to Tanaka [4], ξ can be expressed as a function of σ and θ during the transformation process, i.e.,

$$\xi = \Xi(\sigma, \theta). \tag{4}$$

The above equality also gives the starting condition of the transformation. The function $\Xi(\sigma, \theta)$ takes different forms for the martensitic and the reverse transformations, say, $\Xi_M(\sigma, \theta)$ and $\Xi_R(\sigma, \theta)$. We employ

$$\left.\begin{aligned} \Xi_M(\sigma, \theta) &= 1 - \exp\{b_M c_M(\theta_{Ms} - \theta) + b_M \sigma\}, \\ \Xi_R(\sigma, \theta) &= \exp\{b_A c_A(\theta_{As} - \theta) + b_A \sigma\}, \end{aligned}\right\} \tag{5}$$

where b_M, c_M, b_A and c_A are assumed to be material constants and θ_{Ms} and θ_{As} denote the temperatures corresponding to M_S and A_S points, respectively. The martensitic transformation proceeds if Ξ_M increases, and so does the reverse one if Ξ_R decreases. Otherwise ξ remains constant and elastic deformation takes place.

Omitting an intermediary derivation, eq.(1) is reduced to

$$C\dot{\theta} - \frac{\theta}{\rho_R}\Theta\dot{\varepsilon} - H\dot{\xi} + \frac{1}{\rho_R}q' = r, \quad C = -\theta\frac{\partial^2 \psi}{\partial \theta^2}, \quad H = -\frac{\partial \psi}{\partial \xi} + \theta\frac{\partial^2 \psi}{\partial \theta \partial \xi}. \tag{6}$$

3. CONSTITUTIVE MODELING OF THE TiNi SHAPE MEMORY ALLOY WIRE

To describe the quantitative material behavior, we assume a simple form for ψ considering the temperature dependence of the elastic constants and the recoverable transformation strain. Only linear dependence is taken into account here. Then the coefficients in eq.(3) are simplified as

$$\left.\begin{aligned} D &= D_0\{1 + \alpha(\theta - \theta_0)\}, \quad \Theta = D_0(\alpha\varepsilon - \pi\varepsilon_R\xi), \\ \Omega &= -D_0\varepsilon_R\{1 + \pi(\theta - \theta_0)\}. \end{aligned}\right\} \tag{7}$$

Material properties such as the electric resistivity reveal variation during the phase change. Here the resistivity r_e is assumed to be written as

$$\left.\begin{aligned} r_e &= (1 - \xi)r_A(\theta) + \xi r_M(\theta), \\ r_A(\theta) &= \begin{cases} r_{A0}\{1 + \kappa_A(\theta - \theta_0)\}, & \theta < \theta_1 \\ r_{A0}\{1 + \kappa_A(\theta_1 - \theta_0)\}, & \theta \geqq \theta_1 \end{cases} \\ r_M(\theta) &= r_{M0}\{1 + \kappa_M(\theta - \theta_0)\}. \end{aligned}\right\} \tag{8}$$

The above constants are determined by experiments. The constants which appear in eqs.(5) and (7) are determined from the measured stress-strain curves, and those in eq.(8) from the resistance measurements during the loading-unloading of the specimen, both under some different constant temperatures. Determined constants are:
$D_0 = 22110$[MPa], $\alpha = 0.0071$[1/K], $\varepsilon_R = 0.0401$, $\pi = 0.0115$[1/K],
$b_M = -0.051$[1/MPa], $c_M = 6.02$[MPa/K], $b_A = 0.077$[1/MPa], $c_A = 6.67$[MPa/K],
$\theta_{Ms} = 278$[K], $\theta_{As} = 330$[K], $\theta_0 = 293$[K], $\theta_1 = 333$[K]
$r_{M0} = 6.2\times10^{-4}$ [$\Omega\cdot$mm], $\kappa_M = 1.82\times10^{-3}$ [1/K],
$r_{A0} = 5.9\times10^{-4}$ [$\Omega\cdot$mm], $\kappa_A = -9.07\times10^{-3}$ [1/K].
The model well describes the experimental data as shown in Figure 1 and Figure 2.

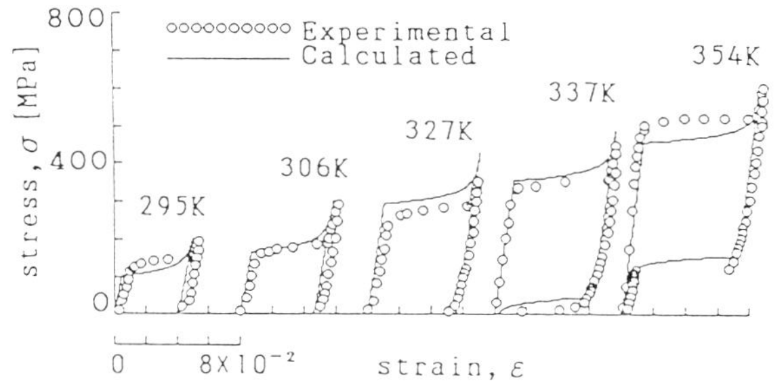

Figure 1. Stress-strain curves.

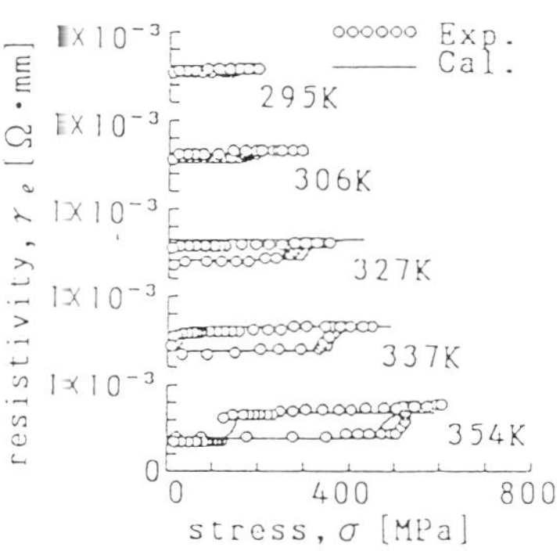

Figure 2. Variation of resistivity with stress.

4.DYNAMIC RESPONSE OF AN ACTUATOR

We consider a simple type of SMA actuators, so called dead-weight bias type. An SMA wire is fixed at one end and loaded by a dead weight. The stress is kept constant. The temperature of the wire rises when heated by the Joule effect and falls when cooled by the heat transfer into the air. The output is the displacement, or the strain, of the wire. When heated the wire contracts due to the shape memory effect, and when cooled it is elongated due to the stress-induced martensitic transformation.

Assume that the state of the wire is uniform and that the heat supply consists of the Joule heat and the heat transfer into the air, then from (6) we obtain

$$C_V\dot{\theta} - \Theta_V\dot{\varepsilon} - H_V\dot{\xi} = \frac{V^2}{R_e} - \tau a(\theta - \theta_r), \quad R_e = r_e\frac{4l}{\pi d^2}, \tag{9}$$

where

$C_V = v\rho_R C$, $\Theta_V = v\Theta\theta$, $H_V = v\rho_R H$

are simply assumed to be constants. And V is the voltage, θ_r the environmental temperature and τ the constant of heat transfer into the air. The geometrical dimension is determined by the length l and the

diameter d, thereby the volume $v = \pi d^2 l / 4$ and the surface area $a = \pi d l$. We can obtain the variation of θ, ξ and ε by solving eqs.(3), (4) and (9) with the condition that σ is constant. Figure 3 shows the variations of the temperature and the strain, when the stepwise voltage is applied to the wire. We have put the constants:
$C_V = 0.012$[J/K], $\Theta_V = 3.0$[J], $H_V = 0.25$[J], $\tau = 0.00023$[J/(mm^2·K·sec)], $\theta_r = 298$[K], $l = 60$[mm], $d = 0.2$[mm], $\sigma = 245$[MPa].
The experimental data are also shown, where the curve is more rounded but shows similar behavior.

5.SIMULATION OF FEEDBACK CONTROL : AN EXAMPLE

To achieve the full performance of the actuator, design of an effective control system is important. We demonstrate an example of the feedback control, where the position (strain) is controlled by manipulating the electric voltage according to the continuously monitored position (strain). The control strategy is simple: for the reference strain ε_r, if ε exceeds $\varepsilon_r + \delta$ then the voltage V is set as V_0, and V is reset to be 0 when ε falls below $\varepsilon_r - \delta$, where δ is the permissible error prescribed as 0.001. A computed result is shown in Figure 4. We have set the reference strain piecewise constant and put $V_0 = 1.2$[V]. By the simple on-off method described above, the figure indicates that the strain can be kept near the reference value, although there exists some amount of delay and the voltage should be supplied periodically.

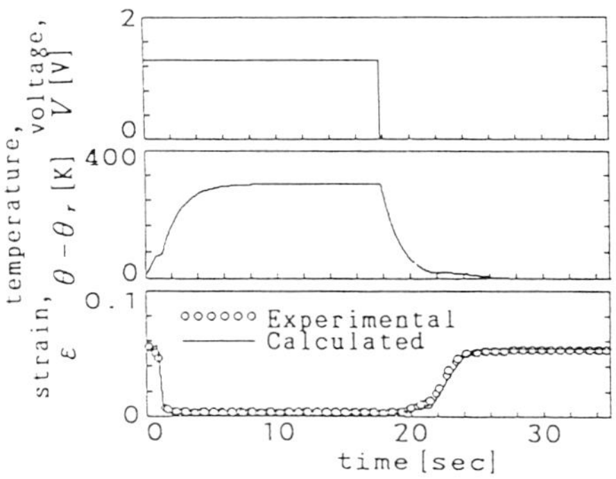

Figure 3. Response to stepwise voltage change.

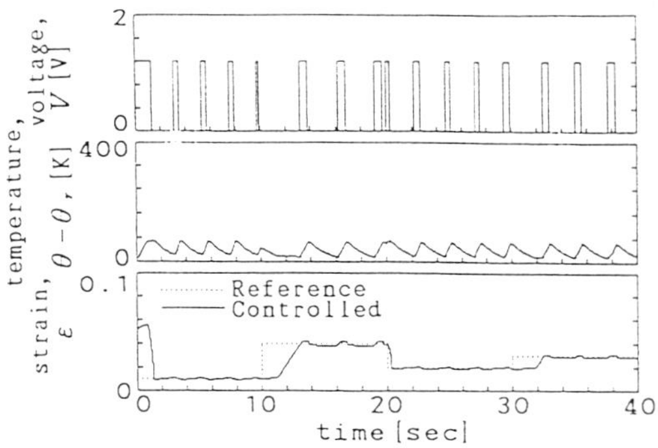

Figure 4. Simulated feedback control.

6.REFERENCES

1 A.Baz,K.Iman and J.McCoy, J.Intell.Mater.Syst.and Struct.,1(1990)105.

2 E.Matsumoto,T.Furuta and T.Shibata, Mechanical Behaviour of Materials-VI(eds.M.Jono and T.Inoue),Vol.3,599,Pergamon Press,1991.

3 C.Liang and C.A.Rogers, Trans.ASME,J.Mech.Design,114(1992)223.

4 K.Tanaka, Res Mechanica,18(1986)251.

Simulation and Design of Applied Electromagnetic Systems
T. Honma (Editor)

The film transformer

S.Hayano, Y.Midorikawa and Y.Saito

College of Engineering, Hosei University, Kajino, Koganei, Tokyo 184, Japan

Abstract

In order to realize the ultra light and small power supplies, we propose here a film shape transformer. This new transformer is constructed by the chemical etching processes. Experimental works show that this film transformer takes about 98 percent coupling factor over 500kHz and operates over 95 percent efficiency at 1.5MHz.

1. INTRODUCTION

With the developments of modern electronic devices such as the book size computer, word processor and codeless telephone, it is essentially required to reduce the size and weight of their electric power supplies. In order to design small and compact electric power supplies, it is required to reduce the size of magnetic devices, e.g. reactor and transformer. One of ways to reduce the size of a magnetic device is to employ high frequency excitation [1,2]. In such a high frequency exciting condition, a serious problem is that the major performance of the device is dominated by the frequency characteristics of the core magnetic materials. To overcome this difficulty, new magnetic materials typically represented by amorphous magnetic material have been exploited and utilized [3,4]. Nevertheless, it is difficult to avoid an essential increase of iron loss in accordance with the rise of operating frequency. Another solution of this problem is to exploit a coreless transformer. Previously, we have succeeded in realizing the coreless transformer with high efficiency and demonstrated its usefulness for the DC to DC converters [5,6].

In the present paper, an ultimate thin and light weight high frequency transformer named the film transformer is proposed. This new transformer is composed of the lamination of thin film conductors. Each of the films is constructed by the chemical etching processes. Basic operating principle is the skin effects similar to those of our previously exploited coreless transformer [5]. This fact was confirmed by comparing their frequency characteristics. Depending on the load conditions efficiency of the film transformer reaches over the 95%. Thus, we experimentally verify that our film transformer is applicable to the practical power supplies.

2. THE FILM TRANSFORMERS

2.1 Principle

Figure 1(a) shows a typical core type conventional transformer which utilizes the magnetic flux linking the conductors wound around the magnetic core. On the other side, our film transformer utilizes the magnetic flux enclosing the current carrying conductors as shown in Fig.1(b). This basic operating principle is the same as those of the twisted coils transformer [1]. However, in order to couple the primary and secondary circuits in a flat surface, we arrange both of the primary and secondary coils in a circular, rectangular or L shape with proper separations.

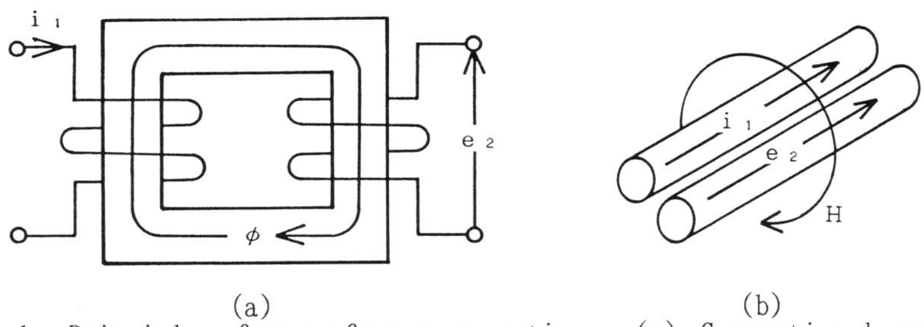

Figure 1. Principle of transformer operation. (a) Conventional core type transformer, and (b) new type transformer.

Figure 2(a) shows a film transformer whose primary and secondary coils are arranged in a coaxially circular shape. When alternating current is flowing through the primary coil, the magnetic flux starts from the center to outer on the top surface and returns from outer to center on the under surface of coaxially circular arranged coils or viceversa. This magnetic flux essentially induces a voltage in a secondary coil because the secondary coil is arranged in a coaxially circular shape on a film base common to primary and secondary coils.

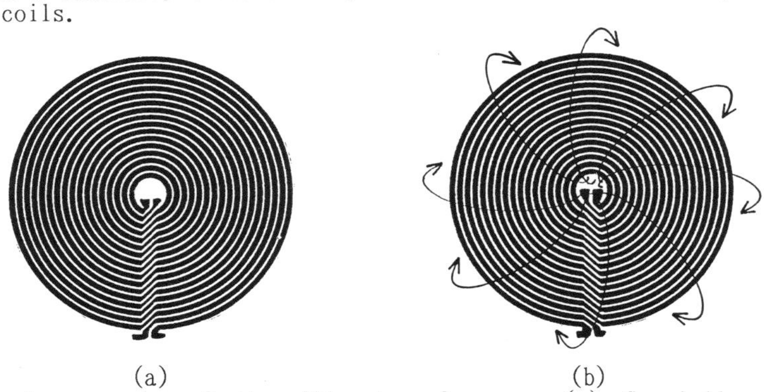

Figure 2. An example of the film transformer. (a) Coaxially circular arranged coils, and (b) a state of magnetic flux flow.

2.2 Practical considerations

In order to realize the film transformer for practical use, at least, three points should be taken into account the design of film transformer.

The first is concerning with the terminal arrangement of the primary and secondary coils. It is desirable to install their terminals at the outside of film. Let us consider the coil arrangement shown in Fig. 2(a), then it can be observed that one of the primary terminals locates at the outside and the other locates at the center of the film base. This means that connection to the electric source should be done crossing over the transformer coils. This condition is the same about secondary connection. To remove this fault, the coaxial circular arranged coils having symmetrical pattern to those of Fig. 2(a) is so laminated as to coincide with the direction of magnetic flux flow, and connected their terminals at the center. Thus, it is possible to install both of the primary and secondary terminals at the outside of film.

The second is concerning with the transformer ratio. As described in the reference [1], the transformer ratio of this type transformer depends on the length of coils, i.e. a transformer ratio corresponds to those of the primary and secondary coils length. If a primary coil is longer than those of secondary, then lower voltage can be obtained at the secondary circuit but leakage flux of the primary coil is essentially increased. Even though a large number of laminations is required, this fault can be removed by changing the circuit connection. For example, let us consider a pair of film transformer. If two primary coils are connected in series and two secondary coils are connected in parallel, then it possible to get a half of primary impressed voltage at the secondary terminals.

The third is concerning with the improvement of coupling factor. Because of an edge effect at the outer side, the magnetic flux does not flow ideally as shown in Fig.2(b) but it may take the closed paths penetrating the base films. Namely, the magnetic flux flow normal to the film surface is caused by the edge effect at the outer side. Obviously, this bypassing magnetic flux makes the coupling factor lower. One of the solutions of this problem is to increase the number of laminations.

2.3 Experimental

We made the three types of trial film transformers by the chemical etching processes. One is the coaxially circular coil pattern shown in Fig.2(a) and the others are shown in Fig.3.

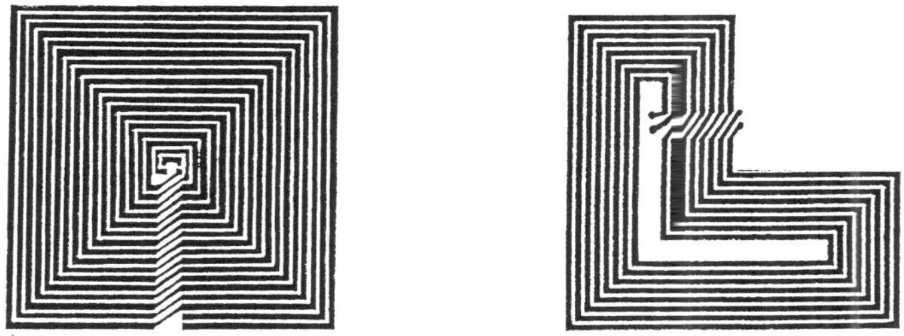

(a) (b)

Figure 3. The trial film transformer patterns. (a) Coaxially square arranged coils, and (b) coaxially L shape arranged coils.

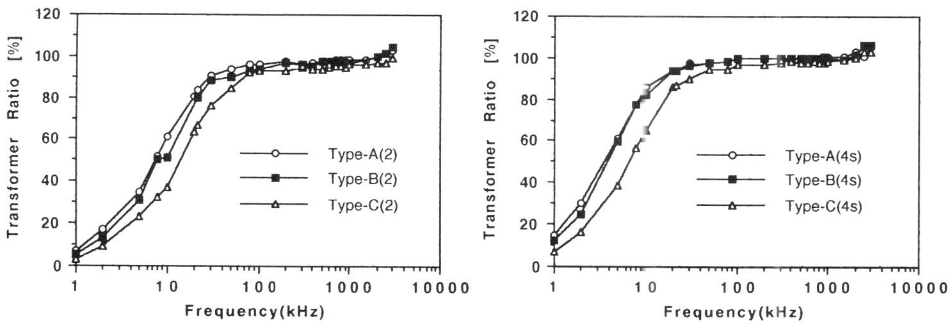

(a) (b)

Figure 4. The frequency characteristic of the transformer ratio, which corresponds to the coupling factor at high frequency (a) Two layers, and (b) four layers (series connected). Types A, B and C are the circular, rectangular and L shape film transformers, respectively.

Figure 4 shows the frequency characteristic of the transformer ratio, which corresponds to the coupling factor at high frequency. In this figure, 4(a) shows a transformer ratio of the film transformers composed of a series connected two films. On the other side, Fig.4(b) show a transformer ratio of the film transformers composed of a series connected four films. The difference between them suggests that the frequency characteristic of the coupling factor can be improved by increasing the number of film lamination.

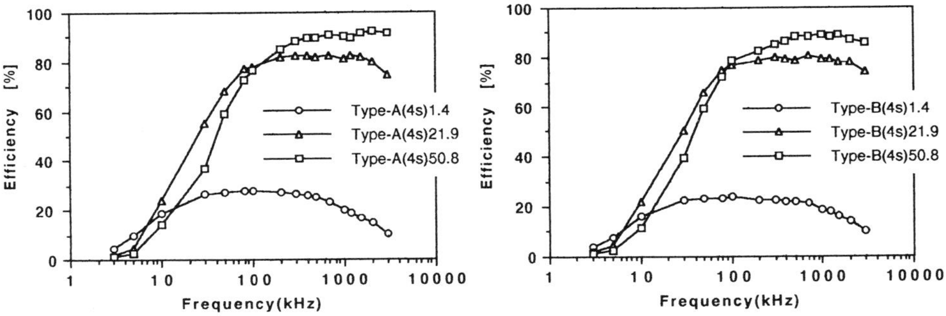

(a) (b)

Figure 5. The frequency characteristics of efficiency. (a) Four layers series connected circular shape film transformer, (b) four layers series connected rectangular shape film transformer. 1.4, 21.9 and 50.8 denote the load resistance value in Ohm.

Figure 5 shows the frequency characteristics of efficiency. The results in Fig.5 suggest that it is possible to realize highly efficient film transformers even though the load and operating frequency conditions are limited. Also, Fig. 5 shows that the optimum operating frequency depends on the load resistance value, and it moves to a higher frequency in accordance with the larger load resistance value. This tendency is similar to the results in the Ref.[5]. Also, the frequency characteristic of transformer ratio in Fig. 4 is similar to those of the Ref.[5]. These similarities suggest that the basic operating principle of film transformers is the skin effects.

3. CONCLUSION

In the present paper, an ultimate thin and light weight high frequency transformer has been proposed. Depending on the load conditions, efficiency of the film transformer reaches over the 95%. Thus, we have experimentally verified that our film transformer is applicable to the practical power supplies.

4. REFERENCES

1 K.Harada et al., Proc. IEEE, Vol.76, April(1988) pp.353-361.
2 F.C.Lee, Proc. IEEE, Vol.76, April(1988) pp.377-390.
3 K.Arai et al., IEEE Trans. Magn., Vol.MAG-20, No.5, Sep.(1984) pp.1463-1465.
4 R.V.Major et al., IEEE Magn., Vol.MAG-20, No.5, Sep.(1984) pp.1415-1416.
5 S.Hayano et al,IEEE Trans. Magn., Vol.27, No.6, Nov.(1991) pp.5205-5207.
6 T.Ogawa et al, Paper on Technical Meeting of Magnetics in IEEJ, MAG-92-132, July (1992).

Simulation and Design of Applied Electromagnetic Systems
T. Honma (Editor)

A Coreless film shape induction motor for the application to small devices

S.Hayano, O.Ishizawa and Y.Saito

College of Engineering, Hosei University, Kajino, Koganei
Tokyo 184, Japan

Abstract

In the present paper, we propose a coreless film shape axial type induction motor for the application to small devices. In accordance with the developments of modern large scale integrated circuits, the smaller mechanical devices, i.e. micromachines, are intensively exploiting for the medical applications. Because of their physical dimensions, the small devices essentially require the the small current but high voltage type actuators. Our new induction motor proposed here has a possibility to realizing the magnetic induction type actuators by employing a large number of magnetic poles and high frequency excitation, even though most of the existing actuators are the electric field types. As an initial test motor, we constructed an axial type induction motor having 7cm diameter and 5mm width. Examination of this trial motor reveals that our induction motor may be applicable to the micromachines.

1. INTRODUCTION

As an application of the technologies associated with the developments of modern large scale integrated (LSI) electronic circuits, the smaller mechanical devices, i.e. micromachines, are now intensively exploiting for the medical applications. Because of their physical dimensions, their electric circuits consists of the extremely small size electric conductors. This means that the mechanical power source of micromachines must be driven by the low current but high voltage source, e.g. piezoelectric actuator and electric motor utilizing electric field energy [1,2]. The electric motor utilizing electric field energy does not require a large currents but its speed and torque control technologies have not been yet established. On the other side, the control technologies concerning with the electric motors utilizing magnetic field energy, i.e. conventional DC and AC electric motors, are well established because they have been used over 100 years. Among the various conventional electric motors, the polyphase induction motor is one of the most reliable and widely used electric motors. With the developments and widespreads of power electronics technologies, continuous speed and torque control of the induction motors are quite easily carried out by means of the power LSI inverters. This replaces a large number of DC motors by the polyphase induction motors. However, conventional polyphase induction motor is always composed of the iron core to control the magnetic flux flows in the machine. In principle, application of the high frequency excitation to the machines equipping a large number of magnetic poles reduces the operation current into small value but increases an iron loss. Furthermore, it is difficult to reduce their size and weight, because their mechanical structure and coil arrangement are too complex.

In the present paper, we propose a coreless film shape axial type induction motor for the application to small devices. Our coreless film shape induction motor can be constructed by a large number of magnetic poles with a high frequency excitation so that it does not require a large operation current. Further, application of the high frequency excitation to our motor does not increase an iron loss because our motor does not have any iron core to control the magnetic flux flow. Our new induction motor proposed here has a possibility to realizing the magnetic induction type actuators by employing a large number of magnetic poles and high frequency excitation. One of the key ideas of our motor utilizes a low magnetic resistance comprising small air gap along the flux path and large area normal to the flux path in the film shape axial type motor. The other key idea is that the rotor is installed on the both sides of the stator coils. This rotor arrangement removes a stator yoke and realizes a light weight motor. Furthermore, the coils of stator as well as rotor can be constructed by the chemical etching processes. This make it possible to realize a short air gap, light weight and cheaper producing cost. Also, in order to fit the micromachines, the size of this new motor can be minimized by means of the lithograph techniques. As an initial test motor, we constructed an axial type induction motor having 7cm diameter and 5mm width. Examination of this trial motor reveals that our induction motor may be applicable to the micromachines.

2. A coreless film shape axial type induction motor

2.1 Basic principle and structure

Figure 1(a) shows a schematic diagram of the new motor. The stator is composed of the lamination of polyphase coils. Each of the polyphase coils is located at $2\pi/3$ different positions in electrical angle. Figure 1(b) shows one of the 2 poles stator coil patterns. Application of the balanced polyphase currents to the stator exciting coils yields the rotating magnetic fields on both sides on the stator. These rotating magnetic fields induce the currents on the rotors. In order to reduce the induced currents on the rotors, the rotors themselves rotate toward the same direction of the stator rotating magnetic fields. Because of the rotor enclosing stator, it is not necessary to install a stator yoke. The stator as well as rotor are constructed by the film shape conductors so that the air gaps along the flux path can be minimized also the surface areas normal to the flux path can be maximized. Thereby, the magnetic resistances in the direction of axis can be reduced to take a small in value. Thus, it is possible to realize our coreless film shape axial type induction motor.

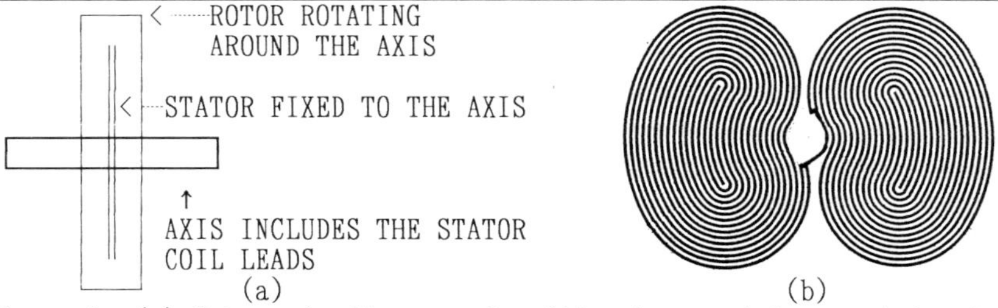

(a) (b)

Figure 1. (a) Schematic diagram of a film shape axial type induction motor. Stator polyphase coils are fixed to the axis and their leads are taken out to the external circuit through the axis. Rotor encloses the stator coils and rotates around the axis. (b) One of the 2 poles stator coil patterns.

2.2 System equation and parameter determination

By means of the method of symmetrical coordinate, it is possible to write a following system equation of the positive phase circuit [3].

$$\begin{vmatrix} E \\ 0 \end{vmatrix} = \begin{vmatrix} R_1 + jX_1 & jX_m \\ jsX_m & R_2 + jsX_2 \end{vmatrix} \cdot \begin{vmatrix} I_1 \\ I_2 \end{vmatrix}, \tag{1}$$

where $E, I_1, I_2, R_1, R_2, X_1, X_2$, X_m and j are the stator phase voltage, stator current, rotor current, stator resistance, rotor resistance, stator reactance, rotor reactance, mutual reactance and imaginary unit $\sqrt{-1}$, respectively. Also, the slip s in (1) is defined by

$$s = (\omega - \omega_m)/\omega, \tag{2}$$

where ω and ω_m are the angular frequency of the exciting current and mechanical angular velocity referred to the electrical angular velocity, respectively.

Defining a transforming matrix

$$A = \begin{vmatrix} 1 & \\ & a \end{vmatrix}, \tag{3}$$

(1) is transformed by

$$A \cdot \begin{vmatrix} E \\ 0 \end{vmatrix} = \left[A \cdot \begin{vmatrix} R_1 + jX_1 & jX_m \\ jsX_m & R_2 + jsX_2 \end{vmatrix} \cdot A^{-1} \right] \cdot A \cdot \begin{vmatrix} I_1 \\ I_2 \end{vmatrix}$$

$$= \begin{vmatrix} R_1 + jX_1 & jX_m/a \\ jsaX_m & R_2 + jsX_2 \end{vmatrix} \cdot \begin{vmatrix} I_1 \\ aI_2 \end{vmatrix}$$

or dividing the second column by a^2

$$= \begin{vmatrix} R_1 + jX_1 & jX_m/a \\ jsX_m/a & (R_2 + jsX_2)/a^2 \end{vmatrix} \cdot \begin{vmatrix} I_1 \\ I_2' \end{vmatrix}. \tag{4}$$

When we select $a = \sqrt{R_2}$, (4) reduces to

$$\begin{vmatrix} E \\ 0 \end{vmatrix} = \begin{vmatrix} R_1 + jX_1 & jX_m/\sqrt{R_2} \\ jsX_m/\sqrt{R_2} & 1 + jsX_2/R_2 \end{vmatrix} \cdot \begin{vmatrix} I_1 \\ I_2' \end{vmatrix}. \tag{5}$$

Let consider the no load condition, i.e. $s \simeq 0$ and $I_2' \simeq 0$, in (5), then the primal resistance R_1 and reactance X_1 can be obtained by

$$R_1 = P_\emptyset/(3 \cdot I_1{}^2) \text{ and } X_1 = \sqrt{(E/I_1)^2 - R_1{}^2}, \tag{6a}$$

where $P_\emptyset$ is the total input power at no load. Also, let consider the lock condition, i.e. $s = 1$, in (5), then following relationships can be derived as

$$\begin{aligned} \alpha_1 &= P_l/(3 \cdot I_1{}^2) - R_1 \\ &= [X_m{}^2/R_2]/[1 + (X_2/R_2)^2], \end{aligned} \tag{6b}$$

$$\begin{aligned} \alpha_2 &= X_1 - \sqrt{(E/I_1)^2 - (\alpha_1 + R_1)} \\ &= [X_m{}^2 \cdot X_2/R_2{}^2]/[1 + (X_2/R_2)^2], \end{aligned} \tag{6c}$$

where P_l is the total input power at lock condition. By means of (6b) and (6c), the other parameters in (5) can be obtained by

$$X_2/R_2 = \alpha_2/\alpha_1 \text{ and } X_m/\sqrt{R_2} = \sqrt{\alpha_1[1+(\alpha_2/\alpha_1)^2]}. \quad (6d)$$

Thus, we can evaluate the parameters by means of the no load and lock tests [4]. Using these parameters, the output mechanical power P_m and torque T_m are respectively calculated by

$$P_m = 3[(1-s)/s]I_{2s}'^2 \quad [W], \quad (7a)$$

$$T_m = 3[p/(s\omega)]I_{2s}'^2 \quad [N\text{-}m], \quad (7b)$$

where p and I_{2s}' are the number of pole pairs and rotor current at an arbitrary slip s.

2.3 Experimental

Figure 2(a) and 2(b) show a picture of the tested motor and the torque vs. speed curve. The experimental torque T_m was obtained by measuring the stator current I_1 and total input power P_{in} at an arbitrary slip s, viz.,

$$T_m = p[P_{in} - 3 \cdot R_1 I_1{}^2]/\omega. \quad (8)$$

As shown in Fig.2(b), the maximum output power of this tested motor was about 2-5[W], further, fairly good agreement between the experimented and calculated results were obtained.

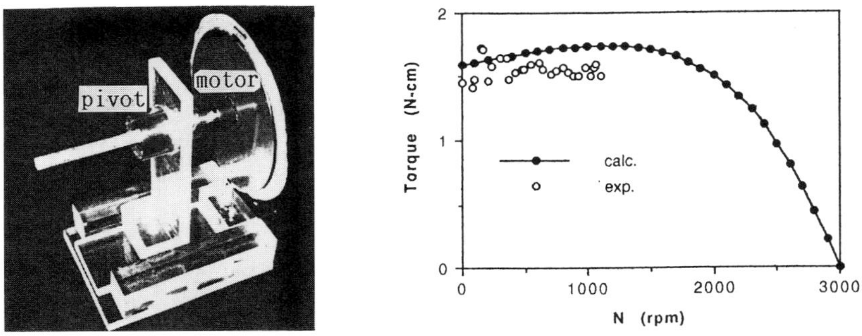

(a) (b)

Figure 2. (a) Tested trial motor. (b) Torque vs. speed characteristic.

3. CONCLUSION

As shown above, we have proposed the coreless film shape axial type polyphase induction motor for the small machines use. The tests of an initial trial motor have suggested that the motor utilizing magnetic field energy may be possible to exploit for the micromachine's mechanical power source use.

4. REFERENCES

1 K.R.Udayakumar et al., Nonlinear Phenomena in Electromagnetic Fields, (Elsevier Science Publishers B.V., 1992)pp.97-100.

2 Pooja C.Sardeshmukh et al., Nonlinear Phenomena in Electromagnetic Fields, (Elsevier Science Publishers B.V., 1992)pp.109-112.

3 H.H.Woodson and J.R.Melcher, Electromechanical Dynamics, Part-1: Discrete Systems (John Wiley & Sons, INC., New York, 1968).

4 Y.Saito et al., Bulletin of the College of Engineering, Hosei University, No.12, Jan.(1976)pp.81-103.

Simulation and Design of Applied Electromagnetic Systems
T. Honma (Editor)

FEM analysis and design of polar shoes in permanent-magnet DC motors

M. Rizzo[a], A. Savini[b] and J. Turowski[c]

[a]Department of Electrical Engineering, University of Palermo, Viale delle Scienze, 90140 Palermo, Italy

[b]Department of Electrical Engineering, University of Pavia, Via Abbiategrasso 209, 27100 Pavia, Italy

[c]Institute of Electrical Machines and Transformers, ul. Stefanowskiego 18/22, 90-924 Lodz, Poland

Abstract

The choice of different materials for permanent-magnets and the shape of the airgap can strongly affect the performance of permanent-magnet small DC motors. The present study is aimed at assessing, by means of the finite element method, the influence of the shape and the size of polar shoes on the magnetic field intensity at the airgap, for a given configuration of rotor and stator.

1. INTRODUCTION

In permanent-magnet small DC motors the distribution of magnetic field is strictly dependent, among other factors, on the nature of the magnet employed.

One of the problems implied by the use of a permanent magnet is the possible demagnetization of the magnet itself because of the contrasting effect of the magnetic field created by the reaction flux. Airgaps with increasing width towards the pole edges as well as polar shoes made with mild steel have been recognized to be expedient in order to protect the magnet against the reaction flux of armature current. Polar shoes can also determine widening or focusing of magnetic flux at airgap [1]. Furthermore, the size of polar shoes is influential on the loss of magnetic energy. Actually, for a slotless armature polar shoes can be made with solid mild steel. Conversely, in a slotted armature the choice of laminated iron for polar shoes is mandatory, if one wants to avoid significant "tooth-ripple" eddy current losses. The latter increase at high speed, so strongly lowering the efficiency of the motor.

2. FEM MODELLING

To the purpose of investigating the influence of both various polar shoes and of airgaps with variable width, a finite element method (FEM) simulation of the magnetic field, in terms of magnetic vector potential **A**, has been carried out in the cross-sections of four small motors. The FEM code employed could take the magnetic nonlinearity due to saturation into account. The magnet considered for all motors was ferrite with $B_r = 0.22$ T and $H_c = 120$ kA/m with radial magnetization [2]. The demagnetization curve of the magnet has been assumed to be linear, even if the code employed was able to consider any recoil line. For the sake of simplicity, effects due to eddy currents as well as end effects have been neglected. The boundary condition **A** = 0 has been introduced for the exterior border. Four motors have been examined, whose cross-sections have been subdivided into first-order triangular elements giving rise to meshes having the same number of elements (1975) and nodes (980).

3. MODELS OF MOTORS

The four types of brushless permanent-magnet DC motors considered are shown in Figures 1 and 2.

They have: i) same radial height of the magnet h = 7.7 mm; ii) same external diameter of stator D = 42.0 mm; iii) same width at airgap $\delta = 0.6$ mm; iv) rotor with small teeth and large slots (copper machine) in the same number.

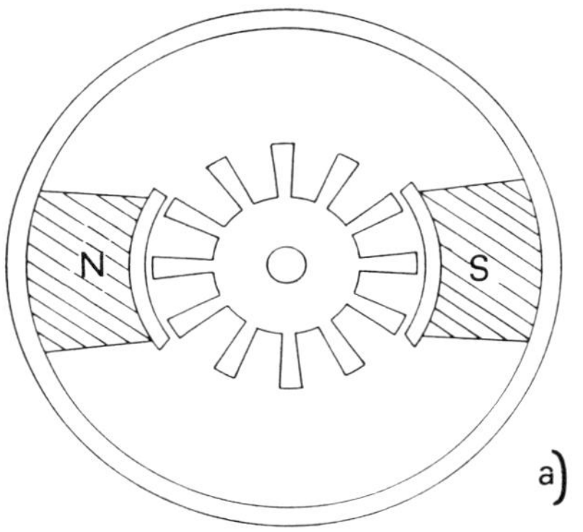

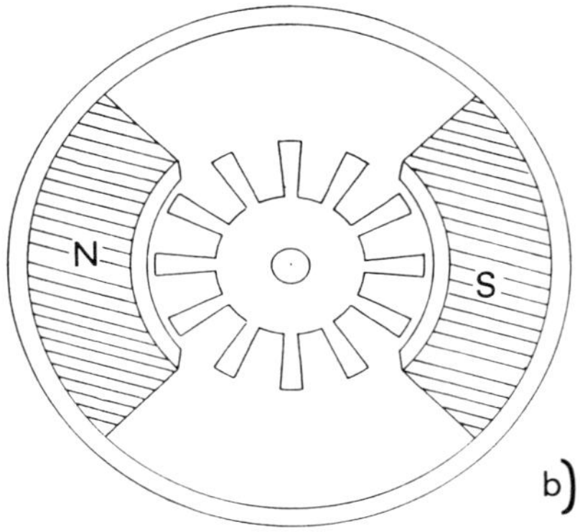

Figure 1. Permanent-magnet DC motors with polar shoe producing: a) extension of flux; b) concentration of flux.

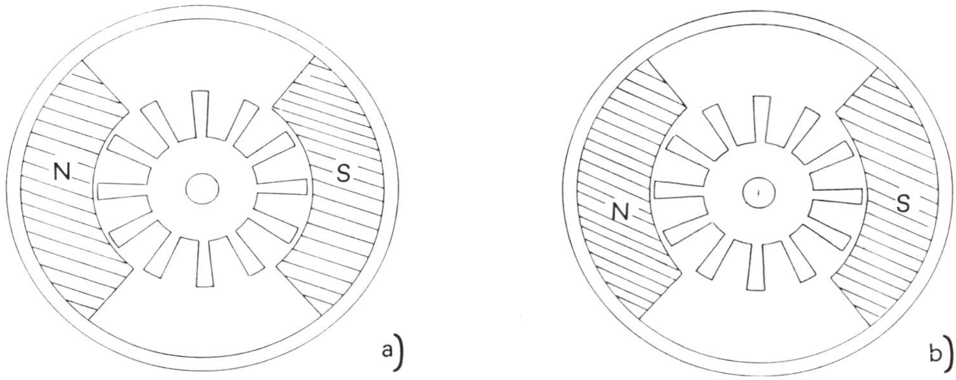

Figure 2. Permanent-magnet DC motors: a) with constant airgap; b) with variable airgap.

All four motors include a stator core made with mild steel. The motors in Figure 1 (a,b) have polar shoes formed by the same material but with different size and shape. The motor in Figure 2 (a) has constant airgap width δ, while that in Figure 2 (b) has a width varying from δ to 2 δ.

3.1 Motors with pole shoes

The waveform of the normal component of flux density at airgap for a pole pitch is depicted in Figure 3 (a,b).

The motor with concentration of flux, compared to that with extension of flux, shows flux densities much higher along the whole pole pitch (peak values, e.g. are 0.26 and 0.17 respectively) as well as remarkably lower distorsion (the zero-point corresponds to 22 degrees for the former and 45 degrees for the latter).

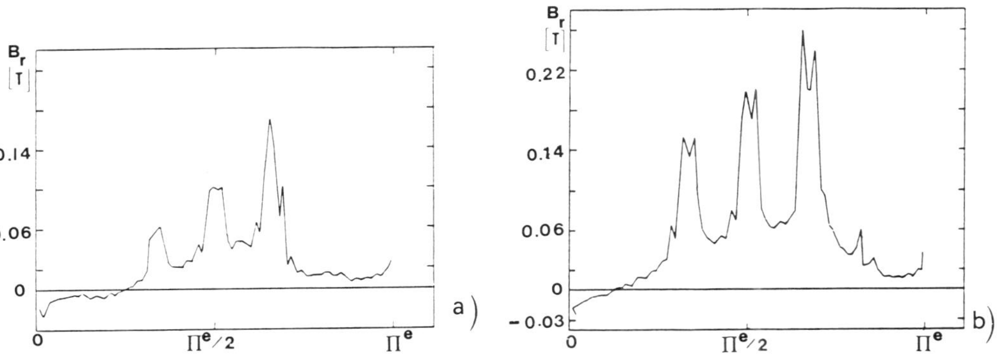

Figure 3. Flux density at airgap in a pole pitch: a) motor with pole shoe and extension of flux; b) motor with pole shoe and concentration of flux.

3.2 Motors with variable airgap

The distribution of the normal component of flux density at airgap for a pole pitch is shown in Figure 4 (a,b).

The two motors exhibit very similar distributions; the values are comparable even at pole edges and the zero-point is coincident.

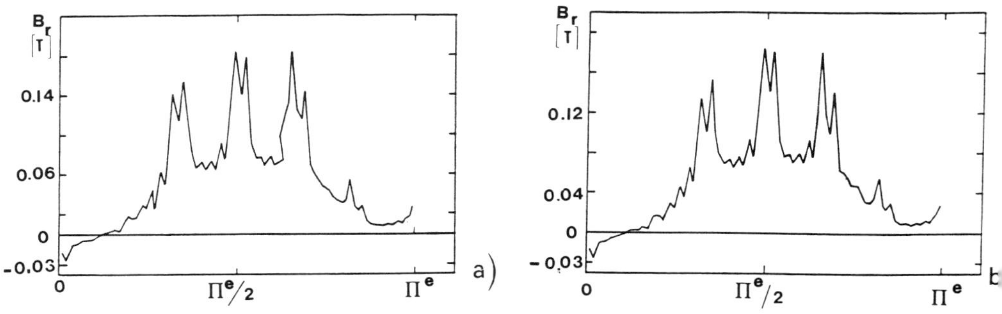

Figure 4. Flux density at airgap in a pole pitch: a) motor with constant airgap; b) motor with variable airgap.

4. CONCLUSION

The analysis of the magnetic field at airgap for the four different configurations of permanent-magnet small DC motors has clearly shown that the presence of a pole shoe and, in particular, of a pole shoe with extension of flux, can improve the performance of the motor. In the absence of this pole shoe the variable airgap does not affect the performance in a negative way so that it can be a rather economical way to avoid the demagnetization of magnet.

5. REFERENCES

1 J. Turowski, K. Komeza, A. Pelikant, S. Wiak, M. Rizzo and A. Savini, Electromagnetic Field Problems in Small Electronically Controlled Motors, Proceedings International Conference ACEMP'92, Kusadasy Turkey, 2/2 pp 430-437.

2 M. Rizzo, A. Savini, J. Turowski and S. Wiak, The Influence of Permanent Magnet on Brushless DC Motor, Fifth Biennal IEEE C.E.F.C., Harvey, USA, 3-5 August 1992.

Superconductors

Simulation and Design of Applied Electromagnetic Systems
T. Honma (Editor)
1994 Elsevier Science B.V.

AC applications of superconductors at 50/60 Hz frequency

I. Hlásnik

Division of Electrical and Computer Engineering, Faculty of Engineering,
Yokohama National University, TEPCO - Chair, Tokiwadai, Hodogaya-ku, Yokohama, 240

Abstract

Paper presents a historical review of the research and development of low loss AC composites with very fine superconducting filaments as well as a description of the actual situation of their applications at 50/60 Hz frequency.

1. INTRODUCTION

Modern practical superconductors (TS) such as NbTi and Nb_3Sn with high critical current density (J_c of the order of $10^3 \sim 10^4$ A/mm^2), with $\rho = 0$ in.DC regime and with high critical magnetic field ($B_{c2} = 5$ to 20 T) are very attractive for the use in electrical engineering. They allow to increase the efficiency and to decrease drastically the size and mass and in some cases also the capital costs of electrical machines and devices.

However their very low working temperatures ($T_w \leq 4.2$ K) and AC losses when exposed to time varying magnetic field have hindered their use at industrial 50/60 Hz frequency.

The AC losses have two components: hysteresis losses and coupling current losses. Their corresponding volumic densities in multifilamentary composite placed in a transverse external AC magnetic field of the amplitude $B_m >> \mu_o dJ_c$ and frequency f are given as

$$p_h = \frac{8}{3\pi} dJ_c B_m f \quad (1) \qquad \text{and} \qquad p_{cc} = \frac{B_m^2 l_p^2 f^2}{2\rho_\perp} \quad (2)$$

respectively,where d is the filament diameter, l_p the twist pitch length and $\rho_\perp$ the effective transverse resistivity of the composite.

These losses have led through the superconductor warming up to an excessive decrease of working currents and in concomitance with the energy needed for refrigeration ($\zeta = 300 \sim 1000$ W at 300 K for evacuation of 1 W from 4.2 K) they made the total energy consumption in superconducting AC devices economically unfavorable. In 1981 to 1983 it has been shown that in modern multifilamentary NbTi composites the AC losses can be sufficiently reduced to eliminate the current degradation and to allow the energy saving in AC windings [1] - [5].

In this paper the basic criterion of energetic feasibility of AC superconducting winding is given. Then the results obtained in the development of low loss AC multifilamentary composites (ACMFC), cables and of superconducting windings including the description of main characteristics of some actually produced ACMFC are presented.

Finally several examples of results obtained in feasibility studies and in experimental research of models of superconducting AC machines and devices (SACMD) are given.

2. CRITERION OF THE ENERGETIC FEASIBILITY OF AC SUPERCONDUCTING WINDINGS

The first criterion about the energetic feasibility of superconducting AC winding has been derived in [2] as

$$\xi_e = \frac{V_{ws}p_{ts}}{V_{wn}p_{tn}} = \frac{\sqrt{2}\,\nu J_n}{\lambda J_c(B_{mm})}\frac{p_{ts}}{p_{tn}} = \frac{\sqrt{2}\,\nu(1+\zeta)}{\rho_n J_n \lambda J_c(B_{mm})}\left[\frac{8d\lambda J_c B_m f}{3\pi} + \frac{B_m^2 l_p^2 f^2}{2\rho_\perp}\right] < 1 \qquad (3)$$

where V_{ws}, V_{wn} and p_{ts}, p_{tn} are the conductor volume and total loss density in the conductors of the superconducting and conventional windings including energy for refrigeration, respectively, J_n is the current density in the normal conductor at the nominal conditions, ν is the fault current overload coefficient, $\lambda J_c(B_{mm})$ being the minimum critical current density in the superconducting winding. For $\xi_e < 1$ i.e. for the energy saving it is needed

$d < 1\ \mu m$, $l_p \cong 1$ mm, $\rho_\perp > 10^{-9}\ \Omega m$, $\lambda J_c \geq 1\ kA/mm^2$ and $\nu \leq 1.5$ to 2. (4)

ACMFC for AC winding is characterized by the normalized AC loss power density $p_s(B_m) / \lambda J_c(B_{mm})$ rather than by its absolute value $p_s(B_m)$ only. Because with increasing B_m $p_s(B_m)$ increases and $\lambda J_c(B_m)$ decreases, ξ_e increases with B_m. Therefore in AC superconducting windings B_m cannot be much higher than 1T in the present state of art.

3. TECHNOLOGY AND CHARACTERISTIC PARAMETERS OF AC MULTIFILAMENTARY COMPOSITES, CABLES AND WINDINGS

At the present time only NbTi ACMFC are used for AC 50/60 Hz applications. To obtain filament diameters $d < 1\ \mu m$ the multiple compaction process is used. Due to the proximity effect there is a lower limit of filament diameter and of the thickness of normal metal between filaments d_n. Actually for NbTi filaments and CuNi matrix $d = 0.1 \sim 0.5\ \mu m$ and $d_n \cong 0.1\ \mu m$. Because of the mechanical and electrical reasons l_p should be higher than 5 times the wire diameter d_w, for needed $l_p \cong 1$ mm d_n is $0.1 \sim 0.3$ mm. $\rho_\perp$ is increased by using high resistive Cu-Ni matrix ($\rho_{CuNi} = 1.3 \sim 3.55 \times 10^{-7}\ \Omega m$), so $\rho_\perp = 10^{-8} \sim 10^{-7}\ \Omega m$. λJ_c can be increased by decreasing d_n as well as by decreasing the number of compaction levels and by increasing J_c. At $B \leq 1$ T $\lambda J_c \geq 1\ kA/mm^2 > 100\ J_n$ is usual. This allows to reduce strongly the volume, mass and price of the winding. The conductor price per 1 kAm at B = 1 T of industrially produced ACMFC is about 7 $ / kAm. It is approximately the same as that of Cu conductors for turbogenerators with $J_n = 10\ A/mm^2$, but 4 times smaller than that of Cu conductors for power transformers with $J_n = 2.5\ A/mm^2$.

Normalize losses $p_s(0.5T) / \lambda J_c(0.5T) \cong 5\ \mu W/Am$ and $p_s(1T) / \lambda J_c(1T) \cong 20\ \mu W/Am$ are about 10 times smaller than corresponding reduced losses in Cu winding in power transformer or in turbogenerator stator winding assuming $\zeta = 500$ W/W and $\nu = 1.5$, respectively. It means that the use of modern ACMFC in both types of these machines could lead to an energy saving of about 90 % of losses in conventional windings.

The resistivity of ACMFC in the normal state can be very high of the order of $10^{-7}\ \Omega m$. This feature allows to use them for superconducting switches and fault current limiters.

Composites with mentioned parameters are produced by several companies.

Fig. 1 shows the cross section view of a typical ACMFC produced by GEC Alsthom Co.

Critical currents of ACMFC are of the order of several 10 A only. For many AC machines and devices with currents of 1 ~ 10 kA fully transposed multi-strand compacted cables are necessary. The Fig. 2 shows different types of fully transposed cables and their combinations. In spite of perfect transposition there is often observed important current degradation in kA - class superconducting cables. Many sources of this degradation have been determined and eliminated. Nevertheless the current degradation of 10 kA - class cables remains the most important issue of ACMFC applications. Some works are in progress to solve this problem by appropriate choice of the cabling lengths and directions at different cabling levels [6], [7].

Technology of AC 50/60 Hz superconducting windings should secure intensive cooling and good fixations of all strands in the winding.

One such technology, very suitable for cylindrical coils uses multi - layer FRP cylinders with spiral and axial grooves on their surface as coil formers (see Fig. 3). The axial grooves, form cooling channels, the spiral grooves prevent strand motion. Moreover these cylinders serve as interlayer and coil end insulation.

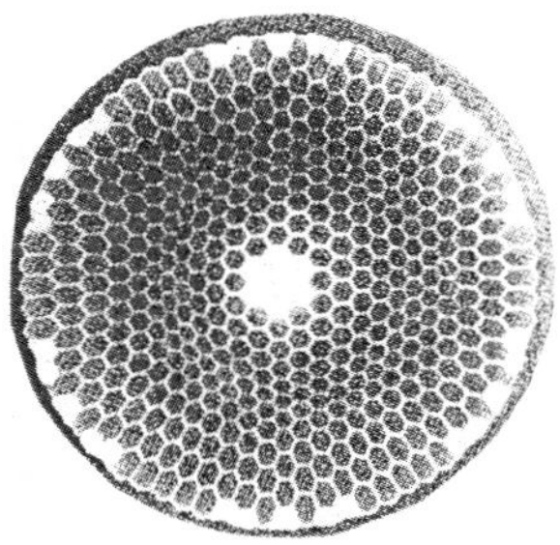

Figure 1. Cross-section of a typical ACMFC (GEC Alsthom).

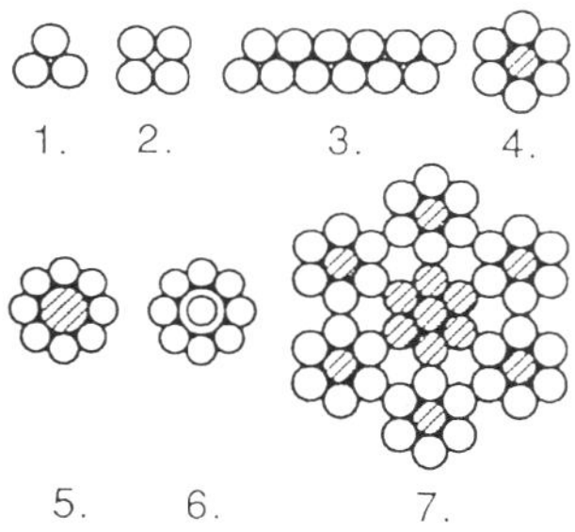

Figure 2. Different types of fully transposed compacted cables.

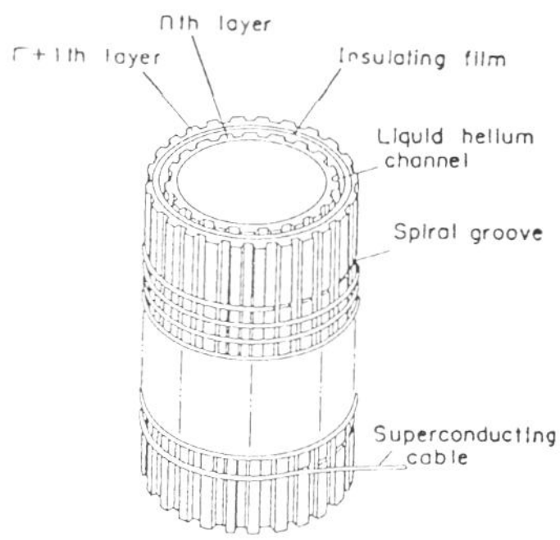

Figure 3. Winding with cylindrical spacers.

For windings with other form than cylindrical good fixation of ACMFC can be achieved only by epoxy - impregnation. However the thickness of the impregnated winding should be sufficiently small (less than about 2 mm) to avoid the current degradation [8].

4. RESEARCH AND DEVELOPMENT OF SUPERCONDUCTING AC MACHINES AND DEVICES (SACMD)

The main advantages expected from these new superconductor applications are the increase of efficiency and the decrease of the size and mass of SACMD. New advanced technical solutions for many acute problems of science and technology such as e.g. the fault current limitation in electric power systems are also expected.

The most promising AC application of ACMFC is **SUPERCONDUCTING FAULT CURRENT LIMITER** (SFCL).

It is based on the rapid automatic transition ($\tau_t \cong 100\mu s$) of an AC superconducting composite from the superconducting state with zero resistivity into the normal state with high resistivity (ρ_n of the order of $10^{-7}\Omega m$) which occurs when the current and/or temperature of the conductor become higher than their critical values. The transition time is much smaller than the time necessary for detection, commutation and current limitation by classical switchgear (> 10 ms).

Schematic configuration of the most developed SFCL is shown in Fig. 4. It consists of a superconducting trigger coil with small impedance and of a limiting reactance or a resistance Z_p connected in parallel. When the trigger coil transits at the fault the limiting reactance or resistance decreases the overvoltage due to high $\partial I / \partial t$ and the inductance of the line. The quench current I_Q of the trigger coil should equal to the required maximum fault current I_f. Its impedance R_N in the normal state together with that of the limiting coil or shunt resistor should secure the required reduction of the current in the electric power grid during the short. Further requirement on the trigger coil is short recovery time, $\tau_r < 1s$, necessary to recover the superconducting state. The SFCL shown in Fig. 5 from [9] was able to limit the fault current to 2 kA and to secure the current overload coefficient $\nu \leq 1.5$ at applied voltages 6.6 kV_{rms}. The total losses of 2.3 W measured at 1000 A_{rms} are small enough to be evacuated with small Gifford - Mc Mahon type refrigerator. Its recovery time is between 1 ~ 10 s and should be still decreased. Fig. 6 shows a conceptual design for a 6 kV / 2 kA three phase limiter in a common cryostat with two stage refrigerator. The SFCL from [10] has similar parameters.

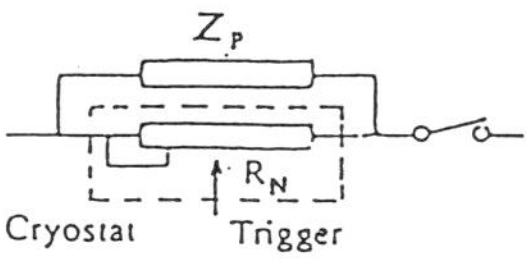

Figure 4. Configuration of superconducting fault current limiter.

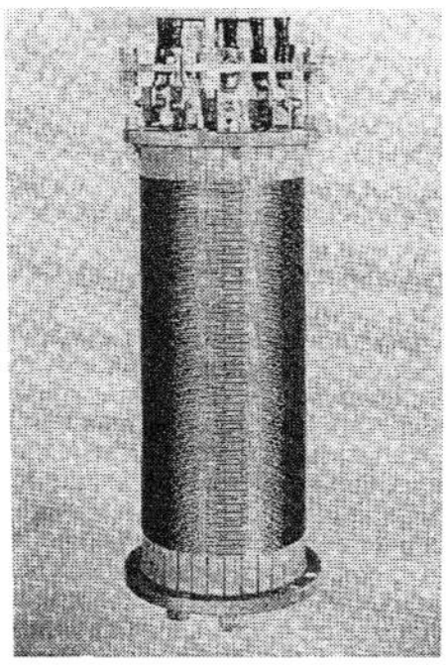

Figure 5. 6.6 kV / 1.5 kA-class fault current limiter [9].

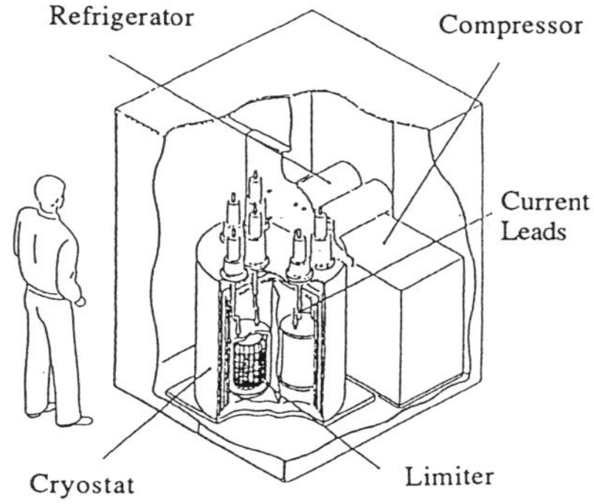

Figure 6. Conceptual design for a 6 kV/2 kA three phase fault current limiter [9].

These models have attained apparent limiting power $P_l = UI_{lim}$ of the order of 10 ~ 20 MVA. The technology of their windings and cryostat is mastered, so that they have good prospects to be inserted as pilot AC device into the electric power system in the near future. They have no conventional counterpart.

There are many feasibility studies of **SUPERCONDUCTING POWER TRANSFORMERS** (SPT). In [11] it has been shown that the total life cycle cost of core, conductor and losses for SPT with rating power $P > 380$ MVA are smaller than those of a conventional one and at $P = 1000$ MVA it is smaller by 30 %. The weight of the core and conductor of such a SPT is smaller by 38 % compared to that of a conventional one.

In [12] it has been shown that for a transformer with $P = 200$ kVA the total weight can be reduced by a factor of 10 when iron core is reduced. General conclusion concerning the iron core of large SPT is that it should be at room temperature.

The largest model of SPT was described by K. Funaki et al. [6] (see Fig. 7). Its design rating parameters are 1000 kVA / 3000 V / 220 V at 60 Hz. At first tests however its secondary winding wound from a fully transposed $6 \times 6 \times 6$ ACMFC cable has attained only 56% of the designed current. After the amelioration of the current transfer between the cable and the terminal plate it is expected that the nominal rating of 1000 kVA will be reached in near future.

Today a technology exists to construct superconducting transformer, in regard to the elementary ACMFC, insulation and mechanical structures.

Additional R&D effort is still required to develop mainly superconducting several kA - class cables with decreased degradation of quench current, practical high voltage cryogenic bushing as well as to master the problems of the use of high pressure supercritical helium as coolant and insulating medium.

FULLY SUPERCONDUCTING GENERATOR (FSG) with both rotor and stator windings superconducting is also being developed. The main expected advantage compared to cryoturbogenerator with only rotor winding superconducting are a decrease of total mass by 15 ~ 25 %, a decrease of the rotor length by about 15 % and an increase of efficiency by about 0.3. %. The models with rating power of the order of 10 kW only have been built until now [13].

Figure 7. 1000 kVA-class superconducting transformer [6].

A **SUPERCONDUCTING LINEAR INDUCTION MOTOR** (SLIM) can provide much higher trust in devices with wide air gap than a conventional linear induction motor because of higher current density in the superconducting winding. The SLIM can have a great impact to many industrial fields especially to a steel making plant. If this application would be realized very important economic effect of the order of 10^9 \$ could be obtained. In actual models high degradation of the quenching current in impregnated winding made from AC cables was observed [13].

To stabilize an electric power line against different short disturbances the possibility of sudden release or sudden storage of the energy from or in a superconducting magnet can be used. The corresponding **SUPERCONDUCTING MAGNETIC ENERGY STORAGE** (SMES) system consists basically of two convertors and of a superconducting magnet. Different models have been verified theoretically and experimentally [14]. An interesting result is that even a relatively small SMES power which is less than 15 % of the generator power is effective to stabilize both DC as well as AC power lines and can be realized by present technology already in the near future.

In the last decade **THERMALLY CONTROLLED SUPERCONDUCTING SWITCHES** (TCSS) with repetition frequency up to 5 Hz for superconducting rectifiers for powering DC superconducting magnets have been developed using ACMFC [15]. The repetition frequency has been increased to 50/60 Hz by the use of thin NbTi ribbon [16]. This led to a decrease of dimensions of the superconducting transformer and of the total price as well as to an increase of the overall efficiency. The use of HT_cS thin films for TCSS with 50/60 Hz frequency seems to be very promising [17].

5. CONCLUSIONS.

Basic AC multifilamentary composites and kA-class compacted cables allowing important energy saving as well as dimension and mass reduction in SACMD have been developed and are commercially available. The technology of AC cylindrical superconducting windings has been mastered and successfully applied to superconducting fault current limiter and power transformer.

Nevertheless 10 kA - class cables and impregnated AC superconducting windings with other than cylindrical geometry are still to be developed.

Actually the most promising and developed application of ACMFC is the superconducting fault current limiter which has good prospects to be inserted as pilot AC device into the electric power system in the near future. Other SACMD are also successfully being developed.

The prospects of the use of HT_cS in this domain are very attractive because of the substantial increase of the refrigeration efficiency at 77 K and of the simplification and enhanced reliability of the needed cryogenic equipment. However mechanical and electrical characteristics of HT_cS should be same as those of low temperature superconductors. At present time it is not yet true, but the technology of HT_cS is progressing.

REFERENCES

1 I. Hlásnik, IEEE Trans. on Mag., MAG - 17, (1981),p. 2261
2 I. Hlásnik, Cryogenics 23, (1983), p. 508
3 I. Hlásnik, E. W. Seibt, J. Appl. Phys. 54, (1983), p. 5479
4 I. Hlásnik, , J. Phys. 45, (1984), p. C1 - 459
5 P. Dubots, et al., J. Phys. 45, (1984), p. C1 - 467
6 K. Funaki et al., IEEE Trans. on Magnetics 28, (1992), p. 394 and Cryogenics 32 (1992) ICEC Supplement, 489
7 N. Amemiya, I. Hlásnik, O. Tsukamoto, Influence of longitudinal magnetic field on thermomagnetic instabilities in a. c. superconducting cables, to be published in Cryogenics
8 O. Tsukamoto et al., Proc.of ICEC 12, Southampton (1988), 724
9 D. Ito et al., Cryogenics 32 (1992) ICEC Supplement, 462
10 T. Verhaege et al., Experimental 7.2 kV / 1 kA_{rms} / 3 kA_{peak} current limiter system, presented as paper LLD - 5 at the Appl. Supercond. Conf., Chicago, (1992)
11 H. Riemersma et al., IEEE Trans. on Power App. and Syst., 100, (1981), p. 3398
12 A. Février et al., Proc. of I.C.E.C.11, Berlin, (1986).
13 O. Tsukamoto, Proc. of Workshop on AC Superconductivity, National Lab. of High Energy Physics, Tsukuba, (1992), p. 44.
14 F. Irie et al., IEEE Trans. on Mag. 28 (1992) 426
15 J. Sikenga et al., IEEE Trans. on Mag. 25 (1989) 1771
16 I. Hlásnik et al., Proc.of the 11th Internat. Conf. on Mag. Tech., Tsukuba, (1989), 431
17 N. V. Markovsky et al., Cryogenics 32 (1992) ICEC Supplement, 624

Simulation and Design of Applied Electromagnetic Systems
T. Honma (Editor)

Applications of electromagnetic phenomena in high temperature superconducting materials

Masato Murakami

ISTEC, Superconductivity Research Laboratory, 1-10-13, Shinonome, Koto, Tokyo, Japan

Abstract

Large persistent currents can flow in the superconductors when they exhibit a large pinning force. The electromagnetic interaction between such persistent currents and magnetic field gives both repulsive and attractive forces. Various applications are now under consideration using such interaction, e. g., magnetic bearing, flywheel, transport system. The pinned superconductor can also either shield or trap magnetic fields by persistent currents. Magnetic shield and superconducting permanent magnets can be made using such properties.

1. INTRODUCTION

The response of the superconductor to magnetic field is attractive for various applications[1]. In magnetic fields below H_{c1}, type II superconductors can prevent magnetic flux penetration absolutely, which is known as the Meissner effect. Under such circumstances, a magnet can levitate above the superconductor and vice versa. However, the repulsive force due to the Meissner effect is small, which limits the applications. On the other hand, a repulsive force can also be obtained above H_{c1}, when the superconductor exhibits pinning [2-4]. Although the external field enters the superconductor, the internal field has a gradient corresponding to

$$\text{rot}\, \boldsymbol{B} = \mu_0 \boldsymbol{J_c}. \tag{1}$$

In one dimension this equation is simplified as $dB/dx = \mu_0 J_c$, where B is the magnetic induction in T, μ_0 the permeability of free space, and J_c is the critical current density in A/m^2. The physical origin of the flux-density gradient is the pinning of the magnetic flux lines by structural defects in the material. In other words, large supercurrents (J_c) can flow in the superconductor by virtue of flux pinning. As will be shown below, the interaction between magnetic field and such large supercurrents can bring various attractive applications of superconductors. In this paper, I will introduce possible applications of electromagnetic forces of high T_c superconductors.

2. MAGNETIC FORCE

The force between a superconductor and a magnet is given by

$$\boldsymbol{F} = \int (\boldsymbol{m} \bullet \nabla) \boldsymbol{H}\, dv \tag{2}$$

In one dimension, this equation is simplified as $F = m\, dH/dx$, where m is the magnetic moment of a superconductor, and dH/dx is a field gradient produced by a magnet . m is equal to Mv where M is the magnetization per unit volume and v is the volume. dH/dx is a constant, provided that we use the same magnet, while M is dependent on J_c and the size of the shielding current loop (r) according to the following relation: $M = A J_c r$, where A is a constant which depends on sample geometry and r is the radius of the shielding current loop. It should be born in mind that, in contrast to conventional magnetic materials, M of a superconductor depends on its size, which is a peculiar characteristic of a superconductor, and is the key to large levitation force.

In order to increase F under the condition that the volume of a superconductor is constant, one must increase M, and therefore either J_c or r. Recently, the development of melt processing of Y-Ba-Cu-O superconductors has enabled us to have high temperature superconductors which exhibit large M and therefore large F[5]. This opened up the possibility of applications of electromagnetic forces of superconductors.

Magnetic force is also given by the following expression[6]:

$$\boldsymbol{F} = \int \boldsymbol{J}_\mathbf{c} \times \boldsymbol{B} \, \mathrm{dv} \tag{3}$$

This is more direct expression to show that the interaction of supercurrent (J_c) and magnetic induction (B) is the source of the magnetic force. In static field, equations (2) and (3) give the same results.

3. APPLICATION

3.1. Levitation

A heavy object can be levitated by a repulsive force between a magnet and MPMG processed Y-Ba-Cu-O superconductors[7]. Various applications using such levitation have been proposed.

3.1.1. *Physics experiment*

Since an object can be levitated without contact, several physics experiments have been proposed. One example is fluid dynamics experiment. If we support an object in midair, flow of the fluid around the object can be analyzed under the ideal conditions.

3.1.2. *Lunar telescope*

The moon is the best site for an astronomical observatory. It is a stable platform with a dark sky, perfect seeing, and offers wavelength not accessible from the earth. Long observing times are possible ranging from 14 days to forever (in the poles). However, the limited payload weight of existing launch vehicles and lack of maintenance were the obstacles to building telescopes on the moon.

Recently, Chen *et al.*[8] has proposed the design of a lunar telescope using a combination of advanced optical fabrication techniques and levitation by oxide superconductors. Melt processed Y-Ba-Cu-O superconductors are used to levitate, steer and position the telescopes.

The moon has several advantages for levitation by the superconductors. The ground temperature is estimated to be 40K if permanently shielded from sunlight. This temperature is low enough for $YBa_2Cu_3O_x$ compounds to be superconducting. It is also important to note that the gravity on the moon is seven times smaller than that on the earth. This means that an object seven times heavier can be levitated on the moon.

3.1.3. *Display*

Levitation itself might be used for various applications. One such example is the display of jewels. Although this is not scientific area, it might be good commercial use. One Japanese department store is interested in displaying precious metals and stones on the turntable levitated by superconductors. Some Japanese hotel is also interested in levitating a married couple as one of the attractions in the wedding party. Levitation of a person has been demonstrated in many exhibitions and fairs, and has attracted attention of a number of people.

3.2. Suspension

It is very interesting to note that a superconductor can be suspended below a magnet, and vice versa. This clearly demonstrates that the suspension is due to flux pinning.

3.2.1. *Transport system on a guide rail*

When we construct a guide rail such that a number of magnets are placed with the same pole facing the same direction, there is no friction for the translational motion of the superconductor. Furthermore, by virtue of flux pinning, the levitated superconductor will not deviate from this magnetic guide.

Recently, Toshiba corporation has succeeded in constructing a transport system of LSI in a clean room[9]. In this system, motion of the LSI container with four bulk superconductors can be controlled electromagnetically.

A change in magnet configuration from NNN to NSN causes friction, and this can be utilized for constructing a brake or a stop.

3.2.2. *Transport system without magnetic guide rail*

It is also possible to construct a non-contact transport system without magnetic guide rail. Here we use a pair of a superconductor and a magnet . If we move the superconductor, the levitated magnet is transported without contact. Such a system has already been constructed by Nippon Sinku[10] for transport of LSI.

3.3. Rotation device

3.3.1. *Magnetic bearing*

Several groups[11-13] have succeeded in fabricating bearings using melt processed bulk Y-Ba-Cu-O. Superconducting bearings are very simple devices; bulk Y-Ba-Cu-O is used as a stator and a permanent magnet for the rotor. The attractive and repulsive forces between the rotor and the stator causes the stator to be suspended in mid-air, where it is thus able to spin freely.

However, in the case of actual devices, the space in which the rotor can move is extremely limited and therefore, the distance between the superconducting stator and a magnetic rotor is found to be too small to obtain a repulsive force enough to suspend the rotor, since in such a case a magnetic field is trapped by the superconductor, resulting in only a small repulsive force. The problem has recently been solved utilizing the attractive force, which becomes largest when the superconductor is placed in contact with the magnet before cooling[13].

3.3.2. *Flywheel system using bulk superconductors*

The flywheel system has long been studied as a candidate for an energy storage system. However, energy storage for a long duration was difficult because of the friction in the mechanical bearings.

Superconducting magnetic bearings will make it possible to store energy for a longer period, since there is no friction. A flywheel system spinning in a vacuum with no friction can store kinetic energy imparted to the disk from outside, and whenever necessary, can discharge the stored energy in the form of electric energy.

The energy (E (J)) stored by the flywheel is given by

$$E = (1/4)\, M\, r^2\, \omega^2 \tag{4}$$

where M is the mass (kg), r is the radius (m), and ω is the angle velocity (rad/s) of the rotating disk. It is clear that it is possible to increase the amount of stored energy by increasing the diameter of the disk.

However, if we simply enlarge the size of a magnet, the strength of magnet is weakened. This problem can be solved by arranging magnet rings concentrically with non-magnetic aluminum spacers between the rings.

It is also important to improve the strength of the rotating disk against the centrifugal force for scaling up. Recently, it is proposed that the problem of the centrifugal force is greatly reduced by constructing a ring flywheel[14]. The beneficial point of the superconducting flywheel is that the wheel can be supported by a large area instead of the axis. In the case of a ring flywheel, the concentration of the force at the rotating axis is no more a serious problem.

In the future, it might be possible to convert a solar energy into electric currents by the solar cells, and the energy is stored in a flywheel, and is discharged when it is necessary.

3.4. Application of flux trapping

It is possible to construct a kind of superconducting permanent magnet using flux pinning. A significant magnetic field can be trapped by a superconductor when it exhibits large flux pinning force. If the flux pinning force is increased, a larger amount of magnetic field can be trapped.

At 5K, a small piece of melt processed Y-Ba-Cu-O sample with dimensions of 0.1x3.3 mm^3 exhibits MH_{max} of 80MGOe, which is larger than that of any conventional magnetic materials[15].

4. SUMMARY

Until now major applications of superconductors have been confined to products in the form of wires and thin films. However, the development of MPMG bulk superconductors which have a very strong pinning force, has opened up the possibility of many different applications in the bulk monolithic form.

A superconducting magnetic bearing, utilizing large repulsive and attractive forces due to flux pinning between MPMG-Y-Ba-Cu-O superconductors and magnets, enabled a 2.4 kg shaft to rotate safely at over 30,000 rpm.

It is also expected that storage of over 10kWh of energy will be achievable in the future superconducting flywheel systems. Possible applications include backup power sources for buildings , supercomputers and load leveling systems for the day and night.

5. REFERENCES

1. D. Wolfstein , T. E. Seidel, D. W. Johnson, Jr. and W. W.Rhodes: *J. Superconductivity* **2** (1989) p. 211.

2. F. C. Moon, M. M. Yanoviak and R. Ware : *Appl. Phys. Lett.* **52** (1988) p. 1534.

3. E. H. Brandt: *Appl. Phys. Lett.* **53** (1988) p. 1554.

4. W. G. Harter, A. M. Hermann and Z. Z. Sheng: *Appl. Phys.Lett.* **53** (1988) p. 1119.

5. H. Fujimoto, M. Murakami, S. Gotoh, N. Koshizuka and S. Tanaka: *Advances in Superconductivity* **2** (1990) p. 285.

6. M. Murakami, T. Oyama, H. Fujimoto, T, Taguchi, S. Gotoh, Y. Shiohara, N. Koshizuka and S. Tanaka: *Jpn. J. Appl. Phys.* **29** (1990) p. L1991.

7. M. Uesaka and K. Miya: *Melt Processed High Temperature Superconductors* (World Scientific, 1993, ed. M. Murakami) p. 284.

8. P. C. Chen, R. J. Oliversen, and H. Hojaji: *Proc. TCSUH Workshop* (World Scientific, 1992) p. 513.

9. H. Ogiwara: *Paper presented at 1992 spring annual meeting of Japanes Applied Physics Society.*

10. N. Minami et al.: *Extended abstract of 47th annual meeting of Japan Cryogenic Society* (1992) 136.

11. F.C.Moon, P.Z.Chang, H.Hojaji, A.Barkatt and A.N.Thorpe: *Jpn.J.Appl.Phys.* **29** (1990) p. 1257.

12. T. Yotsuya, M. Shibayama and R. Takahata: *Advances in Cryogenic Engineering* **37** (1992) p. 519.

13. H.Takaichi, M.Murakami, A.Kondoh, N.Koshizuka, S.Tanaka, H.Fukuyama, K.Seki, T.Takizawa and S.Aihara: *Proc. THIRD INTERNATIONAL SYMPOSSIUM ON MAGNETIC BEARINGS* (1992) P. 10.

14. R. Shimada and H. Chikaraishi: *Paper presented at Electrical Society Meeting* (1989).

15. S. Gotoh, M. Murakami, H. Fujimoto, and N. Koshizuka: *J. Appl. Phys.* **72** (1992) p. 2404.

Simulation and Design of Applied Electromagnetic Systems
T. Honma (Editor)

Techniques Applied to the Magnetic Design of the Superconducting Dipoles for LHC

S. Russenschuck and T. Tortschanoff
CERN, CH-1211 Geneva 23 Switzerland

Abstract

The Large Hadron Collider (LHC) project foresees the installation of superconducting accelerator magnets on top of the existing Large Electron Positron Storage Ring (LEP). CERN aims to develop for this the strongest possible magnets in order to bring two counter rotating proton beams into collision at the highest possible centre of mass energy. The paper describes the design activity for these magnets which is confronted with many different objectives and constraints requiring special care in the selection of field simulation and optimization techniques.

1. Introduction

The LHC project foresees to add in the 26.7 km long tunnel of the existing LEP a new accelerator consisting of superconducting magnets producing the highest possible magnetic bending power. The magnetic components most difficult to produce are the dipoles of this machine. The two beam channels and space limitation lead to the conception of the so-called two-in-one magnets [1]. These magnets have the excitation coils around the two beam channels both placed in one common iron yoke (Fig. 1).

The conductor adopted is based on NbTi multifilamentary strands braided into a Rutherford type cable, key-stoned with an angle of not more than 2 degrees. To reach the highest possible field strength it must be cooled to the temperature of Helium II, 1.8 K. The structure of the coil consists of two layers of conductors arranged in several blocks of turns.

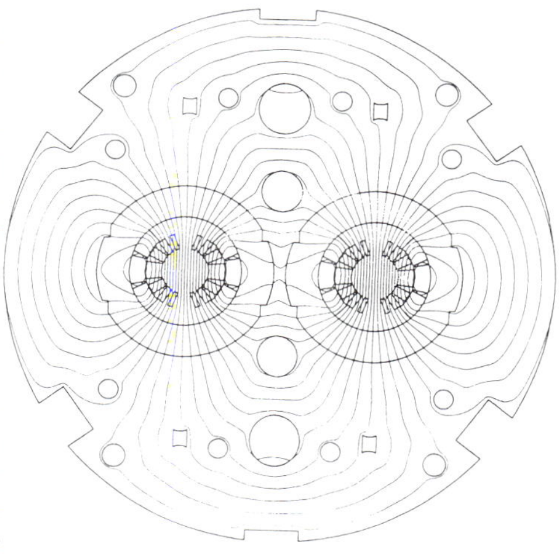

Fig. 1. Cross-section of LHC dipole

2. Magnetic Design

The objective of the magnetic design for the LHC dipole magnets is not only the maximum field in the useful aperture which has to be achieved. All higher order unwanted multipole terms of the field, as obtained from Fourier analysis, must be minimized. Further, the working points of the two layers have to be balanced.

This leads to the so called grading of the coils, which means that their peak fields in the conductor must have about the same margin from the critical current versus critical field curve. This is achieved by using a conductor with a smaller cross-section, implying a different cable with different strands.

In the two-in-one structure a fraction of the return flux of one magnet will pass through the aperture of the other and cause a certain field enhancement and a certain cross-talk. Besides the sextupole component due to iron saturation, the cross-talk manifests itself mainly in the appearance of a field dependent quadrupole component and, to a much lesser extent, an octupole. The design variables are the coil block disposition, the shape of the yoke, the form of the inner surface of the yoke together with its outer radius and a certain number of holes. The fact that the magnets are very long with respect to the transverse dimension of their aperture allows concentration on two dimensional computations for the optimization.

3. Analytical Approach for Coil Optimization and Inverse Problem Solving

Because of their ease of use, the speed of computation and the initially sufficient precision of the results, analytical computations are performed when starting the design of an accelerator magnet. Applying Biot-Savart's law to coils described by a collection of filamentary line currents has the advantage of easily permitting its extension to 3-D computations. Care must be taken to describe the geometry of the coils accurately. The smallest deviation in conductor location may result in changes of the multipole content, which is vital to the correct functioning of an accelerator. In fact, the analytical methods are used for determining the sensitivity to conductor locations with respect to the field quality. The modelling is made more difficult by the key-stoning of the conductor which has as consequence the variation of the current density over each cable cross-section. Also the piling up of the conductors in the blocks of turns must be described precisely if reliable results are expected. Fig. 2 shows a typical dipole coil configuration as used for the analytical computations.

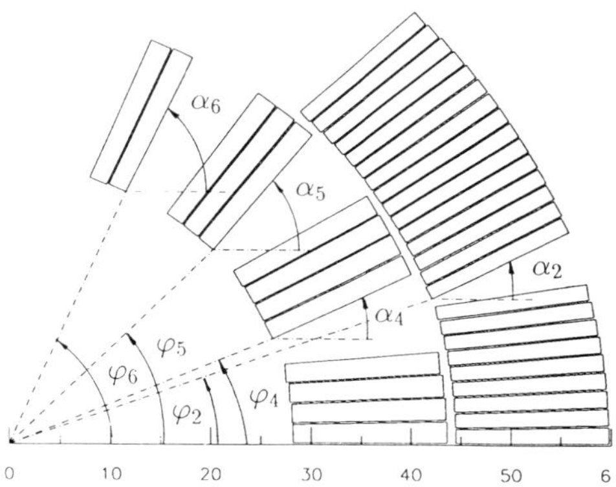

Fig. 2. Dipole coil as used in computation

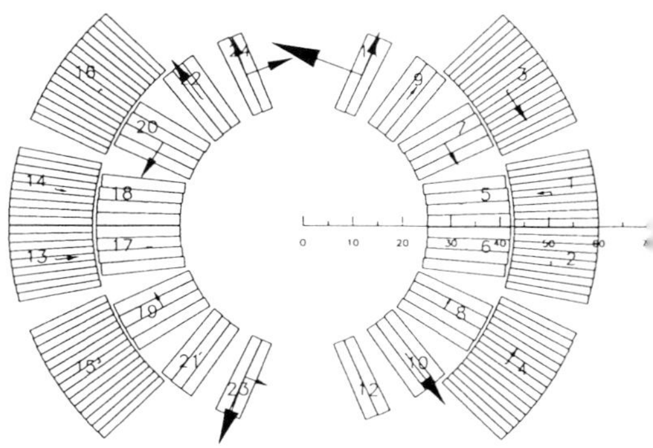

Fig. 3. Result of inverse problem approach

The software routines developped for the analytical computations lend themselves well to the application of optimization techniques used to find the required coil geometries fulfilling best the above mentioned criteria. The design variables for the optimization are the position and inclination of the blocks as shown in fig.2. The problem is formulated as an objective weighting function which is minimized by the search algorithm EXTREM [2].

Once the magnetic measurements of a magnet are perfomed and the multipoles evaluated, one usually is confronted with a set of unwanted terms of which one wished to know the origin. Here the inverse problem approach is indicated. It consists of using least squares minimization routines to find coil geometries which produce exactly the multipole content measured. The inverse problem solving task is more difficul than the initial optimization. The inverse problem has to deal with an objective function which has to contain all multipole terms, which implies all un-allowed terms including the skew ones. On the other hand the number of geometry variables is a multiple of those for the design optimisation, since the coil positioning errors may well be asymmetric. The Levenberg-Marquardt algorithm has been used for solving the least squares minimization problem. Even though the uniqueness is not guaranteed, the results of these computations allow an indication to be obtained of which kind of perturbations may have occurred and of which magnitude they must have been. Fig. 3 shows an example of such a computation. In this case all conductor blocks were left free to move both in radial and azimuthal direction.

4. Numerical Calculations

Once effects like saturation due to permeability are to be computed for a magnet one is constrained to use more complex numerical programs. Most of the modern ones are based on the well-known finite element method.

For the LHC magnets as developped at CERN nowadays, two two-dimensional packages are most in use: the older POISSON package for which a number of improvements have been made and the more modern commercially available, FLUX-2D program, which is characterized by its user-friendly graphical interface and a fast and powerful meshing routine. POISSON in its CERN version was developed with accelerator magnets in mind and was equipped with the multipole analysis from the beginning. POISSON in CERN permits the choice between the Conjugate Gradient Method or the Block Cholesky method for the solution of the matrix equation. FLUX2D uses the ICCG (Incomplete Cholesky Conjugate Gradient) method.

Up to now only the CERN POISSON program has been used in the frame of applying optimization to it for the design of the yoke of the LHC dipole magnet. This required among other changes the parametrization of many of the input data. Once the routines of which POISSON consists have been imbedded into the optimisation program it was possible to perform optimization runs for finding best iron yoke configurations. Since more than a hundred passages may be required before convergence is reached, very powerfull computer configurations must be used. At CERN this is done either on a CRAY X-MP/48 machine or, more recently on a centralized workstation cluster consisting of eight IBM RS6000 RISC system computers.

5. Mathematical Optimization of the Iron Yoke

The design variables may be the collar radius, outer yoke radius the position and dimension of some holes etc. (fig. 4.) The objectives of the optimization are minimized quadrupole and sextupole variations versus the excitation and to achieve a dipole field as high as possible. Usually it is not possible to find a design with all objectives in their individual optimum. The problem is that a further reduction of the quadrupole variation by means of additional holes in the iron insert of the dipole causes a degradation of the dipole field and an increase of the sextupole variation. Such problems are called vector-optimization problems (VOP). Characteristic of VOPs is the appearance of an objective conflict where the individual solutions for each single objective function differ and no solution exists where all the objectives reach their individual minimum. Optimal solutions are therefore characterized by the Pareto-optimality criterion and a solution is found if one objective can only be improved by deteriorating at least one other objective.

It is clear that this yields a set of solutions. One method among others for finding a compromise solution out of the Pareto-optimal solution set, is to define the problem in the objective weighting function, with the user supplied weighting factors. Another way of finding Pareto-optimal solutions is to minimize only one objective while considering the others with constraints. It can be proved that both approaches are equivalent.

In the latter case, in order to consider the nonlinear constraints, a penalty term is added to the objective function which depends on the degree to which the constraints are violated and vanishes if the constraints are satisfied. In order to achieve feasibility of the result the penalty factors have to be chosen relatively large. The final step is then to minimize the penalty function by means of an optimization algorithm for unconstrained problems. The applied optimization algorithm is the search routine EXTREM by Jacob which proved to be very efficient with respect to convergence behavior, accuracy and robustness i.e. the success in obtaining a solution without problem dependent parameters.

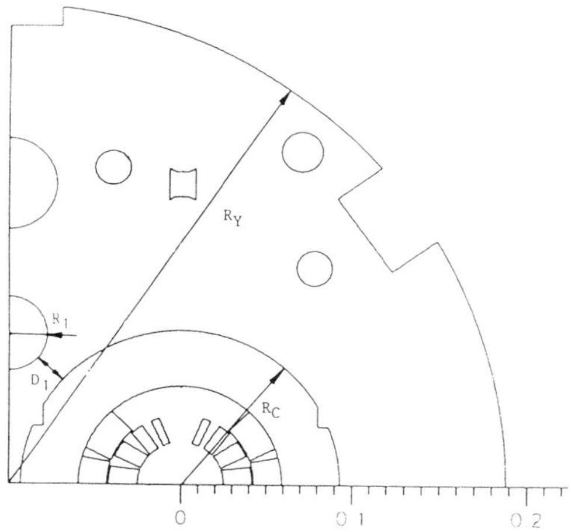

Fig. 4 Iron yoke configuration for optimization

6. References

1 Perin, R.: The Superconducting Magnet System for the LHC, IEEE-Transactions on Magnetics, 27 (1991) 1735-1742

2 Jacob, H.G.: Rechnergestützte Optimierung statischer und dynamischer Systeme, Springer, 1982.

Simulation and Design of Applied Electromagnetic Systems
T. Honma (Editor)

Induced critical current density limit of a Bi-family superconducting oxide tape

H.Ogiwara[a], T.Kitamura[b] and T.Hasegawa[b]

[a] Toshiba Research and Development Center, 4-1 Ukishima, Kawasaki-ku, Kawasaki-shi, Kanagawa, 210 Japan

[b] Showa Electric Wire & Cable Co., Ltd. 2-1-1 Odasakae, Kawasaki-ku, Kawasaki-shi, Kanagawa 210 Japan

Abstract

The contact between a mica-flake type microstructure of a bismuth family superconductor in a silver sheathed tape can have the effect of becoming loose in a high magnetic field when the tape carries a large steady current at a high current density. We investigated the effect of a transport current at a high current density with a high magnetic field on the critical current density of a silver sheathed bismuth family superconducting oxide tape conductor. A silver sheathed bismuth family superconducting oxide conductor is presently the most promising real-use engineering material for applied superconductivity. We have already demonstrated its feasibility by test-fabrication and by testing a series of over one-tesla class superconducting coils. In our previous reports, we have suggested the possible existence of a kind of missing link between our first-size oxide superconducting coils and today's superconducting coils used in the MAGLEV and MRI technologies. The fact that the thinner the superconducting oxide core in a silver sheathed tape is, the higher critical current density can be obtained can prevent from realizing a wire or tape with a large transport current carrying capability. Further, the oxide microstructure adjacent to the silver sheath, carrying current at a high current density, experiences a large electromagnetic force, and local contacts between the mica-flake type microstructure of the bismuth family superconducting oxide can have the effect of being enhanced or weakened depending on the directions of the parallel magnetic field.

1. INTRODUCTION

We have obtained the critical current density of 66,000A/cm^2 with the short sample of an Ag sheathed Bi-2223 tape conductor at 77K, zero external magnetic field[1]. The critical current density of this level can not be attained in the conductor of a coil. However, the magnetic field of about 1.5 Teslas was generated successfully in the bore of the small first-size coil at 4.2K, and 0.16 Teslas at liquid nitrogen temperature[2]. This coil generated successfully 0.7 Teslas at the external magnetic field of 14 teslas to prove feasibility of a small-size laboratory magnet which can generate a high magnetic field. There are many problems in using this kind of conductor for superconducting magnets of MAGLEV train and an MRI system. The magnetic field which a coil generates is determined by magnetmotive force which has the unit of amperes multiplied by turns. A number of fundamental problems, in the basic quantities such as current transport capacity and the length of a conductor needed for constructing a medium size superconducting magnet as above mentioned, for example, are still remaining[3]. In order to obtain a higher critical current density under the present status of R&D, we have to make the thickness of the Bi-2223 core in a silver sheath substrate thinner in the order 10μm[4]. This fact severely reduces the amount of current passing through the conductor. A press/roll process for making an oxide core thinner gives a crystal orientation to the core oxide, and results in a higher critical current density. Moreover, this orientation introduces a strong anisotropy in the current transport property of the conductor. The oxide core consists of lots of oriented mica-flake-like minute Bi-2223 crystals of a thickness around 1μm. Therefore, the mechanical and electrical properties are easily affected when mechanical stress and deformation are applied to the tape conductor. A large electric current at a high current density which flows through conductor under a high magnetic field, produces a large Lorentz force and stress that affects the pile-up structure of the mica-flake-like Bi-2223 crystals.

This report presents the results of a preliminary consideration on the electromagnetodynamic effect to the critical current density which is due to the steady transport current through the microstructure of a Ag sheathed Bi-2223 tape conductor.

2. STRUCTURE OF A TAPE CONDUCTOR

Fig.1 shows schematically the structures of the 1.5 Tesla coil, the cross section of conductors, the cross section of a conductor element, and the microstructure of a Bi-2223 core. The electric current is thought to pass through the tape conductor as illustrated in Fig.1d, following the so-called Brick wall model. Each oxide flake makes contact with adjacent flakes via weak links. The critical current density and the amount of the critical current decrease to the value of from one third to one fifth of the value at zero magnetic field. Therefore, a number of oxide cores are needed for passing hundreds ampere order of a transport current in order to generate a magnetic field of as high as several tesla. When it comes to making multicore systems like metallic fine multicore conductors produces R&D produces but poor conductors with only a small amount of average critical current density due to an unequal press/roll processing effect on each core in the multicore system. This was the reason why we used the conductor structure as in Fig. 1b.

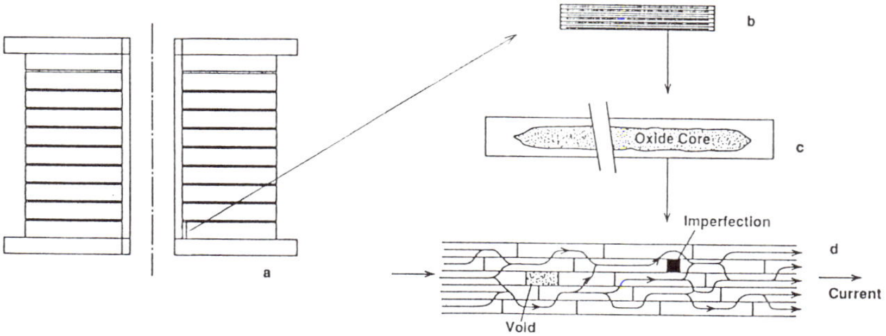

Fig.1 Schematic structures of the coil, cunductor, elementary tape and oxide core

3. SIGNIFICANCE OF BENDING-STRESS CHARACTERISTICS

Fig.2 shows the bending strain effect on the critical current density. We used this result for designing the inner diameter of our coils. This figure indicates the change of the quality of the contact between the flakes. The bending stress finally results in peeling off each flakes.

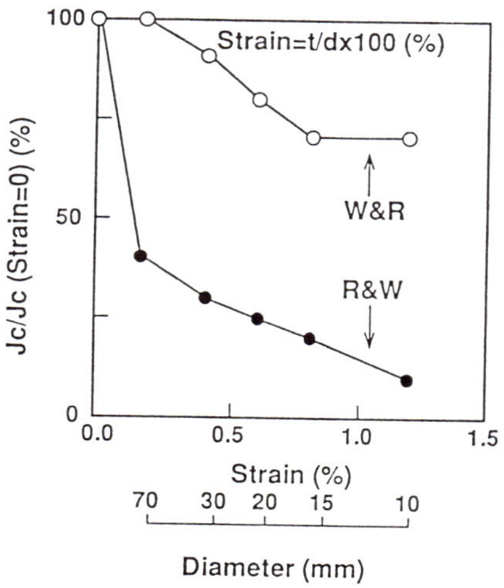

Fig.2 Bending strain chracteristics of the Ag shathed Bi-2223 tape conductor

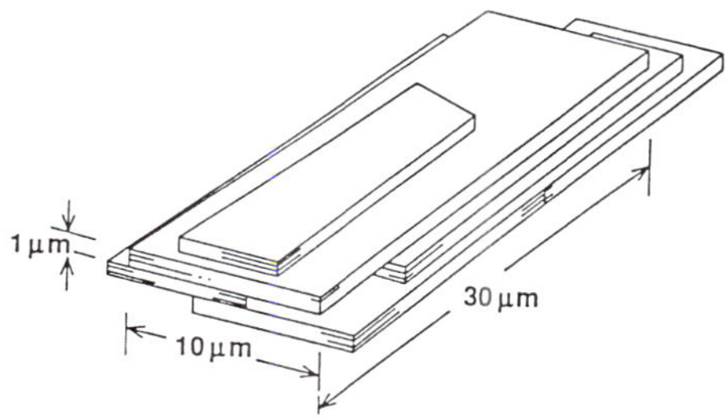

Fig.3 Mica-flake-like microstructure

Here, we consider the model like Fig.3, in which the core of 30μm thickness and 3mm length consists of elementary flakes of 1μm thickness, 10μm width and 50μm length. Only 20μm in 30μm thickness of the oxide core just adjacent to the silver sheath substrate is well orientated and has superconductivity, and the central part of the core of 10μm thickness acts almost like an insulator.

4. STRESS DUE TO STEADY CURRENTS

Electromagnetodynamic forces work on the core structure of the model above mentioned. From the bending stress effect of Fig.2, we can find the effect of these kinds of forces. The poor magnetic properties of the Bi-2223 tape conductor can be recognized as being the results of the stress due to the steady current in the brick wall structure of the Bi-2223 core in part. Possible electromagnetodynamic effects are;

(1) Hoop Stress A hoop stress works on tape carrying a current in a magnetic field.

(2) Propulsive force between the current and its minor image which is created by a silver sheath metal. This force can be significant in case of a single sided vapor deposition tape.

(3) Lorentze force generated by a transport current and a magnetic field. This is again important with a single sided vapor deposition tapes.

(4) Buckling or crimping of the pile-up flake structure with a current passing through a thin flat oxide core.

(5) Imperfection extension with current flow around voids and impurities.

5. CONCLUSIVE REMARK

The mica-flake-like microstructure of the superconducting oxide core of a silver sheathed Bi-2223 tape conductor suffers a number of electromagnetodynamic stresses due to current passing through the conductor. The forces which weaken the contact between the oxide flakes possibly result in poor magnetic property.

6. REFERENCES

1 Y.Yamada, M.Satou, S.Murase, T.Kitamura, Y.Kamisada, presented at ISS92
2 T.Kitamura, T.Hasegawa, H.Ogiwara, presented at ASC-Chicago1992
3 H.Ogiwara, T.Hasegawa, T.Kitamura, presented at 3rd WCS Munich 1992
4 H.Wilhelm, H.W.Newmuller, G.Ries, Physica-C, 2399(1992)

Simulation and Design of Applied Electromagnetic Systems
T. Honma (Editor)

Computer simulations of flux pinning properties in type II superconductors

O. Ichimaru and E. Kusayanagi

Department of Electrical and Electronic Engineering, Faculty of Engineering,
Oita University, 700, Dannoharu, Oita, 870-11 Japan

Abstract

Voltage-current characteristics in type II superconductors with various types of pins are investigated as a function of magnetic field by computer simulation. The magnetic field dependence of the volume pinning force shows a good fit to experimental results for some Nb_3Sn film. It is seen that the addition of small number of point pins degrades the pinning forces by grain boundaries. This unusual degradation of the pinning force can be explained by the softening of the flux line lattice due to the point pins.

1.INTRODUCTION

The interaction between the flux line lattice in type II superconductors and the pinning potential in the sample is difficult to study because the problem is intrinsically non-linear. Experimentally we cannot get the correct informations about pins in the sample. Numerical simulations are very useful for the investigation of such complex systems. In this paper, the flux pinning properties under the co-existence of different types of pins in a two dimensional model system are discussed by computer simulation.

The detailed studies of flux pinning by means of computer simulation have been done since 1980's [1-7]. Brandt [1-4] has investigated the collective pinning theory of Larkin and Ovchinnikov [8], while Jensen et al [5-7] studied the lattice deformations by pins using a molecular dynamics annealing method. Voltage-current characteristics are obtained in the simple case of identical point pins also by Jensen et al [6], but their magnetic field dependences have not yet been studied.

2.MODEL SYSTEM AND METHOD OF SIMULATION

The computer simulations were performed on the following model [1-7]. The system used consists of Nv vortices with variable position $\mathbf{r}_i$, Np point pins with random or regularly arranged position $\mathbf{r}_j^p$ and Ng grain boundaries shown by straight lines in a basic area with periodic boundary conditions (Fig.1). The potential energy of the system is given by

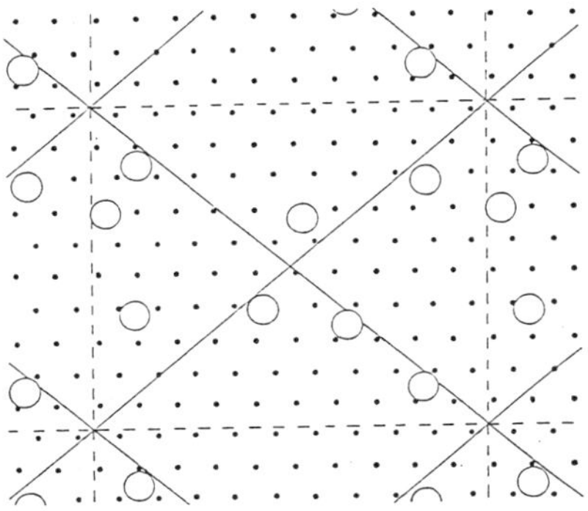

Figure 1. Triangular lattice of Nv = 10x10 = 100 vortices (solid dots), random arrangement of Np=8 attractive point pins (circles) and Ng=2 attractive grain boundaries (solid lines) along the diagonals of the periodicity area (broken lines). This vortex system corresponds to the reduced magnetic field b=0.1.

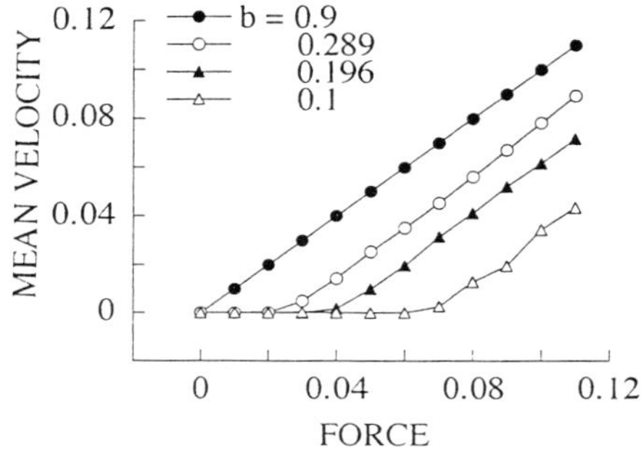

Figure 2. Mean velocity of vortices versus applied force for different values of the magnetic field b. The system parameters are Av=0.3, Apo=Ago=-0.1, Rp=Rg=0.25.

$$U = \sum_{i=1}^{N_v} \left[\sum_{j=i+1}^{N_v} V_v(|\mathbf{r}_i - \mathbf{r}_j|) + \sum_{j=1}^{N_p} V_p(|\mathbf{r}_i - \mathbf{r}_j^p|) + \sum_{j=1}^{2} V_g(|\mathbf{r}_i - \mathbf{r}_{ij}^g|) \right] \tag{1}$$

where $\mathbf{r}_{ij}^g$ is the position vector of the foot of the line perpendicular to the j th grain boundary dropped from the i th vortex. The interaction potentials Vv (> 0) and Vp (< 0) are rotationally symmetric. But Vg (< 0) varies only in the direction perpendicular to the grain boundary. The functional forms are chosen to be Gaussian $v(\rho)=\exp(-\rho^2)$ for numerical convenience.

$$V_v(r) = A_v v(r/R_v),\ V_p(r) = A_p v(r/R_p),\ V_g(r) = A_g v(r/R_g) \tag{2}$$

The unit of the potentials is fixed by choosing Av=0.3. The length is reduced by the spacing of the ideal flux line lattice at Nv=100. The equation of motion is given by

$$\eta \frac{d\mathbf{r}_i}{dt} = -\frac{\partial U}{\partial \mathbf{r}_i} + \mathbf{f} \tag{3}$$

where η is the friction coefficient and **f** the driving force. Time scale is fixed by choosing $\eta = 1$. First the driving force is set to zero and the ideal triangular lattice is relaxed to the pinning potentials using Euler method,

$$\mathbf{r}_i(t+\Delta t) = \mathbf{r}_i(t) + \Delta t \frac{d\mathbf{r}_i}{dt} \tag{4}$$

A constant homogeneous driving force is then applied and the center of mass velocity is measured. This velocity-force curve corresponds to the voltage-current characteristics in type II superconductors. The critical current density Jc is determined by the force at which the mean velocity deviates from zero.

The magnetic field dependences of Jc's are obtained by changing the number of vortices from 100 to 900 in the basic area. The Nv=1000 corresponds to the state at the upper critical field Bc2. We assume the field dependence of the pinning potentials as,

$$A_p = A_{po}(1\text{-}b), \quad A_g = A_{go}(1\text{-}b) \tag{5}$$

where b (=B/Bc2) is the reduced field. The rigidity of the flux line lattice is crucial in determining the correct magnetic field dependence of the volume pinning force Fp (=Jc b). We chose the range Rv of the interaction potentials between flux lines so that the field dependence of the ratio of shear (C_{66}) and compressional modulus (C_{11}) is consistent with the relation [2],

$$C_{66} / (C_{11} - C_{66}) \doteqdot (1 - b)^2 / 10\, b\, \kappa^2 \tag{6}$$

where κ is the Ginzburg-Landau parameter. Here we adopted $\kappa = 34$ as a typical value of Nb_3Sn.

3. RESULTS AND DISCUSSION

Figure 2 shows the mean velocity of vortices versus applied force for different values of magnetic field. These curves show a close resemblance to the experimental ones. The Fp(b) curve are plotted in Figure 3 for three types of pinning system: one contains only the grain boundaries (solid triangles), the others are composed of both the grain boundaries and the random (solid circles) or regular (open circles) arrangement of point pins. It should be noted that the pinning forces of the system with point pins are lowered in low fields than that for the system without point pins. This seems rather unusual results, but can be explained by considering the softening of the shear modulus C_{66} due to the random potential. Jensen et al. showed that the effective shear modulus depends on the pinning strength Ap as $C_{66}^{eff} \propto (1 - \alpha Ap)$, where α is a numerical constant.[6] The Fp(b) curve for only the grain boundaries ($Fp^{GB}(b)$) shows characteristics of the Fp(b) determined by the shearing flow [9], though its exact field dependence is close to not $\sqrt{b}\,(1\text{-}b)^2$ but $b^{3/4}(1\text{-}b)^3$. Here we assumed that Fp(b) is proportional to C_{66}^{eff} and recalculated the Fp(b) for the pinning system with the random point pins ($Fp^{GB+PP}(b)$) from $Fp^{GB}(b)$ as $Fp^{GB+PP}(b)=Fp^{GB}(b)(1\text{-}\alpha Ap)$. The results are shown in Figure 3 by dotted lines. They shows fairly good fit to the Fp(b) curve by the simulation for the case of random point pins. Figure 4 shows a comparison of the reduced $Fp^{GB}(b)$ curve with the experimental results [10] for Nb_3Sn film with a thickness of ~ 0.5 μm formed on a tungsten filament. Note that the $Fp^{GB}(b)$ by the simulation agrees well with the experimental one. The Fp's are normalized by $(3/16\pi^2)^{1/4}Bc_2^2/\mu_o\xi\kappa^2$, where ξ is the coherence length, μ_o the permeability of free space. As for the maximum value of Fp(b), we obtained $Fpmax^{simu}$ ~1.1×10^{11}(N/m^3) assuming Bc2 =24 (T). The experimental one is estimated as $Fpmax^{exp}$ ~ 3.2×10^{10} (N/m^3). The grain size $D=(30\pi^{1/2})\xi$ in simulation gives about 197(nm). The experimemtal D is not given. But it may be estimated to be smaller than the simulation value, since the standard sizes of grains in Nb_3Sn are considered to be 50~200(nm)[11]. Therefore the difference between $Fpmax^{simu}$ and $Fpmax^{exp}$ would be caused by the fact that the elementary pinning force of the grain boundaries Ag/Rg is assumed a little too large.

This work is partly supported by Saneyoshi Scholarship Foundation.

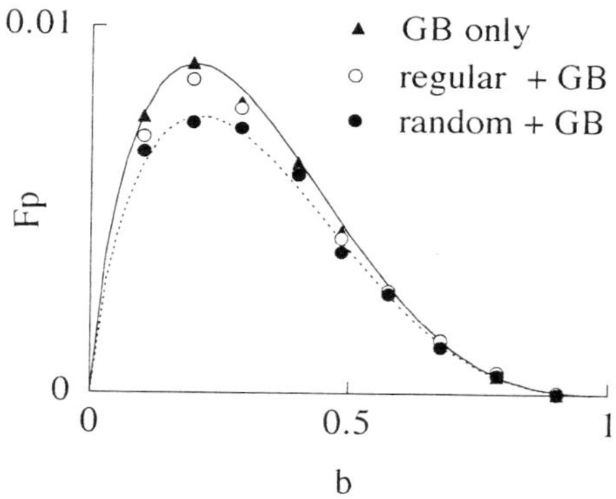

Figure 3. Pinning forces versus magnetic field for different types of pinning system. Solid and broken lines denote $C_1b^{3/4}(1-b)^3$ and $C_1b^{3/4}(1-b)^3(1-\alpha Ap)$ with C_1=0.059, α=2.0, respectively.

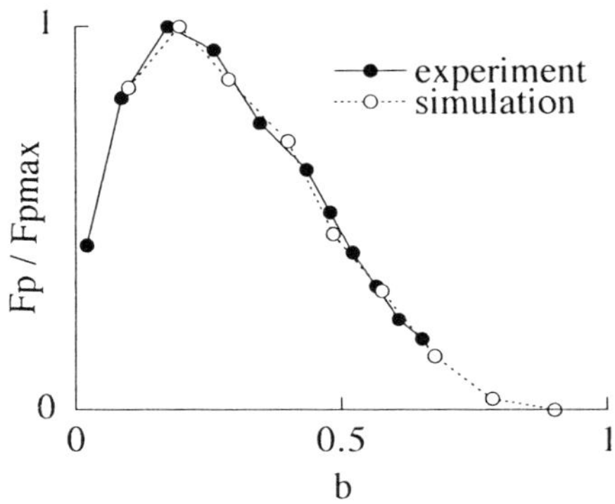

Figure 4. Comparison of the magnetic field dependence of the $Fp^{GB}(b)$ (open circles) with the experimental results (solid circles) from Ref. 10.

4.REFFERENCES

1 E.H.Brandt, Phys. Rev. Lett. 50 (1983)1599.
2 E.H.Brandt, J. Low Temp. Phys. 53 (1983) 41.
3 E.H.Brandt, J. Low Temp. Phys. 53 (1983) 71.
4 E.H.Brandt, Phys. Rev. B 34 (1986) 6514.
5 H.J.Jensen, A.Brass and A.J.Berlinsky, Phys. Rev. Lett. 60 (1988) 1676.
6 H.J.Jensen, A.Brass, Y.Brechet and A.J.Berlinsky, Cryogenics 29 (1989) 367.
7 H.J.Jensen, Y.Brechet and A.Brass, J. Low Temp. Phys. 74 (1989) 293.
8 A.I.Larkin and Y.N.Ovchinnikov, J. Low Temp. Phys. 34 (1979) 409.
9 E.J.Kramer, J. Appl. Phys. 44 (1973) 1360.
10 H.Tateishi, K.Arai and K.Agatsuma, Paper of Technical Meeting on Static Apparatus and Rotating Machinery, IEEJ, SA-92-61(1992).
11 R.M.Scanlan, W.A.Fietz and E.F.Koch, J. Appl. Phys. 46 (1975) 2244.

Simulation and Design of Applied Electromagnetic Systems
T. Honma (Editor)

Improvement of numerical scheme to calculate electromagnetic fields in type-II superconductors (quasi 3-D analysis)

H. Hashizume and S. Toda

Department of Nuclear Engineering, Tohoku University, Aramaki, Aoba, Sendai, 980, Japan

Abstract

Quasi three-dimensional formulation is performed to evaluate current distributions in type-II superconductors when both transport currents and magnetic field parallel to the currents exist. Further the numerical scheme to calculate electromagnetic fields in the superconductors is improved in order to reduce the CPU time required for the quasi three-dimensional analysis.

1. Introduction

Recently, numerical analyses to treat directly superconductor have been well made after the author demonstrated possibility and performance of the constitutive equation and its numerical scheme in terms of electric field and superconducting current based on the critical state models[1-6]. The proposed scheme, however, needs iterative procedure and then their analyses were performed based on 2-D or axisymmetric codes due to limitation of computational capability. Therefore effect of existence of the transport current on a.c. losses is discussed only when both the transport and shielding currents induced by magnetic filed exist on a same plane[4,5]. In this paper quasi three-dimensional analysis to treat these currents is performed together with introducing an improved numerical scheme to reduce the required CPU time.

2. Formulation

The governing equation based on the current vector potential method is given by[1,4]

$$\nabla \times \frac{1}{\sigma} \nabla \times \mathbf{T} = -\frac{\partial(\mathbf{B}_0+\mathbf{B}_e)}{\partial t} \quad (1) \qquad \text{, and} \qquad \mathbf{J} = \nabla \times \mathbf{T} \quad (2)$$

where

$$\mathbf{B}_e(P) = \frac{\mu_0}{4\pi}\int \mathbf{J}(P')\times\nabla'\frac{1}{R(P,P')}\,dV(P') = \frac{\mu_0}{4\pi}\int \nabla'\times\mathbf{T}(P')\times\nabla'\frac{1}{R(P,P')}\,dV(P') \quad (3)$$

In this paper, transport current flowing in z-direction is taken into account as shown in fig.1. To perform quasi three dimensional formulation, the variable, T, is assumed independent of z coordinate and then eqs.(2) and (3) are calculated as follows;

$$\nabla \times \mathbf{T} = \left(\frac{\partial T_z}{\partial y}, -\frac{\partial T_z}{\partial x}, \frac{\partial T_y}{\partial x} - \frac{\partial T_x}{\partial y}\right) \quad (4)$$

$$\mathbf{B}_e(P) = \frac{\mu_0}{2\pi}\int \nabla'\times\mathbf{T}(P')\times\frac{\mathbf{R}(P\leftarrow P')}{R(P,P')^2}\,dx(P')dy(P') \quad (5)$$

In deriving eq.(5), integration in terms of z is performed for the range of $-\infty < z < \infty$. Since eq.(2) or (4) gives no unique solution in terms of T, the Colulomb gauge condition is imposed in the numerical analysis by introducing a reduced integration method. It was confirmed through the numerical analysis that without the gauge condition the system matrix becomes singular and therefore the gauge condition is necessary in the numerical analysis.

The following boundary conditions in terms of T are selected;

$T_z = 0$, (6)

$T_y = T_0$ on A-B , $T_y = -T_0$ on C-D lines (as seen in figure 1) , (7)

$T_x = -T_0$ on B-C , $T_x = T_0$ on D-A lines . (8)

The constant value, T_0, is determined due to the amount of transport current flowing in z-direction.

3. Constitutive equation and improvement of numerical scheme

The following constitutive equations to treat zero resistivity in type-II superconductors have already been proposed and their validity has also been verified through numerical calculations [1,3,4,5,6];

$$\mathbf{J} = \sigma_S \mathbf{E},\ \sigma_S = \frac{J_c}{|\mathbf{E}|}\quad (|\mathbf{E}| \neq 0) \qquad (9)\ ,\quad \text{and} \quad \frac{\partial \mathbf{J}}{\partial t} = 0 \quad (|\mathbf{E}|=0) \qquad (10)$$

where J_c means critical current density. In the numerical calculations, however, iterative procedure is necessary to satisfy eqs.(9) and (10) , as follows[1,3,4];

1) initial value of σ_S is assumed sufficiently large at each time step

2) if $|\mathbf{J}| > J_c$,then $\sigma_S(n=i) = \frac{J_c}{|\mathbf{J}|}\ \sigma_S(n=i-1) = \frac{J_c}{|\mathbf{E}|}$ (n: iteration number) (11)

Therefore in each iteration step, a dense system matrix of the linear algebra equations must be solved using a direct matrix equation solution procedure such as the LU decomposition method. This procedure requires relatively large CPU time.

In this study a new algorithm is developed which employes both the LU decomposition and backward substitution methods. The governing equation given by eqs.(1) and (5) is modified as follows;

$$\nabla \times \frac{1}{\sigma_S(n=i_0)} \nabla \times \mathbf{T}(n=i) + \frac{\mu_0}{2\pi}\frac{\partial}{\partial t}\left(\int \nabla' \times \mathbf{T}(n=i) \times \frac{\mathbf{R}}{R^2}\, dx(P')dy(P')\right)$$
$$= -\frac{\partial \mathbf{B}_0}{\partial t} - \left(\nabla \times \frac{1}{\sigma_S(n=i)} \nabla \times \mathbf{T}(n=i-1) - \nabla \times \frac{1}{\sigma_S(n=i_0)} \nabla \times \mathbf{T}(n=i-1)\right) \qquad (12)$$

where i_0 corresponds to the iteration number when the LU decomposition is performed. In the numerical scheme, the LU decomposition is performed with putting $i_0=i$ when the change of σ_S is relatively large, while the backward substitution is done when it is small. The validity of this procedure will be verified as follows. When the change of σ_S is small, the first term on the left hand side of eq.(12) well approximates the term replacing $\sigma_S(n=i_0)$ with $\sigma_S(n=i)$. Then the left hand side of eq.(12) does not change, which makes the backward substitution available. On the other hand, when the change becomes large, the backward substitution method does not work well because the solution usually diverges.

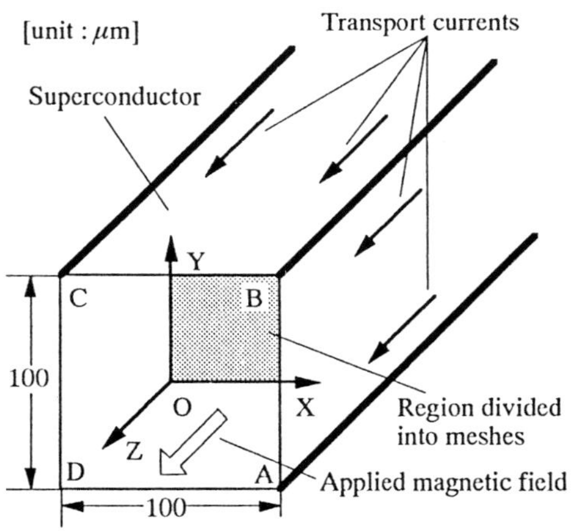

Figure 1. Model used in numerical analysis

4. Numerical results

Equation (12) is solved based on the finite element method using eight-node isoparametoric elements. Figure 1 shows a model of superconductor used in the numerical analysis, where transport current flowing in z-direction exists. One quarter of the cross section of the superconductor($50[\mu m] \times 50[\mu m]$) is divided into 4×4 meshes taking its symmetry into account. Magnitude of the critical current density, J_C, is assumed $2.0[GA/m^2]$.

In table 1, results of the schemes using the backward substitution procedure is compared with those without the procedure when the applied magmatic field increases from 0[T] to 0.06[T] by 0.02[T] without the transport current . The number of iteration at each time step does not differ in both schemes. The CPU time used in the proposed scheme, however, is reduced to about one quarter of the CPU time required in the calculation without the backward substitution procedure.

Table 1 Comparison of schemes with and without backward substitution

	Scheme without backward substitution			Proposed Scheme		
Time step	1	2	3	1	2	3
Total iteration number	366	74	160	366	74	160
Number of LU decomposition	366	74	160	46	23	38
Number of backward substitution	-	-	-	320	51	122
CPU time (sec)	164			37		

Figure 2 shows distribution of shielding current when the longitudinal magnetic field, B_Z, is 0.06[T]. Penetration depth is estimated 23.9 [μm] by the equation; $P=B_Z/\mu_0 J_C$, while the numerical results indicates almost the same value as seen in the figure. Hysteresis loss, Q, calculated by the code is $1373[J/m^3]$, which coincides with the value obtained by the following equation[7];

$$Q = \frac{B_Z^2}{8\mu_0}\left(\frac{2\beta}{3} - \frac{\beta^2}{3}\right) = 1388\ [J/m^3] \quad \left(\beta = \frac{B_Z}{B_Z^*}\right) \tag{13}$$

where B_Z^* is an amplitude of magnetic field just penetrating to the center of the conductor and is calculated 0.127[T] in the geometry used in this analysis.

In figure 3, distribution of transport current flowing in z-direction is shown when the longitudinal magnetic field is not applied. Total transport current , I_{trans}, is 12[A], which means that average of the transport current density is 1.2 $[GA/m^2]$ and less than the critical current density. As seen in the figure, the distribution of transport currents is localized near the surface.

In figure 4, distributions of both transport and shielding currents are shown when the longitudinal field is applied to the superconductor in which the transport current distribution shown in figure 2 is achieved. In this analysis, an absolute value of J, ($|J|=\sqrt{J_X^2+J_y^2+J_Z^2}$), is assumed equal to J_C. As seen in figs.4(a) and (b), both the shielding and transport currents invade further into the center of the superconductor. Due to this larger penetration depth, the hysteresis loss is calculated to be $2832[J/m^3]$, which is almost twice larger than the value without the transport current.

5. Conclusions

Through the present study the following conclusions are obtained;

(1)Quasi three-dimensional formulation is successfully performed based on the T-method.

(2)Using the improved iterative scheme to treat the type-II superconductor, CPU time is reduced to one quarter in comparison with the previous scheme

(3)Numerical results indicate that existence of transport current induces twice larger hysteresis loss when the longitudinal field is applied.

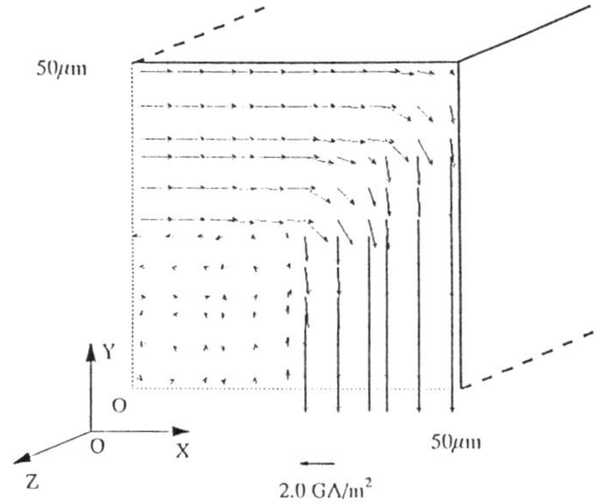

Figure 2. Shielding current distribution without transport current (T_Z=0.06T)

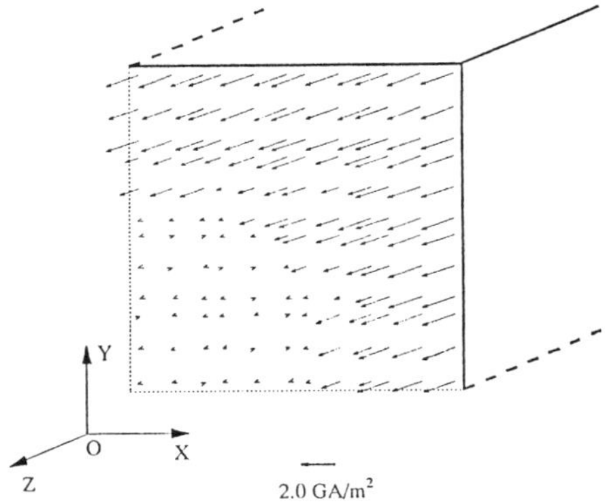

Figure 3. Transport current distribution without longitudinal magnetic field (I_{trans} = 12A)

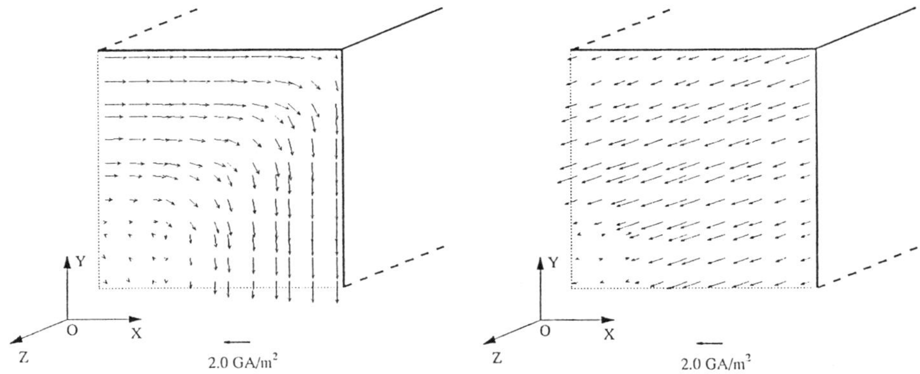

(a) Shielding current distribution (x and y components)

(b) Transport current distribution (z component)

Figure 4. Current distribution when both transport and shielding currents exist (B_Z=0.06 T, I_{trans} = 12A)

References

[1] H. Hashizume et al, Cryogenics, 1991, vol.31, pp.601-606.
[2] T. Sugiura et al, Int. J. of Appl. Ele. Mag. in Mat., 1991, vol.2, pp.183-196.
[3] H. Hashizume et al, IEEE Trans. on Mag.,1992, March, pp.1332-1335.
[4] H. Hashizume et al, Int. J. of Appl. Ele. Mag. in Mat., 1992, vol.3, pp.205-213.
[5] T. Kurusu et al, The 5th CEFC, Claremont, USA, August, 1992.
[6] M. Uesaka et al, Nonlinear phenomena in ele. mag. fileds, Elsevier, pp.289-292, 1992.
[7] M. N. Wilson, Superconducting Magnets, Clarendon Press, Oxford, 1983.

Simulation and Design of Applied Electromagnetic Systems
T. Honma (Editor)

Numerical analysis on magnetic shielding effects of superconducting plate

A. Kamitani, M. Miyauchi and A. Satoh

Department of Electrical and Information Engineering, Faculty of Engineering, Yamagata University, Johnan 4-3-16, Yonezawa, Yamagata 992, Japan

Abstract

The magnetic shielding effects of the superconducting plate are investigated by use of the numerical simulation. A numerical code to solve the open boundary problems of the magnetostatic field is developed and, by means of the code, the shielding effects of axially symmetric plates and partially overlapped plates are investigated. The results of computations show that the shielding effect of the axially symmetric plate is improved with the decrease of ellipticity and with the increase of triangularity. In case of overlapped plates, leakage of the magnetic flux through a gap is shown to decrease with the increase of overlapped length.

1. INTRODUCTION

In detecting the magnetic fields produced in human body, even the earth magnetism and other magnetic signals resulting from electric instruments become noise. In this sense, the magnetic shielding is an essential problem for the biomagnetic measurement. Recently it has been proposed that high T_c oxide superconductors should be used to construct a magnetic shielding apparatus[1-3]. The reason for this is that they are easy to keep the superconducting state by use of liquid nitrogen or a mini refrigerator. On the other hand, it is well known that oxide superconductors are too brittle to be shaped into large plates. In order to avoid these difficulties, many small plates should be joined into a large shielding apparatus.

The purpose of this paper is to develop the numerical code for analyzing the magnetostatic field around axially/slab symmetric superconducting plates and to investigate their magnetic shielding effects by use of the code. In the next section, we investigate the influence of the plate shape on the magnetic shielding by using the axially symmetric model. The dependence of the shielding effect on magnetic permeability of the plate is also studied there. In the third section, we investigate the magnetic shielding effects of two plates with partially overlapped region by means of the slab symmetric model. Conclusions are presented in the final section.

2. SHIELDING EFFECTS OF AXIALLY SYMMETRIC PLATE

We assume, at first, the shielding plate is axially symmetrical. Let us use the cylindrical coordinate system (z,r,ϕ) and take the symmetry axis as z-axis. The plate surface is assumed as

$$\left(e\frac{z}{b} - \delta\frac{r^2}{b^2}\right)^2 + \left(\frac{r}{b}\right)^2 = 1 \quad , \tag{1}$$

where e and δ are ellipticity and triangularity, respectively. As the source of the magnetic field, we place an 1-turn coil at $(-5b/6, b/30)$. In the following, $\partial\Omega_s$ denotes the plate surface and the region bounded by $\partial\Omega_s$ and z-axis is represented by Ω_1. In addition, Ω_2 denotes the region outside Ω_1 and contains the 1-turn coil in it.

Under the above assumptions, the behavior of the magnetic field in Ω_m is represented by

$$c^{(m)}(z_i, r_i)\, \psi^{(m)}(z_i, r_i) = \int_{\Omega_m} \mu^{(m)} j_\phi^{(m)} w^* dzdr + \int_{\partial\Omega_s} \left(\frac{w^*}{r} \frac{\partial\psi^{(m)}}{\partial n} - \frac{\psi^{(m)}}{r} \frac{\partial w^*}{\partial n} \right) dl \quad , \tag{2}$$

where $j_\phi^{(m)}$ and $\mu^{(m)}$ denote the ϕ-component of the current density and magnetic permeability, respectively. In addition, $\psi^{(m)}$ is the magnetic flux function for the axially symmetric model and is associated with magnetic induction **B**. Further, $c^{(m)}(z_i, r_i)$ denotes the so-called shape coefficient and $\partial/\partial n$ stands for the normal derivative. The function w* is now given by

$$w^*(z, r\, ;\, z_i, r_i) = \frac{(rr_i)^{1/2}}{2\pi k} \left[(2-k^2)K(k) - 2E(k)\right] \quad , \tag{3}$$

where K(k) and E(k) are complete elliptic integrals of the first and the second kind, respectively, and k is defined by

$$k^2 = \frac{4rr_i}{(z-z_i)^2 + (r+r_i)^2} \quad .$$

As the interface conditions, we impose the conditions that the tangential component of magnetic field and the normal one of magnetic induction be continuous across the plate surface $\partial\Omega_s$.

Following the standard manner of the boundary element method, we can solve eq.(2) under the above interface conditions and determine the values of $\psi^{(m)}$ and **B** at an arbitrary location. As the measure of the magnetic shielding, we adopt the decay factor defined by $\alpha = 20 \log (|\mathbf{B}|/|\mathbf{B}_n|)$. Here, $\mathbf{B}_n$ is magnetic induction in case of no shielding plate.

First, we investigate the influence of the relative magnetic permeability μ_s on the magnetic shielding. Figure 1 shows the dependence of the decay factors α on the relative magnetic permeability μ_s. We see from this figure that though the value of α is small when $\mu_s \cong 1$, it decreases with the decrease of μ_s. In the case of $\mu_s \leq 10^{-4}$, the value becomes nearly constant.

Next, we also investigate the effect of the plate shape on the magnetic shielding. In Figure

2, we show the decay factors as functions of triangularity δ. The results of computations show that the decay factor α on z-axis decreases with the decrease of e and with the increase of δ.

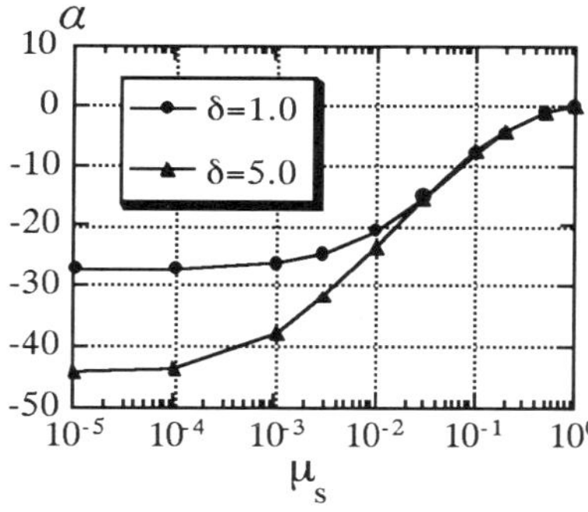

Figure 1 The dependence of the decay factors α on the relative magnetic permeability μ_s (z/b=0.5, e=6.0).

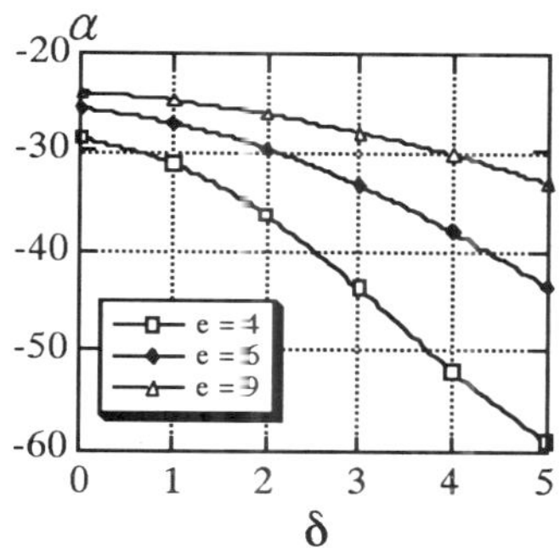

Figure 2 The dependence of the decay factors α on triangularity δ (z/b=0.5, $\mu_s=10^{-4}$).

3. SHIELDING EFFECTS OF PARTIALLY OVERLAPPED PLATES

Although the previous model is realistic for the axially symmetric shielding plate, it is impossible to change the location of the coil center along the plate. In order to investigate the dependence of the shielding effect on the coil location, we introduce the second model. In this model, we assume that the shielding plate has a slab symmetry and we take the symmetry axis as z-axis. Under these assumptions, the governing equation is transformed into

$$c^{(m)}(x_i,y_i)\,\overline{\psi}^{(m)}(x_i,y_i) = \int_{\Omega_m} \mu^{(m)} j_z^{(m)} \overline{w}^* dxdy + \int_{\partial\Omega_s} \left(\overline{w}^* \frac{\partial\overline{\psi}^{(m)}}{\partial n} - \overline{\psi}^{(m)} \frac{\partial\overline{w}^*}{\partial n} \right) dl \ , \tag{4}$$

in which $\overline{\psi}^{(m)}$ is the magnetic flux function for the z-independent model and $\overline{w}^*$ is given by

$$\overline{w}^* = -\frac{1}{2\pi} \log \sqrt{(x-x_i)^2+(y-y_i)^2} \ . \tag{5}$$

The above boundary integral equation (4) can be solved in a similar way to the previous model.

Let us investigate the magnetic shielding effects of two plates with partially overlapped region. In actual experiments, an adhesive tape is used to join two plates and its relative magnetic permeability is nearly equal to unity. Therefore, we might assume that two plates are placed with a gap. In Figure 3, we show the cross section of the adopted model. Two shielding plates of width $3a+y_0/2$ and thickness a are placed with a gap of width a/5, and their overlapped length is equal to y_0. As the source of the magnetic field, we place two wires at (-2a,y-a/2) and (-2a,y+a/2), and apply currents, -I and I. Under the above assumptions, decay factors are calculated at (2a, y). Figure 4 shows the dependence of decay factors on coil location y. In this figure, decay factors are also depicted for various values of overlapped length. We see from this figure that three curves take the almost same values near both ends of the plates. In addition, the minimum decay factor for case of $y_0/a=6$ is about twice as large as that for case of $y_0/a=1$. The reason for this is explained as follows. Since the magnetic flux leaks out through a gap, the magnetic shielding is improved with the increase of overlapped length.

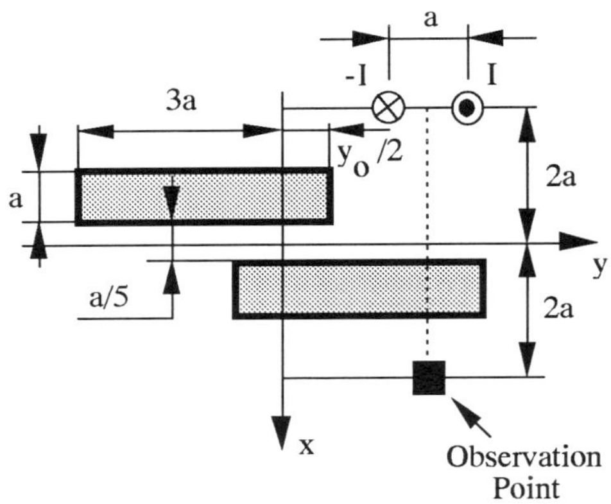

Figure 3 Two shielding plates with a gap.

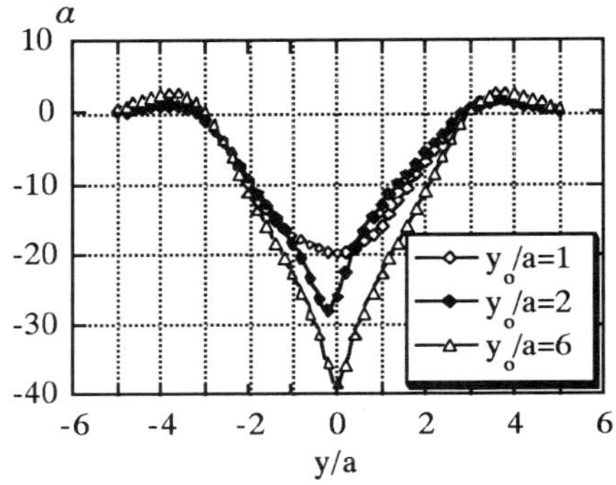

Figure 4 Decay factors as functions of coil position y/a in case of $\mu_s=10^{-3}$.

4. CONCLUSIONS

We have investigated the magnetic shielding effects of the superconducting plates by use of the numerical simulation. Conclusions obtained in the present study are as follows. 1) The shielding effect of the axially symmetric plate is improved with the decrease of ellipticity and with the increase of triangularity. 2) In case of overlapped plates, leakage of the magnetic flux through a gap decreases with the increase of overlapped length.

REFERENCES

1 S. Ohshima and K. Okuyama, Jpn. J. Appl. Phys. **29** (1990) 2403.
2 A. A. Hussain and M. Sayer, Cryogenics **32** (1992) 64.
3 A. Satoh, A. Kamitani, S. Ohshima and M. Katagiri, Applied Electromagnetics in Materials and Computational Technology, Hokkaido University Press, Sapporo, 1992, pp 175-180.

Simulation and Design of Applied Electromagnetic Systems
T. Honma (Editor)

Evaluation of dynamic magnetic force of high T_c superconductor with flux flow and creep

Y. Yoshida, M.Ukibe, M.Uesaka and K.Miya

Nuclear Engineering Research Laboratory
Faculty of Engineering, The University of Tokyo
Tokai-mura, Ibaraki 319-11, Japan

Abstract

This paper describes the evaluation of dynamic magnetic force of high T_c superconductor based on the flux flow and creep model. Numerical analyses and experiment were performed in order to investigate the dynamic magnetic force characteristics. Measured magnetic force of an MPMG processed YBaCuO superconductor shows fast growth and decay as a result of motion of fluxoids. A computer code based on the macroscopic theory shows good agreement with experimental results of dynamic magnetic force. Especially, it was confirmed that the source of the fast growth and decay is the flux flow.

1. INTRODUCTION

Recently an application of high T_c melt grown YBaCuO superconductors[1] to a superconducting levitation system becomes promising. An estimation of both static and dynamic magnetic force is one of the most important subjects. Several numerical methods have been proposed based on the critical state model for the static magnetic force [2][3].

High T_c superconductors show flux flow and creep phenomena due to which the shielding current and magnetic force grow and decay with time. Though these phenomena affect characteristics of the dynamic magnetic force in superconducting systems, numerical analysis has not been done so far due to the strong nonlinearity. Therefore the purpose of this paper is to develop a three dimensional numerical method taking into account the flux flow and creep based on the current vector potential method and thereby to evaluate the dynamic characteristics of the magnetic force.

2. FLUX FLOW AND CREEP

The flux flow and creep are the phenomena due to the motion of fluxoids which induces the electric field. The Joule loss due to flux flow and creep makes the magnetic force decay in time.

According to the Kim-Anderson theory[4], the constitutive equation between the current density and electric field in the flux creep region is represented as follows;

$$E = 2\rho_c J_c \sinh(\frac{U_0}{k\theta}\frac{J}{J_c})\exp(-\frac{U_0}{k\theta}) \tag{1}$$

where E is the electric field, J the current density, θ the temperature, U_0 the pinning potential, k the Boltzmann constant, ρ_c the creep resistivity, J_c the critical current density without thermal activation. For the flux flow region, the following expression is well-accepted [5];

$$E = E_c + \rho_f J_c(J/J_c - 1) \tag{2}$$

where $E_c = \rho_c J_c$ and ρ_f is the flow resistivity.

Above mentioned constitutive relations are shown with the critical state model in Fig.1.

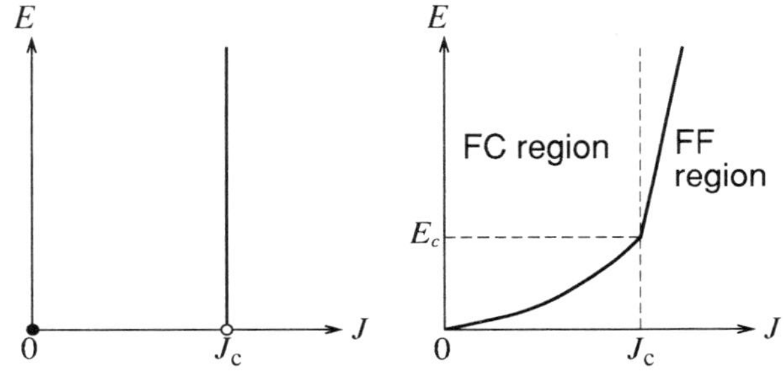

(a) Critical state model (b) Flux flow, flux creep model

Fig.1 Constitutive relation between electric field and current density

3. NUMERICAL METHOD

It is assumed in the numerical analysis that an MPMG processed sample comprises thin plates parallel to the a-b plane where supercurrents between the plates are negligibly small. This assumption is based on the anisotropy of J_c values. Thus, the thin plate approximation can be applied to computation of screening currents in a superconductor in order to reduce the number of degrees of freedom[3].

A governing equation for normal component of current vector potential, T, takes the following form;

$$\frac{1}{\sigma}\nabla^2 T - \mu_0 \dot{T} - \frac{\mu_0}{4\pi}\boldsymbol{n}\cdot\int \dot{T}\nabla'\frac{1}{R}dS = \dot{B}_0 \tag{3}$$

where $\boldsymbol{J} = \nabla T \times \boldsymbol{n}$.

In order to treat the nonlinear constitutive relation such as eq.(1), the following artificial conductivity is introduced to linearize the governing equation;

$$\sigma = J/f(J) \tag{4}$$

where the constitutive relation is written as $E=f(J)$. An iterative method was applied to get a converged solution.

4. EXPERIMENT

The experiment was performed using an MPMG processed cylindrical YBaCuO superconductor of 18mm in diameter and 2.5 mm in thickness to investigate the growth and decay of the magnetic force for different applied magnetic field changes. Schematic drawing of the experiment is shown in Fig.2. The cylindrical permanent magnet of 25 mm in diameter and 22.5 mm in thickness was fixed at the free end of the cantilever beam. The XY stage drives the magnet from 25 mm to 0.5 mm in gap between the magnet and the superconductor at a specified constant speed. The magnetic force was measured by the strain gage attached to the clamped end of the beam.

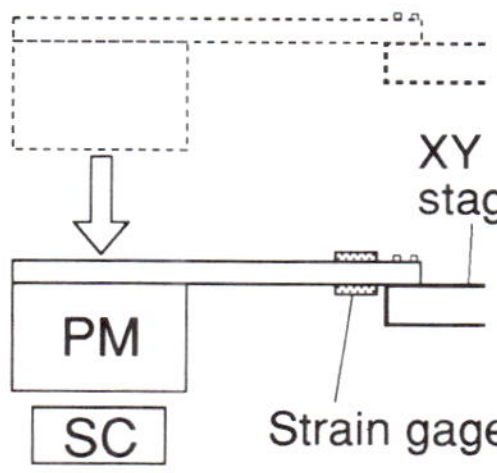

Fig.2 Experimental model

5. RESULTS AND DISCUSSION

Fig.3 shows experimental and numerical results of time dependence of the magnetic force. In the calculation we used both the flux flow and creep model and the critical state one. The fast growth and decay are observed in the time domain of 0–3sec in the experimental results and its amplitude is far larger than the numerical results with the critical state model where J_c=1.3×10^{7}A/m^2. In the flux flow and creep model, we choose J_c=1.3×10^{7}A/m^2, U_0=92meV, ρ_f=7.62× 10$^{-10}\Omega$m, E_c=0.1mV/m in order to make the calculated curve coincide with the measured one. Good agreement between them confirms that the fast growth and decay occur primarily due to the flux flow.

Time-variation of the electric field in the superconductor is shown in Fig.4. These results were obtained from computation based on the flux flow and creep model. The electric field penetrates from the radially outer elements in the superconductor. As the electric field becomes larger, the faster decay occurs.

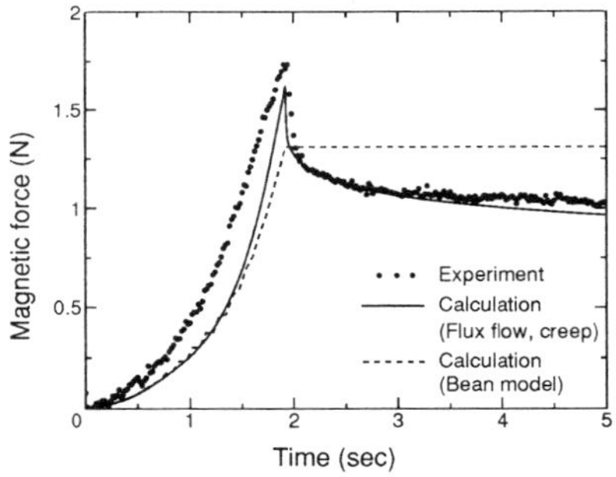

Fig.3 Time-variation of magnetic force

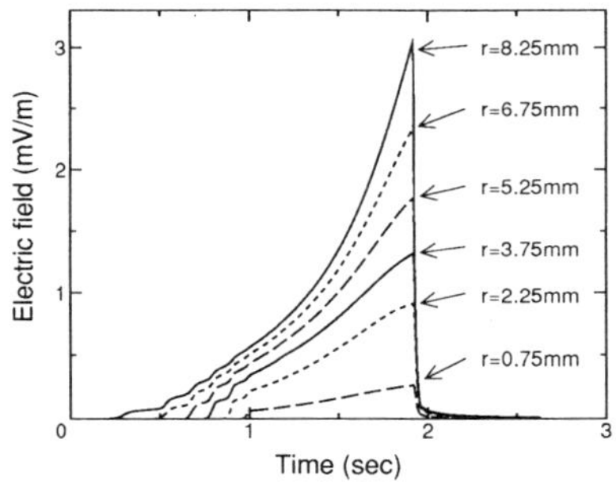

Fig.4 Time-variation of electric field

Fig.5 shows the experimental and numerical results of the magnetic force in 3 cases of different speeds of motion of the magnet. From these results we can find the dependence of the magnetic force on the speed of the applied magnetic field change. When the field change is faster, the peak value of the magnetic force becomes larger.

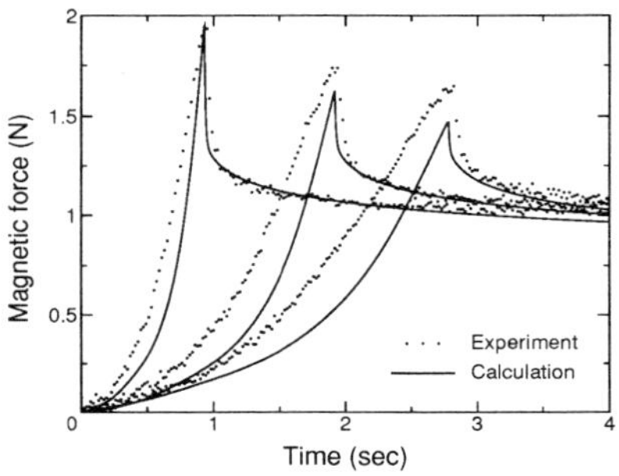

Fig.5 Dependence of force on Time

6. CONCLUSIONS

- Experimental results of the magnetic force of the MPMG processed YBaCuO superconductor show the fast growth and decay which are attributed to the flux flow.
- The numerical method to evaluate the dynamic magnetic force of the superconductor was developed based on the flux flow and creep model.
- Using the computer code developed here, the dynamic characteristics of the magnetic force has been analyzed quantitatively taking into account the flux flow and creep.

References

[1] M.Murakami et al., Jpn. J. of Appl. Phys., Vol.29, No.11, pp.L1991-L1994, 1990
[2] N.Takeda et al., Elsevier Studies in Appl. Elec. Mag. in Materials 3, 1992, pp.125-128
[3] M.Uesaka et al., Int. J. Appl. Elec. Mag. in Materials, Vol.4, No.1, 1993, in press
[4] P.W.Anderson, Phys. Rev. Lett., Vol.9, No.7, pp.309-311, 1962
[5] K.Yamafuji et al., Cryogenics, Vol.32, 1992, pp.569-577

Simulation and Design of Applied Electromagnetic Systems
T. Honma (Editor)

Inverse analysis of current distribution in high-T_c superconductor

N. Takeda, M. Uesaka and K. Miya
Nuclear Engineering Research Laboratory,
Faculty of Engineering, The University of Tokyo

Tokai-mura, Ibaraki 319-11, Japan

Abstract

This paper describes the inverse analysis to estimate superconducting shielding current (supercurrent) paths and grain boundaries of a melt-grown high-T_c superconductor by using measured magnetic field data. For the inverse analysis, the following two optimizing method are used: Least Squares Method (LSM) and Genetic Algorithm (GA). Through numerical experiments, supercurrent paths and grain boundaries have been successfully reconstructed by the two methods. Especially, it has been found that GA is more robust for random noises than LSM.

1. INTRODUCTION

Most of large melt-grown high-T_c superconductors contain many cracks and grain boundaries. When they are magnetized, supercurrents are distorted or localized by the cracks and grain boundaries. However, it is often assumed that a superconductor consists of a single grain in numerical analysis of magnetization and magnetic force[1]. This discrepancy is the major source of disagreement between experimental and numerical results[2]. Therefore, it is very important to develop an inspecting method to estimate the supercurrent paths and grain boundaries.

The purpose of this study is to develop an inversion technique to estimate supercurrent paths and grain boundaries by using measured magnetic field data. For the inverse analysis, we use the following two optimizing methods: Least Squares Method and Genetic Algorithm. We have carried out numerical experiments on estimation of supercurrent paths and grain boundaries. The results given by these two methods are compared with each other and the effectiveness of these method are discussed.

2. INVERSE ANALYSIS

2.1. Description of inverse analysis

We suppose the following experiment. A cylindrical superconductor of 18mm in diameter and 2.5mm in thickness has grain boundaries as shown in Figure 1. After the permanent magnet approaches the superconductor from the gap of 25mm to 10mm, the shielding current flows in the superconductor. The shielding current was obtained by using the three dimensional computer code based on the current vector potential method[2]. Figure 1 shows the contour map of normal component of current vector potential T_n. The contour-lines correspond to the shielding current paths. In the inverse analysis, the distribution of T_n in the superconductor is estimated by using computed magnetic field data on the plane located from 2mm to 5mm over the superconductor. The relation between the normal component of flux density B_n and T_n can be written by[3]

$$B_n = \frac{\mu_0}{4\pi}\boldsymbol{n} \cdot \int T_n \nabla \frac{1}{R} dS, \tag{1}$$

where $\boldsymbol{n}$ is the unit normal vector of the surface. We also carry out the numerical experiment by using computed field data with 10% random noise.

2.2. Linear Least Squares Method

From Eq.1, we can get the system equation: $[A]\{x\} = \{y\}$ where $[A]$ is the Jacobian matrix($n \times m$), $\{x\}$ is the unknown vector ($m \times 1$) and $\{y\}$ is the input vector ($n \times 1$). We can solve this system equation for x using the linear Least Squares Method(linear LSM) with the modified Gram-Schmidt orthogonalization.

2.3. Genetic algorithm

Introduction of Genetic Algorithm Genetic Algorithm is a kind of optimizing methods and it is based on the mechanics of natural selection and genetics[4]. GA is different from other optimizing method in the following ways: (1) GA works with a coding of the parameter set, that is called "chromosome" or "string," (2) GA searches from multiple point, (3) GA uses probabilistic transition rules. Because of these features, GA is efficient comparing to other optimizing method to solve the problem whose objective function is multiple-peak and/or is discontinuous. In other words, GA is a robust method.

Procedure of GA is composed of three processes: Reproduction, Crossover, and Mutation. Reproduction is the process in which individual "strings" are copied according to their objective function values. Biologists call this function as the fitness function. Crossover and Mutation are the processes in which individual "strings" are changed. In Crossover, two "strings" exchange arbitrary parts of them and become two new strings. In Mutation, an arbitrary character in a "string" is changed into another character. In this study, the mutation ratio is 1%.

Analysis model for GA If the system equation can be expressed in the above linear equation, linear LSM is clearly more effective and efficient to solve it than other probabilistic optimization methods such as GA. In order to utilize GA effectively, we propose the following simplified non-linear model.

First, consider a group of artificial cylindrical grains with the conical T_n distribution in a superconductor as shown in Fig. 2. Its slope corresponds to the critical current density: $|\nabla T_n| = J_c$. T_n distribution in the whole superconductor is regarded as the superposition of these conical distributions. Every "cone" (grain) has four parameters: the coordinate of the center (x_c, y_c), T_n value on the center T_c, the radius of the "cone" r_c. By the above modeling, the estimation problem of T_n distribution is transformed into the optimizing problem of the locations of these "cones." Grain boundaries can be recognized as the lines along local minima. In this study, the number of "cones" is 20.

Coding into "string" To apply GA, we must code parameters of a system into a "string." In this study, we code a sequence of parameters of the "cones" into the "string," regarding figures in the sequence as characters in the "string." The number of these "strings" is 320.

Fitness function Optimizing methods need the objective function or the fitness function which is the criterion of optimization. In this study, we define the fitness function as

$$(Fitness\ function) = 1 / \left\| \frac{\boldsymbol{B}_o}{\|\boldsymbol{B}_o\|} - \frac{\boldsymbol{B}_a}{\|\boldsymbol{B}_a\|} \right\|, \tag{2}$$

where $\boldsymbol{B}_o$ is the observed vector whose components are flux density data obtained at each input (computed) point and $\boldsymbol{B}_a$ is the calculated vector from the assumed T_n distribution and $\|\cdot\|$ is Euclidean norm. The problem is transformed into the maximizing problem of the fitness function.

3. RESULTS AND DISCUSSION

First, we applied linear LSM and investigated the dependence of the solution on the condition number. The condition number is a parameter which indicate a sensitivity of a solution to input data and expected to be less than 10.

When the numbers of input and model points are 198 and 81, respectively, the condition number becomes 8.8 and an accurate solution can be obtained. However, in case of 100 input points, the condition number becomes more than 10^3 and it is hard to recognize grain boundaries. We tried to solve the problem with four different location of input points: 2, 3, 4, 5mm over the superconductor. Among the above locations, we can get a reasonable solution only in case of 4mm as shown in Fig. 3.

Furthermore, in the case of a large condition number, the solution is sensitive to noise in input data. Figure 4 shows the result in case of the input data with 10% random noise. In this case, the inverse analysis is failed.

The number of input data must be large so as to decrease the condition number theoretically. However, it is expected to be as small as possible from the viewpoints of time and cost of experiment and calculation. For example, the computation time is about 10 minutes (with SPARCstation ELC) in case of 100 input points while in case of 196 input points it is about 30 minutes.

On the other hand, GA can solve the problem effectively even if the problem is rather ill-condition. Figure 5 shows the result with the same condition as Fig. 3. Furthermore, GA can get similar results in other cases than 4mm and give a reasonable solution although the input data have random noise. Figure 6 shows the result with the same condition as Fig. 4. These results indicate that GA is very robust for the problem.

In addition to the above advantage of GA, the computation time of GA is comparable with that of linear LSM. GA spends about 6 seconds on one iterative step; 100 iterative steps, which correspond to about 10 minutes, are enough to get an almost reasonable result. T_n distribution is sufficiently converged after 10 minutes although fitness function is still increasing.

4. CONCLUSION

Using linear LSM and GA, we have carried out the numerical experiment on estimation of shielding current in a melt-grown superconductor. The linear LSM is effective if observed data do not have random noise, but it is very weak against the noise in the input data. On the other hand, it has been confirmed that GA is very robust for both the noise and ill-condition of the problem.

REFERENCES

[1] N.Takeda et al., Elsevier Studies in Appl. Elec. Mag. in Materials 3 (1992)
[2] M.Uesaka et al., Int. J. Appl. Elec. Mag. in Materials, Vol.4, No.1 (1993) in press
[3] K.Miyata et al., Proc. of 9th International Conference on Nondestructive Evaluation in the Nuclear Industry (1988)
[4] D.E.Goldberg, "Genetic Algorithms in Search, Optimization and Machine Learning," Addison-Wesley (1989)

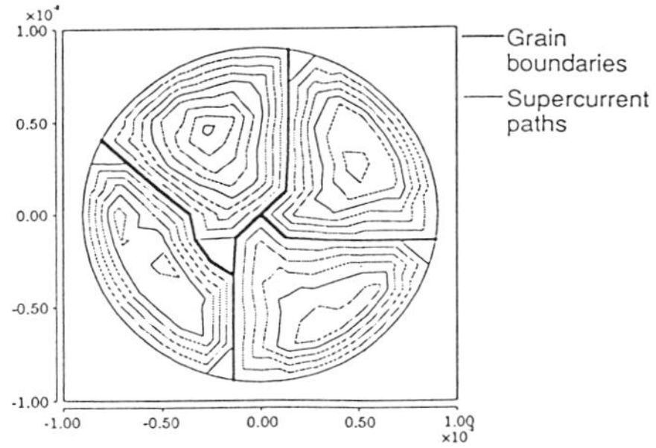

Fig. 1 Distribution of T_n in a superconductor with grain boundary

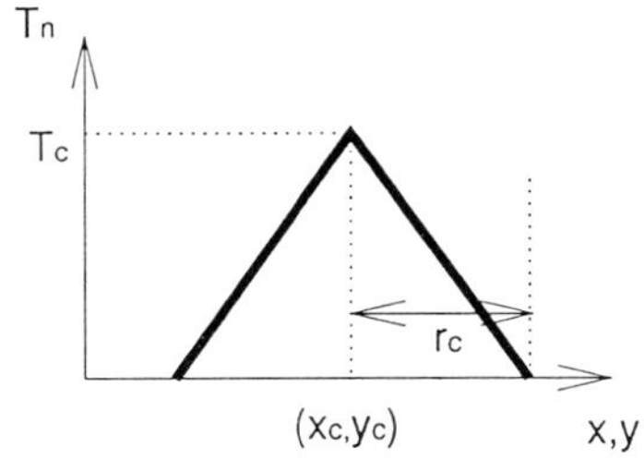

Fig. 2 Conical T_n distribution of artificial grain

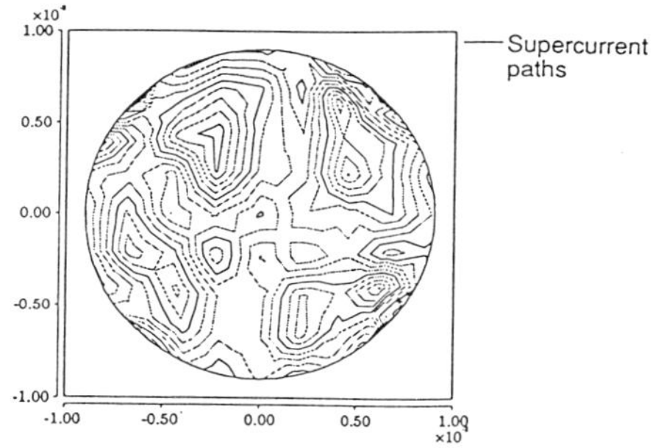

Fig. 3 Result of inversion by LSM without noise

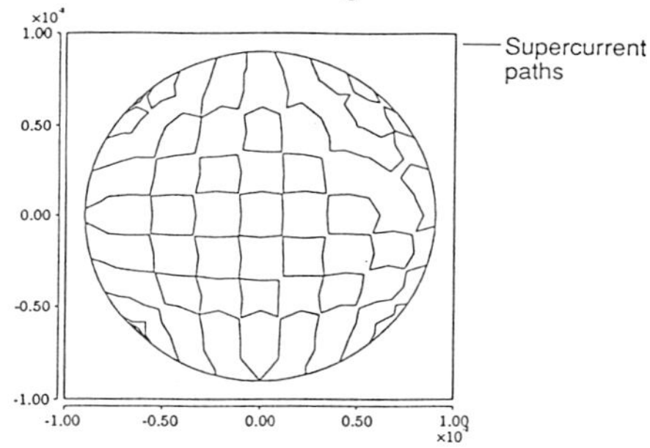

Fig. 4 Result of inversion by LSM with 10% random noise

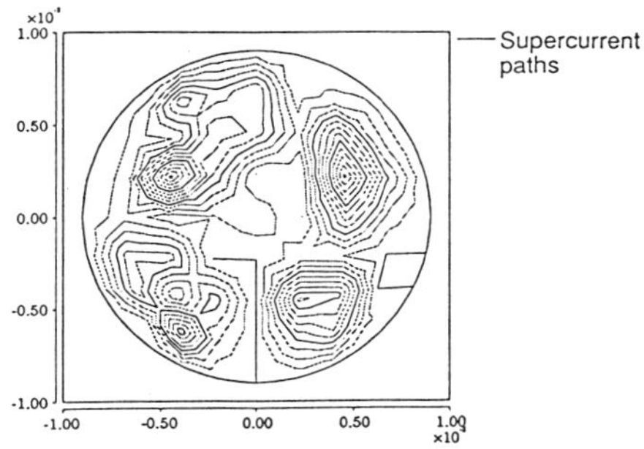

Fig. 5 Result of inversion by GA without noise

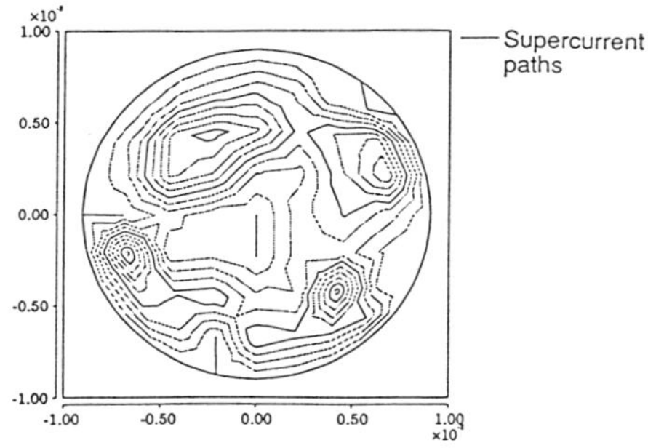

Fig. 6 Result of inversion by GA with 10% random noise

Simulation and Design of Applied Electromagnetic Systems
T. Honma (Editor)

Effects of a grain boundary on levitation force of a high T_c superconductor

M. Tsuchimoto, H. Takeuchi and T. Honma

Department of Electrical Engineering, Faculty of Engineering,
Hokkaido University, Kita 13, Nishi 8, Kita-ku, Sapporo 060, Japan

Abstract

Levitation force of a high T_c superconductor (HTSC) is evaluated numerically by using a model with a grain boundary. Effects of the grain boundary on the levitation force are discussed from a view point of equivalent critical current density. It is shown that existence of grain boundaries is one of reasons of the difference between the measured critical current density and the equivalent critical current density.

1. INTRODUCTION

Large levitation force and a stable equilibrium are obtained with a permanent magnet and the Melt-Powder-Melt-Growth processed HTSC [1]. Evaluation of the levitation force is important for many applications, such as magnetic bearing and flywheel. It is said that grain boundaries and cracks in the HTSC disturb screening current distribution and reduce the levitation force. The problem is studied numerically by using models with and without grain boundary.

2. FORMULATION

The numerical method to analyze the electromagnetic phenomena in Type-II superconductors was developed by Sugiura and Hashizume, et.al. [2-4]. In the present paper, the multi-layers 2D code with the current vector potential method (T-method) [3, 4] is applied to the analysis because of anisotropic structure of the HTSC. Governing equation of the T-method is obtained from Maxwell equations and Biot-Savart's law :

$$\nabla\times\mathbf{E}=-\frac{\partial\mathbf{B}}{\partial t},\ \nabla\times\mathbf{B}=\mu_0\mathbf{J},\ \nabla\cdot\mathbf{B}=0\,,\ \mathbf{J}=\sigma\mathbf{E},\ \mathbf{B}=\frac{\mu_0}{4\pi}\int\mathbf{J}\times\nabla\frac{1}{R}dV\,,\ \mathbf{J}=\mathbf{J}_0+\mathbf{J}_e\,, \tag{1}$$

where μ_0 and σ are magnetic permeability in air and conductivity, and $\mathbf{J}_0$, $\mathbf{J}_e$ are external current and screening current densities. When screening current distribution is assumed constant through a thickness of a layer, the following scalar equation is obtained with normal component T of the current vector potential defined by $\nabla\times\mathbf{T}=\mathbf{J}_e$ [3-4] :

$$\frac{\partial}{\partial x}\left(\frac{1}{\sigma}\frac{\partial T}{\partial x}\right)+\frac{\partial}{\partial y}\left(\frac{1}{\sigma}\frac{\partial T}{\partial y}\right)-\mu_0\frac{\partial T}{\partial t}-\frac{\mu_0}{4\pi}\int_S\frac{\partial T}{\partial t}\frac{\partial}{\partial z}\frac{1}{R}\,dS=\frac{\partial B_z}{\partial t}. \tag{2}$$

Nonlinear constitutive relations between screening current $\mathbf{J}_e$ and electric field $\mathbf{E}$ are derived from a critical state model [2-5].

$$\mathbf{J}_e=J_c(|\mathbf{B}|)\frac{\mathbf{E}}{|\mathbf{E}|}\quad if\ |\mathbf{E}|\neq 0\ ,\quad \frac{\partial \mathbf{J}_e}{\partial t}=0\quad if\ |\mathbf{E}|=0\ . \tag{3}$$

The Bean model (critical current density J_c=constant) is applied to the following procedure for a simple calculation. At first, conductivity of the HTSC is set very large value assuming that the HTSC is a very good conductor. Then the conductivity is corrected as follows on each elements in iterative scheme until the maximum screening current is converged to J_c:

$$\sigma_{new}=\sigma_{old}\frac{J_c}{J}\quad if\ J>J_c,\quad \sigma_{new}=\sigma_{old}\quad if\ J\leqq J_c\,. \tag{4}$$

3. NUMERICAL RESULTS AND DISCUSSION

The present HTSC is composed from several grains. A numerical model is divided into half with a grain boundary as shown in Fig. 1. The HTSC disc is 18.0[mm] in a diameter and 10.0[mm] in thickness, and that of the magnet is 25.0[mm], 22.5[mm], respectively. The model is divided four layers with 270 elements in each layer.

When the magnet is moved toward and away from the HTSC, the levitation force between the magnet and the HTSC shows hysteresis property and repulsive force often changes to attractive force [1-5]. Since grain boundaries and cracks exist in the HTSC, equivalent critical current density is applied to the numerical analyses [4]. Figures 2 show numerical levitation forces, where dash line and solid line show solutions of the model with and without the grain boundary. Figure 2 (a) shows solutions with the same equivalent critical current density (J_c= 3.1×10^7[A/m^2]) for the both models. It is shown that the grain boundary reduces the levitation force and levitation force is larger for the model without grain boundary. Figure 2 (b) shows the levitation force of the model with the grain boundary by using the duplicate critical current density (J_c=6.2×10^7[A/m^2]). The measured critical current density of a small sample of the HTSC is

2.5×10^8[A/m^2]. These results show existence of grain boundaries is one of reasons of the difference between the measured critical current density and the equivalent critical current density in the numerical analysis.

Figure 3 shows current distributions of the first and the fourth layers of the model with the grain boundary for $J_c = 3.1 \times 10^7$[A/m^2]. Screening currents penetrate from the top and side surfaces as the magnet moves toward HTSC in Figures 3 (a), (b). The repulsive force in Figure 2 is generated by the screening currents. When the magnet is moved close to 1.0 [mm] and removed away from the HTSC, the reversed currents are obtained as shown in Figure 3 (c). The electromagnetic force between the magnet and the reversed current is the attractive force. The different current distributions at the same gap length explain the hysteresis property in Figure 2.

4. CONCLUSION

Effects of the grain boundary for the levitation force of the HTSC are discussed from a view point of equivalent critical current density. Existence of grain boundaries is one of reasons of the difference between the measured critical current density and the equivalent critical current density.

5.REFERENCES

1 M. Murakami, T. Oyama, H. Fujimoto, T. Taguchi, S. Gotoh, Y. Shiohara, N. Koshizuka and S. Tanaka, *Jpn. J. Appl. Phys.*, 29, (1990), L1991-1994.

2 T. Sugiura, H. Hashizume and K. Miya, *Int. J. Appl. Electromagn. in Materials*, 2, (1991), 183-196.

3 H. Hashizume, T. Sugiura, K. Miya, Y. Ando, S. Akita, S. Torii, Y. Kubota, and T. Ogasawara, Cryogenics, 31, (1991), 601-606.

4 Takeda, N., Tsuchimoto, M., Uesaka, M. and Miya, Proc. of Int. Symp. on nonlinear phenomena in Electromagn. Field, Elsevier, (1992), 125-128.

5 M.Tsuchimoto, T. Honma, N. Takeda, M. Uesaka and K. Miya, *Engineering Analysis with Boundary Elements*, Vol.11, in press, (1993).

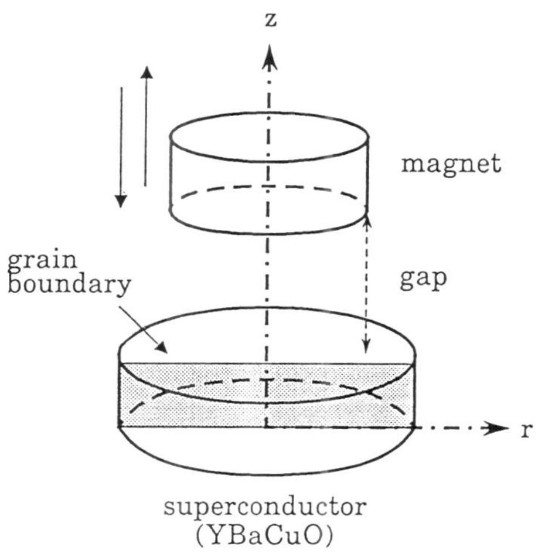

Figure 1. Numerical model of the HTSC with a grain boundary

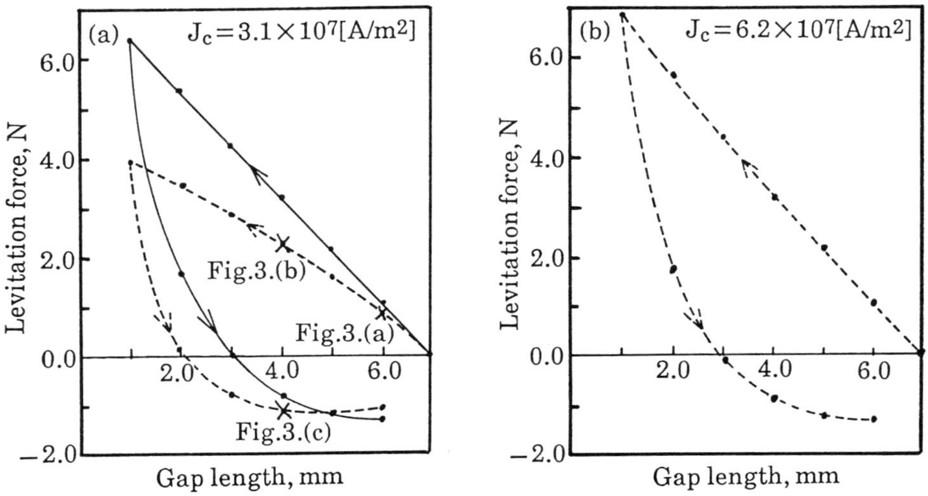

Figure 2. Numerical levitation forces, -- : with grain boundary, — : without grain boundary

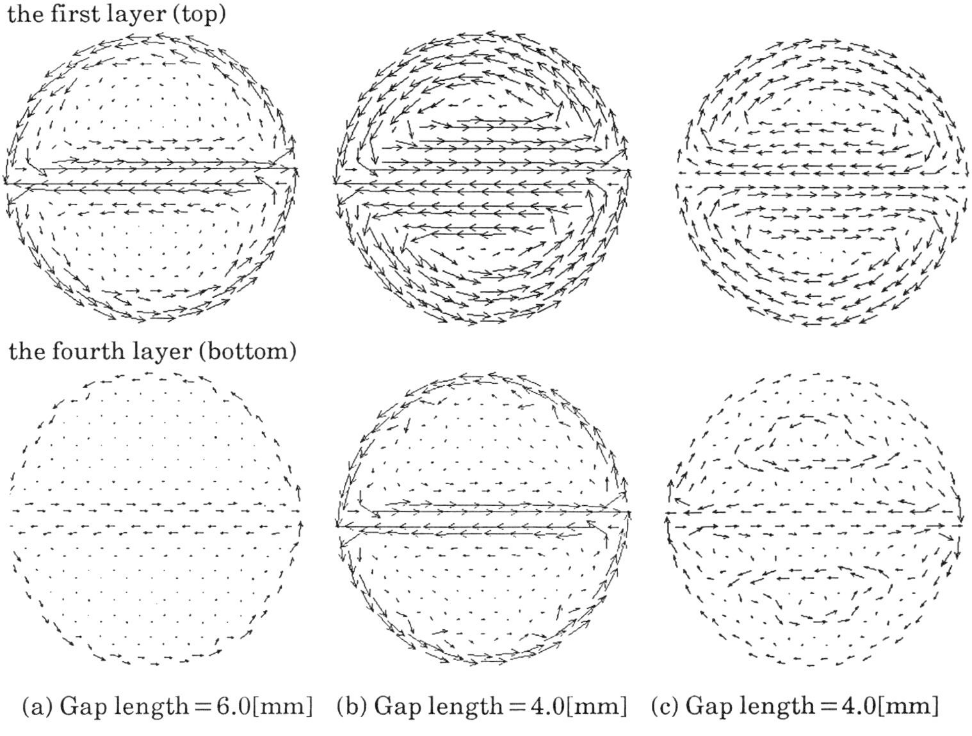

(a) Gap length = 6.0[mm] (b) Gap length = 4.0[mm] (c) Gap length = 4.0[mm]

Figure 3. Screening current distributions the first and the fourth layers

Electromagnetic Materials

Simulation and Design of Applied Electromagnetic Systems
T. Honma (Editor)

Ferromagnetic resonance–an unusual but powerful method for the study of magnetic anisotropy and magnetization reversal processes

K.A. Hempel

Institut für Werkstoffe der Elektrotechnik, Aachen University of Technology, Templergraben 55, 5100 Aachen, Germany

Abstract

The ferromagnetic resonance frequency of a material with uniaxial crystal anisotropy shows hysteresis in dependence of the applied dc magnetic field. Consequences and applications of this effect are discussed.

1. HYSTERESIS OF THE RESONANCE FREQUENCY OF A SINGLE CRYSTAL

The behaviour of an atomic moment in a magnetic dc field H can be described in a classical way by a precessional motion around the direction of the dc field because of the angular momentum connected to the magnetic moment. The frequency of the precession is proportional to the magnitude of the dc field and plays the role of the resonance frequency in an electron spin experiment usually performed in the microwave region. So the response of a ferromagnetic material to a microwave magnetic field is determined by the resonance frequencies of the system of magnetic moments which are strongly coupled by the exchange and stray field interactions. Even in the simple case of an ideal single–domain crystal in the shape of an ellipsoid of rotation and with a positive uniaxial magnetocrystalline anisotropy the dependence of the resonance frequency of the uniform precessional mode on the external magnetic dc field is rather complicated. In Figure 1 the reduced resonance frequency

$$\Omega_0 = \frac{\omega_0}{\gamma H'_A} \quad ; \quad H'_A = \frac{2K_1}{\mu_0 M_s} + \left(N_\perp - N_\parallel\right) M_s \tag{1}$$

is presented for different angles ϑ_0 between the direction of the dc field and the magnetically easy axis when the dc field is continously varied from positive to negative saturation [1]. Here γ is the gyromagnetic ratio, H'_A is the effective anisotropy field strength determined by the first order anisotropy constant K_1 (assuming $K_2 = K_3 = \cdots = 0$), the saturation magnetization M_s and the demagnetization factors $N_\perp$ and $N_\parallel$ perpendicular and parallel to the rotational axis, respectively and $\omega_{nat} = \gamma H'_A$ the natural resonance frequency. When the dc field is varied from $H = 0$ to negative values (from remanence to point B, see inset of Figure 1) the resonance frequency decreases for all values of ϑ_0 and reaches the value 0 exactly at the value of H at which the magnetization switches irreversibly to a new stable orientation (follow e.g. the curve for $\vartheta_0 = 0°$ as a simple example). Simultaneously, the resonance frequency also switches to a "stable" value ($\Omega = 2$

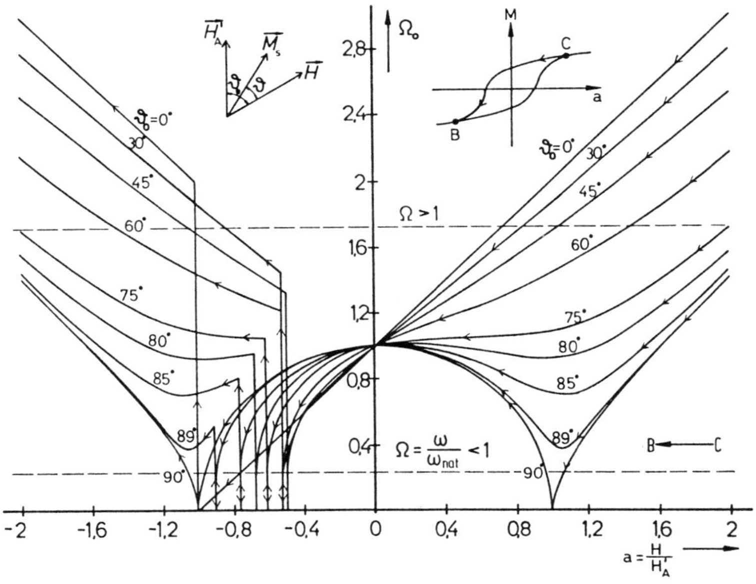

Figure 1. Reduced resonance frequency of an uniaxial single–domain particle

for $\vartheta_0 = 0°$) and then increases for increasing negative values of H. Within the model assumed the resonance frequency shows a hysteresis as does the magnetization according to the Stoner–Wohlfarth model.

2. MICROWAWE ABSORPTION OF A POWDER SAMPLE

Assuming a sample of non–interacting single–domain particles with random orientation of the easy axes the absorption in an ac magnetic field (which is in most cases perpendicular to the dc field) can be calculated at a given reduced measuring frequency $\Omega = \frac{\omega}{\omega_{nat}}$ as follows.

For any value of the dc field we have to sum up the imaginary parts of the ac susceptibility for all the values of ϑ_0 according to Figure 1 taking into account a finite value of the absorption line width corresponding to the damping of the resonance. Especially at dc–field values at which $\Omega = \Omega_0$ holds we expect large values of absorption. For any value of $\Omega < 2$ the absorption is different for positive and negative values of H following the course $C \to B$ in Figure 1. The absorption shows a hysteresis which is especially strong for values $\Omega < 1$ where the absorption at positive dc–field is mainly caused by *the particles with* $\vartheta_0 \approx 90°$ whereas at negative dc–fields *all the particles* contribute to the absorption. With the same method we can calculate the ac absorption along secondary hysteresis loops of various kind taking into account the appropriate values of the resonance frequency of the particles depending on whether or not they have already performed the irreversible jumps.

3. EXPERIMENTAL RESULTS

A sound test of the theory can be performed if we have to a good approximation a powder sample of non–interacting ideal particles. Very small barium ferrite particles with dimensions about $0.1\mu m$ produced by chemical coprecipitation [2] which are placed into a nonmagnetic matrix to reduce the magnetic interaction are well suited for that purpose.

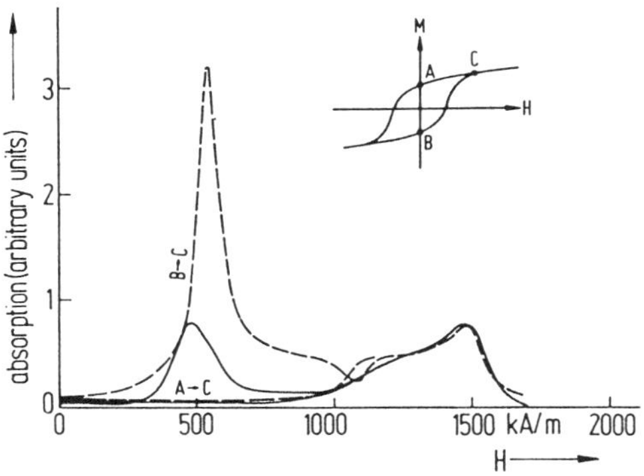

Figure 2a. Calculated (– –) and measured (—–) hysteresis of microwave absorption at f = 22.6 Ghz; $p = 91\%$, $H'_A = 1234$ kA/m.

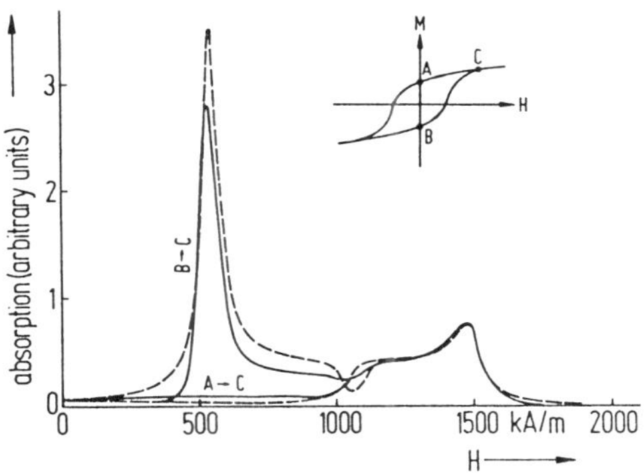

Figure 2b. Calculated (– –) and measured (—–) hysteresis of microwave absorption at f = 22.6 Ghz; $p = 9.3\%$, $H'_A = 1228$ kA/m.

In the Figures 2a and 2b the absorption of two such samples with a packing density p of 91% and 9.3%, respectively, along the primary hysteresis curve is shown [3]. The height and shape of the irreversible peak is extremely sensitive against variations of the magnetic interaction produced by variation of the packing density, quite in contrast to the "normal" absorption. The reason is that for larger values of p the switching field does no longer follow the simple Stoner–Wohlfarth model so that the calculation of the absorption becomes incorrect in the field region where the magnetization jumps occur.

The formation of the barium ferrite from the precipitate during the various stages of the process can also be studied by ferromagnetic resonance [4]. In Figure 3 the microwave absorption at 19.6 GHz of coprecipitated samples quenched at different temperatures is shown. Up to 700°C the absorption exhibits no hysteresis. At higher temperature where

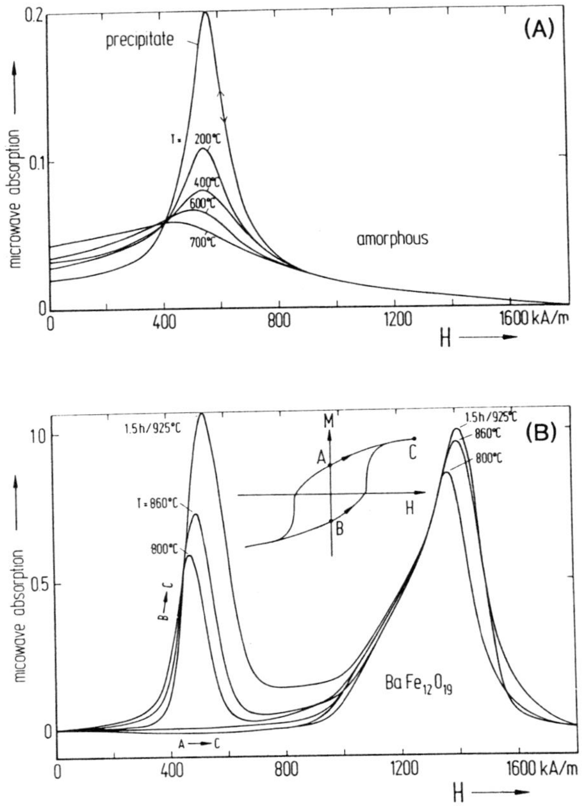

Figure 3. Microwave absorption (arbitrary units) of samples quenched at (A) temperatures up to 700°C (samples are amorphous) and (B) higher temperatures (starting material is converted to $BaFe_{12}O_{19}$ and absorption shows hysteresis typical of samples consisting of single–domain particles).

the formation of the ferrite begins the hysteresis grows up where the height of the "irreversible" peak corresponds to the coercivity of the samples.

REFERENCES

1 P. Grohs, K.A. Hempel, F. Schumacher, H. Hibst, Advances in Ceramics Vol.16, Ed. F.F.Y. Wang, American Ceramic Society (1985)
2 K. Haneda, C. Miyakawa, H. Kojima, J. Amer. Cer. Soc., 57, 354 (1974)
3 P. Grohs, K.A. Hempel, IEEE Trans. Mag., Vol. MAG–20, 5, 1633 (1984)
4 W. Roos, J. Amer. Cer. Soc., 63, 601 (1980)

Simulation and Design of Applied Electromagnetic Systems
T. Honma (Editor)

Sinusoidal, distorted, rotational fluxes and related loss phenomena in soft magnetic materials: from measurement to prediction

F. Fiorillo

Istituto Elettrotecnico Nazionale Galileo Ferraris and INFM-GNSM, 10125 Torino, Italy

Abstract

Magnetic power losses under complex flux waveforms can be accurately measured by means of improved techniques and quantitatively assessed through novel theoretical approaches. Advances are largely due to extensive application of digital methods in experiments and of statistical concepts in theory. Some recent achievements concerning alternating and rotational regimes in soft magnetic laminations are briefly reported and discussed in this paper.

1. INTRODUCTION

The design of magnetic cores is generally based upon the standard material data provided by the manufacurer, which are obtained under conventional controlled exciting conditions (e.g. 50 Hz power losses under sinusoidal flux and Epstein method). But actual cores very often operate under complex conditions. Flux waveform distortion is ubiquitous in devices, especially when relatively high magnetic induction levels are attained, and rotational fluxes are always present in cores of motors, generators and three-phase transformers. Core loss prediction, although crucial to optimum design in all these cases, is often hindered by lack of theory and one has to rely on empirical methods of calculation. Progress towards improved and versatile measuring methods and advances in theoretical predictions have been made in recent times. In this paper we shall outline some significant achievements obtained in theory and measurement of power losses in soft magnetic laminations under alternating and rotational fields. It will be shown that power losses under generic flux waveform can be accurately predicted on the basis of standard data obtained under sinusoidal induction, while, in the fundamental case of non-oriented laminations, rotational losses can quantitatively be related to alternating ones, a result pointing to improved core loss prediction in rotating machines.

2. ALTERNATING LOSSES

The measurement of 50 Hz - 60 Hz power losses in toroidal or Epstein frame samples is currently made under sinusoidal flux, according to existing standards. Recent advances in this field were chiefly focussed on the development of digital procedures, that have considerably improved measuring accuracy, sensitivity and long term stability [1]. But the increasing use of magnetic devices under a variety of exciting frequencies and induction waveshapes calls for the elaboration of wide ranging, flexible and accurate apparatus, emulating the actual working conditions of magnetic cores. Central to such apparatus is a tight control of the induction waveform, under the widest possible range of magnetizing frequencies. Such a control is usually accomplished in modern electronic wattmeters by means of analog feedback circuitry. This is howeverfrequently plagued by auto-oscillations, making difficult or impossible the material characterization above few kHz [2]. It has recently been shown that such basic drawback can be completely overcome, in a wide range of magnetizing frequencies (0.5 Hz - some hundred kHz), by means of digital feedback [3]. The working principle of the measuring circuit, which relates to a low frequency setup proposed by Birkelbach et al. [4], can be summarized as follows. A computer generated waveform is fed into an arbitrary function generator (HP 3245 A) and, *via* a wide band power amplifier (NF 4015), supplied to the primary circuit. The voltage induced on the secondary winding is suitably amplified and sampled, together with the voltage drop produced by the primary current on a calibrated resistor, by means of a fast digital oscilloscope (10 bit 150 MHz Le Croy 9430). Primary current and secondary voltage waveforms are then transferred and stored into the PC, where further elaboration is performed. Feedback control is accomplished by means of an iterative procedure, where the induction $B_i(t)$ and the applied field $H_i(t)$ at the i-th iterative step are taken as parametric representation of the $B(H)$ hysteresis curve. By imposing the desired time dependence of induction $B(t)$ (e.g. sinusoidal, trapezoidal, distorted, etc.) in place of the sampled one $B_i(t)$, the ensuing field function $H_{i+1}(t)$ at the $(i+1)$-th step is determined and the waveform to be programmed into the generator is computed, taking into account the impedance of the primary circuit. Iteration is stopped at step n, characterized by a $B_n(t)$ function satisfactorily emulating the desired one $B(t)$, an objective rule being provided by comparison of the form factors of $B_n(t)$ and $B(t)$. This method of induction waveshape control is independent of the operating frequency and in principle free of any problem regarding auto-oscillations.

The basic concept of loss separation in magnetic materials, where the total *average* specific power loss P is defined as the sum of the hysteresis P_h, excess P_e, and classical P_c loss components, has been fruitfully exploited in recent times by G. Bertotti in the elaboration of his statistical theory of losses, by far the most exhaustive and successfull approach to this challenging subject [5]. Instrumental to the development of such theory is the assumption that any change of magnetization in a ferromagnetic sample occurs, at a given instant of time t, through the simultaneous reversal of "magnetic objects" (MOs). If the magnetic domains are widely spaced, like in grain-oriented materials, the MOs are identified with the individual Bloch walls, while in finely spaced domain materials they are associated with groups of strongly interacting walls. The statistics of the local coercive fields, defining the threshold for reversal of magnetization by the single MOs, dominates the magnetization process and determines the macroscopic response of the sample to a cyclically varying applied field. A single parameter V_o, defining such a distribution, is assumed to incorporate the information on the material microstructure. Borrowing on such a theory, Fiorillo and Novikov [2] postulated that, if $\dot{B}(t)$ is the induction derivative in the sample, an *instantaneous* excess specific power loss $P_e(t) = (\sigma G V_o S)^{1/2}|\dot{B}(t)|^{3/2}$ can be defined, with σ the material conductivity, $G = 0.1357$ and S the sample cross-sectional area. As discussed in Ref. [2], combining this assumption with the standard calculation of the classical component P_c and averaging over the magnetization period $T = 1/f_m$, one can make explicit the loss separation equation under a generic periodic time dependence of induction $B(t) = \Sigma_n B_n sin(2\pi n f_m t + \phi_n)$, being B_n and ϕ_n the amplitude and phase of the n-th Fourier component. It turns out the following general expression for the total power loss in a lamination of thickness d, under an induction waveshape described through the previous Fourier development

$$P(f_m) = P_h(f_m) + P_c(f_m) + P_e(f_m) = P_h(f_m) + (\sigma\pi^2 d^2 f_m^2/6)\cdot\Sigma_n n^2 B_n^2 + (\sigma G V_o S)^{1/2}\cdot(1/T)\int_0^T |\Sigma_n 2\pi n f_m B_n \cos(2\pi n f_m t + \phi_n)|^{3/2}\, dt \qquad (1).$$

The hysteresis component $P_h(f_m)$ is independent of the particular $B(t)$ waveform and can be precisely determined by means of a quasi-static measurement. The value of V_o can be obtained by means of a single measurement under sinusoidal induction (e.g. at 50 or 60 Hz). It can be concluded that mere knowledge of two standard quantities suffices to predict the power losses under generically distorted induction. It should be remarked that the present treatment overcomes the ambiguities and the substantial inaccuracy of previous literature treatments, which could not account for the role of excess losses [6]. Figure 1 provides an example of the dependence of power losses on the degree of distortion in a $Co_{71}Fe_4B_{15}Si_{10}$ amorphous ribbon, magnetized at a frequency f_m = 1 kHz and peak induction B_p = 0.5 T. $B(t)$ is made of fundamental and third harmonic component, with ratio B_3/B_1 up to 0.3 and phase shift ϕ_3 up to 30°. While the prediction by Equation (1) of $P(f_m)$ vs. distortion can accurately reproduce the experiments (continuous lines), the classical literature model (dashed lines) fails to provide a correct outcome.

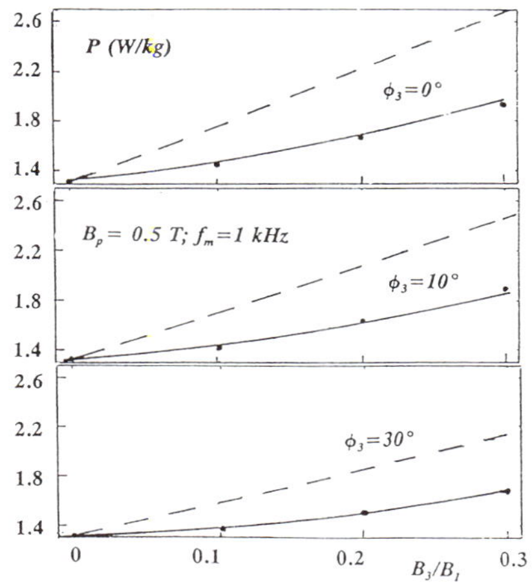

Figure 1. Power loss vs. degree of distortion in amorphous ribbons. Points: experiments. Continuous lines: prediction by Equation 1. Dashed lines: prediction by the classical theory.

3. ROTATIONAL LOSSES

The measurement of losses under rotational fields poses special problems, related to the vectorial character of the quantities involved. At present it largely remains a laboratory technique, although efforts are being made to assess the whole matter, in order to possibly produce a standard. A survey on the main measuring methods has been recently provided by Sievert [7]. A typical measuring arrangement consists in placing a circular or square shaped sample inside the gap of a two- or three-phase laminated yoke, in the plane of the magnetic field lines. Control of the circular (or elliptical) induction **B** is achieved either by means of feedback or exploiting the sample demagnetizing effect. The modulus of induction is determined in general by measuring its components through orthogonal windings threaded along holes drilled in the central region of the sample. Orthogonal point contacts can also be used [8]. Loss is then measured either by detection of the field components, by means of flat orthogonal H-coils, subsequent multiplication **H**·d**B**/dt and time averaging or by determining the rate of rise of the sample temperature. Simultaneous application of these two independent methods, when achieved, represents a good test of the measuring accuracy [9].

The experiments show that the rotational energy loss per cycle $W_r = P_r/f_m$, while exhibiting a same non-linear dependence on f_m as alternating losses, always passes through a maximum with increasing peak induction B_p. The hysteresis component W_{hr}, in particular is shown to vanish at saturation [10], a fact which correlates with the disappearance of the domain walls. Recently, the author has shown that the relationship between rotational and alternating hysteresis losses in isotropic laminations can be quantitatively assessed postulating three basic mechanisms of magnetization rotation: 1) linear combination of local unidirectional hysteresis loops at low inductions (Rayleigh region); 2) domain rearrangements between easy directions at intermediate inductions; 3) coherent lossless spin rotation towards saturation [11, 12]. The ratio between rotational and alternating losses W_{hr}/W_{ha}, a monotonically decreasing function of B_p, is expressed according to the equation

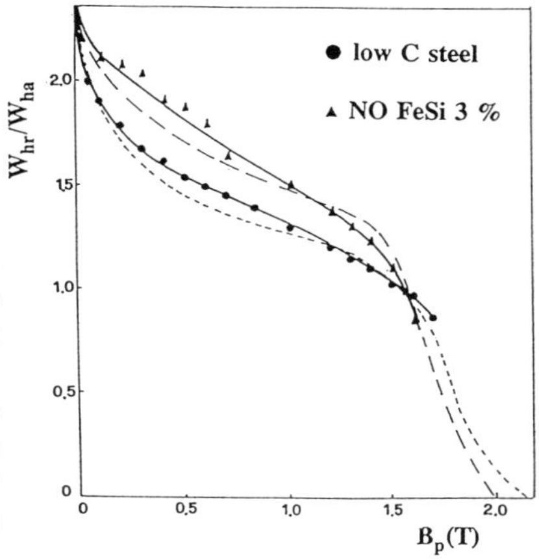

Figure 2. Ratio of rotational to alternating hysteresis losses in non-oriented laminations. Dashed lines are predicted by Equation (2).

$$W_{hr}/W_{ha} = n_w \cdot W_{wr}/W_{ha} + n_s \cdot W_{sr}/W_{ha} \quad (2),$$

where n_w and n_s represent the fractional sample volumes where processes 1) and 2) respectively occur, giving rise to the associated rotational hysteresis losses W_{wr} and W_{sr}. All the quantities appearing in Equation (2) can basically be induced from standard experiments and, provided the material is non-oriented, are rather independent of microstructural details. This formulation thus appears to provide quite a general theoretical approach to rotational losses. Figure 3 shows excellent agreement between theory and experiments in the case of low C steel and FeSi 3 % non-oriented laminations.

4. REFERENCES

1 H. Ahlers and J.D. Sievert, PTB-Mitteilungen 94 (1984) 99.
2 F. Fiorillo and A. Novikov, IEEE Trans. Magn., 26 (1990) 2559.
3 G. Bertotti, E. Ferrara, F. Fiorillo, and M. Pasquale, J. Appl. Phys., to appear.
4 G. Birkelbach, K.A. Hempel, and F.J. Schulte, IEEE Trans. Magn., 22 (1986) 505.
5 G. Bertotti, IEEE Trans. Magn., 24 (1988) 621.
6 R.A. Newbury, IEEE Trans. Magn., MAG-14 (1978) 263.
7 J. Sievert, J. Magn. Magn. Mater., 112 (1992) 50.
8 W. Brix, K.A. Hempel, and F.J. Schulte, IEEE Trans. Magn., MAG-20 (1984) 1708.
9 F. Fiorillo and A.M. Rietto, IEEE Trans. Magn., 24 (1988) 1960.
10 K. Narita and T. Yamaguchi, IEEE Trans. Magn., MAG-11 (1975) 1661.
11 F. Fiorillo, PTB Bericht, E-43 (1992) 11.
12 F. Fiorillo and A.M. Rietto, J. Appl. Phys., to appear.

Simulation and Design of Applied Electromagnetic Systems
T. Honma (Editor)

The Torque on Elliptical (110)[001] FeSi-Discs

H. Hauser and P. Fulmek

Institute of Materials Science in Electrical Engineering,
University of Technology in Vienna, Gußhausstraße 27–29/355, A-1040 Wien, Austria.
Financial support was provided by the FWF (Project Nr. P8483-TEC).

Abstract
Experimental results of torsion magnetometry at strong magnetic fields are compared with theoretic calculations on an energy model for magnetic crystal anisotropy and shape anisotropy. Showing torque characteristics of a symmetric four-pole magnet, discs of certain axis relations and orientations could be used as rotors in small synchronous motors.

1. INTRODUCTION

Ferromagnetism is in the most cases an effect of a special crystal structure and therefore an anisotropic phenomenon. Grain oriented transformer steel sheets with Goss-texture take advantage of the crystal anisotropy and provide high permeability and small losses for magnetizing processes in the rolling direction [001].
The anisotropy of the energy of demagnetizing fields is dependent on the geometrical shape. As grain oriented FeSi sheets have almost single–crystalline magnetic properties, the torque at saturation was calculated by superposition of crystal energy and shape energy. Two interesting aspects have been revealed: Almost circular discs oriented in [1$\bar{1}$0] direction show the torque characteristic of a 4–pole rotor for small synchronous motors without the disadvantage of distinct geometrical poles. The torque characteristics of elliptical discs with their long axis in the direction of most hard magnetization, allow to determine the orientation of the ferromagnetic crystallites in the sheet plane. This effect can be applied for quality control of grain oriented transformer steel sheets.

2. ANISOTROPIC ENERGIES AND TORQUE

We are using an energy model to determine the magnetic state of the material. A stable state of magnetization is established for minimum energy, only. The anisotropic energy contributions are described in [1] for cubic crystals. Inner magnetostrictive strain energy was eliminated by stress relief annealing, and no external stress had been applied during our experiments. A series expansion of the magnetocrystalline energy density is

$$E_K = K_0 + K_1(\alpha_x^2\alpha_y^2 + \alpha_x^2\alpha_z^2 + \alpha_y^2\alpha_z^2) + K_2(\alpha_x^2\alpha_y^2\alpha_z^2) \quad , \qquad (1)$$

with only two anisotropy constants K_1, K_2, for any direction of $\vec{M}_s = M_s \cdot \vec{\alpha}$.

For special geometric shapes, we can interpret the stray field as product of a demagnetizing matrix **N** and the magnetization $\vec{M}_s$ [2]. Demagnetizing factors are calculable for general ellipsoids [3]. The energy density of the shape anisotropy of ellipsoids is

$$E_F = \frac{1}{2}\mu_0 M_s^2 (N_a \alpha_x^2 + N_b \alpha_y^2 + N_c \alpha_z^2) \ . \tag{3}$$

For a first approximation it is sufficient to simply add the magnetocrystalline energy, the shape energy and the energy of magnetization in the applied field. But as the applied field H_a is assumed to be strong enough to saturate the sample in any direction in the sheet plane, the angle between $\vec{H}_a$ and $\vec{M}_s$ is zero. The relevant total energy E_T is

$$E_T = E_K + E_F . \tag{4}$$

Fig. 1 shows the superposition of the anisotropy energies without applied field. The enormous effect of the shape anisotropy for a circular disc is visualized by "zooming" into the energy area. The influence of crystal energy is almost invisible in the first part of the picture. It is almost impossible to magnetize a sheet in any direction pointing out of the sheet plane as the shape energy builds a huge energy lobe perpendicular to the plane.

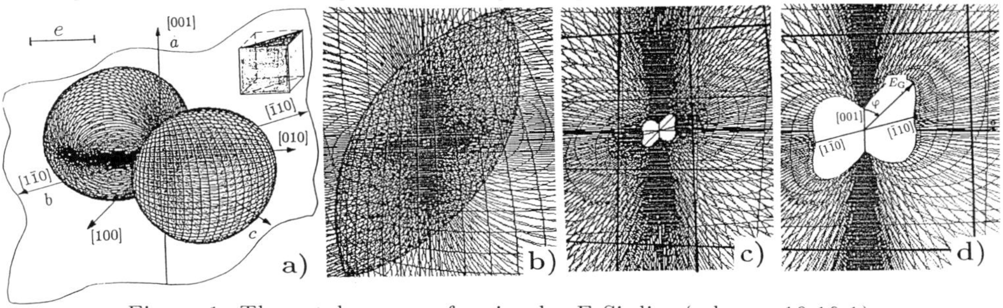

Figure 1: The total energy of a circular FeSi-disc (a:b:c = 10:10:1);

Under two-dimensional conditions, we have to calculate the torque acting on the disc as the first derivative of the total energy Eq. (4) in the sheet plane with respect to the rotation angle φ:

$$T = \frac{dE}{d\varphi} = \sin 2\varphi(\tfrac{1}{4}K_1 + \tfrac{1}{64}K_2) + \sin 4\varphi(\tfrac{3}{8}K_1 + \tfrac{1}{16}K_2) - \tfrac{3}{64}K_2 \sin 6\varphi +$$
$$+\mu_0 M_s^2 (N_b - N_a)\tfrac{1}{4} \sin 2(\varphi - \psi) \ . \tag{5}$$

The angle between the magnetization $\vec{M}_s$ and the rolling direction is φ, the orientation of the long axis of the elliptical disc in respect to the [001]–direction is ψ. Equation (5) assumes that the (110)–plane of the crystallites is identical to the sheet plane. For all other cases, the torque was computed by numerically derivating the energy areas, e.g. the area of Fig.1.

3. EXPERIMENTS AND CALCULATIONS

For our experiments we used Thyssen's Grain–Oriented Electrical Sheet ORSI 89 (3.5% Si; nominal thickness 0.27mm) with the following magnetic specifications [4]:
M_s = 1.62MA/m, K_0 = 0.58kJ/m^3, K_1 = 28.0kJ/m^3, K_2 = −9.14kJ/m^3. Discs of

axial ratios of 5:4, 4:3, 3:2, 2:1, and 1:1 were cut with a CO_2–laser. The discs have been deburred, and stress relief annealing has been carried out to restore the original magnetic properties according to the manufacturer's specifications.
We used a self–made torsion magnetometer [5]. The first measurement was done on a circular disc. In this case no effect of shape anisotropy may occur and the result directly proofs the values of the crystal constants. as the torque characteristic always was symmetrical, we did not plot the second half of the curve ($180° \leq \varphi \leq 360°$) in Figs. 2–4.
The most important aspects influencing our experiments and calculations are:
- The thickness of the sheet does not correspond exactly to the axis–dimension of an ellipsoid of the same volume. But only for exact ellipsoides the demagnetizing field and the magnetization are homogeneous and allow, thus, an exact calculation of demagnetizing factors.
- The cylindrical hole in the center of our discs influenced the homogenity. This hole is necessary to correctly mount the disc in our torsion magnetometer.

3.1. Orientation of the long axis in the [001]–direction ($\psi = 0°$)
Figure 2 shows the total energy of the discs in the (110) plane and the corresponding torque characteristics. Crystal energy and shape energy supply a minimum in the [001] direction. Even for an axial ratio of 5:4 (between the [001] and [1$\bar{1}$0]–directions) only the shape anisotropy determines the sign of the torque.

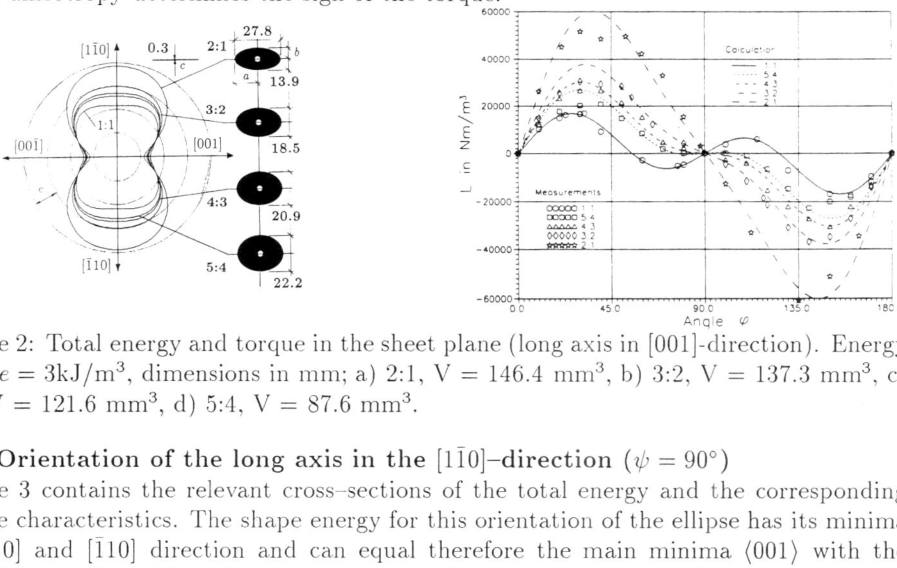

Figure 2: Total energy and torque in the sheet plane (long axis in [001]-direction). Energy scale e = 3kJ/m^3, dimensions in mm; a) 2:1, V = 146.4 mm^3, b) 3:2, V = 137.3 mm^3, c) 4:3, V = 121.6 mm^3, d) 5:4, V = 87.6 mm^3.

3.2. Orientation of the long axis in the [1$\bar{1}$0]–direction ($\psi = 90°$)
Figure 3 contains the relevant cross–sections of the total energy and the corresponding torque characteristics. The shape energy for this orientation of the ellipse has its minima in [1$\bar{1}$0] and [$\bar{1}$10] direction and can equal therefore the main minima $\langle 001 \rangle$ with the secondary minima $\langle 011 \rangle$ of the magnetocrystalline energy in the (110)-plane.

For an axial ratio of 7:6 the positive and the negative amplitude of the torque become equal. A disc of this axial ratio shows the characteristic of a 4–pole rotor of a synchronous motor, without any distinct geometric poles. A rotor utilizing this effect could be manufactured easily and would cause almost no losses due to air turbulences and provide silent run, as it is almost of circular shape.

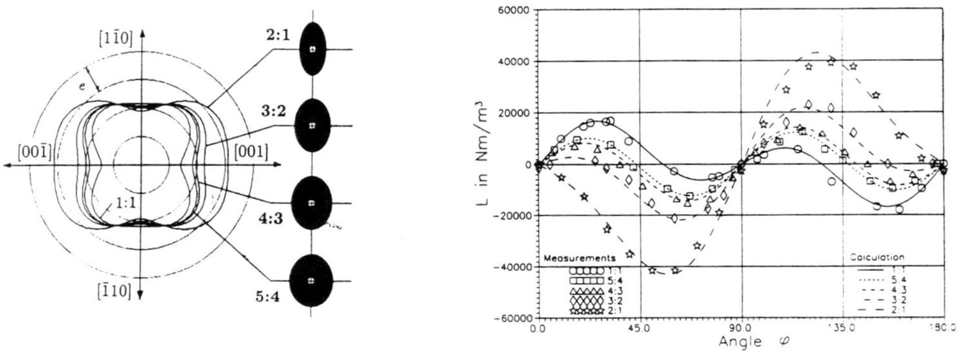

Figure 3: Total energy and torque in the sheet plane (long axis in [1$\bar{1}$0]-direction).

3.3. Orientation of the long axis in the [1$\bar{1}$1]–direction ($\psi = 54.7°$)

The theoretical superposition of crystal energy and shape anisotropy results in small deviations of the minima, in comparison to the minima–directions of the crystal energy alone. The two minima of the shape energy are flattening the maxima of the magnetocrystalline energy in the [1$\bar{1}$1] and [$\bar{1}$11] directions. At axis ratios between 2:1 and 3:2, the total energy does not change approximately between 35° and 55°, so the torque is vanishing in this sector (see Fig.4). This leads to a high sensitivity for the precision of the orientation of the axes of the ellipse and the orientation of the crystallites with respect to the rolling direction. Therefore we can interprete significant differences between calculated and measured torque data as misorientation of the crystallites in the sheet plane. The correspondence between measurement and calculation was attained by assuming an average crystallite–misorientation of 2°.

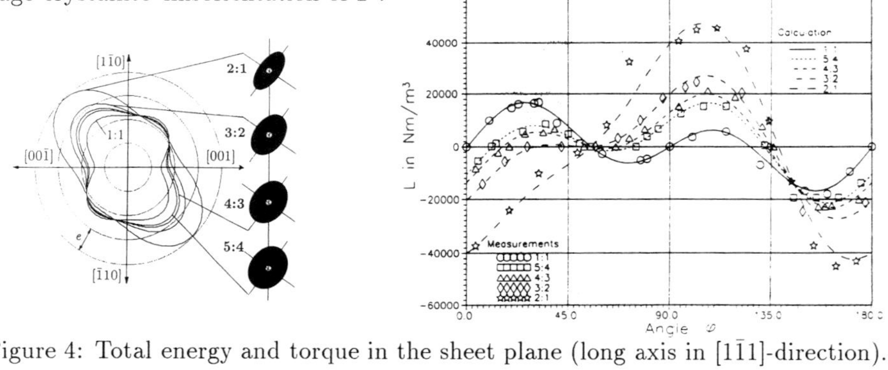

Figure 4: Total energy and torque in the sheet plane (long axis in [1$\bar{1}$1]-direction).

REFERENCES

1 Hauser, H., Fulmek, P.: IEEE Trans. Magn. 28 (3), 1815 (1992).
2 Fulmek, P., Hauser, H.: Z. Metallkd. 83 (12), 853 (1992).
3 Osborne, J.A.: Phys. Rev. 67 (11), 309 (1945).
4 Hauser, H.: Jap. J. Appl. Phys. 31 (10), 3336 (1992).
5 Hauser, H., Fulmek, P.: To be published in Arch. Elektrotech. 76 (1993).

Simulation and Design of Applied Electromagnetic Systems
T. Honma (Editor)

Acoustic Analysis of the Barkhausen Noise in Stressed Magnetic Materials

M. Akita and M. Enokizono

Faculty of Engineering, Oita University, 700 Dannoharu, Oita 870-11, Japan

Abstract

This paper describes the acoustic analysis of the Barkhausen noise on the magnetic materials with thermal fatigue for the detection of its flaws and precognition before the break down. As acoustic analyses, obtaining power of the waveform, zero crossing rate analysis and frequency analysis are performed for this purpose. These results show that the fatigue of the materials and the spectral feature of the noise is closely concerned.

1. INTRODUCTION

The increase of fatigue in magnetic materials is known as the causes of the break down of materials. And the techniques for detecting this fatigue are very important for the flaw detection and precognition before the breakdown of the materials. When the fatigue of the materials happens, it is assumed that the stressing distribution is transformed at the place of flaws. So, detecting the change of the stressing distribution is important for this problem. However, there is no definitive method for the detection of the fatigue until now.

For the detection of the flaws of magnetic materials with nondestructive method, we notice the Barkhausen noise generated by magnetizing ferromagnetic materials. Barkhausen noise is generated when the magnetic materials are magnetized and the flopping of the magnetic domain [1]. So, it is concerned that the features of the Barkhausen noise are varied, if the fatigue of the material affects the motion of the magnetic domain. In this paper, the relationship between the feature of the Barkhausen noise and stressing is mentioned. The acoustic analysis of the Barkhausen noise is investigated and the basic cues for the detection of the fatigue are presented.

2. EXPERIMENTAL MATERIALS

The experiments are made using the stainless steel sheet with magnetism. The fatigue is given these sheets by changing the aging time. The thermal stress of the material increases as the ageing time increases. The temperature of the aging is 450 °C. The size of the specimens for the measurement is 34.8 x 34.8 x 1.04 mm. In the experiment, seven kinds of the specimen with fatigue are prepared in this experiment. The ageing times of these samples are 3, 10, 30, 100, 300, 1000, and 3000 hours. Table 1 shows the chemical component of the specimens.

Table 1 The chemical component of the specimens (%)

C	Si	Mn	P	S	Ni	Cr	Mo	N	Nb	Fe
0.044	1.16	0.62	0.028	0.003	9.10	20.60	2.24	0.049	0.01	Bal.

Under the rotating magnetic field, the Barkhausen noise is measured. By this method, we can always obtain the noise. In this experiment, the rotating magnetic flux density and its frequency are changed. The values of the magnetic flux density B are changed from 0.0145T to 0.0725T at each step by about 0.0097 T. The frequency of the rotating field is changed from 30 Hz to 180 Hz at intervals of 30 Hz.

The generating Barkhausen noise waveforms are sampled with a 1MHz sampling rate, quantized by a 10-bit quantizer. About 260000 points of waveform data is stored for each examination.

3. ACOUSTIC ANALYSIS

We define several parameters for obtaining the cues of the Barkhausen noise. First, the average energy of the noise waveform is calculated for examining the strength of the noise. Next, zero crossing rate of the waveform measured for examining the degree of the flopping of the magnetic domain briefly. These analyses are performed using the beginning 0.13 seconds of the data. As spectral features parameters for the noise waveform, cepstral coefficients calculated by improved cepstral method[2]. This method is performed using repetition algorithm using Fast Fourier Transform. Using this method, the envelope passing through the peak of the power spectrum of the noise is obtained. Analysis is made using 2.048 ms Blackman window with 2.048 ms frame period. In each frame, 64 cepstral coefficients are coded. The variance of the cepstrum is defined for expressing the feature of the Barkhausen noise. We denote the i-th order of the cepstral coefficients as c_i, and express the cepstral variance as follows.

$$V=\sum_{i=0}^{N} c_i^2, \quad (N=32) \tag{1}$$

The parameter obtained at the beginning 8 frames is averaged and used later as the spectral cues. This parameter expresses the broad spectral features of the power spectrum. If the value is small, the feature of the noise is similar to the white noise. And the parameter is getting large, as the noise is colored.

4. EXPERIMENTAL EVALUATION OF THE PARAMETERS AND DISCUSSION

Figure 1 shows the zero crossing rate of the noise on the single specimen in case that the magnetic field changed. In this experiment, we use the specimens having the aging time 3 and 3000 hour. In the experiments, the frequency of the rotating field is changed from 30 Hz to 100 Hz at intervals of 10 Hz. The result shows that the relationship between the fatigue and the parameter exists. So we examine the relationship between the acoustic parameter and the fatigue under the condition of the constant magnetic field.

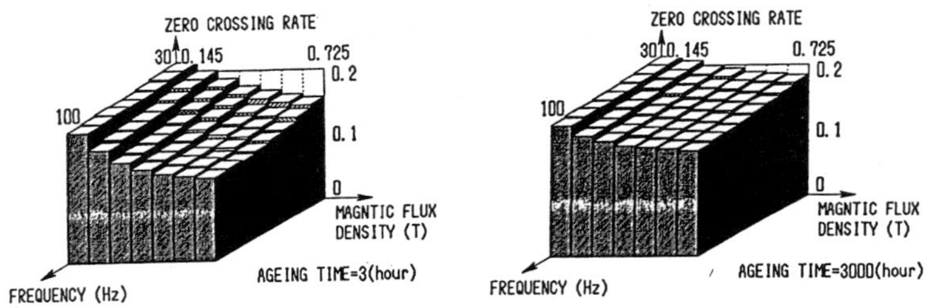

Fig. 1 The zero crossing rate of the noise ageing time = 3 and 3000 hour

Figure 2 shows the relationship between the average power of the noise waveform and the ageing time of the samples. Figure 3 shows the relationship of the power, the frequency of the magnetic field and the ageing time at B = 0.0339T. It shows that the power of the noise increases as the magnetic field is getting stronger. And the power of the noise increases as the frequency of the field increases. However, there is no relation between the power and the ageing time. So, it is considered that this parameter is not the cue of the fatigue.

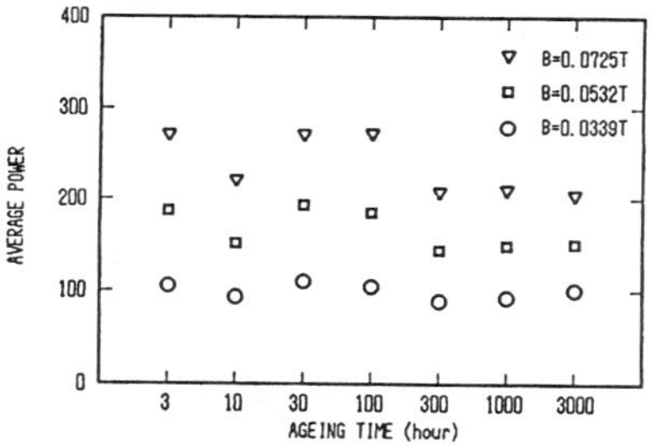

Fig. 2 The power of the Barkhausen noise field frequency = 60 Hz

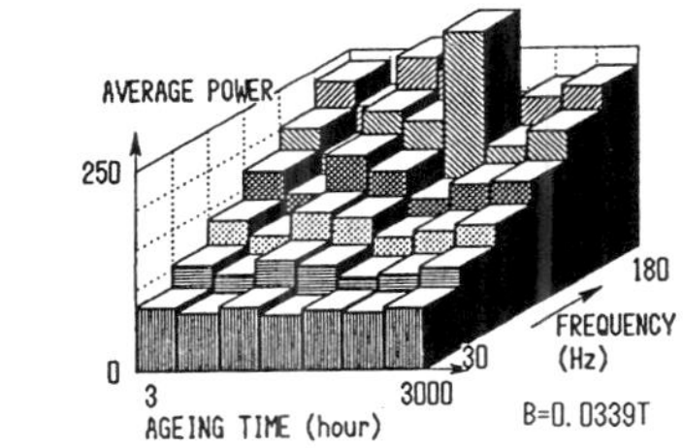

Fig. 3 The power of the Barkhausen noise B = 0.0339T

Figure 4 shows the relationship between the zero crossing rate of the noise waveform and the ageing time. Figure 5 shows the relationship of the zero crossing rate, the frequency of the magnetic field and the ageing time at B = 0.0339T. These figures show that the zero crossing rate increases as the ageing time increases in case of constant magnetic flux density. This result means that the high frequency components of the noise waveform increase as the fatigue of the materials increases.

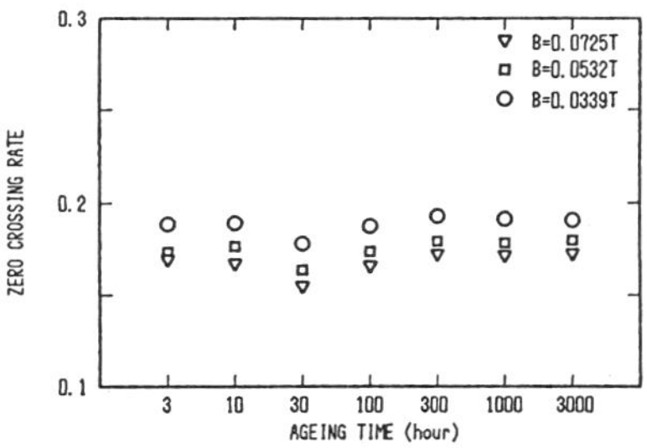

Fig. 4 The zero crossing rate of the noise field frequency = 60 Hz

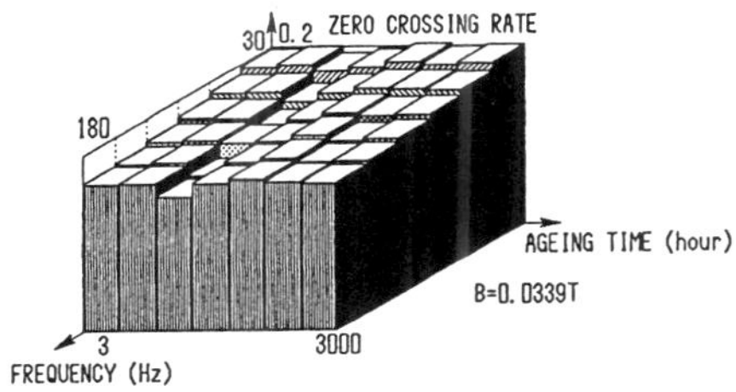

Fig. 5 The zero crossing rate of the noise B = 0.0339T

To make this result clearly, the relationship of the cepstral variance of the noise waveform and the ageing time is plotted. In this experiment, the first 32 cepstral coefficients are used. Figure 6 shows the relation between the order of the cepstral coefficients and the cepstral variance. The high order cepstral coefficients decays, so we cut off the coefficients at 32th order. Figures 7 and 8 show the relation between the cepstral variance and the ageing time.The result shows clearly that the cepstral variance decreases as

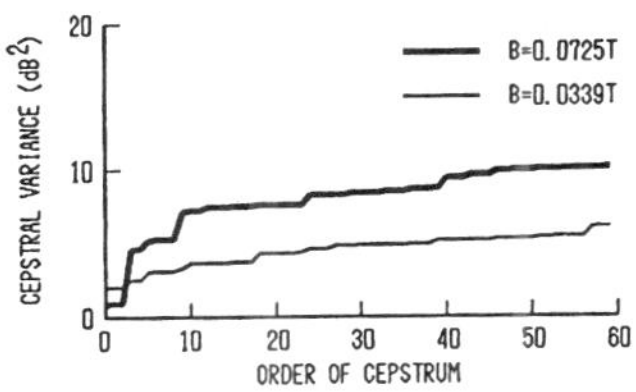

Fig. 6 Relationship between the cepstral variance and the order of the cepstrum

the fatigue increases in case that B is constant. This result show that the Barkhausen noise gets similar to the white noise with the increasing of the fatigue.

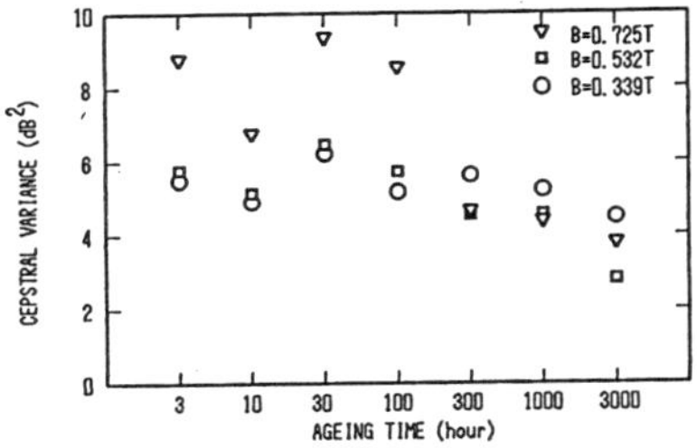

Fig. 7 The cepstral variance of the noise
field frequency = 60 Hz

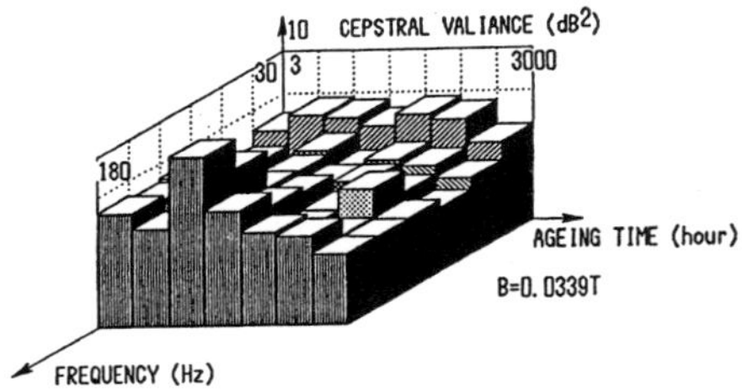

Fig. 8 The cepstral variance of the noise
B = 0.0339T

Figure 9 shows the cepstral variance of the noise is calculated with the single material under the same condition as the experiment on Fig.1. In this figure, the changing pattern of the value on two specimens is similar. The results of experiments show that the cue of the fatigue is the frequency features of noise waveform. That is, the Barkhausen noise by the specimens increasing the thermal fatigue is whitened. It is considered that the Barkhausen noise is generated by the movements of the magnetic wall with magnetization of the magnetic materials. So, in this result, it is supposed that the magnetization patterns of the materials are changed by the thermal fatigue, namely, the fatigued materials are difficult to magnetize and the amplitude of the Barkhausen jump with magnetization decreases and the time of the jump increases. The future problem is to clear the influence of the examination.

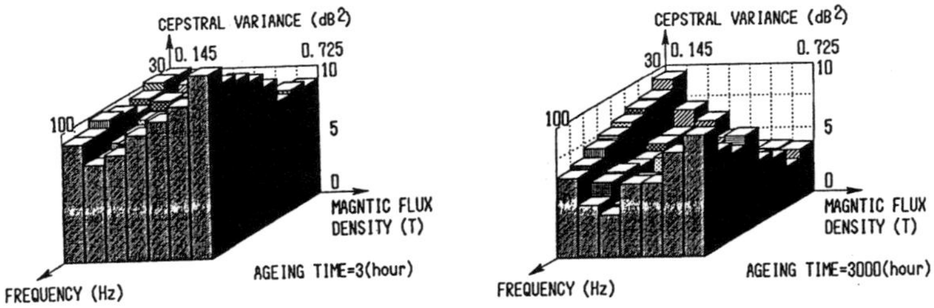

Fig. 9 The cepstral variance of the noise ageing time = 3 and 3000 hour

5. CONCLUSION

The considerations of the present paper lead to the following main conclusions:

1. The principal cue of the fatigue is not the strength of the noise waveform but the frequency feature of the noise.
2. The clear cue of the fatigue is cepstral variance of the noise waveform.
3. The Barkhausen noise gets similar to the white noise with increasing the fatigue of the specimen.

6. REFERENCES

1. R.Ranjan J. Appl. Phys. 61-8 (1987) 3196
2. S.Imai and Y.Abe Trans. IECE Vol.62-A No.4 (1979) 217 (In Japanese)

Simulation and Design of Applied Electromagnetic Systems
T. Honma (Editor)

Numerical analysis of domain structure in magnetic thin film

M. Enokizono, T. Todaka, and Y. Midou

Dept. of Electrical and Electronic Engineering, Faculty of Engineering, Oita University, 700 Dannoharu, Oita 870-11, Japan

Abstract

This paper gives a numerical analysis method of domain structure in a magnetic thin film. A stable domain structure of thin film in external magnetic field is analyzed by the iterative method that minimizes its total domain energy. In this analysis, we assumed that the total domain energy is sum of the magnetostatic energy, the anisotropy energy, the domain wall energy, and the elastic energy. In particular, we applied the Biot-Savart formula and the Gauss quadrature formula on surface integral to the calculation of the magnetostatic energy. The advantage of this method is that the calculation points in each domain are considerably reduced in comparison with the conventional method.

1. INTRODUCTION

The analysis of domain stricture is usually carried out by the minimization of its total domain energy that consists of the magnetostatic energy and the anisotropy energy and the domain wall energy and the zeeman energy. In the conventional analysis, because the calculation of the magnetostatic energy in a domain is very difficult, it was usually assumed to be zero or to calculate with approximate formula. However the consideration of the magnetostatic energy is very important in the common application problems. The numerical technique to evaluate the magnetostatic energy is first reported by Saka, 1988 [1]. In this method, the domain is divided into points that has a unit area, and it is assumed that the sum of the energy of each calculation point is equal to be the magnetostatic energy in the domain. More than 800 calculation points have been required to evaluated the energy accurately.

In this paper, we applied the Biot-Savart formula and the Gauss quadrature formula on surface integral to the calculation of the magnetostatic energy. The magnetostatic energy therefore can be calculated accurately and valuated at the integral points. The advantage of this method is that the calculation points in each domain and the CPU time are considerably reduced in comparison with the conventional method. In the example of this method, we assumed that the domain energy is composed of the Block type component and Neel type component. The elastic energies are taken into account in addition. First, we set the number of domains so that the domain energy without the external field becomes minimum. Next, the stable domain structure in an applied field is computed iteratively by using the Newton-Raphson method. The vertex coordinates of the domains and the direction of magnetization in each domain are treated as an unknown variable in the minimization. The conditions used in this analysis are as follows: 1) the magnetization is uniform in each domain; 2) the magnetization is parallel to the film surface; 3) the easy axis is parallel to x-axis; 4) the magnetization does not change in direction of the thickness; 5) the number of domains in the area under consideration is constant; 6) the thickness of film is very thin.

2. FORMULATION

From the Biot-Savart formula and the above condition 6, the demagnetization $\mathbf{H}_d$ is obtained as follow:

$$\mathbf{H}_{d_i} = \int \frac{\mathbf{M}_j \cdot \mathbf{n}}{4\pi r_{ij}^3} r\, \delta\, d\Gamma \tag{1}$$

Where, $\mathbf{M}$ is the magnetization, $\mathbf{n}$ the outward unit normal vector, δ the thickness of the film, $\mathbf{r}$ the vector from the source point "i" (the integral point of the Gauss quadrature formula for surface integral) to the point under consideration "j" (the integral points on the wall), and $d\Gamma$ the line element on the wall (which is pass of the line integral). Using the Gauss quadrature formula and eq. (1), the magnetostatic energy E_s is written as

$$E_{s_i} = -\frac{\mu_0}{2} \iint_{-1}^{1} (\mathbf{H}_{d_i} \cdot \mathbf{M}_i)\, \delta\, |J|\, d\eta\, d\zeta \tag{2}$$

where, η and ζ are the normal coordinates, and $|J|$ is the Jacobean of its transformation. The anisotropy energy E_a per domain is defined by [1-2]

$$E_a = K_u \sin^2\theta_1 \tag{3}$$

where K_u is the anisotropy constant, and θ_1 the angle between the easy axis and the magnetization vector. The Zeeman energy E_m per domain is defined by [1-4]

$$E_m = -\mu_0 M \cdot H_e \cos\theta_2 \tag{4}$$

where $\mathbf{H}_e$ is the external magnetic field applied to the film, and θ_2 the angle between the applied field and the magnetization vector. The wall energy E_w and the elastic energy E_e per domain are written as [1-3]

$$E_w = \left(E_b \sin^2\left(\frac{\theta_3}{2}\right) + E_n\left(1 - \cos\frac{\theta_3}{2}\right)^2\right), \tag{5}$$

$$E_e = \frac{9}{8}\lambda^2 C_{ii} \tag{6}$$

Where, E_b is the Bloch wall energy in 180 degrees wall, E_n the Neel wall energy in 180 degrees wall, θ_3 the angle between the magnetization vectors of each neighboring domain, C_{ii} the elastic modulus, and λ the magnetostriction constant. E_b and E_n are written as follows:

$$E_b = \frac{a}{2} K_u + \frac{\pi^2 A_e}{a} + \frac{a^2\mu_0 M_s^2}{4(a+\delta)}, \tag{7}$$

$$E_n = \frac{b}{2} K_u + \frac{\pi^2 A_e}{b} + \frac{b\delta\mu_0 M_s^2}{4(b+\delta)}. \tag{8}$$

Where "a" or "b" is the width of the domain wall and A_e is the exchange constant. Minimizing the E_b and E_n independently, the width a and the width b can be obtained analytically. The analysis of domain structure can be carried out by the minimization of the total domain energy, which is sum of the above energies. In this calculation, the vertex coordinates of the domains and the direction of magnetization in each domain are modified iteratively by using the Newton-Raphson method to get a stable state.

3. RESULTS AND DISCUSSION

Figure 1 shows the distribution of the demagnetization field, which is calculated by eq. (1). It is evident that the demagnetization field becomes large near the wall, and its distribution is nonuniform. For the accurate calculation of the magnetostatic energy, it is necessary to evaluate its high-order distribution. The conditions used in this analysis are: C_{ii}

$= 2.41 \times 10^{11}$ J/m^3; $\lambda = 2 \times 10^{-5}$; $A_e = 8.3 \times 10^{-12}$ J/m; $\delta = 5.0 \times 10^{-8}$ m; $K_u = 4.2 \times 10^4$ J/m; and $M_s = 1.719 \times 10^6$ A/m. Figure 2 shows the relationship between the magnetostatic energy calculated by eq. (2) and the inclination angle of the triangular domain θ (see Fig. 2). It is found that the magnetostatic energy becomes negative value with θ being smaller than 45 degrees. Because it becomes zero at $\theta = 45$ degrees, we therefore set the initial inclination angle θ at 45 degree. Figure 3 shows the relationship between the above energies and the number of the triangular domains N (N = 1 ~ 5). In this case, the zeeman energy and the magnetostatic energy are equal to be zero. Since the total energy becomes minimum when the number of the triangular elements is equal to be three, this structure can be regarded as an initial structure.

Figure 4 shows the domain structure in each external field condition: (a) Hx = 120 A/m; (b) Hx = 170 A/m; (c) Hx = 200 A/m; (d) Hy = 120 A/m; (e) Hy = 170 A/m; and (f) Hy = 200 A/m. As shown in these figures, the domains are grown when its magnetization vector has the same component at the applied field. The magnetization vector turned toward the applied field, slightly. As a result, it can be said that this method is very useful for the analysis of magnetic domain structure in common application problems.

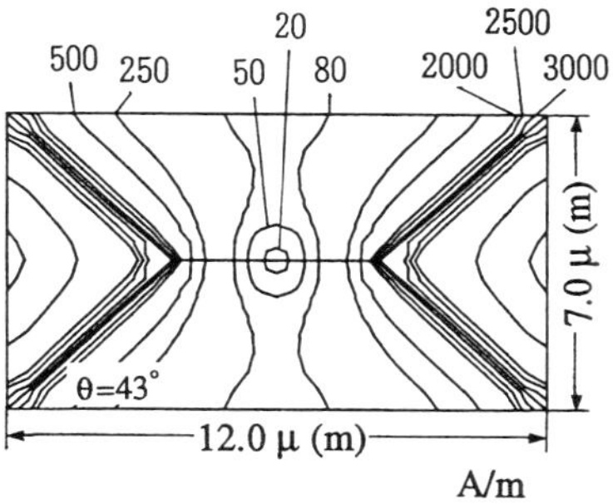

Fig. 1 Distribution of demagnetization field.

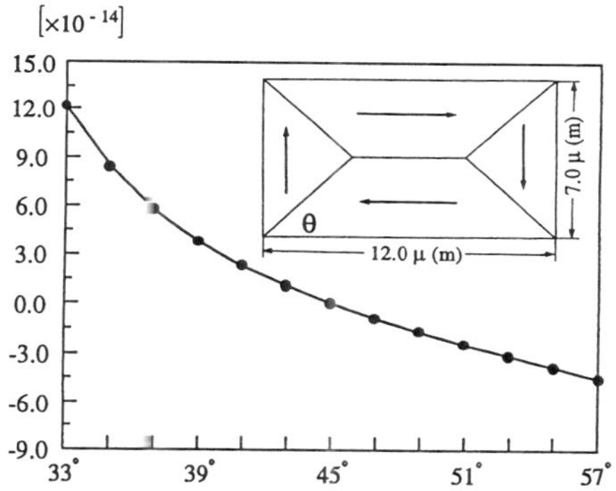

Fig. 2 Magnetostatic energy for the angle θ.

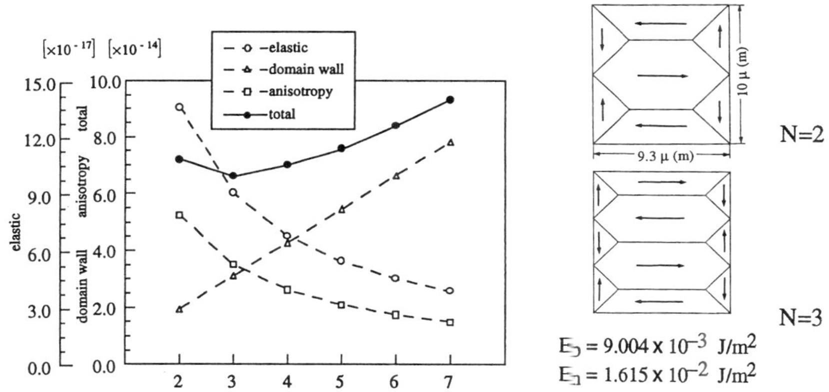

Fig.3 Relationship between the energies and the number of triangular domain N.

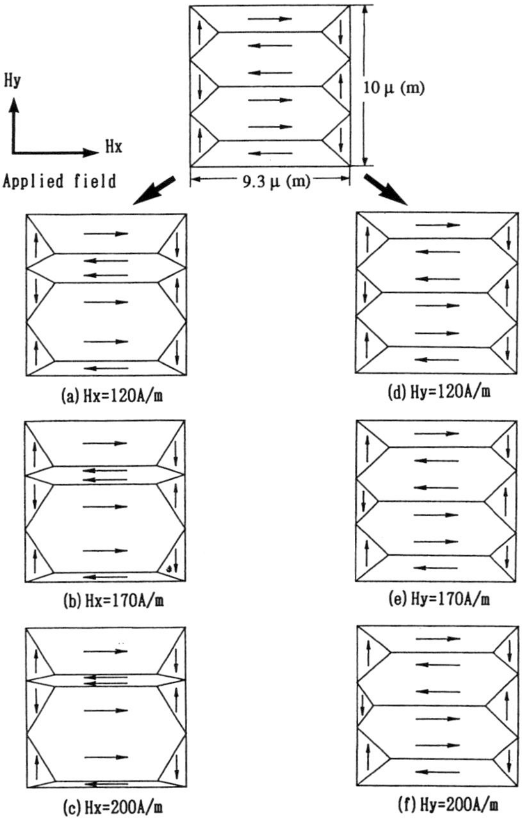

Fig.4 Movement of the domain wall

4. CONCLUSION

In this paper, we have applied the Biot-Savart formula and the Gauss quadrature formula on surface integral to the calculation of the magnetostatic energy in the domain structure analysis. The advantage of this method is that the calculation points in each domain and the CPU time are considerably reduced in comparison with the conventional method. In the example of this method, it was shown that our method is very useful for the analysis of magnetic domain structure.

5. REFERENCES

[1] C. Saka, K. Shiiki and K. Shinagawa, IEICE Tech. Report, MR88-38 (1988) pp. 25-32
[2] S. Iida, et al., Magnetic Film Engineering, Maruzen, Tokyo (1977)
[3] S. Chikazumi, Physics of Magnetism, Joho Wiley & Sons, Inc., New York (1964)
[4] R.F.Sooho, IEEE Trans. on Magn. , Vol. 18, No. 6 (1982) pp. 1128-1130

Simulation and Design of Applied Electromagnetic Systems
T. Honma (Editor)

Magnetoresistive Sensors - Past, Present and Future

A.J. Moses and P.C.Y. Ling

Wolfson Centre for Magnetics Technology, School of Electrical, Electronic and Systems Engineering, University of Wales Cardiff, P.O. Box 917, Newport Road, Cardiff, United Kingdom.

Abstract

This paper highlights the development of magnetoresistive (MR) sensors from the discovery of the MR effect in 1857 to its utilisation in the 1970's and the ultimate breakthrough in the discovery of the Giant Magnetoresistive (GMR) effect. Applications of MR effect in data retrieval systems, anti-fraudualent devices and non-destructive testings are described.

1. INTRODUCTION

The magnetoresistive (MR) effect was discovered over a hundred years. Thompson [1] in 1857 discovered significant changes in anisotropic magnetoresistance of ferromagnetic materials when they were placed in a uniform magnetic field However, the applications of MR effect were not fully utilised because the change in magnetoresistance was restricted by demagnetisation. Better understanding of the effect of demagnetisation was achieved using thin-film technology. A thin layer of ferromagnetic wafer (≈ 300Å) was produced to minimise the effect.

This paper reviews some applications of MR sensors in data retrieval, magnetic stripe cards, anti-fraudulent devices and Non-Destructive Testing (NDT) systems.

The Giant Magnetoresistive effect, its present development and future prospect are also discussed.

2. FABRICATION, CHARACTERISTIC AND BIASING TECHNIQUES

A typical MR sensor element is shown in Figure 1. This consists of a thin-film MR element 2mm long, 50μm wide and 300Å thick. The alloy 83%Ni, 17%Fe by weight is commonly used although various compositions of Ni, Fe and Co are used for more specific customised applications. The fabrication of MR sensors and transducers is largely by vacuum evaporation. Photolithography and etching methods on glass subtrates permit batch fabrication. Evaporation of an alloy is then commenced in a vacuum of 10^{-5} Torr and in the presence of a magnetic field of about 7kA/m in order to induce an uniaxial anisotropy

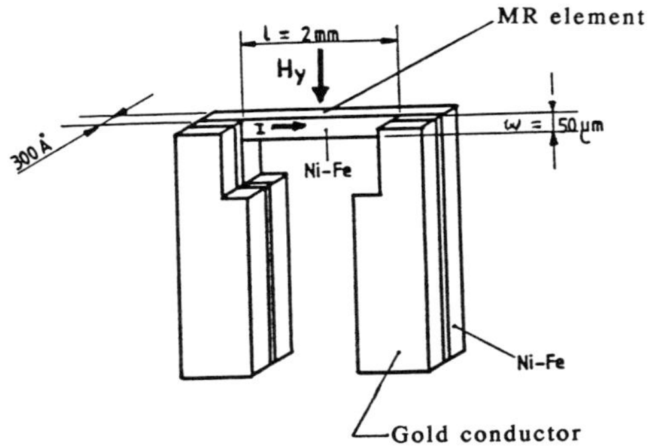

Figure 1 A typical MR element NiFe thin-film.

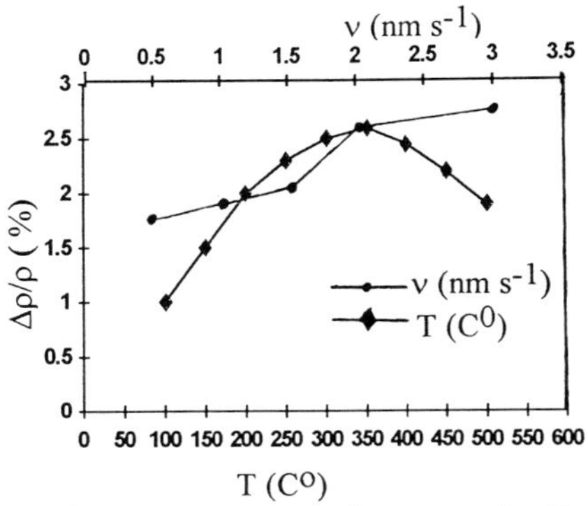

Figure 2 Dependence of MR response of $Ni_{83}Fe_{17}$ alloy on deposition temperature and rate.

axis. The ultimate $\Delta\rho/\rho$ response is dependent on the deposition temperature and rate. Figure 2 shows that in order to yield highest MR response the deposition temperature should be about 350°C and the deposition rate should be as high as possible. However large deposition rate requires a high deposition temperature hence the MR response is always limited to about 2.5% for $Ni_{83}Fe_{17}$ alloy.

The main advantage of MR field detection is that it responds directly to the magnetic field 'H' and not its time derivative 'dH/dt' as in the case of conventional inductive read/write heads. Equation 1 describes a generalised MR transfer characteristic when a constant current 'I' passes the MR element and a bias field, H_b, is applied. Equation 1 is independent of time. 'K' is a constant for the magnetic material used.

$$V = K\, I \int_0^y \int_0^x \frac{h_y(y,x)}{w\, l}\, dy\, dx \qquad (1)$$

Four biasing techniques shown in Figure 3 are commonly used in the MR element and may be fabricated in addition to the normal stages. These are (a) Permanent-magnet chip, (b) Conductor, (c) Anisotropic and (d) Barber-pole structure. The provision of a small permanent magnet at a suitable distance from the MR element can give the required bias field (Figure 3a). Similary a d.c. bias field may be produced by passing a current through a conductor close to the MR element (Figure 3b). Anisotropic biasing can be achieved by inducing the anisotropy axis into the MR element during deposition to provide an initial magnetisation rotation (Figure 3c). The barber-pole biasing arrangement is achieved by rotating the sense current so that the sensing current itself produces a bias field (Figure 3d).

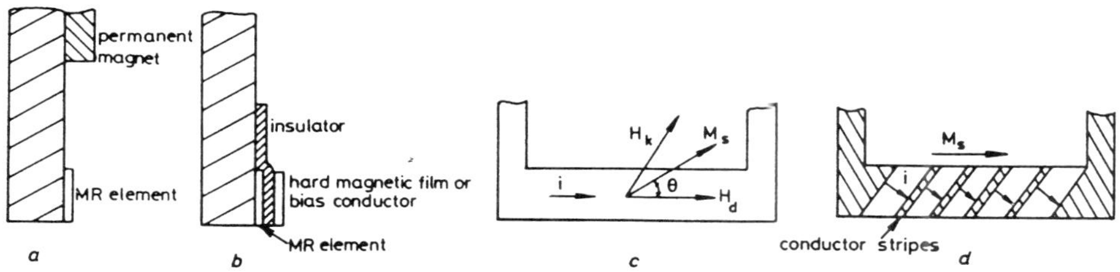

Figure 3 Biasing techniques of (a) permanent-magnet,(b) conductor, (c) anisotropic and (d) barber pole.

3. APPLICATIONS IN BRIEF

Simple unbiased MR heads have found immediate applications in many low density data retrieval systems (≈500 bit per inch) such as the magnetic stripe card (MSC) system with data recorded on a low coercivity magnetic medium (≈24kA/m). The format and configuration of MSCs are governed by ISO standards 7810-7813 [2].

MSCs have found usage as a stored value card using high coercive magnetic media [3]. One powerful fraud deterrent known as 'Magnetic Watermark' was patented by the Thorn E.M.I.(U.K.) Group [3]. The complex Watermark consists of structuring the coating magnetic orientation by using varying computer driven fields. The final product has an intrinsic and non-erasable magnetic data-carrying pattern. The magnetic watermark pattern can be scanned by arrays of MR sensors, decoded by a computer and ultimately the identity of the stored value card is reviewed.

Because of their small size and high sensitivity MR sensors have many potential applications in non-destructive inspection of steel surfaces. Figure 4 shows a novel system for scanning the surface of electrical steel sheet in order to map the localised power loss distribution. The sensor element in this case is 2mm x 10μm x 400Å and is situated 0.05mm above the sheet. Readings of localised power loss are made at 0.34mm steps by multiplying the signals proportional to the field and rate of change of flux density. This technique can give valuable information on grain structure, orientation and domain patterns to users of electrical steel [4].

4. GIANT MAGNETORESISTIVE EFFECT (GMR)

The discovery of the GMR was made in 1988 when Baibich et al [5] reported that $\Delta R/R$ changes of as much as 50% occured in $(Fe/Cr)_n$ multilayer ultrathin films at room temperature. For the GMR effects to appear, changes in the relative orientation of the magnetisation in adjacent magnetic metal layers must occur and the thickness of each ultrathin film must be a fraction of the mean free path of an electron in the multilayer array. However the effect still lacks quantitative and predictive understanding.

The GMR head offer greater sensitivity and lower noise ratio than its inductive counterpart. The main drawback of the GMR sensors is the need for a large field to switch the array from parallel to antiparallel configuration. This large field (≈1.6MA/m) is six times higher than the coercive field of the storage media and certainly causes data corruption.

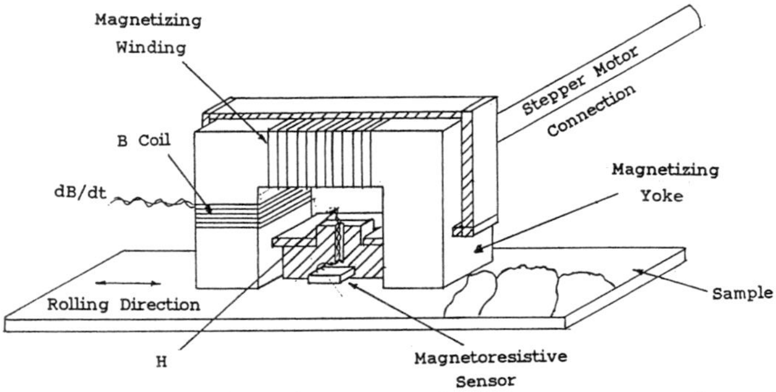

Figure 4 Schematic diagram of MR sensor system for investigating localised losses in electrical steels.

5. CONCLUSION

The MR effect is finding many more applications ranging from magnetic field sensors to data retrieval systems but it is used mostly in magnetic stripe card readers. GMR has dramatically increased the $\Delta R/R$ response but the underlying phenomenon of the electron scattering between the ultrathin multilayers is not fully understood at present.

6. REFERENCES

1 Thompson,W.,"Magnetoresistive effect in iron", Royal Society, London, 1857.

2 A.J. Moghaddam and A.J. Moses, "A non destructive automated method of localised power loss measurement in grain oriented 3% silicon iron", J.M.M.M., 112, pp132-134, 1992.

3 International Standard Organisation (ISO), "Magnetic stripe cards", 7810-7813, 1979.

4 Harrop, P., "The electronic purse", IEE Review, pp. 227-231, June 1992.

5 Bailbich, M.N. et al.,"Giant magnetoresistance of (001)Fe/(001)Cr magnetic superlattices", Phys.Rev.Lett.61, p.2472, 1988.

ACKNOWLEGEMENT

The authors would like to acknowledge reference to the work of A.J. Moghaddam, D. So and P. Koutsopoulos at the Wolfson Centre for Magnetics Technology.

The symmetry in Eqs.(2) ($P_{-(1-\nu)} = P_{-\nu}$) allows us to write

$$V_r = (r - r_o)(K - r - r_o)P_{-r/K}/Z, \ I_r = (r - r_v)(K - r - r_v)P_{-r/K}/Z \tag{4}$$

The mechanical wave-field in the system is following ($\gamma^{\pm}$ are the SAW reflection coefficients from left, and right half-infinite systems of strips)

$$\begin{aligned} &\frac{U_0}{U_{-1}} = \frac{R_1 - k - \chi}{R_0 - k - \chi^*}, \ \gamma^- = \frac{X + Q - \chi}{X - Q - \chi^*}, \ \gamma^+ = \frac{X + Q - \chi^*}{X - Q - \chi} \\ &\text{- for } r = r_o: \ U^+(x) = U_0 e^{-jr_o x} + U_{-1}e^{-j(r_o - K)x} = U_0(1 + \gamma^+ e^{jKx})e^{-jr_o x} \\ &\text{- for } r = K - r_o: \ U^-(x) = U_0 e^{-j(K-r_o)x} + U_{-1}e^{jr_o x} = U_{-1}(1 + \gamma^- e^{-jKx})e^{jr_o x} \end{aligned} \tag{5}$$

3. GENERATION AND DETECTION OF SAW

Equations (2) give us the Fourier transform of two 'discrete functions - the sets of strip potentials and strip currents. The strip potentials are known, thus

$$V_m = \frac{1}{K}\int_0^K V(r)e^{-jrm\Lambda}\,dr \tag{6}$$

allows us to evaluate α_0, which introduced to the corresponding equations, yields

$$\begin{aligned} &I_n = j\omega C_{n-m}V_m + YV_m e^{-jr_o|n-m|\Lambda} \\ &C_k = \frac{\varepsilon_e}{\pi}\left(\frac{1}{4} - k^2\right)^{-1}, \ Y = 4\omega\varepsilon_e\frac{\Delta v}{v}\gamma\left(\frac{\sin \pi r_o/K}{P_{-r_o/K}}\right)^2, \ \gamma = \frac{(1+\gamma^+)(1+\gamma^-)}{1 - \gamma^+\gamma^-} \\ &U_0(x > m\Lambda) = -j\kappa 2\sqrt{\varepsilon_e}\frac{\sin \pi r_o/K}{P_{-r_o/K}}\frac{X - Q - \chi}{K - 2r_o}V_m e^{-jr_o(x - m\Lambda)} \\ &U_{-1}(x < m\Lambda) = -j\kappa 2\sqrt{\varepsilon_e}\frac{\sin \pi r_o/K}{P_{-r_o/K}}\frac{X - Q - \chi^*}{K - 2r_o}V_m e^{-j(K - r_o)(x - m\Lambda)} \end{aligned} \tag{7}$$

which, after introducing certain 'A-amplitudes' defined by

$$A^+_{>m}e^{-jr_o(n-m)\Lambda} = j\sqrt{\frac{\omega}{2\gamma}}(1+\gamma^+)U_0(n\Lambda), \ A^-_{<m}e^{jr_o(n-m)\Lambda} = j\sqrt{\frac{\omega}{2\gamma}}(1+\gamma^-)U_{-1}(n\Lambda) \tag{8}$$

for $n > m$ and $n < m$ correspondingly, can be rewritten in following simple form

$$\begin{aligned} &I_n = j\omega C_{|n-m|}V_m + YV_n + \sqrt{2Y}(A^+_n + A^-_n) \\ &A^+_{>n} = A^+_n + \sqrt{\frac{Y}{2}}V_n, \ A_{n+1} = A^+_{>n}e^{-jr_o\Lambda}, \ A^-_{<n} = A^-_n + \sqrt{\frac{Y}{2}}V_n, \ A_{n-1} = A^-_{<n}e^{-jr_o\Lambda} \end{aligned} \tag{9}$$

where A^+_m and A^-_m are "A-amplitudes" of SAWs at the center of m-th strip which are generated by the other sources in the system. It is assumed that the generation takes place just after the strip center, looking in the propagation direction.

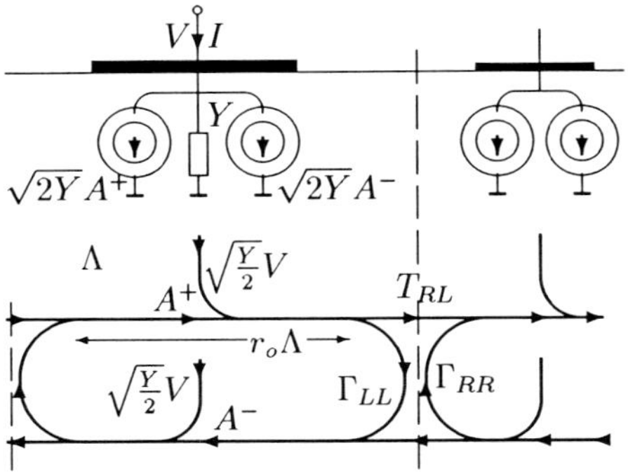

Figure 1. P-matrix model of a single section of almost periodic strips.

4. A JUNCTION OF TWO SYSTEMS OF STRIPS

There are four different waves involved in equations (7)
- a composed SAW propagating right (U^+ in (2)), its forward-propagating, and backward--propagating components are

$$\frac{1}{1+\gamma^+}A^+e^{-jr_o x} + \frac{\gamma^+}{1+\gamma^+}A^+e^{-j(r_o-K)x}$$

- and similarly a composed SAW propagating left (U^-); in the above relation we should change indices + to −, and wave-numbers r_o to $-r_o$, and $r_o - K$ to $K - r_o$.

The wave-field continuity requires that wave-components having close wave-numbers should be continuous. Let us consider two different systems of periodic strips having close wave-numbers $K_L \approx K_R$ which edge strips are placed at x_R and $-x_L$ on both sides of $x = 0$. The above-mentioned wave-field continuity results in

$$A_L^- = \Gamma_{LL}A_L^+e^{-j2r_Lx_L} + T_{LR}A_R^-e^{-j(r_Lx_L+r_Rx_R)} \tag{10}$$

$$\Gamma_{LL} = \frac{1+\gamma_L^-}{1+\gamma_L^+}\frac{\gamma_R^+e^{-jK_Rx_R} - \gamma_L^+e^{jK_Lx_L}}{1-\gamma_R^+\gamma_L^-e^{-j(K_Lx_L+K_Rx_R)}},\quad T_{LR} = \frac{1+\gamma_L^-}{1+\gamma_R^-}\frac{(1-\gamma_R^-\gamma_R^+)\sqrt{\gamma_R/\gamma_L}}{1-\gamma_R^+\gamma_L^-e^{-j(K_Lx_L+K_Rx_R)}}$$

where indices L and R mark quantities valid in left, and right systems of strips. Simultaneous changes of indices $+ \Leftrightarrow -$ and $R \Leftrightarrow L$ gives us the second set of equations.

Eqs. (7), (9) and (10) are the P-matrix for a single section of a system of almost periodic strips, Fig.1. It helps analyzing IDTs, and more complicated structures [2], [3].

Acknowledgment: This work was sponsored by KBN under Grant 312129101

5. REFERENCES

1 T. Aoki, K.A. Ingebrigtsen, IEEE Trans., SU-24 (1977), 167–178
2 M. Takeuchi, K. Yamanouchi, 1988 IEEE Ultras. Symp. Proc., 57–63
3 E. Danicki, 1992 IEEE Ultras. Symp. Proc., to appear

Simulation and Design of Applied Electromagnetic Systems
T. Honma (Editor)

Convenient orientations for SPUDT applications forced on piezoelectric crystal by external biasing stress

D. Gafka and E. Danicki

Institute of Fundamental Technological Research PAS,
Świętokrzyska 21, 00-049 Warsaw, Poland

Abstract
The paper presents results of convenient lithium niobate and quartz orientations useful for single phase unidirectional transducer (SPUDT) applications. The piezoelectric substrate is subjected to external and homogeneous biasing stress. The stress influence on the existing unidirectionality effect was numerically tested. The external stress changes the material properties, so that unidirectionality effect can appear in crystal orientations, where it does not exist without biasing stress. The stressed material is described by effective material tensors and nonlinear problem of wave propagation in such a media is numerically solved. Results can be useful for designing unidirectional transducers.

1. INTRODUCTION

Unidirectional transducer (UDT) finds important applications in signal processing. Many ideas were utilized in designing such interdigital transducer (IDT), which can generate surface acoustic wave (SAW) only in one direction. Well-known technical realizations are the three-phase UDT, combined IDT with SAW coupler with folded strips, and IDT with internal reflections caused by the difference in thickness of metalization, using mass loading effect or floating electrodes, called single phase UDT (SPUDT) [2,4,5]. All of the mentioned approaches have some disadvantages (complex matching network, complex strips pattern, or difficulties in designing or fabrications). An other approach has been proposed by Wright [1] and is called natural SPUDT (NSPUDT), because takes advantage of natural properties of special crystal orientations, for which simple uniform IDT exhibits unidirectionality [1,3,6,7]. Such orientations were found for SiO_2, $LiNbO_3$, $LiTaO_3$, GaAs, but are not frequent. Applying the external homogeneous stress to the crystal changes the SAW propagation conditions, that gives one more degree of design freedom. The paper presents some results concerning the idea of *natural unidirectionality* forced by an external stress. This new approach gives some interesting possibilities in SAW technique: improvement of beam steering, usage of typical orientations of substrates for SPUDT applications, corrections of fabrication errors on natural cuts, control of direction of wave generation (left, right) or control of power volume in a given direction.

2. ANALYTICAL MODEL OF PIEZOELETRIC CRYSTALS UNDER BIAS

The problem of the stress influence on the propagation of waves in piezoelectrics was studied by many authors. Different methods were used to solve the problem (e.g. variational methods). This work is based on the paper [8], where a number of references for this problem is given. Derivation of constitutive equations for electroelastic media upon mechanical or electrical bias is given in [8], basing on general rotationally invariant nonlinear electroelastic equations (energy balance equation and bilinear Gibbs function expansion). As a result, nonlinear constitutive relations for large biasing stress, and linear but parametric equations of state for small fields (surface waves) written in material coordinates were given. It can be written for SAW as follows

$$T_{ij} = c^{eff}_{ijkl} U_{k,l} + e^{eff}_{mij} \Phi_{,m} \qquad D_i = e^{eff}_{ikl} U_{k,l} - \epsilon^{eff}_{ij} \Phi_{,j} \tag{1}$$

where T, U, D, Φ are stress, displacement, electrical induction, and potential, respectively, of propagating surface wave. Effective material tensors are functions of biasing stress

$$c^{eff}_{ijkl} = c_{ijkl} + 2\underline{c}_{ijklmn}\tilde{U}_{m,n} \qquad e^{eff}_{mij} = e_{mij} + \underline{e}_{mijkl}\tilde{U}_{k,l} \qquad \epsilon^{eff}_{ij} = \epsilon_{ij} + \underline{l}_{ijkl}\tilde{U}_{k,l}. \tag{2}$$

$\tilde{U}$ is a displacement vector of biasing stress and

$$\begin{aligned} \underline{c}_{ijklmn} &= c_{inkl}\delta_{jm} + 0.5c_{ijnl}\delta_{mk} + 0.5c_{ijklmn} \\ \underline{e}_{mijkl} &= e_{mil}\delta_{jk} + e_{mijkl} \\ \underline{l}_{ijkl} &= l_{ijkl} - \epsilon_0(\delta_{il}\delta_{jk} + \delta_{ik}\delta_{jl} - \delta_{kl}\delta_{ij}), \end{aligned} \tag{3}$$

where c, e, ϵ, l are material tensors of elasticity, electroelasticity, permittivity and electrostriction of proper order given in tables of crystals.

Propagation of waves in the biased substrate is described by equations of motion

$$T_{ij,j} = \rho\ddot{U}_i \qquad D_{i,i} = 0 \tag{4}$$

with boundary conditions

$$T_{ij}n_j = 0, \quad n_j \text{ versor normal to the surface} \tag{5}$$

and

$$D_i n_i = 0 \quad \text{or} \quad E_i \times n_i = 0, \tag{6}$$

with the assumption of fulfilling the constitutive relations (1)-(3).

The system of equations (1)-(6) can be solved only numerically. Results of computation, with much attention pay to unidirectionality effects, are presented in the next section.

3. UNIDIRECTIONAL ORIENTATIONS OF SiO_2 AND $LiNbO_3$

The condition for the unidirectional behavior of IDT is the phase difference between the displacement and potential on the surface of piezoelectric halfspace for the propagating SAW ($\Delta\phi = 45^o$). Such a result was the subject of research for two crystals: $LiNbO_3$ and SiO_2, because only for these substrates third order material data are available. Electromechanical coupling coefficient ($\Delta V/V$) and beam steering angle (α), as important for applications, were also computed. All parameters are dependent on three Euler's angles (λ, μ, θ), describing cut and direction of SAW propagation, and also six components of symmetric stress tensor. The vector function $[\Delta\phi, \Delta V/V, \alpha] = f(\lambda, \mu, \theta, T_1, ..., T_6)$ was computed for two piezoelectric crystals. Only chosen results for $LiNbO_3$ Y cut are presented in Fig.1-3 to show the influence of the external stress on the phase difference $\Delta\phi$ for main direction of biasing stress. Additionally, some results are collected in the Table, describing cuts and proper stresses for which unidirectionality effect appears.

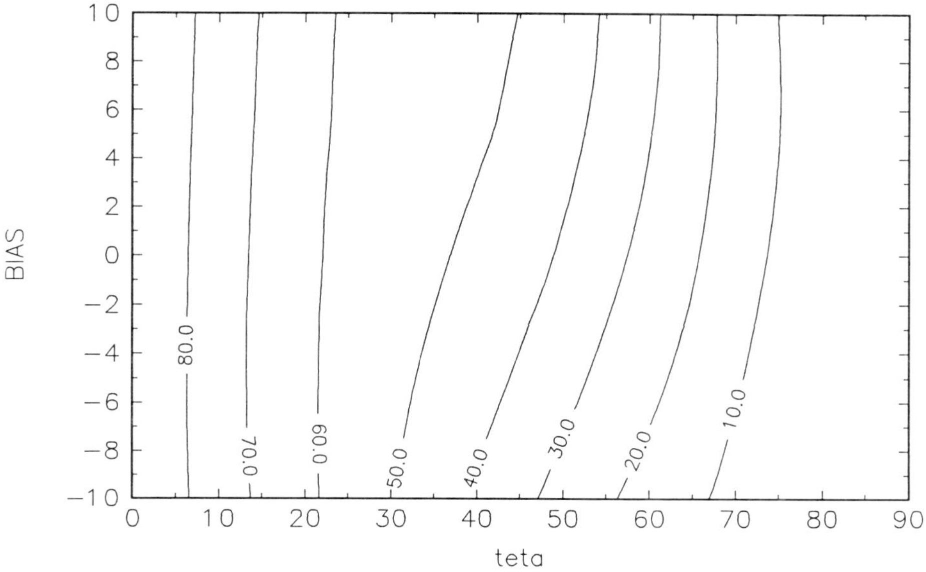

Figure 2: Contour lines of $\Delta\phi[deg] = const.$ in $\theta[deg] - T_2[GPa]$ plane.

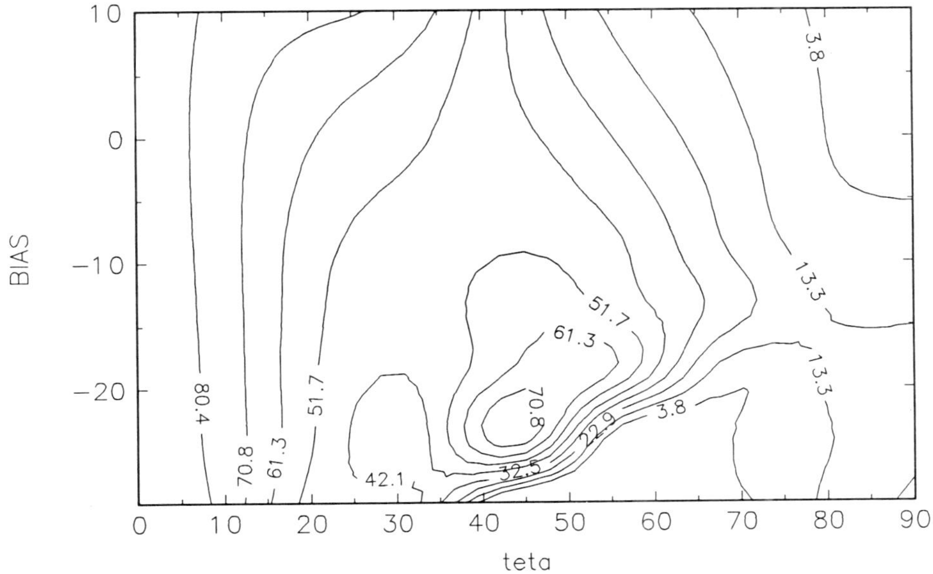

Figure 3: Contour lines of $\Delta\phi[deg] = const.$ in $\theta[deg] - T_3[GPa]$ plane.

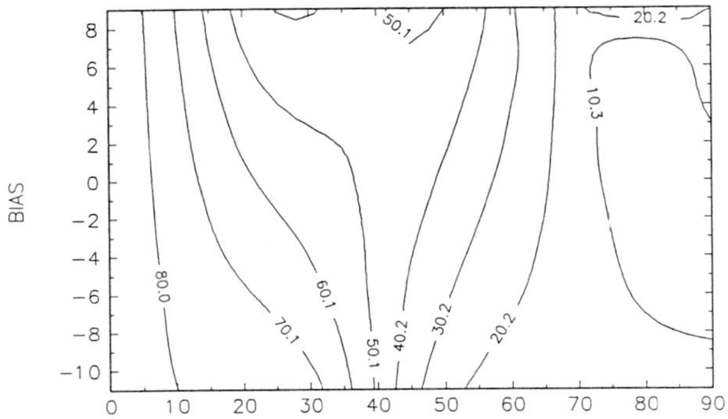

Figure 1: Contour lines of $\Delta\phi[deg] = const.$ in $\theta[deg] - T_1[GPa]$ plane.

Crystal	Euler's Angles	Velocity [km/s]	$\Delta V/V$	α [°]	$\Delta\phi$ [°]	Stress [GPa]
SiO_2	0. 132.75 24.69	3.2463	0.00065	5.1	-45.0	0
SiO_2	0. 132.75 23.53	3.7286	0.00033	0.0	-45.0	$T_{11} = -18$
SiO_2	0. 132.75 27.26	3.5333	0.00087	0.0	-45.0	$T_{22} = -7$
SiO_2	0. 132.75 26.60	3.0878	0.00078	0.0	-45.0	$T_{33} = -10.1$
SiO_2	0. 0. 5.52	3.2782	0.00009	6.4	-44.9	0
SiO_2	0. 0. 69.41	4.0945	0.00046	-12.2	-44.9	0
$LiNbO_3$	0. 0. 44.10	3.6379	0.0063	-3.7	-45.0	$T_{11} = 10$
$LiNbO_3$	0. 0. 43.75	3.8516	0.0027	-4.5	-45.0	$T_{13} = 3.3$
$LiNbO_3$	0. 0. 18.00	3.7287	0.0053	1.4	-45.0	$T_{12} = 2.9$
$LiNbO_3$	0. 90. 43.81	3.5375	0.0140	-7.3	-45.0	0

4. CONCLUSIONS

As one can see it is possible to make corrections of beam steering, usually appeared for natural unidirectional cuts by applying external bias (compare row 2,3,4 to 1 in the Table). Some other cuts, for which unidirectionality effect appears, were also found. The disadvantage is the stress magnitude, necessary to obtain the effect. It is very large, so the implementation possibility is rather uneasy in the considered cases. The only interesting application of stress in these cases is a correction of designing and fabrication errors on natural cuts. This conclusion is based only on two tested materials, lithium niobate and quartz. In some new materials the discussed effect may be stronger. It can be tested after linear and nonlinear material data will be published.

4. REFERENCES

1 P.V. Wright, 1985 IEEE Ultrasonics Symp. Proc., 58-63 (1985).
2 K. Yamanouchi, Z.H. Chen, T. Meguro, 1985 Ultrasonics Symp. Proc., 74-90 (1985).
3 T. Thorvaldsson, 1989 IEEE Ultrasonics Symp. Proc., 91-96 (1989).
4 E. Danicki, 1990 IEEE Ultrasonics Symposium Proc., 113-116 (1990).
5 C. Kappacher, O. Maenner, W. Ruile, R. Dill, 1991 Ultras. Symp. Proc., 1-4 (1991).
6 D.F. Thomson, P.V. Wright, 1991 IEEE Ultrasonics Symp. Proc., 181-184 (1991).
7 E. Danicki, 1991 IEEE Ultrasonics Symp. Proc., 145-148 (1991).
8 D. Gafka, J. Tani, J. Appl. Physics **70**, 6679-6686 (1991).

Simulation and Design of Applied Electromagnetic Systems
T. Honma (Editor)

A three-dimensional finite element analysis of microwave distribution in a cavity

K. Yasutake[a] and Y. Kagawa[b]

[a]Production Technology Laboratories, Dept. 4, SHARP Corp.
Ichinomoto-cho, Tenri City, Nara, 632, Japan

[b]Dept. of Electrical and Electronic Engineering, Okayama University
Tsushima-naka, Okayama, 700, Japan

Abstract

Three dimensional finite element program for electromagnetic field problem is developed. Some numerical solutions are examined compared with experimental results for the microwave strength distributions in some cavities. The both results agree well, the electric field is evaluated by the electric current induced in the small loop antenna and by the temperature distribution over the dielectric loss inserted in the field.

1. Introduction

It is important to understand the precise wave behavior for the design of microwave devices and appliances. Exact solution to Maxwell equations is not always possible, but numerical approach is now available [1], [2]. The finite element method (FEM) is known to be a powerful technique for electromagnetic field analysis, but spurious solutions are sometimes observed in the eigenvalues, to which some remedies have been proposed [3], [4].

We developed a FEM program for three-dimensional inhomogeneous electromagnetic field solutions. To verify the validity, some experiments for the field distribution in a cavity are carried out and the results are compared with the FEM solution. The program is accompanied with pre / post processors to serve as a part of CAE system.

2. Governing Equation and Discretization

Time-harmonic and source-free three dimensional electromagnetic fields satisfy the vector Helmholtz equation.

$$\nabla^2\Phi + \omega^2\varepsilon\mu\Phi = 0$$

, where Φ is an electric field vector **E**=(Ex, Ey, Ez) or a magnetic field vector **H**=(Hx, Hy, Hz), ω is an angular frequency, ε is a permittivity and μ is a permeability. We formulate the discretized equations using Nedelec elements [5] with the constraint of Lagrange multipliers [6] to avoid spurious solutions. For the boundary conditions, the walls are assumed to be of perfect electric conductors. Namely, the condition

$$\mathbf{E} \times \mathbf{n} = 0$$

is satisfied on the boundary wall, where **n** is normal vector. The field is excited by a thin and short antenna.

3. Numerical Examples and Experiments

The Electric field distribution is measured by two methods. The first one is that the induced current is measured by a loop probe placed in the electric field. And the second is that the electric field distribution is measured by a thermo-graphy in terms of the temperature distribution over a dielectric thin film placed in the field, as the absorbed energy causes the temperature increase.

3.1 Electric field strength and an electric current

Fig. 1 shows a rectangular cavity with a feeding waveguide which is here called a small oven. It has many holes in line over the conducting wall and the microwave with a frquency of 2.45 GHz is fed from the antenna in the end of waveguide. As shown in Fig. 2, the electric field inside is searched by the loop probe through the hole. Fig. 3 shows the current distribution in the vicinity of the inside surface of the small oven, for which only the half is shown.

It is supposed that electric current is proportional to the electric field strength in the direction of the probe axis. Therefore, calculation is also made for the electric field **E** = (Ex, Ey, Ez). Fig. 4 is the distribution of a normalized electric field strength |Ez| (filled line) by a FEM solution and the normalized electric current j (dashed line) by a probe method along the central line of the cavity. The numbers and the position of the peaks and dips are both in good agreement.

3.2 Electric field and temperature distribution

Fig. 5 shows another rectangular cavity coupled with the same waveguide which is here called a large oven. A Polycarbonate thin film (0.5 mm thick) is placed in the central part of the cavity. Electric field energy is absorbed to heat the film due to its own dielectric loss. The temperature of the film reaches to stationary state in a few seconds. As shown in Fig. 6, the temperature distribution over the surface is measured remotely by means of the thermo-graphic technique.

Calculation is made for the electric field intensity, that is the square of the electric field strength.

$$|\mathbf{E}|^2 = E_x^2 + E_y^2 + E_z^2.$$

Fig. 7 and Fig. 8 show the distribution of the numerical solution and the measured temperature, respectively. The similar patterns of the distributions are obtained. Thus, the validity of the numerical solutions is confirmed .

4. REFERENCES

[1] P. Silvester, " High-order finite element waveguide analisys ", IEEE Trans. Microwave Theory Tech., MTT-17, pp. 651-653, Aug. (1969)

[2] A. Konrad, " A direct three-dimensional finite element method for the solution of electromagnetic fields in cavities ", IEEE Trans. Magn., MAG-21, pp. 2276-2279, Nov. (1985)

[3] M. Hano, " Finite-element analysis of dielectric-loaded waveguides ", IEEE Trans. Microwave Theory Tech., MTT-32, pp. 1275-1279, Oct. (1984)

[4] M. Koshiba, K. Hayata, and M. Suzuki, " Finite-element formulation in terms of the electric-field vector for electromagnetic waveguide problems ", IEEE Trans. Microwave Theory Tech., MTT-33, pp. 900-905, Oct. (1985)

[5] J.C.Nedelec," Mixed finite elements in R^3 ", Numer.Math.,35, pp315-341(1980)

[6] Kikuchi. F, " Mixed and penalty formulations for finite element analysis of an eigenvalue problem in electromagnetism ", Computer Meth.Appl.Mech.Eng., 64, pp509-521(1987)

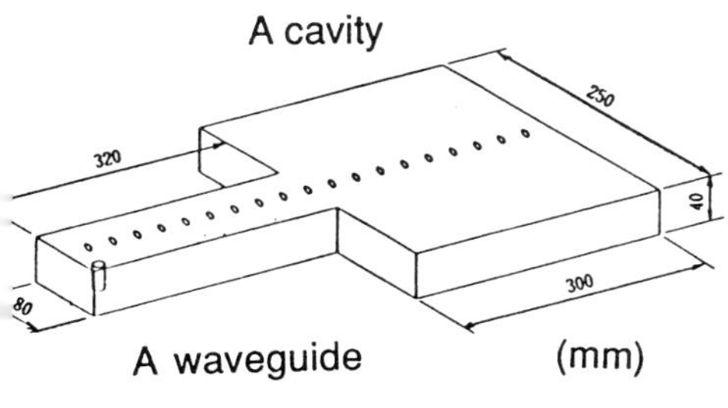

Fig. 1. A small oven: A cavity, a waveguide and an antenna.

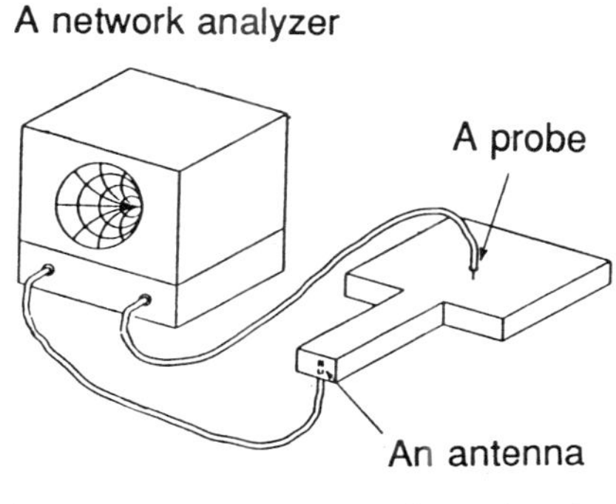

Fig. 2. A measurement of an electric current by a probe method.

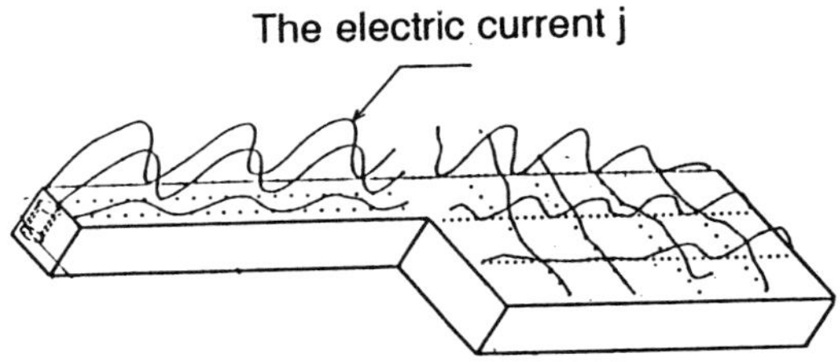

Fig. 3. The electric current distribution in the small oven (Experimental).

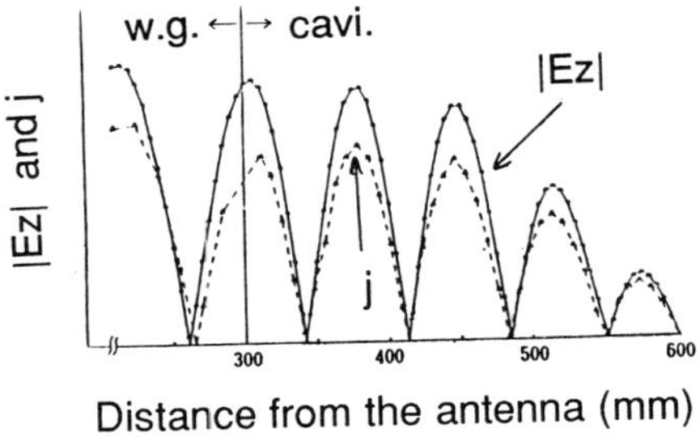

Fig. 4. The electric field component |Ez| (Numerical) and the electric current j (Experimental).

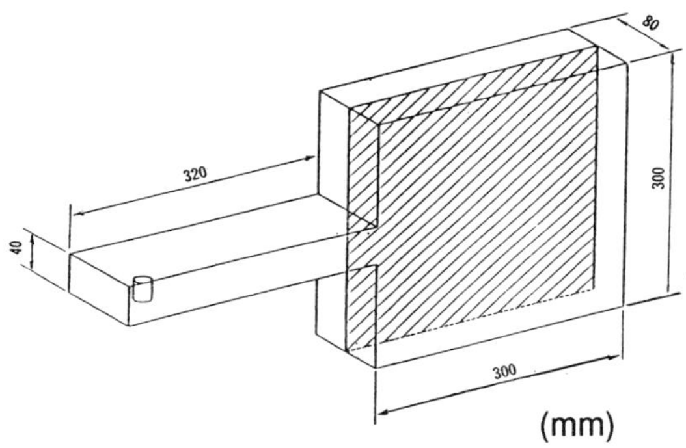

Fig. 5. A large oven and a film inside a cavity.

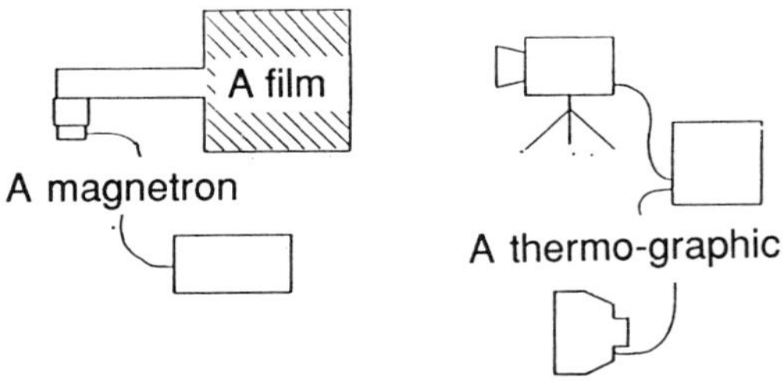

Fig. 6. A measurement of the temperature distribution by a thermo-graphic method.

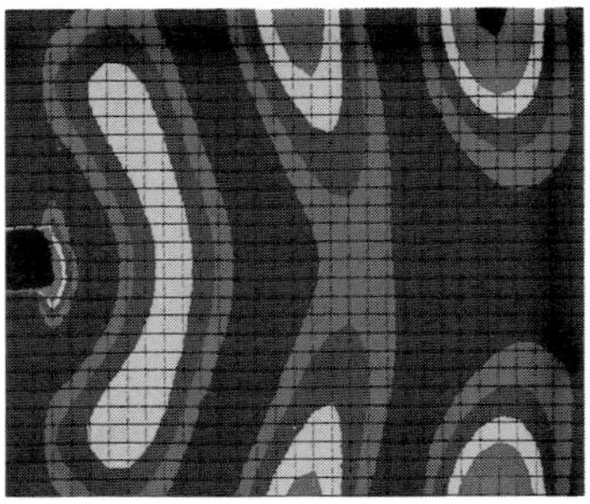

Fig. 7. The distribution of the electric field energy $|\mathbf{E}|^2$ (Numerical).

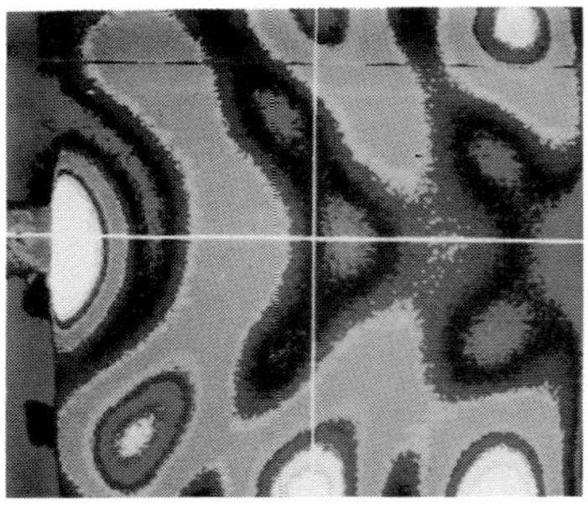

Fig. 8. The distribution of the temperature (Experimental).

Simulation and Design of Applied Electromagnetic Systems
T. Honma (Editor)

Removing spurious solutions in boundary elemental method for multimedia problems using virtual boundary dividing method

Satoshi Tomioka[a], Takeaki Enoto[a] and Ichiro Fukai[b]

[a]Dept. of Nuclear Engineering, Faculty of Engineering, Hokkaido University.
[b] Dept. of Electrical Engineering, Faculty of Engineering, Hokkaido University.
Kita 13, Nishi 8, Kita-ku, Sapporo, 060, Japan

Abstract
In the analysis of multi-media problems by the boundary element method (BEM), we have found that spurious solutions superimpose onto the true solutions. To remove the spurious solutions, we propose a new method named the "Virtual Boundary Dividing Method". The virtual boundary dividing method is one of the region or boundary selecting methods for multi-media problems. It obeys Huygens' principle as long as optical paths don't cross another medium. We show the validity of the virtual boundary dividing method for removing of spurious modes.

1. Virtual boundary dividing method

The BEM in electro-magnetic wave problems is understood by Huygens' principle. The waves propagate from their origins to secondary points, and the summation of these waves at each secondary point becomes the next origin. In uniform-medium problems we don't need to consider secondary points, but in multi-media problems we must consider them at the interface of each media. Thus , in conventional BEM , the boundaries are set at the interface of each region with different physical constants , and the number of regions and media are equal.

Consider the case of a medium within the other medium, some waves cross into inner medium. (Fig.1(a)). In order to exclude these crossing paths, the virtual boundary is set in the same physical constant medium (Fig.1(b)). In the BEM formulation, the region integral can be replaced with the boundary integral by Green's integral theorem. The difference between two methods is shown in Fig.2 as the regional operation of the outer medium.

We can derive the next equation for observation point i from the BEM basic equations (Helmholtz equation, Green's integral theorem and Green's function) under the assumption that, in the boundary elements, the potential u_i and its derivatives q_i are constant or known functions (e.g., linearly).

$$c_i u_i + \sum_j^N h_{ij} u_j = \sum_j^N g_{ij} q_j \tag{1}$$

$$h_{ij} = \int_{\Gamma_j} q^*(\vec{r_j};\vec{r_i})\, d\Gamma \quad , \quad g_{ij} = \int_{\Gamma_j} u^*(\vec{r_j};\vec{r_i})\, d\Gamma$$

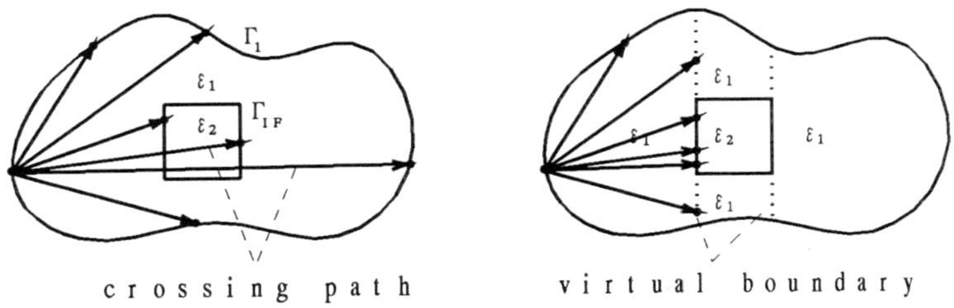

(a) The conventional method (b) The virtual boundary method

Figure 1: The optical path

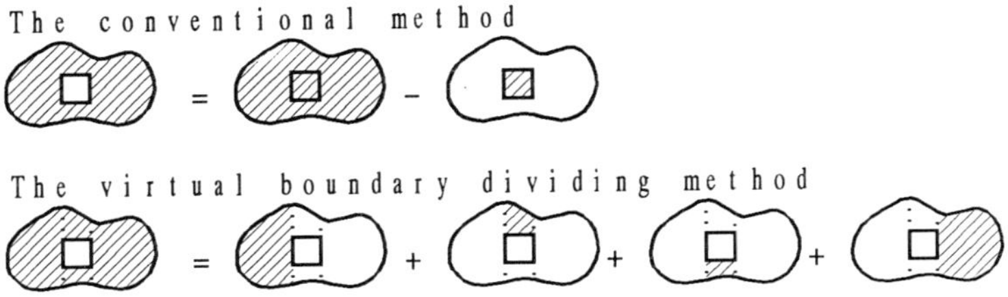

Figure 2: The regions integral operation for outer region

To explain the cause of spurious solutions, which will be discussed in later sections, it is necessary to introduce here the system of equations by the conventional method. In the conventional method the boundary Γ is separated into two boundaries for the outer region. One is the original boundary Γ_1, and the other is the interface boundary Γ_{IF}. For the inner region , the boundary is only Γ_{IF}. Using the interface boundary conditions, which are the continuity of potential u and its derivative q , where the significance of the outer region's derivative q has opposite to the inner region's. For every point i on the boundary the Eq.(1) becomes the Nth set of simultaneous equations as follows:

$$\begin{pmatrix} \mathrm{G}_1^1 & -{}^*\mathrm{G}_{IF(1)}^1 \\ \mathrm{G}_1^{IF(1)} & -{}^*\mathrm{G}_{IF(1)}^{IF(1)} \\ 0 & \mathrm{G}_{IF(2)}^{IF(2)} \end{pmatrix} \begin{pmatrix} q_1 \\ q_{IF(2)} \end{pmatrix} = \begin{pmatrix} \mathrm{H}_1^1 & -{}^*\mathrm{H}_{IF(1)}^1 \\ \mathrm{H}_1^{IF(1)} & -{}^*\mathrm{H}_{IF(1)}^{IF(1)} \\ 0 & \mathrm{H}_{IF(2)}^{IF(2)} \end{pmatrix} \begin{pmatrix} u_1 \\ u_{IF(2)} \end{pmatrix} \tag{2}$$

where, the subscripts and the superscripts indicate the boundary and nodal points' identifiers, and the asterisk means reverse order of the matrixes.

In the virtual boundary dividing method, we can construct the simultaneous equations similarly.

2. Analysis model

Fig.3 shows the analysis model and the virtual boundaries. The Fabry-Pérot resonator (FPR) has two mirrors, one of which is connected to the waveguide excited TE_{10} mode.

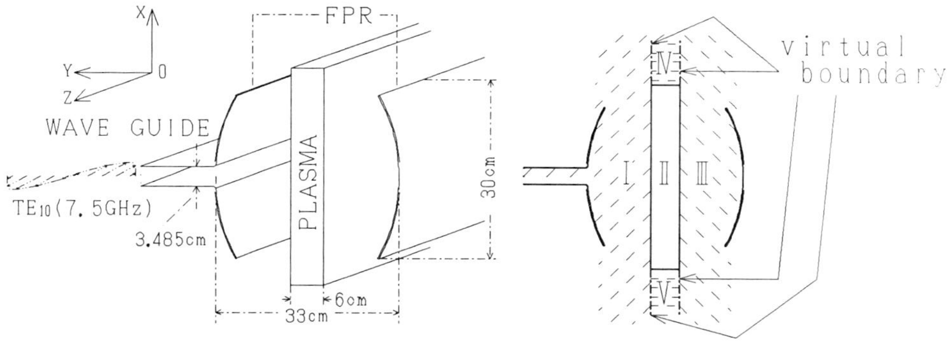

Figure 3: The FPR model

This FPR contains two media, one is a vacuum and the other is a plasma. Since the plasma has uniform density distribution, may be regarded as a homogeneous dielectric medium which has frequency dispersion. The FPR walls are perfect conductors.

The virtual boundaries are set in a vacuum medium to exclude the paths crossing the inner medium.

3. Condition number

The a set of simultaneous equations are given as follows.

$$\mathrm{A\,x=b} \tag{3}$$

Whether or not these equations can be easily solved depends on the singularity of matrix A. In order to evaluate the singularity, we introduce the condition number.

$$\eta(\mathrm{A}) = \left(\frac{V}{V_0}\right)^{\frac{1}{N-1}} = \left(\frac{\det(\mathrm{A})\det(\mathrm{A})^*}{\prod_i^N \sum_j^N a_{ij}a_{ij}^*}\right)^{\frac{1}{2(N-1)}} \tag{4}$$

where the asterisk means complex conjugate. In Eq.(3), if we choose row vector a_i of matrix A as the basis, then the component of the basis becomes vector x when constructing vector b in oblique vector space. The denominator of the right side of Eq.(4) is recognized as the square volume V_0 of an N-dimensional rectangular solid when all of $\{a_i\}$ are orthogonal to each other. The numerator is recognized as the square volume V of an N-dimensional parallelepiped. Thus, the condition number indicates the one-dimensional orthogonality.

4. Result

For our calculation, the distance between adjacent nodes is nearly 1/10 of the minimum wavelength, and we used linear boundary elements. Shown in Fig.4, the frequency characteristics of the complex reflection coefficients in the waveguide, analyzed using two

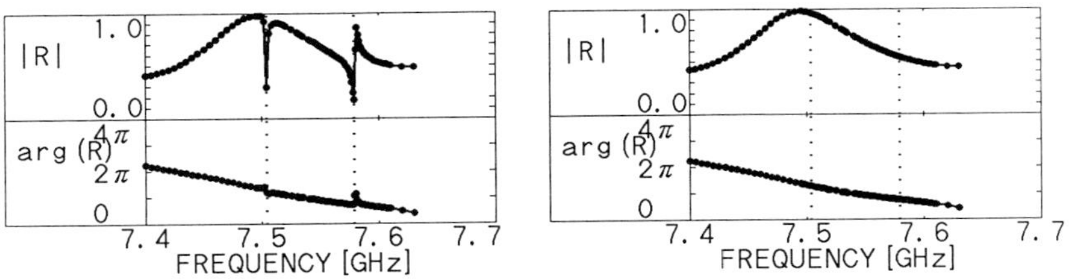

Figure 4: The complex reflection coefficients

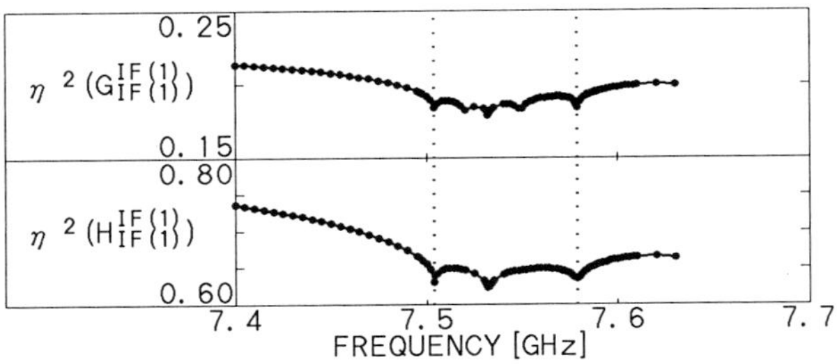

Figure 5: The condition numbers

methods, respectively. The differences in the results of each method are recognized as the spurious solutions.

Fig.5 shows the frequency characteristic of the condition numbers of sub-matrixes $G_{IF(1)}^{IF(1)}$ and $H_{IF(2)}^{IF(2)}$ which are square matrix in Eq.(2), as analyzed by the conventional method. We find the local minima in Fig.5, and their frequencies correspond to spurious solutions.

5. Conclusion

We discovered the existence of the spurious solutions by the conventional method of BEM. Using the virtual boundary dividing method, we can remove these solutions. While these methods have the difference of subtraction and addition in the region integral, they are same meaning mathematically. Thus spurious solutions will appear due to numerical error. Evaluating the condition numbers, we conclude that the cause is the singularities in sub-matrices which are used in subtracting the inner region from all regions included in the outer region's counter-integral.

Acknowledgement

One of the authors, Prof. Fukai passed away on the 11th December 1992. During his life, he contributed a great deal to this research. The success of this research is largely thanks to him. We offer tribute to him for his brilliant scientific achievement, and we wish him peaceful rest.

Simulation and Design of Applied Electromagnetic Systems
T. Honma (Editor)

Some Problems of Microstrip Antenna Input Impedance Calculations

T. Rakó and Gy. Veszely

Department of Microwave Telecommunications, Technical University of Budapest,
1111 Budapest, Goldmann Gy. tér 3, Hungary

Abstract

Two spectral domain integral equation techniques for calculating the input impedance of a microstrip patch are discussed and compared. The solutions are based on the moment method and the reaction equation respectively. A numerical investigation is presented to determine how the critical parameters of the solutions should be chosen. The convergence of the input impedance is increased by using stationary formulas.

1. INTRODUCTION

The present work aims at examining two analysis methods for calculating the input impedance of microstrip patches. The first solution is based on spectral domain integral equations solved by moment method. [1,2] This model applies an open-ended stub in order that gap excitation can be used. [1] However the parameters of the stub are not definite which introduces ambiguity in the solution. This paper presents considerations based on numerical investigations for choosing the stub. The convergence of the results is increased by using a stationary formula or the radiated power for the input impedance calculation.

The second method applies reaction equation. [3] This solution allows us to model both coaxial probe and microstrip line feeds. It leads to the same matrix equation as the moment method technique. However the excitation vector is different which results faster convergence.

2. MOMENT METHOD SOLUTION

This technique formulates a set of coupled integral equations for the patch current density $\overline{J}^p$ by forcing the total tangential electric field $\overline{E}^t$ to vanish on the conducting patch. The total field is the sum of the incident field $\overline{E}^i$ and the electric field produced by the patch current. The latter can be conveniently formulated with the help of the dyadic Green's function $\underline{\underline{G}}$.

$$0 = \overline{E}^t = \overline{E}^i + \iint_{patch} \underline{\underline{G}}(\bar{r},\bar{r}')\bar{J}(\bar{r}')d\bar{r}' \tag{1.}$$

The Green's function can be easily derived in spectral domain. Then the current density components on the patch are expanded into sets of rooftop basis functions B^i. Taking inner products with the weighting functions chosen similarly to the basis functions leads to a matrix equation. [1,2]

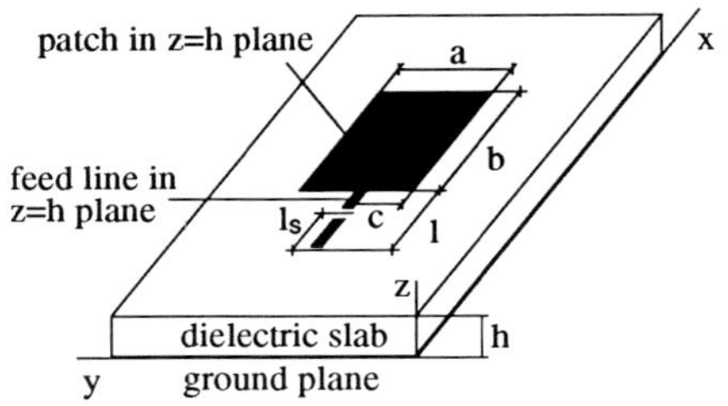

Figure 1. Model of the problem.

$$-\underline{U} = \underline{\underline{Z}}\,\underline{I} \tag{2.}$$

where $\underline{\underline{Z}} = \frac{1}{4\pi^2}\int_{-\infty}^{\infty}\int_{-\infty}^{\infty} \underline{\underline{\tilde{G}}}\,\underline{\tilde{B}}\,\underline{\tilde{B}}^* dk_x dk_y$, $\quad \underline{U} = \iint_{patch} \overline{E}\,\underline{\tilde{B}} d\bar{r}$,

and the tidle denotes the Fourier transform.

Once the above equation is solved for the unknown coefficients of the basis functions in $\underline{I}$, the current density is known. The most convenient way to determine the input impedance is to use gap excitation by applying an additional open-ended stub and follow the two-step procedure suggested by Splitt. [1]

Table1

Variation of the calculated input resistance for a resonant patch. (Measured value : 230Ω)

Stub Length [mm]	2dx	3dx	4dx	5dx	6dx
Input Res. [Ω]	231.73	249.5	271.85	292.79	333.78

a=55.6166mm, b=46.5502mm
c=3a/7, dx=b/7 *, dy=a/7 *
h=1.57mm, ε_r=2.2
N_{fft}=32, f=2.07GHz

* Lengthes of subsections in the x and y directions respectively.

This method provides different input impedances for different stub-lengthes as it can be seen from Table 1. The reason of this ambiguity is that the patch currents have an effect on the electric field on the stub, therefore the calculated input impedance of the stub for the 'stub-antenna' problem is different from that of the 'stub-stub' problem.

Considering Fig. 2 it can be concluded that the numerical solution is unstable at resonant stub-lengthes. Therefore such lengthes should be avoided. According to Table 1 short stubs are preferred. The stub-width has to be also minimazed so that the characteristic impedance of the line can increase and can get closer to the input impedance resulting thus lower standing wave ratio on the stub.

The convergence of the calculated input impedance can be increased by applying the following stationary formula.

$$Z_{in} = -\iint_{patch} \overline{E}(\overline{r})d\overline{r} \Big/ I_{feed} \tag{3.}$$

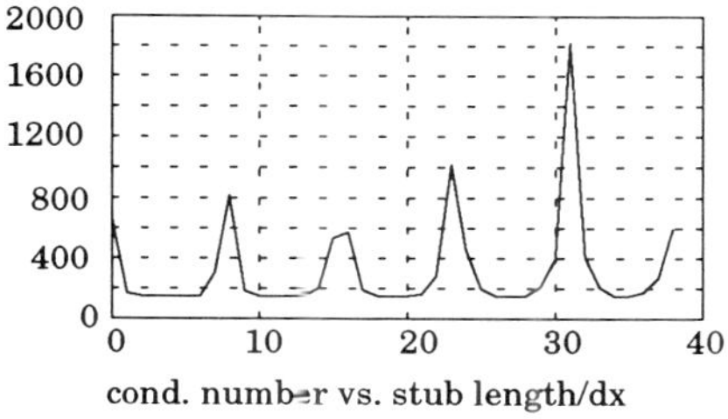

Figure 2. Condition number of the matrix equation for the patch examined in Table 1.

This technique attempts to compensate for the numerical errors in the calculation which lead to a spred electric field instead of a single impulse in the gap.

Better convergence can be also achieved by computing the radiation resistance from the radiated power.

$$R_r = 2P_r / I_{feed}^2 \tag{4.}$$

The above two formulas have similar effects as it is shown in Table 2.

Table 2

Comparison of input impedances computed by different formulas

Impedance Calculation	U/I	Stationary Formula	Radiated Power
Z_{in} [Ω]	292.79	265.96	257.31

l=33.25mm
See Table 1 for other parameters.

3. REACTION EQUATION SOLUTION

This technique is suitable for modeling both a microstrip feed line and a coaxial probe excitation. The excitation is replaced by different current sources in the two cases. [3] If $\overline{E}^p$ is the electric field due to a test current assumed to be present on the patch and $\overline{E}^i$ is the electric field produced by the excitation, then applying the reaction equation one obtains an integral equation with the patch current density $\overline{J}^p$ as an unknown variable.

$$-\iiint_{source} \overline{J}^i \overline{E}^p dV = \iiint_{patch} \overline{J}^p \overline{E}^i dV \tag{5.}$$

The right side of the above equation leads to the same matrix as the moment method solution. However the excitation vector is different (see left side) and, therefore, the convergence is faster as it is suggested by Table 3.

Table 3
Comparison of input resistances computed by different methods.

N_{fft}	32	64	128	See Table 1 for parameters.	
R_{in} [Ω]	271.85	265.43	270.2	***Gap Type Feed (U/I)***	l=4dx
R_{in} [Ω]	219.64	233.45	237.06	***Gap Type Feed (Stationary Formula)***	l=4dx
R_{in} [Ω]	210.99	276.82	247.39	***Coaxial Probe***	
R_{in} [Ω]	197.54	273.21	247.46	***Microstrip Feed Line***	

One can conclude from Table 3 that reaction equation solutions provide similar input impedances. The best convergence is obtained in case of a microstrip feed line. The convergence of the moment method technique is less rapid. However it can be greatly improved by applying a stationary formula. In case of reaction equation solutions the deviation from the measured value is less then 8 percent for $N_{FFT} \geq 128$. It must be also mentioned that by using this technique there is no need to apply an additional stub and, therefore, no inconveniences arise in the solution.

4. CONCLUSIONS

Two methods were presented for calculating the input impedance of a microstrip patch. It was pointed out that reaction equation solutions provide the most favorite convergence and that this technique allows us to avoid the ambiguities in the input impedance calculation which arise in case of the moment method solution. Two formulas were suggested to improve the convergence of the moment method technique. A numerical investigation provided hints for chosing the additional stub needed in the latter solution.

6. REFERENCES

1. G Splitt, Rectangular Electromagnetically Coupled Microstrip Antennas in Multilayered Structures, 18th European Microwave Conf., Sweden, Stockholm, (1988) 1043
2. T. Rakó, Comparison between the Scattered Fields Computed by Geometrical Theory of Diffraction and Spectral-Domain Integral Equation Solution, 2nd Japan-Hungary Joint Seminar on Appl. Electromagn. in Materials and Computat. Techn., Japan, Sapporo (1992) 238
3. M. D. Deshpande and M. C. Bailey, Input Impedance of Microstrip Antennas, IEEE Trans. Ant. Prop., vol. AP-30, no. 4 (1982) 645

Simulation and Design of Applied Electromagnetic Systems
T. Honma (Editor)

Electromagnetic Inverse Scattering Problem for Perfectly Conducting Rough Surface

K. Sato and M. Tanaka

Department of Electrical and Electronic Engineering, Oita University
700 Dannoharu, Oita 870-11, Japan

Abstract

This paper deals with an electromagnetic inverse scattering problem for a one-dimensional perfectly conducting rough surface. We present a simple inversion algorithm for reconstructing the rough-surface profile based on the modified Newton-Kantorovich method. The numerical results confirm that the algorithm provides high-quality reconstructions even for cases where the scattered field data contains the measurement error. Furthermore, the results demonstrate fast convergence properties of the iterative procedure.

1. INTRODUCTION

Electromagnetic scattering by a rough surface has been of great interest and practical importance owing to potential applications in remote sensing of land and sea [1]. In the far zone, the scattered field may be obtained based on the Kirchhoff approximation and the stationary-phase formulation [1,2]. Thus the total field at a point on the rough surface is equal to the incident field plus the field specularly reflected by an infinite plane tangent to the point [1]. The scattered field is then independent of the state of polarization of the incident field.

In this paper, we present a simple inversion algorithm for reconstructing a one-dimensional perfectly conducting rough surface with the knowledge of an incident field and the resulting scattered far-field data. The inverse scattering problem may be formulated as the solution of a nonlinear integral equation for the rough-surface profile f. Applying the modified Newton-Kantorovich method [3] to the integral equation, one can obtain an iterative formula for getting f. While the formula is a linear integral equation of the first kind, the rough-surface profile may be recovered without employing the usual regularization techniques. Setting an initial guess of the rough-surface profile to zero, we can get the n-th estimate of f by directly performing the inverse Fourier transform of its spectrum, which is bandlimited within a low-frequency band. For a sinusoidal surface and a surface consisting of a fundamental undulation and a higher harmonic, computer simulations are made to show the validity of the proposed algorithm. From the numerical results, one can see that our procedure yields high-quality reconstructions even if the scattered field data contains the measurement error. It is also shown that the iterative algorithm has the properties of fast convergence.

2. PROBLEM FORMULATION

Consider a plane electromagnetic wave obliquely incident from free space upon a one-dimensional perfectly conducting rough surface $z=f(x)$, as shown in Fig. 1. In this figure, the incident and scattered angles are denoted by θ_i and θ_s, respectively. The inverse scattering problem discussed here can be formulated as the solution of the following nonlinear integral equation:

$$\mathbf{A} f(x) = 0. \tag{1}$$

The nonlinear operator $\mathbf{A}$ may be written as

$$\mathbf{A} f(x) = -W(\theta_s; \theta_i) + \frac{2[1 + \cos(\theta_s + \theta_i)]}{\cos\theta_s + \cos\theta_i} \int_{-L/2}^{L/2} \exp\{j[\alpha x + \beta f(x)]\}\, dx, \tag{2}$$

where

$$\alpha = k_0 (\sin\theta_s - \sin\theta_i), \quad \beta = k_0 (\cos\theta_s + \cos\theta_i), \tag{3}$$
$$-\pi/2 < \theta_s, \ \theta_i < \pi/2. \tag{4}$$

Here W, k_0, and L indicate a pattern function of the scattered far field, the wavenumber of free space, and the extent of the rough surface, respectively. Note that the scattered field is obtained based on the Kirchhoff approximation and the stationary-phase formulation [1,2]. The rough-surface profile is assumed to fall to zero at the ends $x = \pm L/2$. From Eqs. (3) and (4), we have

$$\theta_s = \cos^{-1}[\pm(\sqrt{\alpha^2 + \beta^2}/2k_0)] + \tan^{-1}(\alpha/\beta), \tag{5}$$
$$\theta_i = \cos^{-1}[\pm(\sqrt{\alpha^2 + \beta^2}/2k_0)] - \tan^{-1}(\alpha/\beta), \tag{6}$$
$$|\alpha| \le \sqrt{4k_0^2 - \beta^2}, \quad k_0 < \beta < 2k_0. \tag{7}$$

In Eqs. (5) and (6), the upper and lower signs are in order.

Applying the modified Newton-Kantorovich method [3] to Eq. (1), we get

$$\mathbf{A}'(f_0)[f_{n+1}(x) - f_n(x)] = -\mathbf{A} f_n(x), \qquad n = 0,1,2,\cdots. \tag{8}$$

Here f_n indicates the n-th estimate of f, and $\mathbf{A}'(f_0)$ denotes the Fréchet derivative of $\mathbf{A}$ at an initial guess f_0. It should be remarked that $\mathbf{A}'(f_0)$ is a linear operator. Substituting the expression for the Fréchet derivative and Eq. (2) with $f = f_n$ into Eq. (8) leads to an iterative formula for getting f_{n+1} from f_n. While the formula is a linear integral equation of the first kind, the rough-surface profile may be recovered without employing the regularization techniques. Setting $f_0 = 0$, the integral equation becomes

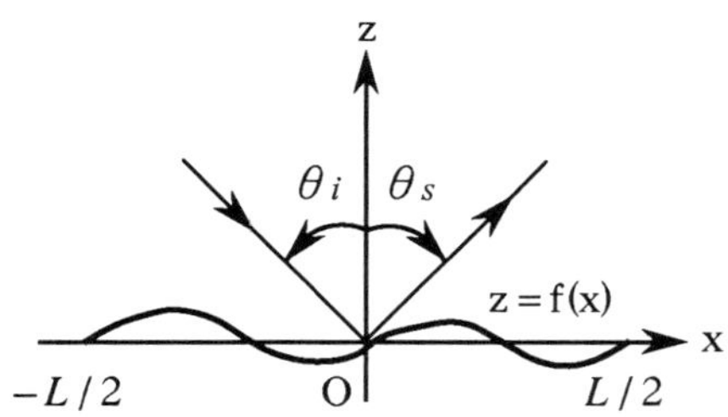

Figure 1. Scattering geometry for rough surface.

$$F_{n+1}(\alpha) = \int_{-L/2}^{L/2} f_{n+1}(x) \exp(j\alpha x)\, dx$$
$$= F_n(\alpha) - j\frac{W(\theta_s; \theta_i)}{2k_0[1 + \cos(\theta_s + \theta_i)]}$$
$$+ \frac{j}{k_0(\cos\theta_s + \cos\theta_i)} \int_{-L/2}^{L/2} \exp[j\beta f_n(x)] \exp(j\alpha x)\, dx. \tag{9}$$

From Eq. (9), we can obtain f_{n+1} by directly performing the inverse Fourier transform of its spectrum F_{n+1}. As is evident from Eq. (7), F_{n+1} is bandlimited within a low-frequency band.

3. NUMERICAL RESULTS

Computer simulations are made for the following rough-surface profile:

$$f(x) = h_1 \sin(2\pi x / \Lambda) + h_2 \sin(2\pi M x / \Lambda), \tag{10}$$

where Λ and M denote the rough-surface period and the order of the harmonic. Now consider the parameters: $L / \lambda_0 = 400$, $\Lambda / \lambda_0 = 20$, and $\beta / k_0 = 1.1$. Here λ_0 is the wavelength in free space.

In the iteration process of reconstruction, the imaginary part of f_n is set to zero with *a priori* knowledge of the *real* profile function. In Eqs. (5) and (6), we choose the upper (plus) signs. The iteration terminates if the relative mean-squared error is less than the convergent criterion 10^{-8}.

Figure 2 presents the reconstructions of two rough-surface profiles from the scattered field data without noise. Figure 2 (a) shows the reconstructed result of the sinusoidal rough-surface profile described by Eq. (10) with $h_1 / \lambda_0 = 0.15$ and $h_2 / \lambda_0 = 0$. The final convergent solution after 11 iterations and the true profile are, respectively, denoted by the circle points and the solid curve. Figure 2 (b) illustrates reconstructed result of the rough-surface profile consisting of a fundamental undulation and a higher harmonic, which is described by Eq. (10) with $h_1 / \lambda_0 = 0.12$, $h_2 / \lambda_0 = 0.03$, and $M = 5$. The final convergent solution is the result from 10-th iteration. From Fig. 2, we can see that the reconstructions based on our algorithm show excellent agreement with the true profiles.

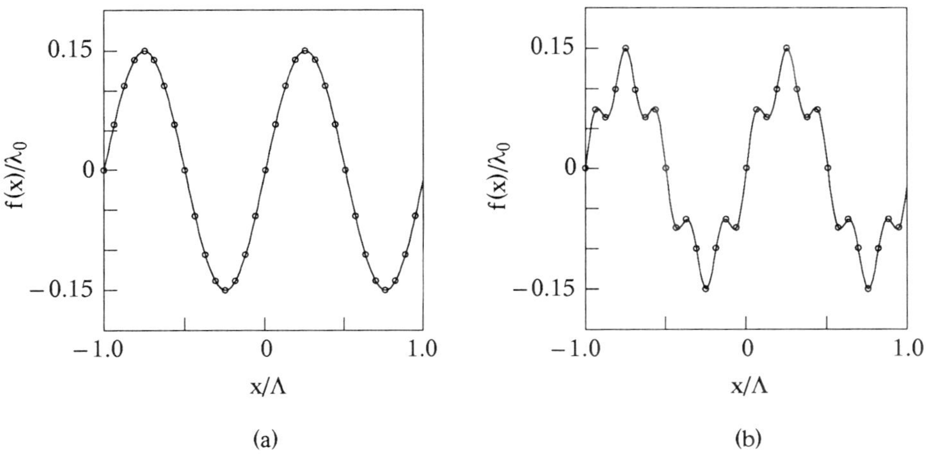

Figure 2. Reconstructions of rough-surface profiles from the scattered field data without noise. (a) Sinusoidal surface. (b) Surface consisting of a fundamental undulation and a higher harmonic. The circle points and the solid curves show the results based on the present method and the true profiles, respectively.

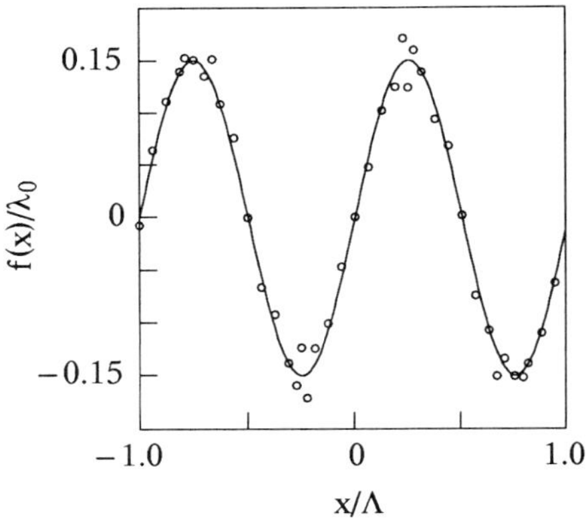

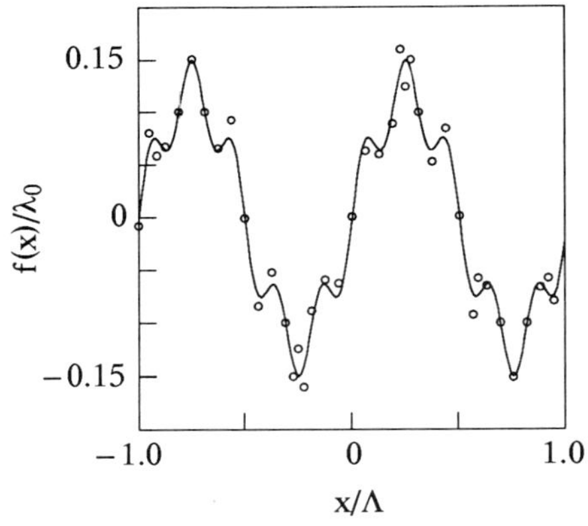

Figure 3. Reconstructions of rough-surface profiles from the scattered field data with signal-to-noise ratio 20 dB. (a) Sinusoidal surface. (b) Surface consisting of a fundamental undulation and a higher harmonic. The circle points and the solid curves show the results based on the present method and the true profiles, respectively.

Figure 3 shows the reconstructions of the same surface profiles as in Fig. 2 from the scattered field data with signal-to-noise ratio 20 dB. The circle points and the solid curves denote the results based on the present method and the true profiles, respectively. In Figs. 3 (a) and 3 (b), the final convergent solutions are the results from 13-th and 12-th iterations. From Fig. 3, it can be concluded that our procedure yields high-quality reconstructions even if the scattered field data contains the measurement error.

4. CONCLUSION

A simple inversion algorithm for reconstructing a perfectly conducting rough surface based on the modified Newton-Kantorovich method has been presented. From the numerical results, it is seen that our procedure provides high-quality reconstructions even for cases where the scattered field data contains the measurement error. It is also shown that the iterative algorithm has the properties of fast convergence.

REFERENCES

[1] F. T. Ulaby, R. K. Moore, and A. K. Fung, Microwave Remote Sensing Active and Passive, Vol. II, Addison-Wesley Publishing Co., London (1982) 922.
[2] R. J. Wombell and J. A. DeSanto, J. Opt. Soc. Am. A, Vol. 8, No. 12, (1991) 1892.
[3] L. V. Kantorovich and G. P. Akilov, Functional Analysis, 2nd ed., translated by H. L. Silcock, Pergamon Press, New York (1982) 524.

Simulation and Design of Applied Electromagnetic Systems
T. Honma (Editor)

Design Considerations for Polarization Conversion/Matching Filters Using Stratified Chiral Slab

A. Kusunoki and M. Tanaka

Department of Electrical and Electronic Engineering, Oita University
700 Dannoharu, Oita 870-11, Japan

Abstract

This paper presents design considerations for polarization conversion/matching filters using a stratified chiral slab, which is located on a dielectric substrate. The perpendicular and parallel power components and the Stokes parameters of the transmitted and reflected waves are computed for the incident wave of linear and circular polarizations. The numerical results demonstrate that the slab acts as efficient filters active at a wide range of frequency.

1. INTRODUCTION

Chiral media have been known in optics under the more common name of optically active materials since the early part of the nineteenth century [1,2]. In recent years, the media have received considerable attention in electromagnetics' community owing to novel applications in the optical, millimeter wave, and microwave regimes. It is well known that [2] the media exhibit interesting properties of polarization birefringence and circular dichroism. Many researchers have studied the interaction of electromagnetic waves with chiral structures [3-6]. The subject continues to be of great interest and practical importance in modern electromagnetics.

It is the purpose of this paper to present design considerations for polarization conversion/matching filters using a stratified chiral slab, which is located on a dielectric substrate. In the analysis, the incident field is assumed to be a plane monochromatic wave with any arbitrary polarization. Note that [1,2] the electric field inside a chiral medium is expressed as a sum of left- and right-circularly polarized (LCP and RCP) plane waves of different phase velocities. Imposing the boundary conditions at the interfaces of the slab, we can obtain the electric field components of the transmitted and reflected waves. For the incident wave of linear and circular polarizations, the perpendicular and parallel power components and the Stokes parameters [7] of the two waves are computed to show the effects of chirality on depolarization properties of the stratified chiral slab. Considering the numerical results, it is seen that the slab acts as efficient filters active at a wide range of frequency.

2. FORMULATION OF THE PROBLEM

Consider a plane monochromatic wave with any arbitrary polarization incident from free space upon a stratified chiral slab, as shown in Fig. 1. The slab is located on a dielectric substrate with material properties ε_s and μ_s. The m-th dielectric chiral layer (m=1, 2, ···, N) of the slab is described by the following constitutive relations [2,3]:

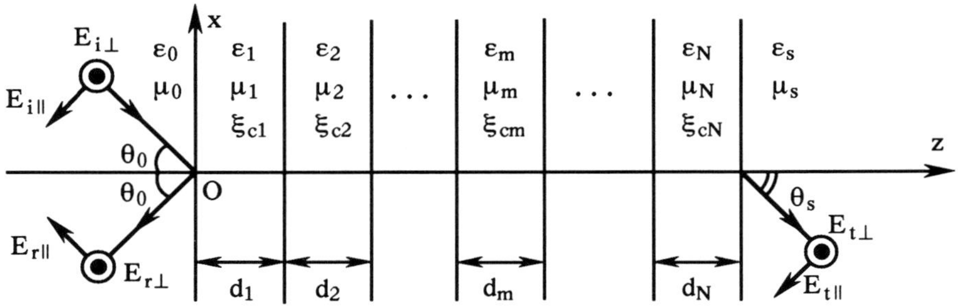

Figure 1. Geometry of stratified chiral slab.

$$\mathbf{D}_m = \varepsilon_m \mathbf{E}_m + j\,\xi_{cm}\,\mathbf{B}_m\ , \tag{1}$$

$$\mathbf{H}_m = j\,\xi_{cm}\,\mathbf{E}_m + (1/\mu_m)\,\mathbf{B}_m\ , \tag{2}$$

where ε_m, μ_m, and ξ_{cm} are the permittivity, permeability, and chirality admittance of the layer, respectively. Note that ξ_{cm} denotes the degree of chirality [2].

The incident electric field with any arbitrary polarization can be written as

$$\mathbf{E}_i = [-E_{i\parallel}(\cos\theta_0\,\mathbf{u}_x + \sin\theta_0\,\mathbf{u}_z) + E_{i\perp}\,\mathbf{u}_y]\exp[jk_0(z\cos\theta_0 - x\sin\theta_0)]\ , \tag{3}$$

where $\mathbf{u}_\alpha$ (α = x, y, z) and k_0 are the unit Cartesian vectors and the wave number of free space, respectively. Note that the subscripts $\perp$ and $\parallel$ refer to the perpendicular and parallel components . The incident power is now normalized to unity, thus $|E_{i\perp}|^2 + |E_{i\parallel}|^2 = 1$.

Lets us denote by $\mathbf{E}_t$ and $\mathbf{E}_r$ the transmitted and reflected electric fields, respectively. $\mathbf{E}_t$ and $\mathbf{E}_r$ are similarly expressed in terms of their perpendicular and parallel field components. It is well known that [1,2] the electric field inside a chiral medium may be represented as a sum of the LCP and RCP plane waves of different phase velocities. Thus there are four waves in the m-th chiral layer, two propagating toward the right interface and the other two propagating toward the left interface. Using a 4×4 matrix [6], which relates the four amplitudes of the LCP and RCP electric fields in the (m+1)-th layer to those in the m-th layer, one may readily obtain the field components of the transmitted and reflected waves.

The transmitted and reflected powers are, respectively, expressed as

$$P_t = P_{t\perp} + P_{t\parallel} = \frac{\eta_0\cos\theta_s}{\eta_s\cos\theta_0}\left(|E_{t\perp}|^2 + |E_{t\parallel}|^2\right), \tag{4}$$

$$P_r = P_{r\perp} + P_{r\parallel} = |E_{r\perp}|^2 + |E_{r\parallel}|^2\ , \tag{5}$$

where $\eta_0 = \sqrt{\mu_0/\varepsilon_0}$ and $\eta_s = \sqrt{\mu_s/\varepsilon_s}$ are the intrinsic impedances of free space and the substrate, respectively.

Note that [7] the Stokes parameters uniquely describe every possible state of polarization of a plane wave. The Stokes parameters are defined as

$$S_0 = |E_\perp|^2 + |E_\parallel|^2, \quad (6a) \qquad S_1 = |E_\perp|^2 - |E_\parallel|^2, \quad (6b)$$

$$S_2 = 2\,\mathrm{Re}\left(E_\perp E_\parallel^*\right), \quad (6c) \qquad S_3 = 2\,\mathrm{Im}\left(E_\perp E_\parallel^*\right). \quad (6d)$$

Here Re and Im are the real and imaginary parts, and the asterisk denotes complex conjugation. It is easily seen from Eq. (6) that $S_1^2 + S_2^2 + S_3^2 = S_0^2$. It should be remarked that S_3 determines the rotation sense of the wave [7]. The plus (minus) sign of S_3 corresponds to left (right) elliptic polarizations, and $S_3 = 0$ to linear polarizations. Especially, the left-handed (right-handed) circular polarization is denoted by $S_3 = S_0$ ($S_3 = -S_0$). In the following, let us denote the Stokes parameters of the transmitted and reflected waves as $S_{t\beta}$ and $S_{r\beta}$ ($\beta = 0, 1, 2, 3$), respectively.

3. NUMERICAL RESULTS

Numerical results are presented for the incident wave of linear and circular polarizations. Only the transmission characteristics are shown here on account of space consideration. The refractive index of the dielectric substrate is now set as $n_s = \sqrt{\varepsilon_s / \varepsilon_0} = 1.52$. Let us consider the thirteen-layered structure :

1.0 | LMMLHHLHHLMML | 1.52,

where we employ the following parameters : $n_L = \sqrt{\varepsilon_L / \varepsilon_0} = 1.35$, $n_M = \sqrt{\varepsilon_M / \varepsilon_0} = 3.5$, $n_H = \sqrt{\varepsilon_H / \varepsilon_0} = 4.0$, $\xi_{cL} = \xi_{cM} = \xi_{cH} = 4.54 \times 10^{-4}$ S , $n_L d_L = n_M d_M = n_H d_H = \lambda_c / 4$, and $\mu_L = \mu_M = \mu_H = \mu_s = \mu_0$. Here λ_c denotes the wavelength in free space at the central frequency f_c.

For the incident wave of perpendicular polarization, i.e., $E_{i\parallel} = 0$, the power components and the Stokes parameters of the transmitted wave are shown in Fig. 2 as a function of the angle θ_0 of incidence. In this case, $P_{t\parallel}$ and $P_{t\perp}$ correspond to the cross- and co-polarized power components, respectively. Here we consider the case of $f/f_c = 1.0$. From Fig. 2, one can see that the stratified chiral slab passes only the cross-polarized power component of the transmitted wave for θ_0 less than 15°. Thus the stratified chiral slab acts as a polarization-

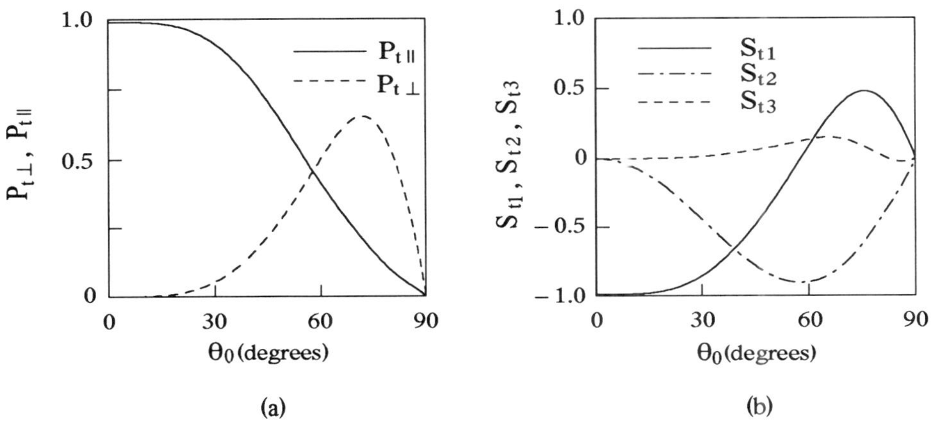

Figure 2. Power components and Stokes parameters of transmitted wave for the incident wave of perpendicular polarization. (a) Power components. (b) Stokes parameters.

conversion transmission filter and an antireflection filter. Furthermore, the cross-polarized power component is dominant over a wide range of the angle of incidence. It is also shown that S_{t3} nearly equals to zero independent of θ_0, and then the transmitted wave is almost linearly polarized.

Figure 3 illustrates the same parameters as in Fig. 2 versus frequency for the incident wave of left-handed circular polarization. Now we consider the case of normal incidence. From Fig. 3, it is seen that the transmitted wave is decomposed into the perpendicular and parallel field components of the same amplitude. Then the slab may be used as a circular-to-linear polarization conversion/matching element at the frequency region around $f/f_c = 1.0$. In addition, the polarization state of the transmitted wave remains left-circularly polarized since $S_{t0} = S_{t3} > 0$.

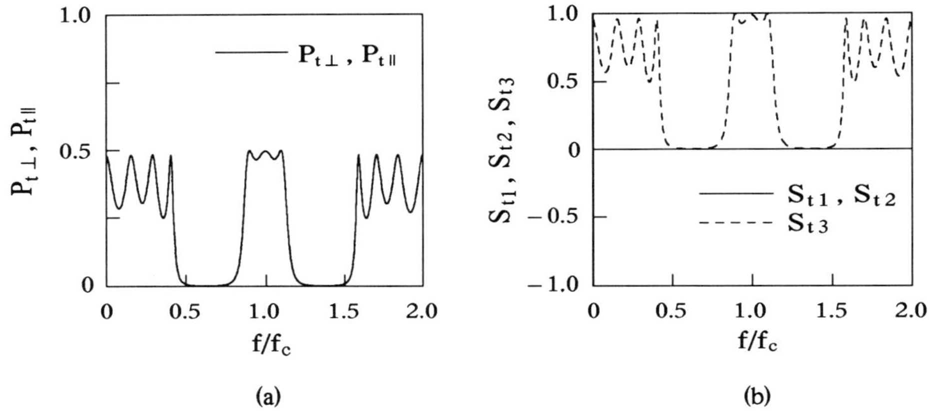

Figure 3. Power components and Stokes parameters of transmitted wave for the incident wave of left-handed circular polarization. (a) Power components. (b) Stokes parameters.

4. CONCLUSION

Design considerations for polarization conversion/matching filters using a stratified chiral slab have been presented. The perpendicular and parallel power components and the Stokes parameters of the transmitted and reflected waves are computed for the incident wave of linear and circular polarizations. The numerical results demonstrate that the slab acts as efficient filters active at a wide range of frequency.

REFERENCES

[1] A. Lakhtakia, V. K. Varadan, and V. V. Varadan, Time-Harmonic Electromagnetic Fields in Chiral Media, Springer-Verlag, Berlin (1989).
[2] N. Engheta and D. L. Jaggard, IEEE AP-S Newslett., Vol. 30, No.5 (1988) 6.
[3] S. Bassiri, C. H. Papas, and N. Engheta, J. Opt. Soc. Am. A, Vol. 5, No. 9 (1988) 1459.
[4] A. Lakhtakia, V. V. Varadan, and V. K. Varadan, J. Opt. Soc. Am. A, Vol. 6, No. 1 (1989) 23.
[5] S. H. Yueh and J. A. Kong, J. Electro. Waves Appl., Vol. 5, No. 7 (1991) 701.
[6] M. Tanaka and K. Sato, Trans. IEICE, Vol. J75-C-I, No. 10 (1992) 677.
[7] H. Mott, Polarization in Antennas and Radar, John Wiley & Sons, New York (1986).

Simulation and Design of Applied Electromagnetic Systems
T. Honma (Editor)

Numerical Analysis of a Microwave Field in the Chamber with Multilayer Walls of Microwave Absorbing Materials

S. Shirakawa ,* M. Koizumi* and H. Yokoyama**

*Energy Research Laboratory, Hitachi Ltd. 7-2-1 Omika-cho, Hitachi-shi, 319-12, Japan

**Hitachi Engineering Co. Ltd. 3-2-1 Saiwai-cho, Hitachi-shi, 317, Japan

Abstract

The electromagnetic field in an anechoic chamber depends on the geometrical shape of the chamber, the arrangement of the absorbing material and impedance of the antennas. This electromagnetic field is analyzed using the moment method assuming that the wedged wall in anechoic chamber is approximated as a flat wall giving the reflection coefficient of the wedged microwave absorber. The numerical results show that this formulation is useful if the calculation point is at a sufficient distance from the wall.

1. Introduction

To measure electromagnetic noise from electrical machinery, many microwave anechoic chambers have been built. The electromagnetic field in an anechoic chamber depends on the geometrical shape of the chamber, arrangement of the absorbing material and impedance of the antennas. Moreover, since the absorbing rate on the walls decreases with longer wave length, reflections from walls must be considered for accurate calculation of the electromagnetic field at low frequencies. The conventional moment method[1] can give accurate results to problems with an ideal conducting medium, but not to problems involving a medium with loss. On the other hand, computational schemes based on geometrical optics can treat reflection on a lossy medium, but applicability is limited to a chamber with highly absorbing walls. This is because these schemes exclude self-consistent evaluation of antenna impedance, arising from electromagnetic interference which must be taken into account at low frequencies, and it is difficult to give sufficient accuracy in calculating many traces of the reflecting rays until the strength of the rays is sufficiently reduced[2]. In this paper, we propose a formulation of the moment method which is applicable to calculation of the microwave field in a chamber enclosed with a lossy medium assuming that the wedged wall in the anechoic chamber is approximated as a flat wall giving the reflection coefficient of the wedged microwave absorber. This formulation has an advantage of considering the self-consistent antenna impedance and infinitely reflecting waves between walls, which cannot be calculated by a ray tracing method. Finally, we compare the calculated microwave field in the anechoic chamber with measurements.

2. Model of Microwave Absorber's Wall

When the wave length is sufficiently longer than the height of microwave absorbers in an anechoic chamber, the wedges of the absorber can be approximated as a flat multilayer of the lossy medium as shown in Fig.1. Then, each layer's electrical parameter is given by

$$\varepsilon_n = \frac{\text{(absorber cross section)}}{\text{(total cross section)}}(\varepsilon - \varepsilon_o) + \varepsilon_o \quad (1), \qquad \mu_n = \frac{\text{(absorber cross section)}}{\text{(total cross section)}}(\mu - \mu_o) + \mu_o \quad (2).$$

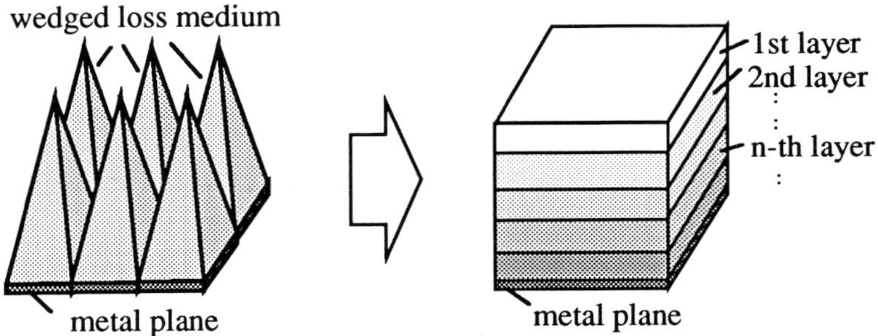

(a) Wedged microwave absorber (b) Flat multlayer with lossy medium

Fig.1 Model of microwave absorber wall

where ε_o is the permittivity and μ_o, the permeability in the vacuum; ε is the permittivity and μ, the permeability in the absorber; ε_n is the permittivity and μ_n, the permeability in the n-th layer. From these electric parameters, characteristic impedance of each layer and reflection coefficient on surface can be written as follows:

Impedance:

$$Z_{TEn}=\frac{Z_0\mu_0\mu_n}{\sqrt{\varepsilon_n\mu_n-\varepsilon_0\mu_0\sin\theta_n}} \quad (3) \qquad Z_{TMn}=\frac{Z_0\sqrt{\varepsilon_n\mu_n-\varepsilon_0\mu_0\sin\theta_n}}{\varepsilon_0\varepsilon_n} \quad (4)$$

$$Z_{n-1}=Z_{TEn\text{ or }TMn}\frac{Z_n+Z_{TEn\text{ or }TMn}\tanh j\,k_nd_n}{Z_{TEn\text{ or }TMn}+Z_n\tanh j\,k_nd_n} \quad (5) \qquad k_n=\omega\sqrt{\varepsilon_n\mu_n-\varepsilon_0\mu_0\sin\theta_n} \quad (6)$$

Reflection coefficient:

$$R_{TE}=\frac{Z_1-Z_o/\cos\theta_1}{Z_1+Z_o/\cos\theta_1} \quad (7) \qquad R_{TM}=\frac{Z_1-Z_o\cos\theta_1}{Z_1+Z_o\cos\theta_1} \quad (8)$$

where $Z_{TMn\text{ or }TEn}$ is the characteristic impedance of the n-th layer against the TE or TM wave; Z_0, the characteristic impedance of the free space, Z_n, the surface impedance of the n-th layer; k_n, the propagation constant in the n-th layer; $j=\sqrt{-1}$; ω, the angle frequency of the microwave, θ_n, the incident angle into the n-th layer and $R_{TM\text{ or }TE}$, the reflection coefficient on the first layer's surface against the TE or TM wave. The $\theta_{n(2\cdots N)}$ is calculated by Snell's law of lossy medium. The Z_1 in Eqs. (7) and (8) is calculated from a recurrence formula Eqs. (5).

3. Formulation of Moment Method

For a perfect conductor, the tangential component of electric intensity is zero. With an imperfectly conducting medium, the difference **E**' between tangential components of incident and reflective electric intensity is not zero. This electric intensity **E**' is absorbed into the lossy medium. This difference can also be expressed by a reflection coefficient and incident electric intensity. Then, the boundary condition on the surface of the loss medium is given by

$$\mathbf{E}_{inc}+\mathbf{E}_{rad}=(1+R)\mathbf{E}_{inc} \quad (9)$$

where $\mathbf{E}_{inc}$, $\mathbf{E}_{rad}$, the tangential components of incident and reflective electric intensities and R is the reflection coefficient. The difference between tangential components of incident and reflective electric intensities drives currents on the surface of the loss medium. Then, using this difference as driving voltage, a moment method matrix equation is written by

$$[-R_{ij}][Z_{ij}][I_j]=[V_i] \tag{10}$$

where $[Z_{ij}]$ is the impedance matrix; $[I_j]$, the unknown current vector; $[V_i]$, the driving voltage vector; and $[R_{ij}]$, the reflection coefficient matrix. The reflection coefficient R_{ij} is calculated by R_{TM}, R_{TE} and TE and TM components of the incident wave. We note that when the unknown current number i=j, then $R_{ij}=-1$. In computation of the electromagnetic field in the chamber with wedged lossy walls, we use the reflection coefficient of Eqs. (7) and (8).

4. Numerical Results and Discussion

We tested this formulation by comparing numerical and measured results for an anechoic chamber with a wedged microwave absorber. Figure 2 shows the shapes and size of the anechoic chamber and the position of the transmitting antenna. The height of the transmitting antenna is 1 meter. The receiving antenna moves in over a distance of 1 to 4 m. The ds is distance between the wall A to the transmitting antenna. The absorber is composed of a wedged absorber joined to a ferrite plate. The parameters for the wedged absorber and ferrite plate used in anechoic chamber are summarized in Table 1. Figure 3 shows the rate of reflected to incident amplitude on absorbers calculated by Eqs. (7)-(8). It is found that the pseudo Brewster angle which is the angle at the minimum rate of reflected to incident amplitudes appears at TM wave. We simulate the chamber covered with absorbers except for its conductive floor. Figure 4 shows the attenuation at the receiving point in the the chamber, where the attenuation is the ratio of transmitting power to receiving power. Then, the ds is 2.85 m. From Fig.4, comparing the numerical results with measured ones shows each pattern is similar, but the difference between them increases near the ceiling. This is because the electromagnetic field near the ceiling strongly depends on the geometrical shape of the wedged absorber. Consequently, this formulation is useful if the calculation point is at a sufficient distance from the wall.

5. Conclusion

In this paper, we analyzed the microwave field in the anechoic chamber having walls approximated as a flat surface and giving the reflection coefficient of the wedged microwave absorbers by using the moment method. From the numerical results, the pattern of the attenuation was seen to be similar to the measured one, but the difference between them increased near the ceiling. We consider that this formulation is useful if the calculation point is at a sufficient distance from the wall.

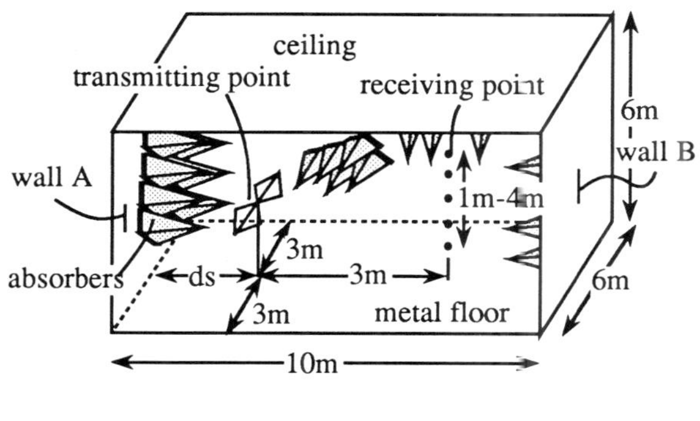

Fig.2 Size of numerical model

Table 1 Physical parameters

parameter	wedged absorber	ferrite plate
height(mm)	1500	8
permittivity($\varepsilon / \varepsilon_0$) at 90MHz	11 - j17	15 - j0.2
permeability(μ / μ_0) at 90MHz	1.0	18 - j83

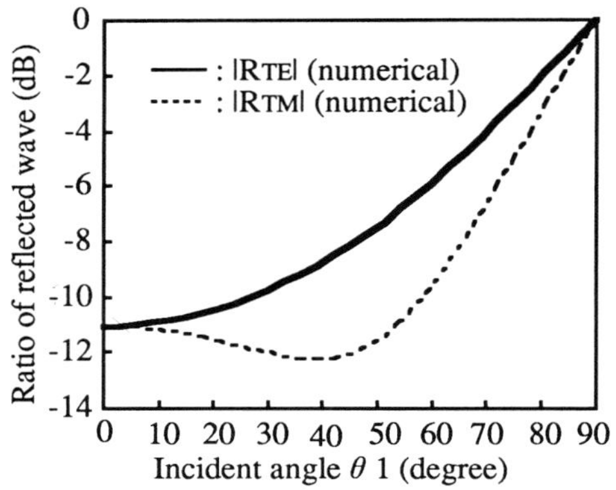

Fig.3 Ratio of reflected wave on absorber

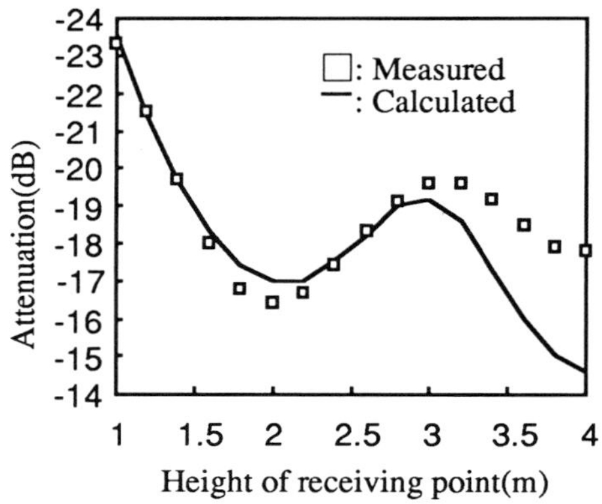

Fig.4 Attenuation in chamber
(Frequency: 90MHz
Transmitting antenna: Horizontal)

References

[1] J. Moore and R. Pizer, "Moment Method in Electromagnetics", Research Studies Press, Letchworth, Hertfordshire, England (1983)

[2] K. Nishimura, et al., "Numerical Simulation of Anechoic Chamber Characteristics for EMI", IEICE Japan, EMCJ(in Japanese), Vol.88, No.5, pp21-27, (1988)

Simulation and Design of Applied Electromagnetic Systems
T. Honma (Editor)

High speed shuttering characteristics of proximity focused image intensifiers

Brian Landberg[a], Makoto Kagata[a], Satoshi Tomioka[a], Takeaki Enoto[a] and Ichiro Fukai[b]

[a]Dept. of Nuclear Engineering, Faculty of Engineering, Hokkaido University. Kita 13, Nishi 8, Kita-ku, Sapporo, 060, Japan

[b] Dept. of Electrical Engineering, Faculty of Engineering, Hokkaido University. Kita 13, Nishi 8, Kita-ku, Sapporo, 060, Japan

Abstract

In view of the increasing importance of recording and measuring the processes of instantaneous reactions such as laser fusion, the precision of existing shuttering technology is no longer sufficient. In this paper, a system including a proximity focused image intensifier and an external transparent electrode is proposed. Using this apparatus, a shuttering rate of 100ps-order is realized and a comparison of the new design with that of the photoelectric plate-based shuttering system is included. Also, two methods of mathematical analysis, FD-TD and R-Line representation, are used to successfully simulate the shuttering process.

1. ABOUT HIGH SPEED CAMERAS

There are two kinds of high-speed cameras currently being used to record instantaneous phenomena. These are the framing (cine) style camera and the streak camera. The latter uses elongated, fixed strips of film across which the camera itself does the moving, thus forming narrow (1-dimensional) images that are continuous in time. The former type, the framing camera, is the one used in our research. It follows the same logic as cameras used in every-day life, except that the extremely high framing rate demands a more responsive shuttering mechanism. Since, mechanical devices are not physically capable of supporting nanosecond-order framing, we use an electronic pulse to control the shuttering. The required specifications for our ultra high-speed camera are as follows. Spacial resolution: 10 line- pairs/mm. Light spectrum range: visible light or x-rays. Shuttering pulse voltage: 600V. Shuttering time (per frame): 100ps - 1ns. Note that it is extremely difficult to produce a high-voltage, high frequency pulse as necessary. However, with current available technology, it is possible to produce a 100ps pulse at up to approximately 1000V. In the case of high-speed framing cameras, there are three basic shuttering designs, which are listed as follows: Internal mesh; Low R photocathode; and External mesh. In the internal mesh design, the shuttering pulse is applied to the input surface of the photocathode. Since the photocathode has a high resistance that impedes the smooth propogation of a fast pulse, the Low R design was put together using a photocathode consisting of alternative less resistant materials. The new materials, however, are thicker and not completely transparent, so that the resolving ability of the apparatus is insufficient

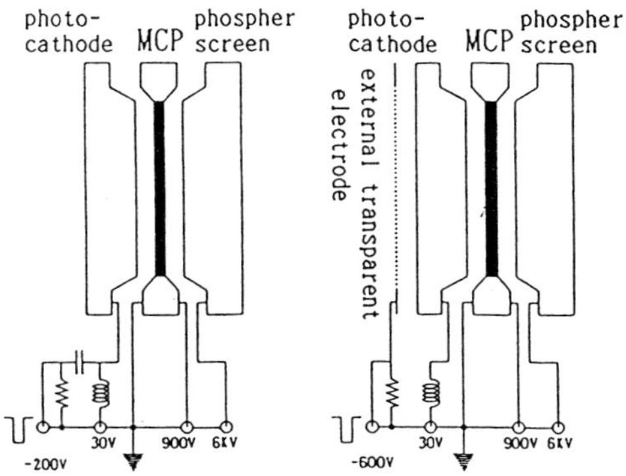

Figure 1: Diagram of two proximity-focused image intensifier designs. In the left design (Internal mesh), the shuttering pulse is applied to the input surface of the photocathode. The right design (External mesh) incorporates an external transparent electrode to allow for smoother pulse propogation and improve resolution by eliminating shadows.

and was not used in our experiments. Additionally, a shadow cast on the MCP by the internally positioned electrode in both designs acts to further worsen resolution. Thus, the third, External mesh assembly was designed. In this assembly, the pulse is applied to an externally positioned electrode which allows for smooth propogation of a fast pulse and does not weaken resolution.

2. PRINCIPLE OF THE ULTRA HIGH-SPEED CAMERA

The camera apparati used in our experiments employ a proximity- focused image intensifier to enable a shuttering rate of nanosecond order. The principle set-ups of the two shuttering systems used are depicted in fig.1. As shown, while the input surface of the MCP (micro-channel plate) is grounded, the photocathode can be supplied with a positive (+30V) voltage bias, leaving the image intensifier cut-off. Thus, by applying a negative, reverse-biased pulse to the photocathode, it is possible to control the shuttering with nanosecond precision. However, if the limitation of slow pulse propogation through the photocathode plate (apparatus on left) is avoided, the shuttering speed can be further increased. By affixing a low-resistance, external transparent electrode to the front surface of the image intensifier (apparatus on right), we are able to circumvent this delay to accomodate sub-nanosecond framing speeds.

Placing the shuttering control externally also serves to avoid shadows and practically eliminate the "iris-effect" that had previously made sub-nanosecond framing difficult. Note that the use of the external electrode places the distance between the shutter mechanism and the MCP at three times that of former designs, such that the applied pulse must also be increased from -200V(left-design voltage) to -600V(right-design voltage). Further, in the newly designed apparatus the external electrode can be divided into parts, resulting in a compound structure of two hemi-intensifiers or conceivably, of four quadrant-intensifiers. Thus, with staggered pulses, shuttering speeds of 100ps-order are feasable.

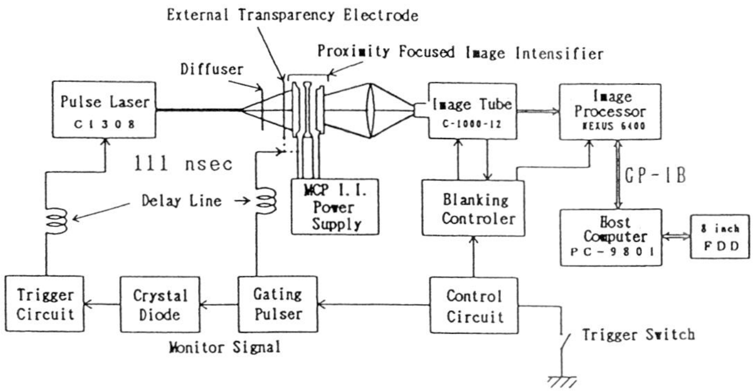

Figure 2: Experimental set-up for measuring shuttering characteristics of the high-speed shuttering system. Ultra-short single laser pulses (50ps) are shot through the apparatus, one at a time, as the source light. Reflected images display the light intensity distribution throughout the electrode at various steps in the shuttering process.

3. MEASURING SHUTTERING CHARACTERISTICS

The experimental set-up for measuring the shuttering characteristics for the ultra high-speed camera is shown in fig.2. These results were obtained by the following experimental procedure. According to the gating control of the 2.2 nanosecond FWHM pulse, the shuttering mechanism alternately opens and closes. In addition to this pulse, a helium neon laser is employed to produce single pulses of even higher frequency (approximately 50ps). These pulses, when shined upon the proximity focused image intensifier, reflect where the shutter is closed and pass through where the shutter is open. Thus, the images that come back to the computer image processor exhibit the shuttering characteristics at various instants in the 2.2 ns gating pulse's propogation.

4. RESULTS

As shown in fig.3, experimental results using the external transparent electrode exhibit both marked improvement over those gained using previous proximity-focused image intensifier systems and good agreement with the results obtained by Finite Difference computer analysis. Since the two-dimensional analysis ignores any interference of the photocathode with the shuttering pulse, a separate, three-dimensional FD-TD computer-simulated analysis is also implemented to include possible effects from the photocathode. Displayed in fig.4, these results agree well with both experimental and Finite Difference analysis results. In previous experiments it is evident that the pulse first reaches the edge of the photocathode and moves radially toward the center (the "iris-effect"). However, in experiments using the external electrode-controlled gating, the "iris-effect" is nearly eliminated due to the pulse's rapid propogation through the low-resistance electrode. Additionally, the use of an internal electrode in previous experiments resulted in a shadow, cast upon the MCP when the source traversed the electrode. The new external electrode design removes this shadow-effect, and thus significantly improves the resolving ability of the camera system.

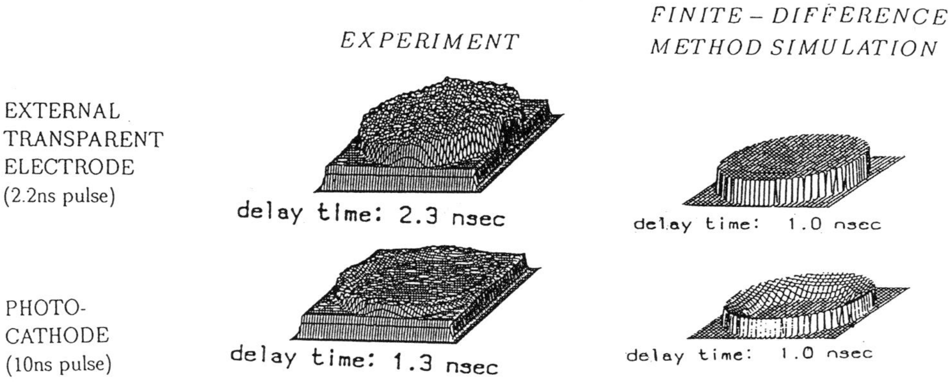

Figure 3: Comparison of experimental and Finite Difference results. Experimental results are measurements of light intensity distribution in the electrode, while analytical results are measurements of the electrode's electric field. These two quantities are directly proportional to each other. Note that the 2-D, finite difference results are based upon the simplified assumption that the electric field is completely unaffected by the centrally positioned photocathode. Axial symmetry is also assumed.

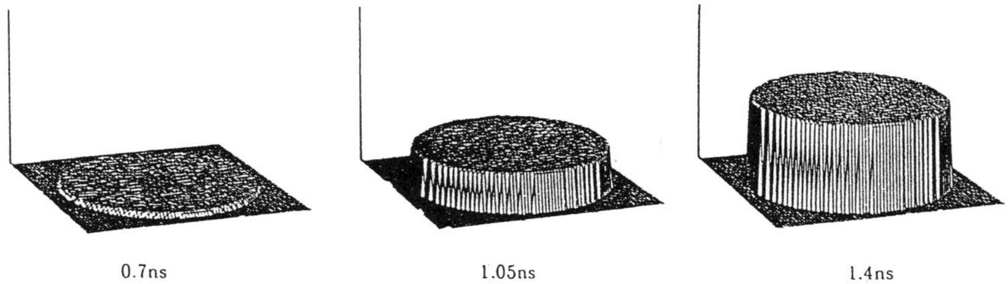

Figure 4: FD-TD analysis results. This is a 3-dimensional calculation which does not assume axial symmetry (it is found that the validity of such symmetry is lost with an input pulse of less than 1ns). The FD-TD analysis also takes into account any influence of the photocathode upon the electric field, although this influence is found to be negligible.

Simulation and Design of Applied Electromagnetic Systems
T. Honma (Editor)

Numerical analysis of ultra high-speed shuttering characteristics using finite-difference time-domain method

Makoto Kagata[a], Brian D.Landberg[a], Satoshi Tomioka[a],
Takeaki Enoto[a] and Ichiro Fukai[b]

[a]Dept. of Nuclear Engineering, Faculty of Engineering, Hokkaido University.

[b]Dept. of Electrical Engineering, Faculty of Engineering, Hokkaido University.
Kita 13, Nishi 8, Kita-ku, Sapporo, 060, Japan

Abstract

In the study on laser driven fusion, the most important problem is to obtain high-resolution images of the plasma implosion process. It is thus necessary to use a high-speed framing camera system. This paper presents a numerical analysis of ultra high-speed shuttering characteristics using the FD-TD method. These analytical results show good agreement with experimental results, and using this method, we determine the cause of leakage of shuttering fields from one of the divided electrodes to the other. Thus, it is clear that the FD-TD method is valid and useful for analysis of shuttering characteristics of high-speed camera systems.

1. Introduction

Shuttering of the proximity focused image intensifier(PFII) is achieved by applying a voltage pulse to the electrode, so that shuttering characteristics depend upon propagation characteristics of the pulse through the control electrode.

But, the photocathode has such high resistivity that a short pulse can hardly propagate, which makes shuttering characteristics worse. So, we proposed adding a less resistive, external transparent electrode(ETE). This drastically improves shuttering characteristics.

Shuttering triggered by applying a pulse to the ETE has another advantage. If we divide the ETE into parts, and apply staggered, delayed pulses to these divided ETEs, then we get multiple images to fire alternately. Shuttering using PFII cannot operate at a high repetition rate because of the long relaxation time of the phospher plate and limited capability of the data processing system. However, it is easy to split the optical source beam into two, and divide the ETE. Thus, by applying staggered delayed pulses to these divided ETEs, we can get sequential images of the targets development in time. Fig.1 shows a conceptual diagram of the multiframe camera system and geometry of the divided ETE. In previous work, we have treated the electrode as radially symmetrical, but this

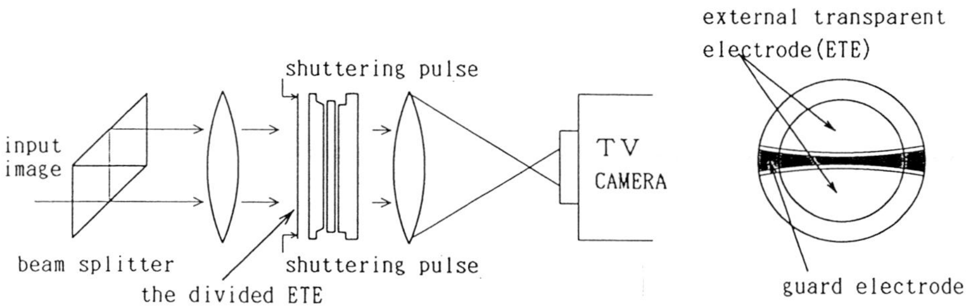

Figure 1: concept of multiframe camera system and divided ETE

assumption does not hold when we use the divided ETE. Furthermore, in order to include the effects that result from the actual 3-dimensional distribution of the control electrode, we employed the 3-dimensional FD-TD method to analyze the pulse propagation.

2. Analytical model

The finite-difference time-domain(FD-TD) method was first proposed by Yee[1]. It is a popular technique that solves Maxwell's curl equations by converting them into difference equations,which are solved in sequential time steps by alternately calculating the electric and magnetic fields at nodal points in the computational domain[1].

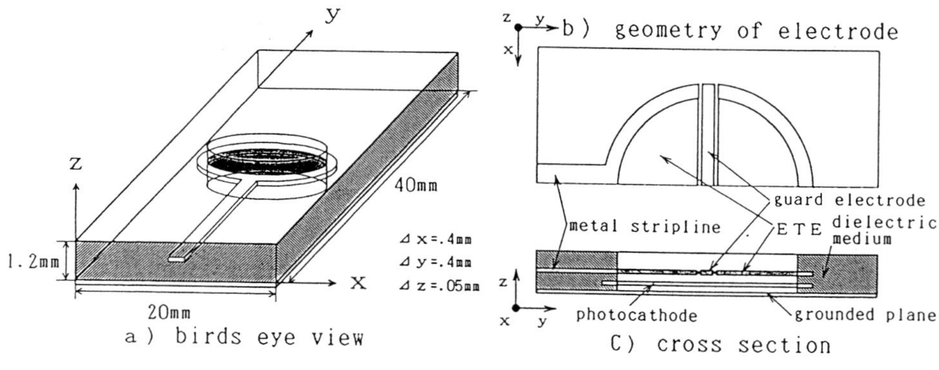

Figure 2: schematic diagram for computational domain

Fig.2 shows the computational domain for this calculation. It consists of a grounded plane, a dielectric medium, and a metal stripline(embedded in the dielectric medium). The shuttering process takes place in the field between the photocathode and the microchannel plate surface plane(MCP_in), and since the shuttering pulse is applied to the ETE, the computational domain includes the photocathode, ETE and MCP_in. The applied pulse propagates along the microstripline, reaches the ETE, and forms the shuttering field. The center of the model is layered as shown in Fig.2(c), and the electrodes are supported by the dielectric medium.

FD-TD formulas include the constitutive parameters μ, ϵ and σ at all points. Yet we don't know the thicknesses of the photocathode and ETE, which are necessary in calculating these values. We can, however, measure their surface resistivities. Thus, using a reasonable estimate of the thicknesses(Δz) we are able to determine the parameters, which would yield the measured resistivity.

All tangential components of the electrical field and normal components of the magnetic field vanish on the metal stripline because of assumption that it is a perfectly conducting surface. This also holds for the grounded plane.

We operate PFII in open space, therefore, artificial boundaries must be employed to truncate the mesh. We employ an absorbing boundary condition[2] as the artificial boundary condition. The absorbing boundary condition is classified into first-order and second-order approximations. We used the first approximation to calculate boundary values in this research.

3. Numerical results

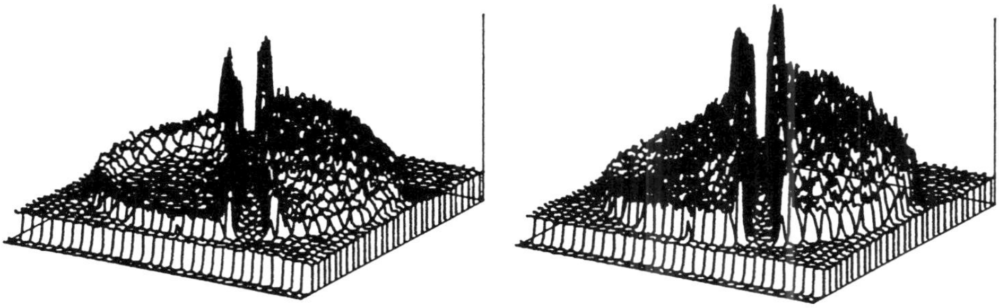

t=4.1nsec. t=4.5nsec.

Figure 3: experimental results

Fig.3 and Fig.4 show experimental results[3], and numerical results, respectively. In experiment, we divided the ETE as shown in Fig.1, and applied a 2.2nanosecond FWHM pulse to one of these parts. The other side of the divided ETE and the guard electrode were biased to prevent accerelating photoelectrons. Then we fired an ultra-short laser pulse synchronized to the shuttering pulse. By staggering the former to coincide with several sequential instants in the shuttering cycle, it is possible to record and measure the shuttering process. The distribution of output light intensity with change in time is shown in Fig.3.

The shuttering pulse was applied to only the right electrode in this figure, however, output light appeared on the left side. In other words, there was undesirable leakage of the shuttering field.

Fig.4 shows the shuttering field distribution with change in time when a 2.2nanosecond FWHM pulse is applied to the right side of the divided ETE. In these results, we note that

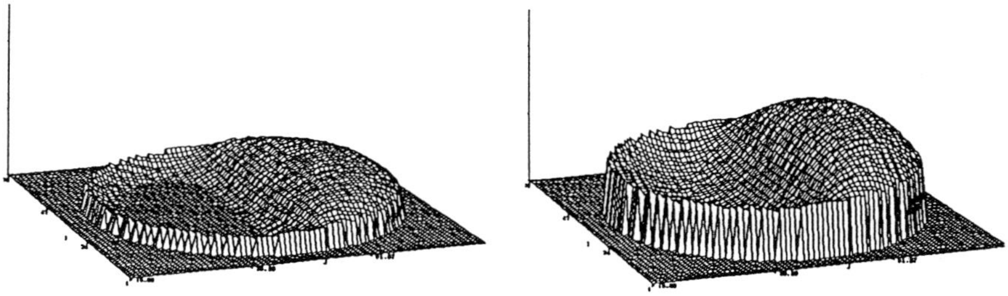

t = 1 . 7 2 n S E C   t = 2 . 3 0 n S E C

Figure 4: numerical results

there is leakage of the shuttering field to the reverse-biased part. Leakage characteristics both on the experimental and numerical results are similar to the iris effect that appears in when the shuttering pulse is applied to the photocathode. In simulation, which neglects conductivity of the photocathode (and its periphery electrode), there is no leakage. Thus, we found that the shuttering field leaks by way of the photocathode (especially its periphery electrode).

4. Conclusion

Shuttering characteristics of ultra high-speed cameras with a divided ETE were analyzed by the FD-TD method. Numerical results show good agreement with experimental ones. From numerical results, we found that the leakage of the shuttering field is a result of the photocathode. It has been shown that the FD-TD approach is a valid and useful method for calculating the transient response of ultra high-speed cameras.

References

[1] Kane S. Yee. "Numerical Solution of Inital Boundary Value Problems Involving Maxwell's Equations in Isotropic Media", *IEEE TRANS. ON ANTENNAS AND PROPAGATION*, VOL. AP–14,NO. 3,pp. 302–307

[2] Gerrit Mur "Absorbing Boundary Conditions for the Finite–difference Approximation of the Time–Domain Electromagnetic–Field Equation", *IEEE TRANS. ON ELECTROMAGN. COMPAT.* VOL. EMC-23,NO. 4,pp.377–382

[3] Takeaki Enoto and Kazuo Haruyama "High Speed Gating of the Proximity Focused Channel Plate Intensifier with an Additional External Electrode and MTF Characteristics" *KAKUYUGO KENKYU* VOL. 62,NO. 6,pp.467–477

Simulation and Design of Applied Electromagnetic Systems
T. Honma (Editor)

Unified formulation for vector and scalar potential fields for Lorentz gauge condition by using equivalent circuits

N. Yoshida

Department of Electrical Engineering, Faculty of Engineering,
Hokkaido University, Sapporo 060, Japan

ABSTRACT

In this paper, an unified time-dependent formulation for both the vector potential and the scalar potential is proposed by using the Lorentz gauge condition including loss term. To obtain the equivalent spatial networks of the scalar field, in addition to the electric or magnetic field variable as the current variable, 'F' and 'F * ' functions are newly defined as the voltage variables for electric and magnetic scalar fields, respectively, based on the analogy of the scalar potential field to the velocity potential field in the fluid dynamics.

1. INTRODUCTION

In the analysis of electromagnetic fields, the vector potential has important roles especially when current sources exist. But when true charges also exist, the combined analysis considering scalar fields becomes essential. We have recently proposed a new numerical method for vector analysis of three-dimensional electromagnetic fields in the time domain. This is known as the Spatial Network Method (SNM)[1]. I have already shown that SNM can be expanded to the vector potential fields with the Coulomb's gauge condition by defining a new set of fundamental equations, which are similar to the Maxwell's equation, by using not only the magnetic vector potential but also the electric vector potential [2-3].

In this paper, an unified time-dependent formulation for both the vector potential and the scalar potential is proposed by using the Lorentz gauge condition including loss term [4]. To obtain the equivalent spatial networks of the scalar field, in addition to the electric or magnetic field variable as the current variable, 'F' and 'F * ' functions are newly defined as the voltage variables for electric and magnetic scalar fields, respectively, based on the analogy of the scalar potential field to the velocity potential field in the fluid dynamics.

2. UNIFIED FORMULATION OF SCALAR AND VECTOR POTENTIAL FIELDS BY LORENTZ GAUGE

It is proved that in the spatial network shown in Fig. 1 the wave field for the vector potentials can be simulated by taking the scalar potential fields into consideration. The characteristic equation for the magnetic vector potential "A" and the electric vector potential "S" in loss field is defined as follows. In Fig. 2. the fundamental connection in the network is illustrated in detail.

$$\nabla \times A = \sigma^* S + \mu_0 \frac{\partial S}{\partial t} \quad (\equiv B) \tag{1a}$$

$$\nabla \times S = -\sigma A - \varepsilon_0 \frac{\partial A}{\partial t} \quad (\equiv D) \tag{1b}$$

Here, σ^* is the conductivity for the hypothetical magnetic currents. The following abbreviations are used ; $\Delta d = \Delta x = \Delta y = \Delta z$, $\Delta_t = \partial / \partial t$, $\Delta_t^2 = \partial^2 / \partial t^2$ $\Delta_x = \partial / \partial x$, $\Delta_y = \partial / \partial y$, $\Delta_z = \partial / \partial z$.

The following difference equation can be obtained for node A_1 in Fig. 3.

$$\begin{aligned}
&\{V_y(A_2) + V_y(A_3) - 2V_y(A_1)\} \\
+\ &\{V_y(A_4) + V_y(A_5) - 2V_y(A_1)\} \\
-\ &[\{V_x(D_4) - V_x(D_2) - V_x(D_3) + V_x(D_1)\} \\
-\ &\{V_z(E_4) - V_z(E_2) - V_z(E_3) + V_z(E_1)\}] \\
=\ &4\Delta d^2(\sigma^*/2)\ \{I_{z1} - I_{z2} + I_{x1} - I_{x2}\}_{A1}/\Delta d \\
&+ 4\Delta d^2(\mu_0/2)\ \Delta_t\ \{I_{z1} - I_{z2} + I_{x1} - I_{x2}\}_{A1}/\Delta d \\
&+ 4\Delta d^2(\sigma^*\varepsilon_0/4)\ \Delta_t\ \{V_y(B_1) + V_y(B_2) + V_y(C_1) + V_y(C_2)\} \\
&+ 4\Delta d^2(\varepsilon_0\mu_0/4)\ \Delta_t^2\ \{V_y(B_1) + V_y(B_2) + V_y(C_1) + V_y(C_2)\}
\end{aligned} \tag{2}$$

Each spatial difference of the left hand side in Equ.(2) is expressed by a differential operator when the spatial difference Δd becomes sufficiently small.

$$\frac{\partial^2 V_y}{\partial x^2} + \frac{\partial^2 V_y}{\partial z^2} - \frac{\partial}{\partial y}\left(\frac{\partial V_x}{\partial x} - \frac{\partial V_z}{\partial z}\right) = \frac{\partial^2 A_y}{\partial x^2} + \frac{\partial^2 A_y}{\partial z^2} - \frac{\partial}{\partial y}\left(\frac{\partial A_x}{\partial x} + \frac{\partial A_z}{\partial z}\right) \tag{3}$$

The terms in the parenthesis in equ.(2) are transformed as follows by considerating 'div A' from the cubic shown in Fig. 1 (to be referred to in Appendix I).

$$-\frac{\partial}{\partial y}\left(\nabla \cdot A - \frac{\partial A_y}{\partial y}\right) = -\frac{\partial}{\partial y}\nabla \cdot A + \frac{\partial^2 A_y}{\partial y^2} = \frac{\partial}{\partial y}(\mu_0 F) + \frac{\partial^2 A_y}{\partial y^2} \tag{4}$$

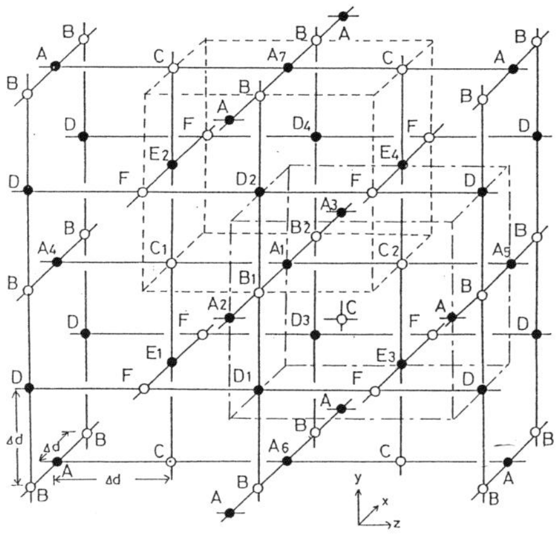

Figure 1. Three-dimensional lattice network for vector potential fields.

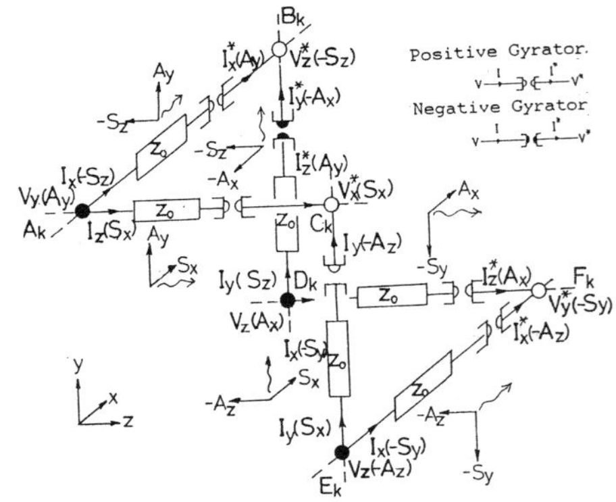

Figure 2. Fundamental Connection in the network.

Then, the left hand side of Equ.(2) is given as follows;

Left hand side $= \nabla^2 A_y + \mu_0 \nabla_y F$

On the other hand, when the spatial difference Δd becomes sufficiently small and the equivalent voltage variables are also rewritten by the corresponding component of the vector potential, the right hand side of Equ.(4) is given as

Right hand side =

$$\sigma\sigma^* A_y(A_1) + (\mu_0\sigma + \varepsilon_0\sigma^*)\frac{\partial}{\partial t}A_y(A_1) + \varepsilon_0\mu_0\frac{\partial^2}{\partial t^2}A_y(A_1) \tag{5}$$

The vectorial expression for all the components is given as follows.

$$\nabla^2 A + \mu_0 \nabla F - \varepsilon_0\mu_0\frac{\partial^2 A}{\partial t^2} - (\mu_0\sigma + \varepsilon_0\sigma^*)\frac{\partial A}{\partial t} - \sigma\sigma^* A = 0 \tag{6}$$

Based on the Lorentz gauge condition including the loss term, the combined wave equation for both the magnetic and electric vector potentials is given as follows by using the current source as presented in Appendix Ⅱ.

$$\left\{\nabla^2 - \varepsilon_0\mu_0\frac{\partial^2}{\partial t^2} - (\mu_0\sigma + \varepsilon_0\sigma^*)\frac{\partial}{\partial t} - \sigma\sigma^*\right\}\left[\frac{A}{S}\right] = -\left[\frac{\mu_0 J}{\varepsilon_0 J^*}\right] \tag{7}$$

(J^* : Magnetic current density)

3. CONCLUSION

The unified spatial network for both the vector potential field and the scalar potential field based on the Lorentz gauge condition with the loss term is presented. This formulation will have the advantage for analyses of quantum effect devices because of the direct evaluation of the Hamiltonian in the Schrodinger equation by both potential variables.

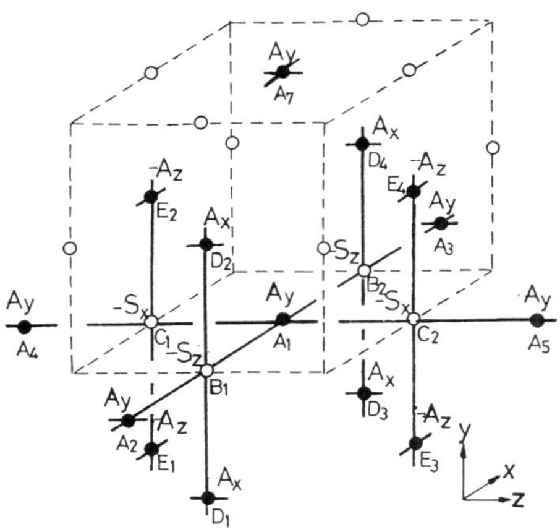

Figure 3. Arrangement of nodes arround node A_1 .

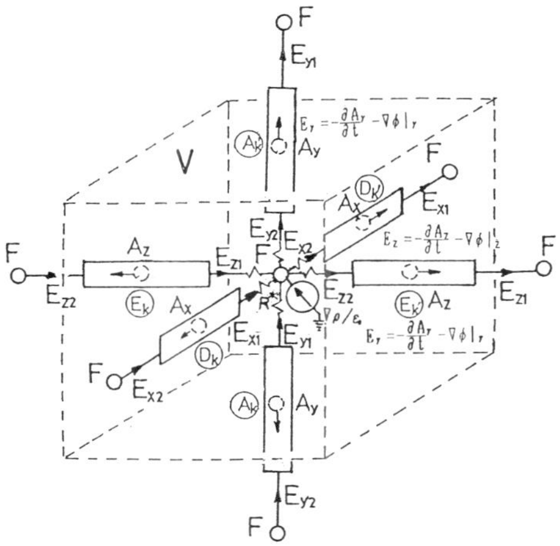

Figure 4. Equivalent circuit for electric scalar field.

4. REFERENCES

1 E. Yamashita;editor, Analysis Methods for Electromagnetic Wave Problems, Artech House, Boston (1990).
2 N. Yoshida, Report of E/M Field Theory Group in IEE Japan, EMT-88-84 (1988).
3 N. Yoshida, Radio Science, (1991) 259.
4 N. Yoshida, Proc. of 1992 URSI Int. Symp. on Electromagnetc Theory (1992) 24.

Appendix I

Based on the Lorentz gauge condition for loss field, the equivalent circuit expression for the scalar field is derived by using both the 'F' function and the electric field 'Es'. The 'F' function is defined as follows.

$$\nabla \cdot A = -\varepsilon_0 \mu_0 \frac{\partial \phi}{\partial t} - \mu_0 \sigma \phi = -\mu_0 F \qquad (A_I\ 1)$$

$$\therefore \quad F = \varepsilon_0 \frac{\partial \phi}{\partial t} + \sigma \phi \qquad (A_I\ 2a)$$

ϕ is the electric scalar potential. The static electric field E_S is given as;

$$E_S = -\nabla \phi \qquad (A_I\ 2b)$$

Relation between F and E_S is given by the following characteristic equations,

$$-\nabla F = \varepsilon_0 \frac{\partial E_S}{\partial t} + \sigma E_S \qquad (A_I\ 3a)$$

$$-\nabla \cdot E_S = \mu_0 \frac{\partial F}{\partial t} - \frac{\rho}{\varepsilon_0} \qquad (A_I\ 3b)$$

Wave equation for 'F' is obtained for the equivalent circuit shown in Fig.4 as;

$$\nabla^2 F - \varepsilon_0 \mu_0 \frac{\partial^2 F}{\partial t^2} - \mu_0 \sigma \frac{\partial F}{\partial t} = \nabla \cdot J_S \qquad (A_I\ 4)$$

Appendix II

Processing the gradient operator to both sides of Equ.(A_I 1) gives;

$$\mu_0 \nabla F = \varepsilon_0 \mu_0 \frac{\partial}{\partial t} (\nabla \phi) + \mu_0 \sigma \nabla \phi \qquad (A_I\ 1)$$

On the other hand, the divergence of the laminar component of current J_S and the scalar electric field Es are defined as follows;

$$\nabla \cdot J_S = -\frac{\partial \rho}{\partial t} - \frac{\sigma \rho}{\varepsilon_0} \qquad (A_I\ 2)$$

$$\nabla \cdot E_S = \frac{\rho}{\varepsilon_0} \qquad (A_I\ 3)$$

After substituting Equ.(A_I 3) for Equ.(A_I 2) and multiplying μ_0 to both sides of Equ.(A_I 2), integrating this equation side by side gives the next equation.

$$\mu_0 J_S + \mu_0 J_R = -\varepsilon_0 \mu_0 \frac{\partial E_S}{\partial t} - \mu_0 \sigma E_S \qquad (A_I\ 4)$$

, Here J_R is the solenidal component of current density. Since the right sides of Equ.(A_I 1) and Equ. (A_I 4) are identical, the next relation can be derived.

$$\mu_0 \nabla F = \mu_0 J = -\varepsilon_0 \mu_0 \frac{\partial E_S}{\partial t} - \mu_0 \sigma E_S \qquad (A_I\ 5)$$

, where $J = J_S + J_R$

Simulation and Design of Applied Electromagnetic Systems
T. Honma (Editor)

A complementary variational analysis of waveguides

H. Igarashi and T. Honma

Department of Electrical Engineering, Faculty of Engineering,
Hokkaido University, Kita 13, Nishi 8, Kita-ku, Sapporo, 060 Japan

Abstract

The cutoff frequencies of a rectangular waveguide are analyzed using the finite element method based on the complementary variational principle. The present method is shown to provide the lower bounds of the cutoff frequencies of the waveguide whereas the conventional FEM based on the Rayleigh quotient gives the upper bounds.

1. INTRODUCTION

The cutoff frequencies of hollow metallic waveguides are obtained by solving the two-dimensional Helmholtz equation. The finite element method (FEM) based on the Rayleigh quotient has been shown to give the accurate solutions to those problems. However, it is known that the FEM provices only the upper bounds of the cutoff frequencies. It has been, therefore, difficult to evaluate the error bounds of them because of lack of the lower bounds.

For the potential problems, the FEM based on the complementary variational principle [1] has been shown to give the lower bounds of the exact energy of the field [2]. Moreover, in 1989, Ladeveze *et al.* introduced the complementary variational method which provided the lower bounds of the eigenfrequencies of an elastic structure [3]. In this paper, we report an analysis of the cutoff frequencies of a rectangular waveguide by the FEM based on the complementary variational principle.

2. FORMULATION

We consider here the two dimensional scalar Helmholtz equation

$$\nabla^2\phi + \lambda\phi = 0, \quad in\ \Omega, \tag{1}$$

under the boundary conditions

$$\phi = 0 \quad on\ \partial\Omega_1, \quad \frac{\partial\phi}{\partial n} = 0 \quad on\ \partial\Omega_2, \tag{2}$$

where n denotes the outward normal vector of the boundaries $\partial\Omega_i (i=1,2)$. The upper bounds of the eigenvalues of this problem λ_i ($i=1, 2, 3...$) can be obtained by minimizing the Rayleigh quotient

$$R(\phi) = \int_\Omega \nabla\phi \cdot \nabla\phi \, d\Omega \Big/ \int_\Omega \phi^2 \, d\Omega \, . \tag{3}$$

On the other hand, the quotient R'

$$R'(\phi) = \int_\Omega \phi^2 \, d\Omega \Big/ \int_\Omega \int_\Omega \phi(\mathbf{r}) \, G(\mathbf{r}\,;\mathbf{r}_0) \, \phi(\mathbf{r}_0) \, d\Omega_0 \, d\Omega \tag{4}$$

yields the integral equation corresponding to (1),

$$\phi(\mathbf{r}) = \lambda \int_\Omega G(\mathbf{r}\,;\mathbf{r}_0) \, \phi(\mathbf{r}_0) \, d\Omega_0 \, , \tag{5}$$

at its stationary point under the boundary conditions (2) [4], where $G(\mathbf{r};\mathbf{r}_0)$ denotes the Green function satisfying the equation

$$\nabla^2 G(\mathbf{r}\,;\mathbf{r}_0) + \delta(\mathbf{r} - \mathbf{r}_0) = 0 \, , \tag{6}$$

under the boundary conditions (2). Hence, one can also obtain the eigenvalues λ_i from variation of (4).

By use of the solution $\Psi(\phi)$ to the Poisson equation

$$\nabla^2 \Psi = -\phi \, , \tag{7}$$

one can rewrite the quotient (4) in the form

$$R'(\phi) = \int_\Omega \phi^2 \, d\Omega \Big/ \int_\Omega \nabla\Psi(\phi) \cdot \nabla\Psi(\phi) \, d\Omega \, , \tag{8}$$

where $\Psi(\phi)$ is expressed in terms of $G(\mathbf{r}, \mathbf{r}_0)$ as follows :

$$\Psi(\phi) = \int_\Omega G(\mathbf{r}\,;\mathbf{r}_0) \, \phi(\mathbf{r}_0) \, d\Omega_0 \, . \tag{9}$$

The lower bounds of λ_i will be obtained below by applying the complementary variational principle to (8).

We express the exact and approximate solutions to the scalar Helmholtz equation (1) as ϕ^* and ϕ, and define the error as $\varepsilon = \phi^* - \phi$. Assuming the error ε is perpendicular to the approximate solution ϕ, we have the inequality [3]

$$\frac{1}{R'(\phi)} + \frac{1}{R'(\varepsilon)} - \frac{1}{R'(\phi^*)} \geqq 0 \tag{10}$$

From (10), the lower bound of the eigenvalue λ^* for the exact solution ϕ^* can be expressed in terms of the eigenvalue λ for the approximate solution ϕ as follows :

$$\lambda^* \geqq \frac{\lambda}{1 + \lambda / p} \, , \tag{11}$$

where the constant p denotes the lower bound of $R'(\varepsilon)$. One can, therefore, evaluate the lower bound of the exact eigenvalue by (11) providing that the constant p is known.

In general, it is difficult to obtain the solution $\Psi(\phi)$ involving the Green function $G(\mathbf{r};\mathbf{r}_0)$ for an arbitrary geometry of the region Ω. In order to overcome this difficulty, following Ladeveze *et al.*, we introduce here an alternative expression of (8) through the adoption of the complementary variational principle for static electric fields. To do this, (7) is divided into the following canonical equations

$$\nabla \cdot \mathbf{u} = \phi , \tag{12}$$

$$-\nabla \Psi = \mathbf{u} . \tag{13}$$

Applying the complementary variational principle to (12) and (13), we see that the inequalities

$$\int_\Omega \mathbf{u}\cdot\mathbf{u}\, d\Omega \geqq \int_\Omega \nabla\Psi(\phi)\cdot\nabla\Psi(\phi)\, d\Omega \geqq 2\int_\Omega \Psi\phi d\Omega - \int_\Omega \nabla\Psi\cdot\nabla\Psi\, d\Omega . \tag{14}$$

hold [1]. Using (8), (12) and (14), the lower bound of the quotient $R'(\phi)$ is evaluated by $R_C(\mathbf{u})$ as follows :

$$R'(\phi) = \frac{\int_\Omega \phi^2\, d\Omega}{\int_\Omega \nabla\Psi(\phi)\cdot\nabla\Psi(\phi)\, d\Omega} \geqq \frac{\int_\Omega (\nabla\cdot\mathbf{u})^2\, d\Omega}{\int_\Omega \mathbf{u}\cdot\mathbf{u}\, d\Omega} \equiv R_C(\mathbf{u}) . \tag{15}$$

Since the quotient $R_C(\mathbf{u})$ does not now include the Green function, its stationary value may be easily obtained by FEM.

The lower bound of the constant p in (11) can be evaluated in terms of the maximum diameter ρ_{max} of the finite elements as follows [3, 5] :

$$p \geqq \pi^2 / \rho_{max}^2 . \tag{16}$$

(The diameter of a triangle element corresponds to the maximum length of its sides.) Hence, one can easily evaluate the lower bound of p by (16).

If we find the stationary points of $R_C(\mathbf{u})$ under the condition that all the components of $\mathbf{u}$ are independent variables, solenoidal solutions appear in numerical solutions. The equilibrium element used frequently in structural analyses is employed here to avoid this spectrum pollution.

3. NUMERICAL RESULTS AND CONCLUSION

We apply the FEM based on the variational principle mentioned in the previous section to the analysis of the cutoff wavenumber k of a rectangular waveguide of wide 1.0 and height 0.5.

The numerical results are shown in Figure 1, in which "Complementary" and "Rayleigh" denote the cutoff wavenumbers obtained from the quotients $R_C(\mathbf{u})$ and (3), respectively. We see that, as expected, the wavenumbers by the complementary variational FEM approaches from under the exact ones as the maximum diameter ρ_{max} of the finite element decreases. The use of the conventional FEM together with complementary FEM may allow to evaluate the error bounds of cutoff frequencies of hollow metallic waveguides of arbitrary cross-sections.

4. REFERENCES

1 A. M. Arthurs, Complementary Variational Principles, Chap. 4, Clarendon Press, Oxford, 1970.

2 J. Penmann, J. R. Fraser, J. R. Smith and M. D. Grieve, IEEE Trans. on Mag., 19 (1983) 2288.

3 P. Ladeveze and J. P. Pelle, Int. J. for Numerical Methods in Eng., 28 (1989) 1929.

4 P. M. Morse and H. Feshbach, *Methods of Theoretical Physics*, Part II, Chap. 9, McGraw-Hill, New York, 1953.

5 L. E. Payne and H. F. Weinberger, Arch. Rational Mech. Anal., 5 (1960) 286.

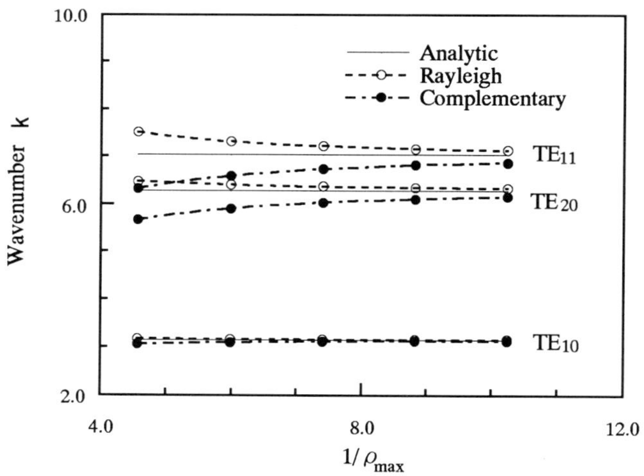

Figure 1. Convergence of the eigenvalues $k=\lambda^{1/2}$ for a rectangular waveguide of 1.0×0.5. The abscissa ρ_{max} denotes the maximum diameter of the triangle elements.

Biological Systems

Simulation and Design of Applied Electromagnetic Systems
T. Honma (Editor)

Modeling and analysis of magnetic field sensory system extracted from sockeye salmon

Y. Sakaki[a], N. Daikuhara[a] and M. Ogura[b]

[a]Dept. of Electrical and Electronics Engineering, Chiba University, 1-33, Yayoi-cho, Inage-ku, Chiba-shi, 263 Japan

[b]Far Seas Fisheries Lab., Fisheries Agency, 5-7-1, Orido, Shimizu-shi, 424 Japan

Abstract:A model of magnetic sensory system for migration of salmon is proposed on the basis of the organs extracted from the skull of sockeye salmon. The organ is elliptical and is encircled by a lot of small magnetic particles, then the model is set up as a closed chain of small spherical magnets arranged elliptically. The behavior of model under application of static magnetic field is analyzed using FEM technique. The result shows that the maximum torque is generated on the magnetic particles, the magnetization of which is orthogonal to the applied field and the position of maximum torque changes according to the field direction.

1. INTRODUCTION

It would seem that salmon as well as other fish which migrate over long distances, must have navigational abilities in a huge ocean. For survival of fish the abilities should be a fail-safe system which is a combination of several functions, i.e. olfactories, solar compass, electro-orientation, magnetic compass and so on.

Recently the hypothesis that salmon migrates depending on their magnetic sense organ has become reliable, since chains of magnetic particles and organelles encircled by magnetic particles are found in the skull of sockeye salmon[1,2,3]. The authors presented a model of magnetic field sensory system based on the discovered organell and discussed qualitatively the mechanism sensing the geomagnetic field directions[3]. In this paper the mechanism is analyzed quantitatively by using finite element method(FEM) on the basis of the model presented previously. The result shows that the responded part of the organell changes according to the direction of applied magnetic field.

2. MODELING OF THE SENSE ORGAN

The candidate of magnetic sense organ is extracted from the head of salmon[3]. There are a couple of elliptical organells encircled by a lot of small magnetic particles linking as chain which is not straight as that of

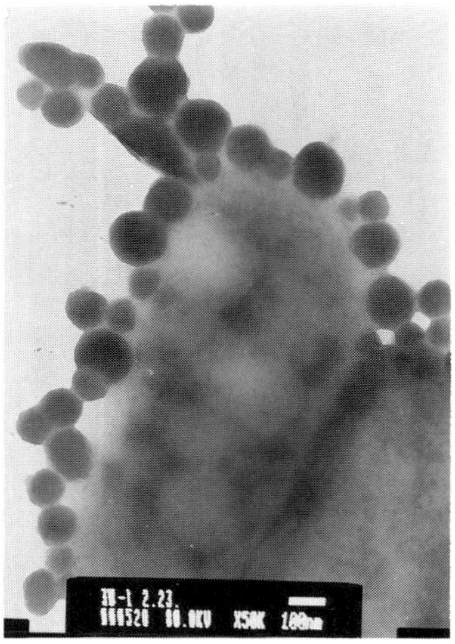

Fig.1 A candidate of magnetic sense organ extracted from the skull of sockeye salmon[3]

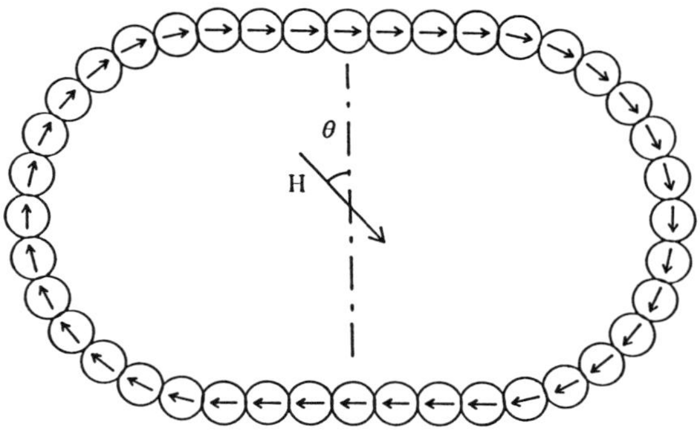

Fig.2 Model of magnetic sense organ

bacterial magnetite[4]. The minor and major axes of the elliptical organells are 0.65 to 0.75 μ m and 1.3 to 1.8 μ m, respectively. The diameters of magnetic particles range from 50 to 100nm. These sizes correspond to the single domain magnet. EDX analysis shows that the main component of the particles is iron. Based on previous works[1,2] we believe that these particles are composed of single crystal magnetite. Each particle is covered by very thin transparent organic film which prevents the aggregation of particles due to their magnetic attraction force. The film also sticks the particles tightly to the organell.

On the basis of Fig.1, we propose a model of the magnetic sensory system as shown in Fig.2 where a chain made of 40 magnet spheres is arranged elliptically. The diameter of sphere is 75nm which is the average value of particles observed. As the material and properties of elliptical organ is not known, we only analyze the behaviors of magnetic particles, then the elliptical organ itself will not be shown hereafter. The arrow in each sphere shows the direction of magnetization in the absence of magnetic field. The intensity of magnetization is 0.592[T] which is the saturation magnetization of magnetite and the relative recoil permeability μ s=1.1. The base axis is the vertical dashed line and the direction of external field is taken by angle θ in anticlockwise direction from the axis. As the FEM analysis is performed by a two dimensional program, the magnet spheres are replaced by magnet cylinders with infinite length in the following analysis.

3. ANALYSIS OF MECHANISM SENSING MAGNETIC FIELD DIRECTION

The region for finite element analysis including the model is divided into about 4000 triangle meshes. Then we calculate the torque acting on each magnet cylinder with unit length under application of magnetic field 50[μT] which is nearly the same as the geomagnetic field intensity and convert them into the ones of magnet rod with 75nm height. In this system there are leakage flux from the magnets especially located at the curved parts of the model and the field due to the flux affects the magnets themselves and exerts torque on the system even in the absence of magnetic field. The torque acting on each magnet cylinder are, therefore, calculated under the conditions with and without applied magnetic field and the net torque **T** are introduced as the difference between the torque values under both conditions. The distribution of net torque and the resulting tangential forces **f** are shown in Fig. 3 for θ = 0 and in Fig.4 for θ = 120°. In these figures the length of arc in each cylinder means the intensity of torque **T** and the length of arrow along each cylinder means the intensity of tangential force **f**. The reference torque and force are given in the figures.

The torque **T** is maximum for the cylinders, the magnetization direction of which is orthogonal to the applied field. The tangential force **f** arises from a cylinder, the magnetization direction of which is parallel to the applied field, increases along the cylinders where the magnetization direction is

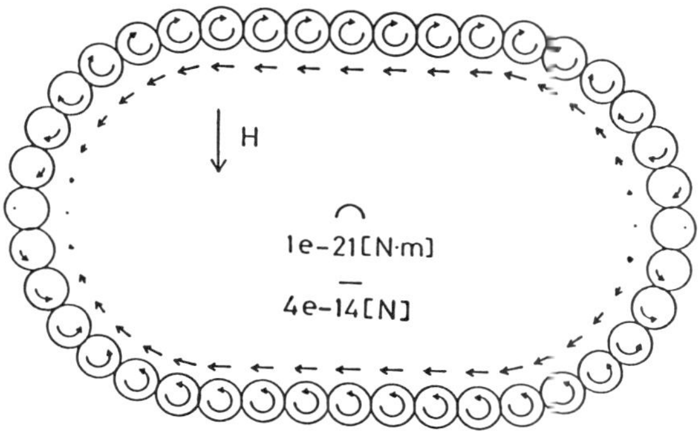

Fig.3 Torque and tangential force acting on each magnet. Applied field is 50μT and θ is 0°.

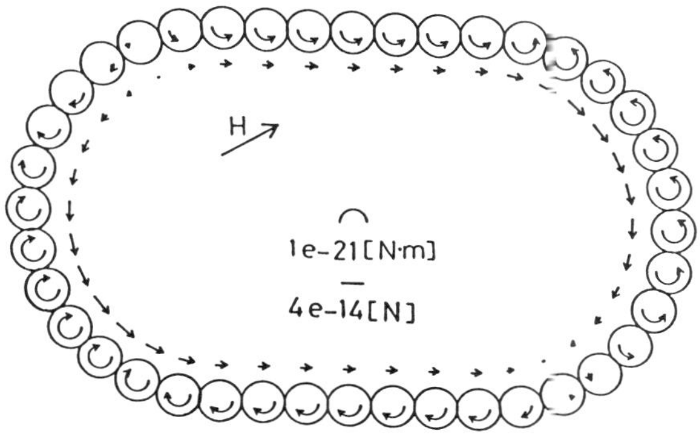

Fig.4 Torque and tangential force acting on each magnet. Applied field is 50μT and θ is 120°.

orthogonal and terminates to a cylinder, the magnetization direction of which is antiparallel. The position of maximum torque and tangential force acting on the cylinder changes according to the direction of magnetic field. This force may stimulate the elliptical organ mechanically. The sensing mechanism controlling navigational bearings during migration in a huge ocean is a matter of conjecture. In order to sense the generated torque, the deformation of organell is necessary, however, the viscous damping constant and elastic constant of organ and internal organell are not known and a quantitative discussion has not been carried out yet. From the results of this study we only presume that the small organell in the elliptical organ may sense the force generated from the interaction of geomagnetic field and the magnet spheres and converts the signal into electrical one and sends it to the brain through nervous system. The discussion on the detailed mechanism will be given in future.

4. CONCLUSION

A model of magnetic field sensory system of salmon is proposed and the mechanism sensing the direction of external magnetic field is analyzed using FEM technique. The force and torque exerting on the organ changes according to the applied field direction. The discussion on the mechanism sending the force signal to brain is not clear, however, this model will present an impact to the research of biogenic sense organ in salmon.

Acknowledgement

The authors thank Prof. Y.Saito of Hosei University for his suggestions on the modeling of the magnet sphere and FEM analysis. They also thank the support by Grant-in Aids from the Ministry of Education, Science and Culture, and from Biocosmos Program of the Ministry of Agriculture, Forestry and Fisheries of Japan.

REFERENCES

[1]J.L.Kirschvink, M.M.Walker, S-B.Chang, A.E.Dizon and K.A.Peterson,"Chains of single-domain magnetite particles in chinook salmon, Oncorhynchus Tshawytscha", J. Comp. Physiol. **A 157**, pp.357-381, 1985.

[2]S.Mann, N.C.Sparks, M.M.Walker and J.L.Kirschvink,"Ultrastructure, Morphology and Organization of Biogenic Magnetite from Sockeye Salmon, Oncorhynchus Nerka: Implication for Magnetoreception", J.exp. Biol. **140**, pp.35-49, 1988.

[3]Y.Sakaki, T.Motomiya, M.Kato and M.Ogura, "Possible mechanism of biomagnetic sense organ extracted from sockeye salmon", IEEE Trans. Magn., **MAG-26**, pp.1554-1556, 1990.

[4]R.B.Frankel, R.P.Blakemore and R.S.Wolfe,"Magnetite in Freshwater Magnetotactic Bacteria", Science, **203**, pp.1355-1356, 1979.

Simulation and Design of Applied Electromagnetic Systems
T. Honma (Editor)

Looking inside the human brain with magnetoencephalography

M. Sams
Low Temperature Laboratory, Helsinki University of Technology, 02150 Espoo, Finland

1. INTRODUCTION

Currents flowing in a group of parallel cortical pyramidal cells produce tiny magnetic fields which can be detected outside the head by means of SQUID (superconducting quantum interference device) sensors. The recording of these fields, the magnetoencephalography (MEG), opens an entirely noninvasive window to the functions of the human brain. The excellent time resolution and good spatial resolution make MEG a powerful tool for cognitive neuroscience to study the cortical basis of mental functions which changes quickly in time and space.

The 20-year history of MEG has seen the development of instruments from those with a single channel instruments to those containing 24–37 channels. The latter instruments cover an area of the head large enough for a quick determination of the magnetic field pattern necessary for calculation of the corresponding neural generator in a case of a simple local source. However, the study of the information flow in and between different parts of the cortex has been tedious under natural information processing conditions. Such measurements can now be carried out with a new 122-channel magnetometer, accomplished in our laboratory.

The new device has 122 planar first-order gradiometers located at 61 measurement sites. Each sensor unit contains a pair of gradiometers measuring the orthogonal tangential derivatives $\partial B_Z/\partial x$ and $\partial B_Z/\partial y$ where Bz is the field component normal to the bottom of the dewar at the measurement site. The average channel separation is 44 mm. The channels are located at the helmet-shaped surface which fits the heads of about 95% of the adult population. The magnetometer is a prototype developed by Neuromag Ltd., in close collaboration with the brain research group of the Low Temperature Laboratory, and it has been operational since June, 1992.

This instrument allows the simultaneous comparison of the activities generated in the left and right auditory cortex. This is advantageous, *e.g.*, in studies of cortical processing of speech.

In the present paper we show examples of results obtained with our previous 24-channel and the new 122-channel instruments.

2. 24-CHANNEL STUDIES OF ATTENTION AND AUDIO-VISUAL SPEECH PERCEPTION

The effect of selective attention on activity of the right human auditory cortex was studied in two conditions [1]. In the first, a randomized tone sequence was delivered to the left ear with a constant interstimulus interval (ISI) of 0.4 s. Subject's task was to discriminate infrequent (15%) 150-ms tones among 50-ms 'standards' of 1 or 3 kHz. When the auditory evoked magnetic field (AEF) to the attended standards was subtracted from that to the ignored ones, an attention effect, called magnetic difference, Md, was found (Fig. 1A). Md peaked at 195–220 ms and its field distribution could be satisfactorily explained with an equivalent current dipole in the supratemporal auditory cortex. The source was about 1 cm anterior to the generator site of the N100m deflection and could not be separated from the source of the P200m.

In another condition, tones were led randomly to opposite ears with an ISI of 240–300 ms. In addition to the Md around 200 ms, a separate 'early Md', adding up to N100m deflection was observed, starting already at 30–40 ms (lower part of Fig. 1A). The source areas of N100m and early Md overlapped, suggesting similar generators for both deflections.

The results show that attention can modify the activity of two different areas in the supratemporal auditory cortex. We suggest that the earlier effect reflects enhanced activity in the neurons underlying N100m to attended tones and the latter effect a modification of P200m to ignored tones.

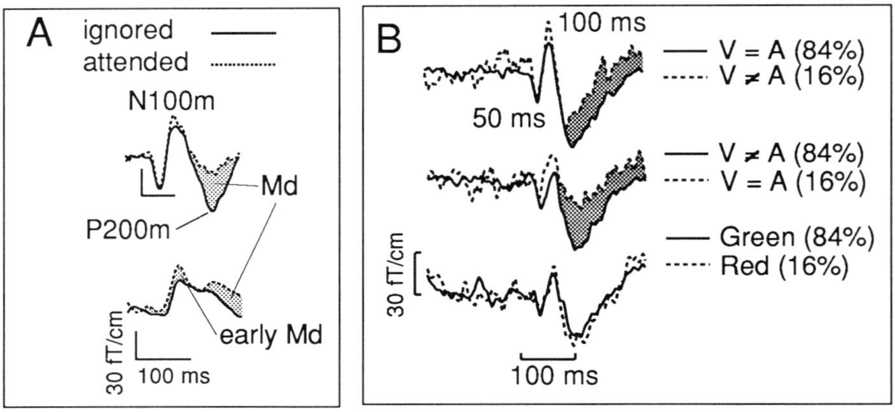

Figure 1. (A) Responses to ignored and attended auditory stimuli. The signals in the upper part of the figure were obtained when the stimuli were delivered to the same ear. In the lower part, the two stimuli were delivered in opposite ears. The shadowed areas show the selective-attention effect. (B) Magnetic responses of one subject to the /pa/ syllables with congruent (V=A) and incongruent (V≠A) articulation. The shadowed area indicates the effect of the visual stimulus which is not seen in the control condition with light stimuli.

Seeing the articulatory movements of a speaker's face supports speech understanding; this is particularly clear in a noisy environment or when hearing is defective. A dramatic example of the audiovisual integration is the "McGurk effect": when an auditory /ba/ is associated with a videotaped face articulating /ga/, the auditory perception is altered by the visual input and the subject hears /da/ or /ga/ [2].

Up to now there has been no information about the actual neuroanatomical area where the visual information affects the auditory perception. We used neuromagnetic measurements to answer this question [3]. The auditory stimulus was the natural syllable /pa/ which was dubbed to the visual /ka/ or /pa/ articulation. Combinations where the visual and auditory stimuli were in concordance (V=A, 84% of the stimuli) and where they were discordant (V≠A, 16% of the stimuli) were joined to a film of a speaker articulating one or the other of the syllables about once every seconds. In some subjects, the probabilities of the audiovisual stimuli were also reversed. In a control condition the face was replaced by a green (84%) or red (16%) light stimulus.

The magnetic responses to infrequent V≠A stimuli had similar 50-ms and 100-ms deflections as those to the V=A stimuli. However, starting at about 180 ms, the two responses differed (Fig. 1B). A rather similar difference waveform was elicited by infrequent V=A stimuli among frequent V≠A stimuli. However, the signals to syllables preceded by frequent green and infrequent red light stimuli were identical. The response differences could be explained by changed activity of the supratemporal cortex, showing that visual information from the articulatory movements may have an entry into the human auditory cortex.

3. 122-CHANNEL STUDIES ON CHANGE DETECTION IN THE AUDITORY CORTEX

Infrequent deviations in an otherwise monotonous auditory stimulus sequence elicit both in EEG and MEG a specific 'mismatch response', which is not generated by infrequent stimuli presented alone without the intervening standards. We have recorded mismatch fields (MMFs) to changes in, *e.g.*, tone frequency, intensity, and duration. MMFs are elicited also by more complex deviations, such as the direction of the frequency glide, periodicity pitch, and synthetic as well as natural syllables. The various MMFs have their neural origin in the supratemporal auditory cortex.

We used the 122-channel neuromagnetometer [4] to study the cortical sources of MMF elicited by different types of deviants. The 50-ms 1-kHz standard stimuli were presented at 0.6-s interstimulus interval (ISI). The deviants were (i) of higher frequency (1.1 kHz), (ii) of longer duration (100 ms), or (iii) similar to standards but presented with a shorter ISI (0.4 s). The standards and different deviants were in the same sequence and they were delivered to both ears. The responses to standards and frequency deviants presented to the left ear are shown in Fig. 2.

The standard stimuli elicited a clearly larger N100m responses over the contra- than ipsilateral hemisphere. The deviants, on the other hand, elicited a right-hemisphere-dominant MMF. The same was true for right-ear stimuli. Interestingly, both

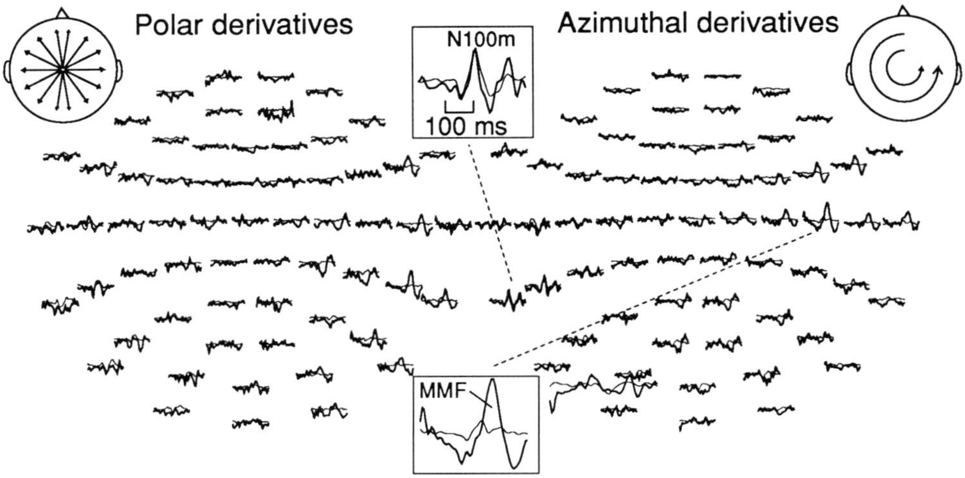

Figure 2. Responses of one subject to standards (thin lines) and frequency deviants (thick lines) recorded with the 122-channel device. Stimuli were presented to the right ear. The polar and azimuthal gradients are shown separately. The passband is 2–30 Hz. Unpublished data from the Low Temperature Laboratory.

standards and deviants elicited equally large N100m over the left hemisphere. These preliminary data show a clear dissociation of the N100m and MMF responses as well as the right hemisphere dominance of the frequency MMF. MMFs to duration and ISI deviations also tend to be larger in the right hemisphere, suggesting its special role in the auditory change detection.

4. ACKNOWLEDGEMENTS

This study was supported by the Academy of Finland

5. REFERENCES

1 J. Rif, R. Hari, M. Hämäläinen and M. Sams, Electroenceph. Clin. Neurophysiol., 79 (1991) 464.

2 H. McGurk and J. MacDonald, Nature 264 (1976) 746.

3 M. Sams, R. Aulanko, M. Hämäläinen, R. Hari, O.V. Lounasmaa, S-T. Lu and J. Simola, Neurosci. Lett. 127 (1991) 141.

4 A. Ahonen, M. Hämäläinen, M. Kajola, J. Knuutila, P. Laine, O.V. Lounasmaa, J. Simola, C. Tesche and V. Vilkman, Proc. 14th Ann. Int. Conf. IEEE Eng. Med. Biol. Soc., November 2–4, 1992 Lyon, France, 16.

Simulation and Design of Applied Electromagnetic Systems
T. Honma (Editor)

Changes of alpha wave distribution when subjects perceive binocular rivalry

Tetsuo Kobayashi[a], Minoru Morita[b], Toru Owada[a] and Shinya Kuriki[a]

[a]Research Institute for Electronic Science, Hokkaido University, N12, W6, Kita-ku, Sapporo, 060 Japan
[b]Hokkaido Institute of Technology, 7-15 Maeda, Teine-ku, Sapporo, 060 Japan

Abstract

To elucidate the alternation process of binocular rivalry, electroencephalographic spectra during subjects perceive binocular rivalry were analyzed at 12 electrode positions on a whole head according to the international 10-20 system. Real and illusory triangles (white, 10 cd/m^2) on a black screen were used as the visual stimuli in the experiments. Attenuation distribution of alpha wave during binocular rivalry was analyzed to estimate the region in the brain where the neural activities related to binocular rivalry are taking place. The most prominent feature is the significant difference of the attenuation between rivalry and fusion in the occipital and posterior-temporal regions.

1. INTRODUCTION

Binocular rivalry is the process whereby different visual stimuli from left and right eyes are recognized alternately. This psychological process is thought to represent a fundamental property of the human binocular visual system. Although the binocular rivalry has been studied for years, its mechanism has been only partly clarified[1-4].

The aim of the present paper is to investigate whether any changes appear in the EEGs obtained while subjects perceive binocular rivalry. It is well known that some psychological processes cause changes in event related potential (ERP). The ERP is usually measured to investigate the cognitive process. However, ERP triggered by the alternation of the perceived visual stimuli during binocular rivalry can not be utilized in the present study because binocular rivalry is the stochastic process and it is not possible to determine the exact time when the perceived dominant stimulus alternates. Instead, spontaneous EEGs while the subjects perceive binocular rivalry were measured and their spectra were analyzed. The spectra of EEGs while the subjects perceive a fused image (this condition is called as binocular fusion) and the subjects close their eyes were also analyzed for the comparison. Several features of measured EEGs were discussed by the point of alpha attenuation. The topographical distribution of the attenuation was analyzed to estimate the region in the brain where the neural activities related to binocular rivalry are taking place.

2. METHODS

Twelve volunteer students participated in the experiments. They were between 18 and 27 years of age and all had normal or corrected to normal visual acuity. The apparatus of the experiment is schematically shown in Fig.1. A 32 bit personal computer and a 15" high resolution color display (analog RGB, 1120x750 dots) were used to present visual stimuli. To make subjects perceive binocular rivalry, two different pairs of real and illusory triangles shown in Fig.2, both of which were about 1°x1° of visual angle, white, 10 cd/m^2, on the black screen (0.085 cd/m^2) were given to the left and right eyes separately through a stereoscope. The EEGs when subjects perceive binocular rivalry were measured at 12 different locations (Fp1, Fp2, F7, Fz, F8, C3, C4, T5, T6, Pz, O1, O2 of International 10-20 system) with both ears grounded connection. The EEGs were also measured when the subjects were given the same figure to both eyes (this is the state of binocular fusion) and when the subjects closed their eyes. The measurement time in each condition was for 180 sec separated by 60 sec rest. Successive elementary 8 sec spectra of the EEGs at each location were obtained using a spectrum analyzer after the experiment. Sixteen elementary spectra were used to obtain the average spectrum. The spectrum analysis was based on the FFT (sampling points are 1024, sampling frequency is 128Hz).

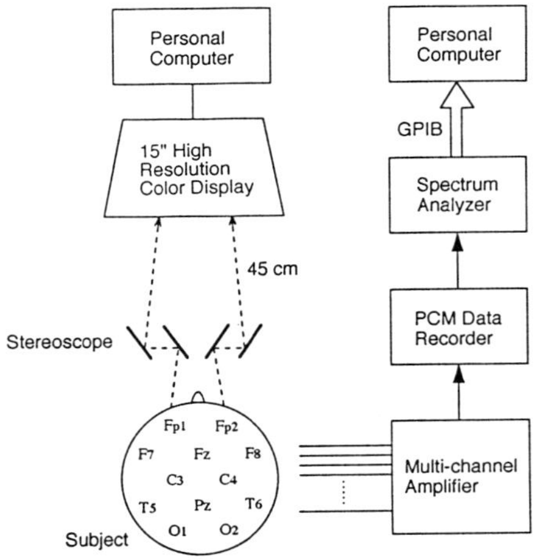

Figure 1. Arrangement of experimental equipment. An analog RGB 15" high resolution color display is used to provide visual stimuli.

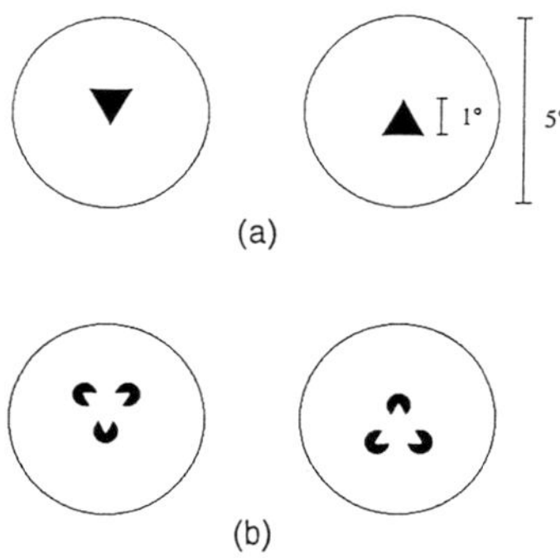

Figure 2. Two pairs of stimuli of (a) real triangles and (b) Illusory triangles used in the experiments.

3. RESULTS AND DISCUSSION

When subjects were closing their eyes, alpha wave predominated in the spontaneous EEG. Alpha wave decreased while the subjects perceived binocular rivalry and fusion. This can be considered as one of the phenomena called "alpha attenuation" [5]. The alpha attenuation is generally thought to be related to the increase of neural activities of the cortex [6].

Although the stimulus strength given to both eyes during rivalry and fusion is considered to be the same, there was a difference in the amount of the attenuation. This difference of the attenuation is thought to reflect the cortical activities related to binocular rivalry.

In order to examine the spatial difference quantitatively, we analyzed the normalized attenuation. The normalized attenuation in rivalry (and in fusion) is defined as the ratio of alpha wave magnitudes in rivalry (and in fusion) to the one in eyes-closed condition.

Figure.3 shows the topographical maps of the grand average of alpha attenuation in rivalry and fusion. The percentage of attenuation is more than 61% at O_1, O_2, and gradually decreases toward the frontal region in binocular rivalry. The spatial uniformity of attenuation of alpha waves in each condition was examined using statistical t-test. The attenuation was

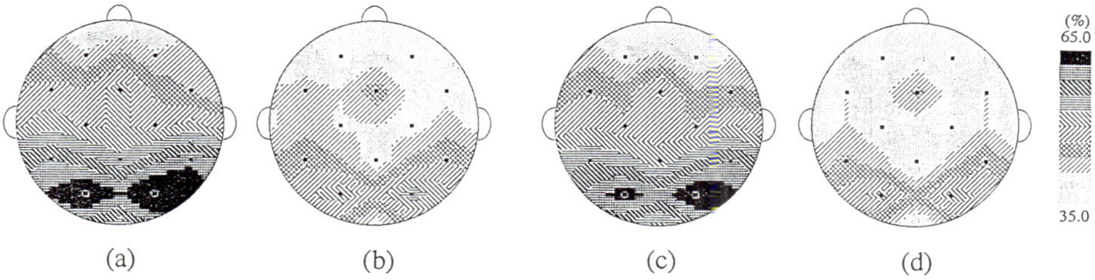

Figure 3. Topographical maps of alpha attenuations during (a) binocular rivalry and (b) binocular fusion for real triangles, and during (c) binocular rivalry and (d) binocular fusion for illusory triangles.

Table 1
Mean and S.D. of alpha attenuation of 12 subjects normalized by the magnitude of alpha wave in eye-closed condition.

	Real triangle		Illusory triangle	
	Binocular Rivalry	Binocular Fusion	Binocular Rivalry	Binocular Fusion
Leads	*Mean ± S.D. (%)*	*Mean ± S.D. (%)*	*Mean ± S.D. (%)*	*Mean ± S.D. (%)*
F_{p1}	41.2 ± 14.6	36.5 ± 13.6	39.3 ± 12.3	36.4 ± 11.1
F_{p2}	41.2 ± 13.8	35.8 ± 11.9	37.7 ± 12.7	35.7 ± 11.2
F_7	46.6 ± 14.2	41.8 ± 17.3	45.7 ± 13.4	40.7 ± 17.5
F_z	50.0 ± 11.5	45.1 ± 15.0	48.8 ± 11.7	44.2 ± 15.1
F_8	43.0 ± 17.4	39.9 ± 18.6	44.0 ± 17.0	40.7 ± 17.0
C_3	47.9 ± 12.1	40.3 ± 17.3	46.4 ± 13.2	38.4 ± 17.8
C_4	47.3 ± 11.9	39.4 ± 16.2	46.0 ± 12.6	37.1 ± 18.0
T_5	56.7 ± 9.0	45.4 ± 14.4 *	53.6 ± 8.2	44.1 ± 13.7 *
P_z	51.5 ± 9.7	38.4 ± 16.5 *	49.3 ± 10.7	35.6 ± 16.3 *
T_6	58.7 ± 9.1	46.1 ± 11.8 **	57.2 ± 8.3	45.8 ± 11.8 **
O_1	62.4 ± 7.6	50.5 ± 14.1 **	61.0 ± 8.0	49.7 ± 13.3 *
O_2	64.4 ± 11.3	52.7 ± 16.1 *	63.0 ± 11.0	50.2 ± 16.8 *

* $P < 0.05$
** $P < 0.01$

significantly larger at O_1, O_2 (p<0.01), T_5 and T_6 (p<0.05) than the ones at remaining positions in binocular rivalry. The normalized attenuation in binocular fusion at O_1 and O_2 was slightly larger, but the difference was not significant.

Table 1 shows the mean and S.D. of the normalized attenuation at all electrode positions. The attenuations in binocular rivalry are significantly larger than those in fusion at T_5, T_6, Ps, O_1 and O_2 for both real and illusory triangles. Figure.4 shows topographical maps of the difference of alpha attenuations between binocular rivalry and binocular fusion. There is no significant difference at the remaining 7 positions.

The alpha wave attenuates largest at the occipital region in all cases. This seems reasonable because visual stimuli may primary activate the visual cortex. It is noticeable that the alpha wave attenuates greatly at the posterior-temporal region compared with our previous results for the rivalry caused by two different oriented lines [7]. The difference of alpha attenuations for real and illusory triangles is not significant.

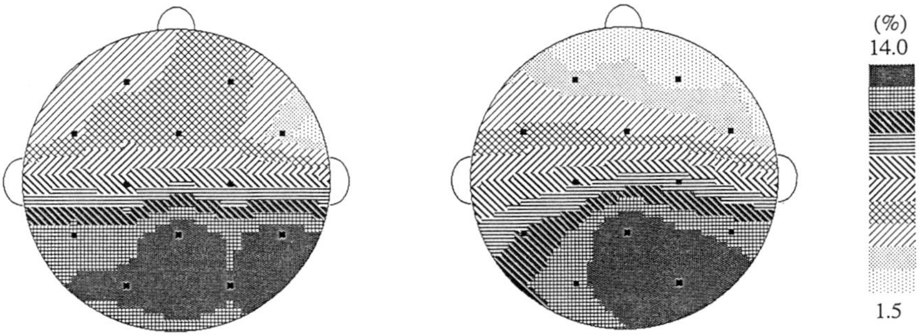

(a) (b)

Figure 4. Topographical maps of the difference of alpha attenuations between binocular rivalry and binocular fusion. (a) and (b) represent the differences for real triangles and for illusory triangles, respectively.

4. CONCLUSION

The levels of EEG activity in alpha band detected over the occipital and posterior-temporal regions reduced significantly when subjects perceive binocular rivalry caused by real and illusory triangles compared with the ones in binocular fusion. This result suggests that neural activities related to the binocular rivalry may be taking place in the occipital and posterior-temporal regions.

In a recent study, Kaufman et al. [8] suggested that it is possible to make use of cortical maps of magnetoencephalogram (MEG) reflecting differences in levels of spontaneous activity as functional images similar to those generated by PET. We think that the topography of alpha attenuation of EEG may also be used as one of the functional brain imaging techniques.

5. REFERENCES

1. W.L.M. Levelt, The alternation process in binocular rivalry. Brit J Psychol, 57 (1966) 225-238.
2. J.M. Wolfe, Stereopsis and Binocular Rivalry. Psychological Review, 93 (1986) 269-282.
3. T.J. Mueller and R. Blake, A fresh look at the temporal dynamics of binocular rivalry. Biological Cybernetics, 61 (1989) 223-232.
4. N.K. Logothetis and J.D. Schall, Neuronal correlates of subjective visual perception. Science, 245 (1989) 761-763.
5. IFSECN, A glossary of terms commonly used by clinical electroenphalographers. Electroenceph. clin. Neurophysiol., 37 (1974) 538-548.
6. W.J. Ray and H.W. Cole, EEG alpha reflects attentional demands, and beta activity reflects emotional and cognitive processes. Science, 228 (1985) 750-752
7. T. Kobayashi, M. Morita, and S. Kuriki, Analysis of alpha attenuation distribution related to binocular rivalry. Proc. Int. Conf. IEEE-EMB, (Paris, 1992) 2742-2743
8. L. Kaufman, J.H. Kaufman, and J.Z. Wang, On cortical folds and neuromagnetic fields. Electroencephalo. clin. Neurophysiol., 79 (1991) 211-226

Simulation and Design of Applied Electromagnetic Systems
T. Honma (Editor)

Source localization of K-complexes during sleep and auditory-evoked responses using MEG measurements

K. Iramina and S. Ueno

Department of Electronics, Kyushu University, Fukuoka 812, Japan

Abstract

This study focuses on source estimation of spontaneous MEG activity during sleep. Sources of K-complexes are investigated by magnetoencephalograph (MEG) and electroencephalograph (EEG) measurements. The MEG suggests that the sources of the K-complexes can be modeled by two current dipoles. The relationship between the sources of K-complexes and the sources of auditory- evoked responses was also investigated. The dipoles for the K-complexes are estimated at locations 5 mm apart from the sources of N100 components of auditory evoked responses.

1. INTRODUCTION

During the past decade, substantial advances in the understanding of the functional organization of the human brain have been made through a technique of MEG topographic mapping. Most of these investigations were concerned with estimation of localization in evoked magnetic fields. Compared with many reports on source estimation of evoked magnetic fields [1-3], only a few neuromagnetic studies on MEG activities during sleep have been reported [4-7].

Sources of K-complexes in sleep stage two and sources of N100 components of auditory-evoked responses are investigated. Although these activities have been studied by many groups using EEG methods, sources of these activities within the brain are poorly understood. We analyzed these waveforms based on both topographic EEG maps and magnetic fields measurements. The K-complex is most readily evoked by auditory stimulus during sleep stage 2. The relationship between K-complexes and auditory-evoked responses was investigated using dipole localization methods. The dipoles were estimated at locations 5 mm apart from the sources of N100 components of auditory evoked responses.

2. METHODS

Both MEG and EEG activities were recorded simultaneously. The magnetic field components perpendicular to the surface of the head were measured using a 7 ch DC-SQUID with a first order gradiometer. The MEG data were obtained mainly from the temporal and frontal regions of the head. EEG data were recorded from 14 locations over the surface of the head, corresponding to the points defined by the international 10-20 system.

In a computer simulation, the head was represented by a homogeneous spherical volume conductor with a radius R=0.1m. Normal components of magnetic fields on the surface of the sphere can be determined by the Biot-Savart law. The potentials were obtained by solving a simultaneous equation [8,9].

In the recording of auditory evoked responses, auditory stimuli were presented to both ears. Tone bursts of 1 kHz with a 100 ms plateau were delivered randomly every 1-2 seconds.

The sources of K-complexes and auditory evoked responses were estimated by the least square method using a homogeneous spherical model.

3. RESULTS AND DISCUSSION

Figure 1 shows the EEG waveforms of the K-complexes measured at 14 locations and the topographic EEG map. The spatial pattern demonstrates that the K-complex is maximal over the frontal region, and minimal in the occipital region. Figure 2 shows the simultaneous measurement of K-complexes using MEG and EEG. The flux reversal appeared between the second and third channels, and between the fourth and fifth channels. As a first approximation, a single dipole, which is located along the midline in the head, is assumed to be an electric source of a K-complex. Although the single dipole model simulates the EEG topography well, the model is not sufficient to simulate phase relationships of MEG K-complexes measured at different points as shown in Fig. 3.

Figure 4 shows the distribution of the polarities of the magnetic fields. The symbols x indicate locations where magnetic components emerge from the head, whereas the solid dots indicate locations where flux enters. Magnetic field polarities of K-complexes change between C3 and T3, and between C4 and T4. Based upon the results of the MEG measurements, particularly, according to the distribution of the magnetic fields polarities, the source of the K-complexes can be modeled by two current dipoles which are located in at both sides of the midline. Figure 5 shows the spatial patterns of EEG and MEG simulated by two dipoles. This source model explains the phase patterns of the magnetic fields of the K-complex well.

The K-complexes are most readily evoked by auditory stimulus during sleep stage 2. Sources for K-complex and N100 components of auditory evoked response were estimated by the least square method to investigate the relationship between K-complexes and auditory-evoked responses. Figure 6 shows the auditory-evoked potentials recorded at 14 locations, and

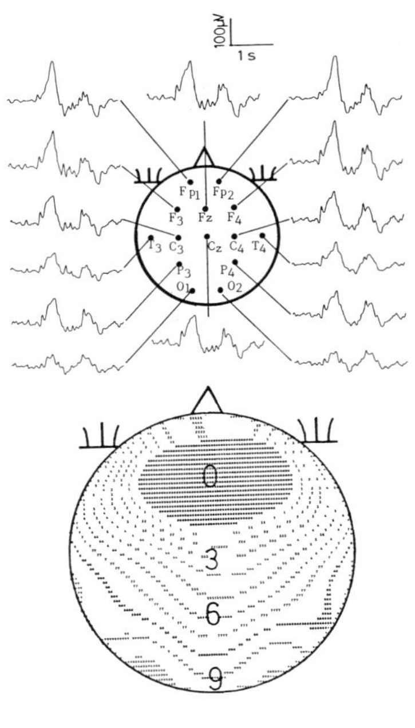

Fig.1 The EEG topography of K-complexes obtained from a normal adult subject in sleep stage 2. The potentials are quantized into 11 levels from 0 to X instead of 10

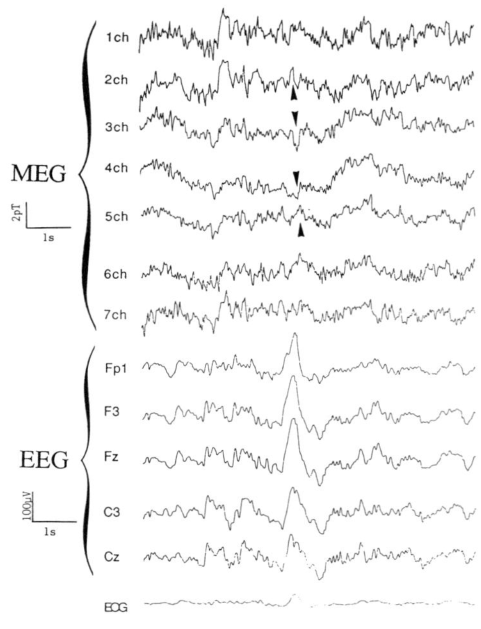

Fig.2 Simultaneous measurements of K-complexes using MEG and EEG

the topographic EEG map of N100 component. The N100 components are negative potentials that appear about 100 ms after stimulus. The spatial patterns demonstrate that the N100 components are maximal in the front-central region, minimal in the occipital region. This topographic map shows a similar spatial pattern of K-complexes. Figure 7 shows the auditory-

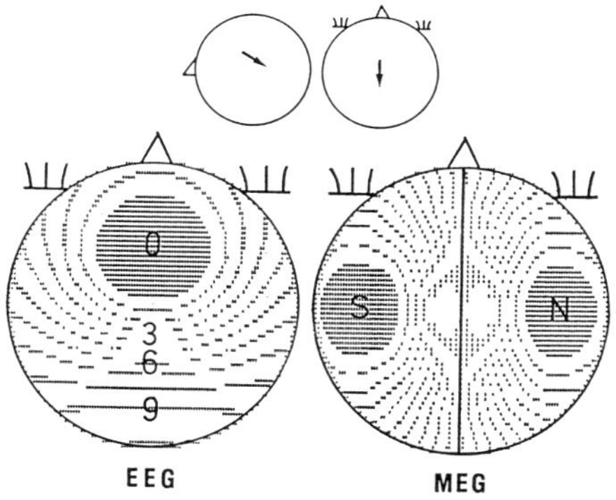

Fig.3 Spatial patterns of EEG and MEG over the surface of the sphere generated by a single dipole

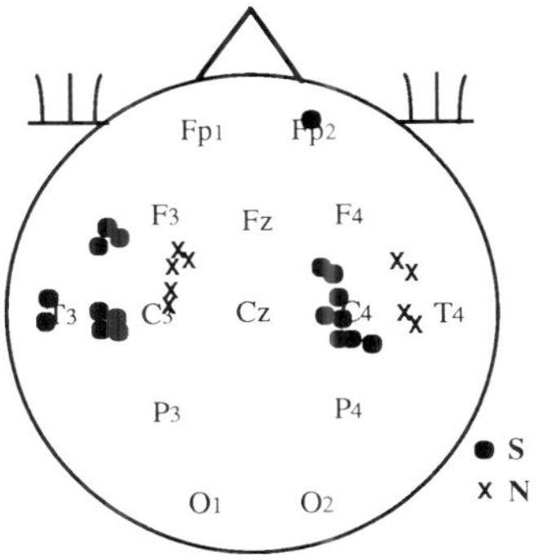

Fig.4 Distributions of polarities of magnetic fields of K-complexes over the surface of the head (: S pole, x: N pole)

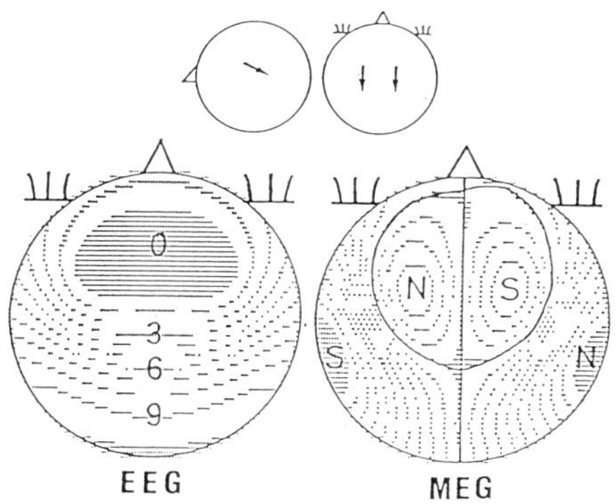

Fig.5 Spatial patterns of EEG and MEG generated by two dipoles which are positioned in parallel and in the same direction in the sphere

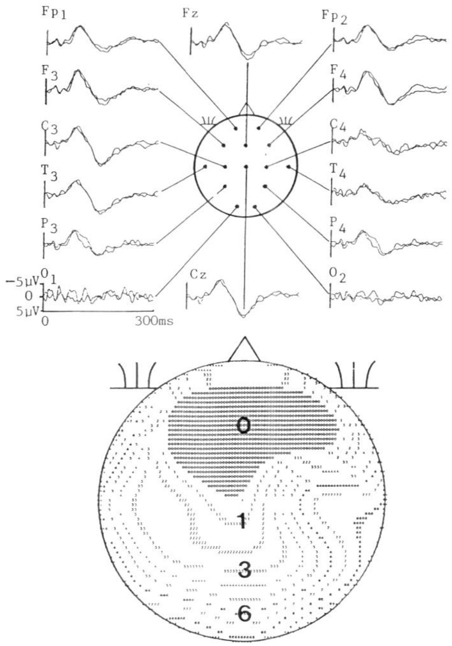

Fig.6 Auditory evoked potentials and spatial pattern of N100 components of evoked potentials

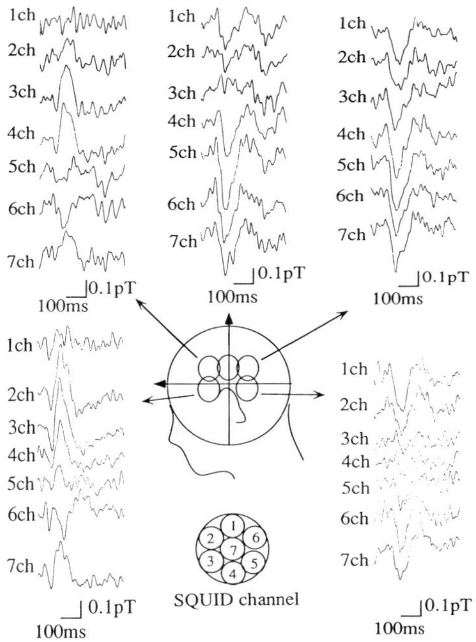

Fig.7 Auditory evoked magnetic fields

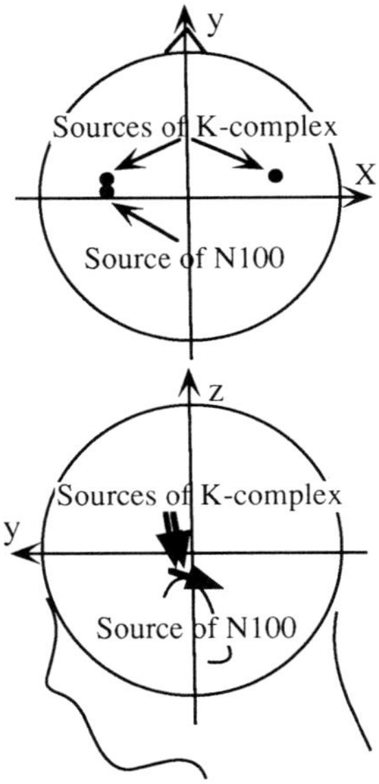

Fig.8 Sources of the K-complexes estimated by EEG data and source of N100 components of auditory evoked magnetic field

evoked magnetic fields and spatial distributions of N100 components of magnetic field over the right hemisphere. Positions of sources of EEG N100 components coincide with positions of sources of MEG N100 components. The dipoles are estimated in the right and left auditory areas . These locations are 5 mm apart from the dipoles of K-complexes estimated by EEG data as shown in Fig.8. That is, the mechanism of K-complexes is closely related to auditory evoked responses.

This work was supported in part by the grants 03505002 from the Ministry of Education Science and Culture, Japan

4. REFERENCES

1 G.L. Romani, S.J. Williamson and L. Kaufman. Science 216 (1982) 1339
2 S.J. Williamson and L. Kaufman. IEEE Trans. Magn. MAG-19 (1983) 835
3 R.L Rogers, A. C. Papanicolaou, S. Baumann, C. Saydjari, H.M. Eisenberg. In: S.J. Williamson, M. Hoke, G. Stroink and M. Kotani (eds.) Advances in Biomagnetism. Ple num Press, New York (1989) 105
4 J.R. Hughes, D.E. Hendrix, J. Cohen, F.H. Duffy, M.L.Scholl and B.N.Cuffin. Electroenceph. Clin. Neurophysiol.; 40 (1976) 262
5 N. Nakasato, K. Hatanaka, K. Seki, T. Yoshimoto and H. Kado. In: S.J. Williamson, M. Hoke, G. Stroink and M. Kotani (eds.) Advances in Biomagnetism. Plenum Press, New York (1989) 265
6 S. Ueno and K. Iramina. Brain Topography Vol.3 (1990) 151
7 S.-T. Lu, M. Kajola, S.-L. Joutsiniemi, J. Knuutila and R. Hari. Electroenceph. clin. Neuro physiol., 82 (1992) 182
8 S. Ueno and Y. Fukui. Memoirs of the Faculty of Eng. Kyushu Univ. 38 (1978) 207
9 S. Ueno, H. Wakisako and S. Matsuoka. Il Nuovo Cimento 2D (1983) 558

Simulation and Design of Applied Electromagnetic Systems
T. Honma (Editor)

Gustatory Evoked Electric and Magnetic Fields of the Human Brain

T. Yoshiura[a], S. Ueno[b], K. Iramina[b], and K. Masuda[a]

[a]Department of Radiology, Faculty of Medicine, Kyushu University, Fukuoka 812, Japan

[b]Department of Electronics, Faculty of Engineering, Kyushu University, Fukuoka 812, Japan

Abstract

The magnetic and electrical sources of the gustatory responses in the human brain were analyzed using magnetoencephalography (MEG) and electroencephalography (EEG) . We stimulated a human tongue with NaCl, acetic acid, and distilled water. Gustatory evoked potentials were measured at 14 points on the surface of the head, and gustatory evoked magnetic fields were measured with a dc SQUID system over the left hemisphere of the subject's brain. Slow electrical responses, dominant in the frontal region, were observed as responses to the taste stimuli. Magnetic fields evoked by gustatory stimuli were observed in the left temporal region. On the basis of the experimental results, dipole models for the magnetic and electrical gustatory responses were proposed.

1. INTRODUCTION

The functions of auditory, visual, and somatosensory systems of the human brain have been studied well by EEG and MEG methods. However, very little attention has been paid to the brain functions associated with the sense of taste [1, 2]. Although it is known that the primary taste area is located at the anterior insula and the frontal operculum in the brain of monkeys [3-5], the process of gustatory information in the human central nervous system is poorly understood. In the present study, we stimulated a human tongue with taste solutions and measured gustatory evoked potentials and gustatory evoked magnetic fields. We also proposed dipole models for the magnetic and electrical gustatory responses on the basis of the experimental results.

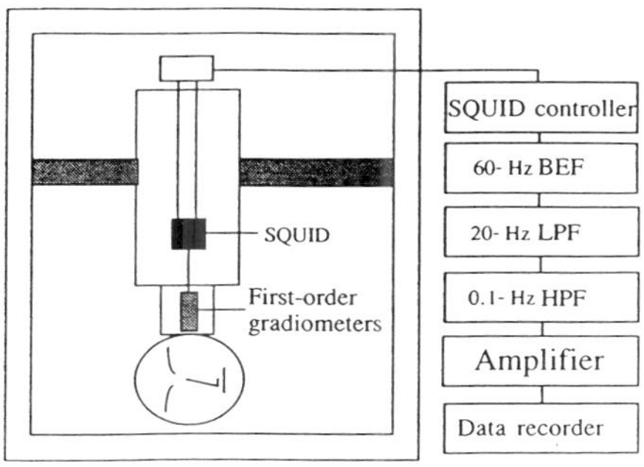

Fig.1 MEG measurement system using a dc SQUID

2. METHODS

We prepared NaCl solutions for studying

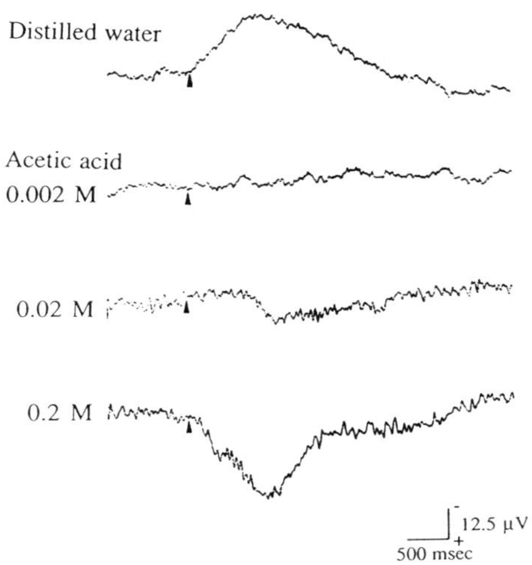

Fig.2 Electrical brain responses to the stimulations of distilled water and acetic acid

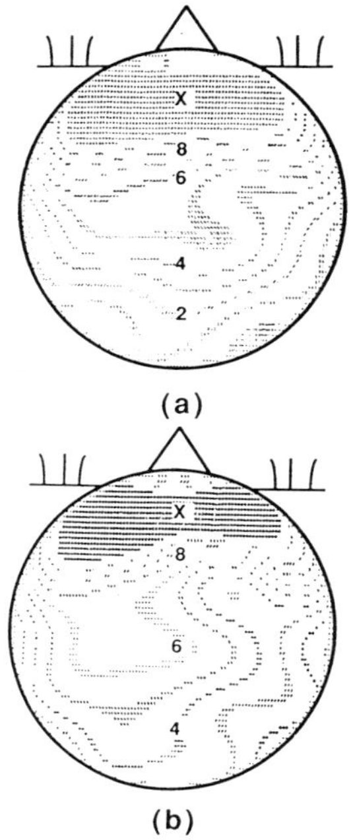

Fig.3 EEG topographies of gustatory evoked potentials: (a) distilled water; (b) 1 M NaCl

salty taste and acetic acid solutions for studying sour taste. We also prepared distilled water for a control study. 1.0ml of each test solution was delivered into the oral cavity to stimulate the tongue.

Evoked potentials over the surface of the head were recorded at 14 points of the 10-20 system, and the topographical maps [6] of the potentials were obtained. The amplified signals (bandpass 0.1Hz-30Hz) were recorded. Twenty trials were averaged for each test solution. Gustatory evoked magnetic fields were measured using a dc SQUID system with first order gradiometers over the left hemisphere of the subject's head (Fig.1). The stimulation method was the same as that used in the experiments of gustatory evoked potentials. The data were filtered (bandpass 0.1Hz-20Hz) and averaged. Twenty trials were averaged for each test solution.

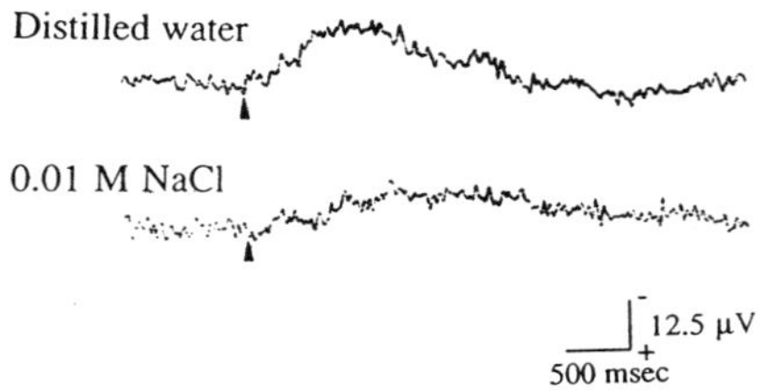

Fig.4 Brain's electrical responses to the stimulations of (a) distilled water and (b) 0.01 M NaCl

Based on the experimentally obtained topographical patterns of EEG and MEG, we carried out a computer simulation to estimate the location of the brain electrical sources for the gustatory responses, solving a forward problem in a spherical head model [7, 8].

3. RESULTS AND DISCUSSION

Fig.2 shows the electrical brain responses to the stimulations of distilled water and various

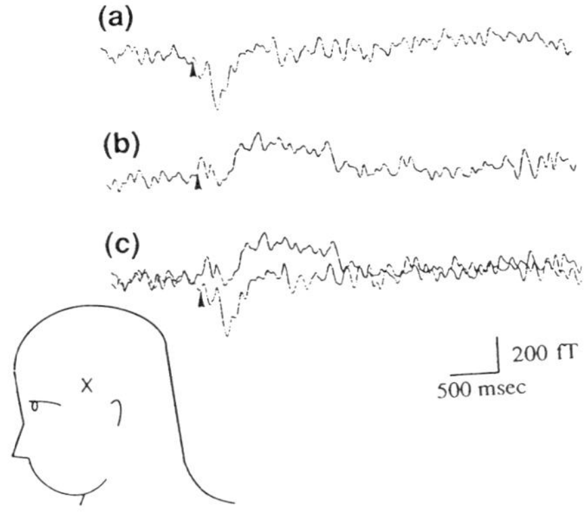

Fig.5 Brain's magnetic responses to the stimulations of (a) distilled water and (b) 1 M NaCl. (a) and (b) are superimposed in (c)

concentrations of acetic acid. A slow negative evoked potential was observed as a response to the stimulation of distilled water. On the other hand, a slow positive evoked potential was observed as a response to the stimulation of acetic acid. As the figure shows, the amplitude of the evoked potentials grew larger with the increasing concentration of acetic acid. Fig.3 shows the EEG topographies of the gustatory evoked potentials at the latency of 1000msec induced by distilled water and 1M NaCl. In each topography, the amplitude of the evoked potentials is displayed in 11 levels (0-9 and X). It is observed that gustatory evoked potentials induced by both distilled water and NaCl are dominant in the frontal region. As shown in Fig.4, the negative electrical brain response to the stimulation of distilled water disappeared when the tongue was stimulated by 0.01M NaCl. It is known that human saliva contains 0.01M NaCl. It is suggested that this negative electrical brain response to the stimulation of distilled water is caused by the decrease spontaneous activities of the neurons in the gustatory cortex when taste receptors on the tongue habituated to NaCl in saliva are washed out with distilled water.

To estimate the sources of the gustatory brain

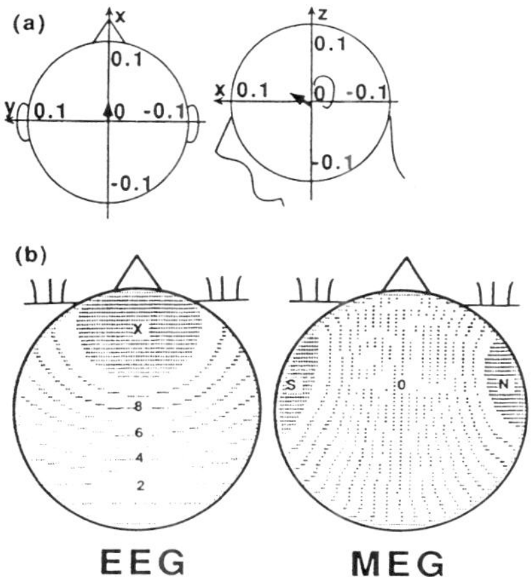

Fig.6 Forward problem of EEG and MEG: (a) one-dipole source model for magnetic and electrical gustatory brain responses; (b) spatial patterns of EEG and MEG derived from the model in (a)

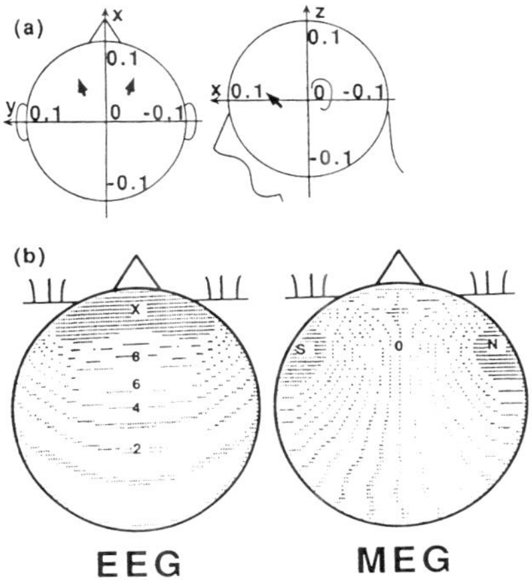

Fig.7 Forward problem of EEG and MEG: (a) two-dipole source model for magnetic and electrical gustatory brain responses; (b) spatial patterns of EEG and MEG derived from the model in (a)

responses more accurately, gustatory evoked magnetic fields were also measured. Fig.5 shows the gustatory evoked magnetic fields recorded in the temporal region of the subject's head. Fig.5(a) shows the waveform of the magnetic brain response to the stimulation of distilled water, and Fig.5(b) shows the waveform of the magnetic brain response to the stimulation of 1M NaCl. In Fig.5(c), the waveforms in (a) and (b) are superimposed. Monophasic gustatory evoked magnetic fields similar to the gustatory evoked potentials shown in Fig.2 were observed. No reversal was observed in the polarity of the gustatory evoked magnetic fields over the left hemisphere of the subject's head.

On the basis of these experimental results, we introduced brain source models for gustatory evoked responses using the homogeneous spherical conductor model. The radius of the model sphere is 0.1m. Fig.6(a) provides a simple example of a dipole source model that is consistent with the EEG topography shown in Fig.3(a). In this model, a single current dipole is located in the midline at (0.01, 0.00, 0.00) in the coordinate system shown in Fig.6(a), and the dipole is oriented at a 30-degree rise from the x-y plane. Fig.6(b) shows the spatial patterns of EEG and MEG derived from the model in Fig.6(a). The calculated spatial pattern of MEG is compatible with the experimental result that no reversal was observed in the polarity of the evoked magnetic fields over the left hemisphere of the subject's head. However, if we assume that the source is located in the cerebral cortex, this model is inconsistent with the anatomical knowledge that the human cerebral cortex is divided into two hemispheres.

Fig.7(a) shows a two-dipole source model for gustatory evoked brain responses. In this model, the dipoles are located at (0.04, 0.03, 0.00) and (0.04, -0.03, 0.00), in the region equivalent to the bottom of the anterior cranial fossa in the actual head. Each dipole is oriented 17 degrees lateral to the midline and at a 45-degree rise from the x-y plane. Fig.7(b) shows the spatial pattern of EEG and MEG derived from the model in Fig.7(a). It is understood that the experimentally obtained spatial patterns of gustatory evoked electrical and magnetic fields can be explained by brain source models with two bilaterally located current dipoles. In the present study, the tongue was stimulated bilaterally. Therefore, the two dipole model is more suitable for the gustatory evoked brain responses than the one-dipole model shown in Fig.6(a).

4. REFERENCES

1 M. Funakoshi and Y. Kawamura, Electroencephalography and Clinical Neurophysiology, vol. 30, pp. 205-209, (1971)
2 G. Kobal, Electroencephalography and Clinical Neurophysiology, vol. 62, pp. 449-454, (1985)
3 T.C. Pritchard, R. B. Hamilton, J. R. Morse and R. Norgen, The Journal of Comparative Neurology, vol. 244, pp. 213-228, (1986)
4 S. Yaxley, E. T. Rolls and Z. J. Sienkiewicz, Journal of Neurophysiology, vol. 63, pp. 689-700, (1990)
5 T. R. Scott, S. Yaxley, Z. J. Sienkiewicz, and E. T. Rolls, Journal of Neurophysiology, vol. 63, pp. 876-890, (1986)
6 S. Ueno, S. Matsuoka, T. Mizoguchi, M. Nagashima and C. Cheng, Memoirs of the Faculty of Eng., Kyushu Univ., vol. 34, pp. 165-209, (1975)
7 S. Ueno and Y. Fukui, Memoris of Faculty of Eng., Kyushu Univ., vol. 38, pp. 207-217, (1978)
8 S. Ueno and K. Iramina, Brain Topography, vol., 3, pp. 151-165, (1990)

Simulation and Design of Applied Electromagnetic Systems
T. Honma (Editor)

An algorithm of eye movements in biological vision

Tohru Yagi[a], Kazutoshi Gouhara[b], and Yoshiki Uchikawa[c]

[a,c]Dept. of Electronic-Mechanical Engineering, Nagoya University, Nagoya 464-01, Japan

[b]Dept. of Electrical Engineering, Chubu University, Kasugai 487, Japan

Abstract

We propose an algorithm of eye movements in the human vision. Quantitative analysis of the computed eye movements and the human eye movements revealed they showed very similar eye movements. This implies that the algorithm emulates the human foveation which is applicable to an image extraction processing.

1. INTRODUCTION

Since humans have much higher resolution in the focal visual area than the periphery [1], eye movements are required to focus the object on the fovea [2]. This visual behavior called *foveation* seems to be processed at the lower level of the visual system like the retina [3]. In the retina, the visual signals are transferred topologically from the photo receptor layer to the retinal ganglion cell layer. Each ganglion cell connects several photo receptor cells through several kinds of cells and makes a receptive field. Neighboring receptor fields are overlapped to process the visual signals [4]. We assume that non-uniform resolution is caused by the following specific features of the retinal structure. One is the distribution of cells in the retina. Cells are located more densely in the fovea than the peripheral region [5]. The other is the size of the receptive field. The size of receptor fields in the central region is relatively small and increasingly larger towards the periphery of the retina [6].

Although the retina has various kinds of cells, in this paper we look closely at photo receptor cells and retinal ganglion cells only, pay attention to the latter feature, and assume the uniform distribution of cells. Then we make a mathematical model of the signal processing in the retina. Using this model we will make an algorithm of eye movements in foveation, and then simulate it in the computer. On the other hand, the actual human eye movements will be examined in a cognitive experiment. Finally, both results in the computer simulation and the experiment will be compared to evaluate the algorithm.

2. COMPUTED EYE MOVEMENTS

2.1. The architecture of an algorithm

In our algorithm the image is processed with a mathematical model of the retinal structure (Figure1). The model consists of two layers. Assume that the first layer corresponds to a layer of photo receptor cells, and the second to ganglion cells. Since the connection strength between two layers can be expressed as the spatial filters, this model is denoted as a mathematical equation.

Define $I(x,y)$ as the input visual signals, $F_{x,y}(\xi,\eta)$ as the spatial filters, and $A_{x,y}$ as the integral area. Hence the processed image $O(x,y)$ is as follows:

$$O(x,y)=f\left[\int\int_{A_{x,y}} F_{x,y}(\xi,\eta)\cdot I\,(x+\xi,y+\eta)d\xi d\eta\right]$$

$A_{x,y}$ is the group of ξ and η which suffices the equation, $|\xi^2+\eta^2|\leq R(L)$.

After the image is processed, the position of the pixel that has the maximum signal value among the pixels in the processed image is detected. Afterwards, the fovea is relocated on that position, and the relocation process is completed.

To prevent the model from continuing to detect the same objects, we delete circuler areas which are centered on the detected position. If the adequate size to cover the whole object is chosen, and the detected position is precisely on the center of the object, the object will be deleted completely.

2.2. Computer simulation using this algorithm

In the computer simulation of the algorithm, we used pillbox function as a spatial filter $F_{x,y}(\xi,\eta)$. This function had a cylindrical shape. The radius of the cylinder denoted with "R(L)", was increased proportionally to the square root of the distance "L". As displayed images, we used very simple computer graphic images. The images were black and white and consisted of seven to ten objects that were identical in brightness, shape, and size.

Figure 2 shows the simulation results in four different images. Symbol "S" indicates the starting position of the eye movements, symbol "□" as the fixation position, and lines between fixation points as the locus of the fovea. The fovea moved towards the object, and each object was detected one by one.

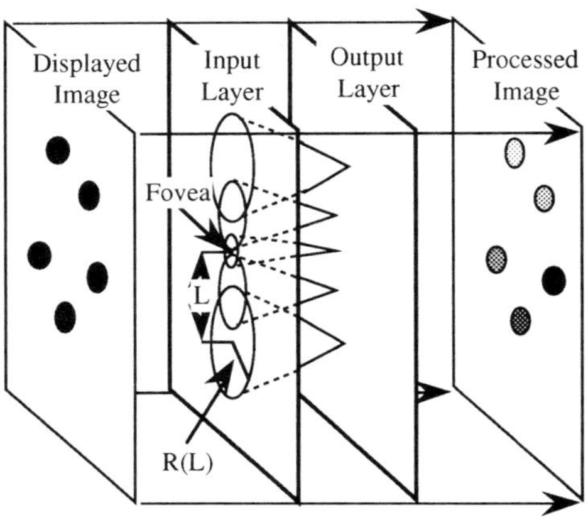

Fig.1 The structure of Retinal Model

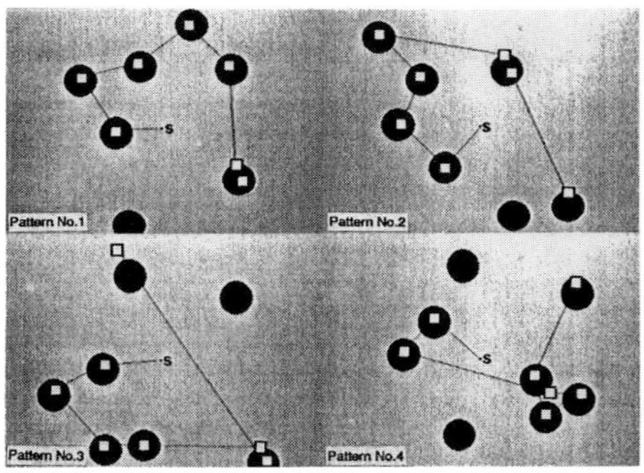

Fig. 2 Computed eye movements

3. HUMAN EYE MOVEMENTS

3.1. A cognitive experiment

To record the actual eye movements, we make a cognitive experiment using ophthalmography which is produced by NAC Co. The experiment is performed in the dark room. Figure 3 shows the facilities for the experiment. The screen is set 1.5m away from the test person. The computer graphic image is displayed on the screen from an overhead projector and a liquid crystal display that are located behind the test person.

The experimental procedure is as follows:

Before displaying the image a test person is told to stare at the center of the screen. The image, which is the same as the one in the computer simulation, is displayed for only a few short seconds since the aim is to prevent the test person from recognizing the objects in the displayed image. During the display of the image, the position of the eyes is recorded as the x-y spatial data every 1/600th of a second. After the experiment, the locus of eye movements is obtained from the data and imposed into the displayed image. It enables us to observe the eye movements on the displayed image.

3.2. Experimental results

Figure 4 is one of the results. The eye movements of one person are indicated as the locus on the displayed image. Some parts of the locus make the blotch in the figure. We found out in another investigation which is not presented here that this is caused by the eyes pausing momentarily at one point before proceeding to the next.

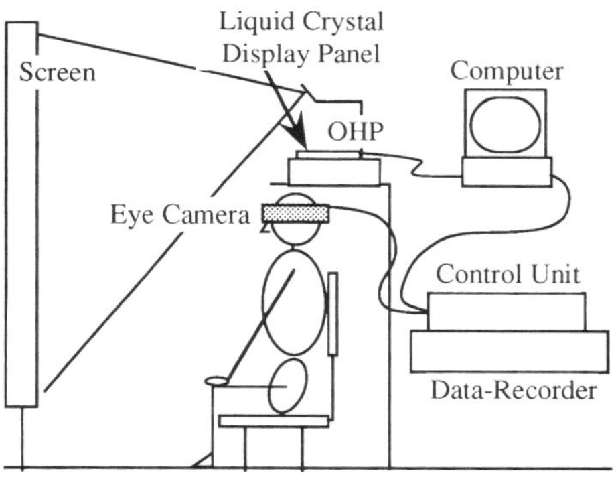

Fig. 3 Experimental facilities in the cognitive experiment

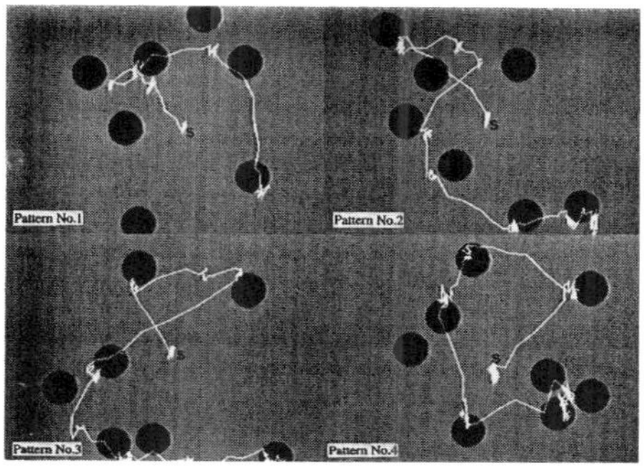

Fig. 4 Human eye movements

4. EVALUATION OF RESULTS

Both results in the computer simulation and the experiment showed quite similar eye movements. They were alike in three points:

1. Fixation points were on or close to the objects.
2. Only one eye movement was required to accomplish movement between fixation points.
3. Eye movements occurred towards the closest object from each fixation point.

In this paper we shall discuss the third feature. To analyze this tendency we need to evaluate the proximity between each fixation point. L, Lmax and Lmin indicate the distance between each fixation point, the maximum and the minimum distance between the present fixation point and the objects, respectively. We define the proximity index "p" as the following:

$$p=\frac{L_{max}-L}{L_{max}-L_{min}}$$

Figure 5, for example, indicates the proximity index from one fixation point to each object. The fixation point is adjusted to the center of the closest object. If eyes move towards the object "B" from the fixation point "A", the proximity index "p" is 0.590 on this eye movement. Likewise, all simulated and experimental eye movements were analyzed in the same way, then the frequencies of each value were graphed as a histogram in Figure 7. This denotes that most of the eye movements in the simulation and experiment occurred towards the object that is closely located from the present fixation point.

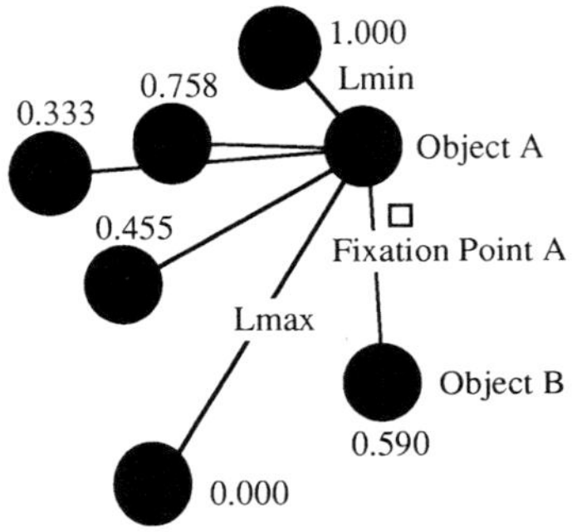

Fig. 5 Proximity index between fixation points

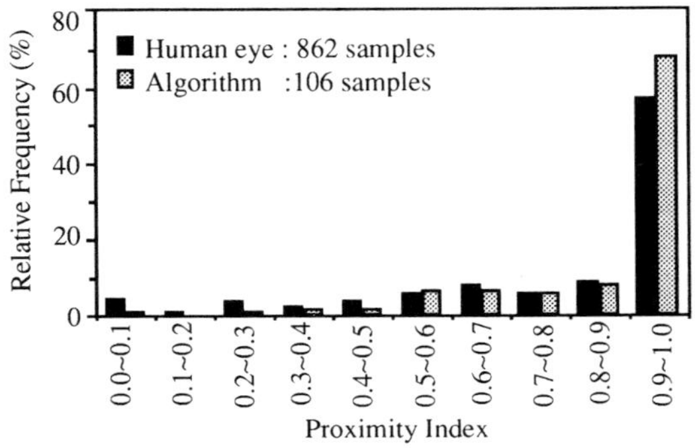

Fig. 6 Proximity index betweem a fixation point to each object

5. CONCLUSIONS

In this paper we made a mathematical model of the retinal signal processing. The size of the receptor field was small in the fovea and gradually became larger towards the periphery. Using this model we proposed an algorithm of eye movements in foveation. Its computer simulation showed that each object in the image was detected one by one using simulated eye movements. This result was compared with the actual human eye movements that were recorded with ophthalmography in a cognitive experiment. Comparison revealed that both results were basically identical. It implies the algorithm proposed in this paper realized eye movements in foveation.

It has been clear that the ability of the proposed algorithm varies depending on the two mathematical functions: the function which causes the size of the receptive field to change, and the function which defines the spatial distribution of the connecting value between input and output layers. These functions, as well as the procedure to delete every detected object in the image, will be investigated in a further study.

6. REFERENCES

[1] Wertheim T. : "Uber die indirekte Sehscharfe", *Z.Psychol. Physiol.Sinnesorg*, 7, 172-187, 1894

[2] Noton D. and Stark L. W. : "Scanpaths in eye movements during pattern perception", *Science*, Vol. 171, pp. 308-311, 1971

[3] Schneider G. E. : "Two visual systems", *Science*, Vol. 163, pp.895-902, 1969

[4] Marr D. *ision*. W.H.Freeman and Company, 1982

[5] Lindsay P. H. and Norman D. A., *Human Information Processing : An Introduction to Psychology, 2nd Edition*, Academic Press, 1977

[6] Sterling P., Freed M. and Smith R.G. : "Microcircuitry and functional architecture of the cat retina", *Trends in Neuro Science*, Vol. 9, No. 5, pp. 186-192, 1986

[7] Volkmann F. C., Schick A. M. L. and Riggs L. A. : "Time course of visual inhibition during voluntary saccades", *J. Opt. Soc. Am.*, 58, pp. 562-569, 1968

Simulation and Design of Applied Electromagnetic Systems
T. Honma (Editor)

The optimum configurations of SQUID sensors for the magnetocardiogram and magnetoencephalogram

H. Saotome and Y. Saito

College of Engineering, Hosei University, 3-7-2 Kajino, Koganei, Tokyo 184, Japan

Abstract

This paper deals with optimum SQUID sensor configurations for biological system applications. It is shown that the optimum SQUID sensor configurations make it possible to establish locally orthogonal coordinate systems at the measurement surfaces. This leads to a faster field source searching algorithm based on the sampled pattern matching method. Some test examples demonstrate that current distribution in the heart can be estimated at a speed of about ten times faster than the conventional sampled pattern matching algorithm.

1. INTRODUCTION

Obtaining the current signal distribution in a human body from electromagnetic field data measured over the body takes a very important role in medical diagnosis. This process is reduced to solving the inverse problem. In order to identify the current signal distributions in the human heart from a magnetocardiogram (MCG) and in the human brain from a magnetoencephalogram (MEG), we have successfully developed the sampled pattern matching (SPM) algorithm [1-5]. Moreover, this algorithm has been applied to non-destructive testing [6].

In the previous studies, the Cartesian coordinate system was used to estimate current element (or dipole) vectors. In this paper, we propose the locally orthogonal coordinate systems, such as the cylindrical and spherical coordinate systems, for MCG and MEG analyses, respectively. With the result of employing the locally orthogonal coordinate systems, we reveal that the optimum SQUID sensor configurations make it possible to implement the SPM method for fast medical diagnosis with MCGs and MEGs.

2. FASTER SPM METHOD UTILIZING THE LOCALLY ORTHOGONAL COORDINATE SYSTEMS

The MCG and MEG analyses can be reduced to searching for distribution of the source current density **J**. This problem is expressed by the following integral equation:

$$\mathbf{H} = \nabla \times \int_{v} G \mathbf{J} \, dv, \tag{1}$$

where **H** is the magnetic field intensity and G denotes the Green function. In (1), the product **J** dv corresponds to the current dipole vector. The measured magnetic field pattern (MCG or MEG) is caused by current dipole vectors. Therefore, the identification of current distribution is carried out by estimating not only current dipole positions but also their directions. Identifying the current dipole direction requires considerable CPU time in the Cartesian coordinate system. For example, 72-time SPM processes are necessary at the same spatial position when the angular resolution is 5 degrees [3-5]. The locally orthogonal coordinate systems, i.e. the cylindrical and spherical coordinate systems proposed in this paper, can remove the angular subdivision, so that an efficient SPM implementaion can be carried out. This is because the magnetic field pattern on the measurement surface of these coordinate systems can be decomposed into two orthogonal field pattern components whereas it can not be done in the Cartesian coordinate system.

With these locally orthogonal coordinate systems, the system equation for the MCG and MEG analyses is derived from (1). By discretizing a target volume into m small subdivisions, we get

$$\mathbf{u} = \sum_{j=1}^{m} (\alpha_{pj}\, \mathbf{d}_{pj} + \alpha_{qj}\, \mathbf{d}_{qj}); \quad p=r,\ q=\theta \quad \text{or} \quad p=\theta,\ q=\phi; \tag{2}$$

where $\mathbf{u}$ shows a MCG or MEG pattern, α_{pj} and α_{qj} are p and q components of a current dipole vector at a position j. When we have n measurement points, the n-dimensional vectors $\mathbf{d}_{pj}$ and $\mathbf{d}_{qj}$ show the estimated magnetic field patterns due to α_{pj} and α_{qj}, respectively. In the proposed coordinate systems, the orthogonality between $\mathbf{d}_{pj}$ and $\mathbf{d}_{qj}$

$$\mathbf{d}_{pj}{}^{T} \cdot \mathbf{d}_{qj} = 0, \tag{3}$$

is satisfied as shown in Fig. 1 for the cylindrical coordinates and Fig. 2 for the spherical coordinates. Making the most of (3), the pattern matching rates of $\mathbf{d}_{pj}$ and $\mathbf{d}_{qj}$ with respect to the measured MCG or MEG $\mathbf{u}$ can be independently evaluated by

$$\gamma_{tj} = \mathbf{u}^{T} \cdot \mathbf{d}_{tj} \,/\, (\|\mathbf{u}\|\, \|\mathbf{d}_{tj}\|); \quad t=p,\ q. \tag{4}$$

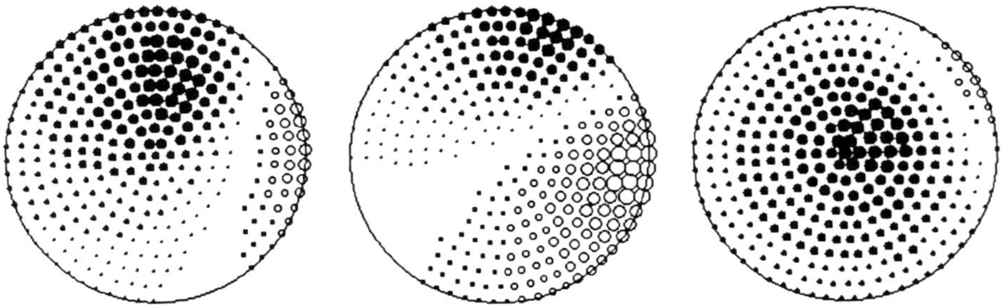

(a) $\mathbf{u}$ pattern (b) $\mathbf{d}_r$ pattern (c) $\mathbf{d}_\theta$ pattern

Figure 1. Magnetic field patterns normal to a circular measurement surface of the radius r=1 in the cylindrical coordinate system when the current dipole vector $(\alpha_r, \alpha_\theta)=(1,1)$ is located at $(r,\theta,z)=(0.8,\pi/6,0.8)$. The circles ● and ○ indicate N and S poles, respectively. Obviously, $\mathbf{u}=\mathbf{d}_r+\mathbf{d}_\theta$ and $\mathbf{d}_r{}^T\cdot\mathbf{d}_\theta=0$ are held.

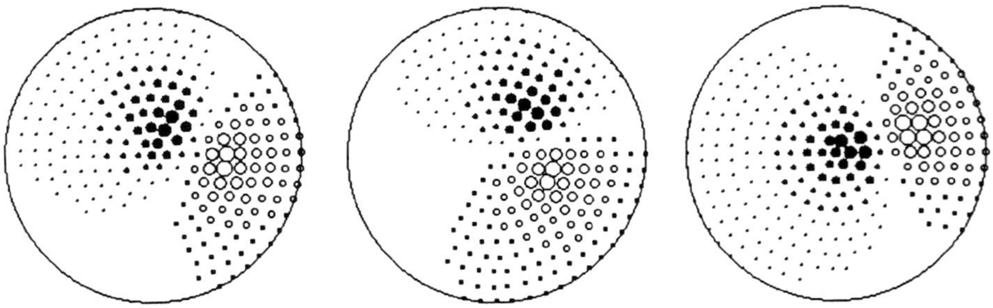

(a) $\mathbf{u}$ pattern (b) $\mathbf{d}_\theta$ pattern (c) $\mathbf{d}_\phi$ pattern

Figure 2. Magnetic field patterns normal to a spherical measurement surface of the radius r=1 and the angle $\theta \leqq \pi/4$ in the spherical coordinate system when the current dipole vector $(\alpha_\theta, \alpha_\phi)=(1,1)$ is located at $(r,\theta,\phi)=(0.8,\pi/12,\pi/12)$. The circles ● and ○ indicate N and S poles, respectively. $\mathbf{u}=\mathbf{d}_\theta+\mathbf{d}_\phi$ and $\mathbf{d}_\theta{}^T\cdot\mathbf{d}_\phi=0$ are held.

From (4), it is possible to obtain the amplitude of the normalized current dipole spectrum [3,4], γ_j, and its spatial angle on the p-q plane, ψ_j:

$$\gamma_j = \sqrt{\gamma_{pj}^2 + \gamma_{qj}^2}, \tag{5}$$

$$\psi_j = \tan^{-1}(\gamma_{qj} / \gamma_{pj}), \tag{6}$$

where ψ_j is determined by taking account of the signs of γ_{pj} and γ_{qj}. After finding out the first current dipole vector, we continue with the procedures similar to (4)-(6) up to the peak of γ [1-6].

Therefore, employing the proposed coordinate systems for the MCG and MEG analyses removes the repetitive process for determining the angle of a current dipole vector. As a result, we suggest the optimum configurations of SQUID sensors as shown in Figs. 1 and 2, rather than mounting them in a square or rectangular grid.

3. EXAMPLES

Figure 3 shows the normalized current dipole vector spectrum distributions obtained from three MCGs of reference [7]. From the comparison of the results obtained by the conventional and proposed algorithms, it is revealed that the

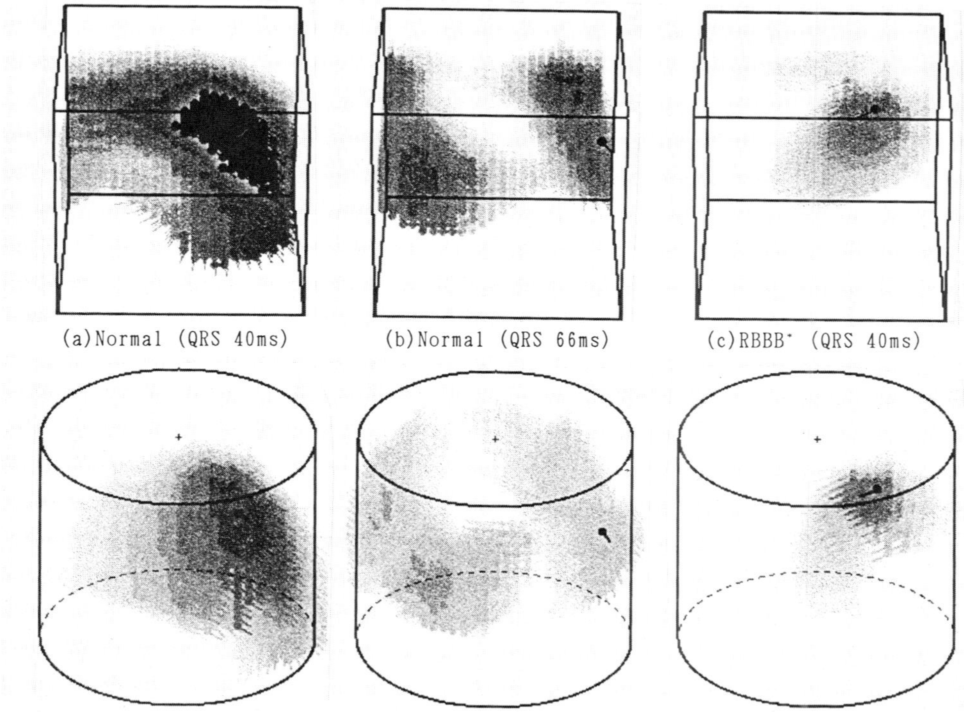

(a)Normal (QRS 40ms) (b)Normal (QRS 66ms) (c)RBBB* (QRS 40ms)

(d)Normal (QRS 40ms) (e)Normal (QRS 66ms) (f)RBBB* (QRS 40ms)

Figure 3. Comparison of the estimated current dipole vector spectrums (γ>0.8) obtained by the conventional and proposed algorithms. (a)-(c)Obtained by the conventional algorithm with the Cartesian coordinate system; m=13225, n=36, $\Delta\psi$=5[deg]. (d)-(f)Obtained by the proposed algorithm with the cylindrical coordinate system; m=14150, n=32. *Right Bundle Branch Block syndrome.

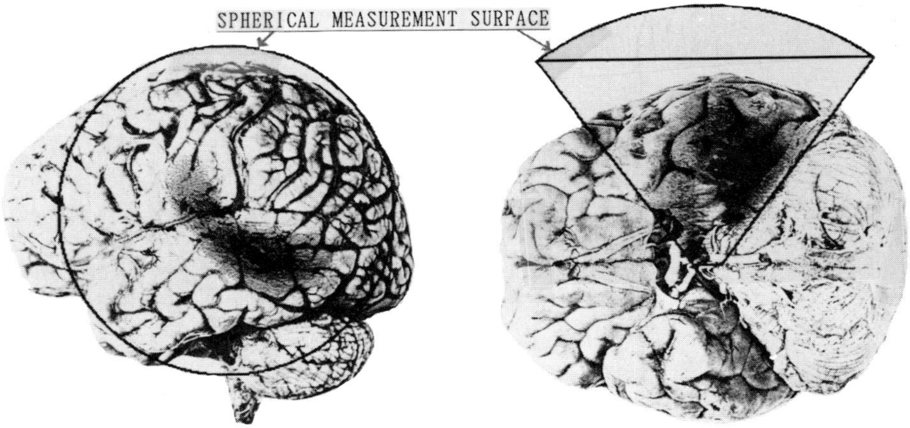

(a)Profile (b)Bottom view

Figure 4. Current dipole vector spectrum distribution (γ>0.9) obtained by applying our new algorithm to the MEG at 110ms after hearing the voice of "ah"; m=75221, n=37. The target region is enclosed by solid lines.

proposed method provides the comparable results to the conventional ones even if the angular subdivision for current dipole vectors has not been carried out. The results of Figs. 3(d)-(f) were obtained more than ten times faster compared with those of Figs. 3(a)-(c).

Using a 37-channel SQUID of which the sensors are mounted on a spherical surface having the radius of curvature, 122mm, Prof.Kuriki's group in Hokkaido University measured a MEG when a person heard a voice of "ah" [8]. The application of our new method to the MEG at 110ms after hearing the voice provided the current dipole distribution in his/her brain as shown in Fig. 4.

4. CONCLUSION

We have proposed the locally orthogonal coordinate systems for the MCG and MEG diagnoses based on the SPM method, as well as the optimum SQUID sensor configurations. Thus, our novel algorithm proposed in this paper can provide medical doctors with current distributions in the heart and in the brain faster than the conventional method.

5. REFERENCES

1 Y.Saito, et al., J. Appl. Phys., 67, No.9, (1990), pp.5830-5832.
2 H.Saotome, et al., Trans. IEE, Japan, Vol.112-A, No.4, (1992), pp.279-286.
3 H.Saotome, et al., Trans. IEE, Japan, Vol.113-C, No.1, (1993), pp.101-108.
4 H.Saotome, et al., Electromagnetic Field Source Searching from the Local Field Measurement, Int. J. Appl. Electromag. Matrls, (1993), in Press.
5 H.Saotome, et al., A Neural Behavior Estimation by the Generalized Correlative Analysis, IEEE Trans. Magn., March, (1993), in Press.
6 H.Saotome, et al., Crack Identification in Metallic Materials, IEEE Trans. Magn., March, (1993), in Press.
7 K.Watanabe, et al., (K.Atsumi et al., Ed., Biomagnetism' 87, Tokyo Denki University Press, Japan, 1988), pp.346-349.
8 S.Kuriki, et al., J. Japan Biomag. Bioelectromag. Society, Vol.5, No.1, June (1992), pp.20-23.

Simulation and Design of Applied Electromagnetic Systems
T. Honma (Editor)

EVALUATION OF ION-CURRENT EXPOSURE ON ANIMAL BODIES USING BOUNDARY ELEMENT METHOD

M.Yamashita[a], K.Shimizu[b], T.Kobayashi[b] and G.Matsumoto[a]

[a]Hokkaido Institute of Technology, 7-15, Maeda, Teine-ku, Sapporo, Japan

[b]Hokkaido University, Kita13-Nishi8, Kita-ku, Sapporo, Japan

Abstract

For the safety assessment of DC power transmission lines, a technique has been developed to evaluate the ion-current exposure on animal bodies. This technique is based on the Boundary Element Method and the space charge terms were processed under an assumption of lumped source of ion-current. Calculation errors were examined and found to be small enough for our applications. Using this technique, the ion-current exposure was evaluated with a human bodies. The field distribution along the body surface were obtained. Their dependence on the ion density and the insulating conditions were clarified.

1. INTRODUCTION

With the recent surge in the demand for electric power, the ultrahigh-voltage direct-current power transmission has been planned to be used in the near future. Before the DC transmission becomes in operation, the safety of the electromagnetic field under the transmission lines has to be assessed. Air ions are generated around the DC power lines and the flow of the air ions, or the so-called ion-current, is observed under the power lines[1]. In regard to the biological effects of the ions in air, numerous studies have been carried out. Either useful or harmful effects have been reported which depend on the polarity and the amount of the ions[2]. For the safety assessment of the transmission lines, the quantitative evaluation of the human and animal exposure to the ion-current is essential.

This paper presents an numerical analysis technique which has been developed to analyze the 3 dimensional field around an animal body with space charges.

2. METHOD OF NUMERICAL ANALYSIS

The object region and its conditions are shown in Fig.1. To model a windless space near the ground surface under a transmission line, a uniform ion-current field is considered.

The governing equation in this region is given by the following Poisson equation:

$$\nabla^2 u - p = 0 \qquad (p = -\rho / \varepsilon) \qquad (1)$$

where u is the potential function, ρ is the space charge density, and ε is the permittivity of air.

A numerical analysis technique based on the boundary element method (BEM) was applied. Equation (1) is expressed as,

$$B_i + \sum_{j=1}^{n} H_{ij}u_j = \sum_{j=1}^{n} G_{ij}q_j \quad , \text{ where } B_i = \int_\Omega p u^* d\Omega \qquad (2).$$

According to the method proposed by Brebbia[3], B_i is obtained by dividing the region into 3 dimensional cells. However, with this division, the important advantage of the BEM is lost. The BEM enables the computation by means of the boundary conditions alone. Also, in the case of 3 dimensional bodies with complex shapes such as animal bodies, a substantial amount of labor is required for the cell division and computation.

In the present study, an approximate method[4] is applied to the ion-current field. As shown in Fig.1 with a dashed line, a planar lumped source of ion-current is assumed horizontally at $z=z_0$. It can be expressed by means of the delta function as

$$p = K\,\delta(z-z_0) \qquad (3)$$

where K is a constant. When eq.(3) is substituted into eq.(2),

$$B_i = \int_\Omega K u^* \delta(z-z_0) d\Omega \qquad (4).$$

Hence, the integration over the entire region is reduced to the one only on the boundary elements so that the advantage of the BEM is maintained.

To apply this method to an animal body, we choose the animal surface as a part of the boundary which is divided finely into constant elements. Then we can derive the electric field value at each element of the body surface.

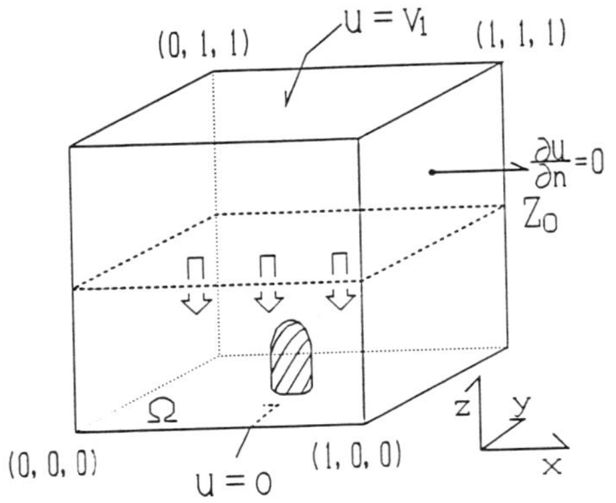

Fig.1 Region and conditions for numerical calculation.

3. EVALUATION OF THE ERROR

The accuracy of the present technique was analyzed with a uniform field shown in Fig.2. The computational conditions were as follows. The boundary conditions V_1 and V_0 were 10 kV/m and 0 V, respectively. The errors were less than 30 % in the worst case and were ten several % at the center point where the animal bodies are placed. The errors were small enough for our applications.

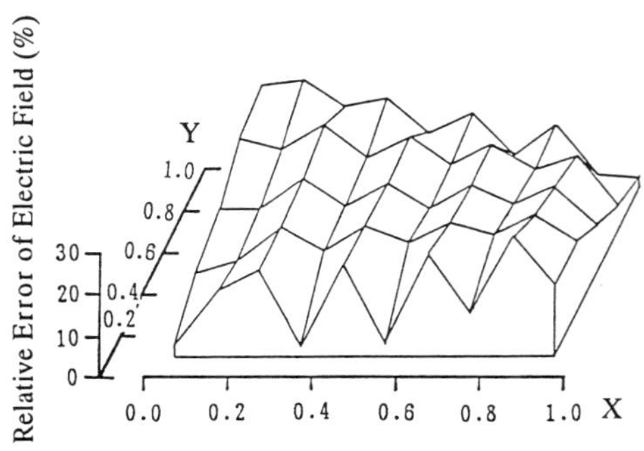

Fig.2 Distribution of electric field error on the ground plane. ($\rho = 10^{-7}$ C/m^3)

4. EXPOSURE ANALYSIS WITH ANIMAL BODY

The present technique was applied to the exposure analysis with a human body in an ion-current field. Fig.3 shows the manner of model division and the calculated results of the surface electric field intensity. The computational conditions were as follows. The boundary conditions V_1 and V_0 were 10 kV/m and 0 V, respectively. The potential of the model was set to be the ground potential (0V). The intensity and the direction of the electric field on the body surface are indicated in a vector. Field concentration can be observed at a head and shoulders.

In the safety assessment of electromagnetic field, the enhancement factor (E.F.) is one of the basic parameters. The E.F. is defined as E_m/E_o, where E_m and E_o are the maximum field at a body surface and unperturbed field on the ground in the Laplace field. The dependence of the E.F. on the space charge density was investigated. Fig.4 shows the result. The influence of the charge density on the E.F. increases rapidly as the density increases from 10^{-8} to 10^{-7}C/m^3.

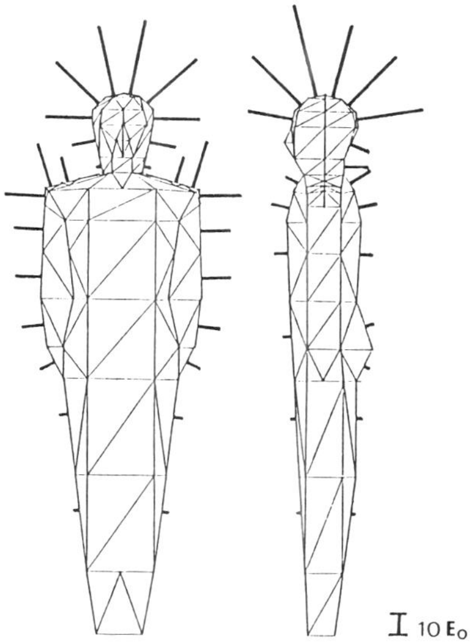

Fig.3 Electric field distribution on the body surface analyzed by BEM. ($\rho = 10^{-7}$C/m^3)

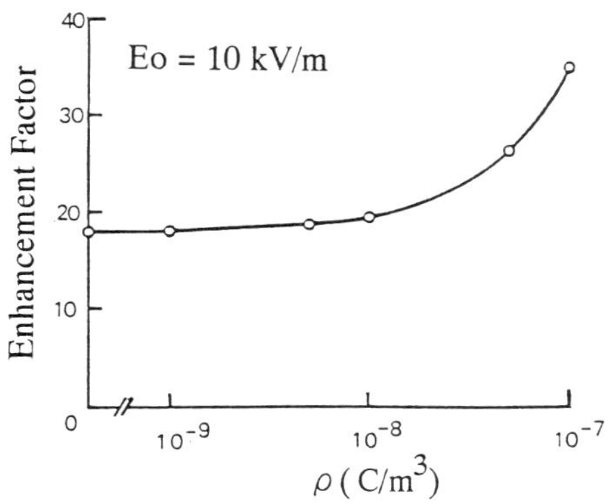

Fig.4 Dependence of enhancement factor on charge density.

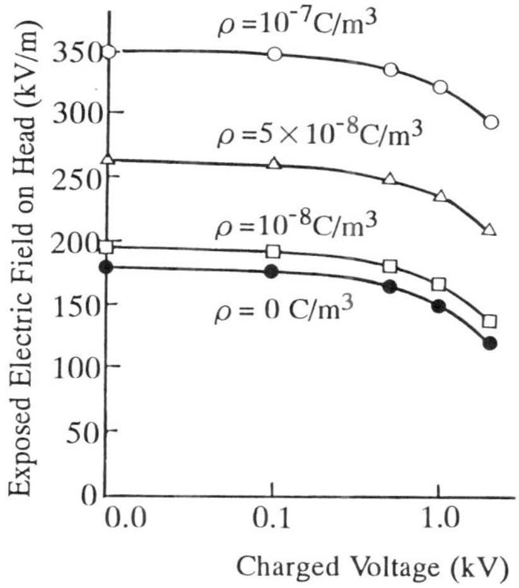

Fig.5 Dependence of human exposure on charged voltage.

Finally, the influence of the charged voltage on the field exposure was investigated with a human model. Fig.5 shows the result. As the charged voltage increases, the concentration of electric field was relieved. Hence, ion-current flowing into the body surface is reduced when the charged voltage increases. On the other hand, the rise of the charged voltage results in the increase of the discharge current, when the human touches a grounded object. Therefore, for the safety, both of these factors have to be taken into account.

5. CONCLUSIONS

For the safety assessment of ion-current field, a technique was developed which evaluates the field exposure on an animal body. It can be applied to the general 3 dimensional object with a complex shape such as an animal body. After the numerical accuracy of the present technique was confirmed, it was applied to a human body.

The distribution of electric field concentration on the human body under the power line was obtained. The dependence of the field concentration was investigated on the space charge density and on the charged voltage of a human body. It was found that the concentration is strongly affected by the existence of the space charges as well as the charged voltages. Hence these factors should not be neglected in the safety assessment of electric field under high-voltage transmission lines.

The present work was supported in part by the grant from Ministry of Education, Science and Culture of Japan(04202133).

6. REFERENCES

1 E.L.Carstensen, Biological effects of transmission line fields,Elsevier,New York, 1987.
2 H.L.König et al., Biologic effects of environmental Electromagnetism, Springer - Verlag,New York,1981.
3 C.A.Brebbia, Fundamentals of Boundary Element Method for Engineers, Pentech Press, London,1978.
4 M.Yamashita et al., Electronics and communications in Japan,3,Vol.74,No.7, 1991.

Simulation and Design of Applied Electromagnetic Systems
T. Honma (Editor)

Mathematical modelling of the problem of micro-currents generated in living bodies by power lines

A. Bossavit

Électricité de France, 1 Av. du Gal de Gaulle,
92141 Clamart

Abstract

We set the mathematical expression of the finite-element problem that has to be solved in order to predict micro-currents in the body of an electrical worker suspended below a power-cable. The distinction between "inductive" and "capacitive" effects is emphasized. Numerical results are briefly discussed.

1. INTRODUCTION

Since the first claims of a possible link between human cancer and the exposure to "extremely low frequency" (ELF) electromagnetic fields [17], and the animated debate they generated [1, 2, 9, 11, 15], it has become important to have reliable estimates of the levels of power deposition in living tissues due to such fields.

Numerical simulation may be used to precisely relate effective levels of magnetic field, induced current, etc., within living bodies, with the ambient levels, for specified sources, and this will help in two respects: 1°- Assess the more or less stringent and realistic character of safety standards to be proposed in the future, 2°- Check whether such standards are *consistent* (for instance, are ceilings of 500 μT for magnetic fields and 10 mA/m^2 for induced currents consistently set?). Such studies have recently been performed by the 'Trifou' team [6] and some results will be presented here. (Other recent numerical studies include [7, 8, 12, 13, 16]. Cf. also [10].)

This paper deals with mathematical modelling only. Our main goal is to present the relevant system of equations, in the simplest possible case (Section 2). We briefly report on recent numerical simulations (Section 3).

2. WHAT TO SOLVE FOR

We shall consider the situation of Fig. 1 as typical. It concerns a worker sitting under a power cable in an insulated carriage (not represented). The question is to compute average current densities in the body. The fine structure of living tissues is ignored, and an average anisotropic conductivity σ is assumed, but its variations with the nature of the tissues are to be taken into account. (Typical values, in MKSA units, are: 0.7 for blood, 0.1 for liver, muscle, lung, 0.03 for fat, 0.01 for bone.) Variants of the situation concern the case of a bystander standing on the ground (on slightly conducting soles), of a person operating a hair-dryer, etc. A rationale for the proposed model can be given by distinguishing two kinds of effect, as follows.

First, there are *inductive* effects, i.e., eddy-currents induced by the 700 A or so alternating current that is carried by the cable. Let us call C the conductive region of interest, in this case the worker's body. Knowing the source-current density j^s (a complex-valued vector field defined over a definite region of space, that we call C^s), one may compute a source-field h^s, such that $\text{rot}\, h^s = j^s$ and $\text{div}\, h^s = 0$. This can be done either by Biot and Savart's formula, or by first computing a vector potential a^s by the integral formula

$$a^s(x) = (4\pi)^{-1} \int_{C^s} dy\, |x - y|^{-1}\, j^s(y),$$

and then setting $h^s = \text{rot}\, a^s$. This source-field generates an electric field e, that obeys Faraday's law, and thus is such that $\text{rot}\, e = -\, i\omega\mu_0\, h^s$. By Ohm's law, this field drives a current-density $j = \sigma e$, which must be conservative, and therefore will satisfy the equation $\text{div}\, j = 0$ in C (which, if taken in the appropriate "weak" sense, includes transmission conditions at material interfaces) and the boundary condition $n \,.\, j = 0$ on the boundary S of region C, where n denotes the outer unitary normal field.

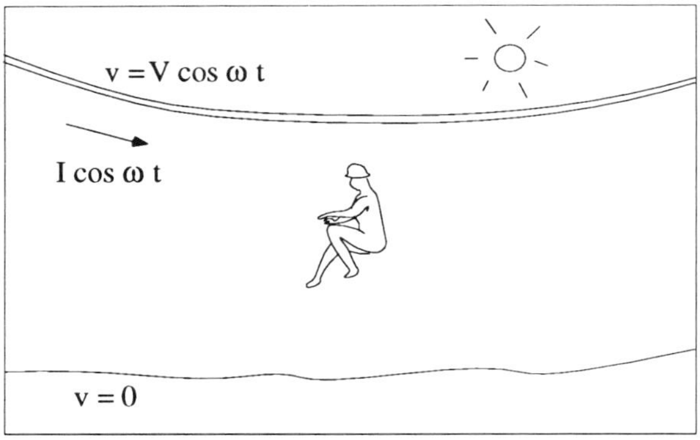

Figure 1. A typical situation, as far as mathematical modelling is concerned.

Now, this induced current will modify the magnetic field, hence a new, corrective inductive effect. But in the present situation, the penetration depth is very large with respect to the body's dimension, so we may ignore this correction. The induced current is thus given, with very good approximation, by the equations

$$\text{rot}\, e = -\, i\omega\mu_0\, h^s, \quad j = \sigma e, \quad \text{div}\, j = 0 \quad \text{in C}, \tag{1}$$

$$n \,.\, j = 0 \quad \text{on S}. \tag{2}$$

There are also *capacitive* effets, that are especially interesting, because they are not predicted by the standard eddy-currents model. They are due to the high value of the electric potential on the cable with respect to the Earth (some 400 kV). The strong electric field that results is deviated towards the body, and is very close to zero inside it. Let us call e_0 this "primary" electric field, which is, in principle, easily calculated: one looks for it as $e_0 = -\, \text{grad}\, v$, where $v = 0$ on the ground surface, $v = V$ on the cable, and $v = c$, an unknown and to be determined constant, in C. There are several possible numerical approaches to this field, by boundary integral methods as in [7], or by finite elements in a large enough region, with some artificial boundary condition at large distance.

What counts in this preliminary computation is the normal derivative $\partial v/\partial n$ on S: the product $i\omega\varepsilon_0\, \partial v/\partial n$ can be interpreted as the complex representation of the time-varying electric charge (conveyed by Maxwell's displacement currents) that enters the body. This charge (the total of which is 0) flows through the body, and thus generates a current j, which obeys Ohm's law, and the corresponding electric field $e = \sigma^{-1} j$ is curl-free. (This assertion is supported by an asymptotic analysis. Cf. [4].) So we have, for this part of the micro-currents,

$$\text{rot } e = 0, \quad j = \sigma e, \quad \text{div } j = 0 \quad \text{in } C, \tag{3}$$

$$n \cdot j = i\omega\varepsilon_0 \, \partial v/\partial n \quad \text{on } S. \tag{4}$$

Let us finally add the two effects. We thus obtain the final model, *find vector-fields* e *and* j *in* C *such that*

$$\text{rot } e = -\, i\omega\mu_0 \, h^s, \quad j = \sigma e, \quad \text{div } j = 0 \quad \text{in } C, \tag{5}$$

$$n \cdot j = i\omega\varepsilon_0 \, \partial v/\partial n \quad \text{on } S. \tag{6}$$

This is a *conduction problem*, but an unconventional one, because there are two kinds of source-terms: a flow of boundary charge, and a distributed electromotive force.

The most natural method to solve this problem seems to be *finite elements* in C with respect to a scalar electric potential, as follows. Let us look for the electric field e of (6) in the form $e = -\, i\omega\mu_0 \, a^s + \text{grad } \psi$, where a^s is the above-defined vector potential. Then, $\text{rot } e = -\, i\omega\mu_0 \, h^s$, and one deals with the other equations by demanding that

$$\text{div}(\sigma \text{ grad } \psi) = \text{div}(i\omega\mu_0\sigma \, a^s) \quad \text{in } C, \tag{7}$$

$$\sigma \, \partial\psi/\partial n = i\omega \, [\varepsilon_0 \, \partial v/\partial n + \sigma\mu_0 \, n \cdot a^s] \quad \text{on } S. \tag{8}$$

A convenient weak form of this problem is, *find* ψ *in* Ψ *such that*

$$\int_C \sigma \text{ grad } \psi \cdot \text{grad } \psi' = \int_C i\omega\mu_0\sigma \, a^s \cdot \text{grad } \psi' + \int_S i\omega\varepsilon_0 \, \partial v/\partial n \, \psi' \tag{9}$$

for all ψ' in Ψ, where the space of test-functions Ψ is made of all complex-valued functions whose gradient is square-integrable over C. From this point on, the application of Galerkin's method is straightforward.

3. RESULTS

In the case when σ is constant in C, the only source terms are in the right-hand side of (8), and a comparison of their orders of magnitude will show whether capacitive or inductive effects dominate. It seems often assumed in the relevant literature that capacitive effects do dominate (inductive effects are neglected, or not even mentioned). Capacitive effects, on the other hand, were considered to be well known, due to earlier studies [14], when we began this one. So we concentrated on inductive effects.

A reference finite-element tetrahedral mesh of one half of a human body was made, by using in conjunction several software systems: GEOMOD™, SUPERTAB™, SIMAIL™, etc. By parameterization, it can accommodate various positions (subject sitting, standing, etc.). It contains 49599 tetrahedra. Inductive effects were computed with the eddy-current code 'Trifou' [3, 5]. (The large ratio of the penetration depth to the characteristic length was not taken advantage of, contrary to what has been done in the above models.) Trifou uses a mix of edge- variables (54463 here) and nodal variables on the conductor's surface (3746). Current-densities of a few mA per sq. meter (230 $\mu A/m^2$ in the case of a pedestrian on the ground, on the basis of a field of 8 A/m there) were found. They are too low to have any "conventional" effect [1]. We refer to [4] for displays of current densities in the case of Fig. 1, and to [6] for detailed figures.

Overall, our numerical results so far do not corroborate the idea that capacitive effects dominate: inductive effects have been found larger.

4. CONCLUSION

The traditional wisdom, according to which displacement currents can be neglected at low frequencies (i.e., when the wavelength is large with respect to characteristic dimensions of the system), is challenged by what precedes. Although 50 Hz is a low frequency, "capacitive effects", for which the term $-i\omega\varepsilon_0$ appears to be responsible, could not be neglected in the case under study. But it is true that they appear as a perturbation [4]: they are accounted for by the *second* term in an asymptotic expansion the first term of which accounts for standard *inductive* effects (eddy-currents). Whether they can be neglected or not depends on the relative magnitude of this second, corrective term: it can be large at low frequencies, when there is a large electric field *outside* conductors, and in that case, capacitive effects are noticeable.

6. REFERENCES

[1] R.K. Adair: "Constraints on biological effects of weak extremely-low-frequency electromagnetic fields", **Phys. Rev. A, 43**, 2 (1991), pp. 1939-48.

[2] F.S. Barnes: "The Effects of Time Varying Magnetic Fields on Biological Materials", **IEEE Trans., MAG-26**, 5 (1990), pp. 2092-97.

[3] A. Bossavit: "The computation of eddy-currents, in dimension 3, by using mixed finite elements and boundary elements in association", **Mathl. Comput. Modelling, 15,** 3-5 (special issue on Boundary Element Methods), 1991, pp. 33-42.

[4] A. Bossavit: "A Theoretical Approach to The Question of Biological Effects of Low Frequency Fields", **IEEE Trans., MAG-** , - (1993), pp. - (to appear).

[5] A. Bossavit, J.C. Vérité: "A mixed FEM-BIEM Method to Solve Eddy-Current Problems", **IEEE Trans. on Magnetism, MAG-18**, 2 (1982), pp. 431-35.

[6] J . Cahouet: "Simulations numériques des champs électromagnétiques induits dans un être humain en présence de sources de champ magnétique", Int. Rept. 881-07, EdF (92141 Clamart, Fr.), 1992.

[7] K.-M. Chen, H.-R. Chang, C.-J. Lin: "Quantification of Interaction Between ELF-LMF Electric Fields and Human Bodies", **IEEE Trans., BME-33**, 8 (1986), pp. 746-55.

[8] A. Chiba, K. Ikada, M. Kitagawa, Y. Yokoi, T. Matsuo, M. Nagata, "Application of finite element method to analysis of induced current densities inside human model exposed to 60-Hz electric field", **IEEE Trans., PAS-103**, 7 (1984), pp. 1895-1902.

[9] M. Coleman, V. Beral: "A Review of Epidemiological Studies of the Health Effects of Living Near or Working with Electricity Generation and Transmission Equipment", **Int. J. Epidemiology, 17**, 1 (1988), pp. 1-13.

[10] D.W. Deno: "Currents induced in the human body by high voltage transmission line electric field — Measurement and calculation of distribution and dose", **IEEE Trans., PAS-96**, 5 (1977), pp. 1517-27.

[11] J. Jauchem: "Alleged Health Effects of Electromagnetic Fields: Misconceptions in the Scientific Literature", **J. Microwave Power and Electromag. Energy, 26**, 4 (1991), pp. 189-95.

[12] N. Orcutt, O.P. Gandhi: "A 3-D Impedance Method to Calculate Power Deposition in Biological Bodies Subjected to Time Varying Magnetic Fields", **IEEE Trans., BME-35**, 8 (1988), pp. 577-83.

[13] B.J. Roth, L.G. Cohen, M. Hallett, W. Friauf, P.J. Basser: "A Theoretical Calculation of the Electric Field Induced by Magnetic Stimulation of a Peripheral Nerve", **Muscle & Nerve, 13**, 8 (1990), pp. 734-41.

[14] K.H. Schneider, H. Studinger, K.H. Weck, H. Steinbigler, D. Utmishi, J. Wiesinger: "Courants de déplacement vers le corps humain causés par le champ diélectrique sous les lignes de transport d'énergie", **CIGRÉ** (112 Bd. Haussmann, 75008 Paris), 36-04 (1974), pp. 1-15.

[15] S. Shulman: "Cancer risks in electro-magnetic fields", **Nature, 345** (June 1990), p. 463.

[16] T. Takuma, T. Kawamoto, K. Isaka, Y. Yokoi: "A Three-Dimensional Method for Calculating Currents Induced in Bodies by Extremely Low-Frequency Electric Fields", **Bioelectromagnetics, 11** (1990), pp. 71-89.

[17] N. Wertheimer, E. Leeper: "Electrical wiring configurations and childhood cancer", **Am. J. Epidemiol., 109**, 3 (1979), pp. 273-84.

Simulation and Design of Applied Electromagnetic Systems
T. Honma (Editor)

Simulation of a magnetic curtain based on molecular dynamics

S.Ueno, M.Iwasaka, H.Eguchi, and T.Kitajima

Department of Electronics, Kyushu University, Hakozaki, Higashi-ku
Fukuoka, 812, Japan

Abstract

The present study focuses on the mechanism of the formation of the "magnetic curtain" which appears in an air atmosphere under gradient magnetic fields. It is assumed that the magnetic curtain is a wall of air made by gradient fields and paramagnetic oxygen. Gas flow experiments were carried out. A model based on molecular dynamics was proposed to explain the experimental results. In this study, trajectories of gas flows were calculated by computer, and the threshold value of magnetic field strength needed to block nitrogen gas flow was compared with the experimental results.

1. INTRODUCTION

We have observed a phenomenon in which candle flames are pressed down by magnetic fields of high intensities.[1] We have also observed a phenomenon in which the flow of gases such as carbon dioxide and oxygen is blocked by magnetic fields.[2] A model called a "magnetic curtain" has been introduced to explain these phenomena.

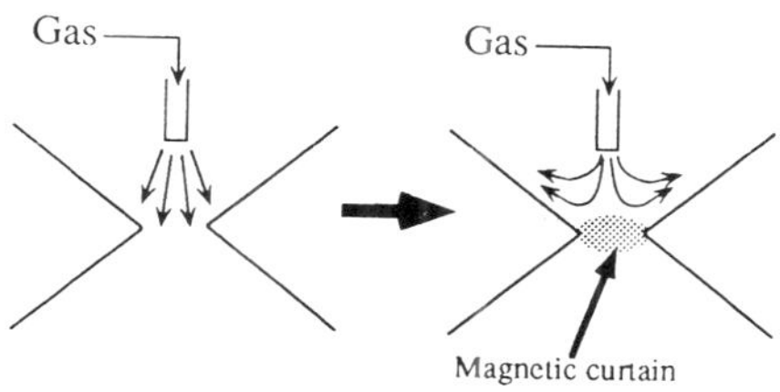

Fig.1 Formation of the magnetic curtain

It is assumed that the magnetic curtain is a wall of air caused by magnetic fields. We have demonstrated that a candle flame can be smothered by a magnetic field, since the magnetic curtain intercepts oxygen.[3] The interception of oxygen by the magnetic curtain quenches flames. Gas flow experiments in air and nitrogen atmos-phere have shown that oxygen as a paramagnetic molecule plays an important role in formation of the magnetic curtain.[4] This study focuses on the mechanism of the magnetic curtain. The magnetic curtain is simulated on the basis of molecular dynamics.

2. GAS FLOW EXPERIMENTS

We observed the blocking of nitrogen gas flow in an air atmosphere. As shown in Fig.2, flows of nitrogen gas were supplied to the airgap between magnetic poles, and the differential pressure was measured with a manometer. Fig.3 shows the change of the differential pressure against the magnetic field strength. The gas flow was clearly blocked at about 1T. The field gradient was about 120 T/m. Let us assume that the magnetic field exists only in x-direction and that the field gradient depends only on a position along the z-axis. Under this assumption, magnetic force acting on 1 m^3 of paramagnetic object is given by

$$F = \frac{\chi}{\mu_0} B \frac{dB}{dz}, \qquad \text{(in z-direction)} \qquad (1)$$

where χ is the magnetic susceptibility of a particle, B is the magnetic flux density, and μ_0 is magnetic permeability of the vacuum. For oxygen molecules, $\chi = 3449 \times 10^{-6}$ at 300 K. The magnetic force acting on 1 mol of oxygen in a 1T and 120 T/m field is approximately 4.2 N, and 7×10^{-24} N per one molecule.

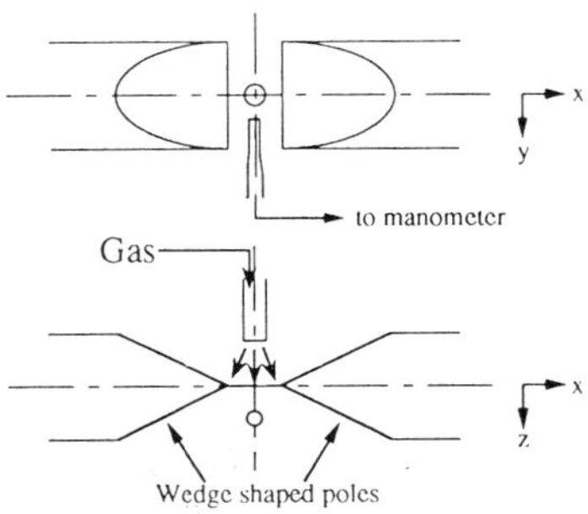

Fig.2 The apparatus of the gas flow experiment

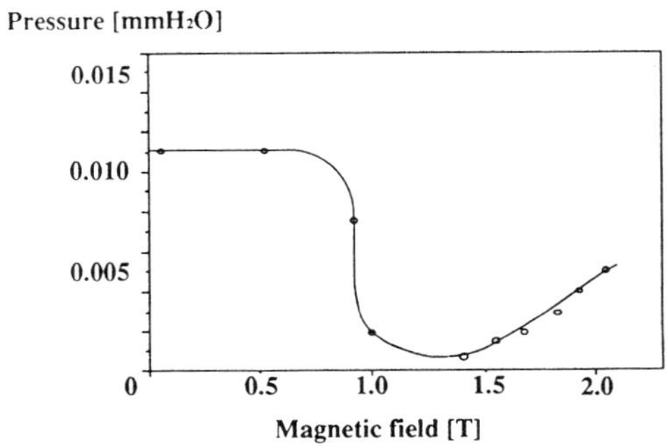

Fig.3 The differential pressure of nitrogen gas flow in various magnetic fields

3. COMPUTER SIMULATION OF THE GAS FLOW IN THE GRADIENT MAGNETIC FIELD

We introduced a model based on the molecular dynamics to explain the experimental results. In this model, gas molecules were treated as rigid spheres obeying the laws of motion for gas molecules. Collisions between molecules were treated as perfectly elastic collisions. The following equation was used to compute the average time between collisions τ from the mean free path for molecules L and the average velocity $\bar{u}$.

$$\tau = \frac{L}{\bar{u}} \quad (2)$$

τ is approximately 1.6×10^{-10} sec for oxygen molecules, and 1.4×10^{-10} sec for nitrogen molecules. According to the molecular dynamics, the velocities of gas molecules obey the Maxwell-Boltzmann velocity distribution. However, the equation for this distribution is complicated and not easily manipulated, so here we use a trigonometric function to approximate it. Velocities were assigned such that the totality of molecules conforms to this distribution. In this simulation, we assumed wedge-shaped poles like those in Fig.1; consequently the magnetic force distribution is as shown in Fig.4. The figure shows the force acting in the positive z-direction, so that oxygen molecules always feel a force tending toward the center of the field gradient. Acceleration in the velocities of oxygen molecules due to the magnetic force was calculated from the potential energy caused by the magnetic field as follows:

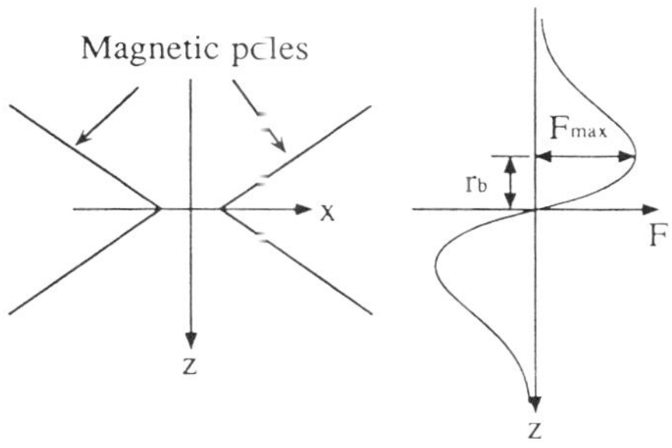

Fig.4 Distribution of magnetic force

$$\Delta v = \sqrt{\frac{2U}{m}} \quad (3)$$

where U is the potential energy, m is the mass of an oxygen molecule. Velocity changes calculated from equation (3) were added to the randomly assigned velocities of oxygen molecules. As a result, flows of nitrogen molecules are affected by the magnetic force acting on oxygen, because the nitrogen molecules move colliding with oxygen molecules. Molecules involved in collisions were chosen at random with a probability of 4:1 for nitrogen : oxygen, based on the composition of air.

Initially velocities are assigned to 10 nitrogen molecules, such that the flow was downward. A molecule was moved for time τ at this velocity, and then another molecule (nitrogen or oxygen) was chosen to collide with it, and the change in velocity was calculated assuming a perfectly elastic collision, and the result added to the flow. The molecule was then moved again for time τ at the new velocity. This was repeated to compute the trajectories of gas molecules.

4. RESULTS AND DISCUSSION

Figures 5(a), (b) and (c) show simulation results for nitrogen molecules. The vertical length of the region shown is 80mm, and the distance from the center of the magnetic field to the peak of the magnetic force, is 20mm. Each molecule moved through 10^8 collisions. The time elapsed is roughly 0.02 sec or so. As the magnetic force is increased, changes occur in the gas flow, and at the maximum magnetic force of 1×10^{-23} N the gas flow clearly is blocked.

Fig.1 Formation of the magnetic curtain

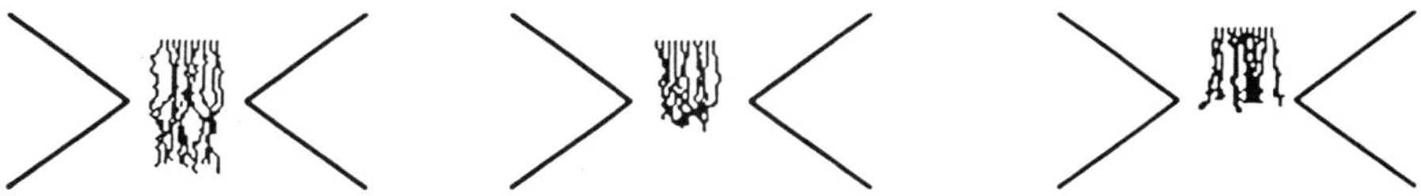

(a) (b) (c)

Fig.5 Trajectories of nitrogen molecules in gradient magnetic fields
(a) Fmax = 1×10^{-24} N (B = 0.14 T) (b) Fmax = 7×10^{-24} (B = 1 T) (c) Fmax = 1×10^{-23} N (B = 1.4 T)

5. REFERENCES

[1] S. Ueno and K. Harada, "Experimental difficulties in observing the effects of magnetic fields on biological and chemical processes", IEEE Trans. Magn. MAG-22 (1986), pp. 866-873

[2] S. Ueno and K. Harada, "Effects of magnetic fields on flames and gas flow", IEEE Trans. Magn. MAG-23 (1987), pp. 2752-2754

[3] S. Ueno, "Quenching of flames by magnetic fields", J. Appl. Phys. 65 (1989), pp. 1243-1245

[4] S. Ueno and M. Iwasaka, "Properties of magnetic curtain produced by magnetic fields", J. Appl. Phys. 67 (1990), pp. 5901-5903

Simulation and Design of Applied Electromagnetic Systems
T. Honma (Editor)

Effects of magnetic fields on fibrin polymerization and dissolution of fibrin

M. Iwasaka, S. Ueno[a] and H. Tsuda[b]

[a]Department of Electronics, Faculty of Engineering, Kyushu University Fukuoka 812, Japan

[b]Clinical Laboratory, Kyushu University Hospital, Kyushu University Fukuoka 812, Japan

Abstract

Effects of magnetic fields on fibrin polymerization and Fibrinolysis were investigated. The present study focuses on the role of ferromagnetic particles on the magnetic orientation of fibrin polymers. Rod-like magnetic particles were mixed into the solution of fibrinogen. The polymerization process of fibrin was exposed to a magnetic field at 8 T. Magnetic particles were agglomerated to form chains of clusters which were oriented parallel to magnetic fields. We observed no oriented fibrin near the chains. In contrast, oriented fibrins were observed where no magnetic particles existed. We have also investigated a possible effect of magnetic fields on the fibrinolytic process. The fibrin dissolution was observed by measuring the absorbance changes at 500 nm.

1. Introduction

Fibrin polymers are diamagnetic materials that are oriented in a magnetic field. Fibrinogen changes to fibrin monomer by the action of protease thrombin, and fibrin monomers are polymerized. In the course of the polymerization process, when a magnetic field of intensity 10 to 20 T is applied, the fibrin fibers orient parallel to the magnetic fields. When no magnetic field is applied, fibrin fibers become entangled in mesh, and no orientation is observed.

This phenomenon was first discovered by Torbet et al. in 1981 [1], and thereafter was reproduced by Yamagishi et al. with an 8 T field [2], and also by the authors with a 6 T field [3]. The present study focuses on the role of ferromagnetic particles in the magnetic orientation of fibrin polymers. A possible effect of magnetic fields on the fibrinolytic process is also investigated.

2. Magnetic orientation

Fibrin monomer has a molecular weight of 340,000 dalton and a size of 450 Å x 90 Å, which are polymerized by electrostatic bonds, and then are stabilized by the covalent bonds between the neighboring fibrin molecules introduced by factor XIII.

The polymerization of fibrin was started by adding thrombin (0.03 NIH units) to 1 ml of 4.8 mg/ml fibrinogen dissolved in 0.05M Tris-HCl buffer, pH 7.4 containing 0.10 M NaCl and 0.5mM EDTA, which was completed within 30~ 60 min. The polymerization process of fibrin was exposed to magnetic fields up to 8.0 T using a superconducting magnet. Magnetic orientation of the fibrin was observed by a scanning electron microscopy; it was observed that the fibrin polymers were oriented parallel to the magnetic fields.

To investigate the effect of ferromagnetic particles on magnetic orientation of fibrin, rod-like magnetic particles (ferrimagnetism γ-Fe_2O_3, 0.3μm x 0.03μm, saturation magnetization 77 emu/g and coercive force 624 Oe) were mixed into the solution of fibrinogen. The polymerization process of fibrin was exposed to a magnetic field at 8.0 T.

As shown in Fig.1, oriented fibrin polymers were observed when the polymerization process of fibrin was exposed to a magnetic field.

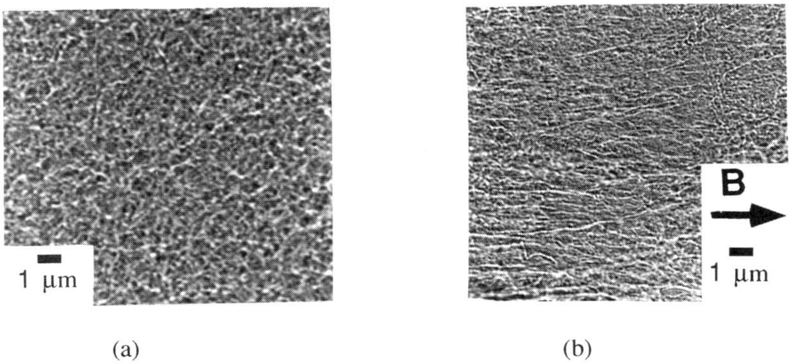

Fig.1 Optical micrographs of fibrin polymers. (a) control ; (b) formed in a magnetic field at 8.0 T.

Magnetic particles were agglomerated to form chains of clusters which were oriented parallel to magnetic fields as shown in Fig.2. The chains of magnetic particles themselves agglomerated and polymerized into a large fiber as shown in Fig.3. We observed no oriented fibrin near the chains. In contrast, oriented fibrins were observed in the large fibers of chains. It seems that the mechanism of fibrin polymerization in the large fiber of magnetic particles was different from the area where small chains of magnetic particles existed.

It is mentioned that magnetic orientation of fibrin is caused by the diamagnetic anisotropy of fibrin monomers [4]. In the magnetic field at 8.0 T, the magnetic energy of fibrin monomer is small compared with thermal energy kT, therefore a fibrin monomer can not be oriented. It is possible for fibrin polymer to have enough magnetic energy and to be oriented by a magnetic field at 8.0 T.

Because magnetic orientation of fibrin was observed when the polymerization process of fibrin was exposed to a magnetic field at 8.0 T, it is estimated that the number of fibrin monomers in the oriented polymer is about 10^7, and the size is 0.1μm ~ 100μm in length. The gaps of oriented chains of magnetic particles were ~1μm as shown in Fig.2.

There is a possible concept to be considered. Right after exposure to a magnetic field, magnetic particles formed chains of clusters that were parallel to a magnetic field, hence the rotative motion of fibrin polymers could be disturbed by chains of magnetic particles.

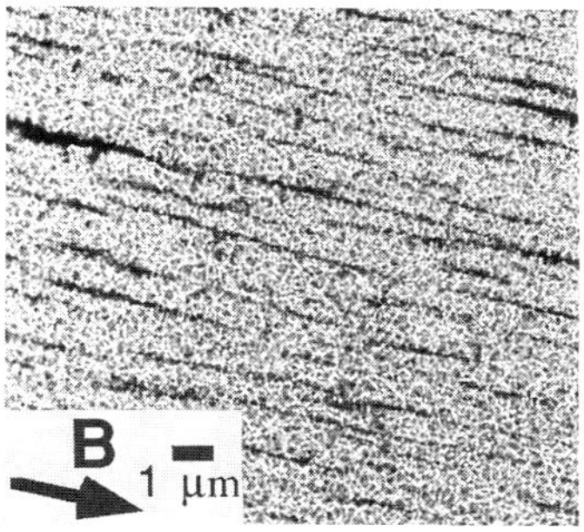

Fig.2 Fibrin and chains of magnet particles. Chains oriented parallel to field are magnetic particles. Fibrin fibers are entangled in mesh in the space between chains.

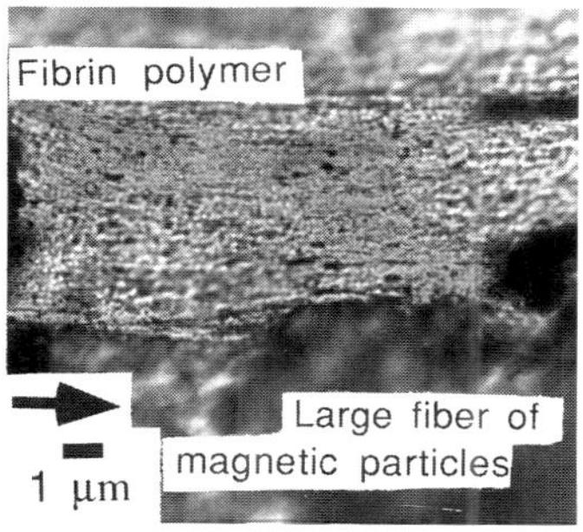

Fig.3 Oriented fibrin polymers in a large fiber of magnetic particles.

Let us compare the magnetic energy of fibrin polymers with that of a cluster of magnetic particles. Magnetic energy of a cluster of ferromagnetic particles is given by

$$U_p = -\frac{1}{2} M H \tag{1}$$

where H is magnetic field, and M is magnetic moment . When the size of the cluster of magnetic particles (saturation magnetization 77 emu/g) was 10μm x 0.1μm, the magnetic moment was estimated to be $\sim 10^{-10}$ emu. In contrast, magnetic energy of fibrin polymer, which is a diamagnetic material, is given by

$$U_f = -\frac{1}{2}\{\chi_\perp + (\chi_{//} - \chi_\perp)\cos^2\theta\} N H^2 \tag{2}$$

where $\chi_{//}$ is diamagnetic susceptibility in the direction of the long axis, $\chi_\perp$ is that of radius, θ is the angle between the direction of a magnetic field and the long axis of fibrins, N is the number of fibrin monomers in a fibrin polymer that orient parallel to a magnetic field ($\sim 10^7$) [2].

When $|\chi_\perp| >> |\chi_{//}|$, Up / Uf is given by

$$\frac{U_p}{U_f} = \frac{M}{\chi_\perp \sin^2\theta \, N H} \tag{3}$$

($\chi_\perp \cong \Delta\chi \times 666 = 4.9 \times 10^{-27}$; $\Delta\chi = 7.3 \times 10^{-29}$ is anisotropy of diamagnetic susceptibility in planer peptide [2,5]) , and the value is estimated to be $\sim 10^4$. It was confirmed that chains of magnetic particles have enough energy to disturb the orientation of fibrin polymer.

3. Dissolution of fibrin in magnetic fields

Coagulation and fibrinolysis occur in vessels in a mutually compensating and balanced state. Fibrin gels are dissolved when polymerized fibrins are degradated to fragments by plasmin. To investigate the effect of magnetic fields on the fibrinolytic process in thrombus and ways to dissolve the fibrin that have oriented in a magnetic field, we carried out the following experiment.

Fibrinogen was transformed to fibrin by incubation with thrombin and plasmin at room temperature for 45 min in a cuvette at room temperature (25 °C), either with magnetic fields or without magnetic fields. After 45 min, samples were removed from magnetic fields and incubated at 37 °C. Then the fibrin dissolution was observed by measuring the absorbance changes at 500 nm. The result is shown in Fig.4. The absorbance of fibrin gels at 500 nm decreased after they were incubated at 37 °C. There were no differences in absorbance between the control group and fibrin gels formed in a magnetic field at 8.0 T.

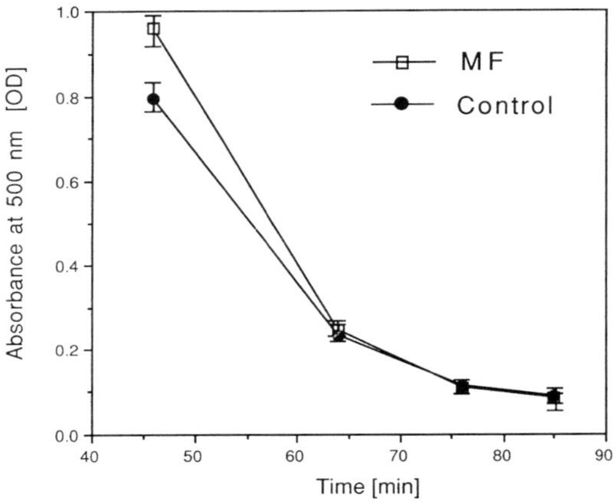

Fig.4 Absorbance at 500 nm in fiblinolitic process. Fibrins (9.6 mg/ml) were incubated with thrombin (0.6 NIH units/ml) and plasmin (0.021 casein units/ml). MF: prepared in 8-T magnetic fields. Control: control group.

4. References

[1] J. Torbet, M. Fryssinet and G. Hudry-Clergeon, Nature, 289, 91, 1981

[2] A. Yamagishi, T. Takeuchi, T. Higashi and M. Date, J. Phys. Soc. Jpn., 58, 2280, 1989

[3] S. Ueno, M. Iwasaka and H. Tsuda, Papers Tech. Meeting Magn., IEEJ, MAG-89-4, 1989

[4] J.M.Freyssinet, J.Torbet, G.Hudry-Clergeon: Biochimie, vol. 66, 81 ,1985

[5] D.L.Worcester : Proc. Natl. Acad. Sci. USA, vol. 75, 5475, 1978

Magnetic Fluids

Simulation and Design of Applied Electromagnetic Systems
T. Honma (Editor)

Gas bubbles in magnetic fluids

V.V.Gogosov[a], G.A.Shaposhnikova[a], A.Ya.Simonovskii[b]
[a] *Moscow University, Institute of Mechanics, 117192 Moscow, Michurinskii Pr. 1, Russia*
[b] *Stavropol Agricultural Institute, 355014 Stavropol, Zootechnicheskii Side-Street 10, Russia*

In this paper some effects of the nonuniform magnetic field on the formation and motion of gas bubbles in magnetic fluids are discussed.

1.Experimental study of the frequency of vapor bubble formation during the boiling of magnetic fluids.

In order to measure the frequency of vapor bubble formation in a magnetic fluid that is boiling in a nonuniform magnetic field the experimental set-up is created. The scheme of this set-up is shown on the Fig.1.The magnetic fluid used in experiments is a colloidal suspension of magnetite in water. The saturation magnetization is 20.4 kA/m.

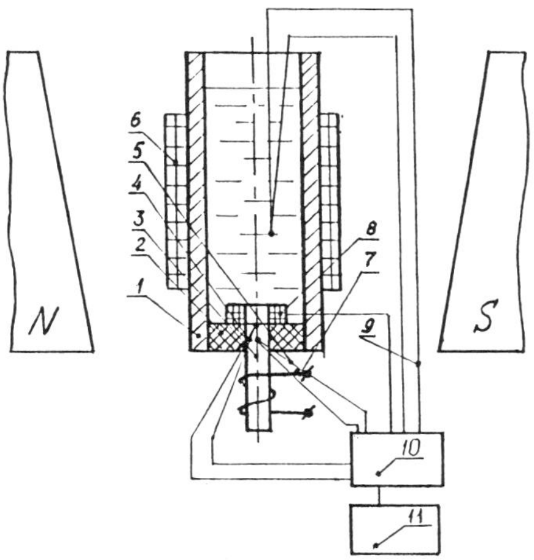

1 – the cylindrical container
2 – the fluoroplastics bottom of the container
3 – the steel rod
4,5,9 – the thermocouples
6,7 – the heaters
8 – the inductive coil
10 – the oscillograph
11 – the amplifier

Fig.1. Experimental set – up scheme

The nonmagnetic container 1 with the fluid is brought into the interpole region of the electromagnet. The electromagnet creates the magnetic field increasing along the gravity direction.

The magnetic fluid boiling and vapor bubble formation take place on the upper end of the rod 3 – on the heat releasing surface. To register the frequency of the bubble formation, the inductive coil 8 is used. The temperature of the magnetic fluid, T_l, the temperature of the heat releasing surface, T_w, and the temperature inside the rod at the distance 3 mm from the end , T_v, are measured with the thermocouples 9,4 and 5.

The temperatures of the magnetic fluid, T_l, and the heater surface, T_w, increase with time. The temperature of the magnetic fluid remains to be less than the temperature of boiling, T_s.

The typical parts of the oscillograms obtained from inductive coil 8 as the result of measurements are given on the Fig 2 a,b,c.

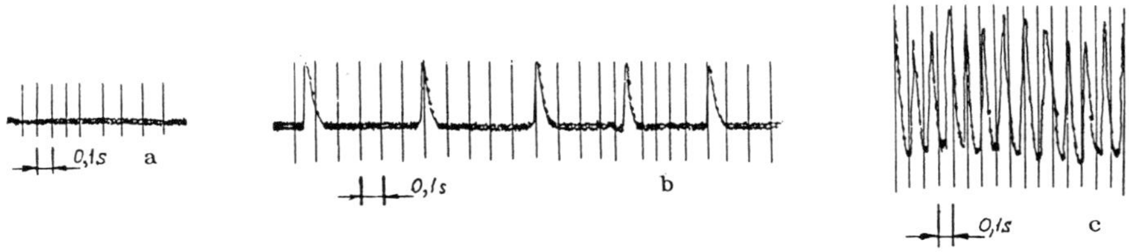

Fig.2. The typical oscillogram ; a – the initial part , b – the beginning of the boiling, c – the boiling near the saturation temperature

As long as the magnetic fluid temperature less than 90°C and the temperature of the heat releasing surface less than 120°C there are no any pulses on the oscillogram, Fig.2a. It means that the bubbles do not form at these temperatures .

At the temperature of the heat releasing surface close to 120°C the inductive pulses appear on the oscillogram, Fig.2b. The generation of these pulses is caused by the bubble formation. The temperature of the magnetic fluid in the volume corresponding to the first stage of the vapor bubble formation on the heat releasing surface ranges up to 90°C.

As the magnetic fluid is heated and the temperature of the heat releasing surface increases, the frequency of the pulses increases too. The amplitude of the pulses grows simultaneously, Fig.2c. With the growth of the temperatures of the fluid kern and the heater wall the frequency of vapor bubble formation and their size increase.

The Fig.3 displays the graphs, showing the dependencies of the frequency of the vapor bubble generation at the boiling of the magnetic fluid in the nonuniform magnetic field. The curve 1 corresponds to the initial stage of boiling, the large values of $\Delta T = T_S - T_1$ (ΔT is of order 8^0C). The curve 2 corresponds to the small values of ΔT (ΔT is of order 1^0C). The magnetic induction B measured in tesla is plotted along the abscissa axis. The curve 1 corresponds to the part of oscillogram shown on the Fig.2b, the curve 2 corresponds to the part on the Fig.2c

It should be noted that the magnitude T_W does not change with variation of B for the constant value of ΔT within the limits of experimental error.The magnitude f was averaged over 12 measurements.

At magnetic field values less than 120 kA/m the detachment frequency of bubbles grows as the magnetic field increases. For small values of ΔT the effect of the magnetic field is more pronounced (the curve 1 on the Fig.3 is more inclined). At the magnetic field values more than 120 kA/m the curves 1 and 2 coincide within the scale chosen to plot the graphs and then go down abruptly. The frequency of bubble formation decreases in one order of magnitude. The frequency f is nearly invariable with the subsequent increase of the magnetic field to 144 kA/m .

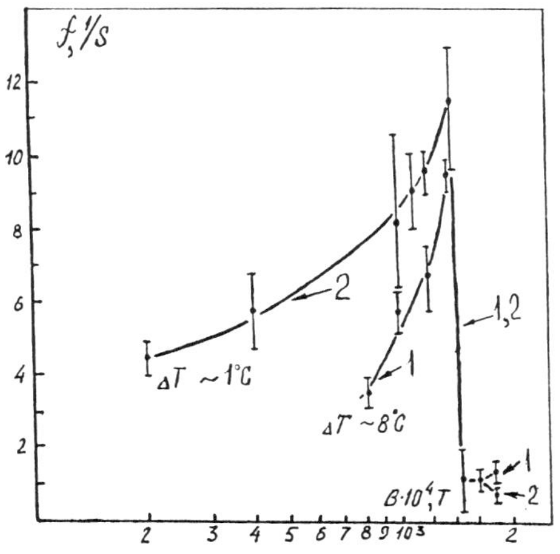

Fig.3. The dependence of the frequency f of the bubbles formation on magnetic induction B

2. The motion of the gas bubbles in the magnetic fluid generated by oscillating magnetic field gradient.

The peculiarity of the behavior of gas bubbles in magnetic fluid is demonstrated with an example of the one-dimensional problem on the motion of gas bubbles in conditions of the microgravitation in the immobile magnetic fluid which is exposed to a magnetic field varying in the following manner

$$H = H_0 + A\,x \sin \omega t, \quad H_0 = \text{const}\ , \quad A = \text{const} \qquad (2.I)$$

For definiteness, let us choose the x-axis to be directed from bottom to top, H_0 is taken to be directed from the top down, $H_0 < 0$. The magnetic field H_0 is strong and the fluid is magnetized to saturation. The case of low-frequency approximation is considered. In this approximation the time derivatives in the equations may be neglected, and the process of pressure - dilation of the gas in bubbles may be assumed isothermal.

In this case the set of equations describing the bubble motion consists of two algebraic equations

$$v_b = -\frac{2a^2}{9\eta_{mf}} M^S_{mf} A \sin \omega t\ , \qquad \frac{a^3}{a_0^3} = \frac{p_b^0}{p_b^0 - M^S_{mf} A x \sin \omega t} \qquad (2.2)$$

where v_b , a are the velocity and radius of bubbles ; η_{mf}, M^S_{mf} are the dynamic viscosity and magnetization of magnetic fluid ; a_0, p_b^0 are the radius of bubble and pressure in it at the initial point in time.

From the equations (2.2) it follows that the velocity of bubble during its upward motion is less than the velocity of its downward motion. As a consequence the bubbles move in the direction of the constant component of magnetic field H_0, executing the oscillating motion . The trajectory of bubble motion is shown on the Fig.4 .

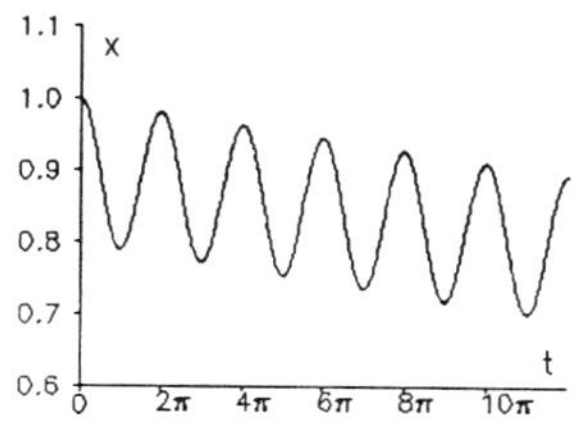

Fig.4. Trajectory of bubble.

The same effect is observed in the presence of the gravitation. When the magnetic field gradient is harmonically varying with time the bubbles may be caused to sink by directing the constant component of the magnetic field H_0 along the force of gravitation.

Simulation and Design of Applied Electromagnetic Systems
T. Honma (Editor)

Constitutive equation of magnetization for magnetic fluids

Y. Ido and M. Kiya

Department of Mechanical Engineering, Hokkaido University, Kita 13, Nishi 8, Kita-ku, Sapporo 060, Japan

Abstract

A new equation of magnetization for magnetic fluids is proposed. This equation is determined by the thermodynamical method that is based on the free energy and the dissipation function.

1. INTRODUCTION

Many researchers have studied about a complete set of basic equations for magnetic fluids, e.g., Shliomis[1], Berkovsky, et al.[2], Shizawa and Tanahashi[3,4], and authors[5]. However, the system of basic equations for magnetic fluids has not been established yet. When we determine the constitutive equations, most of difficulties arise in the treatment of magntization relaxation phenomena and behavior of suspended particles.

In this paper, we determine the equation of magnetization by the thermodynamical method that is based on the free energy and the dissipation function. In this way, the principle of material frame indifference and the principle of maximal dissipation rate[6] are taken into account. The magnetic flux density and the electric displacement are treated as variables of electromagnetic field instead of the electric field and the magnetic field.

2. THERMODYNAMICAL RELATIONS

The energy equation of conducting magnetic fluids is given[5] by

$$\rho \frac{\mathrm{D}u^{(e)}}{\mathrm{D}t} = -\rho p^{(e)} \frac{\mathrm{D}v}{\mathrm{D}t} + \boldsymbol{T}^{(v)} : (\nabla \boldsymbol{v} - \boldsymbol{\epsilon} \cdot \boldsymbol{\Omega}) + \boldsymbol{\Lambda} : \nabla \boldsymbol{\Omega}$$

$$+ \rho \boldsymbol{E} \cdot \frac{\mathrm{D}}{\mathrm{D}t}\left(\frac{\boldsymbol{D}}{\rho}\right) + \rho \boldsymbol{H} \cdot \frac{\mathrm{D}}{\mathrm{D}t}\left(\frac{\boldsymbol{B}}{\rho}\right) - \boldsymbol{M}\boldsymbol{H} : \boldsymbol{\epsilon} \cdot \boldsymbol{\Omega} + \boldsymbol{j}^{*} \cdot \boldsymbol{E}^{*} - \nabla \cdot \boldsymbol{q} + \rho r, \qquad (1)$$

where the notations here are as follows; ρ : the density, $u^{(e)}$: the internal energy density per unit mass, $p^{(e)}$: the pressure, v : the specific volume $= 1/\rho$, $\boldsymbol{T}^{(v)}$: the viscous part of the stress tensor, $\boldsymbol{v}$: the velocity vector, $\boldsymbol{\Lambda}$: the couple stress tensor, $\boldsymbol{\epsilon}$: Levi-Civita's symbol, $\boldsymbol{\Omega}$: the volume averaged angular velocity of suspended particles, $\boldsymbol{E}$: the electric field, $\boldsymbol{D}$: the electric displacement, $\boldsymbol{H}$: the magnetic field, $\boldsymbol{B}$: the magnetic flux density, $\boldsymbol{M}$: the magnetization, $\boldsymbol{j}$: the electric current density, $\boldsymbol{q}$: the heat flux density, r : the intensity of heat sources. The superscript (e) indicates the thermodynamical quantities when electromagnetic field exists. The symbols with an asterisk are the quantities in the moving coordinate system with velocity $\boldsymbol{v}$ against the stationary coordinate system.

The first law of thermodynamics for a magnetic fluid is expressed by

$$\rho T \mathrm{D}s^{(e)}/\mathrm{D}t = \Phi - \nabla \cdot (\boldsymbol{q}/T) + \rho r, \tag{2}$$

where $s^{(e)}$ is the entropy per unit mass and Φ is the dissipation function defined by

$$\Phi = \boldsymbol{T}^{(v)} : (\nabla \boldsymbol{v} - \boldsymbol{\epsilon} \cdot \boldsymbol{\Omega}) + \boldsymbol{\Lambda} : \nabla \boldsymbol{\Omega} + \rho \boldsymbol{H}^{(d)} \cdot \frac{\mathrm{J}}{\mathrm{J}t}\left(\frac{\boldsymbol{B}}{\rho}\right) + \boldsymbol{j}^* \cdot \boldsymbol{E}^* - \frac{\boldsymbol{q}}{T} \cdot \nabla T. \tag{3}$$

The quantity with superscript (d) is the dissipative part of the constitutive equation. The corotational derivative $\mathrm{J}/\mathrm{J}t$ is defined by

$$\mathrm{J}/\mathrm{J}t = \mathrm{D}/\mathrm{D}t - \boldsymbol{\Omega} \times . \tag{4}$$

In the most of previous theories, independent variables of electromagnetic field are the electric field and the magnetic field. Here we treat the electric displacement and the magnetic flux density as variables of electromagnetic field. Substitution of Eq.(3) into Eq.(2) yields the Gibbs equation

$$\mathrm{d}u^{(e)} = -p^{(e)}\mathrm{d}v + \boldsymbol{E} \cdot \mathrm{d}(v\boldsymbol{D}) + \boldsymbol{H}^{(q)} \cdot \mathrm{d}(v\boldsymbol{B}) + T\mathrm{d}s^{(e)}. \tag{5}$$

The Helmholtz free energy is defined by

$$\psi = u^{(e)} - Ts^{(e)}, \tag{6}$$

then the Gibbs equation (5) is rewritten

$$\mathrm{d}\psi = -p^{(e)}\mathrm{d}v + \boldsymbol{E} \cdot \mathrm{d}(v\boldsymbol{D}) + \boldsymbol{H}^{(q)} \cdot \mathrm{d}(v\boldsymbol{B}) - s^{(e)}\mathrm{d}T. \tag{7}$$

We can derive the quasi-conservative parts of the constitutive equations from the Gibbs equation while the dissipative parts of them are determined from the dissipation function by the principle of maximal dissipation rate[6].

3. CONSTITUTIVE EQUATION

The free energy and the dissipation function must satisfy the principle of material frame indifference, i.e., they are independent of coordinate system. Thus, they can be expressed by polynominals of integrity bases composed of respective variables. We assume that phenomenon is compound and the electric polarization is neglected here. Under these assumptions, the free energy is represented by the following form;

$$\psi = a\boldsymbol{B} \cdot \boldsymbol{B} + \tilde{\psi}(T,\ v,\ \boldsymbol{D}) \tag{8}$$

From the Gibbs equation, a partial differential equation

$$\boldsymbol{H}^{(q)} = \rho \left(\frac{\partial \psi}{\partial \boldsymbol{B}} \right)_{T,\ v,\ \boldsymbol{D}} \tag{9}$$

is derived and substituting Eq.(8) into it, the quasi-conservative part of the magnetic field is determined by

$$\boldsymbol{H}^{(q)} = 2\alpha\rho\boldsymbol{B}, \tag{10}$$

where α is a constant.

Since the dissipation function satisfies the principle of material frame indifference, it can be expressed by a polynominal of invariants composed of its variables. From the principle of maximal dissipation rate[6], the dissipative part of the magnetic field is determined by

$$\boldsymbol{H}^{(d)} = \beta \frac{\partial I_H}{\partial [\mathrm{J}(v\boldsymbol{B})/\mathrm{J}t]}, \tag{11}$$

where β is a constant and I_H is the invariant. Under compound process, I_H is given by $I_H = [\mathrm{J}(v\boldsymbol{B})/\mathrm{J}t] \cdot [\mathrm{J}(v\boldsymbol{B})/\mathrm{J}t]$. Substituting it into Eq.(11), the dissipative part of the magnetic field

$$\boldsymbol{H}^{(d)} = 2\beta \frac{\mathrm{J}}{\mathrm{J}t} \left(\frac{\boldsymbol{B}}{\rho} \right), \tag{12}$$

is obtained.

The constitutive equation is given by a sum of the quasi-conservative part and the dissipative part. Thus, the constitutive equation of magnetic field is given by a sum of Eq.(10) and (12), i.e.,

$$\boldsymbol{H} = 2\alpha\rho\boldsymbol{B} + 2\beta \frac{\mathrm{J}}{\mathrm{J}t} \left(\frac{\boldsymbol{B}}{\rho} \right). \tag{13}$$

Equation (13) should include the term related to the relaxation phenomena of magnetization. Thus, when the concentration of suspended particles in a magnetic fluid is increased and suspended particles are fixed under steady electromagnetic field, the equation of magnetization should be the relaxation equation

$$\partial \boldsymbol{M}/\partial t = -(\boldsymbol{M} - \boldsymbol{M}_0)/\tau, \tag{14}$$

where $\boldsymbol{M}_0$ is the equilibrium magnetization. Taking account of this and the equation

$$\boldsymbol{B} = \mu_0 \boldsymbol{H} + \boldsymbol{M}, \tag{15}$$

the constitutive equation (13) can be rewritten

$$\rho \frac{\mathrm{D}}{\mathrm{D}t}\left(\frac{\boldsymbol{B}}{\rho}\right) = \boldsymbol{\Omega} \times \boldsymbol{B} - \frac{1}{\tau}(\boldsymbol{M} - \chi_m \boldsymbol{H}), \tag{16}$$

where χ_m is the magnetic susceptibility and τ is the time corresponding to the relaxation time of magnetization. This equation can represent the effect of magnetization change due to the density gradient. Equation (16) includes the Shliomis relaxation equation[1]. It is equivalent to the Shliomis relaxation equation under assumptions as follows; (i) fluid is incompressible, (ii) the corotation derivative of the magnetic field equals to zero, i.e., $\mathrm{J}\boldsymbol{H}/\mathrm{J}t = \boldsymbol{0}$.

Equation (16) leads the equation of magnetization

$$\boldsymbol{M} = \frac{\chi_m}{\mu_0 + \chi_m}\boldsymbol{B} - \frac{\rho\mu_0\tau}{\mu_0 + \chi_m}\left[\frac{\mathrm{D}}{\mathrm{D}t}\left(\frac{\boldsymbol{B}}{\rho}\right) - \boldsymbol{\Omega} \times \frac{\boldsymbol{B}}{\rho}\right], \tag{17}$$

or

$$\boldsymbol{M} = \chi_m \boldsymbol{H} - \tau\left[\rho\frac{\mathrm{D}}{\mathrm{D}t}\left(\frac{\boldsymbol{B}}{\rho}\right) - \boldsymbol{\Omega} \times \boldsymbol{B}\right]. \tag{18}$$

The above equations are corresponding to the constitutive equation of magnetization. The first term of right-hand side of Eq.(18) is the equilibrium magnetization, the second term represents the effect of density gradient and this term is related to unsteadiness of electromagnetic field, and the last term is related to the relaxation by the rotation of suspended particles. We do not assume incompressibility of fluid here, thus, it is reasonable that the magnetization depends on the density gradient in magnetic fluids. The Shizawa-Tanahashi equation of magnetization [4], derived by the thermodynamical method, cannot describe the magnetic fluid flow behavior under rotating magnetic field, i.e., the flow is independent of the frequency of rotating magnetic field. However, the above equation (18) can analyze the flow under rotating magnetic field.

4. CONCLUDING REMARKS

We proposed the new equation of magnetization. It includes the Shliomis relaxation equation. They are identical under these assumptions; (i) fluid is incompressible, (ii) the corotational derivative of the magnetic field equals to zero. And this equation can describe the relaxation phenomena of magnetization when the density gradient exists.

5. REFERENCES

1 M.I. Shliomis, Sov.Phys.Usp., 17 (1974) 153.

2 B. Berkovsky, A.N. Vislovich and B.E. Kashevsky, IEEE Trans. Magn., MAG-16 (1980), 329.

3 K. Shizawa and T. Tanahashi, Bull.JSME, 28-243 (1985) 1942.

4 K. Shizawa and T. Tanahashi, Bull.JSME, 29-255 (1986) 2878.

5 Y. Ido and T. Tanahashi, J.Phys.Soc.Jpn., 60 (1991) 466.

6 H. Ziegler, An Introduction to Thermomechanics, 245, North-Holland, Amsterdam ·New York·Oxford, 1983.

Simulation and Design of Applied Electromagnetic Systems
T. Honma (Editor)

Theoretical analysis of magnetic fluid flow in consideration of nonuniform dispersion of magnetic particles

K. Shimada and S. Kamiyama

Institute of Fluid Science, Tohoku University, 2-1-1 Katahira, Aoba-ku, Sendai 980, Japan

Abstract

Theoretical analysis is conducted by taking into account nonuniform distribution of magnetic particles. The basic equations are derived by an analogy with a gas molecular motion in the fluidized bed. The equations are numerically solved in the case of one dimensional steady pipe flow in a strong nonuniform magnetic field to clarify the effect of magnetic field on the pipe flow characteristics.

1. INTRODUCTION

The theoretical analysis of magnetic fluid flow have mostly conducted with an assumption of uniform dispersion of magnetic particles [1,2]. However, the experimental results of unsteady pipe flow characteristics in a strong magnetic field showed large deviation from the calculated results [3]. It has also been pointed out that the pipe flow characteristics are influenced strongly on the aggregation of particles near the pole pieces [4]. It is, therefore, necessary to analaze the pipe flow characteristics by using a theory in which the nonuniform distribution of magnetic particles is considered.

In the present paper, the theoretical analysis is conducted by taking into account nonuniform distribution of magnetic particles. The basic equations are derived by an analogy with a gas molecular motion in a fluidized bed [5]. The equations are numerically solved in the case of one dimensional steady pipe flow in a strong nonuniform magnetic field to clarify the effect of magnetic field on the pipe flow characteristics.

2. THEORETICAL ANALYSIS

Now we consider magnetic fluid flowing in a horizontal pipe in an applied magnetic field. In the theoretical analysis, the magnetic fluid is devided into two-phases of magnetic particles and carrier liquid. The conservation equations of mass, momentum and energy of each phase are derived as follows.

The mass conservation equations:
for particles phase,

$$\frac{\partial c^*}{\partial t^*} + \nabla \cdot (c^* \boldsymbol{v}^*) = \frac{1}{Re} \nabla \cdot (\frac{1}{\boldsymbol{Sc}} \nabla c^*), \tag{1}$$

for liquid phase,

$$\frac{\partial}{\partial t^*} c_1^* + \nabla \cdot (c_1^* \boldsymbol{v}_l^*) = 0 \tag{2}$$

where all variables are nondimensionalized by using the representative values as $c^* = c/c_o$, $t^* = tv_o/r_o$, $\boldsymbol{v}^* = \boldsymbol{v}/v_o$, $r^* = r/r_o$, $z^* = z/r_o$, c the volmetric ratio of particles phase per unit volume of magnetic fluid, c_o the volume concentration of magnetic particles without magnetic field, t the time, r, z the coordinates in radial and axial direction, $\boldsymbol{Sc}(= \mu_l/\rho_l \boldsymbol{D})$

the Schmidt number, $Re(=\rho_l v_o r_o/\mu_l)$ the Reynolds number, ρ_l the density of solvent, v_o the velocity of magnetic fluid, r_o the radius of pipe, μ_l the viscosity of solvent, $\boldsymbol{D}$ the mass diffusivity and $\boldsymbol{v}^*$ the mean velocity of liquid and particles phases which is defined as,

$$\boldsymbol{v}^* = \frac{c_1^*}{c_1^* + \tilde{\rho}_o \tilde{c}_o c^*} \boldsymbol{v}_l^* + \frac{\tilde{\rho}_o \tilde{c}_o c^*}{c_1^* + \tilde{\rho}_o \tilde{c}_o c^*} \boldsymbol{v}_p^*, \tag{3}$$

where $c_1^* = (1 - c_o c^*)/(1 - c_o)$, $\tilde{c}_o = c_o/(1 - c_o)$, $\tilde{\rho}_o = \rho_p/\rho_l$, ρ_p the density of magnetic particles, $\boldsymbol{v}_l^*$, $\boldsymbol{v}_p^*$ the mean velocity of liquid and particles phase respectively.

The equation of motion:
for particles phase,

$$\frac{\partial(c^* \boldsymbol{v}_p^*)}{\partial t^*} + \nabla \cdot (c^* \boldsymbol{v}_p^* \boldsymbol{v}_p^*) = \left(\frac{1}{E_1} + \frac{1}{E_2}\right)\left[-\nabla(c^* T_p^*) + Kn T_p^* c^* \left\{\frac{1}{3}\nabla(\nabla \cdot \boldsymbol{v}_p^*) + \Delta \boldsymbol{v}_p^*\right\}\right] + c^* \boldsymbol{f}_1^* + c^* \boldsymbol{f}_2^*, \tag{4}$$

for liquid phase,

$$\frac{\partial}{\partial t^*}(c_1^* \boldsymbol{v}_l^*) + \nabla \cdot (c_1^* \boldsymbol{v}_l^* \boldsymbol{v}_l^*) = \frac{1}{Re}\left[-\nabla p^* + \frac{1}{3}\nabla\left\{\nabla \cdot (c_1^* \boldsymbol{v}_l^*)\right\} + \Delta(c_1^* \boldsymbol{v}_l^*)\right] - \tilde{\rho}_o \tilde{c}_o c^* \boldsymbol{f}_1^* - \frac{Gr}{Re^2} c_1^* T_l^* \boldsymbol{e}, \tag{5}$$

where $E_1(= v_o^2 v_p \rho_p / k(T_w - T_\infty))$ and $E_2(= v_o^2 v_p \rho_p / k T_\infty)$ are the Eckert number, v_p the volume of a particle, k the Boltzmann constant, T_w the temperature at wall of pipe, T_∞ the temperature at infinity, $T_p^*(= (T_p - T_\infty)/(T_w - T_\infty))$ the mean translational kinetic temperature of particles phase, $T_l^*(= (T_l - T_\infty)/(T_w - T_\infty))$ the mean temperature of liquid phase, $Kn(= v_o/\nu r_o)$ the Knudsen number, ν the characteristic collision frequency, $\boldsymbol{f}_1^*(= (\boldsymbol{v}_l^* - \boldsymbol{v}_p^*)/t_q^*)$ the drag of particles phase suffered from liquid phase, $t_q^* = t_q v_o/r_o$, t_q the characteristic time for momentum transfer between particles and liquid phase, $\boldsymbol{f}_2^*(= (H_o L \nabla H^* - 1/Fr^2)\boldsymbol{e})$ the external forces, $\boldsymbol{e}$ the unit vector, $H_o = \mu_o m H_{max}/v_p \rho_p v_o^2$, μ_o the permeability in vacuum, m the magnetic moment, H_{max} the maximum of magnetic field, $L(= \tanh \xi - \xi^{-1})$ the Langevin function, $\xi = \mu_o m H / k T_s$, H the component of magnetic field, $H^*(= H/H_{max})$ the magnetic field, T_s the mean temperature of particles phase, $Fr(= v_o/\sqrt{r_o g})$ the Froude number, g the gravitational acceleration, $p_\infty^*(= p_\infty r_o/\mu_l v_o)$ the pressure of liquid phase corrected for effect of gravity, $Gr(= g\beta(T_w - T_\infty) r_o^3 \rho_l^2/\mu_l^2)$ the Grashof number and β the coefficient of volume expansion of solvent.

The energy equation:
for particles phase,

$$\frac{\partial}{\partial t^*}(\frac{3}{2} c^* T_p^*) + \nabla \cdot (\frac{3}{2} c^* T_p^* \boldsymbol{v}_p^*) = -\tilde{T} T_p^* c^* \nabla \cdot \boldsymbol{v}_p^* + Kn \Big[\tilde{T} T_p^* c^* \Phi_p^*$$
$$+ \frac{5}{2}\left(\frac{1}{E_1} + \frac{1}{E_2}\right)\left\{T_p^* \Delta(c^* T_p^*) + (\nabla T_p^*) \cdot \left\{\nabla(c^* T_p^*)\right\} - T_p^{*2} \Delta c^*\right\}\Big] - \frac{3\tilde{T}}{t_q^*} c^* T_p^* \tag{6}$$

and

$$\frac{\partial(c^* T_s^*)}{\partial t^*} + \nabla \cdot (c^* T_s^* \boldsymbol{v}_p^*) = \frac{Kn}{E_1} \nabla \cdot (c^* T_p^* \nabla T_s^*) + \frac{c^*}{t_T^*}(T_l^* - T_s^*), \tag{7}$$

for liquid phase,

$$\frac{\partial}{\partial t^*}(c_1^* T_l^*) + \nabla \cdot (c_1^* T_l^* \boldsymbol{v}_l^*) = \frac{1}{Pe} \Delta(c_1^* T_l^*) - \tilde{\rho}_o \tilde{c}_o \tilde{c} \frac{c^*}{t_T^*}(T_l^* - T_s^*) + \frac{E}{Re} \Phi_l^*, \tag{8}$$

where $t_T^* = t_T v_o/r_o$, t_T the characteristic time for heat transfer between particles and liquid phase, $\tilde{T} = 1 + T_\infty/(T_w - T_\infty)$, $T_s^* = (T_s - T_\infty)/(T_w - T_\infty)$, $Pe(= c_p r_o v_o \rho_l/\lambda)$ the Peclet number, c_p the specific heat at constant pressure of liquid phase, λ the thermal conductivity of liquid phase, $\tilde{c} = c_s/c_p$, c_s the specific heat of magnetic particles, $\Phi_p^*(= -2(\nabla \cdot \boldsymbol{v}_p^*)^2/3 + 2\boldsymbol{D}_p^*:\boldsymbol{D}_p^*)$ the dissipation function of particles phase, $\boldsymbol{D}_p = (\boldsymbol{v}_p^*\nabla + \nabla \boldsymbol{v}_p^*)/2$, $\Phi_l^*(= -2(\nabla \cdot (c_1^* \boldsymbol{v}_l^*))^2/3 + 2\boldsymbol{D}_l^*:\boldsymbol{D}_l^*)$ the dissipation function of liquid phase, $\boldsymbol{D}_l = ((c_1^* \boldsymbol{v}_l^*)\nabla + \nabla(c_1^* \boldsymbol{v}_l^*))/2$ and $E(= v_o^2/c_p(T_w - T_\infty))$ the Eckert number.

Now we consider one dimensional steady pipe flow as a special case. The length of the pipe is infinitely long and the temperature at wall of the pipe is constant. And it is assumed that variation of the mean translational kinetic temperature T_p is far from Curie point and that the mean temperature of liquid and particles phases are the same as the wall temperature and are constant everywhere in the pipe flow. That is, thermal energy of particles phase is converted to kinetic energy of particles phase.

Equations (1) $\sim$ (8) are solved numerically by Runge-Kutta-Gill method under the given initial condition at the inner of a pipe and $T_w = 305$K and $T_\infty = 295$K. The maximum value of a nonuniform magnetic field applied longitudinally to the axis of the pipe which is installed horizontally is 150kA/m. We use a water based magnetic fluid which has 20% mass concentration of magnetite particles.

3. RESULTS AND DISCUSSION

Figure 1 shows the theoretical results of each variable, that is, v_p^*, $c* - 1$ and $p*$. The calculating region is $z = -80$mm $\sim$ $z = 80$mm in which the magnetic field is applied. The zero point of the pressure is set at $z = -80$mm.

The parameter Kn expresses uniformity of magnetic particles distribution and particles are uniformly distributed at large Kn. On the other hand, Sc expresses dispersion of magnetic particles influenced by the flow and variation of concentration c^* is small at small Sc. The figure is in the case of small Kn of 1.0×10^3 and large Sc of 1.0×10^{-4}.

As the parameter t_q^* is small, drag between particles and solvent phase becomes large and then variation of $c^* - 1$ along direction of pipe axis appears.

The effect of magnetic force on p^* and v_p^* is large, but on $c^* - 1$ is small. This is based on the one dimensional flow analysis at the condition that wall temperature is isothermal and length of pipe is long infinitely and with the assumption that the mean temperature of each phase is the same as the wall temperature. However, if we consider two or three dimensional flow or the exchange of thermal energy between particles and liquid phases, we can grasp more precisely the inner phenomena of magnetic fluid flow in a magnetic field. In such case it is necessary to solve the equations under the boundary condition on the pipe wall, that is, jump condition of particles at the pipe wall.

4. CONCLUSIONS

A theoretical analysis is made to clarify the dispersion of magnetic particles in a nonuniform magnetic field. As a special case of pipe flow problems a steady one dimensional pipe flow problem in a strong nonuniform magnetic field is solved by using the derived equations. The effect of magnetic particles on the flow characteristics are clarified.

5. REFERENCES

1 S. Kamiyama, JSME Int. J., 35-2, (1992), 131.

2 S. Kamiyama and K. Shimada, J. Jpn. Soc. Powder Powder Metal., 36, (1989), 794.

3 S. Kamiyama and K. Shimada, Trans. Jpn. Soc. Mech. Eng., 57-534, B (1991), 526.

4 K. Shimada, T. Oyama, H. Nishiyama and S. Kamiyama, Int. J. Appl. Electromag. Mat., 2-4, (1992), 301.

5 S. Morioka and T. Nakajima, J. Theo. Appl. Mech., 6-1, (1987), 77.

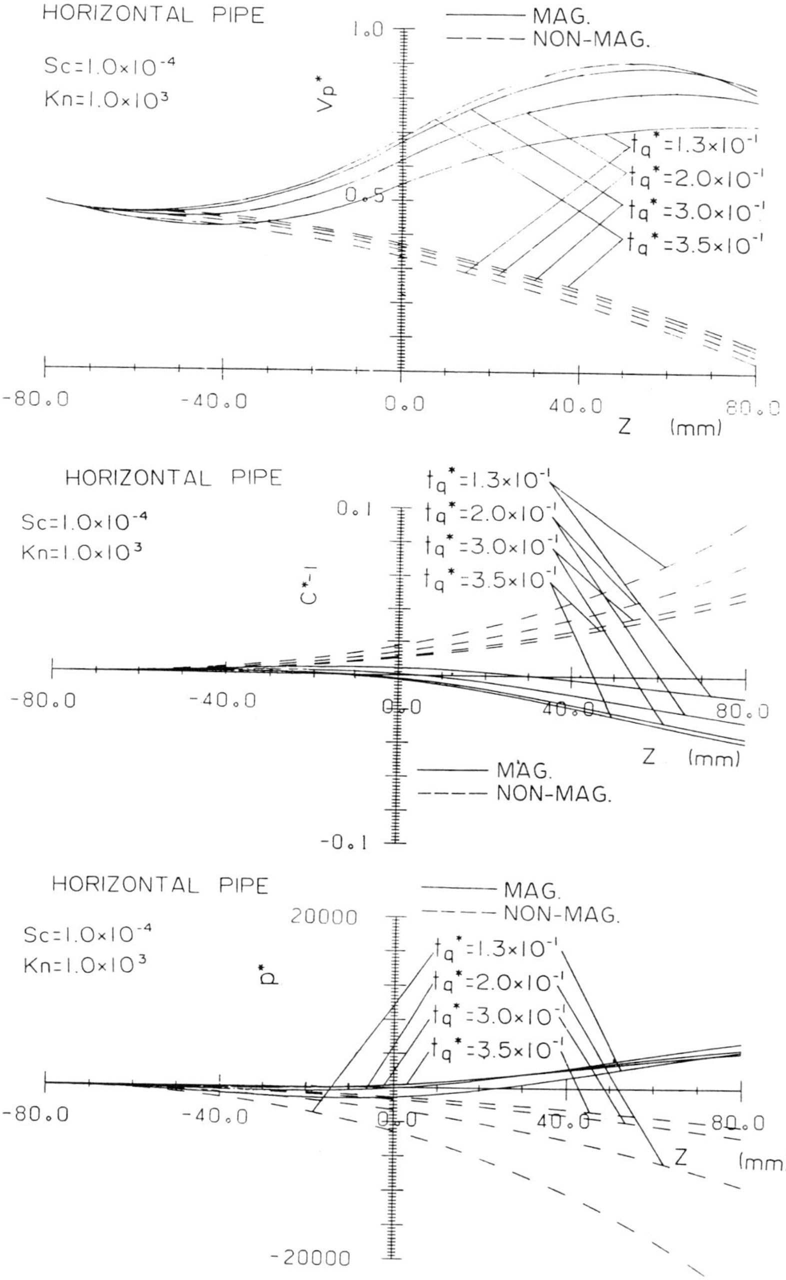

Fig.1 Variation of variables v_p^*, $c^* - 1$ and p^* along z in a magnetic field.

Simulation and Design of Applied Electromagnetic Systems
T. Honma (Editor)

Magnetic Levitation in Magnetic Fluids — Definite Expressions for the Buoyancy Force

Katja Henjes

Department of Mechanical Engineering, Keio University,
3-14-1 Hiyoshi, Kohoku-ku, Yokohama 223, Japan

Abstract

The total levitation force on a body submerged in a Magnetic Fluid is derived. The resulting formula is definite and unambiguous. It does not depend on whether one starts from a theory in terms of $\vec{H}$, or in terms of $\vec{B}$: The correct evaluation of the contribution $\int dV\,[-\nabla p]$ resolves the apparent contradiction between the force density expressions $\mu_o M \nabla H$, or $M \nabla B$.

1 Introduction

Magnetic Fluids are one of the most complex systems that ever have been tried to model: They are fluid, allowing arbitrary internal motion; they are highly magnetizable, with the magnetization in most cases non-uniform and non-linear; and they are temperature-dependent in their magnetization as well as in their density, adding the temperature distribution to the set of variables almost always necessary for a complete description.

Already the quiescent state is not easy to understand: The magnetic field imposes a preferred direction onto the otherwise isotropic fluid. As a consequence, the scalar pressure p is no longer continuous at interfaces, and the pressure gradient $[-\nabla p]$ in the Navier-Stokes equation cannot be integrated across such interfaces. Accordingly, the pressure distribution usually adds to the net force.

Here, the total levitation force on a (magnetic or unmagnetic) body submerged in a Magnetic Fluid is derived. This has to be done in several steps:

1. Find p, the pressure balancing the ponderomotive force in the quiescent background liquid.

2. Solve the magnetostatic problem of the body in the liquid, using the static Maxwell equations, the magnetization law $M(H)$ for the liquid, and the magnetization law $M(H)$ for the body; and compute the ponderomotive force density for every point within the liquid, and within the body.

3. Add the pressure gradient from step 1), and the force density from step 2), according to the Magnetic-Fluid Navier-Stokes equation, and integrate them over the volume of the body, paying due respect to the jump conditions at the interface.

2 The Ferrofluid Navier-Stokes equation

The basic equation to be employed is the ferrofluid Navier-Stokes equation. It is unambiguously determined, but can be stated in two different formulations [1, 2]: The "$\vec{B}$ – formulation" reads

$$\rho \frac{d\vec{v}}{dt} = -\nabla p + M_j \nabla B_j \tag{1}$$

where $p = p(\rho, T, \vec{B})$ is defined through the boundary condition it has to obey at interfaces:

$$\left[p - \frac{\mu_o}{2} M_t^2\right]^{(medium\ 1)} = \left[p - \frac{\mu_o}{2} M_t^2\right]^{(medium\ 2)} \tag{2}$$

Since $M_j \nabla B_j = \mu_o M_j \nabla H_j + \nabla \frac{\mu_o}{2} M^2$, the same equation of motion can be stated also in the form ("$\vec{H}$ – formulation") [3]

$$\rho \frac{d\vec{v}}{dt} = -\nabla p + \mu_o M_j \nabla H_j \tag{3}$$

where $p(\vec{H}) = p(\vec{B}) - \frac{\mu_o}{2} \vec{M}^2$, and obeys:

$$\left[p + \frac{\mu_o}{2} M_n^2\right]^{(medium\ 1)} = \left[p + \frac{\mu_o}{2} M_n^2\right]^{(medium\ 2)} \tag{4}$$

3 The Background Pressure

If a body of a different magnetizability is submerged in a ferrofluid, the system is no longer in equilibrium: The pressure is not able to stably support this alien body. It will be attracted towards, or expelled from, regions of high field intensity, depending on whether its magnetization is higher or lower than that of the background ferrofluid. In this non-equilibrium situation, the magnetic Bernoulli equation [3] does not hold, and the pressure needs to be determined in a completely new way.

Since mass and heat flow are slow processes, ρ and T may be assumed to have still the same values as in the undisturbed quiescent state. But magnetic relaxation takes place much faster. Hence, the static Maxwell equations yield the magnetic field and the magnetization distribution, which reflect the shape and magnetizability of the submerged body: Mechanical equilibrium is not yet reached, but magnetic equilibrium is.

These observations lead to the equations that determine the pressure in the background liquid [2] :

- $\vec{B}$ – formulation:

$$\nabla p \quad = \quad M_j^{old} \nabla B_j^{old} \quad + \quad \nabla \rho^2 \left(\frac{\partial \delta \Phi}{\partial \rho} \right) \tag{5}$$

where $\delta \Phi \quad = \quad - \int_{B^{old}}^{B^{new}} db \;\; M(b)/\rho$ accounts for the change in magnetostriction during the insertion of the body.

- $\vec{H}$ – formulation:

$$\nabla p \quad = \quad \mu_o M_j^{old} \nabla H_j^{old} \quad + \quad \nabla \rho^2 \left(\frac{\partial \delta \phi}{\partial \rho} \right) \tag{6}$$

where $\delta \phi \quad = \quad - \mu_o \int_{H^{old}}^{H^{new}} db \;\; M(b)/\rho$.

Note that $\rho^2 \left(\frac{\partial \phi}{\partial \rho} \right) \; = \; \rho^2 \left(\frac{\partial \Phi}{\partial \rho} \right) - \frac{\mu_o}{2} \vec{M}^2$:

Generally, magnetostriction must not be neglected. It is of the order $\vec{M}^2$ — very small in most gases or liquids but well observable in magnetic fluids. The overall identification of the pressure without the magnetic field $p_0(\rho, T)$ with the pressure in the magnetized liquid p, as done e.g. by Shliomis [4], is not justified.

4 Integration of the Navier-Stokes equation

Let us call "force" the total center-of-mass acceleration of the submerged body:

$$\vec{F} \quad := \quad \int_V dV \; \rho \frac{d\vec{v}}{dt}$$ where V is the volume of the body.

- $\vec{B}$ – formulation

The equation to be integrated is eq.(1), in conjunction with eq.(2). Generally, neither p nor $\vec{B}$ are continuous at the interface. Hence, ∇p or $\partial_i B_j$ in eq.(1) are ill-defined near the surface of the submerged body. To regularize the integral, two additional surfaces are introduced inside and outside the body, separated from it by the infinitesimal distance $\pm \varepsilon$ [2]. Then, $\vec{F}$ is replaced by an integration up to the outer ε-surface, to fully include the effect of the discontinuities. Next, $\vec{F}$ is split into 2 integrals covering different regions:

$$\vec{F} \quad = \quad \int_{V^-} dV \; [\ldots] \quad + \quad \int_{\varepsilon^-}^{\varepsilon^+} dV \; [\ldots] \tag{7}$$

The first integral in eq.(7) covers the volume completely within the body; it can be evaluated without any difficulty. The second integral, covering the narrow sheet around the interface of thickness 2ε, can be rewritten as

$$\vec{F}_2 = \oint d\vec{f} \int_{-\varepsilon}^{\varepsilon} d\vec{R} \left[-\nabla \left(p - \frac{\mu_o}{2} M_t^2 \right) + M_n \nabla B_n + \mu_o M_t \nabla H_t \right] \tag{8}$$

Here, $d\vec{R}$ accounts for the integration accross the layer. A subscript t refers to the tangential, and a subscript n to the normal component of the magnetic field variables. Now, $(p - \frac{\mu_o}{2} M_t^2)$ is continuous at interfaces in the $\vec{B}$ - formulation (eq.2). Hence, all gradients in eq.(8) exist, the integral is well-defined, and one may safely take the limit $\varepsilon \to 0$. It is then easy to calculate the final formula for $\vec{F}$:

$$\vec{F} = - \oint_{S+} d\vec{f} \; p(\vec{B}) + \int_{V-} dV \; M_j \nabla B_j + \oint d\vec{f} \; \Delta \frac{\mu_o}{2} M_t^2 \tag{9}$$

where $\Delta(...) = (...)^{out} - (...)^{in}$. So, the result for the total magnetic force acting on a body submerged in a magnetized fluid consists of 3 contributions: (i) the pressure surface integral, taken at the outer side of the interface, i.e. in the background fluid (eq.5); (ii) the magnetic ponderomotive force density $M_j \nabla B_j$, integrated over the volume of the body; and (iii) the jump $\Delta \frac{1}{2} M_t^2$ in the tangential component M_t of the magnetization, integrated over the entire interface. The pressure within the submerged body does not contribute.

- $\vec{H}$ - formulation

The calculation starting from eq.(3) proceeds quite similar to the one outlined above, and yields

$$\vec{F} = - \oint_{S+} d\vec{f} \; p(\vec{H}) + \int_{V-} dV \; \mu_o M_j \nabla H_j - \oint d\vec{f} \; \Delta \frac{\mu_o}{2} M_n^2 \tag{10}$$

In eq.(10), the jump $\Delta \frac{\mu_o}{2} M_n^2$ in the normal component M_n of the magnetization, integrated over the entire interface, contributes; together with the ponderomotive force, and the pressure in the background fluid (eq.6).

Insertion of $p(\vec{H}) = p(\vec{B}) - \frac{\mu_o}{2} \vec{M}^2$, and some algebra in $\vec{M}^2 = M_n^2 + M_t^2$ confirms that eqs.(9) and (10) are indeed identical:

$$\vec{F}(\vec{B}) = \vec{F}(\vec{H}) \tag{11}$$

References

[1] K. Henjes and M. Liu, Ann. Physics 222 (1993); K. Henjes, Ann. Physics 222 (1993)

[2] K. Henjes, "Buoyancy Forces in Magnetic Fluids", preprint (1993)

[3] R.E. Rosensweig, Ferrohydrodynamics (Cambridge Univ. Press, Cambridge, 1985)

[4] M.I. Shliomis, Sov. Phys. JETP 34 (1972) 1291

Simulation and Design of Applied Electromagnetic Systems
T. Honma (Editor)

Effect of Magnetic Field on Natural Convection of Magnetic Fluids

T. Sawada, H. Kikura and T. Tanahashi

Department of Mechanical Engineering, Keio University
3-14-1 Hiyoshi, Kohoku-ku, Yokohama 223, Japan

Abstract

This paper presented experimental results on natural convection of two magnetic fluids in a cubic enclosure with isothermal hot and cold walls under magnetic fields. Temperature distributions on a front wall was visualized by thermosensitive liquid crystal. The horizontal temperature distributions on the central cross-section and Nusselt numbers were measured. Effect of uniformed and non-uniformed magnetic fields on the heat transfer and natural convection flow were discussed in detail.

1. INTRODUCTION

Natural convection of a magnetic fluid is driven by the buoyancy and magnetic forces. Since the magnetization of a magnetic fluid is changed by the temperature, the thermomagnetic convection is also arisen. Therefore it is expected to control convection of a magnetic fluid by changing an applied magnetic field. Bashtovoy, et al.[1] investigated the effect of the magnetic field on convective heat transfer in a cavity with square cross section. Sawada, et al.[2] and Kikura, et al.[3] presented that it was possible to control the natural convection of a magnetic fluid, changing direction and intensity of a non-uniformed magnetic field.

In the present paper, influences of the vertical uniformed and non-uniformed magnetic fields on the distribution of wall-temperature and Nusselt numbers were investigated precisely and characteristics of the heat transfer of a magnetic fluid were discussed.

2. EXPERIMENTAL APPARATUS AND PROCEDURE

The test section is a cubic enclosure 30mm×30mm×30mm filled with a magnetic fluid (Fig. 1). The two side walls consist of copper plate whose thickness is approximately 1mm. Two thermostats are fitted on both sides of the cubic enclosure to control the temperature of side walls. The water of constant temperature is supplied by a temperature control unit.

The temperature of both thermostats were kept at 20°C in the beginning. The experiment was started by supplying the hot water to the left thermostat. The left side wall was rapidly heated from 20°C to 30°C and its temperature was held there. Then the clockwise flow was induced by natural convection. The flow in the enclosure is regarded as laminar, gravitational acceleration acts parallel to the isothermal walls, and radiation effects is taken to be negligible.

The vertical magnetic fields are applied non-uniformly by cylindrical permanent mag-

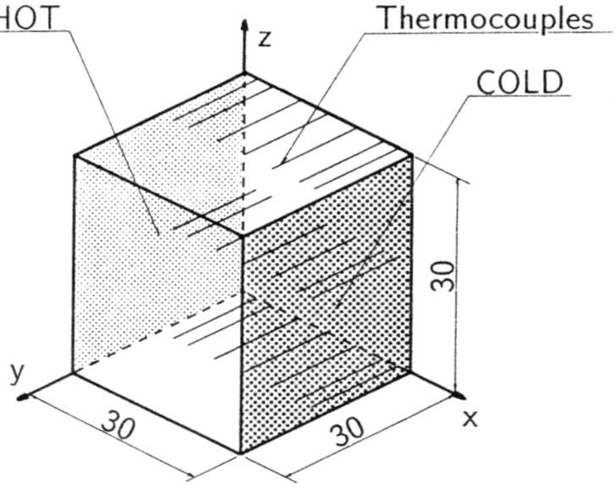

Fig. 1 Test section

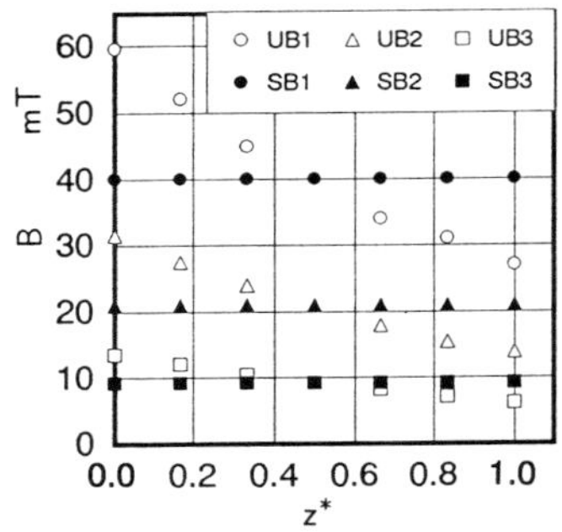

Fig. 2 Distributions of magnetic induction

nets. Its diameter is 115mm. The range of magnetic induction B is 9mT $\sim$ 40 mT. An electromagnet is also used to obtain uniform magnetic fields: B=9mT $\sim$ 400mT. The hole piece of the electromagnet is 120mm in diameter. Figure 2 shows the distributions of the magnetic induction. Here, SB1, SB2 and SB3 indicate uniform magnetic fields, and UB1, UB2 and UB3 non–uniformed fields. z^* is the dimensionless vertical coordinate. Temperature distributions on a front wall were visualized by a thermosensitive liquid crystal sheet. The horizontal temperature distributions on the central cross–section(y^*=0.5) of the cavity were measured by 0.2mm diameter copper–constantan thermocouples. They were inserted straight from the back wall. Two different working fluids were used: kerosene based–magnetic fluid (HC–50, $Pr \approx 128$) and water–based magnetic fluid (W–40, $Pr \approx 108$). Here Pr is Prandtl number. Experiments were mainly carried out for the following four cases:

Case1: There was no magnetic field.

Case2: The magnet was placed under the cavity. The directions of the magnetic and gravitational forces were the same.

Case3: The magnet was placed above the cavity. The direction of the magnetic force was opposite to that of the gravity.

Case4: The uniformed magnetic field was applied.

3. RESULTS AND DISCUSSIONS

Figure 3 shows temperature visualization of the front wall and horizontal dimensionless temperature distributions on the central cross–section(y^*=0.5) along the top wall(z^*=1), the central line(z^*=0.5) and bottom wall(z^*=0) of the cavity. Here B_0 is the magnetic induction at the center of the cavity. These results indicate the steady state.

When there is no magnetic field(case 1), developments of ordinary natural convection are observed and high temperature part grows upward to the cell from the hot plate. In case 2 the magnet was placed under the cavity, the magnetic field gradient is established in the same direction of the gravity. Therefore a magnetic force adds to a buoyant force. Wall–temperature distribution is similar to that of case 1. But the thermal boundary layer near the hot plate is thinner than in case 1 and the horizontal temperature gradient near the walls is larger. When the magnetic force is established in the reverse direction of

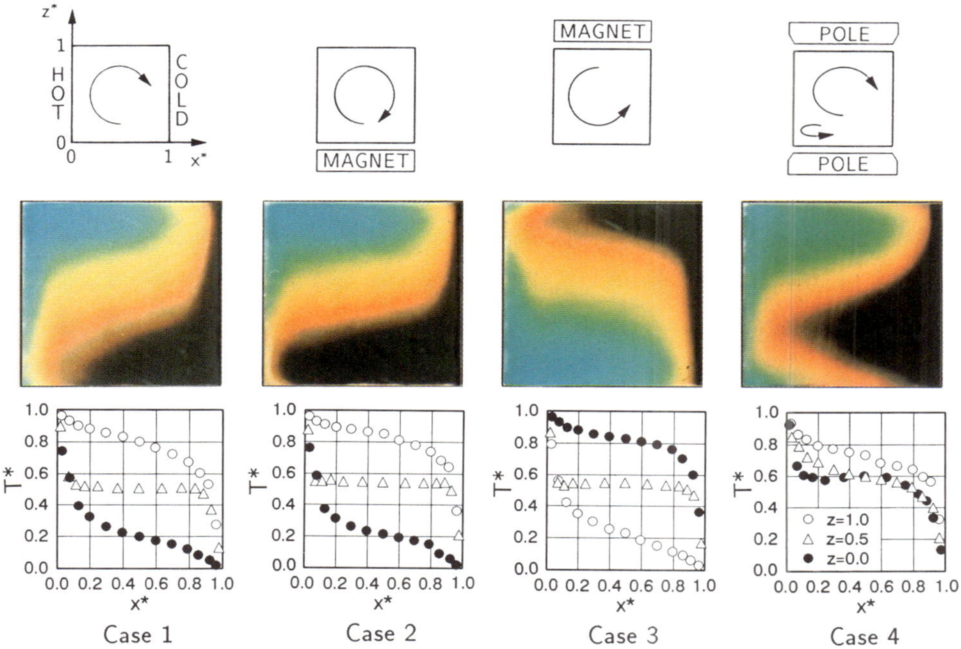

Fig. 3 Temperature visualization and horizontal temperature distributions for each case (HC-50, B_0=40mT)

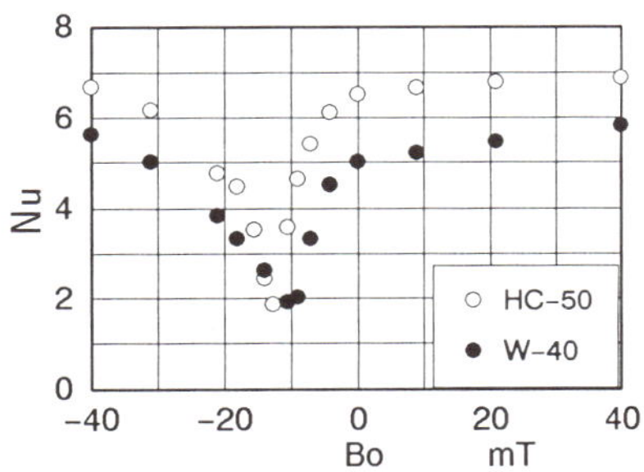

Fig. 4 Nusselt number for cases 1, 2 and 3

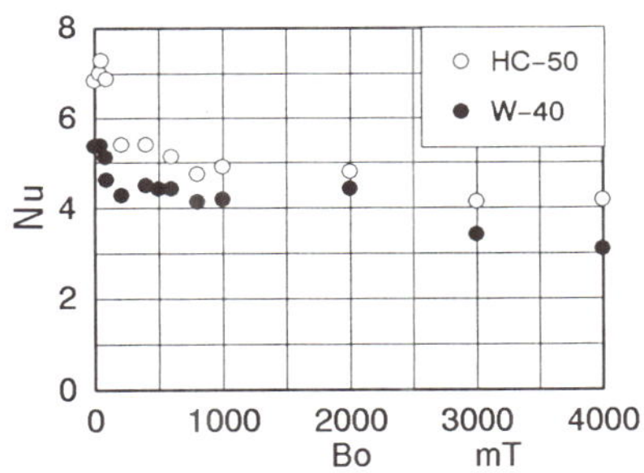

Fig. 5 Nusselt number for case 4

the gravity and its intensity is sufficiently greater than the gravity, the wall-temperature distribution and horizontal temperature distribution are upside down in comparison with that in previous cases as shown in Fig.3(Case 3). In this case the flow is counterclockwise.

Figure 4 shows the relation between the intensity of magnetic field and Nusselt number under non-uniformed magnetic field. $B_0 < 0$ and $B_0 > 0$ correspond to case 3 and case 2, respectively. When the intensity of magnetic field increases to plus, the convection is activated and heat transfer also increases. In the minus region of B_0 it exists a minimum value of Nusselt number because the magnetic force is balanced with gravity, and thermal gravitational and thermomagnetic convection are ideally turned off. Effect of convection diminishes at this point. When the magnetic intensity is sufficiently greater than the

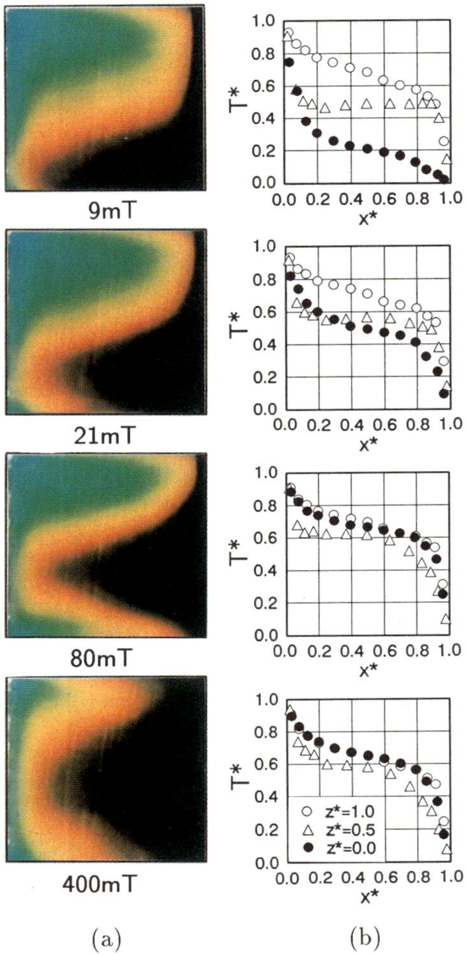

Fig. 6 (a) Temperature visualization and, (b) Horizontal temperature distributions under uniformed magnetic field (HC-50)

gravity, the counterclockwise convection is arisen and Nusselt number increases.

In case 4 there is not essential change of the convection, though it is supposed that the secondary flow yields near the bottom. As shown in Fig.2, the magnetic field is uniform along the z^* axis. If the cubic enclosure filled with a magnetic fluid is set between hole pieces of the electromagnet, the magnetic force lines are not straight and the magnetic field gradient is induced. When the magnetic field is small, the buoyancy is stronger than the magnetic force. As the magnetic field increases, the effect of the magnetic force becomes dominant. Since the magnetic fields near the top and bottom walls are strong, two different convections are induced as shown in Figs.3(case 4) and 6(a). Moreover it is impossible to ignore the effect of the internal spin of the substructure. Microrotaion is suppressed and apparent viscosity increases as magnetic field increases. While, clustering and chaining of the particles of a magnetic fluid is created in presence of magnetic field. These correlation phenomena also prevent the convection from developing. Then the heat transfer is inhibited by the magnetic field. When the intensity of the magnetic field is changed, the thermosensitive liquid crystal shows the variety of temperature distributions for HC-50 as illustrated in Fig.6.

ACKNOWLEDGMENT

This work was partly supported by The Japan Securities Scholarship Foundation and The Grant-in-Aid for Scientific Research of The Ministry of Education, Science and Culture. The authors are grateful to Mr.K.Tsuchihashi of Ebara Corporation for his technical support. The authors express their thanks to Mr.N.Fukuda for his assistance during the experiment.

REFERENCES

1. V.G.Bashtovoy, S.V.Isayev and M.P.Pavlinov, Heat Transfer Soviet Res., 10 (1978), 5.
2. T.Sawada, et al., Experimental Heat Transfer, Fluid Mechanics and Thermodynamics, (1991), 1433.
3. H.Kikura, T.Sawada and T.Tanahashi, J.Magn.Magn.Mat., (1993), in press.

Simulation and Design of Applied Electromagnetic Systems
T. Honma (Editor)

A Basic Study on an Energy Conversion System using a Gas-liquid Two-phase Magnetic Fluid Flow in a Nonuniform Magnetic Field

J. Ishimoto, M. Okubo, H. Nishiyama and S. Kamiyama

Institute of Fluid Science, Tohoku University, 2-1-1, Katahira, Aoba-ku, Sendai 980, Japan

Abstract

The mechanism of the pressure rise in a gas-liquid two-phase pipe flow of magnetic fluid under a nonuniform magnetic field is investigated in detail both theoretically and experimentally. As a result, it is known that the pressure rise is induced mainly by the unbalance of the magnetic body force in the two-phase flow. These fundamental study on the pressure distributions will contribute to the development of the new energy conversion system using a gas-liquid two-phase magnetic fluid flow.

1. INTRODUCTION

An energy conversion system using a gas-liquid two-phase pipe flow of magnetic fluid proposed by one of the authors [1,2] is based on the principle that the flow is accelerated by an additional magnetic driving force which is induced in a two-phase flow of magnetic fluid in a nonuniform magnetic field.

In the present paper, the mechanism of the pressure rise in the two-phase flow of magnetic fluid is investigated in detail both theoretically and experimentally. Firstly, governing equations of one-dimensional gas-liquid two-phase magnetic fluid flow are presented and numerically solved. Also, the pressure distribution in a nonuniform magnetic field region is measured in the cases of two-phase flow, single-phase flow and stationary state using a new experimental apparatus of the flowing system. From the numerical and measurement results, the magnitude of the pressure components which contribute to the total driving force is accurately estimated.

2. THEORETICAL ANALYSIS

2.1. Equations for pressure components

To investigate the mechanism of the pressure rise of the magnetic fluid flow, firstly, the governing equations which represent the two-phase flow characteristics of magnetic fluid are numerically solved. On the physical situation used in the numerical analysis is as well as the experiments. The governing equations are the equation of continuity and motion for gas- and liquid-phases, the energy equation for two-phase flow, the equation of pressure balance between gas- and liquid-phases and ideal gas law. The equation of motion for gas-phase is represented by the translational motion of single gas bubble of equivalent radius $\overline{R}$. We can obtain the distributions of the density of gas-phase ρ_g, the void fraction α, and the velocity of gas- and liquid-phases v_g, v_l from the numerical solutions of those equations. These solutions are utilized to obtain the various pressure components as follows. The pressure distribution ΔP_{lT} is divided into four components as

$$\Delta P_{lT} = \Delta P_{lT(M)} + \Delta P_{lT(BV)} + \Delta P_{lT(C)} + \Delta P_{lT(MR)} \tag{1}$$

$\Delta P_{lT(M)}$ is the pressure component due to the magnetic body force and given by

$$\frac{dP_{lT(M)}}{dz} = (1-\alpha)\mu_0 M_z \frac{dH_z}{dz} \tag{2}$$

where, μ_0 is the permeability in vacuum, M_z the magnetization which is a function of magnetic field H_z and liquid temperature. H_z has a nonuniform distribution along the z-axis.

$\Delta P_{lT(BV)}$ is the pressure component due to the buoyancy force and an ordinary pipe friction force given by

$$\frac{dP_{lT(BV)}}{dz} = \alpha\rho_l g - \frac{32\{\eta_g \alpha v_g + \eta_l(1-\alpha)v_l\}}{D^2} \tag{3}$$

where, η is the viscosity, D the inner diameter of the pipe, the subscript g and l denotes the gas- and the liquid-phase respectively.

$\Delta P_{lT(C)}$ is the pressure component due to the convective acceleration force and given by

$$\frac{dP_{lT(C)}}{dz} = -\rho_g \alpha v_g \frac{dv_g}{dz} - \rho_l(1-\alpha)v_l \frac{dv_l}{dz} \tag{4}$$

$\Delta P_{lT(MR)}$ is the pressure component due to the additional resistance force caused by the magnetic field and given by the following equation based on the experimental result.

$$\Delta P_{lT(MR)} \approx \Delta P_{lS(MR)} = \frac{H_{\max}}{115300.0}(-59.17605 - 57.26076z + 5283.159z^2 + 19987.09z^3) \tag{5}$$

where, the subscript S and T denotes the single- and two-phase respectively. The additional resistance force due to the magnetic field is mainly caused by the particle aggregation due to the magnetic field. And the pressure component due to the single-phase magnetic body force $\Delta P_{lS(M)}$ is defined as $\Delta P_{lS(M)} = \mu_0 \int_0^{H_z} M_z dH_z$.

2.2. Conditions for numerical analysis

The boundary conditions at the inlet section of applied magnetic field region are denoted by the subscript 'in' and at the point where the gas bubble appears are denoted by the subscript '0' and are given as follows. The inlet pressure is $P_{lin} = 104.6$ kPa, the minimum radius of gas bubble $R_0 = R_{\min} = 0.3$ mm, the quality $\chi = 3.7 \times 10^{-4}$, the liquid-phase temperature $T_{lin} = 298$ K, the slip ratio $S_0 = v_{g0}/v_{l0} = 1$, the inlet velocity $v_{lin} = 0.1$ m/s. H_z is given as the approximated polynomial equation by the measurement result of H_z. We use the fluid properties of the kerosene-based temperature sensitive magnetic fluid with 50% mass concentration of manganese-zinc ferrite particles, and the fluid properties is as same as the ones that are used by the experimental study. The other conditions for analysis are as follows. $D = 8.0$ mm, the gas constant is $R_G = 286.8$ J/(kg $\cdot$ K), the surface tension $\gamma = 0.025$ N/m, $\rho_l = 1402$ kg/m^3 and the kinematic viscosity of gas-phase $\nu_g = 15.15 \times 10^{-6}$ m^2/s.

2.3. Results of numerical analysis

Figure 1 shows the calculated result of the pressure distributions of two-phase flow denoted by ΔP_{lT}. In the region; $z \geq 0$ m, the flow becomes two-phase flow so that the void fraction α increases by the appearance of gas-phase. In this region, the magnetic

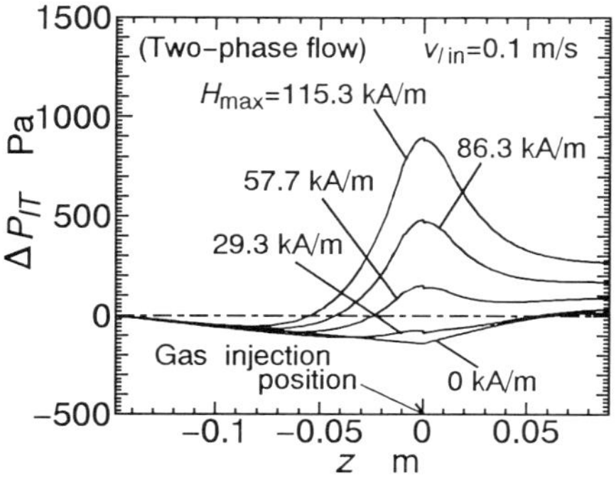

Figure 1. Pressure distributions in two-phase flow (Numerical)

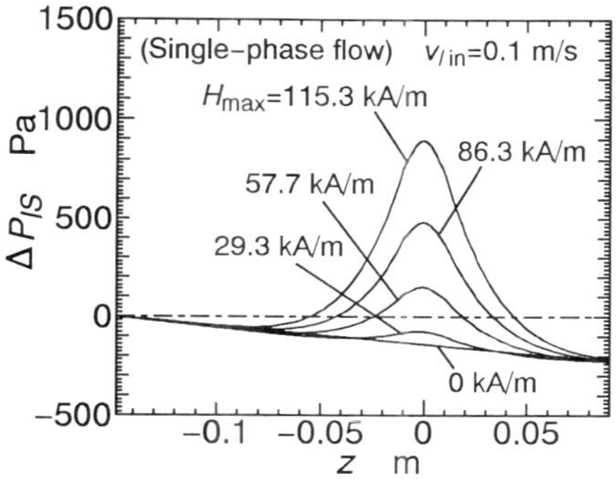

Figure 2. Pressure distributions in single-phase flow (Numerical)

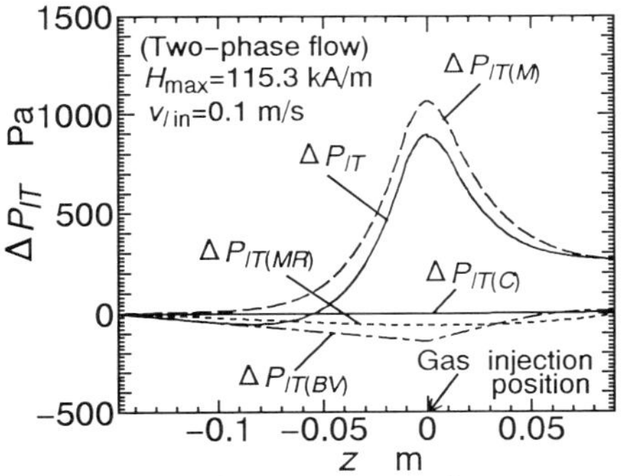

Figure 3. The separated pressure components (Numerical)

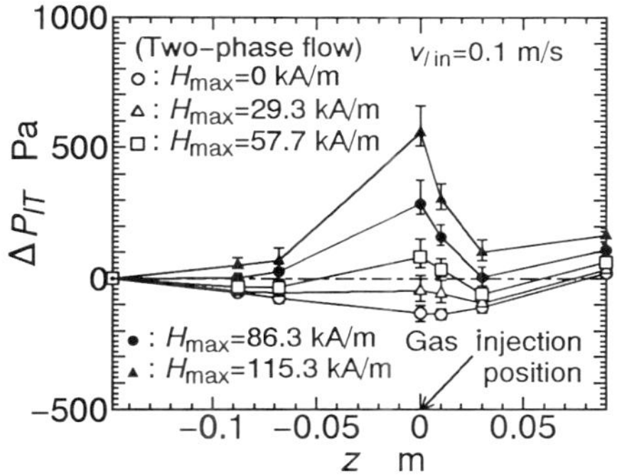

Figure 4. Pressure distributions in two-phase flow (Experimental)

body force are in the negative z direction. However, the force is smaller than one in a single-phase flow in the upstream region ($z < 0$ m) because the apparent magnetization decrease with increase in the void fraction α. As a result, pressure at the exit section of the applied magnetic field region is higher than one in the case of single-phase flow.

Figure 2 shows the pressure distributions of the single-phase flow of magnetic fluid ($\alpha = 0, v_g = 0$) denoted by ΔP_{lS}.

Figure 3 shows the effects of the various separated pressure components which contribute to the total driving force. From this figure, it is clear that $\Delta P_{lT(M)}$ is most dominant to ΔP_{lT}.

3. EXPERIMENTAL STUDY

The experimental apparatus is a small test loop as shown in the previous report [1]. To confirm the validity of the pressure components that are numerically analyzed in the theoretical study, the precise experiment is conducted with this loop to measure the pressure difference of magnetic fluid flow. The gas bubbles are injected into the point where the magnetic field is maximum. Figure 4 shows the pressure distribution of ΔP_{lT}

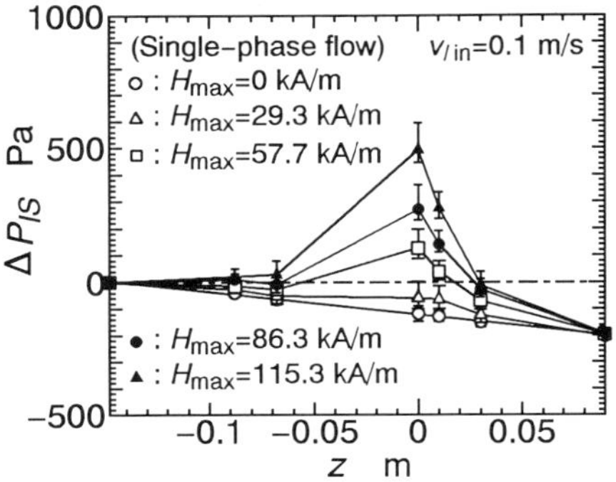

Figure 5. Pressure distributions in single−phase flow (Experimental)

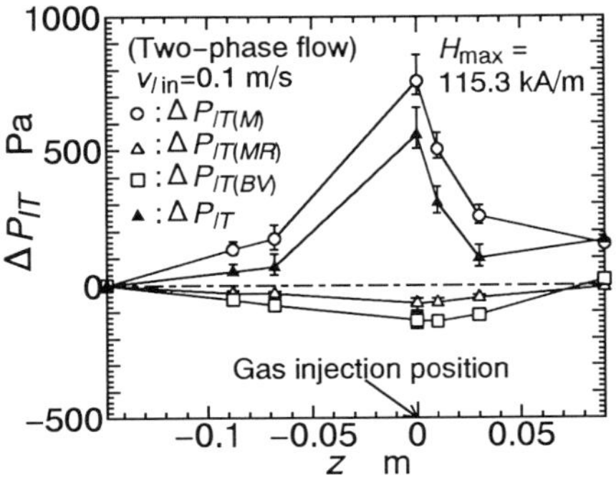

Figure 6. The separeted pressure components (Experimental)

due to gas-injection for the constant mass flow rate $Q_g = 3.7 \times 10^{-4}$ kg/min. Figure 5 shows the pressure distribution of ΔP_{lS}. The inlet velocity on the single-phase flow is controlled to make constant velocity as same as the two-phase flow's one using the pump or the needle valves. The experimental results of ΔP_{lT} and ΔP_{lS} have similar tendency to the numerical ones. By using above measurement results, the method to separate ΔP_{lT} into the pressure components can be written as follows. Firstly, $\Delta P_{lT(MR)}$ is given as $\Delta P_{lT(MR)} \approx \Delta P_{lS(MR)} = \Delta P_{lS[H]} - \Delta P_{lS[H=0]} - \Delta P_{lS(M)}$, where, $\Delta P_{lS(M)}$ is the static pressure which is measured in the applied magnetic field ($\alpha = 0, v_g = v_l = 0$), the subscripts, $[H]$ and $[H = 0]$ denote the values in magnetic field and non-magnetic field respectively. Secondly, $\Delta P_{lT(M)}$ is given as $\Delta P_{lT(M)} = \Delta P_{lT[H]} - \Delta P_{lT[H=0]} - \Delta P_{lT(MR)}$. Finally, $\Delta P_{lT(BV)}$ is given as $\Delta P_{lT(BV)} = \Delta P_{lT[H=0]}$.

Figure 6 shows the results of the various separated pressure components in the experimental study. Each component is reasonable agreement with the numerical ones. Equally to the numerical analysis, it is clear that a large pressure rise in the total driving force is mainly obtained by $\Delta P_{lT(M)}$.

4. CONCLUSION

The pressure distribution in two-phase magnetic fluid flow in a nonuniform magnetic field is investigated both theoretically and experimentally. The pressure rise is induced by using unbalance of magnetic body force between single- and two-phase magnetic fluid flows. It is confirmed that the pressure component due to the magnetic body force is most dominant to the total driving force of magnetic fluid flow.

5. REFERENCES

1 S. Kamiyama and M. Okubo, Gas-liquid Two-phase Flow of Magnetic Fluids, *Proc. Int. Conf. Multiphase Flows*, (1991) 77.

2 S. Kamiyama, T. Kamiya and H. Izu, Effect of Magnetic Field on Gas-liquid Two-phase Pipe Flow of Magnetic Fluid, *Proc. Int. Conf. ISEM Sympo. Appl. Electro. Forces*, (1991) 87.

Simulation and Design of Applied Electromagnetic Systems
T. Honma (Editor)

Subsurface phenomena in magnetic liquid drop collisions

S.Sudo[a], H.Hashimoto[b] and K.Katagiri[b]

[a] Dept. of Mechanical Engineering, Iwaki Meisei University, Iwaki, Japan

[b] Institute of Fluid Science, Tohoku University, Sendai, Japan

Abstract

The subsurface phenomena and the splashing for collisions of magnetic liquid drops against target liquids under both non-magnetic and magnetic fields have been investigated experimentally. It was found that the cavity depth formed by the collision of magnetic drop on the target liquid surface subject to magnetic fields was shallow extremely irrespective of target liquid. The effect of the difference in target liquids on the subsurface phenomena was also found.

1. INTRODUCTION

The subsurface phenomena which occur when a drop of water strikes a water surface have been studied for many years. Worthington studied the process of drop impact on liquid surfaces by means of single-flash photography[1]. Engel[2] and Macklin & Metaxes[3] investigated the maximum radius of the cavity for impacts of drops on liquid layer. Pumphrey & Elmore[4] have recently investigated the entrainment of bubbles by drop impacts. As stated above, the study of drop impact on liquid surfaces has a long history. However, no study has been published on the collisions of magnetic liquid drops on liquid layer except for our investigation[5]. In our previous paper, it was found that the subsurface phenomena by magnetic drop impact on liquid layer under applied magnetic fields differ from ones of water drops; the size of subsurface cavity become smaller in case of the presence of magnetic fields. In this report we describe the details of the subsurface phenomena for impacts of magnetic drops against kerosene and water layers with the applied magnetic fields. The effect of various concentrations of magnetic fluids upon the subsurface phenomena and the difference in target liquids are discussed.

2. EXPERIMENTAL APPARATUS AND PROCEDURES

A schematic diagram of the experimental apparatus is shown in Figure 1. Magnetic liquid drops were formed with a 50-ml burette. The drops were made to fall through the beam of a laser and the hole of a drop cutter. The satellite droplets were excluded by the drop cutter with the drop detector. The drops of magnetic fluid fell into the transparent rectangular tank which contained target liquid. In all experiment, the drop diameter d=2.8mm and the height of fall H=300mm were kept constant. The depth of target liquid was adjusted in the region of L=40mm~5mm. The drop collided with surface of target liquid with an impact velocity of 2.4 m/s under non-magnetic field. The impact velocity under applied magnetic field depended on the depth of target liquid, but the discrepancies in the velocity were slight. For example, in the case of L=20mm and mass concentration of magnetic fluid C=34.4%, the impact velocity was v=2.6m/s. The splashing of the drops was recorded using a high-speed movie camera operated at 2000 frames per second. The impact phenomena were filmed by sidelights illuminated from two 1kW lamps. This method enabled us to recognize two-fluids, that is, drop fluid and target fluid. The subsurface phenomena were analyzed by a film motion analyzer. Sample liquids used in the experiments were water-based ferricolloid W-35 (magnetic fluid) and its dilution with distilled water as drops, and kerosene or distilled water as target liquid. The magnetic drops were magnetized by the magnet installed under the container. Figure 2 shows the distribution of magnetic flux density along the y-direction of the permanent magnet used here. The physical properties of sample magnetic fluids are listed in Table 1.

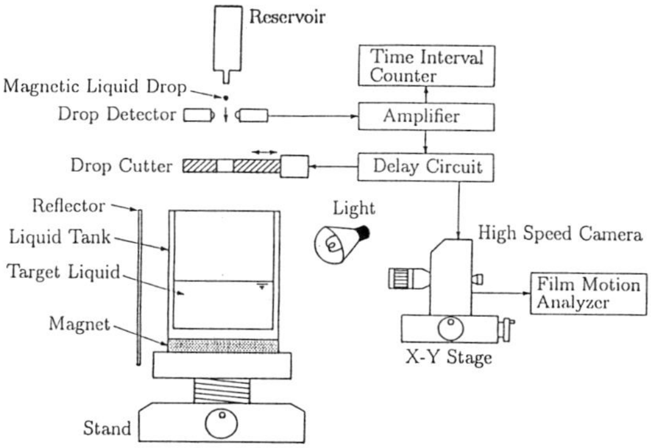

Figure 1. Schematic diagram of experimental apparatus

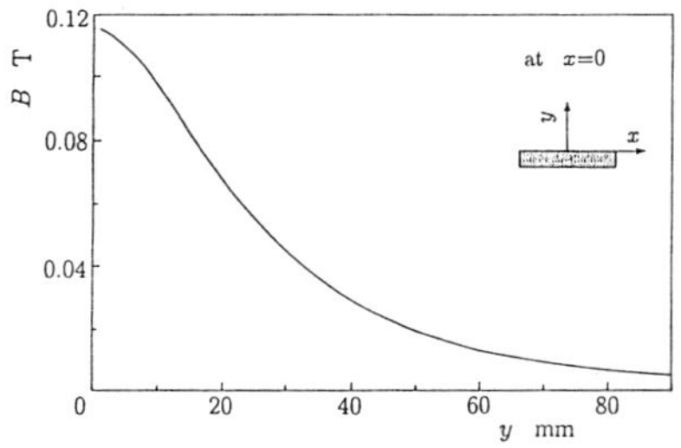

Figure 2. Distribution of magnetic flux density

Table 1
Physical properties of the magnetic fluids (at 20°C)

Mass concentration (%)	Density (kg/m^3)	Viscosity (Pa·s)	Surface tension (N/m)
34.4	1385.0	0.0141	0.0294
20.0	1193.0	0.0026	0.0298
16.1	1150.0	0.0025	0.0302
13.6	1124.0	0.0018	0.0305
9.6	1085.0	0.0014	0.0310
2.7	1022.0	0.0011	0.0316

3. EXPERIMENTAL RESULTS AND DISCUSSION

Without applied magnetic fields, the sequence of events following the impact of a magnetic drop was found to be similar to one of a water drop in the formation of the subsurface cavity and the Rayleigh jet. The sequence of events occurring when a magnetic drop(of mass concentration $C = 16.1\%$) splash into kerosene is shown in Figure 3. The stage, which is nearest to the first instant of contact, is marked as t=0 in Figure 3. The shape of the magnetic drop is almost spherical before the collision, because it is dominated by the surface tension($t = -1$ms). Following the impact, the interior of the cavity is lined by the magnetic fluid which formed the drop, but the magnetic fluid descends to the bottom during the growth of cavity($t = 2$ ~15ms). After the cavity reaches maximum depth, the unstable cylindrical waves appear on the cavity surface(t=25ms). The part of magnetic fluid leaves the cavity with shrinking of the cavity, but the remaining one ascends with the rising of the liquid column($t = 30$ ~50ms). This column is the compound Rayleigh jet.

On the other hand, with applied magnetic fields, the sequence of events is very different from the above. Figure 4 shows detailed photographs of the sequence. It can be seen

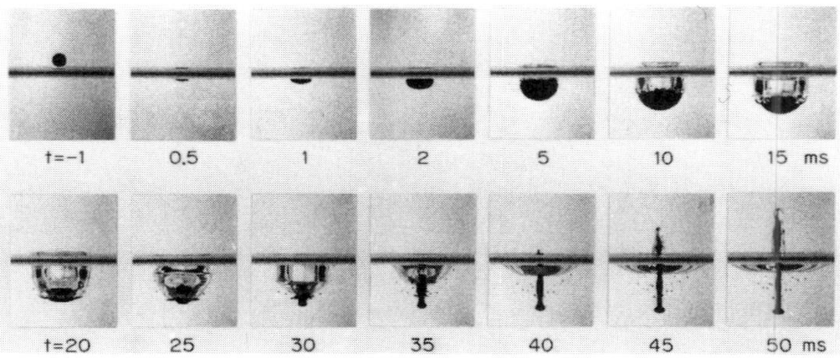

Figure 3. Photographs of drop impact(without magnetic field)

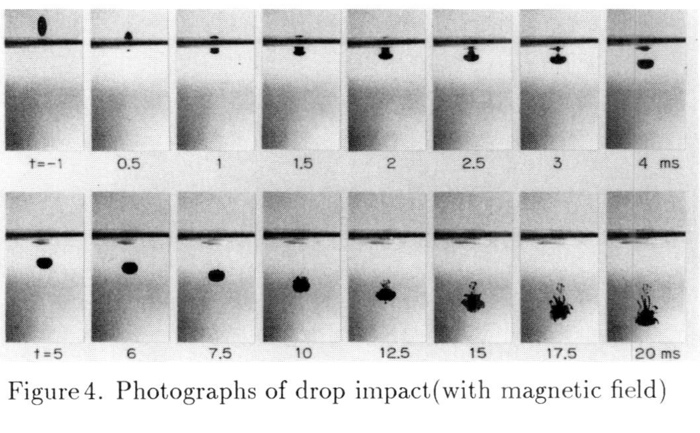

Figure 4. Photographs of drop impact(with magnetic field)

that the magnetic drop is elongated by the action of magnetic field before the impact($t = -1$ms). The magnetic drop collides as if a solid body collides with water, but the upward motion[6] of the leading edge of the kerosene film climbing up along the magnetic fluid surface can not be seen in this experiment($t = 0.5 \sim 1$ms). The drop does not disintegrate through the falling in kerosene for about 10ms after the impact. The cavity depth formed by the impact of magnetic drop on liquid surface is shallow extremely(t=2.5~4ms), that is, the applied magnetic fields have influence on the cavity depth[5]. Figure 5 shows the change of the ratio of the cavity depth S_c to the cavity diameter D_c with time t. The parameter C is mass concentration of magnetic fluid. It can be seen that higher magnetized drop leads S_c/D_c to rapider decrease. In other words, the strength of the applied magnetic fields(the magnetization of drop) have influence on the cavity depth. Experiments were also conducted with water target. The cavity formed by the collision of magnetic drop on the water surface was also extremely small under magnetic fields. Figure 6 shows the relation between the target liquid depth L and the time t_p required for magnetic fluid to reach the bottom. As the value of L increases, the difference of t_p in the target liquid is recognized. The descent velocity of magnetic drop in water was decreased compared with that in kerosene, because of the water-based magnetic fluid diffused in water. Therefore not only did the descent velocity of magnetic fluid depend on the applied field, but it depended on the miscibility.

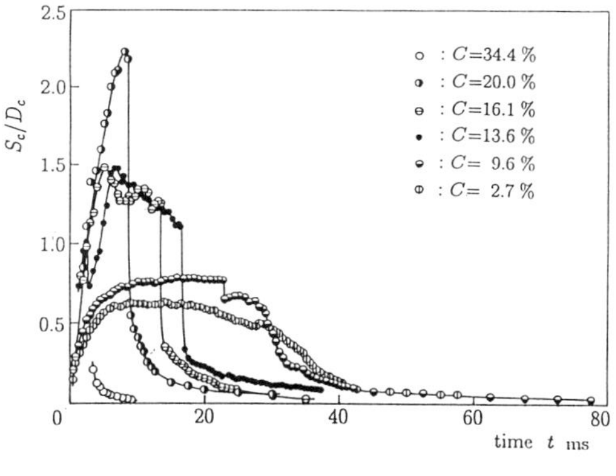

Figure 5. Plots of S_c/D_c against time

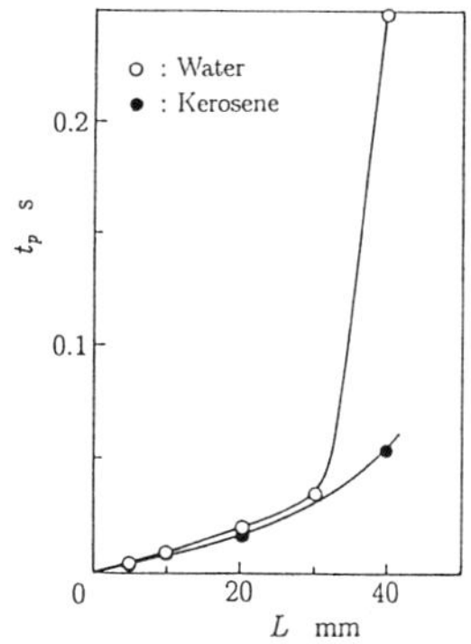

Figure 6. Relation between t_p and L

4. REFERENCES

1 A.M.Worthington, A Study of Splashes, Longman Green and Company, New York,1908.
2 O.G.Engel, J.Appl.Phys., 37(1966) 1798.
3 W.C.Macklin and G.J.Metaxas, J.Appl.Phys., 47(1976) 3963.
4 H.C.Pumphrey and P.A.Elmore, J.Fluid Mech, 220(1990) 539.
5 S.Sudo, H.Hashimoto and K.Katagiri, Proc.ISEM-Sendai,(1992) 131.
6 G.Kuwabara, H.Tanba and K.Kono, J.Phys.Soc. Japan, 56(1987) 2733.

Simulation and Design of Applied Electromagnetic Systems
T. Honma (Editor)

Shear Flow of a Magnetic Fluid between Concentric Annuli

M. Momoi, T. Sawada and T. Tanahashi

Department of Mechanical Engineering, Keio University
3–14–1 Hiyoshi, Kohoku–ku, Yokohama 223, Japan

Abstract

The effect of the substructure of water–based magnetic fluid on the flow between concentric annuli was studied. Power–law and micro polar theories which considered the internal spin of suspensions were used to explain non–newtonian flow properties. The moment acting on the inner cylinder caused by the rotation of the outer one in a rotating magnetic field was discussed, compared with the case under no magnetic field condition.

1. INTRODUCTION

Flow of magnetic fluids are effected by their substructure even if there is no magnetic field. The substructure of magnetic fluids are suspended magnetic particles (diameter is about 10nm) or their aggregations, and the surface of magnetic particles is coated with some dispersion medium. In general, magnetic fluids are divided into two series dependent on the base liquid. One is a water–based magnetic fluid and the other is a kerosene–based. The former has two dispersion layers, and the outer dispersion has hydrophilic groups which yield the aggregation of magnetic particles. So water–based magnetic fluids seem to have higher possibility of being effected by the substructure than kerosene–based.

The rheological property of a water–based magnetic fluid under no magnetic field condition is interesting in the sense that the existence of the substructure makes the flow field non-newtonian. In the present paper, the authors study the shear flow between concentric annuli. Applying power–law theory, the flow appeared to be pseudoplastic. The theoretical relation between the moment acting on the inner cylinder and the angular velocity of the outer one is examined based on micropolar theory[1] which considers the internal spin of the substructure. The authors show that the micropolar effect of the substructure on the shear flow and its influence is obtained experimentally.

A rotating magnetic field is applied by a two pole three–phase induction motor in order to know how large the moment acting on the inner cylinder becomes under the axisymmetric magnetic field.

2. EXPERIMENT

The experimental apparatus is shown in Fig.1. The outer cylinder of radius R=30mm rotates at the constant angular velocity Ω_w up to about 16 rad/s. The inner cylinder of radius κR=27.5mm is hung by the small brass wire whose diameter and modulus of rigidity are 0.5mm and 37.3×10^9Pa, respectively. Here κ is the radius ratio of the inner cylinder to the outer one. The moment acting on the inner cylinder is calculated by the twist angle of the wire. The constant temperature water is supplied in order to maintain the temperature of magnetic fluid at 20°C. Test liquid was a water–based magnetic fluid, shear viscosity $\mu=29.54\times10^{-3}$Pa·s, density ρ=1400 kg/m^3. Rotating magnetic field is induced by a two pole three phase induction motor, changing the supplying current up to 40A and frequency up to 200Hz.

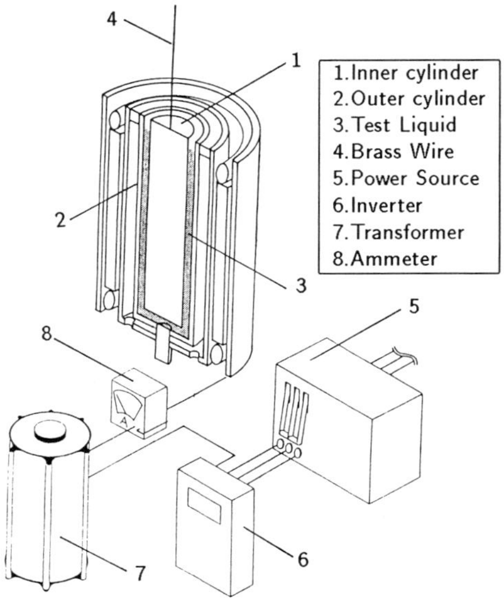

Fig.1 Experimental apparatus

3. RESULTS AND DISCUSSIONS

3.1 No magnetic field condition

Since magnetic fluids have suspended magnetic particles as substructure, its flows should be exactly treated by the micropolar theory. It is an interesting problem that water–based magnetic fluid is treated as newtonian or non–newtonian. The authors apply the power–law theory at first. Constitutive equation of power–law fluid is written by

$$\tau^n = \mu\dot{\gamma} \tag{1}$$

where $\dot{\gamma}$ is the shear strain rate and τ is the shearing stress. When n equals 1, the flow is newtonian. The relation between the moment acting on the inner cylinder per unit length M and the angular velocity of the outer cylinder is

$$\left[\frac{M}{2\pi(\kappa R)^2}\right]^n = \mu\frac{2n\Omega_w}{1-\kappa^{2n}} \tag{2}$$

Taking logarithm of Eq.(2) leads

$$n\log M = \log\Omega_w + \log\frac{2n\mu\left\{2\pi(\kappa R)^2\right\}^n}{1-\kappa^{2n}} \tag{3}$$

n=1.027 was obtained by the experiment. It means that the test liquid was non–newtonian, pseudoplastic. Shear viscosity was measured as 30.78×10^{-3}Pa·s and the relative error was 4.20%. In this point, the power–law theory gives a good approximation.

Next the authors apply the micropolar theory to the same flow. The moment acting on the inner cylinder per unit length is obtained by Sawada–Tanahashi[2]

$$M = 4\pi\mu\Omega_w R^2\left(\frac{\kappa^2}{1-\kappa^2} - D\right) \tag{4}$$

where

$$D=\frac{\kappa}{\Delta(1-\kappa^2)}\Big[(1-\kappa^2)\{I_1(\lambda)K_1(\kappa\lambda)-I_1(\kappa\lambda)K_1(\lambda)\}-\alpha\{I_1(\kappa\lambda)-\kappa I_1(\lambda)\}$$
$$\times\left\{K_0(\lambda)-\kappa^2K_0(\kappa\lambda)\right\}-\alpha\{K_1(\kappa\lambda)-\kappa K_1(\lambda)\}\left\{I_0(\lambda)-\kappa^2I_0\,(\kappa\lambda)\right\}\Big] \quad (5)$$

$$\Delta=\alpha\{K_0(\lambda)-K_0(\kappa\lambda)\}\{I_1(\lambda)-I_1(\kappa\lambda)\}+\alpha\{I_0(\lambda)-I_0(\kappa\lambda)\}$$
$$\times\{K_1(\lambda)-\kappa K_1(\kappa\lambda)\}+\alpha^2(1-\kappa^2)\{I_0(\lambda)K_0(\kappa\lambda)-I_0(\kappa\lambda)K_0(\lambda)\} \quad (6)$$

$$\lambda=kR,\qquad k=\sqrt{\frac{4\mu\mu_1}{\gamma(\mu+\mu_1)}},\qquad \alpha=\frac{\mu+\mu_1}{2\mu_1}\lambda \quad (7)$$

where μ_1 the vortex viscosity, γ the spin viscosity. I_0, I_1, K_0 and K_1 are the modified Bessel functions. D denotes the effect of the substructure on the flow. Tanahashi, et al.[3] determined the vortex viscosity as $\mu_1=\rho i^2/\tau_s$, where i is the volume average of the substructure radius of gyration, and τ_s is the relaxation time of the substructure depending on the fluid rotation. μ_1 is transformed into $\mu_1=3/2\phi\mu_0$, where ϕ is the volume concentration of the substructure and μ_0 is the shear viscosity of carrier liquid. The spin viscosity is given by $\gamma=\mu i^2$[4]. The value of the coefficient D in Eq.(4), which denotes the effect of the substructure on the fluid, was calculated by Eqs.(5)~(7) and D approches to 0 theoretically because λ becomes infinitely large.

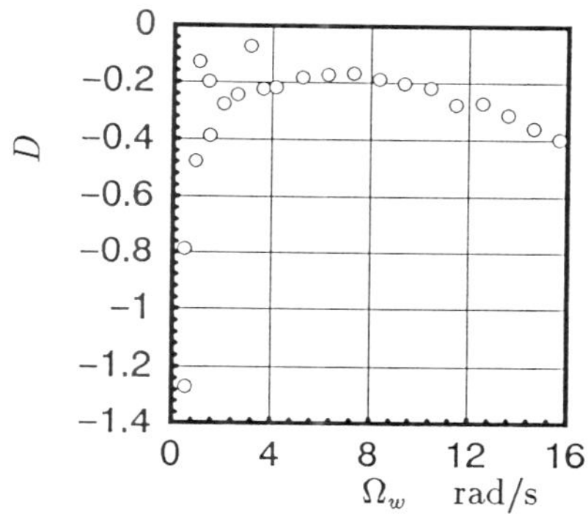

Fig.2 Coefficient D in Eq.(4)

Figure 2 shows the measured value of D. If the test liquid is non–newtonian, D equals 0 for every Ω_w. However, D is not zero as shown in Fig.2. When $D<0$ in Eq.(4), the moment is larger than that of newtonian fluid and it increases in the small region of Ω_w. Difference between theoretical and measured value of D comes from that the authors calculated material constants ignoring the possibility of the aggregation of suspended magnetic particles. That is, the aggregation may be made that the internal spin of the substructure contributed the increase of the moment acting on the inner cylinder. Considering the rotation of a fluid element is $1/2\mid\nabla\times\boldsymbol{v}\mid=\Omega_w/(1-\kappa^2)$ approximately and the Brownian relaxation time is the order of 10^{-6}s, Fig.2 shows the diffusion of the internal spin appeared as the angular velocity between the fluid and the substructure differed.

3.2 Rotating magnetic field

Figure 3 shows the moment M versus the current frequency f when the outer cylinder is fixed, i.e. Ω_w=0. Here B is the averaged magnetic induction. These results agree with the experimental results of Rosensweig, et al.[5] qualitatively in that the moment acted in the same direction as the magnetic field, though experimental conditions are different in some close points, Fig.4 shows the relation between the increase rate of the moment and Ω_w. Here M_0 is the moment in the no magnetic field condition. The moment increases as Ω_w becomes small. This tendency is similar to that in no magnetic field as mentioned

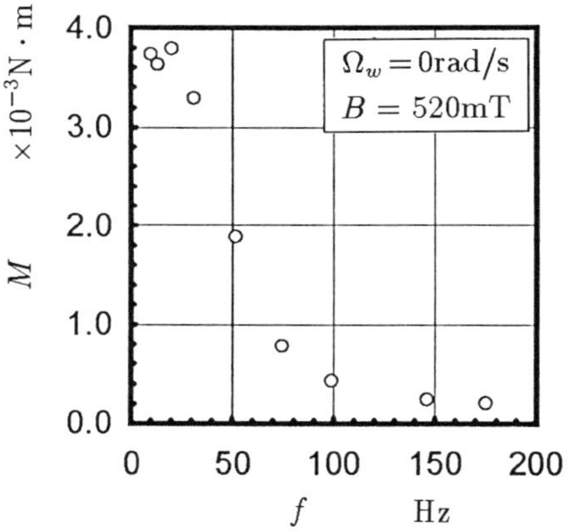

Fig.3 Moment acted on the inner cylinder per unit length.

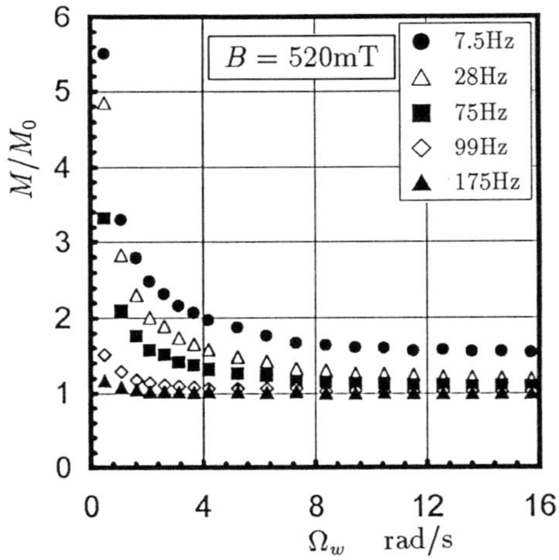

Fig.4 Increase rate of the moment.

in Sec.3.1. Since the aggregation of magnetic particles increases under the rotating magnetic field, the effect of the internal spin becomes remarkable. Dependence of the current frequency on the moment is considered as follows. The difference between the internal spin rate and the rotation of a fluid element becomes large in small f because the internal spin affected by the direction of magnetic lines of force. While this difference decreases with f.

4. CONCLUDING REMARKS

Shear flow of a water–based magnetic fluid was examined by power law and micropolar theories in the no magnetic field condition. Its non–newtonian properties were discussed with experimental results. Effect of the internal spin on the increase of the moment becomes remarkable in low shear strain rate. Under rotating magnetic field, it was shown that the increase of the moment depended on the supplying current frequency.

ACKNOWLEDGMENT

The authors are grateful to Mr.M.Obara of Toyoko Koki Co.Ltd. for his technical support on the experimental apparatus. The authors express their thanks to Mr.K.Kawanaka for his assistance during the experiment.

REFERENCES

1. K.Shizawa and T.Tanahashi, Bull.JSME, 29(1986), 1171.
2. T.Sawada and T.Tanahashi, Bull.JSME, 24(1981), 1778.
3. T.Tanahashi, et al., Bull.JSME, 26(1983), 1509.
4. S.J.Allen and K.A.Kline, Trans.Soc.Rheo., 14(1970), 39.
5. R.E.Rosensweig, J.Popplewell and R.J.Johnston, J.Magn.Magn.Mat., 85(1990), 171.

Simulation and Design of Applied Electromagnetic Systems
T. Honma (Editor)

Dynamic behavior of magnetic fluid in a container subject to vertical vibration

M. Ohaba[a] and S. Sudo[b]

[a]Department of Control System Engineering, Toin-Gakuen University of Yokohama, Kurogane-cho 1614, Midori-ku, Yokohama, Japan

[b]Dept. of Mechanical Engineering, Iwaki Meisei University, Iwaki, Japan

Abstract

This paper presents the results of an experimental study on the dynamic behavior of magnetic fluid surfaces in a vertically vibrated cylindrical container subjected to a magnetic field. Stability boundaries and frequency responses for one-half subharmonic wave motion were investigated under both a non-magnetic and a magnetic field. When a normal magnetic field was introduced, it was found that the resonance frequencies of the surface wave increased and the (0,1) mode wave crest elongated. The predicted resonance frequency by theory of the (1,1) mode wave in a magnetic field agreed with the experimental result.

1. INTRODUCTION

During the course of research in space technology, magnetic fluids were developed, and extensive studies about this fluid have been conducted [1]. In the previous paper [2], we investigated the surface instability and the sloshing phenomena of magnetic fluid in a magnetic field. When a normal magnetic field exceeds a certain critical value, numerous spikes, which have a regular hexagonal pattern, are formed on the fluid surface. That is, the interfacial instability appears on the fluid surface [3]. The purpose of the present study is to examine the nonlinear responses of the magnetic fluid surfaces accompanied with spikes in a cylindrical container vibrated vertically in a magnetic field.

2. EXPERIMENTAL APPARATUS AND PROCEDURES

The experiments were performed with the sinusoidal vertical excitation of the container subject to a magnetic field. The electrodynamic shaker was driven at a given frequency and amplitude. Measurements of the fluid surface amplitude were made using a strobo-vision analyzer. The sample magnetic fluid was water-based ferricolloid W-35. The container partially filled with magnetic fluid was fixed on the vibrating table, and a normal magnetic field was applied to the fluid surface by a disklike permanent magnet which was installed beneath the container. The magnetic flux density was 32 mT at the location equivalent to the fluid surface. The container was made of transparent acrylic plastic, and its inner dimensions were a length of 107 mm, a diameter of 40 mm, and a wall 10 mm thick.

3. EXPERIMENTAL RESULTS AND DISCUSSION

3.1. Stability boundaries

Figure 1 shows the stability boundaries for the mode (1,1) of the 1/2-subharmonic wave motion with the liquid depth h = 30 mm. The solid curves in Fig.1 show the theoretical stability boundaries derived from Mathieu's equation for an ideal fluid in the absence of a magnetic field [4]. The inside region of this stability boundaries is an unstable region, where the subharmonic wave would occur. The incipient points of the wave were obtained by gradually increasing the excitation amplitude Xo from zero at the forcing frequency fo held a constant value. The unstable region was narrowed by the fluid viscous damping. The resonant point of this wave motion may be decided at the point for which the experimental stability boundary most nearly approaches the line Xo = 0.

It is widely known that a normal magnetic field exerts a destabilizing influence on a flat interface between a magnetizable and a non-magnetizable fluid. Hence, when the normal magnetic field was imposed by the magnet, the fluid surface was deformed and had numerous spikes at a liquid depth less than h = 49.5mm. As for the spike motion, it was observed that the spikes on the fluid surface harmonically responded to the vertical excitation in the elongation and contraction of its height outside the unstable region (Fig.1). Similar spike motion like this was also found on a single spike on the vertical vibrating table [5].

The resonance frequency of the fluid motion was shifted towards the higher frequency range in the magnetic field, as shown in Fig.1. The reason for the shift in resonance frequency can be explained by the theory that the effect of the magnetic field is equivalent to that of the apparent gravitational acceleration of Geff, which exerts upon the experimental fluid-container system [6]; Geff = g − (M/ρ) dB/dz, where g the gravitational acceleration, M the intensity of magnetization, ρ the liquid density, B the magnetic flux density, and z the container axis coordinate. On the basis of this theory, the predicted resonance frequency of the mode (1,1), 1/2-subharmonic wave agreed with the experimental result.

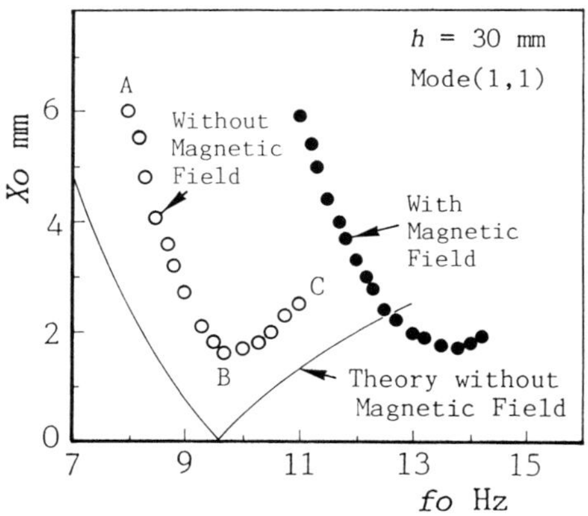

Figure 1. Stability chart for mode (1,1), 1/2-subharmonic.

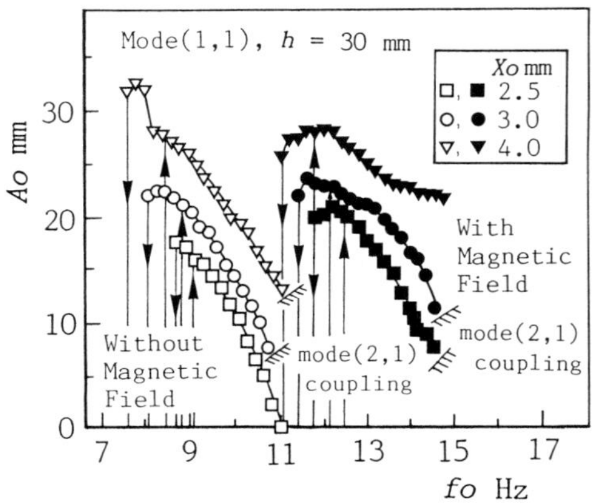

Figure 2. Wave amplitude for mode (1,1), 1/2-subharmonic.

3.2. Fluid amplitude responses for mode (1,1)

Figure 2 shows the fluid amplitude response for mode (1,1), 1/2-subharmonic wave motion, with a liquid depth h = 30 mm, for the different excitation amplitude. The fluid amplitude Ao of this wave mode was measured at the container wall. As the forcing frequency was gradually increased from the lower frequency range of the stability boundary-AB (Fig.1) at a constant excitation amplitude, large amplitude wave was noticed as the frequency fo reached the boundary-AB. The upwards arrows in Fig.2 indicate the generation of this subharmonic fluid motion. On the other hand, as the frequency fo was decreased very slowly from the point on the stability boundary-BC (Fig.1), fluid amplitude continually increased. For this case, the fluid motion existed even at a lower frequency range beyond the stability boundary-AB, but its amplitude was quickly reduced to zero, at the point indicated by the downwards arrows on Fig.2. Such a sudden increase in amplitude and softening in response is the common characteristic behavior for nonlinear dynamic systems.

With the magnetic field experiments shown in Fig.2, the fluid responses shifted towards a higher excitation frequency range, and a nonlinear surface motion similarly occurred. The maximum amplitudes of the wave for Xo = 2.5, 3.0 mm at the mode (1,1) were somewhat increased by the magnetic field. In addition, a swirling sloshing phenomenon was found at the narrow frequency range less than fo = 11.6 Hz for Xo = 4.0 mm.

3.3. Fluid amplitude responses for mode (0,1)

Figure 3 shows the frequency response of the amplitude for the mode (0,1), 1/2-subharmonic wave motion without a magnetic field, for h = 30 mm. The fluid surface amplitude which was measured along the container axis was gradually decreased with the increasing frequency. The solid lines with arrows in Fig.3 indicate the similar nonlinear responses observed at the mode (1,1) fluid motion. Moreover, the wave crest was broken, and liquid drops formed in the dotted-lined frequency range.

In the presence of the magnetic field, the fluid response shifted to a higher excitation frequency range, and the fluid maximum amplitudes became larger than those without a magnetic field (Fig.3), as shown in Fig.4. However, the fluid subharmonic response for Xo = 0.8 mm did not occur. This is attributed to the viscous dissipation of the magnetization

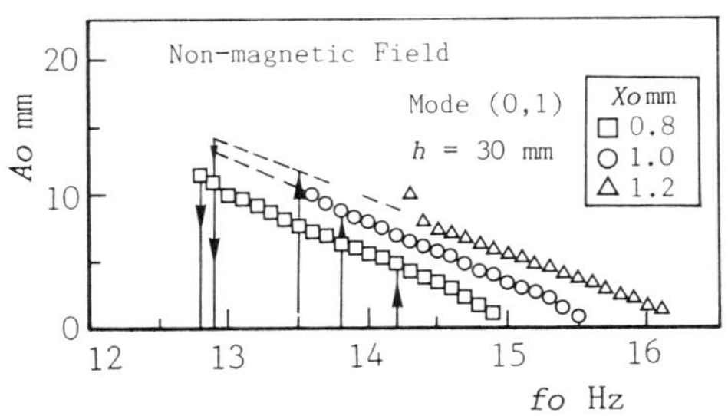

Figure 3. Wave amplitude without magnetic field for mode (0,1).

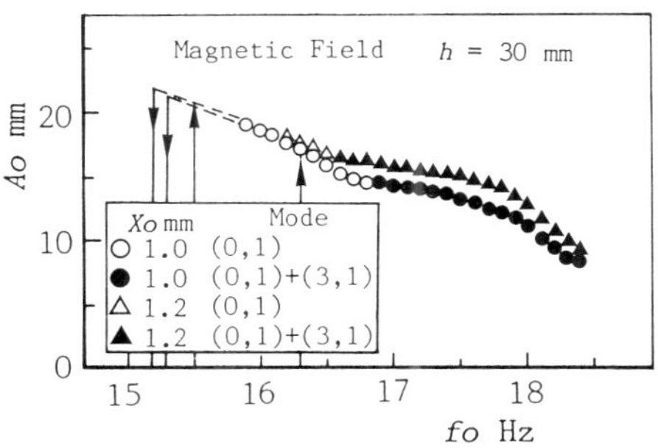

Figure 4. Wave amplitude with magnetic field for mode (0,1).

of the fluid. Such an increase in maximum amplitude of this wave mode was also observed at h = 20, 40 and 49.5 mm, although the magnetic body force acted on the fluid interface. This is because the magnetic flux tends to concentrate at the wave crest, and consequently suppresses the breaking of its crest; the magnetic field along the wave crest provides a stabilizing influence on the unstable disturbance on the fluid interface. At the higher excitation frequency range, the fluid surface motion of the mode (0,1) was coupled with the (3,1) mode wave.

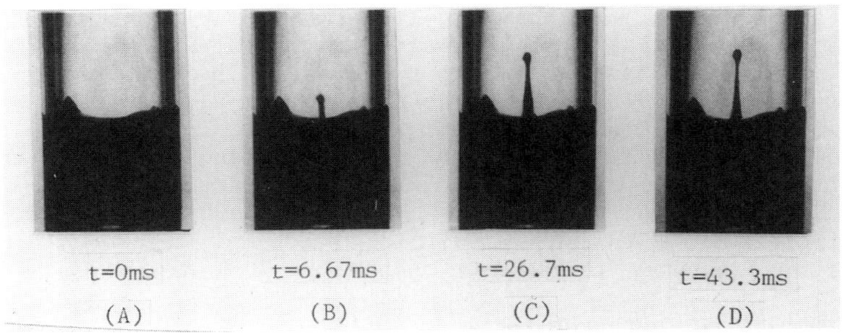

Figure 5. Photographs of 1/2-subharmonic wave for h = 30 mm, fo = 17.0 Hz, and Xo = 1.5 mm, in the normal magnetic field.

3.4. Fluid surface behavior

The photographs in Figure 5 show the fluid surface behavior for the mode (0,1), 1/2-subharmonic wave in the normal magnetic field. The value t in Fig.5 denotes the elapsed time from the appearance of the (0,1) mode response. The 1/2-subharmonic surface response of the mode (3,1) is noticed on the fluid surface in photograph (A). As the fluid response increases, the wave crest of the mode (0,1) is elongated, and then reaches the maximum wave height, as shown in photograph (D) in the magnetic field. Furthermore, the fluid motion was accompanied with a beat motion.

4. CONCLUSION

(1) The 1/2-subharmonic resonance frequency of magnetic fluid was shifted to a higher frequency region due to the normal magnetic field gradient.
(2) The wave crest of the mode (0,1) was elongated by the magnetic field.

5. REFERENCES

1 R.E. Rosensweig, Ferrohydrodynamics, Cambridge University Press, 1985.
2 S. Sudo, H. Hashimoto, A. Ikeda, and K. Katagiri,J. Mag. Mag. Mat.,65 (1987)219.
3 M.D. Cowley and R.E. Rosensweig, J. Fluid Mech., 30, Part 4 (1967) 671.
4 T.B. Benjamin and F. Ursell, Proc. R. Soc., Ser. A, 255 (1954) 505.
5 S. Sudo, M. Ohaba, K. Katagiri and H. Hashimoto, Trans. Japan. Soc. Mech. Eng. Vol.58, No.549 (1992)1434.
6 F.T. Dodge and L.R. Garza, Trans. ASME, Pap. No.71-Vibr-24 (1971) 1.

Simulation and Design of Applied Electromagnetic Systems
T. Honma (Editor)

Some magnetic properties and microstructure of the $Fe_{73.5}Cu_1Nb_3Si_{13.5}B_9$ alloy

J.Zbroszczyk, W.Ciurzyńska, K.Narita[a]

Institute of Physics, Technical University of Częstochowa, Armii Krajowej 19, 42-200 Częstochowa, Poland

[a] Department of Electrical Engineering, Chubu University, Kasugai, Aichi 487, Japan

Abstract

Magnetic properties i.e. coercivity, initial magnetic permeability and hysteresis loops and microstructure for the $Fe_{73.5}Cu_1Nb_3Si_{13.5}B_9$ alloy in the as-quenched state and after annealing at 740 K and 850 K for 0.5 h were investigated.

The distinct improvement of soft magnetic properties was observed after annealing the alloy at 850 k for 0.5 h.

1. INTRODUCTION

It has been found that materials with excellent soft magnetic properties can be prepared by crystallization of amorphous Fe-based alloys with additions of Cu and Nb [1-3]. The $Fe_{73.5}Cu_1Nb_3Si_{13.5}B_9$ alloy is an example of these materials and exhibits a high initial magnetic permeability and a low coercivity after annealing at about 850 K. The crystalline alloy consists of fine grains (about 10 nm) embedded in an amorphous matrix [4,5].

In this paper the magnetic properties such as coercivity and initial magnetic permeability and microstructure of the $Fe_{73.5}Cu_1Nb_3Si_{13.5}B_9$ alloy after different heat treatment are studied.

2. EXPERIMENT

The amorphous ribbon of the $Fe_{73.5}Cu_1Nb_3Si_{13.5}B_9$ alloy was prepared by rapid quenching on a single roller. The width and the thickness of the ribbon were 5 mm and 15 μm, respectively.

The coercivity was measured for the samples of 5 cm in length. However, the initial magnetic permeability and static hysteresis loops were measured for toroidal samples. Using

Mössbauer spectroscopy and X-ray diffraction the microstructure of the alloy was investigated. All investigations were carried out for samples in the as-quenched state and after annealing at 740 K and 850 K in vacuum (p=0.01 Pa) for 0.5 h. The measurements were performed at room temperature.

3. RESULTS AND DISCUSSION.

In Fig.1 the hysteresis loops (ab) and initial permeability versus magnetizing field (c) are shown (f=1.03 kHz). The as-quenched $Fe_{73.5}Cu_1Nb_3Si_{13.5}B_9$ ribbons exhibit rather poor soft magnetic properties (Fig.1a, Table I). At the magnetizing field (6 A/m) lower than the coercive field the magnetization process in the as-quenched alloy takes place by several Barkhausen jumps (Fig.1a).

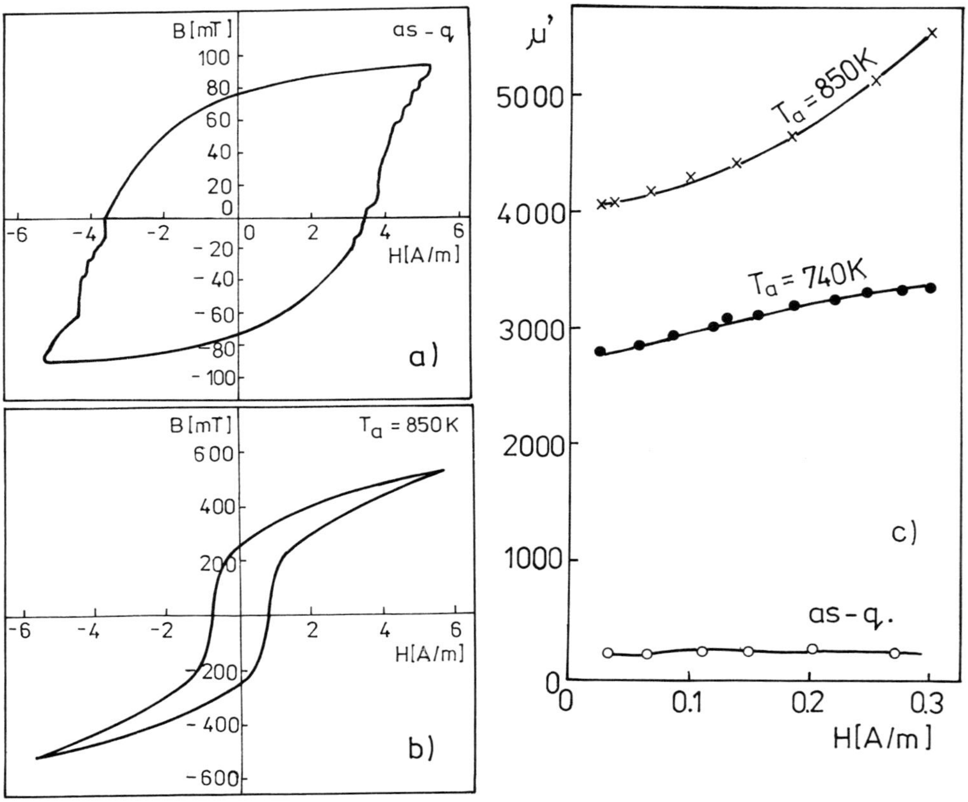

Fig.1 Static hysteresis loops (a,b) and initial permeability versus magnetizing field (c) for the $Fe_{73.5}Cu_1Nb_3Si_{13.5}B_9$ alloy in the as-quenched state (as-q.) and after annealing for 0.5 h at T_a=740 K and T_a=850 K.

The intensity ratios of the six lines of the Mössbauer spectrum (Fig.2a) are about 3:2:1:1:2:3. It corresponds to random orientation of the magnetization in the as-quenched sample. However, the hyperfine field distribution is a superposition of the low and high field components (Fig.2b). It indicates that the composition of the alloy in the as-quenched state is fluctuated. These inhomogeneities influence the magnetic properties of the alloy (Fig.1a, Table I) and they may be the origin of Barkhausen jumps (Fig.1a).

Table 1
Coercivity (H_c) and initial permeability (μ') for the $Fe_{73.5}Cu_1Nb_3Si_{13.5}B_9$ alloy

Sample	H_c[A/m]	μ'
as-quenched	11.94	230
after annealing at 740 K for 0.5 h	2.17	2750
after annealing at 850 K for 0.5 h	1.00	4000

After annealing at 740 K for 0.5 h a few relatively large Barkhausen jumps were observed. This behaviour may be explained by a phase-segregation which takes place during annealing samples at this temperature. It was confirmed by Mössbauer spectroscopy and X-ray diffractometry. The distinct improvement of the soft magnetic properties is obtained after annealing the sample at 850 K (Fig.1b, Table I).

The Mössbauer spectrum for this sample (Fig.2c) can be decomposed into five Zeeman sextets. Four sextets are related to αFe-Si crystalline phase (B_{eff}=31.75 T, 28.71 T, 24.36 T and 19.25 T) and the fifth sextet (B_{eff}=11.51 T) corresponds to low magnetic amorphous matrix. From X-ray studies was concluded that the grain size of αFe-Si phase is about 10 nm and grains are isotropically distributed. The good soft magnetic properties of the $Fe_{73.5}Cu_1Nb_3Si_{13.5}B_9$ alloy after annealing at 850 K for 0.5 h are connected with averaging locally fluctuating anisotropies [5].

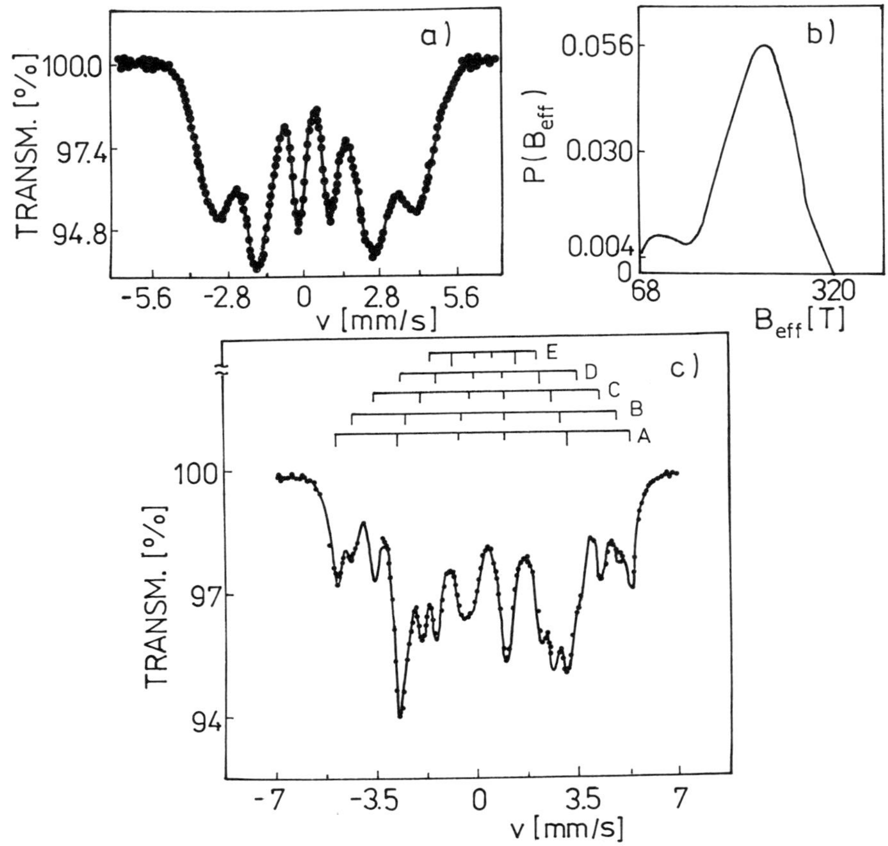

Fig.2 Mössbauer spectra (a,c) and hyperfine field distribution (b) for the $Fe_{73.5}Cu_1Nb_3Si_{13.5}B_9$ alloy in the as-quenched state (a,b) and after annealing at 850 K for 0.5 h (c).

4. REFERENCES

1. Y.Yoshizawa, S.Oguma and K.Yamauchi, J.Appl.Phys.64, 6044 (1988)
2. Y,Yoshizawa and K.Yamauchi, Mat.Trans. J.I.M. 31, 307 (1990)
3. Y.Yoshizawa and K.Yamauchi, J.Jpn.Inst.Metals 53, 241 (1989)
4. K.Hiraga and o.Kohmoto, Mat.Trans.J.I.M.33, 868 (1991)
5. G.Herzer, IEEE Trans.Magn.26, 1397 (1990)

This work is supported by the State Committee for Scientific Research.

Liquid Metals

Simulation and Design of Applied Electromagnetic Systems
T. Honma (Editor)

Levitation melting by duplex configuration of inverters with different frequency

K.Sakuraya[a] T.Watanabe[a] S.Iwasaki[a] A.Fukuzawa[a]
M.Yamazaki[b] T.Take[c] M.Fujita[c]

[a] National Research Institute For Metals,
2-3-12 Nakamegro,Meguro-ku,Tokyo,153,Japan

[b] Electrotechnology Applications R&D Center,Chubu Electric Power Company Inc.,20-1 Sekiyama,Ohdaka cho,Midoriku,Nagoya,459,Japan

[c] Fuji Electric Corporate Research and Development Ltd.,
7 Yawatakaigandori,Ichihara,Chiba,290,Japan

Abstract

Cold crucible induction furnaces are useful for melting metals without contamination, because the molten metals are levitated in the crucible. Stable melting conditions are very important to the successful operation of such equipment. To obtain such conditions, we have designed a new cold crucible that takes advantage of the fact that the levitating forces and joule losses induced in molten metal depend on current frequency. We have designed a new cold crucible with two coils. Higher frequency currents (several tens kHz) flow in the upper coil and lower frequency currents (several kHz) flow in the lower coil.

1.INTRODUCTION

A cold crucible induction furnace is composed of a metal crucible, coils and inverters. A metal crucible is made of many segments, each one cooled internally with water. High frequency currents are made to flow from inverters through coils wrapped around the crucible. Eddy currents are induced in the crucible and the molten metal. Because mutually repulsive forces are induced between the molten metal and the crucible and coils, the molten metal levitates. At the same time, the metal is heated through Joule dissipation. Thus, a chemically reactive metal or alloy, for example, TiAl, can be melted without contamination.

When designing such equipment, it is very important to keep melting conditions stable. To predict melting condition, numerical analyses that consider the magnetic, gravitational and surface tension forces have been studied.[1][2] Because eddy currents flow near the surface of a conductor and a three dimensional analysis is needed, it is difficult to use a numerical method. To get around this problem, we took advantage of the fact that the levitating forces and joule losses induced in the molten metal depend on currents frequency. We designed a new cold crucible apparatus with two coils. Higher frequency (several tens kHz) currents flow in the upper coil and lower frequency (several kHz) currents flow in the lower coil. The molten metal assumed a more perfect shape and had a shorter melting time with a duplex configuration of inverters.

2.TESTING CONDITIONS

2-1 Measurement of the levitating force

Figure 1 shows the experimental apparatus, in which a crucible is closed except for a hole in the center of its bottom, and high frequency currents flow in coils wrapped around the crucible. A copper ball (52 Φ) was suspended in the crucible and levitating forces were measured with a digital balance. The measurement apparatus in the case of a single coil is shown in Fig.2. The frequencies of currents were 3.5, 6.2, 10.0 kHz. The apparatus in the case of two coils is shown in Fig.3. The upper coil has 5 turns and the lower coil has one turn. The shield plate is located between these coils. The current frequency of the upper coil was 30 kHz and that of the lower coil was 3kHz.

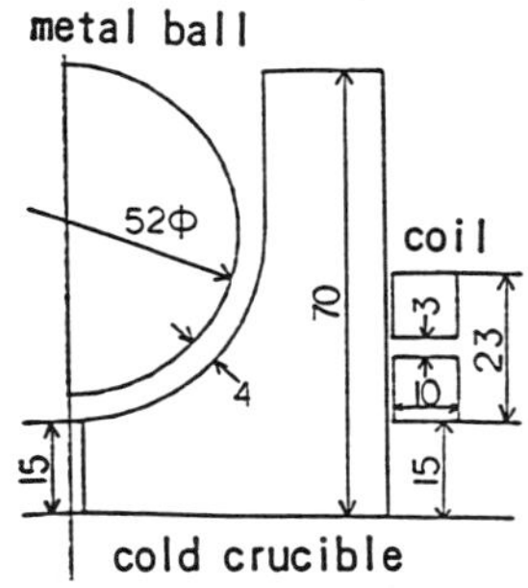

Fig.2 Apparatus with single coil

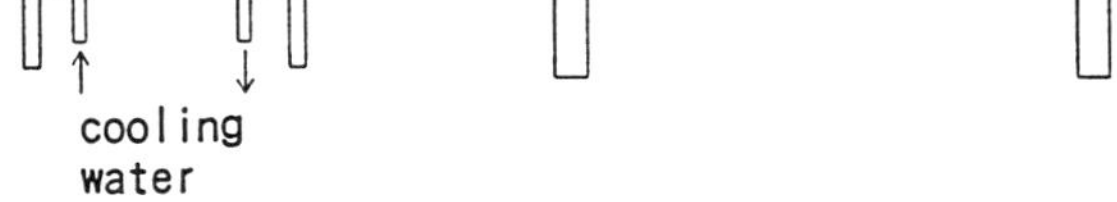

Fig.1 Experimental apparatus

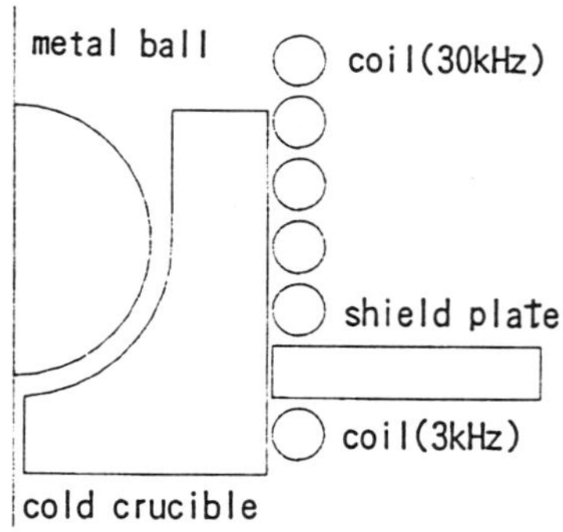

Fig.3 Apparatus with double coils

2-2 Melting test

With the duplex configuration of inverters, power of 3kW, 3kHz was supplied to the lower coil and power of 32kW,30kHz was supplied to the upper coil. In the single power supply drive, the power supplied was 35kW, 30kHz. After the shield plate was removed, 310g of Cu-Al alloy was melted to compare the melting conditions obtainable with the duplex and single power supply drives.

3.TEST RESULTS

3-1 Levitating force

The measured results obtained with the single coil are shown in Fig. 4. The levitating force depends on the frequency of the currents. Induced losses can be obtained with the following equation when eddy currents flow near the surface of a conductor.

$$W = \frac{1}{2\sqrt{2}} \sqrt{\omega \mu \rho} \cdot H m^2 \qquad F = \frac{1}{2} \mu H m^2$$

ω : angle velocity = 2 * π * f
f : frequency(Hz)
ρ : resistivity ($\Omega \cdot$ m)
μ : permeability
H_m: magnetic strength(peak value)
W : induced losses (W/m^2)
F : induced forces (N/m^2)

The relationship between losses and forces can be shown with the following equation.

$$F = W \cdot \sqrt{\frac{2 \cdot \mu}{\omega \cdot \rho}}$$

If W is constant, the induced forces are inversely proportional to frequency. Table 1 shows a comparion of the measured and calculated values mentioned above.

The levitating measured forces with two coils are shown in Fig.5. The levitating forces were induced mainly by the current in the lower coil. Total forces were obtained by adding the forces obtained with each coil.

Table 1 Comparison of measured and calculated values

freq. (kHz)	ratio of forces measured		calculated	
3.5	100		100	
6.2	63	100	71	100
10.0	44	70	55	77

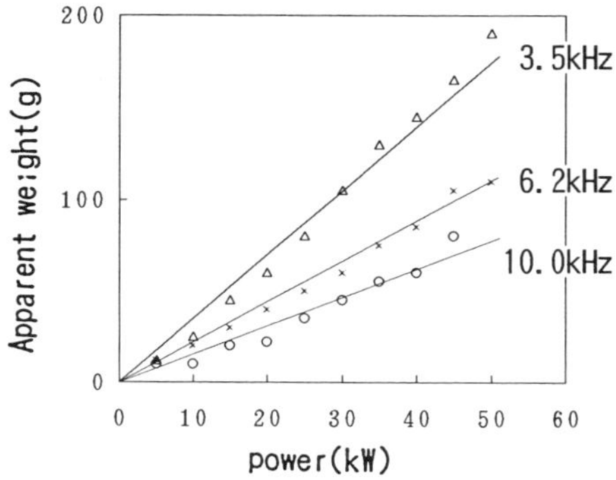

Fig.4 Levitating forces with a single coil

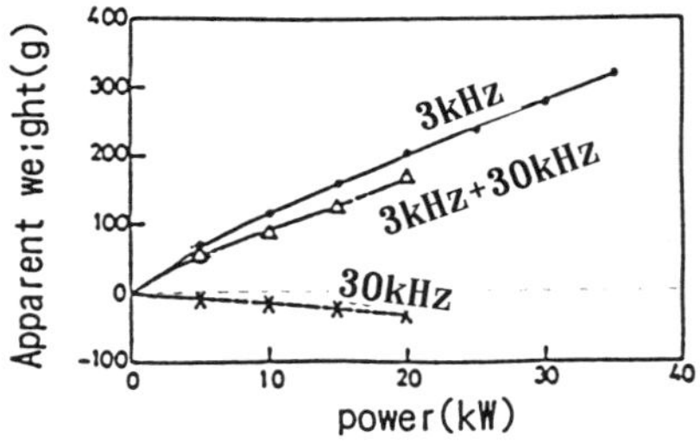

Fig.5 Levitating forces with two coils

3-2 Melting test

The molten metal assumed a more perfect shape with the duplex configuration of inverters, as shown in Fig.6.

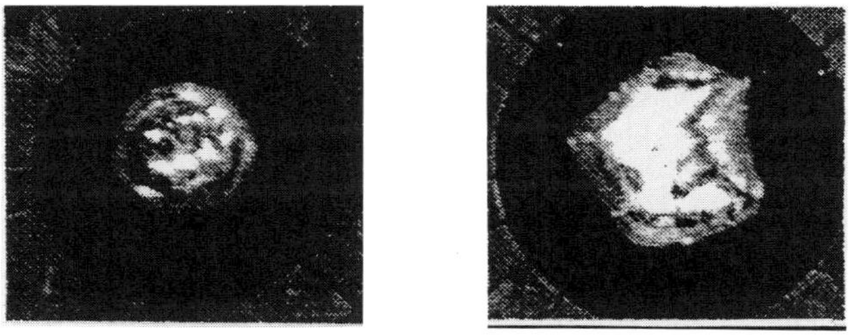

(a)Duplex drive (b)Single drive

Fig.6 Behavior of molten metal

4.CONCLUSION

Our new duplex inverter configuration for a cold crucible allows the heating and levitation of metals to be controlled independentally. This is expected to lead to the successful melting of various kinds and volumes of metals. The authors succeeded in melting 2.3kg of Ti by improving the shape of the crucible and using a duplex configuration of inverters at different frequencies.

5.REFERENCES

1 T.Toh, A.Gagnoud, M.Garnier, CAMP-ISIJ vol.4 (1991)-5
2 O.Iwata, H.Igarashi, T.Honma, The First MAGDA Conf 1992 D-9

Simulation and Design of Applied Electromagnetic Systems
T. Honma (Editor)

Properties of electrorheological fluid dispersing smectite particles

T. Fujita, J. Mochizuki, G. Obinata and K. Yamaguchi

Mining College, Akita University, Akita 010, Japan

Abstract

Hydrophobic fine smectite particles containing adequate amount of moisture were dispersed in alkylnaphthalene. A very stable and transparent fluid was obtained and showed electrorheological (ER) phenomenon. Steady shear viscosity and dynamic viscoelasticity of the fluid were measured and the frequency characteristics were investigated by the experimental damper.

1.INTRODUCTION

There are many desired characteristics for the electrorheological fluid to use in a damping system. To obtain stable ER fluids the ultrafine barium titanate / hydrocarbon suspension [1] and the whisker / silicon oil suspension [2] were reported.

In this report the stable nonsettling condition, large viscosity change under electric field in a wide temperature range, low power requirement and low cost have been investigated for the fluid dispersing smectite. As for dispersants of porous hydrate silicate, zeolite containing water / silicon oil suspension was reported to show ER phenomenon in a wide temperature range [3]. Also the dynamic mechanical study is necessary to use the prepared ER fluid in the damping system. The experimental setup of damper has been assembled and the characteristics have been measured under AC electric field or no field.

2.MATERIALS AND METHODS

2.1 Preparation of fluids

An artificial fine hydrophobic porous smectite covered with surfactant (smectite SAN, produced by CO-OP Chemical CO.LTD, Japan) has the specific surface area of $1\text{-}2 \times 10^3 m^2/kg$ and density of $2670 kg/m^3$. The smectite particles adsorbing some amount of moisture were put into alkylnaphthalen and the suspension was heated at 373-393K for 5min.. After the centrifugation of 300g for 10min., the stable and transparent fluid was obtained and showed ER phenomenon. Much amount of ER fluid could be prepared rapidly by this procedure. Here, as the fluids dispersing high concentration (more than 8wt%) of smectite were prepared, centrifugation was not used.

The desorptions of water from smectite containing moisture and the fluid dispersing its smectite were measured from 293K to 473K by thermogravimetrey (TG) and differential thermal analysis (DTA). Though physically adsorbed water

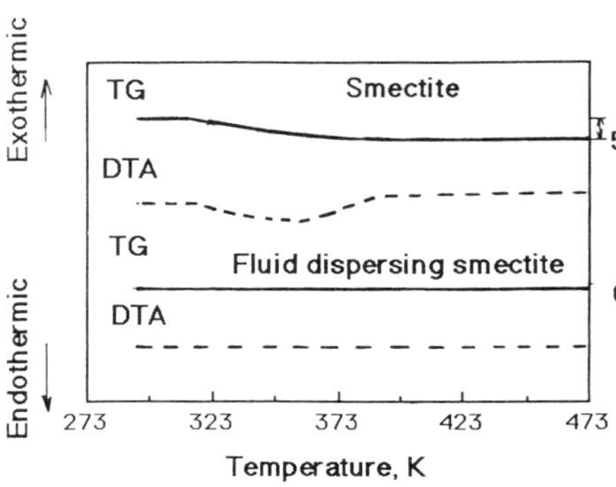

Fig.1 TG and DTA curves of smectite and fluid dispersing smectite.

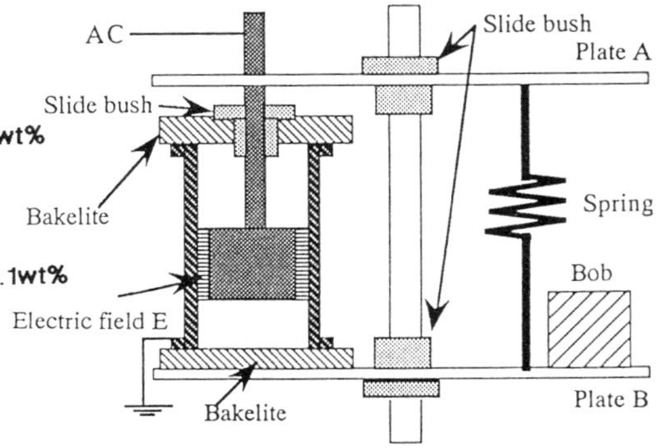

Fig.2 Experimental damper setup

desorbed from smectite, the desorption of water from smectite dispersing in alkylnaphthalene was not almost observed; therefore, ER effect of the fluid is expected under 473K.

The conductivity was $4x10^{-10}$S/m and relative dielectric constant was 2.0 (tanδ 0.18) at 1kHz for the fluid dispersing 9.1wt% of smectite containing 1.8wt% of moisture, which was employed for the dynamic behabior study under electric field.

2.2 Measurements

The rheological properties were measured by the rotation viscometer (Rheometer RM-1, Shimadzu, Japan). The ground outer cylinder of 15.0mm radius was rotated to measure the steady shear viscosity or oscillated to measure dynamic viscoelasticity; on the other hand, the AC high voltage inner cylinder of 13.5mm radius and 70mm height was fixed with spring.

Experimental damper [4] is shown in fig.2. After the prepared fluid was put in the cylinder, plate A was vibrated by the white noise of less than 10Hz under 0 and 2kV/mm electric field and the recorded displacements of plate A and plate B were converted into gain and phase as a function of angular frequency.

3.EXPERIMENTAL RESULTS

3.1 Steady shear viscosity

Flow curves for fluids dispersing 10wt% smectite under 2kV/mm electric field and no field are shown in fig.3 as a function of moisture content in smectite. The optimum moisture content in smectite to increase shear stress difference between on and off electric field was about 5wt%. At larger moisture content, such as 6.2wt%, shear stress difference decreased, comparing that of 5wt% moisture.

Flow curves for fluids dispersing various volume fraction of smectite under 2kV/mm and no field are shown in fig.4. Flow curves for fluids dispersing smectite of 9.1-11.1wt% volume fraction were almost the same; however, shear stress increased exponentially for the fluids dispersing more amount of smectite. At 13.0wt% volume fraction, the yield stress was about 1kPa under 2kV/mm. For each prepared fluid the viscosity increased more than 10 times under 2kV/mm than the one under no field at $10s^{-1}$ shear rate. The electric current flowed about 0.5mA at $10s^{-1}$ shear rate under 3.5kV/mm for the rheometer using fluids shown in fig.4; however, electric current decreased much more as increasing shear rate.

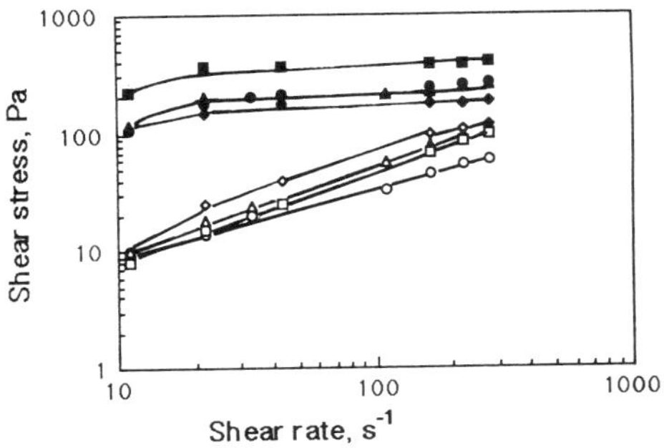

Fig.3 Effect of moisture content in smectite on viscosity of fluid dispersing 10wt% smectite.

Moisture wt%	Electric field 0kV/mm	2kV/mm
1.7	○	●
1.9	△	▲
5.0	□	■
6.2	◇	◆

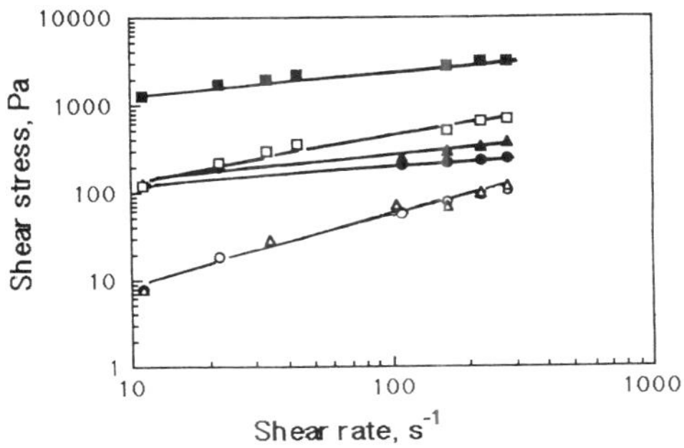

Fig.4 Effect of smectite content in fluid on viscosity of fluid. (moisture content in smectite 1.8wt%)

Smectite content wt%	Electric field 0kV/mm	2kV/mm
9.1	○	●
11.1	△	▲
13.0	□	■

3.2 Dynamic behavior

The dynamic viscoelasticity for the fluid dispersing 9.1wt% of smectite is shown in fig.5. Dynamic rigidity G' and dynamic viscosity η' increased by applying 2kV/mm. While G' was almost constant, η' depended on angular frequency ω, and decreased as ω increased. As the very small strain was used in this experiment, the increase of strain could change G' and η'. In other ER fluids, the larger dynamic stress at larger strain [5] and the dependence of G' and G" (loss modulus) on frequency at different strain [6] were reported.

Frequency characteristics of experimental damper using the prepared fluid is shown in fig.6. Though the resonance was not observed clearly in this system, the gain and phase curves moved to higher frequency region by applying 2kV/mm, because the dynamic rigidity and dynamic viscosity increased comparing those under no field as shown in fig.5. The ratios of estimated spring and damping constants [4] between on and off electric field were the same order as those of dynamic rigidity and dynamic viscosity shown in fig.5. The further investigation of dynamic viscoelasticity at various strains in wide frequency range, and the experiment at the damper where resonance appears sharply should be subject for a future study.

4. CONCLUSION

The suitable ER fluid was prepared by dispersing moisture adsorved smectite in alkylnaphthalen and employed in the experimental damper, whose characteristics could be changed under electric field.

ACKNOWLEDGEMENT

The authors are grateful to CO-OP Chemical CO.LTD. for providing the smectite particles, Mr.T.Takayama (Akita Univ. graduate student) for the experiment of damper and SANEYOSHI SCHOLARSHIP FOUNDATION for the financial support.

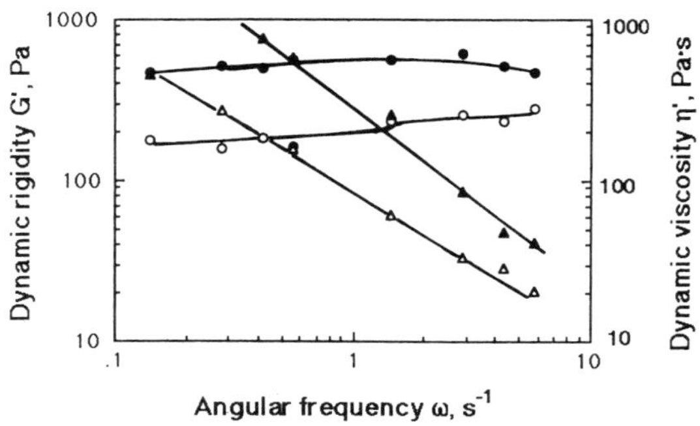

Fig.5 Dynamic rigidity and dynamic viscosity as a function of angular frequency for fluid dispersing smectite.
(smectite 9.1wt%, moisture in smectite 1.8wt%
deflection of outer cylinder 1.75×10^{-2}rad)

	Electric field	
	0kV/mm	2kV/mm
Dynamic rigidity	○	●
Dynamic viscosity	△	▲

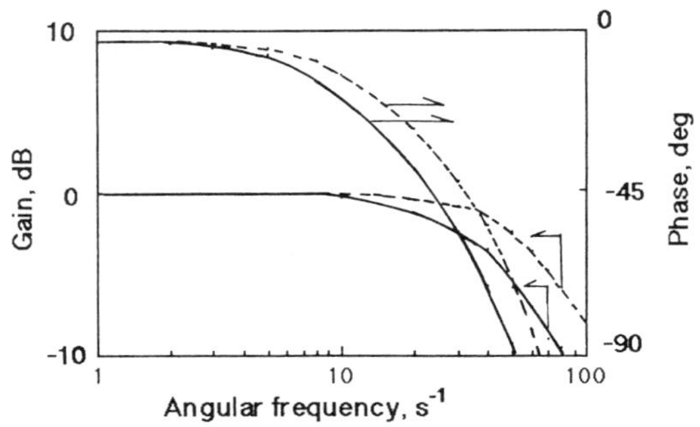

Fig. 6 Frequency characteristics for the experimental damper shown in fig.2.
—— 0kV/mm, ------ 2kV/mm

REFERENCES

1 T.Fujita, I.J.Lin and M.Mamiya, Proc. of the Third Int. ISEM Symposium on Application of Electromagnetic Forces, Sendai, Japan (1991)123.

2 T.Fujita, J.Mochizuki,H.Kongnai and K.Kawata, Proc. of the Fourth Int. ISEM Symposium on Nonlinear Phenomena in Electromagnetic Fields, Nagoya, (1992)157.

3 H.Conrad, A.F.Sprecher, Y.Choi and Y.Chen, J.Rheol. 35 (1991)1393.

4 T.Takayama, G.Obinata, T.Yukawa and T.Fujita, Papers of 7th Symposium on Sendai,Japan (1992)141.

5 W.S.Yen and P.J.Achorn, J.Rheol. 35(1991)1375.

6 Y.Otsubo, M.Sekine and S.Katayama, J.Rheol. 36(1992)479.

Simulation and Design of Applied Electromagnetic Systems
T. Honma (Editor)

Numerical analysis of natural convection of thermo-electrically conducting fluids using the GSMAC finite-element method

Y. Oki, T. Nakai and T. Tanahashi

Department of Mechanical Engineering, Keio University
3–14–1 Hiyoshi, Kohoku–ku, Yokohama 223, Japan

Abstract

In the present paper, the natural convection of thermo-electrically conducting fluids in a square cavity under a constant magnetic field are calculated by using GSMAC-FEM in conjunction with so-called $\boldsymbol{B}$ method. This scheme satisfies both conservation laws of mass and magnetic flux efficiently.

1. INTRODUCTION

Thermo-electrically conducting fluid has been increasingly used in the manufacturing process of semi-conducting material such as silicon crystals. In the present paper, we modify the conventional $\boldsymbol{B}$ method[1] to get the efficient performance of the calculation, keeping the accuracy of the calculation. Hydromagnetic cross helicity is introduced as the new physical quantity which means the effect of interactions between magnetic field and velocity field. Using hydromagnetic helicity, we evaluate the performance between conventional and modified $\boldsymbol{B}$ method quantitatively. In order to verify the present scheme, we discuss the natural convection of molten tin in a square cavity under a constant magnetic field (see **Fig.1**). The numerical results obtained here agreed well with past experimental results[2].

2. GSMAC FINITE-ELEMENT METHOD

In the present numerical analysis, the following assumptions are made to simplify the governing equations : (1) the fluid is Newtonian, (2) the flow is laminar and incompressible, (3) the induced electric field is considered small enough to be negligible, (4) the displacement current is neglected, (5) the bouyancy effects by viscous dissipation and Joule heating are considered small and negligible. Using the GSMAC finite-element method[3], the governing equations can be respectively rewritten in the following element matrix form:

- **Predictor step** ($\boldsymbol{u}^n_\beta, \boldsymbol{B}^n_\beta$, T^n_β are known)

$$\boldsymbol{\omega}^n_e = \frac{1}{\Omega^e}\boldsymbol{C}_\alpha \times \boldsymbol{u}^n_\alpha, \tag{1}$$

$$\boldsymbol{J}^n_e = \frac{1}{\Omega^e}M_{\alpha\beta}\boldsymbol{u}^n_\alpha \times \boldsymbol{B}^n_\beta, \tag{2}$$

$$\overline{M}_{\alpha\beta}\tilde{\boldsymbol{u}}_\beta = \overline{M}_{\alpha\beta}\boldsymbol{u}^n_\beta + \Delta t[\boldsymbol{C}_\alpha H^n_e + M_{\alpha\beta}(\boldsymbol{u}^n_\beta \times \boldsymbol{\omega}^n_e + \frac{Ha^2}{Gr}\boldsymbol{J}^n_e \times \boldsymbol{B}^n_\beta) - \frac{1}{Gr}(D_{\alpha\beta}\boldsymbol{u}^n_\beta - M_{\alpha\beta}T^{n+1}_\beta \boldsymbol{j}) - \boldsymbol{S}_\alpha], \tag{3}$$

$$\overline{M}_{\alpha\beta}\widetilde{\boldsymbol{B}}_\beta = \overline{M}_{\alpha\beta}\boldsymbol{B}^n_\beta + \Delta t\overline{M}_{\alpha\beta}[\boldsymbol{C}_\alpha\Pi^n_e + (\boldsymbol{B}^n_e\cdot\boldsymbol{E}_{\alpha\beta})\boldsymbol{u}^{n+1}_\beta - (\boldsymbol{u}^{n+1}_e\cdot\boldsymbol{E}_{\alpha\beta})\boldsymbol{B}^n_\beta - \frac{1}{Rem}D_{\alpha\beta}\boldsymbol{B}^n_\beta - \boldsymbol{S}^\star_\alpha], \quad (4)$$

· **Simultanious relaxation step** ($\widetilde{\boldsymbol{u}}^{(0)}_\beta = \widetilde{\boldsymbol{u}}_\beta$, $H^{(0)}_e = H^n_e$, $\widetilde{\boldsymbol{B}}^{(0)}_\beta = \widetilde{\boldsymbol{B}}_\beta$, $\Pi^{(0)}_e = \Pi^n_e$ are initial values)

$$\phi^{(k)}_{He} = -\lambda\boldsymbol{C}_\alpha\cdot\widetilde{\boldsymbol{u}}^{(k)}_\alpha, \quad (5)$$

$$H^{(k+1)}_e = H^{(k)}_e + \phi^{(k)}_{He}/\Delta t, \quad (6)$$

$$\overline{M}_{\alpha\beta}\widetilde{\boldsymbol{u}}^{(k+1)}_\beta = \overline{M}_{\alpha\beta}\widetilde{\boldsymbol{u}}^{(k)}_\beta + \boldsymbol{C}_\alpha\phi^{(k)}_{He}, \quad (7)$$

$$\phi^{(k)}_{\Pi e} = -\lambda\boldsymbol{C}_\alpha\cdot\widetilde{\boldsymbol{B}}^{(k)}_\alpha, \quad (8)$$

$$\Pi^{(k+1)}_e = \Pi^{(k)}_e + \phi^{(k)}_{\Pi e}/\Delta t, \quad (9)$$

$$\overline{M}_{\alpha\beta}\widetilde{\boldsymbol{B}}^{(k+1)}_\beta = \overline{M}_{\alpha\beta}\widetilde{\boldsymbol{B}}^{(k)}_\beta + \boldsymbol{C}_\alpha\phi^{(k)}_{\Pi e}, \quad (10)$$

· **Energy equation**

$$\overline{M}_{\alpha\beta}T^{n+1}_\beta = \overline{M}_{\alpha\beta}T^n_\beta + \Delta t[-(\boldsymbol{u}^n_e\cdot\boldsymbol{E}_{\alpha\beta})T^n_\beta - \frac{1}{Ra}(D_{\alpha\beta}T^n_\beta - S'_\alpha) + M_\alpha(\frac{Ha^2Ec}{Gr}\boldsymbol{J}^{n2}_e + \frac{Ec}{Re}\boldsymbol{\omega}^{n2}_e)]. \quad (11)$$

The notations here are as follows. $\boldsymbol{\omega}$: vorticity vector, $\boldsymbol{J}$: electric current density vector, $\boldsymbol{u}$: velocity vector, $\boldsymbol{B}$: magnetic flux density vector, H: Bernoulli function, Π: error function, T: temperature, ϕ: modified potential, λ: relaxation parameter, Gr: Grashof number, Ra: Rayleigh number, Ha: Hartmann number, Rem: magnetic Reynolds number, Ec: Eckert number. The velocity components are interpolated by trilinear functions and the other variables are constant in each element. Subscripts α and β indicate the local-node number in the element and the summation rule is applied. In the present paper, since a four-node isoparametric element is employed, α and β assume the values from 1 to 4. Subscript e indicates that the variable is constant in the element. In order to remove the hourglass numerical instability, the element coefficient matrices and the boundary terms in Eqs. (1), (2), (3), (4) and (11) are analytically calculated as follows:

$$M_{\alpha\beta} = \Omega^e\begin{bmatrix} 1/9 & 1/18 & 1/36 & 1/18 \\ 1/18 & 1/9 & 1/18 & 1/36 \\ 1/36 & 1/18 & 1/9 & 1/18 \\ 1/18 & 1/36 & 1/18 & 1/9 \end{bmatrix}, \quad (12)$$

$$\overline{M}_{\alpha\beta} = \frac{\Omega^e}{4}\delta_{\alpha\beta}, \quad (13)$$

$$M_\alpha = \frac{\Omega^e}{4}, \quad (14)$$

$$\boldsymbol{C}_\alpha = \Delta y\begin{bmatrix} -1/2 \\ 1/2 \\ 1/2 \\ -1/2 \end{bmatrix}\boldsymbol{i} + \Delta x\begin{bmatrix} -1/2 \\ -1/2 \\ 1/2 \\ 1/2 \end{bmatrix}\boldsymbol{j}, \quad (15)$$

$$D_{\alpha\beta} = \frac{\Omega^e}{(\Delta x)^2}\begin{bmatrix} 1/3 & -1/3 & -1/6 & 1/6 \\ -1/3 & 1/3 & 1/6 & -1/6 \\ -1/6 & 1/6 & 1/3 & -1/3 \\ 1/6 & -1/6 & -1/3 & 1/3 \end{bmatrix} + \frac{\Omega^e}{(\Delta y)^2}\begin{bmatrix} 1/3 & 1/3 & -1/6 & -1/6 \\ 1/3 & 1/3 & -1/6 & -1/6 \\ -1/6 & -1/6 & 1/3 & 1/3 \\ -1/6 & -1/6 & 1/3 & 1/3 \end{bmatrix}, \quad (16)$$

$$\boldsymbol{E}_{\alpha\beta} = \Delta y\begin{bmatrix} -1/6 & 1/6 & 1/12 & -1/12 \\ -1/6 & 1/6 & 1/12 & -1/12 \\ -1/12 & 1/12 & 1/6 & -1/6 \\ -1/12 & 1/12 & 1/6 & -1/6 \end{bmatrix}\boldsymbol{i} + \Delta x\begin{bmatrix} -1/6 & -1/12 & 1/12 & 1/6 \\ -1/12 & -1/6 & 1/6 & 1/12 \\ -1/12 & -1/6 & 1/6 & 1/12 \\ -1/6 & -1/12 & 1/12 & 1/6 \end{bmatrix}\boldsymbol{j}, \quad (17)$$

$$\boldsymbol{S}_\alpha = \int_{\Gamma_2^e} N_\alpha(H_e^n \boldsymbol{n} - \frac{1}{Gr}\frac{\partial \boldsymbol{u}^n}{\partial n})d\Gamma_2^e, \quad (18)$$

$$\boldsymbol{S}_\alpha^* = \int_{\Gamma_4^e} N_\alpha(\Pi_e^n \boldsymbol{n} - \frac{1}{Rem}\frac{\partial \boldsymbol{B}^n}{\partial n})d\Gamma_4^e, \quad (19)$$

$$S'_\alpha = \int_{\Gamma_6^e} N_\alpha \frac{\partial T^n}{\partial n} d\Gamma_6^e, \quad (20)$$

where Ω^e is the area of the element, $\boldsymbol{i}$ and $\boldsymbol{j}$ the unit vector in the x and y direction. $\overline{M}_{\alpha\beta}$ is the lumped mass matrix.

3. RESULTS AND DISCUSSION

The finite element mesh consists of 30×30=900 elements and 961 node points. **Figure 2** shows the temperature distribution in the x direction at dimensionless time $t = 9.0 \times 10^4$ for $Ra = 3.659 \times 10^5, Pr = 0.011, Ha = 0.0$. The numerical results obtained here agree well with experimental results[2]. Hydromagnetic cross helicity H_C is introduced as the new physical quantity which means the effect of interactions between magnetic field and velocity field as follows :

$$H_C = \int_\Omega \boldsymbol{B} \cdot \boldsymbol{u} d\Omega \quad (21)$$

Figure 3 (a) and **(b)** show the Hartmann number's dependency ($Ha = 28.3, 452.5, Rem = 10^5$) of hydromagnetic cross helicity between the conventional and present numerical analysis scheme of magnetic field. Under a weak magnetic field, the coupling between the velocity field and magnetic field is so strong that the development of natural convection

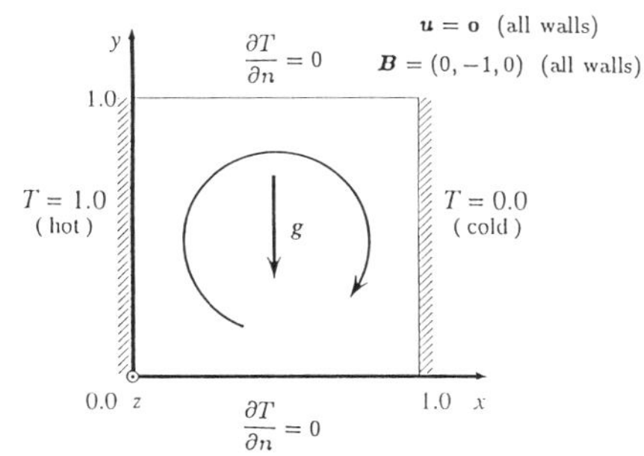

Figure 1. Analytical Model

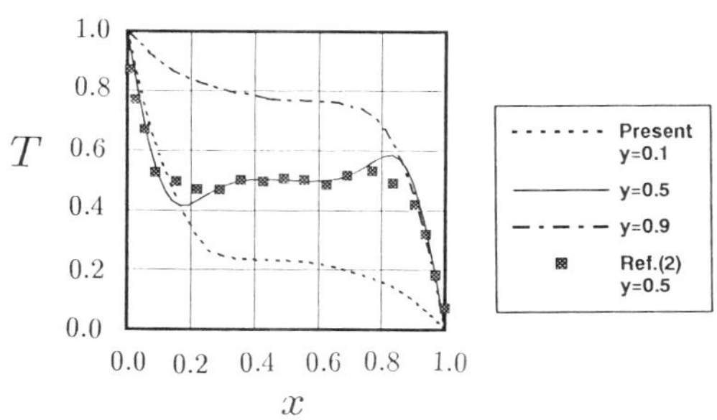

Figure 2. Comparison of Temperature Distribution in the x-Direction

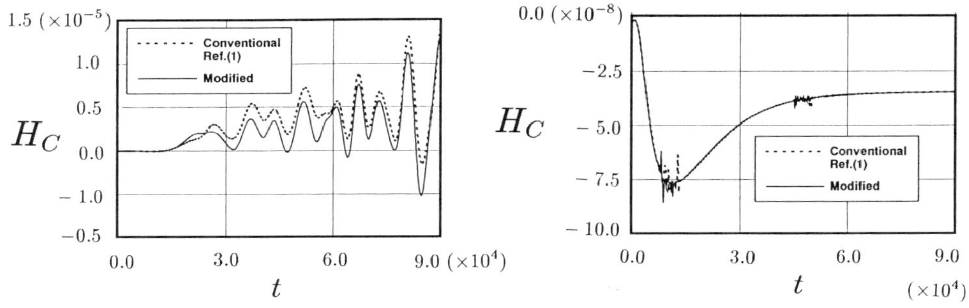

(a) Ha=**28.3**, Rem=**10⁵**   (b) Ha=**452.5**, Rem=**10⁵**

Figure 3. Transient Property of Hydromagnetic Cross Helicity (Magnetic Reynolds Number's Dependency)

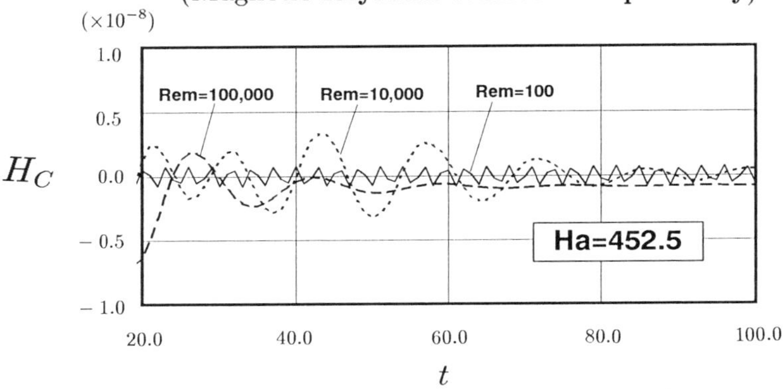

Figure 4. Transient Property of Hydromagnetic Cross Helicity (Hartmann Number's Dependency)

may disturb the magnetic field. On the other hand, under a strong magnetic field, the hydromagnetic helicity approaches zero, so the above-mentioned coupling becomes weaker. Then the hydromagnetic cross helicity is preserved to make the above-mentioned coupling unchanged. Moreover, in the stronger magnetic field, there is no difference between the conventional[1] and present method. **Figure 4** shows the magnetic Reynolds number's dependency ($Ha = 452.5, Rem = 10^2, 10^4, 10^5$) of the hydromagnetic cross helicity in an initial state. The lower is the magnetic Reynolds number, the shorter is the wave length of transient properties for the hydromagnetic cross helicity.

REFERENCES

1. Y.Oki and T.Tanahashi,Trans.Jpn.Soc.Mec.Eng.,59-B(1993) in press (in Japanese).
2. F.Wolff, C.Beckermann and R.Viskanta,Exp.Thermal Fluid Sci.,1(1988) 83.
3. T.Tanahashi, H.Okanaga and T.Saito,Int.J.Numer.Methods Fluids,11(1990) 479.

Simulation and Design of Applied Electromagnetic Systems
T. Honma (Editor)

Effect of resistance variation for Hall measurements in liquid metals

M. Ogita, S. Yasuda, M. Isai, Y. Nakanishi* and I. Tanaka
Coll. of Eng., and RIE*, Shizuoka University, Hamamatsu 432, Japan

Abstract

A new method called ac-dc method is proposed for Hall signal measurements of liquid metals and demonstrated with liquid Hg and Ga, whose Hall coefficient is $5 \sim 7 \times 10^{-5}$ and $7 \sim 9 \times 10^{-5}$ cm^3/coulomb, respectively. It is also confirmed that incorrect signals similar to Hall signals generate due to the resistance variation caused by Lorentz force. This effect is intrinsic for liquid states. The resistance change is found to be proportional to the applied magnetic field B and B^2.

1. Introduction

Hall effect has been widely used in semiconductor physics since it was discovered by E. H. Hall in 1879. For liquid state applications, the two-frequency ac method [1] came to usage since 1960s in order to avoid many undesirable electrode reactions. This method has a high frequency-selectivity and a high sensitivity to detect a very weak Hall signal in liquid states even under noisy and fluctuated circumstances. However, the presence of false signals is the major drawback of this method. This disadvantage arises not only for liquid metals [1], but also for other liquids such as liquid ammonia [2], organic solvents and electrolyte solutions [3]. Therefore, we have proposed a new method [4] called ac-dc method, which uses a dc magnetic field instead of an ac dynamic magnetic field. This method overcomes the nonlinearities arisen at the electrode interfaces. We applied this ac-dc method to liquid state metals such as mercury and gallium as they are the most practical materials in measuring the true Hall signals. Even using this proposed ac-dc method, there is still an ambiguity, hence it may produce some unnecessary signals such as voltage variation which is increasing with the rise of applied magnetic field [4]. The differences in Hall measurements between solid and liquid states of metals, and two-frequency ac method and ac-dc method will be presented in this paper.

2. Experimental procedure

A cell filled with gallium (99.9999%) or mercury (99.9999%) was mounted at the middle point of the air gap of 0 - 0.5 T dc magnetic field or at the middle of an ac magnetic field in half dc strength. This ac magnet is consisted of laminated core made of thin iron sheets of 0.35 mm thick and is in series resonance with a capacitor bank. Temperature of the cell was continuously monitored by a thermistor element and maintained at a constant value. The movements of the sample surface were detected by a position sensitive device (PSD) or a photodiode.

Results and discussion

Hall voltage, V_H, is essentially to be in proportional to the product of applied magnetic field (B_S) and sample current (I_S). However, if there is a nonlinearity in the measuring system or in the test sample, a false signal [4] is easily generated. Our proposed ac-dc method completely avoid false signals, caused by the nonlinearity at the electrode interfaces. The reason is that there is no product of sine function owing to the usage of single frequency ac sample current. In addition, it enables to show a positive or a negative carrier sign corresponding to the polarity, N or S of the applied magnetic field. When this desirable ac-dc method is applied to a liquid sample, a Hall signal appears corresponding to the strength of the applied magnetic field, but there is no correlation for the carrier sign corresponding to N or S fields. This signal, naturally, can not be considered as the true Hall signal. On the other hand, when the same sample is solidified by decreasing the temperature, Hall voltage is revealed clearly having a carrier sign corresponding to N or S fields. Therefore, it is considered that this unnecessary effect comes from the deformation of the liquid sample by the Lorentz force. This effect gives a rise to a faint variation in resistance, which may responsible to make a false signal similar to the Hall signal when the sample is in liquid state.

The resistance change of the liquid metals, Hg and Ga were investigated through Kelvin's double bridge method under an applied magnetic field. A very small change of resistance of 10^{-3} per tesla was observed for these liquid state metals. On the contrary, no change of the resistance was observed when the samples were solidified by decreasing the temperature in fig. 1. It has been also confirmed that the Lorentz force is responsible for the deformation, variation of resistance and movement of the sample when an ac sample current and a dc magnetic field is applied. The movements of liquid samples during the Hall measurements were detected at the surface of liquid mercury using a reflecting beam of He-Ne laser [5]. Moreover it has been found that at higher frequencies the rate of surface movements of the liquid is

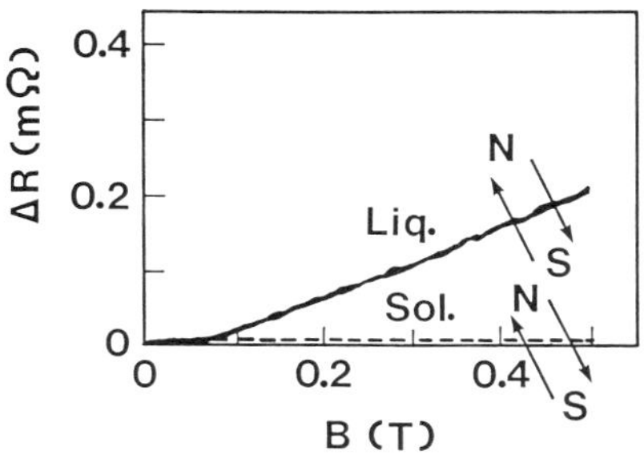

Figure 1. Resistance change of liquid Hg (at 24°C) (solid line) and solid Hg (at -46°C) (doted line).

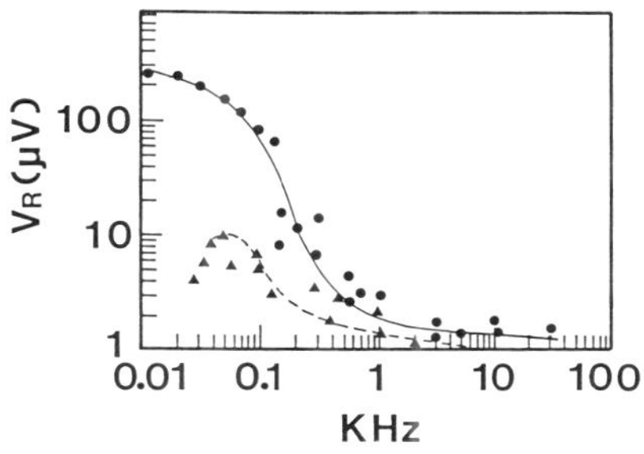

Figure 2. Frequency dependence of variation voltage for liquid Ga (●) and Hg (▲).

lower than the rate of Lorentz force changes. This was further confirmed by means of the four point probe method, where the variation voltage, V_R, caused by resistance change is reduced with increasing the frequency. At the frequencies over 1 kHz, V_R is found to be negligible (see fig. 2). Consequently, we have removed the unnecessary effect with raising the frequency of ac sample current, and successfully detected the Hall signals through the ac-dc method even for the liquid states. The Hall coefficients calculated for Hg and Ga were $5 \sim 7 \times 10^{-5}$ and $7 \sim 9 \times 10^{-5}$ cm^3/Coulomb, respectively. These Hall signals corresponding to the applied N or S field, were found to be proportional to the strength of the magnetic field as shown in fig. 3.

The Hall signals through the two-frequency ac method have also been investigated using a spectrum analyzer in the higher frequency region (see fig. 4a). The signals of $(f_I + f_B)$ and $(f_I - f_B)$ should be the Hall signal. However, there is another outside pair of sidebands of $(f_I \pm 2f_B)$. These extra signals might be due to the intrinsic effect in the liquid state under an applied magnetic field. Subsequently, the sample was solidified by decreasing the temperature. Then, only the Hall signal appears as shown in fig. 4b with one pair of modulated $\pm f_B$ signals. The above outside signals $(\pm 2f_B)$ are accordingly considered due to the resistance change in the liquid states. The variation voltages, ΔV_R, were also observed by means of the four point probe method. A similar spectrum to the Hall voltages which include two pairs of sidebands of $\pm f_B$ and $\pm 2f_B$ were observed at the higher frequency region in the liquid state samples. This also can be attributed to the resistance change in the liquid state. However, there are no peaks except that the signal of sample current, f_I, for the solid state was observed. Consequently, the

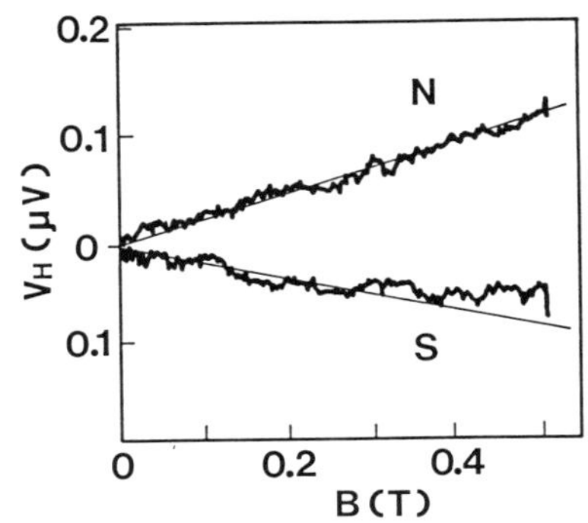

Figure 3. Hall signals for liquid Hg at 23°C through ac-dc method at 7 kHz ac current and dc magnetic field.

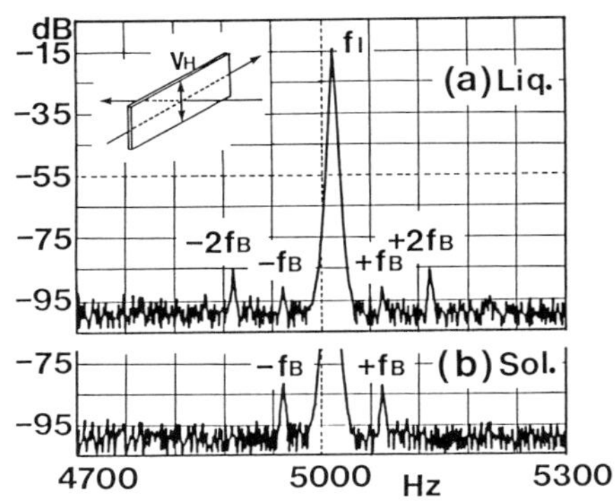

Figure 4. Spectrum of V_H at 5 kHz.
(a): Liquid Ga at 46°C, B=0.23 T.
(b): Solid Ga at 16°C, B=0.23 T.

Lorentz force makes no resistance change even when an external magnetic field is applied to the solid state samples. On the other hand, the outside pair of peaks ($\pm 2f_B$) always shows a variation due to the magnetoresistance corresponding to the applied magnetic field in the liquid states.

Conclusions

In Hall measurements for the liquid state samples, the followings are understood:

There may exist nonlinearities at the electrodes which easily produce false signals for the Hall voltage, V_H.

Variation voltage, ΔV_R, is intrinsic in the liquid state samples under an applied magnetic field. This effect is not based on the nonlinearities at the electrode interfaces because, harmonics trains are not seen in the higher frequency region of the obtained frequency spectrum.

Resistance change, ΔR, is corresponding to B and B^2 for the liquid states, while there is no resistance change in the solid states.

References

[1] N.E. Cusack, et al., J. Sci. Instrum., 42 (1965) 256.
[2] R.D. Nasby and J.C. Thompson, J. Chem. Phys., 49 (1968) 969, 53 (1970) 109.
[3] M.C. Bellisent, et al., J. Electrochem. Soc.,118 (1971) 1944.
[4] M. Ogita and S. Yasuda, J. Electrochem. Soc.,135 (1988) 2547.
[5] M. Ogita, et al., to be published (1993).

Simulation and Design of Applied Electromagnetic Systems
T. Honma (Editor)

Estimation of the critical mass in levitaion melting process

O. Iwata, H. Kawaguchi and T. Honma

Department of Electrical Engineering, Faculty of Engineering, Hokkaido University
Kita 13, Nishi 8, Kita-ku, Sapporo 060 Japan

Abstract

In this paper, an estimation method of the critical mass, which can be levitated, in the electromagnetic levitation melting process, is considered. In this process, the surface tension plays a important role for determination of the metal shape. From this point of view, the estimation method of the temperature and the temperature dependence of the surface tension are discussed.

1. INTRODUCTION

The electromagnetic levitation melting process is a complex system including the electromagnetic force, gravitation, surface tension and fluid motion. The metal shape changes variously depending on these forces. Therefore, the numerical calculation is necessary for theoretical analysis of this process. There exist some numerical methods to analyze this system [1],[2]. The shape of the levitated metal can be calculated very well using these numerical methods.

In this paper, firstly the relation between applied current and the critical metal mass which can be levitated is shown using the Gagnoud's method[1]. Then, it is found that the forces valance around the bottom of the levitated metal is very important to determine the critical mass. On the other hand, there are no electromagnetic force at the bottom. That is, the forces balance between the gravitation and the surface tension should be take into account there. Therefore, the evaluation of the surface tension is very important to consider the forces valance around the bottom, because the gravity effect is large at the bottom. In general, the surface tension is depend on the temperature. These imply that the temperature of the metal influences the critical mass. From this point of view, an estimation method of the critical mass is considered, taking into account the temparature dependence of the surface tension.

2. NUMERICAL SIMULATION METHOD

In this paper, according to the Gagnoud's method, the consideration is done. In this section, assumptions in the consideration and the Gagnoud's method are summarized.

A. Assumptions : It is difficult to consider the electromagnetic force, gravitation, surface tension and fluid motion at the same time. And then, we are interested in the free surface problem on the electromagnetic levitation melting process. So, the following assumptions are used in this paper.

i) The system is axisymmetric.

ii) The frequency of applied currents is so high that the skin depth is much smaller than the metal diameter.

iii) Electromagnetic fields are concentrate to the surface of the metal. So, the fluid dynamic effect can be neglected.

B. Energy minimization principle : Under the above assumptions, the equilibrium shape can be determined by minimizing the following energy functional ϕ,

$$\phi = -\int_{\Omega+V} \frac{|\mathbf{B}|^2}{2\mu_0} dv + \int_V \rho g z \, d\Omega + \int_S \gamma \, dS + \lambda \int_V dv \tag{1}$$

where the first, second and third terms are the magnetic, gravitational and surface tension energies respectively. The last term implies the constraint to keep the metal volume constant. λ is the Lagrange mutiplier for the constraint.

3. NUMERICAL INSTABILITIES

There are some notices should be paid in the numerical simulation of the electromagnetic levitation melting process. In this section, two of most important notices are stated.

A. Increase / decrease of the metal volume

At each iteration step to determine the metal shape, displacement vectors **h** of each element of the surface are determined by the following equation[1],

$$\mathbf{h} = -\varepsilon\left(\frac{|\mathbf{B}|^2}{2\mu_0} + \rho g z + \gamma C + \lambda\right)\mathbf{n} \quad , \tag{2}$$

where ε is the coefficient to adjust the scale of the displacement, C is the mean curvature of the surface and **n** is the normal vector to the surface. Here, if one use the Lagrange mutiplier λ , which is calculated numerically by

$$\lambda = \int_S \left(\frac{|\mathbf{B}|^2}{2\mu_0} + \rho g z + \gamma C\right) dS \Big/ \int_S dS \quad , \tag{3}$$

the metal volume can not be kept constant for the numerical errors (see Fig.2). From this reason, λ should be evaluated exactly as follows. That is to say, the displacement vector **h** can be regarded as the function of λ ($\mathbf{h}=\mathbf{h}(\lambda)$). Then, the metal volume V_l or the volume change δV_l at l - th step in the iteration process becomes the function of λ ,

$$\delta V_l(\mathbf{h}(\lambda)) = V_l(\mathbf{h}) - V_0 = V_l(\lambda) - V_0 \quad . \tag{4}$$

Therefore, one can determine λ which keeps the volume constant exactly to solve the following equation,

$$\delta V_l(\lambda) = 0 \quad , \tag{5}$$

using numerical calculation such as Regula-Falsi method.

B. Unbalanced element size

If you do the iteration without any modifications for the node position on the surface, the element size will be unbalanced (see Fig. 3). One can not express the complex metal shape, if the unbalanced elements occur. Therefore, the mashes should be remade at each iteration step.

Doing the exact calculation of λ and remaking the mashes, one can do many times iteration and can express the complex shape of the levitated melting metal.

In Fig.4, an example of complex shape metals are shown. In this case, the metal can not be levitated because of the weak applied current. For the calculation of this phenomena, many times iterations are needed. Now, considering these two points, one can estimate the relation between the applied current and the critical metal mass which can be levitated for the applied current like the Fig. 5.

4. CONSIDERATION ON METAL TEMPERATURE

It is found from the Fig.4 that the forces balance at the bottom is important to determine if the metal can be levitated. And then, the forces balance at the bottom are determined by only the gravity and the surface tension (there is no electromagnetic force at the bottom). That is, the evaluation of the surface tension is important because the gravity effect is large at the bottom. Considering the temperature dependence of surface tension, an estimation of the metal temperature is important thing in this process. In this section, the estimation method of metal temperature is presented. And then, the temperature dependence of the surface tension in this system is considered.

Now, we shall notice the energy balance in this process. The thermal energy which occurs around the metal surface must be dissipated by radiation if we assume that this system is in vacuum (there are no convection dissipation). Then the following energy balance should be satisfied,

$$\int_V \frac{J^2}{\sigma_c}\, dv - \int_S \sigma_{SB} T^4 \, dS = 0 \tag{6}$$

where σ_c is the electric conductivity, J is the surface current, σ_{SB} is the Stefan-Boltzmann's constant and T is the metal temperature. Using this relation, one can estimate the metal temperature. And then, one can use the modified Eötvös rule[3] for the temperature dependence of the surface tension.

As an example of strong temperature dependence, the Lithium equilibrium shape is shown in the Fig.6. The full line is the equilibrium shape considering the temperature dependence. The broken line is the shape not considering the temperature dependence. One can find the difference at the bottom between both equilibrium shapes. It is predicted that this difference influences the relation between the applied current and the critical metal mass, because the critical mass is sensitive for the forces balance at the bottom.

5. SUMMARY

In this paper, taking notice of the bottom forces balance, an estimation method of the metal temperature and temperature dependence of the metal shape in the electromagnetic levitation melting process have been considered. It has been indicated that the consideration of the temperature dependence is important for the estimation of the critical mass which can be levitated for a applied current.

6. REFERENCES

[1] A.Gagnoud and J.P.Brancher, "Modeling of coupled phenom-ena in electromagnetic levitation", *IEEE Trans. Magn.*, Vol. 21, No. 6. pp. 2424-2427, November 1985 .

[2] A.D.Sneyd and H.K.Moffatt, "Fluid dynamical aspects of the levitation - melting process", *J. Fluid Mech.*, Vol.117, pp. 45-70 1982 .

[3] S.Blairs, "The surface tension and critical properties of liquid metals" J.Colloid and Interface Sci., Vol.67, No.3, pp.548-551, 1978.

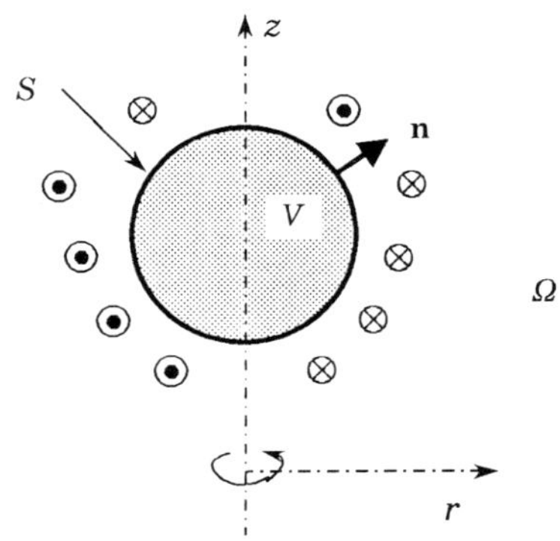

Fig. 1 Configuration of the electromagnetic levitation melting process

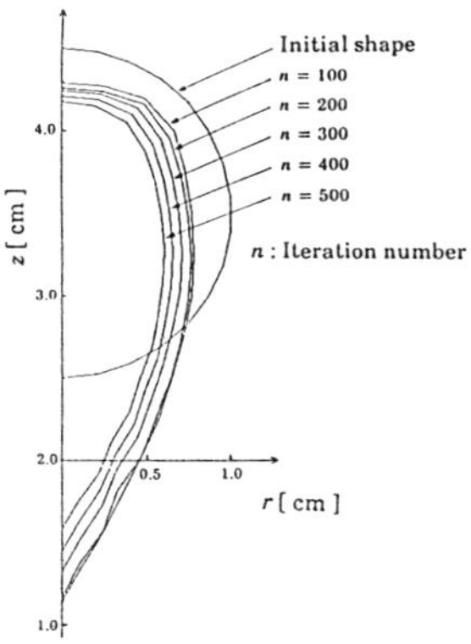

Fig. 2 An example of the metal volume decrease

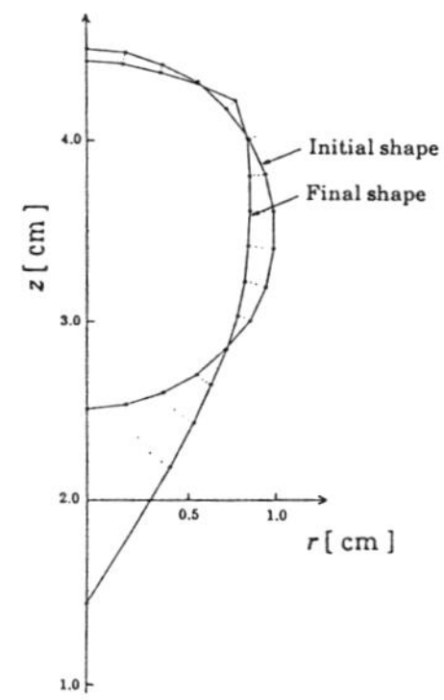

Fig. 3 Trace of each node in the iteration process

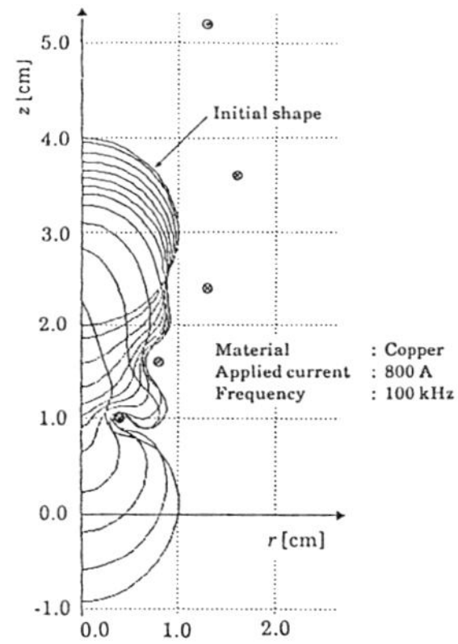

Fig.4 An example of complex shape

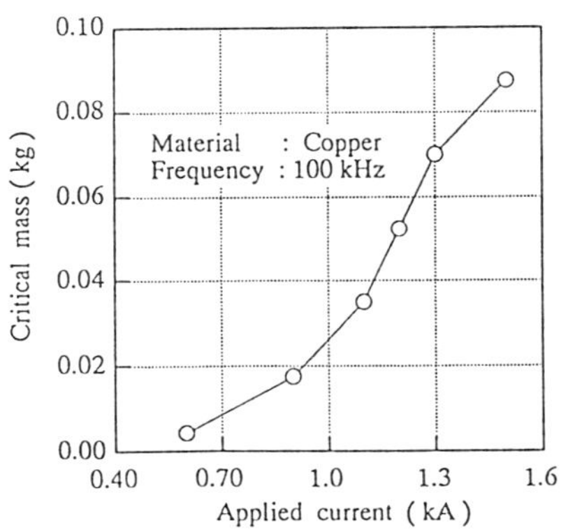

Fig.5 Relation between applied current and critical mass

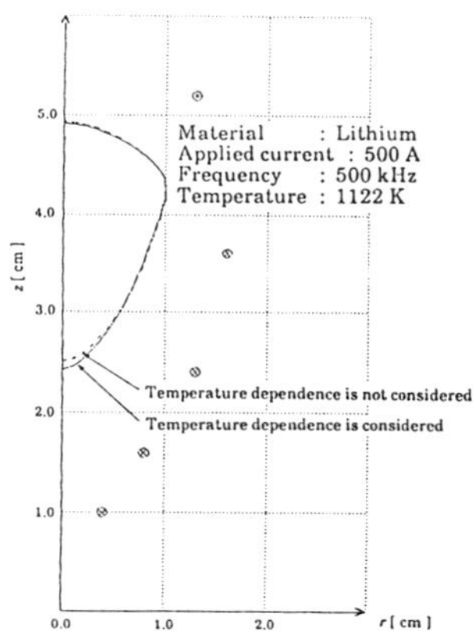

Fig.6 The equilibrium shapes of the Lithium

The second example deals with the mathematical model of the human vocal cords, introduced by Croanjaeger [4]. The vocal cord is modelled as an nonlinear oscillator , with two degrees of freedom, consisting of a flexibly supported point mass. The stiffness and damping qualities of an anisotropic support of a mass correspond to the real properties of the vocal cord. In order to describe the possible contact (impacts) of the vocal cords, an additional elastic element (nonlinear spring) was used in the model. The stiffness of this element was chosen so that for horizontal displacements of the mass approaching the origin the force in the spring increases to infinity. Additionally, it is assumed that flow from the lungs is adiabatic, and the Bernoulli effect is neglected. The lungs and air passage were modelled as a kettle with stiff walls. The real elastic properties of the lungs and air passage were additionally considered in the elastic properties of the air. The escaping air flow is proportional to the horizontal deflection of the vocal cords and the kettle creates approximately identical forces both in the vertical and the horizontal directions of the displacement of the vocal cord.

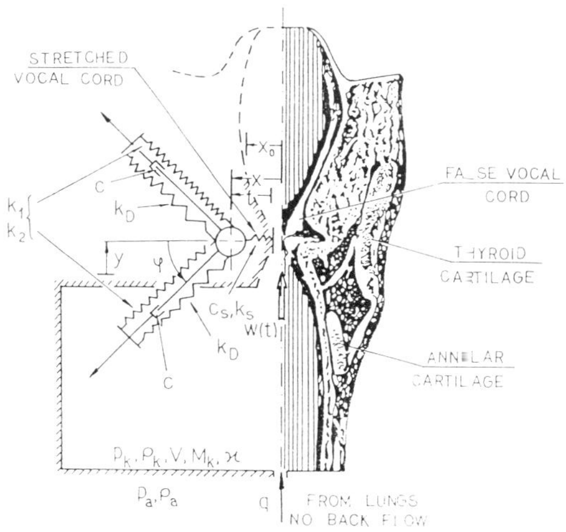

Figure 2. Human vocal cord and its mechanical model.

It is governed by a stiff system of five autonomous nonlinear ordinary differential equations of the first order. First an analysis of the equations of motion based on an analytical approach is given, and then the analytical predictions give a possibility of applying the numerical calculations in order to find fixed points, define their stability and bifurcations. A route from periodicity through quasiperiodicity to chaos is discussed and influence of parametric excitation on the behaviour of a strange attractor is investigated. Many various different conclusions with respect to the real behaviour of the human vocal cords are given.

The third example includes the investigation of nonlinear dynamics of two coupled and externally driven oscillators. A transition between quasiperiodic, strange chaotic and strange non-chaotic attractors are examined in some detail. It is known that chaotic attractors may not be strange. However, there is another class of attractors which are called strange non-chaotic attractors. The word "strange" refers to the complicated geometrical structure of the attractor.

For the investigated system we have discovered strange non-chaotic attractors, as well as quasiperiodic orbits.

3. CONCLUDING REMARKS

The nonlinear behaviour of three different nonlinear mechanical coupled oscillators are analyzed using the numerical strategy which includes the methods based on solving a boundary and an initial value problems. In the case of the four dimensional roll slide oscillator two transition scenarios, one due to period doubling bifurcation and the other due to intermittency, have been detected. Transitional and asymptotic chaos, as well as Rossler band attractors have been demonstrated.

An analysis of the vocal cords oscillations has enabled to draw many conclusions of the influence of control parameters for the behaviour of equilibria and periodic orbits, and their bifurcations. Irregular motion has been found in this case for a small values of damping coefficients which indicates a possibility of the occurrence of chaos during the birds vocal cords oscillations.

Finally, two coupled sinusoidally driven mechanical oscillators have been numerically investigated. We have discovered strange non-chaotic and strange chaotic attractors, as well as quasiperiodic orbits. We have examined the persistence of strange nonchaotic attractors for perturbations in the amplitudes of two exciting forces with incommensurable frequencies.

4. REFERENCES

1 R. Seydel, From Equilibrium to chaos, Elsevier, New York, 1988.

2 J. Awrejcewicz, Bifurcation and Chaos in Coupled Oscillators, World Scientific, Singapore, 1991.

3 F. C. Moon, Chaotic Vibrations, Wiley-Interscience, New York, 1987.

4 R. Cronjaeger, Model of Sound Generation in a Human Larynx, PhD Thesis, Braunschweig, 1978.

Simulation and Design of Applied Electromagnetic Systems
T. Honma (Editor)

Oscillating and steady growth of a dendrite

A.Tanaka[a] and R.Kobayashi[b]

[a]Department of Electrical and Information Engineering, Faculty of Engineering, Yamagata University, Yonezawa, 992, Japan

[b]Department of Applied Mathematics and Informatics, Ryukoku University, Seta, Otsu, 520-21, Japan

Abstract

We have found steady and oscillating growth of a dendrite in a numerical simulation. The resulting shapes are quite similar to real dendrite crystals. We also found the transition from oscillating to steady growth and clarified the importance of anisotropy in the dendritic growth.

1 Introduction

A snow crystal has a beautiful shape as we all know, and some substances exhibit dendritic growth like it. Problems of the dendritic growth, such as sidebranching, pattern selection and so forth, are very interesting but still open. From mathematical point of view, they are good examples of the free boundary problem, and it is difficult to solve them analytically.

Since there are also many difficulties in numerical simulations, several kinds of simplified model have been proposed. The most popular models, such as *Geometrical model* [1] and *Boundary Layer model* [2], are local models. There is a non-local model, but it requires much computation time [3].

We propose a new non-local model called *Phase Field model.* This model has many advantages as compared with others, one of which is its adaptability that enables us to carry out three-dimensional simulations without difficulties [4].

2 Modeling

In the phase field model of two-dimensional crystal growth, two field variables are introduced, i.e. a phase field $p(x,t)$ and a temperature field $T(x,t)$. The former variable p is an ordering parameter, and $p = 0$ means liquid state, and $p = 1$ solid state. The model is derived from the conservation law of enthalpy in consideration of Ginzburg - Landau type free energy given by the following equations with m as a parameter,

$$\Phi[p;m] = \int \frac{1}{2}\varepsilon^2|\nabla p|^2 + F(p;m)dx \tag{1}$$

$$F(p;m) = \frac{1}{4}p^4 - \left(\frac{1}{2}-\frac{m}{3}\right)p^3 + \left(\frac{1}{4}-\frac{m}{2}\right)p^2 \tag{2}$$

where ε is a small value representing the thickness of interface and $|m| < 1/2$. The resulting equations are written as follows,

$$\tau\frac{\partial p}{\partial t} = -\frac{\partial}{\partial x}\left(\varepsilon\frac{\partial\varepsilon}{\partial\theta}\frac{\partial p}{\partial y}\right) + \frac{\partial}{\partial y}\left(\varepsilon\frac{\partial\varepsilon}{\partial\theta}\frac{\partial p}{\partial x}\right) + \nabla\cdot(\varepsilon^2\nabla p) + p(1-p)(p-\frac{1}{2}+m) \tag{3}$$

$$\frac{\partial T}{\partial t} = \nabla^2 T + K\frac{\partial p}{\partial t} \tag{4}$$

where K is dimensionless latent heat. Anisotropy can be taken into account by redefining ε as follows

$$\varepsilon = 1 + \delta_1\cos 4\theta + \delta_2\cos 8\theta. \tag{5}$$

The third term of eq.(5) was omitted in our previous paper, but we added it in agreement with the experimental results of dendrites of NH_4Cl and other substances. NH_4Cl forms a cubic crystalline, which has a 4-fold symmetry, but it also has an anisotropy in the intermediate direction $< 110 >$, and thus we added the term. $\delta_2\cos 8\theta$.

3 Simulation results and discussion

With the above modeling, we carried out numerical simulations using both implicit and explicit schemes. Spatial meshes are 300×300 in most cases, and the numerical parameters are chosen to consist with the physical situation. We varied anisotropy parameters δ_1 and δ_2 only.

As previously reported [5], weak anisotropy results in oscillating tip growth in 4-fold symmetric case(i.e. $\delta_2 = 0$),. When anisotropy become larger, the tip growth becomes steady and the sidebranches grow vertically. The resultant shapes are similar to real dendrites (Fig.1).

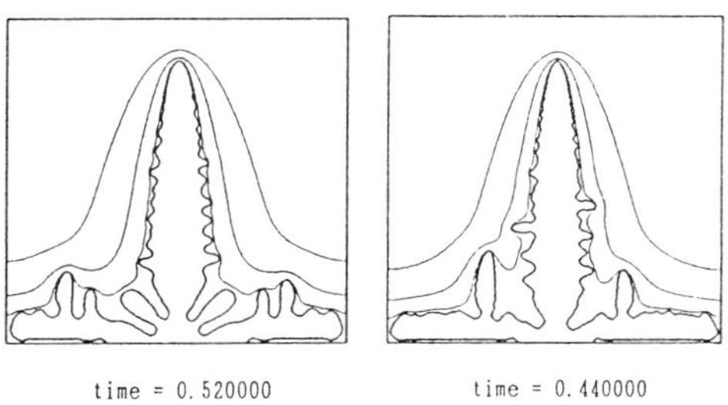

(a) $\delta_1 = 0.05, \delta_2 = 0.01$ (b) $\delta_1 = 0.1, \delta_2 = 0.01$

Figure 1: The steady growth of a main branch.

On the other hand, if we fix δ_1 and increase δ_2, the growth rate of the main branch oscillates and the sidebranches appear by turns on the both sides(Fig.2(a)). It shows good agreement with the experimental results of dendrite whose growth rate oscillates [6] (Fig.2(b)). In the figures we show not only the contour of a dendrite but also the isothermal lines. From Fig.1, we find the growth rate in case (a) is faster than that in case (b). We may say that anisotropy dominates the growth of dendrites.

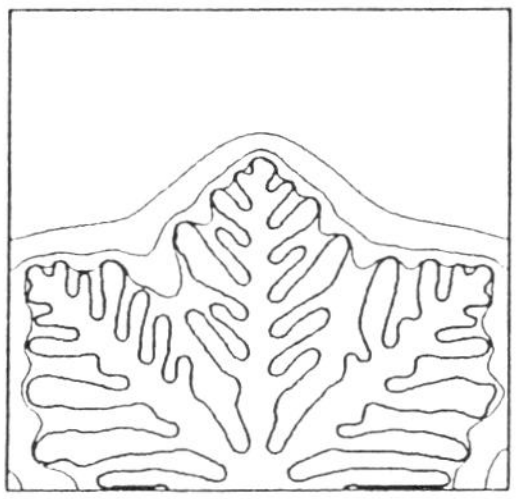

(a)

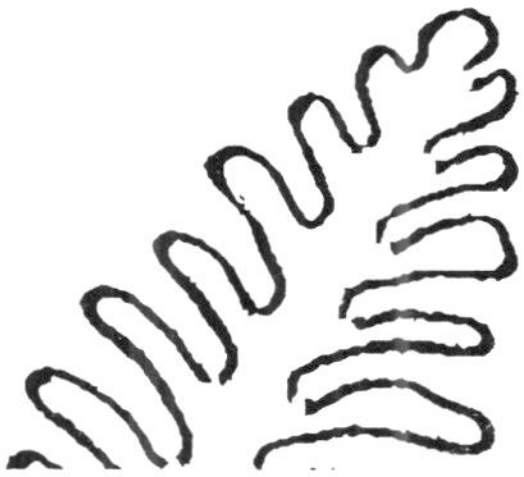

(b)

Figure 2: The oscillating growths of (a) numerical and (b) NH_4Cl dendrites.

Next, we varied both anisotropy parameters δ_1 and δ_2. The resulting phase diagram is shown in Fig.3. We can classify the patterns into six types. Though the classification is rather ambiguous in the case of weak anisotropy the pattern is determined by the ratio δ_1/δ_2.

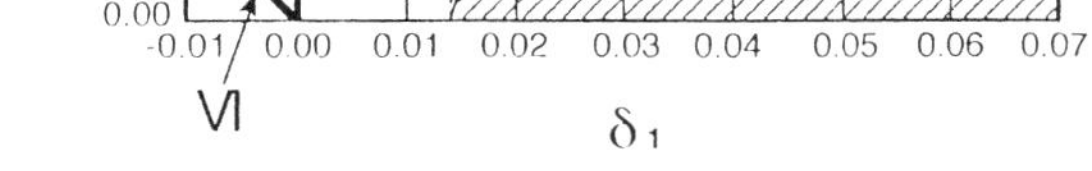

	<100>	<110>
I	steady	non-growth
II	oscillation	non-growth
III	oscillation	oscillation
IV	oscillation	tip-splitting
V	tip-splitting	tip-splitting
VI	non-growth	oscillation

Figure 3: Phase diagram of dendritic patterns.

On the contrary, without anisotropy, we can see an irregular pattern as shown in Fig.4. A real crystal always has anisotropy. This case is not meaningless, since H.Honjo et al. showed that anisotropy of a dendritic crystal can be effectively removed [7]. He used NH_4Cl aqueous solution and the resulting crystal showed a fractal-like structure as in the case of DLA, which is a discrete model of the Laplace equation [8]. Its fractal dimension d_f was 1.67 which shows excellent agreement with that of the two-dimensional DLA.

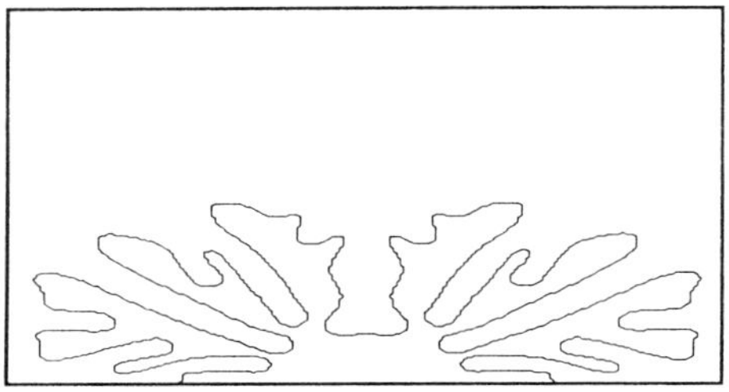

Figure 4: An irregular pattern in an isotropic case

4 Conclusion

We have carried out numerical simulations of dendrites by using the phase field model. In anisotropic cases, the steady and tip-oscillating growth can be seen. The ratio of the degree of anisotropy of 4-fold symmetry to that of 8-fold one determines whether the tip grows steadily or not. On the other hand, in an isotropic case, we find an irregular pattern, and the resulting shape is fractal-like. Therefore it is evident that anisotropy plays an essential role in the pattern formation of the dendritic growth.

References

[1] R. C. Brower, D. A. Kessler, J. Koplik and H. Levine, Phys. Rev. Lett., **51** (1983) 1111.

[2] E. Ben-Jacob, N. Goldenfeld, J. S. Langer and G. Schon, Phys. Rev. Lett., **51** (1983) 1930.

[3] Y. Saito, G. Goldbeck-Wood and H. Müller-Krumbhaar, Phys. Rev. Lett., **58** (1987) 1541.

[4] R. Kobayashi, Pattern Formation in Complex Dissipative System, ed. S. Kai, World Sci. Pub. Singapore (1992) 121.

[5] R. Kobayashi, to appear in Phisica D.

[6] K. Tanaka, Doctoral thesis, Tohoku University (1991)(in Japanese).

[7] H. Honjo, S. Ohta and M. Matsushita, J. Phys. Soc. Jpn., **55** (1986) 2487.

[8] T. A. Wintten and L. M. Sander, Phys. Rev., **B27** (1983) 5686.

Simulation and Design of Applied Electromagnetic Systems
T. Honma (Editor)

Plasma Chaos under oscillatory magnetic Field and its Application to controlling Internal Flux

S. Yatsu, A. Iobe, M. Iwamatsu[a], Y. Aoki, T. Kasahara[b], Y. Abe and N. Kayukawa

Advanced MHD Research Institute, Faculty of Engineering, Hokkaido University, North 13 West 8 Northward, Sapporo, Japan

[a]Railway Technical Research Institute, 2-8-38 Hikari-machi, Kokubunji, Tokyo, Japan

[b]Hokkaido Institute of Technology, 419-2 Maeda Teine Eastward, Sapporo, Japan

Abstract

Radial oscillations of a pinch plasma under an external oscillating magnetic field are analyzed numerically. If the external magnetic field consists of a steady bias field and a periodically oscillating field component, at least two kinds of solution ; divergent and intermittent chaotic mode. Under similar conditions the simulation of a theta pinch plasma is also done and result shows the anomalous growth of internal flux. From these results, a feasible way to control the trapped flux inside of plasma is discussed.

1. INTRODUCTION

When Plasma is compressed by a pinch effect, pressure difference between inside and outside of the plasma excites radial hydromagnetic oscillation with eigen frequencies[1]. We have done a series of numerical analysis to clarify the nonlinear effect derived from a period and an amplitude of applied magnetic field in radial oscillation phenomena[2].

In this report, firstly for a given external field kinds of oscillatory mode are examined. Secondary the simulation results about a theta pinch plasma are shown. Finally a new method to use a nonlinear effect for controlling the trapped flux inside of plasma is presented and briefly discussed its feasibility.

2. NUMERICAL ANALYSIS MODEL

The simulation codes to get both temporal solutions of plasma radial motion and flux change simultaneously were newly developed. The plasma geometry is approximated by an thin annulus with finite width δ. Magnetic field has direction along the axis of plasma annulus, and its distribution is assumed as shown in the Figure 1 [3]. The equations of plasma motion and flux conservation inside of plasma are given by Eqs.(1) and (2).

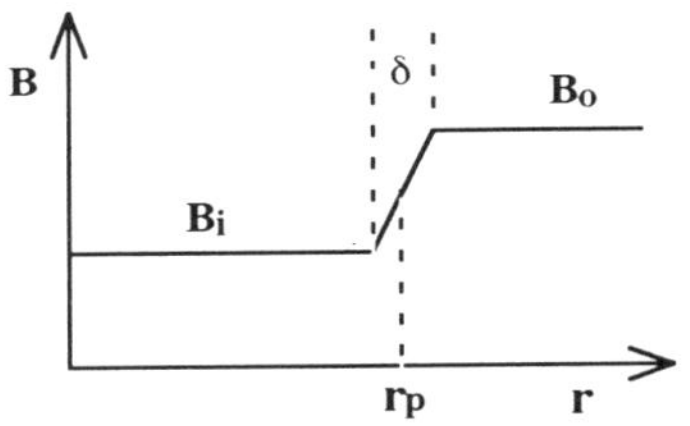

B : Magnetic field
Bo : External applied field
Bi : Internal field
rp : Plasma radius
δ : Plasma sheath thickness

Figure 1. Assumed magnetic field distribution and plasma radius.

$$\frac{d}{dt}\left(M\frac{dr_p}{dt}\right)=\frac{1}{2\mu_0}\left(B_i^2-B_o^2\right)2\pi r_p \tag{1}$$

$$\frac{d\phi_i}{dt}=\frac{2\pi r_p}{\mu_0}\left(\frac{\eta}{\delta}\right)\left(B_o-B_i\right) \tag{2}$$

where M is the plasma line mass and r_p the plasma radius, B_i the internal magnetic field, $\phi_i(=\pi r_p^2 B_i)$ the trapped flux. We assume a single, pressure-less, plasma fluid of finite resistivity η. The externally applied magnetic field B_o is expressed as $B_o = B_b f(t) + B_m e^{-\gamma t}\sin\omega t$.; It is composed of a bias field $B_b f(t)$ and a small alternating field component oscillates periodically with an angular frequency ω and the damping coefficient γ. An angular frequency ratio α/ω and a field amplitude ratio $\varepsilon_b = B_m/B_b$ are used as parameters to condition upon a numerical solution. Where α is the eigen angular frequency of a plasma radial oscillation, and defined as $\alpha^2 = 4\pi B_b^2/\mu_0 M$.

3. RESULTS AND DISCUSSION

3.1. Nonlinear oscillatory modes under a given external field

Firstly kinds of oscillatory mode for a given external field B_o is examined. Plasma sheath thickness δ and resistivity η are included in a damping constant β together with a characteristic length L. β is defined as $\beta = 2\pi\eta / \mu_0 L\delta$. And it is set to be 20kHz. The bias magnetic field is set as a constant(= 1kG), and the alternating field component is given in $2\pi/\omega = 6$ µsec, $\gamma = 0$. Equations are integrated with the initial conditions ; $r_p = 1.01$ cm, $dr_p/dt = 0$, $\phi_i = 0$.

The typical temporal change of plasma radius is shown in Figure 2. The former shows non-periodical intermittence and its Fourier spectrum has a characteristics inversely proportional to square of frequency over the wide range as shown in Figure 3. These features suggests this mode is a chaotic one. This idea is supported by the signs of Lyapunov exponents ; (+, 0, -) shown in Figure 4. The dotted line in Figure 2 shows evanescent decay, but this tendency is true only within time length 1600 µsec which limited by used resources. In larger time scale it may show the intermittence because its Fourier spectrum is almost same as one shown in Figure 3.

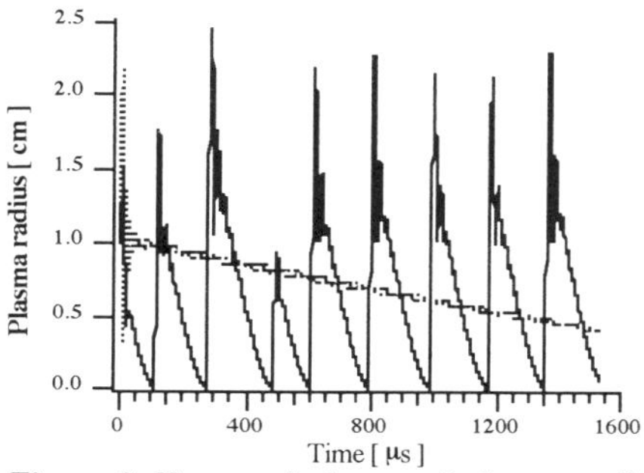

Figure 2. Temporal change of plasma radius.
solid line ; $(\alpha/\omega, \varepsilon_b) = (0.25,\ 0.75)$
dotted line ; $(\alpha/\omega, \varepsilon_b) = (0.50,\ 0.15)$

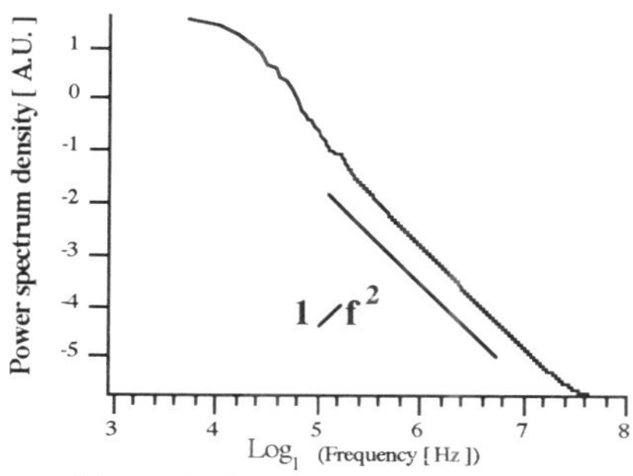

Figure 3. Power spectrum distribution.
$(\alpha/\omega, \varepsilon_b) = (0.25,\ 0.75)$

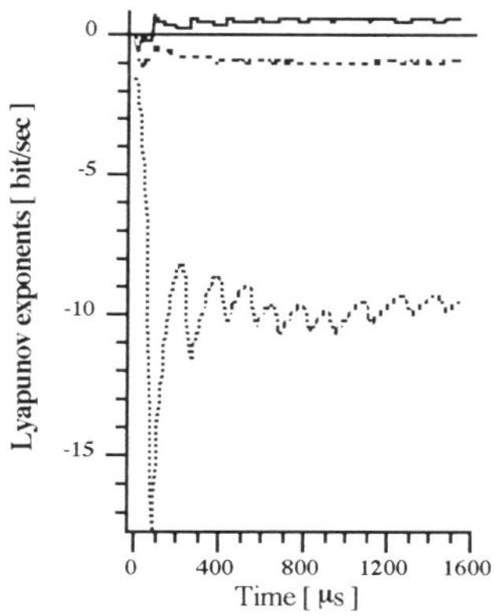

Figure 4. Lyapunov exponents.
$(\alpha/\omega, \varepsilon_b) = (0.25,\ 0.75)$

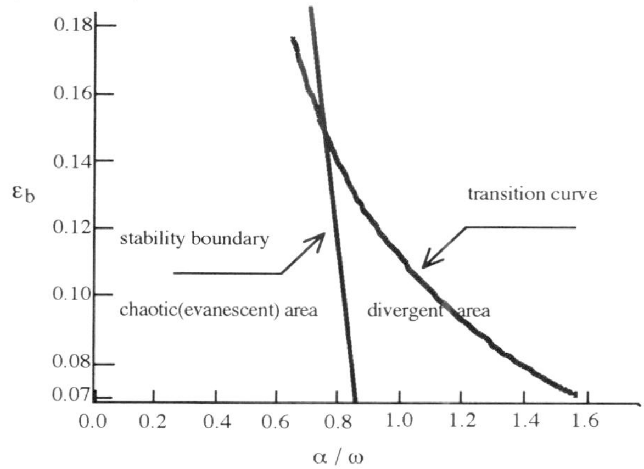

Figure 5. Stability diagram.

In case of a parameter $(\alpha/\omega, \varepsilon_b) = (0.80,\ 0.25)$ the obtained solution is unstable ; oscillatory divergent. The nonlinear resonance between a small alternating external field component and plasma eigen mode may cause this mode. Figure 5 shows each mode and its stability on α/ω-ε_b plane. In the vicinity of the boundary a steady quasi periodic mode is also found.

3.2. Simulation of a theta pinch plasma and a feasible way to control the flux

To simulate the plasma-magnetic field behavior in a theta pinch for He plasma (initial filling pressure 1.33Pa, electron temperature 5eV), the extended equations in which external circuit Equations are added to Eqs.(1),(2) are solved numerically. The temporal change of external field and trapped flux are shown in Figure 6 and Figure 7 respectively.

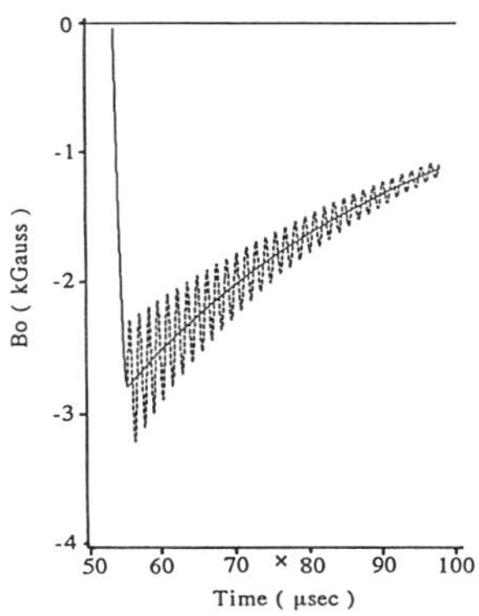

Figure 6. Temporal change of external field. $\omega/2\pi$=750kHz, $1/\alpha$=20μsec, B_m=200Gauss

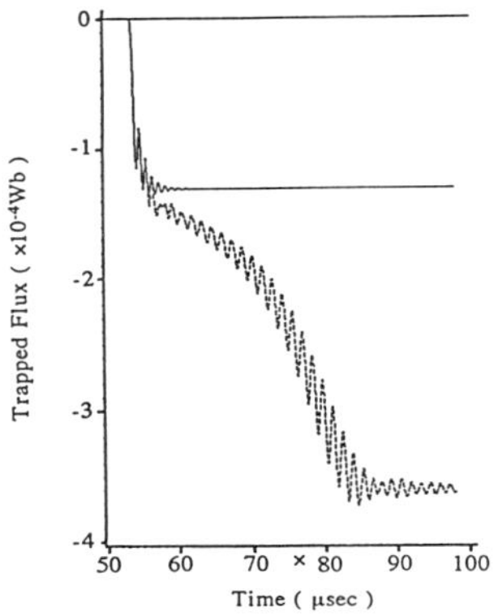

Figure 7. Temporal change of trapped flux. dotted line and solid line correspond to the case with and without small amplitude oscillating field respectively.

In these figures the time when $\alpha/\omega = 1$ is marked by a symbol **x** under the temporal axis. For the case that a small amplitude oscillatory field component is added, the trapped flux increase remarkably before and after the marked time, and at about 85 μsec the quantity of trapped flux reach a saturation level whose value is more than 2.5 times as large as value maintained with only a crowbar field. This anomalous growth of trapped flux is caused by the resonance effect between the external field oscillation and the plasma radial eigen oscillation. The characteristic point (α/ω, ε_b) passes from the lower initial point into the upside on the transition curve displayed in Figure 5 with time. Namely, on the way to chaotic area resonant effect happen and resultant gradual radius expansion and trapped flux growth are caused. After the point arrive in chaotic area, the oscillation change to chaotic stable one and the trapped flux keeps its constant level. In this case, chaos may contribute to form a new stable order.

If above mentioned process is realized by using any hardware a feasible way to control the trapped flux may be obtained. Most important technique is to accomplish both generating resonance and shifting condition to one chaotic mode needs at the same time.

4. CONCLUSION

Nonlinear radial oscillation of a pinch plasma under an external oscillating field is analyzed and divergent mode and chaotic mode are found. Simulation results of a theta pinch plasma can be expressed from a transition between these modes and a new way to control the trapped flux can be considered.

REFERENCES

1 G.B.F. Niblett and T.S. Green, Proc. Phys. Soc.,74(1959)737.1

2 S. Yatsu et al., Proc. of First Japan-CIS Joint Seminar on ElectroMagneto-Mechanics in Structures,(1992)105

3 W.T. Armstrong et al., Appl. Phys. Lett.,38(1981)680

Simulation and Design of Applied Electromagnetic Systems
T. Honma (Editor)

An experimental investigation on chaotic motions of magnetically levitated system

Q. Zhu J. Tani and T. Takagi

The Institute of Fluid Science, Tohoku University
2-1-1 Katahira, Aoba-ku, Sendai, 980, Japan

Abstract

An experimental investigation on chaotic motions of magnetically levitated system with two degrees of freedom is presented. The tested system consists of three cylindrical permanent magnets. One magnet is installed on the vibration exciter and a non-magnet pole is mounted vertically at its center. The other two magnets are taken as movable bodies and can move up and down along the pole. It is observed that the chaotic motions of the system appear near the sum of the first and the second linearized natural frequencies. The method for getting experimental Poincaré map is described in detail.

1. INTRODUCTION

The interest in application of magnetically levitated system has been quite widespread in the recent years. The design of this kind of devices depends in part on the ability of predicting levels of response and force under dynamic conditions. The discovery of chaotic phenomena has demonstrated that for many dynamic non-linear problems, the prediction of the response becomes impossible even though the parameters of the system are known and the inputs are deterministic. It is important to determine the conditions under which chaotic behavior may be expected. Many researchers have been actively trying to understand the dynamic behavior of such kind of system [1]~[2]. Although analyses have been made on certain aspects of the magnetically levitated system, the idealized mechanical models are not relevant to the real device. Thus, an item of present interest is whether chaos is manifested in practical structures. Part of the motivation is to examine the relation between the exciting parameters for chaos and the natural

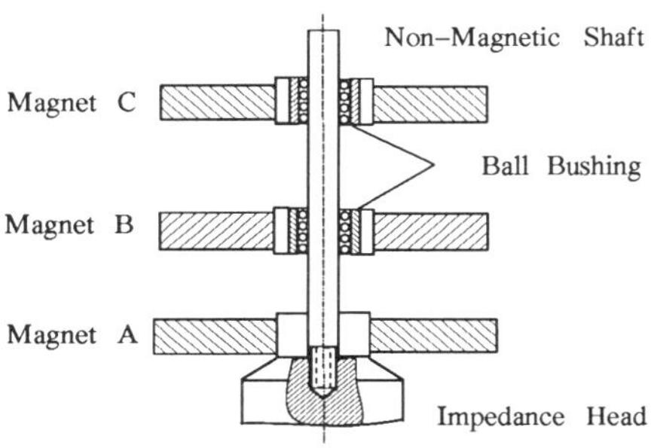

Figure 1. Experimental model

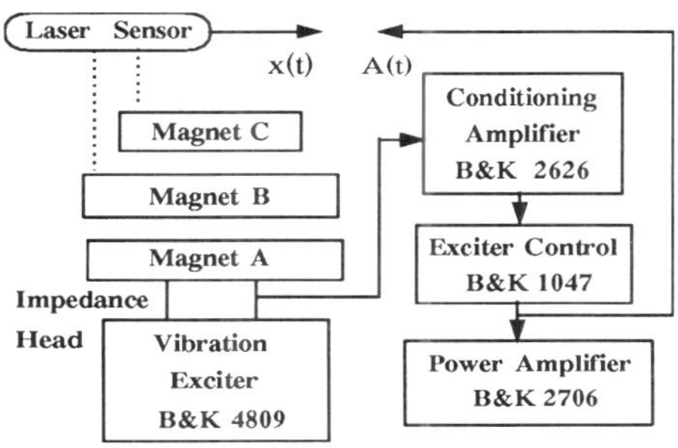

Figure 2. Experimental setup

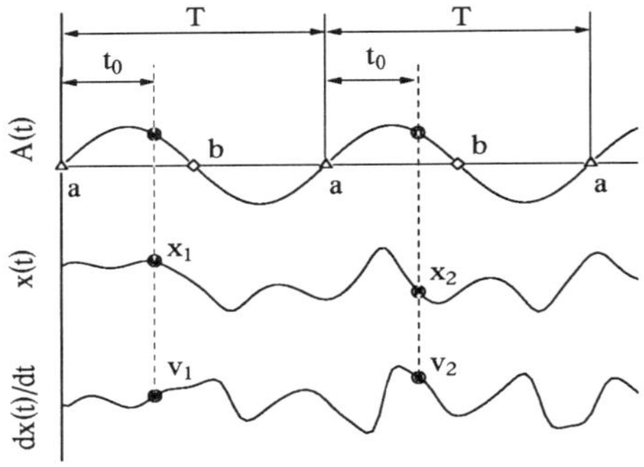

Figure 3. Experimental Poincaré points

frequencies of the system.

2. EXPERIMENTAL EQUIPMENT

Figure 1 illustrates the system used in our investigation. It consists of three cylindrical permanent magnets. Magnet A has the radius 50 mm, thickness 9 mm. Magnet B and C have the same dimension with radius 47 mm and thickness 7 mm. These two magnets can move freely up and down along the non-magnetic pole. The pole has a length of 150 mm and its surface was processed precisely. The bearings constrained the rotation of B and C. The three magnets were separated by their magnetic repulsive forces and constructed a magnetically levitated system with two degrees of freedom. The refined-oil lubricant was injected to the bearings to reduced the friction.

A schematic diagram of the whole test set-up is shown in Fig. 2. The responses of the two movable magnets were picked up by laser sensors and produced a voltage proportional to the displacement of the magnets. The output voltage was fed into an analog differentiator to produce the velocity. The three quantities, displacement, the velocity and the sinusoidal exciting signal were then put into a computer-aided chaos observing system [3] for data processing.

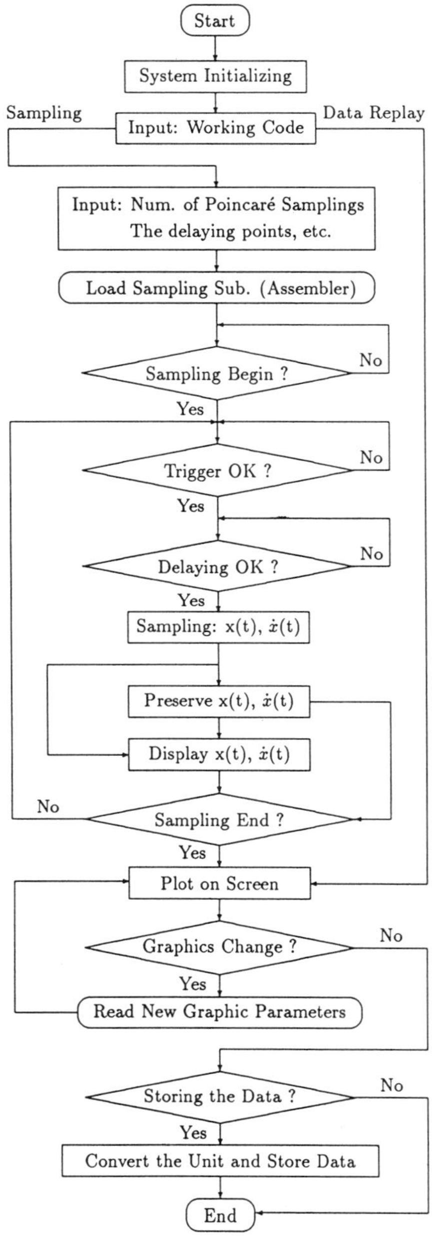

Figure 4. Flow chart of sampling

3. SAMPLING OF POINCARÉ MAP

For a system subject to an excitation $A(t) = A \sin \omega t$, the Poincaré map can be defined in the phase plane as a set of points with $\{x(nT + t_0),\ \dot{x}(nT + t_0)\}$ $(n = 1, 2, \cdots)$, $0 \le t_0 \le T$, where T is the period of forcing motion and t_0 is called initial delaying time that corresponds to the selected phase. The phase changes within $[0,\ 2\pi]$ as t_0 changes within $[0,\ T]$. The ideal of "one point in one cycle" is shown in Fig. 3.

The sampling process in every forcing cycle can be divided into two steps to obtain highly synchronized Poincaré points. In the case of sampling phase $P \in [0°, 180°]$, the time counting of A/D is triggered to start when the forcing wave changes from minus to plus and its voltage level has just reached to 0. The starting points for time counting are marked with character "a". After the delaying time t_0, that is, the phase changes to what we need, the sampling begins. In the situation of the sampling phase $P \in [180°, 360°]$, the time counting starts when the voltage of forcing signal changes from plus to minus and goes down reaching to zero. A simplified logic scheme of the computer program used to obtain or display sampled Poincaré map is shown in Fig. 4.

4. RESULTS AND DISCUSSION

At first, the frequency-response curves of the magnets were obtained by sweeping the frequency range up and down at a small constant level of excitation $A = 0.3$ mm (p-p) and slow sweep rate $S = 0.1$ Hz/s. It was found that the first linear resonant frequency of the system was 4.9 Hz and the second was 12.6 Hz. The dynamic response of the system was investigated in excitation parameter space. The exciting amplitude A ranged from 0.1 mm to 5.0 mm (p-p) and the exciting frequency f ranged from 5.0 Hz to 25.0 Hz. The initial conditions of Magnet B and C were set to be zero for every pair of exciting parameters.

Figure 5 shows the observations made during experiment. It shows the time history and power spectrum of response of the magnets under A=4.0 mm, f=16.2 Hz. The power spectra show discrete peaks in f/n, where $n = 1, 2, 3, 4$. The broad band character around these frequencies suggests that the responses of the magnets are chaotic. To diagnose the chaotic motion, the correlation dimension e was computed from the time historyΩof Magnet B and C. A delay of 0.1 second was chosen for each embedding. In the calculation 20000 data were used. For the embedding dimension of 12, e was found to be 2.8 for Magnetic C and 3.4 for Magnet B. In this case, as seen in experiment, the motion of the two magnets is chaotic.

A representative Poincaré map of an experimental response is shown in Fig. 6 with phase $P = 0°$. To make the system reach steady state, 1000 forcing cycles were allowed to elapse before data was sampled. The exciting parameters were $A = 4.0$ mm (p-p),

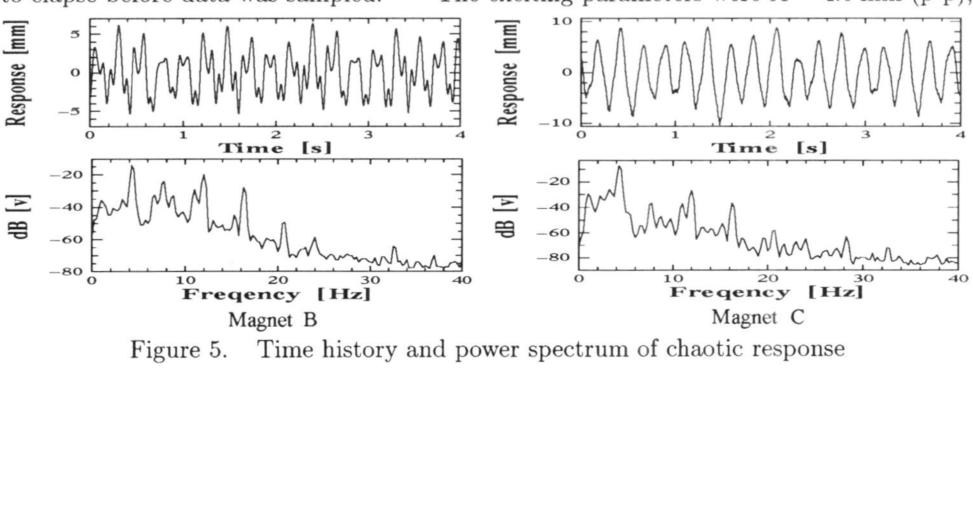

Figure 5. Time history and power spectrum of chaotic response

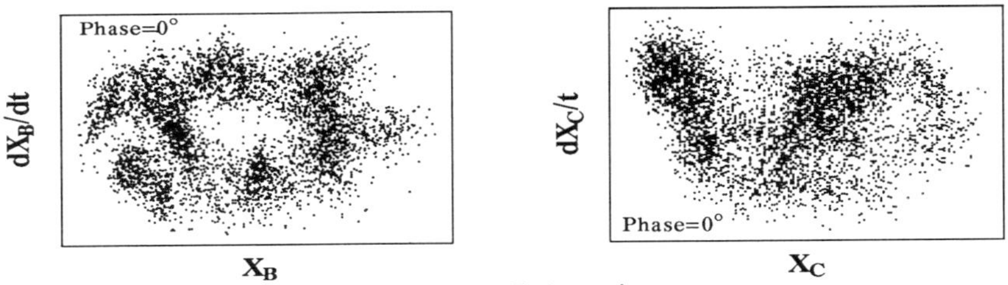

Figure 6. Poincaré map

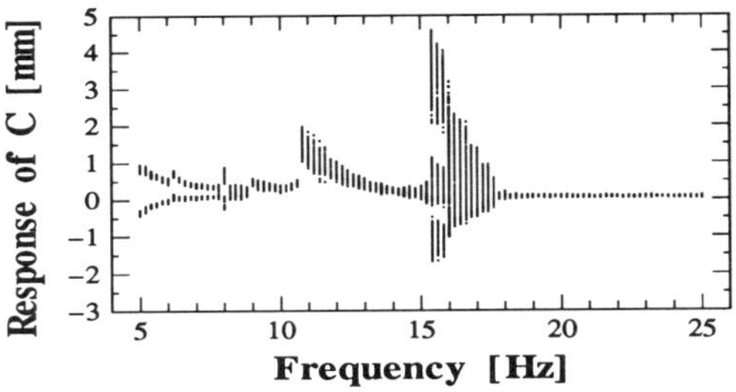

Figure 7. Bifurcation of Magnet C

$f = 16.2$ Hz. Every Poincaré map contains 10000 sampling points.

In the experiment, we observed that the responses of the two magnets were quite complex. Figure 7 shows the bifurcation for Magnet C under the $A = 4.0$ mm and f ranging from 5.0 Hz to 25.0 Hz. For the exciting frequency from 5.0 Hz to 10.6 Hz, the responses of Magnet C are mainly periodic. The response of the magnet became unstableΩwhen the exciting frequency was in the range of 10.8 Hz through 15.2 Hz. The response became chaotic in the frequency 15.2 Hz through 17.8 Hz. This frequency range is near the sum of the two linearized natural frequencies of the system. It should be mentioned that the amplitude of the chaotic response enlarged suddenly when the frequency changes from 15.2 Hz to 15.4 Hz. The same phenomena could also be seen in the Magnet B. This implies that the appearance of chaos could be very sensitive to the change of exciting frequency. It was also observed that the unstable vibrations of the two magnets appeared under the same exciting parameters. When exciting amplitude was small, the responses of the system were not chaotic. The chaos appeared only when the exciting amplitude became large enough.

5. CONCLUSIONS

The results of the experiment show that the magnetically levitated system with two degrees of freedom can indeed give rise to chaos. It concludes that the exciting frequency for inducing large amplitude chaotic response is near the sum of linearized frequencies of the system. This knowledge could be helpful to prevent chaotic response in engineering design of magnetically levitated system with two degrees of freedom.

6. REFERENCES

1 H. Tamura, Electromagnetic forces and Applications, J. Tani and T. Takagi (eds), Elsevier, Amsterdam, (1992) 325.

2 T. Hikihara and Y. Hirane, Int. J. Appl. Electromagnetics in Materials, 3 (1992) 73.

3 J. Tani, Q. Zhu, T. Takagi, and K. Ohtomo, Int. J. Appl. Electromagnetics in Materials, 2 (1992), 317.

Magnetic Levitation

Simulation and Design of Applied Electromagnetic Systems
T. Honma (Editor)

Development of the Test Capsule Falling through the Drop Shaft

Hideyo Sakurai, Sinobu Saito, Takao Azuma and Mi-turu Muto

Ishikawajima-Harima Heavy Industries Co.,Ltd.
6-2, Marunouchi 1-chome, Chiyoda-ku, Tokyo 100 Japan

Abstract

The world's deepest drop shaft facility for microgravity experiment, using former coal mine shaft of 710m in depth was constructed at Kamisunagawa, Hokkaido, Japan in 1991. The rocket shaped capsule in which experimental devices are loaded is falling through the drop shaft and produces microgravity of approximately 1×10^{-5}G for 10 second.

1. Introduction

The drop shaft facility constructed at Kamisunagawa, Hokkaido, Japan in 1991 has a free fall depth of 490m and a total depth of 710m, which is without parallel in the world. Unlike similar facilities in the U.S. and Europe, which have a vacuum shaft, an atmospheric shaft has been chosen, in view of the exceptional depth of the shaft.

A double capsule arrangement has been adopted. Microgravity condition in the inner capsule, in which experimental devices are loaded, is created by floating the inner capsule in the vacuum space of the outer capsule while falling. Two vertical steel rails are oppositely installed in the drop shaft, and the outer capsule is falling down between the rails controlled by magnetic guidance system. The atmospheric drag exerted on the outer capsule is counteracted by an air jet thruster. The motion of the inner capsule (except that in the direction of gravitation) is controlled by electromagnetic bearings for 0.4 second just after released. In this paper, we describe the outline of this system.

2. Basic performance

Basic performance of this facility is as follows ;

(1) Microgravity time : 10 seconds
(2) Microgravity level : approximately 1×10^{-5}G
(3) Deceleration at braking : 8G
(4) Maximum payload weight : 1000kg
(5) Real time data transmission device between falling capsule and control room including colour image.

3. System concept and time sequence for microgravity production

The system concept of this facility is illustrated in fig. 1. The rocket shaped capsule (1.8m in diameter, 8m in length, 5.7ton in weight) is a welded structure of aluminum alloy. Figure 2 shows the capsule, in condition ready for release but still suspended at the top of the shaft. The outer capsule is locked to shaft top by hydraulic clamp, with the inner capsule suspended by electromagnetic clamp to the top of the evacuated interior chamber.

The release of inner and outer capsules follows the time sequence indicated in fig. 3. By decreasing the coil current of electromagnet which suspend the inner capsule, the inner capsule begins to fall. When the force retained by magnetic clamp lowers to 85% of the inner capsule weight, an answer signal indicating "inner capsule released" is transmitted. 35 msec thereafter, the outer capsule release command is generated, and the outer capsule is actually released about 20 msec thereafter.

Upon release, the outer capsule is magnetically guided in its fall between a pair of steel rails laid along the drop shaft. The inner capsule must fall freely in the vacuum space surrounding it without coming into mechanical contact with the outer capsule structure. For this, its movement is restricted by magnetic bearings in the five degrees of freedom excepting downward translation, for 0.4 second just after released.

As the inner capsule is released about 55 msec before the outer capsule is released, the inner capsule displaces downward to dislodge itself from the narrow gap of the magnetic bearings about one second after released. The vertical distance between two capsules is detected by an optical sensor mounted on the outer capsule. The thruster jets are thereafter controlled by feedback loop using the signal of the sensor, to maintain the inner capsule within a prescribed range of distance ; between the lower edge of the magnetic bearings and the bottom shock absorber.

After free fall of 10 seconds, the capsule plunges into the deceleration zone. At this time, the inner capsule is still floating. To minimize the impact of landing to the bottom shock absorber, the distance between the inner and outer capsule is controlled so that, when entering the braking zone, the inner capsule is as close as possible to the bottom shock absorber.

4. Outer capsule magnetic guidance system

The attraction type electromagnetic levitation method has been chosen. Six pairs of electlomagnets are provided on the outer capsule, each pair saddling a guide rail, and their attraction forces are individually controlled by controllers mounted on the capsule. A touchdown bearings (called skid) are provided near the magnets, so as to prevent the pole plates from touching the rail. The magnetic force of each magnet is 300 N max. in Y-direction and 150 N max. in Z-direction. The rail-to-pole plate gap is 8mm, and the rail-to-skid gap is set to 2.5 to 3mm for each side.

The motion of the capsule is controlled in five axes (ie., excepting that in the direction of gravity pull) by six pairs of electromagnets, as shown in fig. 4. Since two pairs are in the Y-direction and four in the Z-direction, the control for the Z-axis may become excessive. For this reason, we have adopted the co-operative controll method for the Z-axis : only three sorts of motion, the translation, the inclination, and the rotation about Z-axis, are picked up in input signals, and are subjected to positive control.

The input signals for control are the gap between magnet and rail, and absolute acceleration in horizontal direction, from which absolute velocity, relative velocity, and relative displacement (relative to the rail) are fed back for controlling. Optimization of each control gain was done in trial-and-error fashion in more than ten drop tests.

5. Motion control of the inner capsule just after released

To maintain the microgravity environment created in inner capsule on its free fall, it is necessary that the inner capsule does not come in touch with the outer capsule. Sice the gap between the two capsules is 75mm on each side, and since such contact can occur when the inner capsule is released with a certain amount of initial velocity, the initial velocity must be reduced to less than 7.5mm/s. To reduce this initial velocity, five sets of magnetic bearings are mounted on the outer capsule. The motion of the inner capsule is controlled for 0.4 second just after released, on a feed back of the relative displacement between the inner and outer capsules. Control gain and control duration time were determined by results of motion analysis of both capsules while falling and then corrected properly by experiments.

6. Braking and recovery of the capsule

After free fall of 10 seconds, the capsule plunges into the braking tube of 1.9m in diameter. Since the bottom of the braking tube is closed, and since the gap between the capsule and braking tube is very narrow, air inside the braking tube is compressed and decelerates the capsule. Fourteen sets of mechanical braking shoes are installed at the bottom part of the braking tube and catch the capsule mechanically. Capsule recovering equipment is hanged down from the top of the shaft by wire rope and the capsule is recovered to the ground in tens of minutes after released.

7. Laser communication

An optical communication system is installed. In this system, laser light is tranmitted through the air between the capsule and the bottom of the shaft. The result of experiments in the falling capsule are observed in real time by this optical transmitter. The command data can be also transmitted in real time from the control room to the capsule by this system.

8. Concluding remarks

Tis drop shaft facility is operated by Japan Microgravity Center, and more than 300 times of drops for microgravity experiment have been successfully performed till the end of 1992.

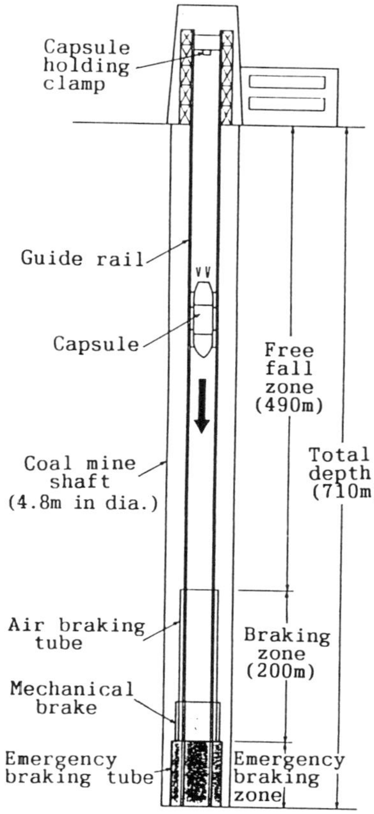

Fig.1 System concept

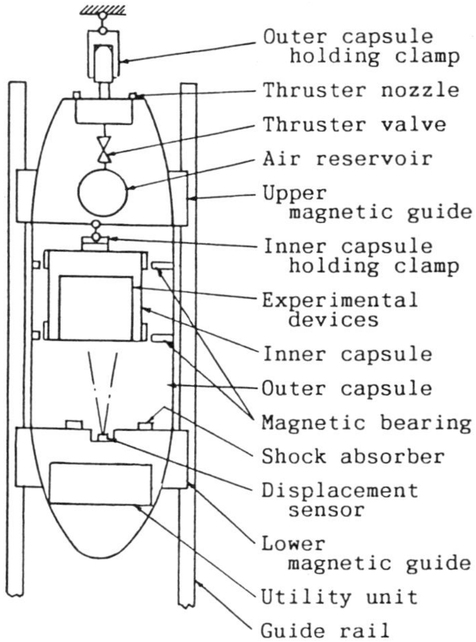

Fig.2 Schematic cross section of capsule

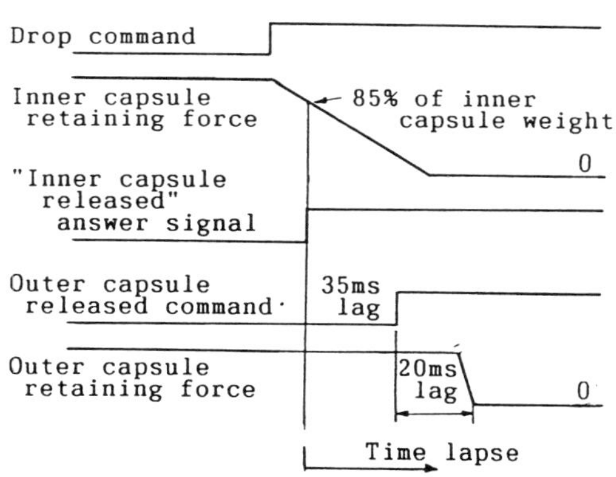

Fig.3 Time sequence of capsule release

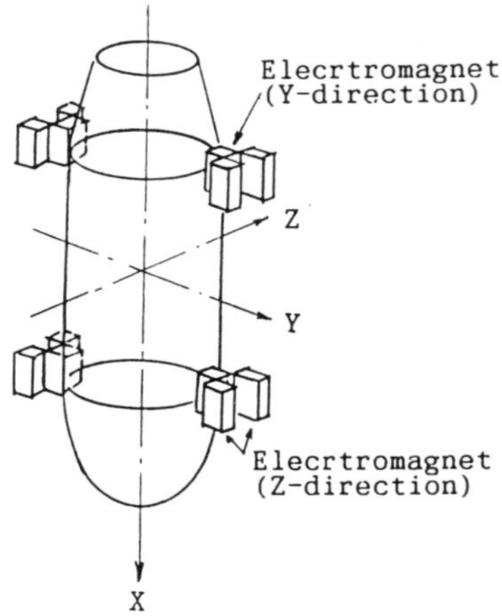

Fig.4 Layout of electromagnet

Simulation and Design of Applied Electromagnetic Systems
T. Honma (Editor)

Optimization of the vertical coils option of the maglev vehicle levitation unit

V.I. Omelyanenko and S.A.Sergeev

Kharkov Politechnic Institute
21 Frunze St., Kharkov, 310002, Ukraine, CIS

Abstract
The optimum characteristics of prospective options of the Levitation Unit (LU) generating the levitation force directed along the gap between the vertical on-board superconducting magnets (SCM) and track coils are compared. Defined are the boundaries of preferable ranges of mechanical load per one vehicle magnet, at which the advantages of single- and double-loop options of track coils are obtained.

1. INTRODUCTION

At least 15 years have passed since the first LU with vertical SCM and track coils was invented [1]. Today we are witnessing a new wave of interest in such devices, the era of normal-flux devices is apparently gone forever. They have been superseded by zero-flux designs with tangentially directed levitation force [2, 3]. The absence of comprehensive information on their characteristics has stimulated this investigation, the aims of which are as follows:
- to compare the optimum characteristics of the LU options with single-loop and double-loop track coils;
- to study the possibility of improvement of the double-loop option by asymmetrization of its upper and lower loops.

2. COMPARISON OF LEVITATION UNIT CONFIGURATIONS

One of the turning points in the history of magnetic levitation transportation was the realization of the fact that the vertical arrangement of the superconducting magnets along the vehicle boards was the most appropriate design. The present day task is to design the track coils ensuring optimal characteristics of the levitation system.

Among the many versions of the LU designs using vertical SCM we have selected three options, which differ in the track coil shape.

They are, firstly, option B, the unit with single-loop track coils (see Figure 1). Second option C, where in each track coil consists of two equal loops located one above another and electrically interconnected. Finally, in the third option G the LU also comprises double-loop "eight"-shaped track coils differing from option C by somewhat smaller dimensions of the upper loop compared with those of the lower one. By reducing the area of the upper loop we increase the SCM magnetic-flux linkage with the track coil and thus increase the levitation force. All three options use the tangential component of the electrodynamic force acting between the track coils and SCM to levitate the vehicle.

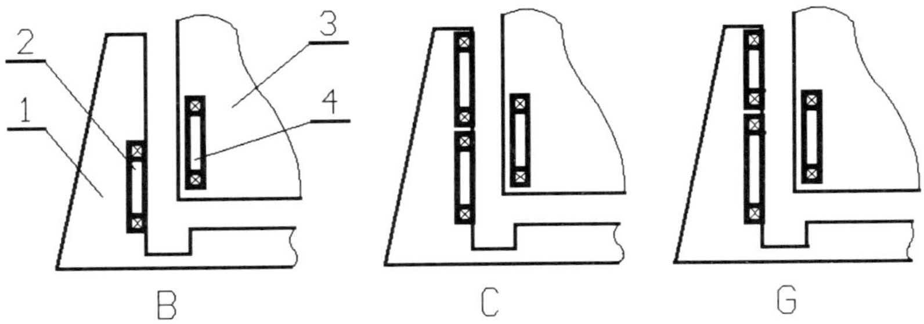

Figure 1. Outline of the compared levitation units
1 - guideway; 2 - track coil; 3 - vehicle; 4 - SCM

3. COMPARISON TECHNIQUE

In developing the mathematical model we have assumed the LU to be composed of one on-board SCM and a chain of shorted track coils. The vector of varied parameters x contains the following variables

$$\mathbf{x} = (\tau, a, b, c, d, y), \tag{1}$$

where τ, a and b are the pole pitch, length and width of the track coil, respectively, and c, d and y are the length, width and vertical displacement of the SCM in the system of coordinates tied to the center of one of the track coils. Magnet velocity v and its MMF I_S, as well as the side clearance z are considered to be constant for all compared versions (v = 100 m/s, I_S = 700 kAT, z = 0.2 m).

The main reason for this comparison was to determine which LU option, while meeting the imposed restrictions as to the value and quality of the levitation force, is able to ensure smaller electric loss E in the track coils.

The proposed technique is based on the comparison of the relative positions of optimum compromise curves min (E/F_L) *versus* F_0 plotted for options B, C and G meeting the following uniform restrictions on criteria and parameters:

1. The LU generates the specified levitation force $F_L = F_0$;
2. The levitation force pulsating factor P does not exceed the level of P_0 = 10 %.
3. The volume consumption w of the track coils conductor per running meter of the guideway is constant.
4. Vectors a and b determining the boundaries of the feasible domain D in the variable space R^N are the same for all options.

Table 1.
Restrictions imposed upon varied parameters

Parameter	τ	a	b	c	d	y
Bottom margin, m	0.44	0.3	0.3	1.0	0.3	0
Top margin, m	1.2	1.0	0.6	1.8	0.6	1

The curve of optimum compromises for each option passes through the points the ordinates of which were found by sequentially solving the problem of criterion vector f minimization

$$f_1(x) = \sum_{i=1}^{N} (\max\{0;\ a_i - x_i\} + \max\{0;\ x_i - b_i\}); \tag{2}$$

$$f_2(x) = |F_L(x) - F_0|;\ f_3(x) = \max\{0;\ P(x) - P\};\ f_4(x) = E(x)\,/\,F_L(x)$$

for various values of $F_0 \in [10, 100]$ kN. So, this curve demonstrates the relation between the value of the required levitation force and the minimum specific loss inherent to given LU. Points below the curve are theoretically unattainable.

Figure 2 shows the summary diagrams combining the most common characteristics of the levitation system: mechanical load per one magnet F_0, SCM's MMF I_S, track design, minimal electric loss E in the levitation system and conductor consumption w. Curve 1 corresponds to option C, curve 2 - to option G, curve 3 - to option B.

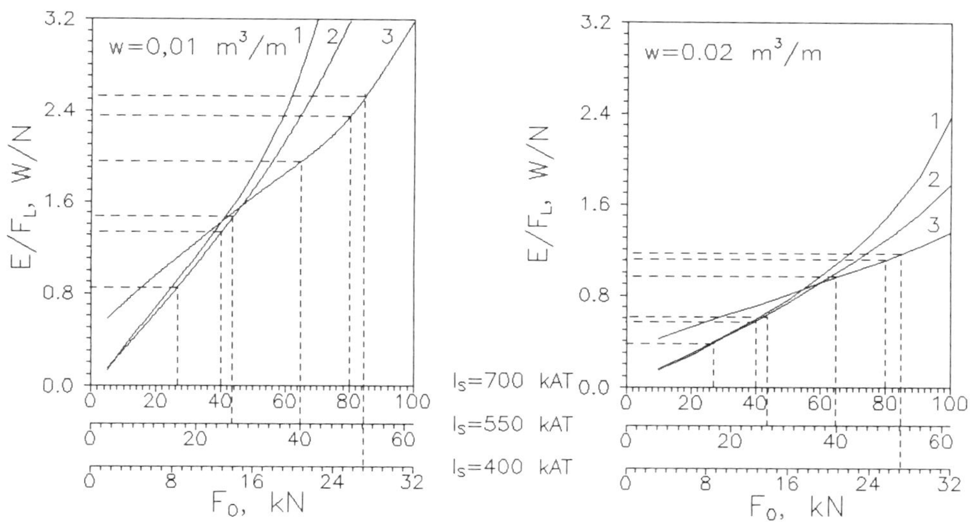

Figure 2. Curves of optimum compromises.

Curves 1 and 3 are obtained by solving the problem in domain $D \in R^6$. As we see, the smallest value of the reduced loss is obtained at the largest conductor consumption and the lowest mechanical load. As the load increases or consumption decreases, the loss increases. The cross-point of curves 1 and 3 separates the expediency ranges of single-loop option and double-loop symmetrical one. To the left from this point the double-loop option has better characteristics, and to the right the single-loop version is more advantageous.

In order to check the efficiency of the track coil loop asymmetrization we used the following technique. Firstly, we complemented vector **x** with two independent variables

x_7 = g, x_8 =h, where g and h are the length and width of the upper loop. The dimensionality N of the variable space R was increased to the eighth order.

Secondly, it should be kept in mind, that the results of option G optimization in the eight-dimensional space, but with additional restrictions g = a and h = b, should be identical to the previously obtained results of option C optimization. Then, starting the optimization process of the option G from the best points of option C, we are able to prove the relative efficiency of option G.

The point is, that the search algorithm [4] used for the calculation program supposes vector x to leave the start point only when there is a possibility of improvement of the criteria vector f values. The opposite is also valid - if there is no possibility to improve the vector of criteria, the vector of variables remains in the start point. Therefore, any displacement of vector x from the initial point during the process of calculation can be interpreted as evidence of the option G superiority over option C.

The optimization results of option G in space R^8 are represented in the diagram by curve 2. Comparing the positions of curves 1 and 2, we can conclude that option G posseses the best characteristics over option C. The greater the required levitation force and the lower the aluminium consumption, the more pronounced is the effect of decreasing loss caused by asymmetrization of the track coil loops.

The chosen system of criteria, except the lifting force only, is invariant with respect to the SCM's MMF. Therefore, introducing a scaling factor $K=(700/I_S)^{0.5}$ for the axis of abscissas we can easily transform the diagram to a form which will correspond to any other desirable SCM's MMF. The relative positions of the curves will thereat remain the same.

Comparing the electric losses in the levitation system E/F_L we have analyzed six lay-out options of the superconducting magnetic system for a vehicle with the mass of 32 t. These options differ in the combination of the MMF and mechanical load per one magnet. We preferred the eight-pole on-board system consisting of the SCM with the dimensions of 1.7 x 0.45 m and MMF 700 kA. Being moderately massive, this system, together with the G-type track coils, would be able to levitate the vehicle with the loss of 200-400 kW per car at the assumed aluminium consumption.

4. CONCLUSION

The comparison of the relative positions of the optimum compromise curves for the LU options with single- and double-loop track coils enabled us to determine the efficiency of each of them.

We have proved that the asymmetrization of the double-loop track coil is the most effective measure to increase the LU operation efficiency.

5. REFERENCES

1. Weh H. Elektrodynamishes Trag - und Fuhrungssystem . - Patentschrift BRD DE 2412221 C3, Int. Cl. B60L 13/00, 14.03.74

2. Fujiwara S. Maglev Trains.-Proc. 2nd ISS'89.-ISTEC.-Tsucuba.-Japan.- pp. 41-46

3. Klimenko E.Yu.,Novikov S.I., Omelyanenko V.I., Sergeev S.A. Superconducting magnet for high speed ground transportation.- Cryogenics. - 1990, N 1, pp.41-45.

4. Omelyanenko V.I., Sergeev S.A., Severin V.P. Multiple Criteria Optimization of the Maglev Vehicle Levitation Unit. - The Proc. of the First Japan - CIS joint seminar on electromagnetomechanics in structures, January 22 - 23, 1992, Tokyo, Japan, pp. 130-133.

Simulation and Design of Applied Electromagnetic Systems
T. Honma (Editor)
1994 Elsevier Science B.V.

Analysis of a Mock-up Levitation Experiment on Superconducting Magnets for Magnetically Levitated Vehicles

Y. Kameoka[a], K. Yoshioka[a], H. Fukumoto[a], T. Takizawa[b], T. Saito[c] and T. Hirata[d]

[a]Energy Research Laboratory, Hitachi Ltd. 1168 Moriyama-cho, Hitachi-shi, 316, Japan

[b]Hitachi Works, Hitachi Ltd. 1-1, Saiwai-cho, Hitachi-shi, 316, Japan

[c]Hitachi Research Laboratory, Hitachi Ltd. 4062 Kuji-cho, Hitachi-shi, 319-12,Japan

[d]Mechanical Engineering Research Laboratory, Hitachi Ltd. 502 Kandatsu-cho, Tsuchiura-shi, 300, Japan

Abstract

A 3D eddy current analysis code was developed and verified through a mock-up levitation experiment on superconducting magnets (SCMs) of magnetically levitated vehicles (MAGLEV). The analytical results of magnetic fields agree experimental one within $\pm 10\%$ on helium (He) container. A coupled electromagnetic and mechanical interaction, under SCM vibrations, is investigated using the code. As a result, the SCM structure with a low resistivity material coating on He container is proposed as a suitable design for reduction of heating loss.

1. INTRODUCTION

Magnetically levitated vehicles (MAGLEV) are planned for the next high-speed mass transport systems. They levitate through interaction of superconducting magnets (SCMs) with the induced magnetic field of the passive coils placed on the tracks, as shown in Fig.1. Each SCM is composed of a vacuum vessel, radiation shield, and helium (He) container.

Since the He container is kept at 4.2K, it is very sensitive to heat penetration. The SCM is exposed to the ripple field of the passive coils, which induces eddy currents on each surface of the components. Eddy currents on the He container may cause significant heat loss. In addition, the eddy current $\mathbf{J}_1$ on the vacuum vessel induces strong vibrations through the $\mathbf{J}_1 \times \mathbf{B}$ force, where **B** is the superconducting magnetic field. This vibration induces a secondary eddy current $\mathbf{J}_2$ on the He container, through $\mathbf{V} \times \mathbf{B}$ emf, where **V** is the vibration velocity. Estimation of heating losses caused by these two types of eddy currents is the principal issue for reliable SCM design.

In this study, a 3D eddy current analysis code, based on a finite element method with thin shell approximation [1] is developed and used for designing a SCM with low eddy current heating loss. The code is verified through a mock-up experiment on the SCMs. Further, a coupled analysis on the electromagnetic and mechanical interaction is presented to find the SCM structure with low heating loss.

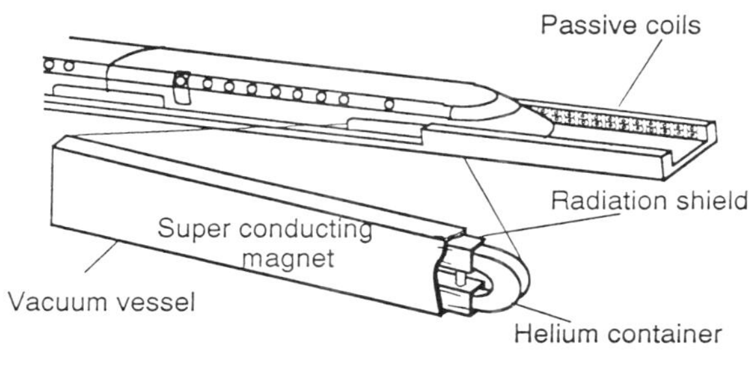

Figure 1. Concept of MAGLEV and superconducting magnet

2. 3D EDDY CURRENT ANALYSIS CODE

The 3D eddy current analysis code is developed in which the SCM is assumed to be composed of non magnetized thin shells. The basic equation is:

$$\frac{\mu_0}{4\pi}\frac{\partial}{\partial t}\int\frac{\mathbf{J}\cdot\mathbf{J}'}{|\mathbf{r}-\mathbf{r}'|}d\mathbf{r}d\mathbf{r}'+\frac{1}{\sigma}\int\mathbf{J}^2d\mathbf{r}+\int\mathbf{J}\cdot\dot{\mathbf{A}}_{ex}d\mathbf{r}=0\cdots(1)$$

where $\mathbf{J}$ and $\mathbf{A}_{ex}$ are the shell surface eddy current density and the magnetic vector potential of the external source, and μ_0 and σ are the vacuum permeability and the conductivity of materials in the SCM. Eq.(1) incorporates conservation of magnetic energy stored in the system. The eddy current $\mathbf{J}$ is presented by a normal component of the current vector potential T_n as,

$$\mathbf{J}=\nabla T_n\times\mathbf{n}\cdots(2)$$

where $\mathbf{n}$ is the unit vector normal to the surface. Discretizing the conductor surface into triangular finite elements, the matrix equationis obtained.

$$\mathbf{M}\dot{\mathbf{T}}+\mathbf{R}\mathbf{T}+\mathbf{V}=0\cdots(3)$$

where $\mathbf{T}=\{T_s\}$, Ts is current vector potential at each node; M and R are the matrixes corresponding to inductance and resistance; $\mathbf{V}$ is the electro motive force due to the externally changing field $\mathbf{A}_{ex}$.

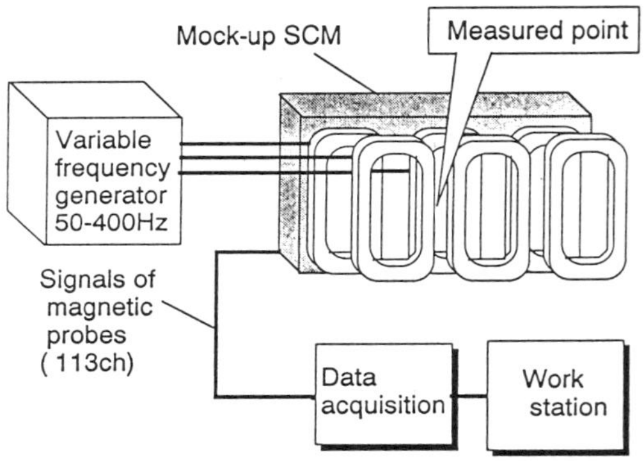

Figure 2. Experimental set up

3. COMPARISON OF ANALYSIS AND MOCK-UP EXPERIMENT

Calculated and experimental magnetic fields for the mock-up experiment are compared. The mock-up SCM and the passive coils system are shown in Fig.2. The passive coils driven by a three-phase variable frequency generator and generates a traveling wave ripple field on the mock-up SCM. The radiation shield and He vessel are cooled down by liquid nitrogen. The magnetic fields are measured by search coils placed on the surfaces of each component. Fig.3 compares typical measured and calculated results. They agree within 20% on the average, although the measured values on radiation shield are somewhat scattered.

After the mock-up experiment, the analysis code is applied to an actual system. The loss of eddy current due to the ripple magnetic filed, which corresponds to that in the Yamanasi experimental line, is calculated at the train velocity of 500km/h. As shown in Fig.4, the eddy patterns reflect the field pattern of the figure-eight passive coils

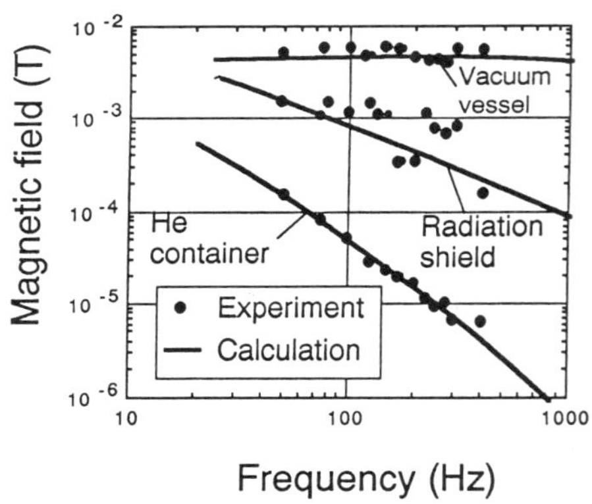

Figure 3. Comparison of calculated andexperimental results from the mock-up experiment

and sweep from the left to the right hand side of the vacuum vessel. The joule loss is 400W on the vacuum vessel, 2W on the radiation shield and 10mW on the He vessel. On the He vessel, the joule loss is sufficiently small for the design limit of 1W.

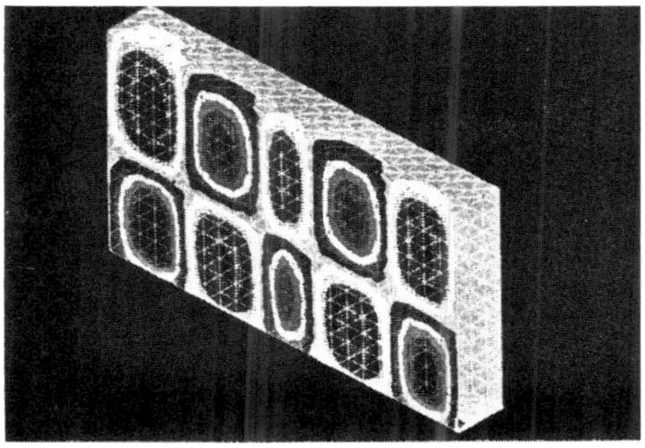

Figure 4. Eddy current pattern on the vacuum vessel due to the ripple magnetic field of passive coils

4. SIMPLE COUPLED ANALYSIS

A simple coupled analysis of electromagnetic and mechanical forces is carried out. This analysis is based on a momentum equation assuming the rigid body displacement of SCM components and circuit equations, as shown in the following equations:

$$m_s\ddot{x} + k_c x = f + \frac{K_0}{g} I \cdots (4)$$

$$L\dot{I} + RI + K_0\dot{x} + M\dot{I}_1 = 0 \cdots (5)$$

$$L_1\dot{I}_1 + R_1 I_1 + M\dot{I} = 0 \cdots (6)$$

$$K_0 = \frac{\partial M_{scm}}{\partial x} \cdot I_{scm} \cdots (7)$$

Structural vibration is described with equivalent mass and stiffness, as shown in Eq.(4). The eddy currents have an eigen flow pattern according to the vibration mode. This is described as a circuit model with equivalent inductances L, M, and equivalent resistance R, as shown in Eqs.(5) and (6). Estimations of L, M and R are made with the eddy current analysis code. Eq.(5) corresponds to the condition on the radiation shield, and Eq.(6), the He container. The vacuum vessel model is excluded, because the eddy current on the vacuum vessel has little effect on the He container. In Eq.(4), the second term of the right side presents the electromagnetic force due to eddy current, and K_0, which is also used in Eq.(5), means the magnetic flux derivative of the SCM. In the equations, m_s is equivalent mass, k_c is stiffness, x is rigid displacement, between the radiation shield and He container, and f is external force. The subscript 1 denotes the He container, and no subscript, the radiation shield.

Fig. 5 shows results of coupled analysis. The frequency responses of (a) eddy current for a unit displacement for various He container resistivities, (b) displacement with and without a coupling consideration, and (c) the joule loss on the He container for various resistivities are shown. In Fig.5(a) , the eddy current per unit displacement for various resistivities of the He container which correspond to 1, 1/10, and 1/100 of resistivity of stainless steel are compared. Using Laplace transformation, from Eq.(4) to Eq.(6) is rewritten as follows:

$$\frac{I}{x} = \frac{K_0 M}{L L_1} \frac{s^2}{(1-\alpha^2)s^2 + \frac{1}{\tau_{eff}}s + \frac{1}{\tau\tau_1}} \cdots (8)$$

where $\tau = \frac{L}{R}$, $\tau_1 = \frac{L_1}{R_1}$, $\frac{1}{\tau_{eff}} = \frac{1}{\tau} + \frac{1}{\tau_1} \cong \frac{1}{\tau_1}$. Eq.(8) indicates that the eddy current on the He container increases in proportion to the square of s, under the threshold frequency f_{th}, which is the inverse of the equivalent L/R time of the eddy current, and it becomes constant above f_{th}. This means that the eddy current is subjected to resistance in the frequency region under f_{th}, but it is subjected to inductance above f_{th}. Decreasing the resistivity of the He container, the eddy current increases below f_{th}, but is constant above it.As shown in Fig.5(b), 'without coupling', resonance vibration without damping is observed, but 'with coupling', the resonance peak is decreased due to eddy current damping and the displacement is suppressed by a factor of 5. As a result, the joule loss shown in Fig.5(c) is reduced if resistivity of the He container is reduced. Coating the He container, which is made of stainless steel, with low resistivity materials such as Cu, or Al is effective for joule loss reduction.

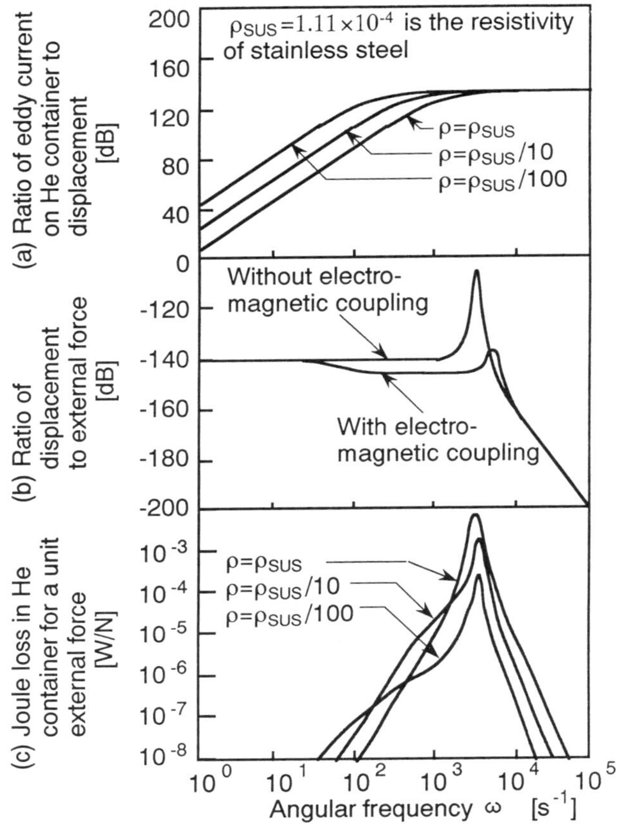

Figure 5. Characteristics of electromagnetic and structural coupled system

According to the complete 3D eddy current simulation, the loss of eddy current on the uncoated He container due to structural vibration between the radiation shield and He container reaches 3W with only a 20 μ m displacement. Under same conditions, the loss of eddy current on the He vessel with low resistivity coating can be suppressed below 0.2W, which is 1/15 of the conventional He vessel.

5. CONCLUSION

In conclusion, the thin shell 3D eddy current analysis code, based on a finite element method, was verified. A simple coupled analysis between electromagnetic and mechanical forces showed that coating with low resistivity material on the He container was effective for reduction of eddy current joule loss.

REFERENCES

[1]H. Fukumoto, et al., "Application of 3D Eddy Current Analysis on Magnetically Levitated Vehicles" IEEE Conf. on Electromagnetic Field Computation, Aug. 1992, Claremont, CA, USA

Simulation and Design of Applied Electromagnetic Systems
T. Honma (Editor)
1994 Elsevier Science B.V.

Safety evaluation test for electromagnetic levitation transportation system H-100

T. Mizuma ,A. Matsumoto and M. Fujino

Traffic Safety Division, Traffic Safety and Nuisance Research Institute 181, 6-38-1 Shinkawa Mitakashi ,Tokyo, Japan
Technology Devision, Chubu HSST Corporation 457, 3-14 Kafukucho Minami-ku Nagoya

Abstract

This paper reports test results on the levitation and guidance system , the propulsion system obtained from basic tests conducted for safety evaluation on the Nagoya test track with electromagnetic levitation vehicle H-100.

1. INTRODUCTION

H-100 transport system is one of the normal conducting magnetic levitation transportation system , supported and guided by magnetic force and driven by linear induction motors . The maximum speed achieved by it's urban transit type is on the order of 100 km/h . Since May 1991 , the transport system has been tested by Chubu HSST Development on the testing track at Ohe , Nagoya.

As the Ministry of Transport of Japan needs to set up ministerial ordinances as technical standards on safety and reliability before this kind of system is put into practical use , it has organized the " Committee for Investigation and Examination of Technical Evaluation Methods for Normalconducting Magnetic Levitation Transport System " (Chaired by Professor Eisuke Masada of the University of Tokyo) . Safety evaluation methods and testing methods are currently examined , while test results are analyzed at the same time for evaluating safety .

This paper reports test results on the levitation and guidance system , the propulsion system obtained from basic tests conducted for safety evaluation.

2. LEVITATION AND GUIDANCE SYSTEM

For the levitation and guidance system , the support force is obtained with the attraction between the electromagnets and the rail , and the guidance force is obtained with the restoring force . The running tests were conducted at up to the speed of 100 km/h on tracks including 100 m radius (100 mR) and 25 mR curved tracks . In order to evaluate the levitation and guidance force, tolerance of the guidance gap , loaded levitation force , and gap characteristics, measurements were taken on the levitation , guidance gap , levitation current, and sliding table movement . Figure 1. shows levitation gap waves measured during running tests at 100km/h .

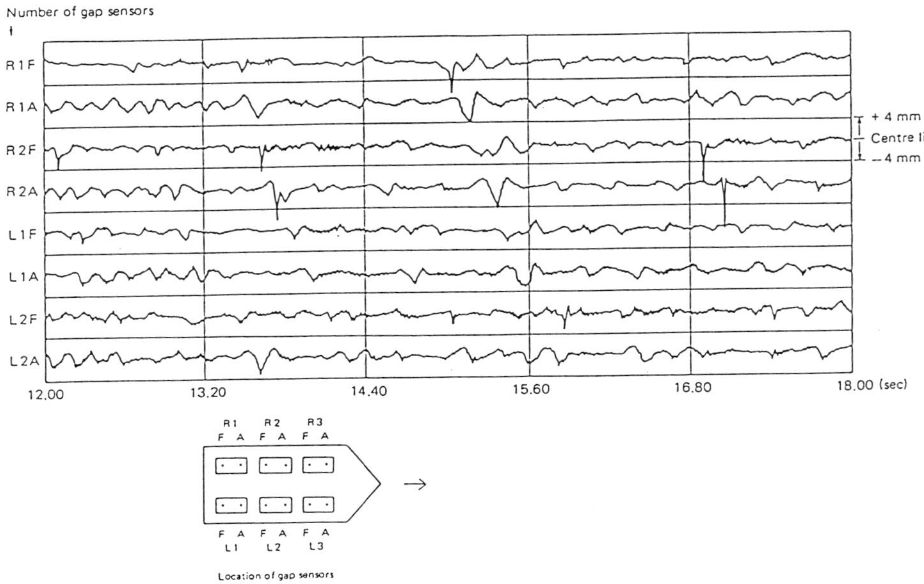

Figure 1. Levitation gap waves measured during running test at 100km/h

The following points were identified after the running tests on the straight track.

1) The levitation gap varied in proportion to the square root of the speed.
2) A maximum variation in the levitation gap was slightly larger with a fully-loaded car : approximately ± 3 mm with an empty car against ± 4 mm with fully loaded car. There was still some room left against the maximum mechanical stroke of ± 6 mm, and the levitation gap was under control.
3) A maximum variation in the guidance gap was approximately 10 mm with an empty car, and 12 mm with a fully-loaded car. With some margain left in the 15 mm maximum allowable variation, no contact was made between the electromagnets and the guidance skids during the normal run. The guidance direction was also found under control.

The following points were identified after the running test on the curved track.

1) A maximum variation in the guidance gap measured on the 100 mR curved track was slightly larger with a fully-loaded train : approximately ± 11 mm with an empty train against ± 13 mm with a fully-loaded train.
2) A maximum variation in the guidance gap on the 25 mR curved track was approximately 31 mm with an empty train, and 35 mm with a fully-loaded train. Estimating the module - rail positions based on the measured gaps, the minimum possible gap between the guidance skid and rail is approximately 2 mm, and no contact should be made.

3) A maximum variation in the levitation gap measured on the curved track was approximately ± 3 mm , and the levitation gap was under control also on the curved track .

Summarizing all test results on levitation and guidance , including other levitation and guidance characteristics values measured at various experiment beams , turnouts or slopes , the levitation control is kept within the designed variation range , and it has been confirmed that there is no problem in obtaining the levitation force and the follow-up function for the variations in the levitation gap .

As for the guidance control , the required force was confirmed , but the guidance gap was found greater than the calculated figure on the 25 mR curved track , and the separation with the rail was sometimes narrowed . The contact, however , was made on the guidance skids , and no problem was found in achieving the strength . The remaining gap left for the contact has to be studied further .

3.PROPULSION

The propulsion system applied here is one of the non-adhesion types obtained with a linear induction motor . The test was designed primarily to check the braking performance at various speed levels up to 100 km/h , and the evaluation was made to check if the required deceleration rate was obtained while stopping the vehicle from a high speed , if electric brakes operated in coordination , and if sufficient provisions were made against regeneration failure .

Figure 2. shows waves describing the inverter control measured with normal braking applied at 100 km/h . When brakes were applied , if the designed electric brakes were provided , the mechanical brakes (driven by hydraulic) did not intervene (thus , not causing initial operation) , and the electric brakes provided all of the braking force all the way down to approximately 42 km/h as regenerative brakes , returning electric power to the power line .

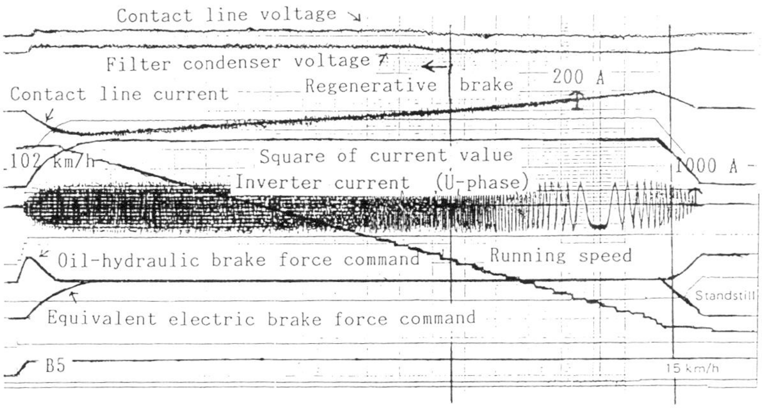

Figure 2. Inverter control measured with normal braking applied at 100 km/h

It shifted to negative-phase brakes , when the vehicle reduced its speed below 18 km/h . The proportion of electric brakes was reduced below 5 km/h , applying the mechanical brakes to stop the vehicle in coordination of the both brakes .

Furthermore , the following points were identified and confirmed in connection with the braking performance .

1) A failure of regeneration immediately triggered the mechanical brakes (with a few hundred seconds of delay) , without extending the braking distance greatly .
2) The electric brakes , when applied at a high speed , performed fairly constant deceleration rate , proving the braking function that was designed taking into consideration end effects .
3) Even when power failed (failure of power supplied from a power substation) the following sequence was confirmed :
   Electric brakes were operated within the speed range where regenerative electric current returned into the power conduct line , and the emergency brake command was issued to apply mechanical brakes to stop the vehicle after the main circuit was cut off with the intervention of alow voltage detection device .

4. CONCLUSION

Some of the test results of the H-100 system obtained on the Nagoya test track have been presented in this paper . Basic safety factors , such as the levitation and guidance system , propulsion system have been proved fairly satisfactory . There remain some behavior particularly on the sharp curved track (25 mR curvature) , reduction of power electric power consumption for propulsion and levitation , selection of the optimum ground system for electric equipment , and improvements in the general riding quality . Furthermore , reliability and durability should be tested .

Therefore , in the future , overall safety and reliability of the H-100 system should be evaluated at the longterm running tests , in which safety against troubles and reliability and durability of the system are to be checked.

5. REFERENCES

1. T. Mizuma , et al. : Safety Evaluation Tests of Model H-100 EMS Maglev Urban Transit System , JREA , February 1992 , Vol. 35
2. T. Mizuma , et al. : Safety Evaluation Tests of Normalconducting Magnetic Levitation Urban Transportation system H-100 (First Report) , Traffic Safety and Nuisance Research Institute , November 1991

Simulation and Design of Applied Electromagnetic Systems
T. Honma (Editor)

Measurement and analysis of eddy currents in guideway structure of superconducting magnetic levitation vehicle system

Atsushi Ichikawa[a], Genpachi Anami[a] and Shigeru Hemmi[b]

[a]Railway Technical Research Institute, 2-8-38 Hikari-cho Kokubunji-shi Tokyo, 185, Japan
[b]CRC Research Institute, Inc., 1-3-D16 Nakase Mihama-ku Chiba-shi, 261-01, Japan

Abstract

The superconducting magnetic levitation vehicle system (MAGLEV) is being developed by JR group in Japan. Eddy currents and/or loop currents are induced in the guideway steel structure by the passage of superconducting magnets (SCM's) . These induced currents cause energy loss and therefore cause magnetic drag force. It is very important to evaluate the magnetic drag force quantitatively before constructing any realistic structure.

Among our experiments, a measurement of the rise in temperature and the magnetic fields near steel plate will be shown. Analysis is also made using our developing computer code named JRMAG in the approximation of the resistive limit.

1. EXPERIMENTAL METHOD

In this case, we have placed a rectangular steel plate at the side wall almost vertically. A view of experimental site is given in Figure 1. The steel plate is either usual ferromagnetic steel or paramagnetic steel (namely, high manganese non-ferromagnetic steel). Dimensions and electrical properties of two rectangular steel plates are summarized in Table 1. Figure 1 also shows SCM's which are passing near the plate.

Then we have measured the rise in temperature at four points near each of four corner (some 5 cm inside of both edges). The temperature rise is measured by thermistors. The steel plate is

Table 1
Dimensions and Electrical Properties of two rectangular steel plates

	Ferromagnetic steel	Paramagnetic steel
Width (mm)	1740	1740
Height (mm)	830	830
Thickness (mm)	3.2	3.2
Electrical Resistivity ($10^{-8}\Omega\cdot m$)	20.0	70.0
Relative permeability (μ_r) maximum	1350	1

covered with adiabatic sheets so that the dissipation of generated heat is small. Magnetic flux density is measured at the center of plate with the help of Hall gaussmeters.

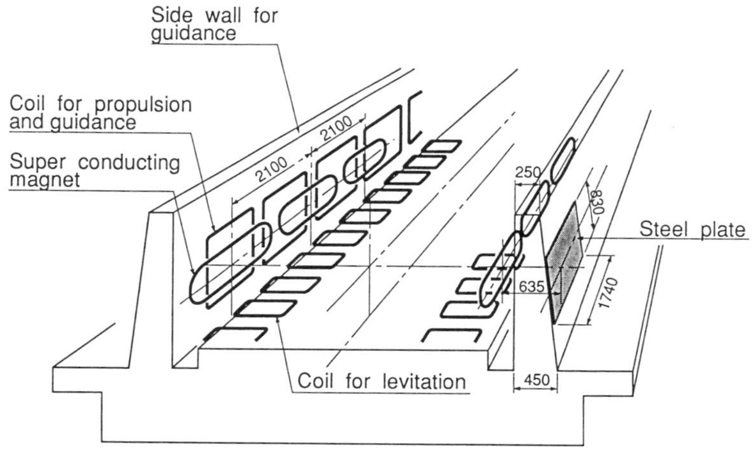

Figure 1 A view of experimental site

2. NUMERICAL METHOD

Unlike many other works[1], our eddy current analysis assumes a resistive limit. The resistive limit is valid for relatively slow speed; but we expect that its approximation will promote efficiency in designing realistic steel structures of the MAGLEV. One justification is that the speed (250 km/h in our case) is small compared to the characteristic velociy[2] (In the case given in Table 1, the characteristic velocity is 1500 km/h for paramagnetic steel plate).

In general, the eddy current density is given by the formula:

$$\boldsymbol{J} = \sigma \left(-\boldsymbol{v}_{\mathrm{SCM}} \times \boldsymbol{B}_{\mathrm{SCM}} - \mathrm{grad}\,\phi - \frac{\partial \boldsymbol{A}}{\partial t} \right) \tag{1}$$

Here, σ is the electric conductivity; $\boldsymbol{v}_{\mathrm{SCM}}$ is the velocity of the vehicle; $\boldsymbol{B}_{\mathrm{SCM}}$ is the magnetic flux density at the conductor induced by the SCM poles; ϕ is the electric scalar potential; $\boldsymbol{A}$ is the vector potential induced by the eddy current and magnetization inside conductor.

In our analysis, the resistive limit approximation neglects the third term of equation (1), and $\boldsymbol{B}_{\mathrm{SCM}}$ of the first term of equation (1) is given by Biot-Savart law. The analysis also implies that ferromagnetic property is not taken into account. Therefore the only unknown variable is the electric scalar potential . To determine the electric scalar potential, the minimum principle of the Joule heat loss is used. As a consequence, the conservation of the electric current is automatic. The FEM implementation is straightforward; it is unnecessary to divide void into meshes. This computer code named JRMAG has been developed. In the JRMAG, a two nodes line element, a four nodes face element and an eight nodes brick element are included. This element library enables analysis of arbitrary shaped realistic steel structures.

So we have simulated our measurements by JRMAG. Three SCM poles are taken into account, though the real vehcle MLU002 at the Miyazaki test center has 3x4 poles[3]. Main feature of the three SCM poles is shown in Table 2. The SCM's are passing at 250 km/h near the steel plate. The distance between a rectangular steel plate and center of SCM is 635 mm.

Table 2
Main feature of SCM (MLU002)

Number of SCM	3 poles 2 line/bogie x 2 bogie
Magnet pole	Opposite magnet pole at right and left
Dimension(Length x Width x Pitch)	1.7 x 0.5 x 2.1 m (Race track shaped)
Magnetomotive force	700 kAT

3. RESULTS

In Figure 2, the measured or computed magnetic flux densities are shown. In this case, the flux densities are dominated by ones the SCM induced and the agreement is satisfactory. In Figure 3, computed profile of eddy currents is given. In Figure 4, computed time history of drag force of paramagnetic steel plate is shown.

Average rise in temperature of our measurement is 0.16°C for ferromagnetic steel plate and 0.02°C for paramagnetic steel plate. Our temperature measurement is shown in Figure 5. Note that measured temperature rise must be substantially small compared to the average value over the plate, because the measured points are located near corners of plates. Therefore, temperature analysis should be made by observing an FEM element which is the nearest of the measured point. Our computed result is 0.16°C for ferromagnetic steel plate and 0.04°C for paramagnetic steel plate. In this case, the rise in temperature of ferromagnetic steel plate can be computed by JRMAG, because the plate is very thin and magnetic flux density in the conductor is over the saturation value which is about 4 kA/m.

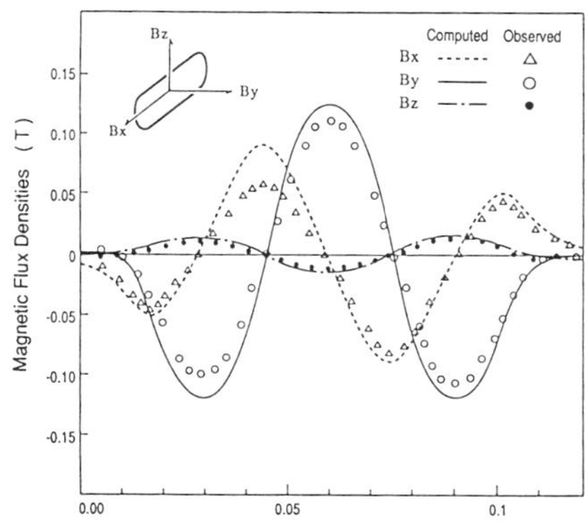

Figure 2 The measured or computed magnetic flux densities

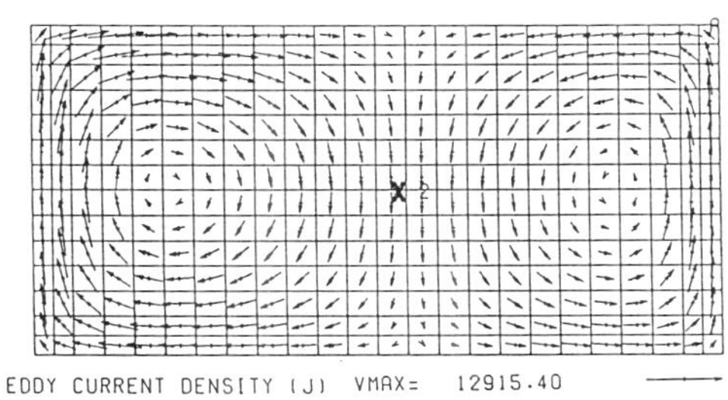

Figure 3 An example of eddy current profile by JRMAG

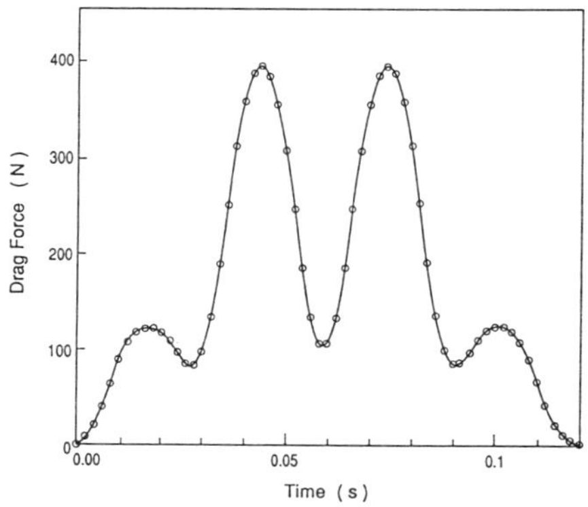

Figure 4 Computed time history of drag force

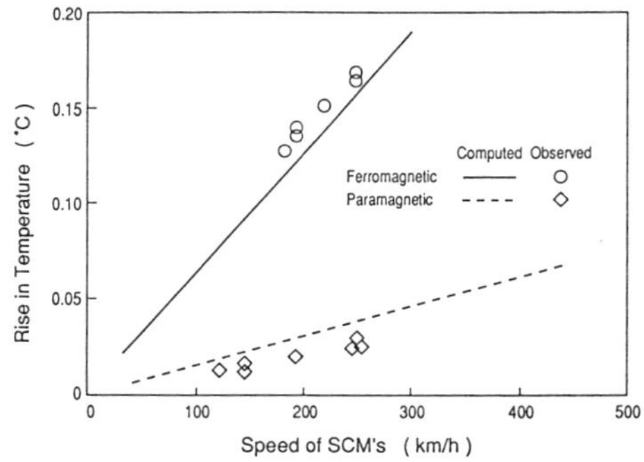

Figure 5 An example of the measurement of temperature rise

4. CONCLUSIONS

As an experimental consequence, characteristic difference between the usual ferromagnetic steel plate and the paramagnetic steel plate has been obtained. For our analysis, a computer code JRMAG has been developed. In the JRMAG, two assumptions are imposed. One is the resistive limit approximation and the other is that structure is built only of the paramagnetic steel. Though our assumptions may look crude at one glance, our computational method is efficient for designing realistic large and complex steel structures.

REFERENCES

1 T.Takahashi and K.Kurita,IEEE Trans. Magn.,24(1988)197.
M.Tsuchimoto, K.Miya, A.Yamashita and M.Hashimoto,IEEE Trans. Magn.,28(1992)1434.
S.Niikura and A.Kameari,IEEE Trans. Magn.,28(1992)1450.
See also the references therein.
2 F.C.Moon, Magneto-solid mechanics, John Wiley & Sons, 1985.
3 T.Fujimoto and S. Fujiwara, Electrodynamic Characteristics of MLU002, Quarterly Report of RTRI, Vol.32, No.2, 1991.

Simulation and Design of Applied Electromagnetic Systems
T. Honma (Editor)

Design of a levitation magnet using Evolution Strategy

H.B.Lee[a], S.Y.Hahn[a], G.S.Cha[b], K.S.Lee[c], P.S.Shin[d], Y.H.Cho[d]

[a] Dept. of Electrical Engr., Seoul National Univ., Seoul 151-742, Korea

[b] Dept. of Electrical Engr., Soonchunhyang Univ., Onyang 336-600, Korea

[c] Dept. of Electrical Engr., Dankook Univ., Seoul 140-714, Korea

[d] Power Apparatus Lab., KERI, Changwon 641-120, Korea

Abstract
This paper deals with the optimal design of an electromagnet for MAGLEV. Effects of eddy current is considered in optimization process. Evolution Strategy is adopted as an optimal design method. Input power and volume of a magnet are chosen as constraints. Levitation forces of the optimized magnet and the actual magnet which is being operated have been compared with. The results shows the optimization of a magnet is more effective at higher speed.

1. INTRODUCTION

The Magnetically Levitated Vehicle(MAGLEV) which is propelled by a Linear Induction Motor(LIM) has to be equipped with magnet and rail for levitation. Generally, the material which is used for the U-shaped magnet and rail has high permeability and conductivity. When the vehicle moves, therefore, eddy current is induced in the rail. Since the eddy current decreases the levitation force of the electromagnet, the effects of the eddy current should be considered in magnet design.

This paper presents the optimal design of an electromagnet with eddy current effects. Evolution strategy is adopted as an optimal design method. Air gap flux by the eddy current is calculated analytically using simplified method. The levitation forces of the optimized magnet and the actual magnet which is being operated have been compared with.

2. OPTIMIZATION ALGORITHM

Evolution strategy proposed by Rechenberg is used as an optimization algorithm[1]. It is based on the probability concept. Evolution strategy is well known as a good measure for the local minimum problem. It also uses objective function only. So it suits well for many cases where the derivatives of objective function with respect to model variables cannot be obtained or is difficult to be obtained.

The (1+1) evolution strategy is used in this paper, where the model variable set of the parent generation and the descendant generation is one[2].

The model variable vector x_c of the descendant generation can be obtained from that of parent generation through the following process.

$$x_{ci} = x_{pi} + \alpha_i \cdot R_i \tag{1}$$

where R_i is a random number which is generated by probability density function with uniform distribution and α_i means step width whose center is x_{pi}.

Next parent generation x_p can be defined by the objective function of parent generation $x_p(F_p)$ and descendant generation $x_c(F_c)$.

$$x_p = \begin{cases} x_c & \text{if } F_c < F_p \\ x_p & \text{if } F_c \geq F_p \end{cases} \tag{2}$$

The step width α is changed during the optimization process. In this paper, step width α is changed by the number of mutation for 10 generations, that is, number of $F_c < F_p$.

$$\alpha = \begin{cases} \alpha \cdot 0.85 & \text{if } \textit{number of mutation} > 10 * N/5 \\ \alpha/0.854 & \textit{otherwise} \end{cases} \tag{3}$$

where N shows the number of design variables.

3. LEVITATION FORCE

Fig.1.a) shows the magnet-rail combination. The exciting current is assumed to be a constant. If the leakage flux outside of the vertical surfaces of the rail is negligible, actual model in Fig.1.a) can be simplified as in Fig.1.b).

In the simplified model, the flux density B consists of the B_0 by exciting current and B_e by the eddy current[3].

$$B(x,z) = B_0 + B_e = B_0 + \sum_{n=1}^{\infty} X_n(x) \cdot \cos \lambda_n z \quad \text{for } 0 < x < L \tag{4}$$

$$B(x,z) = B_e = \sum_{n=1}^{\infty} X_n(x) \cdot \cos \lambda_n z \quad \text{for } x < 0,\ x > L \tag{5}$$

The levitation force by the eddy current in the simplified model is given by

$$F_L = \frac{1}{\mu_0} \int_{-\infty}^{\infty} dx \int_{-a}^{a} dz [B_e(x,z)]^2 \tag{6}$$

$$= F_a^0 \sum_{n=1}^{\infty} c_n^2 f(n) \tag{7}$$

where F_a^0 is the levitation force at standstill.

4. NUMERICAL EXAMPLE

The U-shaped magnet and rail system(Fig.1.a) which is being currently operated in laboratory has been optimized. Nonlinear characteristics of the magnet and rail is considered. Each variable is allowed to be changed with constraints which is given in Table 1.

Table 1
Constraints of each variable

variable	x1[mm]	x2[mm]	x3[mm]	x4[mm]	x5[mm]
upper limit	60	20	10	20	400
lower limit	140	140	60	100	1200

Input power and magnet volume are fixed as constraints. The levitation force at standstill is calculated by FEM, then resultant force with eddy current is obtained by Eq.(7). Initial dimensions before optimization and final dimensions after optimization at several speed are shown in Table 2. Each variable converges well within the upper and lower limit. The length of the vehicle shows the greatest difference between the initial and final value.

Table 2
Dimensions of each variable [mm]

		final value		
variable	initial value	0[km/hr]	50[km/hr]	100[km/hr]
x1	90	77.0	81.2	65.3
x2	90	110.1	68.2	80.9
x3	35	33.5	29.7	28.7
x4	28	33.1	24.1	23.4
x5	716	673.4	1032.2	1090.6

When the speed of a vehicle increases, width of the magnet and rail(x4) becomes narrower to decrease the effect of eddy current. On the contrary, length of the magnet(x5) tends to increase to obey the constraint of constant volume. Fig.2 presents the initial(F_i) and final(F_o) levitation force, where the ratio of two forces(F_o/F_i) increases as the speed becomes higher. It means that optimization of a magnet is more effective at a higher speed.

5. CONCLUSIONS

This paper deals with optimization of a levitation magnet for MAGLEV. Evolution strategy based on the probability concept is chosen as an optimization strategy. Final results show that length of the magnet becomes longer to decrease the effects of eddy current. The effect of optimization becomes more apparent as the speed of vehicle increases.

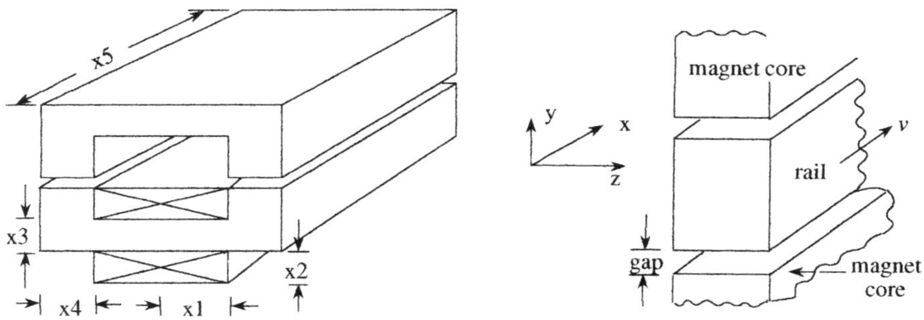

(a) actual model (b) simplified model

Figure 1. Model for analysis

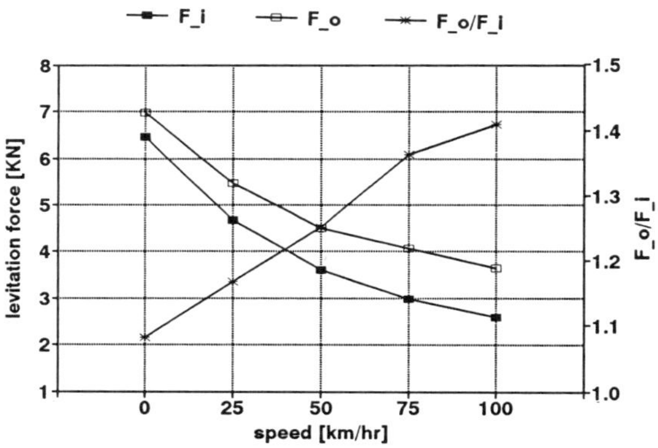

Figure 2. Levitation forces of actual(F_i) and optimized(F_o) magnet

6. REFERENCES

1 I. Rechenberg, Frommann-Holzboog, Stuttgart-Bad Connstatt, 1973.
2 K. Preis and A. Ziegler, COMPEL, 9 (1990) 119
3 S. Yamamura, IEEE Trans. Magn. No. 5 (1975) 1504

Simulation and Design of Applied Electromagnetic Systems
T. Honma (Editor)

Three-Dimensional Design of Hybrid Magnet in Magnetic Levitation System

T.Onuki, Y.Toda, and H.Ogiwara

Department of Electrical Engeneering, Waseda University

3-4-1, Ohkubo Shinjuku-ku, Tokyo 169, JAPAN

Abstract

A magnetic levitation system with both permanent magnets and electromagnets (hybrid magnet) has less powerloss than a conventional attractive-type system. The boundary element method (BEM) is suitable for designing the hybrid magnet in our system because of analyzing very large linear fields. In this paper, we have calculated optimal designs of the hybrid magnet by using both the approximate three-dimensional BEM and the sequential quadratic programming method.

1. Introduction

With the recent advent of high field rare earth magnets, we have utilized this type of magnets in the magnetic levitation system together with ordinal electromagnets. In our levitation system with the hybrid magnets, permanent magnets provide a base value of magnetic motive force (m.m.f) necessary for levitation while electromagnets produce the m.m.f for stabilizing the system. This type of levitation has less powerloss for levitation than a conventional attractive-type with only electromagnets[1].

The boundary element method (BEM) is useful for analizing very large linear fields. It reduces the domain problem to boundary problem, and is particularly suitable for analyzing the infinite space. Furthermore, the required values inside the domain are calculated by using only theoretical expressions.

In our magnetic levitation system, since the magnetic flux density is well below the saturated value because of large air gap length, we can assume that the permeability of iron is constant. In addition, the magnetic flux density $\boldsymbol{B}$ appears in the Maxwell stress tensor $\boldsymbol{T}$ explicitly, from which the attractive force and the guidance force are derived[2][3]. Thus, we adopt the three-dimensional BEM using the tangential components of the magnetic intensity $\boldsymbol{H}$, that is $\boldsymbol{B}/\mu$.

In the process of an optimizing approach, we need huge time to solve simultaneous equations by BEM. Therefore, in this paper, we propose quasi-optimizing approaches, which consists of both BEM and the sequential quadratic programming method (SQP). As examples, we apply this approach to the designs of optimized hybrid magnet, which has the maximum guidance force without changing the stationary attractive force.

2. EXPERIMENTAL MODEL

Fig.1.(a) shows the cross section of our experimental system, and its specifications are denoted in Table I. The model, of which the stationary gap length is 8[mm], is composed of eight magnets, four gap sensors, a digital controller, power amplifiers, batteries, and the secondary conductor for a linear induction motor. The levitation magnet is a hybrid magnet composed of an electromagnet and three permanent magnets as shown in Fig.1.(b). The permanent magnets are arranged on the surface of both the center pole and two edge poles to efficiently provide the flux for the attractive force. These eight hybrid magnets are arranged on four corners of the vehicle. The stagger arrangement on each corner provides the lateral guidance force in addition to the attractive force.

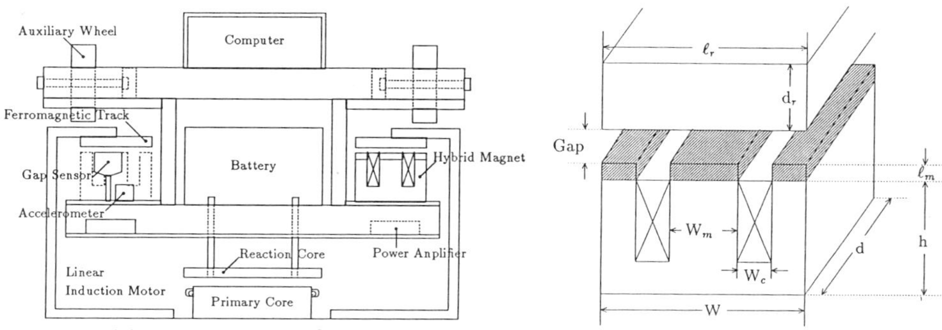

(a) Cross section of system (b) Hybrid magnet

Fig.1 Levitation System investigated

Table.I
Specifications of laboratory model

Vehicle			*Hybrid magnet*		
dimension		8.00×10^{-1} m × $4.60\times10^{-1} m$	dimension	h	5.08×10^{-2} m
weight	M	53.0 kg		W	7.62×10^{-2} m
moment of inertia on	J_θ	0.939 kgm^2		d	6.50×10^{-2} m
moment of inertia on	J_ϕ	2.550 kgm^2		W_c	1.27×10^{-2} m
moment of inertia on	J_ψ	3.030 kgm^2		W_m	2.54×10^{-2} m
Track			hight of PM	ℓ_m	4.40×10^{-3} m
dimension	ℓ_r	7.50×10^{-2} m	No. of coil turns	N	430 T
	d_r	2.50×10^{-2} m	coil's resistance	R	7.26Ω

3. FORMULATION

In the air region, the boundary integral equation derived from the vector Green relationship is

$$C_{p_{air}}\boldsymbol{H}_{p_{air}} = \int_\Omega \boldsymbol{J}\times\nabla\frac{1}{r}d\Omega + \oint_\Gamma \left(\boldsymbol{H}_{q_{air}}\times\boldsymbol{n}_{q_{air}}\right)\times\nabla\frac{1}{r}d\Gamma - \oint_\Gamma \left(\boldsymbol{H}_{q_{air}}\cdot\boldsymbol{n}_{q_{air}}\right)\nabla\frac{1}{r}d\Gamma \qquad (1).$$

Similarly, in the iron region, we can obtain

$$C_{p_{Fe}}\boldsymbol{H}_{p_{Fe}} = \quad 0 \quad + \oint_\Gamma \left(\boldsymbol{H}_{q_{Fe}}\times\boldsymbol{n}_{q_{Fe}}\right)\times\nabla\frac{1}{r}d\Gamma - \oint_\Gamma \left(\boldsymbol{H}_{q_{Fe}}\cdot\boldsymbol{n}_{q_{Fe}}\right)\nabla\frac{1}{r}d\Gamma \qquad (2).$$

The boundary conditions are

$$\left\{ \begin{array}{lcl} \mu_{\mathrm{air}} \boldsymbol{H}_{\mathrm{air}} \cdot \boldsymbol{n}_{\mathrm{air}} & = & \mu_{\mathrm{Fe}} \boldsymbol{H}_{\mathrm{Fe}} \cdot \boldsymbol{n}_{\mathrm{Fe}} \\ \boldsymbol{H}_{\mathrm{air}} \times \boldsymbol{n}_{\mathrm{air}} & = & \boldsymbol{H}_{\mathrm{Fe}} \times \boldsymbol{n}_{\mathrm{Fe}} \end{array} \right. \qquad (3).$$

Utilizing Eq.(3),$\boldsymbol{n}_{air} = -\boldsymbol{n}_{Fe}$ and $\mu_{Fe} = \mu_s \mu_{air}$, we replace $\boldsymbol{H}_{Fe}$ with $\boldsymbol{H}_{air}$ in Eq.(2), then we obtain

$$C_p \boldsymbol{H}_p \times \boldsymbol{n}_p = \frac{1}{1+\mu_s} \int \boldsymbol{J} \times \nabla \frac{1}{r} d\Omega \times \boldsymbol{n}_p + \frac{1-\mu_s}{1+\mu_s} \oint (\boldsymbol{H}_q \times \boldsymbol{n}_q) \times \nabla \frac{1}{r} d\Gamma \times \boldsymbol{n}_p \qquad (4).$$

At inner points in the air region, we have

$$4\pi \boldsymbol{H} = \int \boldsymbol{J} \times \nabla \frac{1}{r} d\Omega + (1-\mu_s) \oint (\boldsymbol{H}_q \times \boldsymbol{n}_q) \times \nabla \frac{1}{r} d\Gamma \qquad (5).$$

By using the Maxwell stress tensor $\boldsymbol{T}$, the attractive force and the lateral guidance force are calculated such as

$$\boldsymbol{F} = \int \nabla \cdot \boldsymbol{T} dv = \oint \boldsymbol{T} \cdot \boldsymbol{n} dS \qquad (6).$$

4. OPTIMIZING METHOD

We are able to calculate the optimum design by using both three-dimensional BEM and SQP method. However, we have to divide the boundary into small elements to obtain accurate characteristics, but it takes huge time to solve a large number of simultaneous equations. Therefore, we apply the following quasi-optimizing approaches to two examples, which enables the maximum lateral guidance force without affecting the attractive force.

1. Permanent magnet height optimization
   (a) We assume the shape of iron body is unchanged.
   (b) The change of the height of permanent magnets replaces with the change of equivalent surface current.
   (c) Subjected constant volume of permanent magnets.
2. Track Shape optimization
   (a) The calculated boundary values on some elements remain almost the same values.
   (b) We fix the calculated boundary values on these elements.
   (c) We solve the simultaneous equations on only others.

As above examples, we attempt to optimize the height of permanent magnets, and the shape of ferromagnetic track, which have the maximum guidance force without changing the stationary attractive force. In case of the permanent magnet heights optimization, the characteristic is shown in Fig.2, and in case of the ferromagnetic track shape one is in Fig.3.

In both cases, the calculation of maximum guidance force is achieved by optimization.

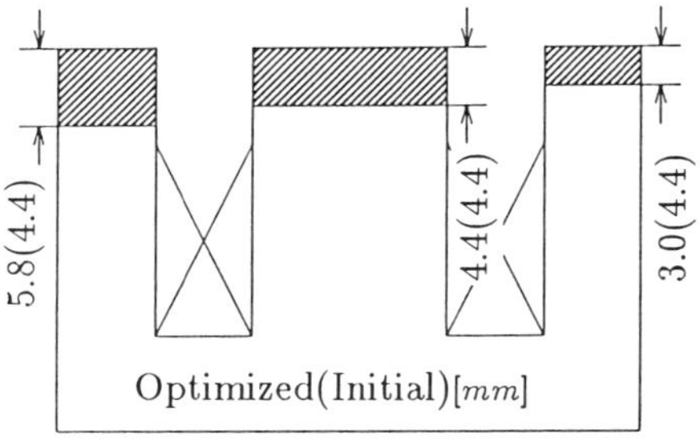

(a) Calculation results

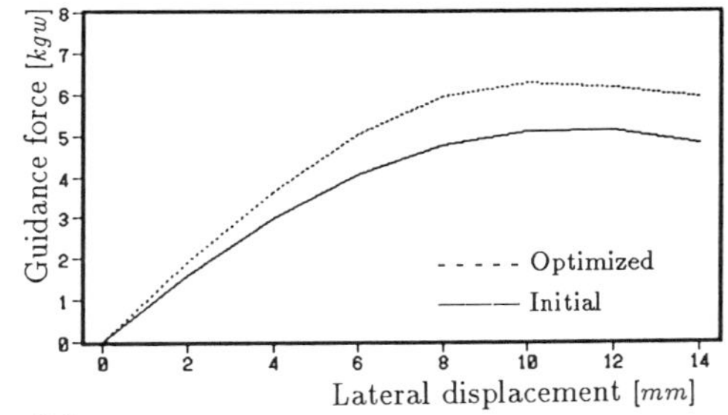

(b) Lateral displacement - Guidance force

Fig.2 PM height optimization

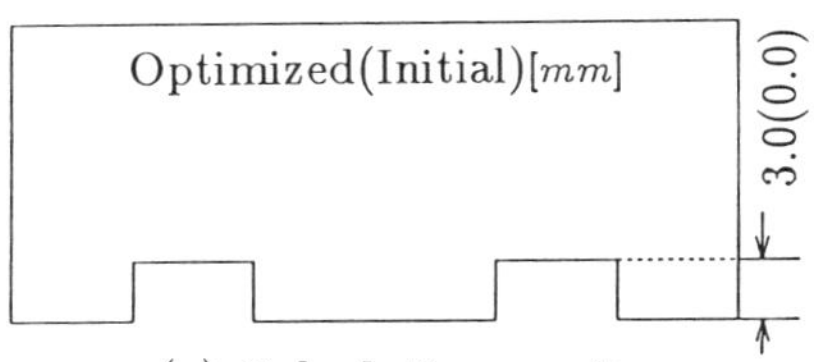

(a) Calculation results

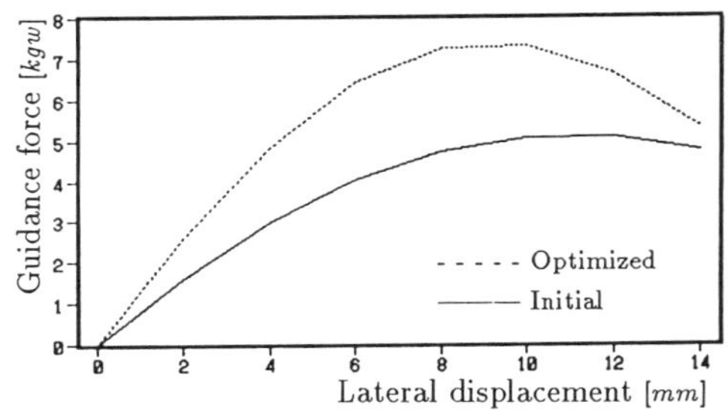

(b) Lateral displacement - Guidance force

Fig.3 Track shape optimization

5. CONCLUSIONS

In this paper, we proposed optimizing approaches for a hybrid magnet, deducing the following conclusions:

1. We analyzed a hybrid magnet by using the three-dimensional boundary element method with the tangential component of magnetic field intensity $\boldsymbol{H}$, because of analyzing domain linearity, showing the validity of the analysis for levitation magnets.
2. We proposed optimizing approaches of designing the hybrid magnet, which had the maximum lateral guidance force assuming that the stationary attractive force is constant, by utilizing both the BEM and the sequential quadratic programming method.

6. REFERENCES

1. M.Morishita et al., "A new Maglev system for magnetically levitated carrier system", *Proc. Int. Conf. of Maglev Transport '86, Vancouver* (1986) pp. 199-204
2. T.Onuki et al., "DESIGN OF THE HYBRID MAGNET IN THE MAGNETIC LEVITATION SYSTEM USING THE BOUNDARY ELEMENT METHOD", *International Journal of Applied Electromagnetics in Materials* (1990) pp.263-269
3. T.Onuki et al., "Optimal Design of Hybrid Magnet in Maglev System with both Permanent and Electro Magnets", *IEEE. Trans. Mag* March,1993 (To be published)

Simulation and Design of Applied Electromagnetic Systems
T. Honma (Editor)

Multi-Objective Optimal Design for Electromagnet with Relative Movement

J. Lee, S.B. Yoon, H.R. Kim, C.J. Ree and D.H. Im

Department of Electrical Engineering, HanYang University, Haeng Dang Dong 17, Sung Dong Gu, Seoul, Korea

Abstract

An efficient optimal design for electromagnet with relative movement is presented. In this problem, there are two conflict design goals which are to minimize drag force and to maximize lift force for MAGLEV. Characteristic equations for lift and drag force are derived from field theory and equivalent magnetic circuit. The trade-off between drag and lift force is performed by Compromised Programming Method. From the results, it shows that the trade-off design between drag and lift force is possible. It also allows flexibility to meet designer's concepts.

1.Introduction

In the field of electrical machine design, the use of optimal design methods, which can be used directly in the stage of design flow, shows that it is very efficient to reduce cost for design. Coincidence with this requirement, optimal design technique for electromagnet has been developed[1]. There are some works [1],[2] for electromagnet at standstill. Since almost electromagnet are not at standstill, it should be considered the dynamic characteristics on the system performance.

An efficient optimal design method for electromagnet with relative movement is presented in this paper. To analyze the dynamic characteristics of drag and lift forces, the characteristic equations are derived from the field theory and magnetic circuit. For the trade-off between drag and lift forces, Compromised Programming Method is used.

From the results of this study, it is shown that the trade-off design between drag and lift forces allows flexibility to meet design needs for electromagnet.

2. Design variables and Characteristic Equations

The considering model for optimal design of elctromanet is shown in Fig. 1. There are relative movement between electromagnet and rail.

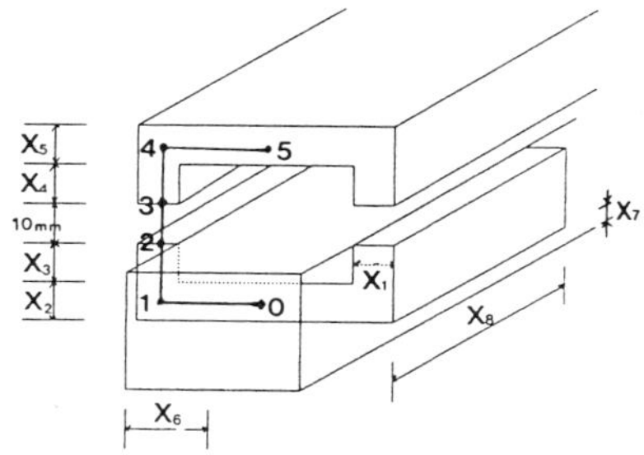

Fig. 1 Considering Model for Optimal Design

Yamamula[3] proposed the analytical method for dynamic characteristics of electromagnet. The equations are eq.(1)– eq.(3)

$$F_a = F_{ao} \sum_{n=1}^{\infty} C_n^2 f(n) \tag{1}$$

$$F_b = F_{ao} \frac{g}{L} \sum_{n=1}^{\infty} [C_n\{ \alpha_n'(e^{\beta nX8}-1) + \beta_n'(e^{-\alpha nX8}-1)\}] \tag{2}$$

where the f(n) can be represented by eq.(3)

$$f(n) = \frac{1}{2X_8} [X_8 + \frac{\beta_n'}{\alpha_n}(2+\beta_n')(1-e^{-\alpha nX8}) + \frac{\alpha_n'}{en}(2-\alpha_n')(1-e^{\beta nX8}) + \frac{2\alpha_n'\beta_n}{\alpha_n+\beta_n}(e^{-\alpha nX8}-e^{\beta nX8})] \tag{3}$$

where the parameters are same as in reference[3]

In eq.(3) it is assumed that the air gap flux density is constant. In this study equivalent magnetic circuit is introduced and the flux density can be calculated. The equivalent magnet circuit is shown in Fig.2

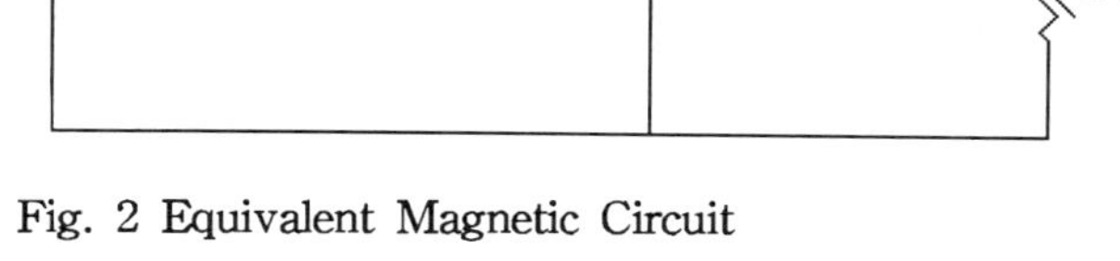

Fig. 2 Equivalent Magnetic Circuit

The flux density can be expressed by eq.(4-a)

$$B_g = \frac{\mathfrak{F}}{\mathcal{R}_t * X1 * X8} \tag{4-a}$$

where $\mathfrak{F}$ is magnetic motive force and $\mathcal{R}_t$ is represented in eq.(4-b)

$$\mathcal{R}_t = [\ (R1+R2+RL)(R3+R4+Rg+RL)-RL^2\ ]^2/RL^2 \tag{4-b}$$

From the derived magnet flux density, characteristic equations for the drag and lift force can be modified as eq.(5), (6).

$$\mathbf{F_a} = \frac{X1*X8\ B_g^2}{\mu_o} * \sum_{n=1}^{\infty} C_n^2\ f(n) \tag{5}$$

$$F_b = \frac{X1*X8\ B_g^2}{\mu_o} * \frac{g}{L} * \sum_{n=1}^{\infty} [C_n\{\alpha_n'(e^{\beta nX8}-1) + \beta_n'(e^{-\alpha nX8}-1)\}] \tag{6}$$

3. Multi-Objective Optimization Technique.

Many techniques for multi-objective optimization are proposed. The compromised Programming Method is used in this study. Here, two objectives between the drag and lift forces are formulated as eq.(7)

$$FSUM(X) = \left| W1^2 * \left[\frac{eq.(5) - FOP1}{FMAX1 - FOP1}\right]^2 - W2^2 * \left[\frac{eq.(6) - FOP2}{FMAX2 - FOP2}\right]^2 \right|^{1/2} \tag{7}$$

where FOP1, FOP2 are desire values , FMAX1, FMAX2 are worst values and W1,W2 are weighting factors.

4. Results and Discussion

The results of optimal design for given model are shown in table 1 with weight factors. As shown in Table 1, the trade-off design conformed to designer's concept is possible with weight factors.

Table 1
Lift and Drag Forces of Optimal Design Model with Weight Factors

Velocity	W1=0.3 , W2=0.7		W1=0.5 , W2=0.5		W1=0.7 , W2=0.3	
	Lift Force	Drag Force	Lift Force	Drag Force	Lift Force	Drag Force
20	3488	182	4576	251	7512	314
40	2828	125	3586	191	5703	261
60	2296	102	2925	145	4202	201
80	1960	92	2655	141	4932	182
100	1612	66	2013	151	4063	181

The optimal design values are shown in Table 2 when the weight factors are 0.5

Table 2
The Optimal Design Values vs. velocity (weight factor : 0.5)

velocity	X1	X2	X3	X4	X5	X6	X7	X8
20	0.154E-1	0.288E-1	0.656E-1	0.69E-1	0.349E-1	0.909E-1	0.602E-1	0.428
40	0.128E-1	0.305E-1	0.727E-1	0.748E-1	0.349E-1	0.907E-1	0.578E-1	0.469
60	0.105E-1	0.297E-1	0.722E-1	0.744E-1	0.349E-1	0.899E-1	0.559E-1	0.500
80	0.107E-1	0.304E-1	0.727E-1	0.751E-1	0.348E-1	0.899E-1	0.557E-1	0.498
100	0.323E-1	0.304E-1	0.727E-1	0.778E-1	0.349E-1	0.927E-1	0.572E-1	0.404

The Fig. 3 shows the drag force vs. iteration and the Fig. 4 shows the lift force vs. iteration.

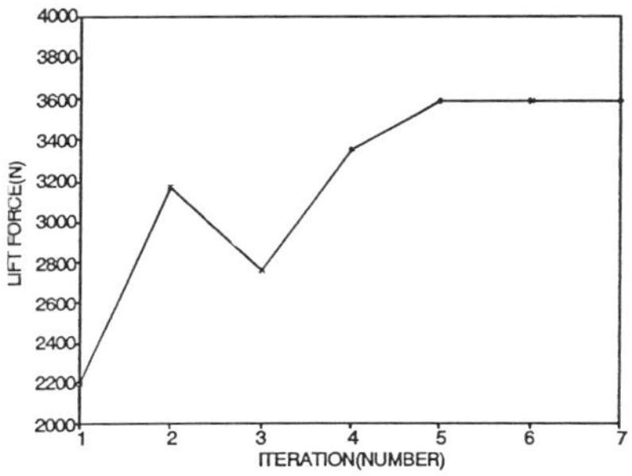

Fig. 3. Lift Force vs. iteration

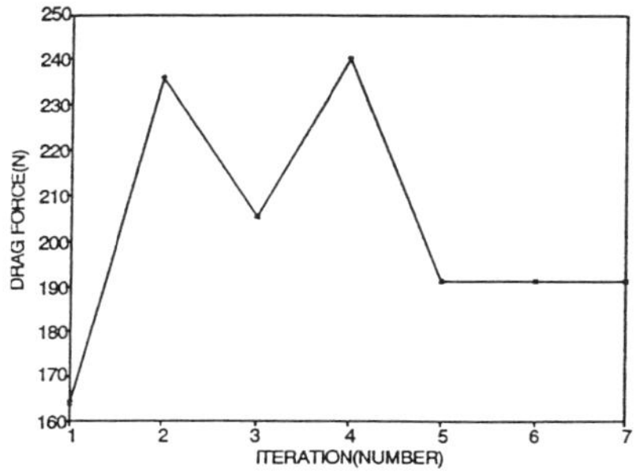

Fig. 4 Drag Force vs. iteration

5. Some Concluding Remarks

The optimal design for electromagnet with relative movement is proposed. For the dynamic analysis, characteristic equations are derived by field theory and equivalent magnetic circuit. For the trade-off design between drag and lift forces, Compromised Programming Method is used. It is shown that the trade-off designs between drag and lift force are possible and that the proposed design method allows flexibility to meet the designer's concept. The present optimization method is also feasible to meet the optimal design of electromagnet and other electric machines.

6.Reference

[1] J. Lee, Y.G. Kim, C.G. Lee, S.M. Jang and D.H. Im, "A Study on the 3D Optimal Design of Electromagnet", Journal of KIEE, 1993. 1.

[2] C.J. Wu, F.C. Lee and R.K. Davis "Minimum Weight EI Core and Pot Core Inductor and Transformer Design", IEEE TRANS. on Mag, Vol. 16, No. 5, Sep. 1980.

[3] S.Yamamula and T.Ito, "Analysis of Speed Characteristics of Attracting Magnet for Magnetic Levitation Vehicles", IEEE TRANS. on Mag., Vol. 11, No. 5, Sep. 1975.

Linear Drive

Simulation and Design of Applied Electromagnetic Systems
T. Honma (Editor)
1994 Elsevier Science B.V.

Design and Testing of a 100 W Class Annular Linear Alternator

Y. Nakamura, S. Ogiwara, and K. Eguchi

Space Technology Research Group, National Aerospace Laboratory

7-44-1 Jindaiji-Higashi, Chofu, Tokyo 182, Japan

Abstarct

The linear alternator driven by a free power piston is an important device in a space Stirling engine. An annular permanent-magnet linear alternator of 100 W class electrical output was designed, fabricated and tested with a future target toward a large-size power model.

In the alternator testing the magnet plunger was directly connected to a motor-driven moving mechanism instead of a power piston. The electrical output of 104 W was obtained for a design operating condition with a 22 mm stroke at 30 Hz. The simple analysis for the alternator performance prediction explained well the experimental characteristics of output voltage and power depending on reciprocating frequency and load damping rate. A suitable capacitor arranged in series to the load resistance attained the output power up to 140 W.

1. INTRODUCTION

Stirling engine is attractive as a space power converter to provide high efficiency even in a comparatively small size. The working gas in the engine varies its pressure by moving it between the high and low temperature operating spaces through the displacer motion, and pushes a power piston oscillatorily against a gas spring. It is desired for the free power piston to directly drive a linear alternator without converting its reciprocating motion into rotating since the conversion is accompanied with mechanical friction to produce erosion and eventually bring about some trouble on the engine endurance.

A Stirling engine model of 125 W has been studied with two-dimensional linear alternator [1]. When a linear alternator is designed to produce a big power, then it requires so large cross-sectional area of moving permanent magnets. The magnet plunger has to be reciprocated with a minimized clearance against the stator. It is difficult for two-dimensional configuration to meet the requirement within the limited stroke. Annular structure will, on the other hand, give the stable configuration for machining, assembling and balancing. We designed a 100 W class annular alternator model. The performance tests on the fabricated model were conducted with a motor-driven piston rod mechanism in order to evaluate its characteristics separately from the Stirling engine and to understand the key design parameters.

2. LINEAR ALTERNATOR

The structure of present baseline linear alternator is shown in Fig. 1. The alternator consists of a plunger with three-staged permanent magnet rings, an inner stator, an outer stator with two cylindrical poles and a power coil. The SmCo magnet rings are radially magnetized and arranged in alternating polarity along their axis. The stator yokes are made of cobalt iron alloy with high saturation flux capability. The radial air gaps between the magnets and each stator, were designed to be 0.35 mm. The main dimensions of the designed alternator are listed on Table 1.

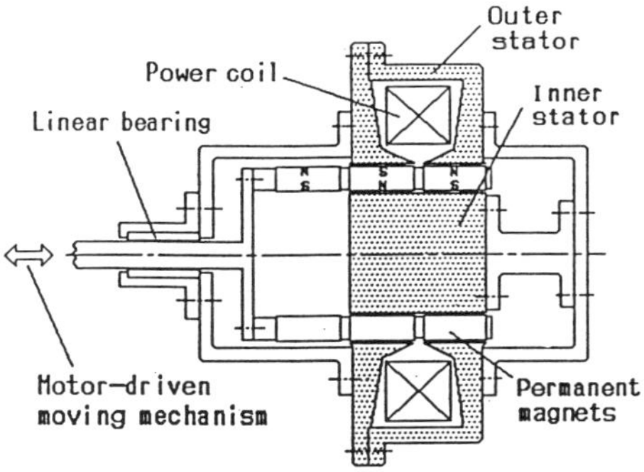

Fig. 1 Structure of the present alternator.

Table 1 Specification of 100 W Alternator Model

Magnet Plunger:	
Magnet Material / Stages	SmCo / 3
Diameter, Inner/Outer (mm)	40 / 56
Stage, Width/Separation (mm)	20 / 2
Stroke (mm)	22
Gap Flux Density (T)	0.38
Outer Stator:	
Diameter, Inner/Outer (mm)	120 / 56.7
Pole, Width/Separation (mm)	20 / 2
Inner Stator:	
Diameter, Inner/Outer (mm)	0 / 39.3
Axial Length (mm)	42
Power Coil:	
Diameter, Inner/Outer (mm)	60 / 108
Cross Sectional Area (cm^2)	6
Winding, Diameter (mm) x Turn	1 x 488
Resistance(Ω) / Inductance(mH)	3.3 / 60

The reciprocating movement of magnets past the poles creates alternating flux through the magnetic circuits surrounding the power coil, thereby generates coil current. The flux changes almost linearly depending on the axial movement of magnet plunger within the stroke. Therefore, if the movement is sinusoidal with the frequency of f, the magnetic flux, Φ, also varies as a sine function of time for the nominal stroke.

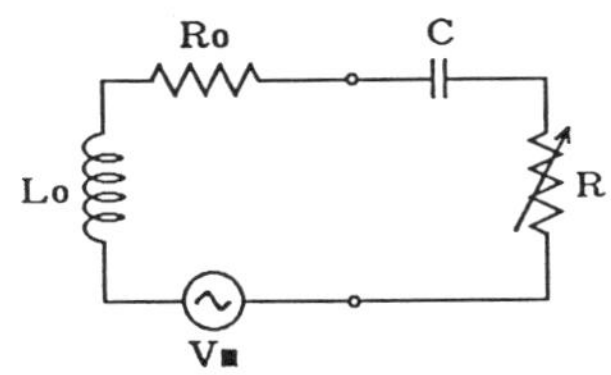

Fig. 2 Equivalent electrical circuit.

(1) $\Phi = B_g \pi D X_s \sin\omega t \quad (\omega = 2\pi f)$

where B_g is the magnetic flux density at the air gap, D is an inner diameter of the magnet and X_s is the stroke. The induced voltage through the power coil with non load is given by

(2) $\dot{V}_m = - Nd\Phi/dt = - \sqrt{2}\, V_{me} \cos\omega t$ (3) $V_{me} = \sqrt{2}\, \pi^2 f N D X_s B_g$

where N is the number of coil winding, and V_{me} is the effective induced voltage.

When the external load is a pure resistance, R, with a tuning capacitor, C in series, the equivalent electric circuit may be shown in Fig. 2, where the resistance and inductance of the power coil are R_o and L_o respectively. The alternate current, $\dot{I}$, and voltage $\dot{V}_m$ of the electric circuits may be expressed in the following relationship.

(4) $\dot{V}_m = (R + R_o + j\omega L_o - 1/\omega C)\, \dot{I}$

The voltage and power output of the load can be written below.

(5) $V_e = V_{me} R / [(R + R_o)^2 + (\omega L_o - 1/\omega C)^2]^{0.5}$

(6) $P_o = V_{me}^2 R/[(R + R_o)^2 + (\omega L_o - 1/\omega C)^2]$

3. EXPERIMENTAL RESULTS AND DISCUSSION

In the experiment, the performance characteristics were obtained by changing the stroke, load resistance and frequency step by step. We used motor driven piston rod mechanism which can change reciprocating frequency at a selected stroke. To make the characteristics clear, we used a pure resistance without capacitor as an external load in most of the present operations. The output voltage increases almost linearly with increases of the frequency as well as the stroke. The output power increases in proportion to the square of the voltage for a constant load resistance as shown in Fig. 3. These variations should be due to the increase of induced voltage expressed in the equations (3) to (6). A higher load resistance makes the voltage higher but the current lower. The output power as a product of these two is maximum at about R = 10 Ω in the present case. The maximum output power at the frequency of 30 Hz was 104 W for the stroke of 22 mm and 113 W for 24 mm stroke.

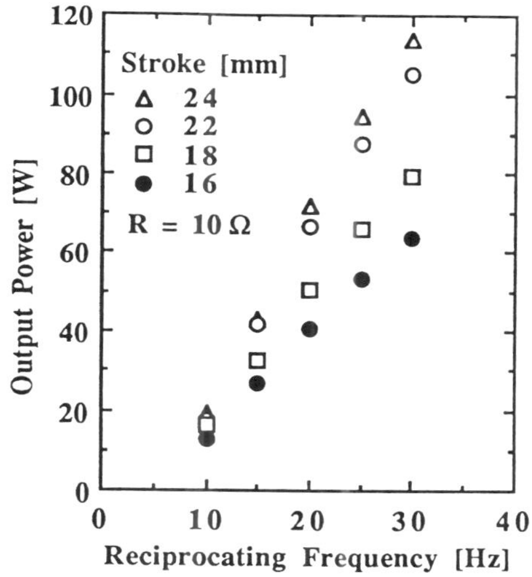

Fig. 3 Output power for different strokes.

Comparison with analytical data

Let us compare the experimental output data with those derived from the previous equations eliminating the capacity term and substituting the physical properties of the present alternator as listed in Table 1. V_{me} at 30 Hz is assumed to be 62 V corresponding to

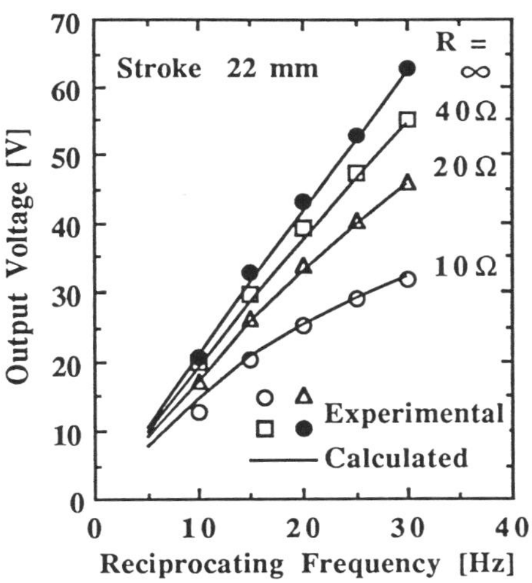

Fig. 4 Output voltage for different resistance.

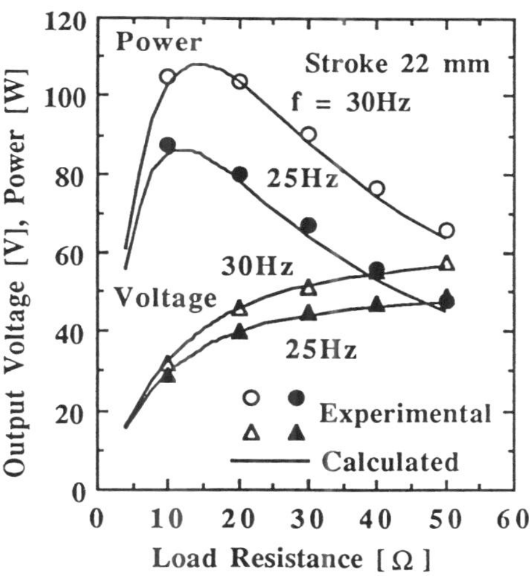

Fig. 5 Output power for different frequencies.

the air gap magnetic flux density of 0.38 T. The gap flux density was estimated to be a higher value of about 0.42 T from the simple permeance method assuming two-dimensional configuration and catalogue demagnetization characteristics on the used permanent magnet. This value may be reduced by some factors. For example, the magnets consist of many elements with different residual magnetic flux density. For the reason, we adopted the above value to fit the experimental induced voltage in the non-load operation. Furthermore, we had to consider the effect of induced current on the magnetization of magnets. The magnets between stator yokes suffer from demagnetization caused by the induced magnetic field. This magnetic field decrement and corresponding variation of induced voltage may be expressed in the following forms,

(7) $\Delta B_g = k_d N I$ (8) $\Delta V = - L_d \, dI/dt$

The equation (8) suggests that this effect is equivalent to the increase of coil inductance from L_o to (L_o+L_d). We calculated the analytical data of the output voltage and power using the above-mentioned values and L_d = 15 mH which is chosen to fit experimental value.

The output voltage vs frequency plots of the operations are shown in Fig. 4 in comparison with the calculations. The output voltage and power are also plotted as a function of the load resistance in Fig. 5 for the frequencies of 25 and 30 Hz. The agreements between the experiment and calculations from Eqs. (5) and (6) are observed in these characteristic curves in all the operated ranges of load resistance and frequency.

Effect of tuning capacity

In the case having tuning capacitor in series to the load resistor, the output power should be much increased by adjusting the capacity to reduce or eliminate the phase difference between the voltage and current. The output power increased with an increase of the capacity as shown in Fig. 6 and reached about 140 W with C = 1,330 μF. This indicates the importance of the tuning capacitor. The power variation depending on the capacitor, however, is different from the expected one from Eq. (6) with the capacity term. This might be caused by the magnet demagnetization due to much increased induced current. We must pay much attention for the capacity selection if the load more complex. Fu, et al. described about the selection criteria of tuning capacitor for the engine/alternator system stability, prevention of magnet demagnetization and power factor correction [2]. These problems should be considered in details on the present device.

4. CONCLUSION

The following findings are concluded from the present study on an annular permanent-magnet linear alternator.

(1) The operating characteristics are understood well with analytical relations of key physical parameters. This will be useful for future scale-up alternator design.

(2) Our research alternator model attained the maximum power of 104 W at the required design point. The addition of tuning capacity also contributed significantly to the increasing of output power up to 144 W.

The precise estimation for the magnetic parameters will be required in the future study; for example the airgap flux density in a complex configuration, and demagnetization of magnet by the induced current, etc. Furhtermore the design considerations available to date have been reflected on an annular alternator configuration for the 300 W Stirling engine.

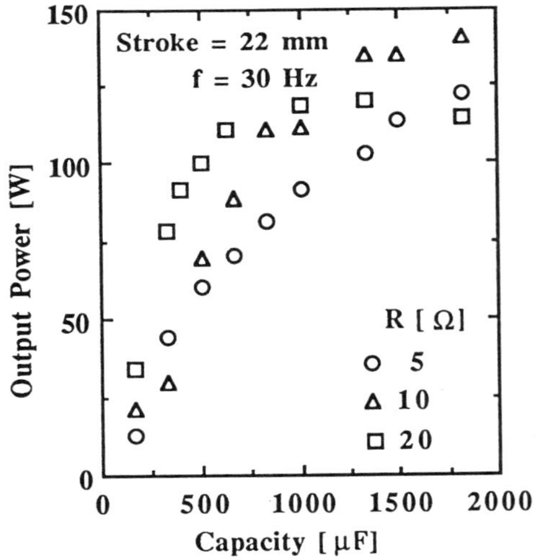

Fig. 6 Tuning capacity effect on power.

5. REFERENCES

1 K. Eguchi, J. Space Power Vol.9 Nos.2/3 (1990) 158.

2. Z. X. Fu, S. A. Maser, and M. Rosswurm, Proc. 27th IECEC No.929262 (1992).

Simulation and Design of Applied Electromagnetic Systems
T. Honma (Editor)

Analysis of magnetic field and dynamic responses of linear dc motor for pen recorder

H. Wakiwaka[a], T. Ninomiya[b], J. Oda[c], T. Morimura[c] and H. Yamada[a]

[a]Faculty of Engineering, Shinshu University, 500 Wakasato, Nagano 380, Japan

[b]Graduate Course, Shinshu University, 500 Wakasato, Nagano 380, Japan

[c]Ohkura Electric Co., Ltd., 2-9-20 Shirako, Wako 351-01, Japan

Abstract

Pen recorders with Linear DC motor (LDM) have already been used in practical applications[1)]. However, an optimum design has not yet been established, considering the dynamic response characteristics of the positioning servomechanism and position sensor, within the unit's limited dimensions.

This paper describes the items listed below. These provide a guideline for optimum design by means of characteristic analysis of pen recorder units, using thin, long-stroke LDM:

(1) Magnetic analysis and position dependency of flux density in gaps of LDM
(2) Position and frequency dependency of time constants
(3) Frequency responses of pen recorder units
(4) Optimum design guideline

1. MAGNETIC ANALYSIS OF LDM FOR PEN RECORDER

LDM thrust F is almost equal to electrodynamic force F , and hence can be obtained from the following formula[2)]:

$$F_e = K_f I \qquad [\mathrm{N}] \tag{1}$$

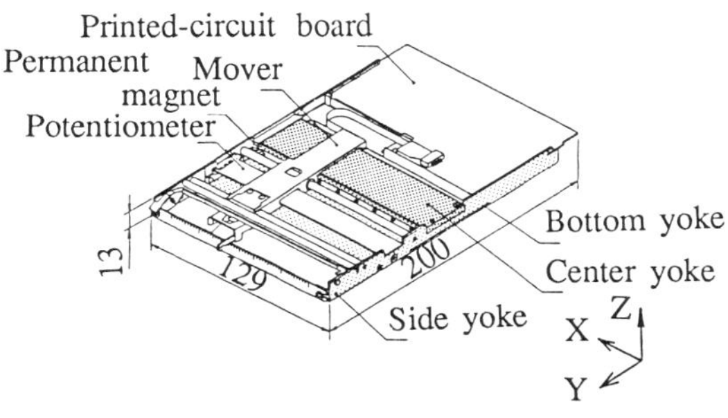

Figure 1. A pen recorder unit with an LDM (units in millimeters).

where $K_f = N\,l\,B$ is thrust constant [N/A], N is number of turns [count], l is length of coil affected by flux density [m], B is flux density acting on the coil [T], I is current in the coil [A].

In this LDM, the flux density acting on the coil is equal to the flux density in the gap. The LDM thrust is proportional to N, l, B and I. N l and I are known as design values. However, B must be determined by design.

Figure 1. shows a pen recorder unit with an LDM. The LDM consists of a mover, a permanent

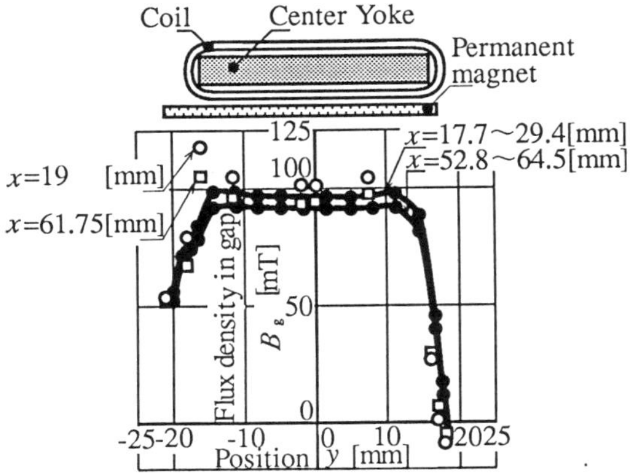

Figure 2. The flux density distribution in the gap in the −Z direction
(o , □ :measured values, —●— :calculated values).

magnet (NdFeB), a center yoke (SUYB-1), a bottom yoke (SPCC-SD), side yokes (SPCC-SD), and a sub-yoke (SPCC-SD). The finite element method (FEM) is used to analyze the three-dimensional magnetic fields of the magnetic circuit comprised of the permanent magnet and yokes.

Figure 2. shows the flux density distribution in the gap in the -Z direction. The actual values are measured by using Hall elements in the LDM.

The flux density acting on the coil of the LDM fluctuates between 90.8 and 98.9 [mT] (8.92%) from the center to the end of the stroke.

2. DYNAMIC CHARACTERISTIC ANALYSIS OF PEN RECORDER WITH LDM

2.1. Time Constants of LDM

The basic parameters used to indicate the LDM response characteristics are electrical time constant T_e and mechanical time constant T_m. They are given by the following formula[2]:

$$T_e = \frac{L}{R} \qquad [s] \tag{2}$$

$$T_m = \frac{(M_c + M_L)R}{(N\,l\,B)^2} \qquad [s] \tag{3}$$

where L is inductance of coil [H], R is resistance of coil [Ω], $M_c + M_L = 0.02$ [kg], M_c is mass of coil [kg], M_L is mass of load [kg].

L and R in the formulas (2) and (3) are found by observing the transfer function of the current for the coil voltage. B uses the result of the magnetic analysis.

Figure 3. shows the positions and the frequency dependency of the time constants of the LDM. The time constants depend on the frequency or position. In the figure, the mechanical time constant changes between 14 and 15.1% from $x = 17.5$ [mm] to $x = 53.5$ [mm].

2.2. Frequency Responses of Pen Recorder Servomechanism

Figure 4. shows a block diagram of the pen recorder servomechanism. Open - loop transfer function G_1 is given by the following formula, according to the figure:

$$G_1 = \frac{K\,K_A\,e^{-st_L}}{s\,N\,l\,B\,(s^2 T_e T_m + s T_m + 1)} \tag{4}$$

where K_A is coefficient of amplifier, K is sensitivity of potentiometer [V/m], t_L is delay of potentiometer and time lost due to free play [s].

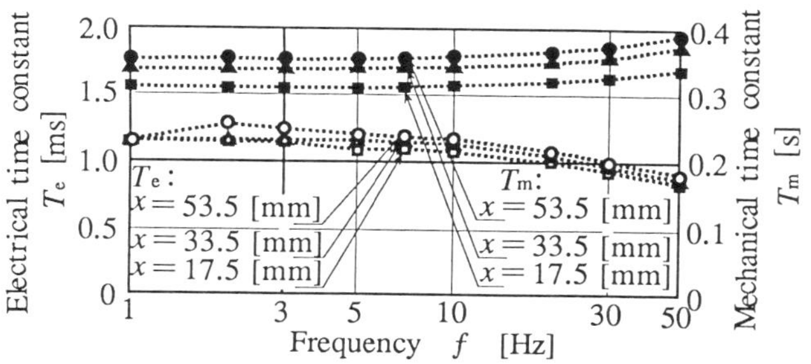

Figure 3. The positions and the frequency dependency of the time constants of the LDM.

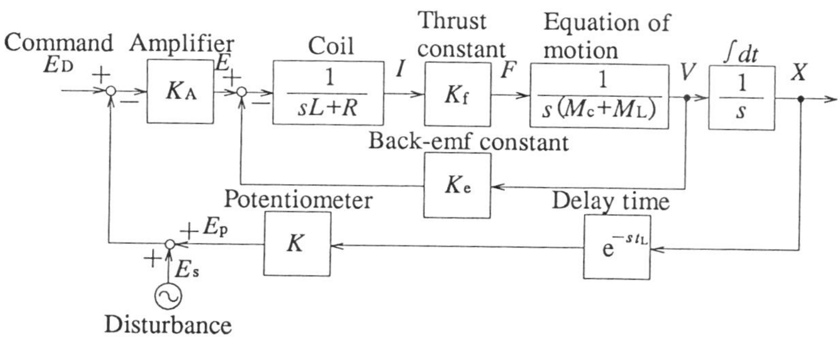

Figure 4. A block diagram of the pen recorder servomechanism.

Table 1 lists the phase margins and gain margins of the pen recorder. According to the table, a stable response is achieved at all positions. Stability depends upon the position.

Table 1
Lists the phase margins ϕ_m and gain margins g_m of the pen recorder

Measuring point x [mm]	Phase margin ϕ_m [°]	Gain clossover frequency f_{cg} [Hz]	Gain margin g_m [dB]	Phase clossover frequency f_{cp} [Hz]
17.5	55.0	3.6	15.3	8.5
33.5	66.3	3.2	18.3	9.5
53.5	85.3	2.4	20.0	9.1
73.5	71.1	3.1	20.0	9.1
93.5	58.7	3.4	15.5	8.2
107.5	66.3	3.4	14.8	8.9

According to Figure 4, the transfer function G_{lc} of position signal E_p for a disturbance signal E_s is found using the following formula:

$$G_{lc} = \frac{G_l}{1 + G_l} \qquad (5)$$

Figure 5. is Nyquist diagrams for an E_p transfer function for E_s in the pen recorder. The figure is obtained from formula (5), using the results of open - loop transfer functions. The frequency has resonant characteristics between 2.6 and 3.7 [Hz] at x = 17.5 [mm]. It decreases greatly at x = 53.5 [mm].

The characteristics at the position or frequency shown in Table 1 and Figure 5 differ, because the time constants depend on the position or frequency, according to formulas (4) and (5).

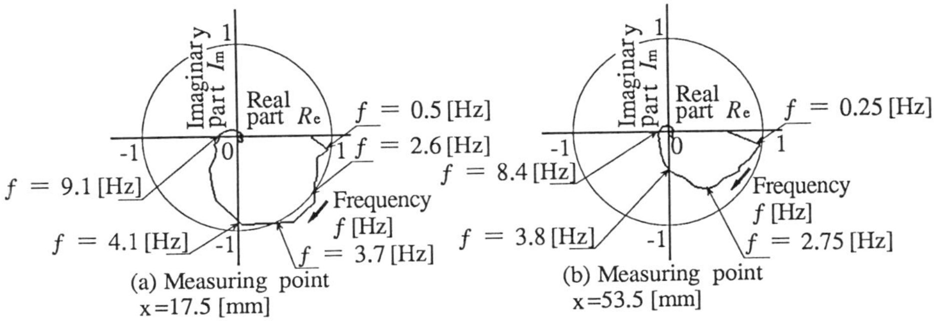

Figure 5. Nyquist diagrams for an E_p transfer function for E_s in the pen recorder.

3. CONCLUSION

This paper has clarified the following items:

(1) The flux density across the center yoke fluctuates between B = 90.8 and 98.9 [mT] (8.92%) from the center (x = 64.5 [mm]) to the end (x = 6 [mm]) of the stroke.

(2) The electrical and mechanical time constants fluctuate, depending on the position or frequency.

(3) In the servomechanism of the pen recorder, stability is apparent at all positions by applying the open - loop transfer function. Stability, however, depends on the position. Resonant characteristics are apparent between 2.6 and 3.7 [Hz] at the initial position (x = 17.7 [mm]) in the frequency response of the position signals for turbulence signals during closed looping.

(4) The response characteristics depend greatly on the position, because of the flux density distribution. Thus, the flux density distribution must be constant across the LDM stroke.

REFERENCES

1) H. Yamada: Industrial Linear Motors, Kogyo Chosakai Pub., pp. 36 - 37 (1981).

2) Magnetic Actuator Technical Committee: Linear motors and their applications, I.E.E. of Japan, pp. 12 - 13 (1987).

Simulation and Design of Applied Electromagnetic Systems
T. Honma (Editor)

Direct vector control scheme of low-speed LIM using primary tapped windings

Dal-Ho Im[a], Byung-Il Kwon[a], Chang-Eob Kim[a,b], Ho Yun[a] and Hyung-Bin Im[a]

[a]Department of Electrical Engineering, Hanyang University,
Haengdang-Dong 17, Sung-Dong Gu, Seoul, Korea

[b]Research and Development Institute, Hyosung Industries Co., Ltd.,
5-4, Dangsan-Dong, Yeong Deung Po Gu, Seoul, Korea

Abstract

This paper presents a novel direct vector control scheme of linear induction motor (LIM). The pole-by-pole d-q model is used to represent the analytical model of LIM, and the primary tapped windings are used to detect the air gap flux. We present the results of a direct vector control of LIM by the pole-by-pole d-q model.

1. INTRODUCTION

Recently, linear induction motors are increasingly being used in the various fields. However, as LIM has a finite length, it has an unbalanced air gap flux at each pole and specific phenomena which is not occured in rotary type induction motors such as end effects, normal force[1]. These phenomena make it difficult to control LIM with high precision, so it is requested to develop the control scheme which can take account of these phenomena.

These phenomena also make it difficult to analyze LIM by analysis over only one pole as rotary inducton machines. Lipo and Nondahl[2-3] analyzed the behavior of LIM by pole-by-pole d-q model. This method can take account of an unbalanced air gap flux due to end effects of LIM. Lipo and Chang[4] proposed the flux sensing method by the stator tapped windings. This method is more robust than other flux sensors such as hall sensor and search coil, and it is insensitive to the primary resistor voltage drop which badly affects on the control performance of motor.

In this paper, we adopt the pole-by-pole d-q model to analyze LIM and the primary tapped windings to detect the flux for direct vector control.

2. VECTOR CONTROL OF LIM BY POLE-BY-POLE d-q MODEL

2.1 Pole-by-pole d-q model[2-3]

The pole-by-pole d-q model is developed on the basis of the assumption that the secondary currents under each pole of primary windings are electrically independent of each other. The d-q axis winding functions are applied to primary winding and secondary windings of each pole. Fig. 1 shows the sketch of pole-by-pole d-q model with winding functions. The portion δ of primary windings is not considered in the analysis as it contains half filled slots.

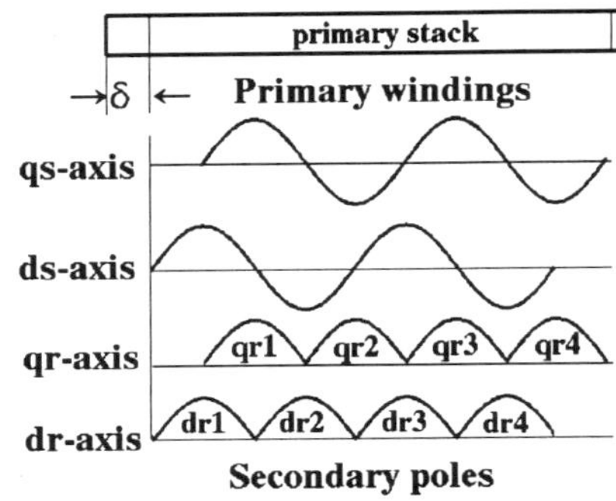

Fig.1 Pole-by-pole d-q model

In the analysis of LIM by pole-by-pole d-q model, the following assumptions are made :

1) Transverse edge effect is taken into account by the Russell and Norsworthy factor[5].

2) Primary and secondary(equivalent) windings are assumed on the surface of the stack and back-iron so that slot effects are neglected and air gap is modified by Carter's coefficient.

3) The primary stack and back-iron is infinitely permeable and the air gap flux is confined to the area directly under the primary at all speeds.

2.2 System matrix and motion equations of LIM[2]

The voltage equations of LIM can be made at each primary and secondary pole, 2 equations for primary and 8 equations for secondary of 4 poles. At each pole of d-axis and q-axis the voltage equations are given in eq.(1) and eq.(2) respectively. The symbols used in equations are same as in ref.[2].

$$V_s = R_s i_s + \frac{d\lambda_s}{dt} + E_s \tag{1}$$

$$0 = R_s i_s + \frac{d\lambda_r}{dt} + E_r \tag{2}$$

The system matrix is given using the speed voltage, flux linkage and equivalent circuit as follows.

$$\mathbb{V} = \mathbb{R}\mathbb{I} + (\mathbb{L}_m + \mathbb{L}_l)\frac{d\mathbb{I}}{dt} + V_r \frac{\pi}{\tau}\mathbb{G}\mathbb{I} \tag{3}$$

where,

$$\mathbb{V} = [\, V_{ds}, V_{qs}, 0, 0, 0, 0, 0, 0, 0, 0 \,]^T$$
$$\mathbb{I} = [\, i_{ds}, i_{qs}, i_{dr1}, i_{qr1}, i_{dr2}, i_{qr2}, i_{dr3}, i_{qr3}, i_{dr4}, i_{qr4} \,]^T$$
$$\mathbb{R} = \mathrm{diag}[R_s, R_s, R_{r1}, R_{r1}, R_{r1}, R_{r1}, R_{r1}, R_{r1}, R_{r1}, R_{r1}]$$

The motion equations in the linear system are given as follows.

$$M\frac{dV_r}{dt} = F_{thrust} - F_{load} \tag{4}$$

$$F_{load} = \mu(9.8 \times M + F_{normal}) \tag{5}$$

$$F_{thrust} = \frac{3}{2}\frac{\pi}{\tau}\mathbb{I}^T\mathbb{G}\,\mathbb{I} \tag{6}$$

$$F_{normal} = -\frac{3}{4g}\mathbb{I}^T\mathbb{G}\,\mathbb{I} \tag{7}$$

Here, M is mass of LIM and load, μ friction coefficient, F_{thrust} thrust, and F_{normal} normal force of LIM.

3. FLUX SENSING USING PRAIMARY TAPPED WINDINGS[4]

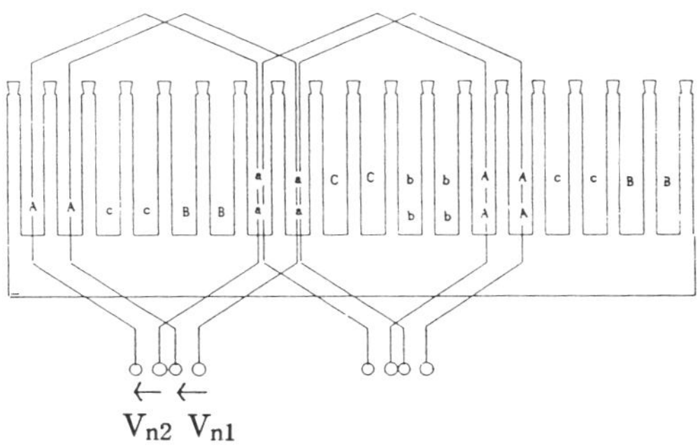

Fig 2. Primary tapped windings of LIM

The principle of flux sensing method for direct vector control of LIM is illustrated in Fig.2.

The air gap flux and secondary flux are calculated by the voltage difference as shown is eq.(8). In the operation of subtraction the resistor voltage drop is eliminated.

$$v_{n2} - v_{n1} = \left(\frac{d\lambda_{2s}}{dt} - \frac{d\lambda_{1s}}{dt} \right) + \left(\frac{d\lambda_{2r}}{dt} - \frac{d\lambda_{1r}}{dt} \right) \tag{8}$$

1) Air gap flux

$$\left. \begin{aligned} \lambda^{s}_{md} &= k_1\left[\int (v_{a2} - v_{a1})dt \right] \\ \lambda^{s}_{mq} &= k_1\left[2\int (v_{b2} - v_{b1})dt + \int (v_{a2} - v_{a1})dt \right] \end{aligned} \right\} \tag{9}$$

2) Secondary flux

$$\left. \begin{aligned} \lambda^{s}_{dr} &= \frac{L_r}{L_m}\lambda^{s}_{md} - L_{lr}\, i^{s}_{ds} \\ \lambda^{s}_{qr} &= \frac{L_r}{L_m}\lambda^{s}_{mq} - L_{lr}\, i^{s}_{qs} \end{aligned} \right\} \tag{10}$$

In low speed LIM since the secondary leakage inductance $L_{lr} \cong 0$ and $L_r = L_{rl} + L_m$, the air gap flux becomes directly the secondary flux[1].

4. SIMULATION AND RESULTS

The simulation is similar to the case of conventional rotary inducton motor. System equation(3) and motion equations(4-7) are used to simulate the vector control of LIM. Motor parameters used in simulation are calculated from the reference[6]. In simulation, hysterisis-current controller is used for current-controlled PWM inverter. Both on-load and off-load condition is assumed in simulation.

The velocity response to the step velocity command is depicted in Fig.3(a), Fig.3(b), Fig.3(c) and Fig.3(d), where load exists in interval of 0~1.0[sec] and friction force and normal force of LIM were considerd. Simulation results show a good control characteristics.

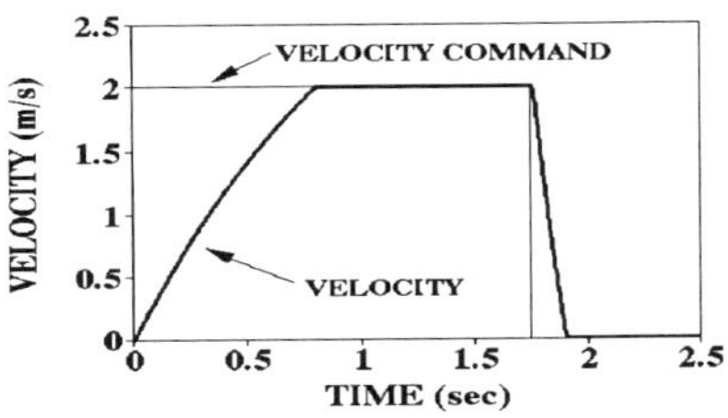

Fig. 3(a) Velocity response

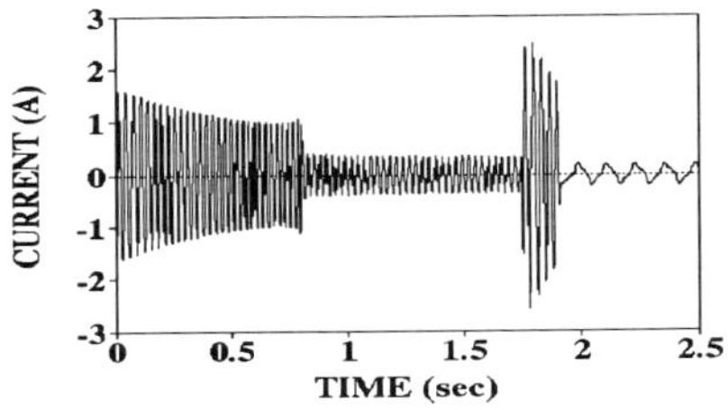

Fig. 3(b) Secondary Current

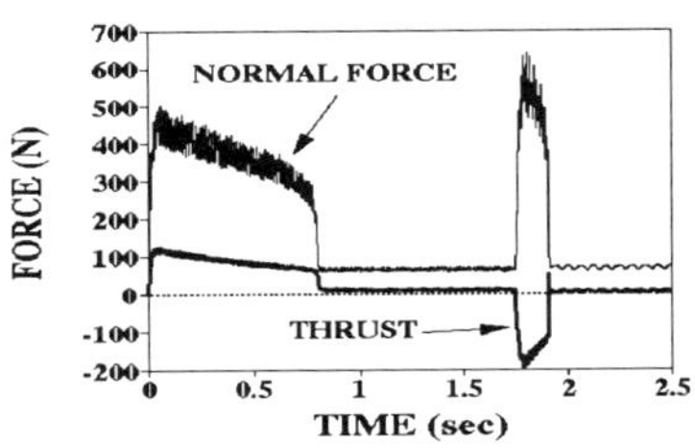

Fig. 3(c) Thrust and Normal force

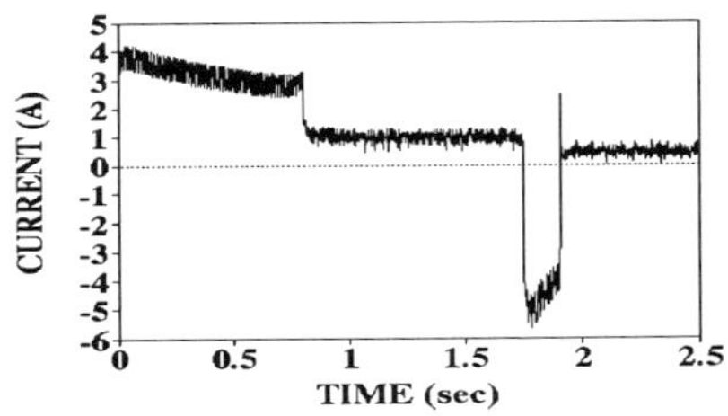

Fig. 3(d) Thrust current

5. CONCLUSION

This paper presents the vector control method of LIM using pole-by-pole d-q model and primary tapped windings. We found that the proposed method were valid for low speed LIM by simulation.

6. REFERENCES

[1] S.A. Nasar and I. Boldea, Linear Motion Electric Machines, John Wiley & Sons, 1976.

[2] T.A. Lipo, et al., "Pole-by-pole d-q Model of A Linear Induction Machine," IEEE Trans. on PAS., Vol. PAS-98, No.2, March/April 1979.

[3] T.A. Nondahl, et al., "Transient Analysis of A Linear Induction Machine using The d-q Pole-by-pole Model," IEEE Trans. on PAS., Vol. PAS-98, No4 July/Aug. 1979.

[4] T.A. Lipo and K.C. Chang, "A New Approach to Flux and Torque Sensing in Induction Machines," IEEE Trans. on Industry Application, Vol. IA-22, July/Aug. 1986, pp.731-737.

[5] R. L. Russell and K. H. Norsworthy, "Eddy Current and Wall Losses in Screened-Rotor Induction Motors," Proc. IEE, Vol. 105A, 1958, pp.163-175.

[6] S. Nonaka, T. Higuchi, "Approximate Equations for Calculation of Characteristics of The Single Sided Linear Induction Motors," JIEE 57-B69 1982, pp.565-572.

Acknowledgement : This work was supported in part by the institute of the Korean Science and Engineering Foundation.(No. 92-47-00-02)

Simulation and Design of Applied Electromagnetic Systems
T. Honma (Editor)

Characteristic analysis of a single-sided linear induction motor for the servo system

D.H. Im[a], C.J. Ree[a], S.C. Park[a], J.W. Im[a], Y.B. Jung[a] Y.H. Cho[b] S.M. Jang[c]

[a]Dept. of Electrical Engineering, Hanyang University, Seoul, 133-791, Korea
[b]Korea Electrotechnology Research Institute, P.O. Box 20, Changwon, Korea
[c]Dept. of Electrical Engineering, Chungnam National University, 302-764, Korea

Abstract

In this paper, the transient characteristics of a single-sided linear induction motor (SLIM) are analyzed at standstill by the finite element method, for the various secondary aluminum depths and mechanical airgaps at the given input frequencies. As a result, $\tau/k_c(g+d_{AL})$ (pole pitch/effective airgap) as a good criterion to design a single-sided linear induction motor for the servo system has been found.

1. INTRODUCTION

Recently, the instantaneous thrust control or the speed control scheme on the basis of the vector control for SLIM have been developed [1-2]. The instantaneous thrust and normal force by the end effect influence on the control schemes because those produce ripples containing double input frequencies [3]. Especially the normal force ripples between the primary and the secondary deteriorate the thrust control in the servo system.

So this paper deals with the transient characteristics of a SLIM at standstill by the finite element method under a constant current drive, for the various secondary aluminum depths and mechanical airgaps at the given input frequencies, i.e., 20, 40, 60 (Hz). From the results, $\tau/k_c(g+d_{AL})$ (pole pitch/effective airgap) and τ/g (pole pitch/mechanical airgap) for the best performance are detererrmined.

Figure 1. shows the experimental apparatus, which has the load cells to measure the thrust and normal force, and the rotary encoder for the position control. The stack width of the primary is 80(mm) and the length of pole pitch is 51(mm), respectively.

Figure 1. Experimental apparatus

2. FINITE ELEMENT ANALYSIS

The governing equation which describes the magnetic vector potential in a SLIM is given by

$$\frac{1}{\mu}\left(\frac{\partial^2 A}{\partial x^2}+\frac{\partial^2 A}{\partial y^2}\right) = -J_o+\sigma\left(\frac{\partial A}{\partial t}+v\frac{\partial A}{\partial x}\right) \tag{1}$$

where σ is the conductivity of the secondary reaction plate, v is the velocity of the primary, respectively. Using the Galerkin method to solve (1), we can write the next equation.

$$\begin{aligned}&\int\int_s \frac{1}{\mu}\left(\frac{\partial N_i}{\partial x}\frac{\partial A}{\partial x}+\frac{\partial N_i}{\partial y}\frac{\partial A}{\partial y}\right)dxdy \\ &-\int\int_s N_i J_o dxdy + \sigma\int\int_s N_i\left(\frac{\partial A}{\partial t}+v\frac{\partial A}{\partial x}\right)dxdy = 0\end{aligned} \tag{2}$$

where N_i is the interpolation function. Manipulaion of equation (2) yields the following equation.

$$([S]+[G])\{A\} = \{F\}J_o - [E]\frac{\partial\{A\}}{\partial t} \tag{3}$$

Equation (3) can be written as follows by using the backward difference method in the time differentiation term [4].

$$\left([S]+[G]+\frac{1}{\Delta t}[E]\right)\{A_{t+\Delta t}\} = \{F\}J_o + \frac{1}{\Delta t}[E]\{A_t\} \tag{4}$$

It is assumed that the primary velocity is zero, so $[G]$ is neglected.

The force acting on volume V equals the surface integral of the Maxwell stress tensor on the surface. So the thrust T_x and the normal force T_y acting on an integral path l are

$$T_x = \int_l \frac{w}{2\mu_o}\{(B_x^2 - B_y^2)n_x + 2n_yB_xB_y)\}dl \tag{5}$$

$$T_y = \int_l \frac{w}{2\mu_o}\{(B_y^2 - B_x^2)n_y + 2n_xB_xB_y)\}dl \tag{6}$$

where n_x and n_y are the unit normal direction vectors, w is the stack height of the primary.

3. RESULTS AND EXPERIMENTS

Figure 2. illustrates the results from the finite element transient analysis under a constant current 3.45 (A). Thrust and normal force have the double frequency components by the end effect, respectively. We can find that the normal force ripple begins to reduce remarkably when the secondary aluminum depth is greater than 2 (mm), but the amplitude of the ripple is almost constant when it is greater than 2.5 (mm).

Table 1. is the results from the following perfomance criteria; 1) the thrust must be sufficient, 2) the normal force ripple must be small. Because the slot and teeth size has already determined, only the secondary aluminum depth and mechanical airgap can be changed as shown in table 1. Carter's coefficient k_c is used to calculate the effective airgap considering the effect of slot.

From the table 1 it is known that the most proper ratio of $\tau/k_c(g+d_{AL})$ (pole pitch/effective airgap) is about 10 and the ratio τ/g (pole pitch/mechanical airgap) is about 25. We think these values to be effective to a large extent.

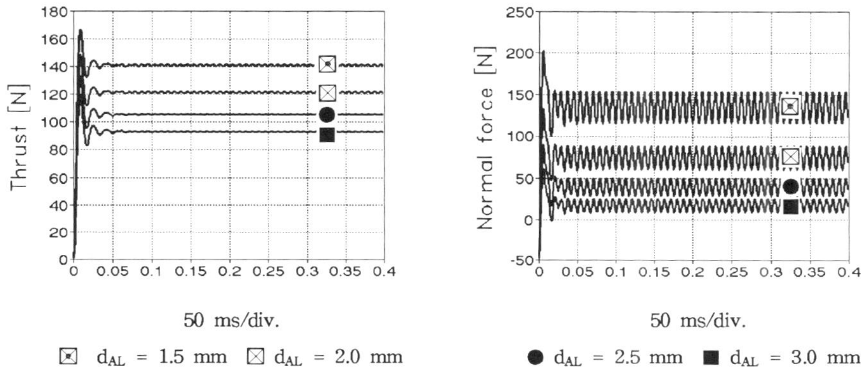

Figure 2. Thrust and Normal force (f=60 Hz, g=2 mm, I=3.45 A)

Table 1.
Performances at the various secondary dimensions and airgaps

Freq. [Hz] 20	40	60	airgap g [mm]	Al. depth d_{AL} [mm]	Carter's coeffi. k_c	Pole pitch/effective airgap $\tau / k_c(g+d_{AL})$
⊠	⊠	⊠	1.5	1.5	1.21	14.05
⊠	⊠	◎		2	1.183	12.319
⊠	⊠	◎		2.5	1.162	10.972
◎	◎			3	1.145	9.897
⊠	⊠	⊠	2	1.5	1.183	12.319
⊠	⊠	◎		2	1.162	10.972
◎	◎	◎		2.5	1.145	9.897
				3	1.132	9.011
⊠	⊠	⊠	2.5	1.5	1.162	10.972
⊠	◎	◎		2	1.145	9.897
◎				2.5	1.132	9.011
				3	1.121	8.271

◎ : sufficient thrust and small normal force ripple
⊠ : excessive normal force or large normal force ripple

Figure 3. shows the experimental results obtained by the apparatus for the servo system supplied with PWM inverter shown in figure 1. It is known that the experimental values are almost agree with the simulation values and contain double-frequency components like in figure 2.

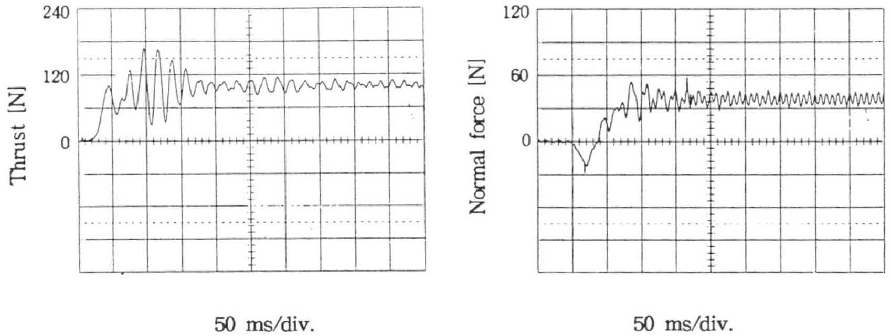

Figure 3. Experimental results (f=60 Hz, g=2 mm, d_{AL}=2.5 mm, I=3.45 A)

4. CONCLUSION

From this study, $\tau/k_c(g+d_{AL})$ (pole pitch/effective airgap) and τ/g (pole pitch/mechanical airgap) which can reduce the normal force ripple between the primary and the secondary that influences on the thrust control in the servo system has been found by the finite element transient analysis under a constant current drive. These results will be very useful especially in the design of a low speed SLIM for the servo system.

5. REFERENCES

1 Takashi Onuki, Hiroyuki Numata, Tsugio Yamamura, Thrust Control of Linear Induction Motor by Detection of Magnetic Field, T. IEE Japan, No. 9, 957 965 (1990)

2 Y. Ogino, k. Abe, K. Ohishi, Y.Takakado, K.Nakano, Super Low Speed Control For LIM, Proceedings of ISPE, 148 152 (1992)

3 Sakae Yamamura, Theory of Linear Induction Motors, University of Tokyo press, 61 (1979)

4 Fnederic Bouillault and Adel Razek, Dynamic Model for Eddy Current Calculation in Saturated Electric Machines, Trans. IEEE, Vol. 19, No. 6, 2639 2642 (1983)

Acknowledgement : This work was supported in part by the institute of the Korean Science and Engineering Foundation.(No. 92-47-00-02)

Simulation and Design of Applied Electromagnetic Systems
T. Honma (Editor)

Characteristics analysis of a tubular linear induction motor for elevators

Dal-Ho Im, Young-Jung Kim, Sang-Yeon Hwang[a] Kyung-Soo Choi[b]

[a] Dept. of Electrical Engineering, Hanyang University, Seoul, 133-791, Korea

[b] Lucky-Goldstar R & D Complex, Anyang, Kyungki-Do, 430-080, Korea

ABSTRACT

This paper presents the analysis method of TLIM for elevators. In analysis method, the transfer matrix to multilayer on cylindrical coordinate is derived from Maxwell equations in terms of modified Bessel function. The penetration depth of rotor core is taken into consideration. The magnetic field distribution in airgap is derived from coupling transfer matrix and complex phasor method.

The validity of the method is verified by comparing the experimental and theoretical results for a experimental machine in locked status.

1. INTRODUCTION

The use of linear motor is currently increasing in many applications such as servo system, conveyer system and elevators. In recent paper[1], the author proposed the complex phasor method of low capacity tubular linear induction motor(TLIM) which was modified from the analysis method of single-sided linear induction motor.

In this paper the analysis method of high capacity TLIM is presented wherein the skin effect is taken into consideration by correction factor. The transfer matrix[2] to the cylindrical multilayer is coupled with the complex phasor method to consider the airgap leakage flux.

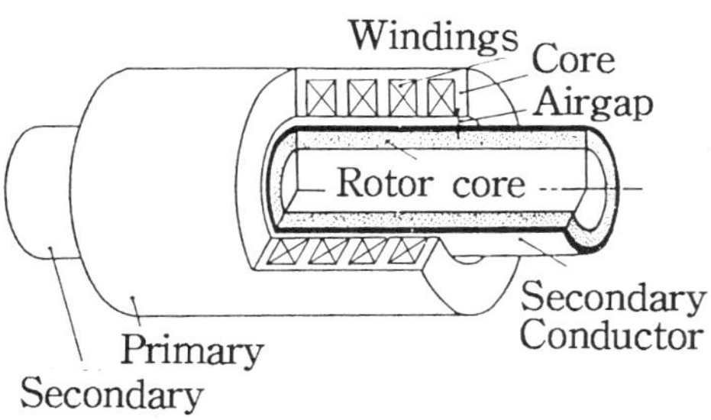

Fig.1 The structure of TLIM

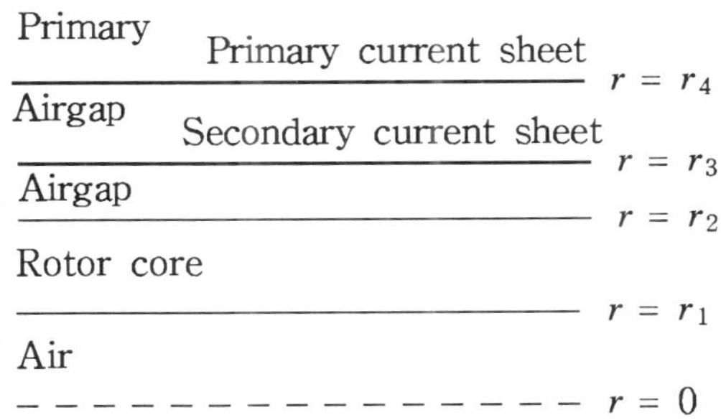

Fig.2 Longitudinal view of multilayer

2. APPLIED THEORY

2.1 Multilayer Model

The 2-D analysis region of the cylindrical multilayer with current sheet, simplified from the actual TLIM in Fig.1 by general assumptions[3], is shown in fig.2. The relation between magnetic airgap g and effective airgap g_e is

$$g_e = K_c\,(1+K_s)\,g \tag{1}$$

where

$$K_s = \frac{\mu_0}{\mu_i\,\delta_i\,g\,K_c\,k^2}\,, \qquad \delta_i = Re\left[\frac{1}{(\,(\tau/\pi)^2 + j\omega_1\mu_i\sigma_i)^2}\right]\,, \qquad k = \pi/\tau\,,$$

τ is pole pitch, K_c carter coefficient. The penetration depth δ_i and saturation coefficient K_s are introduced to consider the effect of conductivity σ_i and permeability μ_i of rotor core[3].

2.2 Transfer Matrix

To calculate the characteristics of TLIM, the flux distribution in airgap must be analyzed accurately. The differential equation of the magnetic field at cylindrical multilayer is modified Bessel equation which can be approximated to the differential equation in terms of exponential function in the case of small airgap machine[1]. Since TLIM, however, has large airgap, the modified Bessel equation must be solved for accurate analysis.

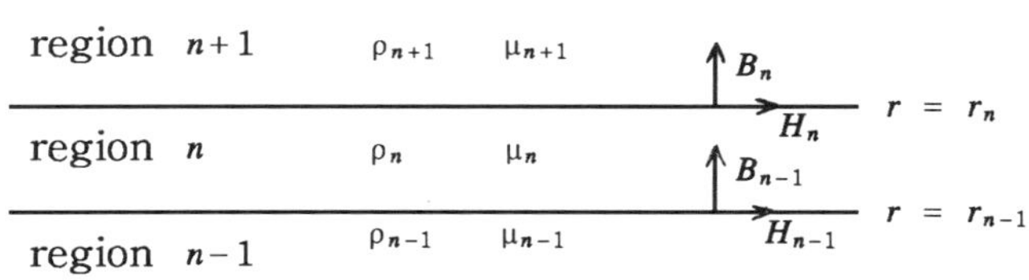

Fig.3 Arbitrary region n

Fig.3 shows the longitudinal view of arbitrary region. The transfer matrix of region n can be defined as

$$\begin{bmatrix} B_n \\ H_n \end{bmatrix} = [T_n]\begin{bmatrix} B_{n-1} \\ H_{n-1} \end{bmatrix} = \begin{bmatrix} T_{na} & T_{nb}/\beta \\ \beta\,T_{nc} & T_{nd} \end{bmatrix}\begin{bmatrix} B_{n-1} \\ H_{n-1} \end{bmatrix} \tag{2}$$

where

$$T_{na} = [\,I_0(\gamma_n r_n)\; b_{n,n-1} - K_0(\gamma_n r_n)\; a_{n,n-1}]\;/\;DET_n\,,$$

$$T_{nb} = [\,-I_0(\gamma_n r_n)\; K_0(\gamma_n r_{n-1}) + K_0(\gamma_n r_n)\; I_0(\gamma_n r_{n-1})]\;/\;DET_n\,,$$

$$T_{nc} = [\,a_{n,n}\, b_{n,n-1} - a_{n,n-1}\, b_{n,n}]\;/\;DET_n\,,$$

$$T_{nd} = [\,b_{n,n}\; I_0(\gamma_n r_{n-1}) + a_{n,n}\; K_0(\gamma_n r_{n-1})\,]\;/\;DET_n\,,$$

$$DET_n = I_0(\gamma_n r_{n-1})\; b_{n,n-1} - K_0(\gamma_n r_{n-1})\; a_{n,n-1}\,,$$

$$a_{n,n-1} = \frac{1}{\gamma_n r} I_0(\gamma_n r_{n-1}) + I_1(\gamma_n r_{n-1})\,,$$

$$b_{n,n-1} = \frac{1}{\gamma_n r} K_0(\gamma_n r_{n-1}) - K_1(\gamma_n r_{n-1}),$$

$$\beta_n = \frac{\pi \gamma_n}{j \tau \mu}, \qquad \gamma = [k^2 + \frac{js\omega_1 \mu_n}{\rho_n}]^{1/2},$$

B is the r component of flux density on boundary, H the z conponent of magnetic field intensity on boundary, ρ conductivity and μ permeability. I_n , K_n are modified Bessel functions[4] and subscripts n, $n-1$ represent outer and inner boundary of the region n respectively.

2.3 Parameters and Thrust

The complex phasor method is used to analyze the transient state[5]. The flux density of airgap can be derived precisely by coupling complex phasor method with transfer matrix, which enables to obtain the leakage flux of airgap. Airgap inductances, indispensable to calculate thrust, can be defined as

$$L_{tt} = (3/\pi)\mu_0(2\pi r_3) N_e^2 [K_{wt}^2 K_{rtt}] \tag{3}$$

$$M = (3/\pi)\mu_0(2\pi r_4) N_e^2 [K_{ws} K_{wt} K_{rst}] \tag{4}$$

where

$$K_{rtt} = \frac{T_{3a}T_{4d} \, [I_0(\gamma r_4)K_0(\gamma r_3) - I_0(\gamma r_3)K_0(\gamma r_4)]}{K_0(\gamma r_3)[T_{3a}T_{4c} - T_{3c}T_{4d}] \, [I_0(\gamma r_4)K_0(\gamma r_3) - I_0(\gamma r_3)K_0(\gamma r_4)]},$$

$$K_{rst} = \frac{T_{3a}(T_{4a}T_{4d} - T_{4c}T_{4b})[I_0(\gamma r_4)k_0(\gamma r_3) - I_0(\gamma r_3)K_0(\gamma r_4)]}{K_0(\gamma r_3)[T_{3a}T_{4c} - T_{3c}T_{4d}] \, [I_0(\gamma r_4)K_0(\gamma r_3) - I_0(\gamma r_3)K_0(\gamma r_4)]},$$

L_{tt} is secondary self inductance, M mutual inductance, N_e the number of equivalent conductors, K_{ws} the primary winding factor, K_{wt} the secondary winding factor. In steady states, eqivalent circuit, omitted in this paper, can be derived from field analysis. Thrust obtained from the cross product of flux density and secondary current distribution is represented in terms of parameters.

$$F_z = 3\frac{\omega_1^2}{V_0} I_s^2 \frac{M^2}{R_2} \frac{s}{1 + \omega_1 s L_{tt}/R_2} \tag{5}$$

where V_0 is synchronous velocity, ω_1 primary angular frequency, I_s the primary phase current, s slip, R_2 equivalent secondary resistance.

3. EXPERIMENTAL RESULTS

TLIM used in experiment was made for the 15-passengers elevator (1000[kgf]). TLIM mounted in the counterweight of conventional elevator system and its specification is shown in fig.4. For elevators need high starting thrust, the characteristics of starting thrust is very important. The computed characteristics of starting thrust are compared with experimental results for the experimental machine in locked status.

Fig.5 shows the characteristics of starting thrusts versus frequency on constant current. Fig.6 shows the starting thrust versus current at the rating frequency of 30[Hz]. Acceptable agreement was shown at frequencies upper than 15[Hz] but the accuracy decreases at low frequencies caused by parasitic thrust in rotor core, as is in Fig.5.

Primary length	108 [mm]
Secondary conductor thickness	4 [mm]
Number of poles	16
Airgap	6.5 [mm]
Tooth width	7 [mm]
Diameter of primary(inside)	330 [mm]
Thickness of rotor core	24 [mm]
Number of slot	48
Slot depth	58 [mm]
Slot Pitch	22.3 [mm]
Pole pitch	67 [mm]

Fig. 4 Experimental apparatus and specification of experimental machine

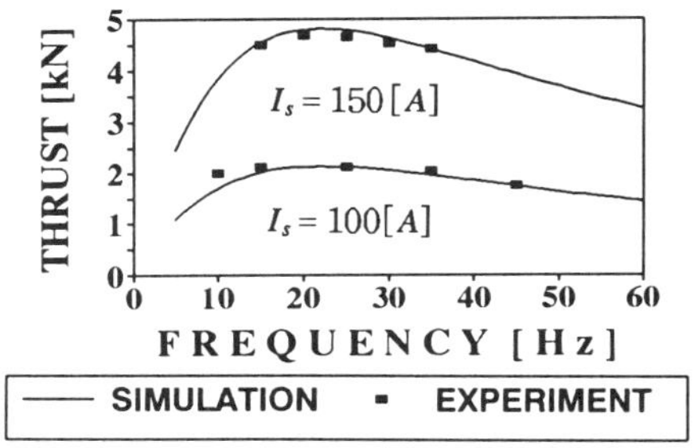

Fig.5 Characteristics of thrust versus frequency

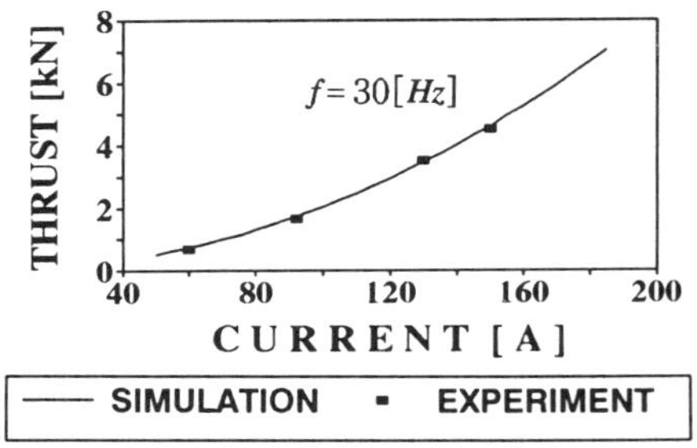

Fig.6 Characteristics of thrust versus current

4. CONCLUSION

The analysis method of thrust characteristics of TLIM was presented. The validity of the method was verified by comparing experimental results with computed values. The method can be applied to design and analysis of the high capacity TLIM with low speed.

REFERENCES

[1] Ir.D.J.de Groot and Pro.C.J.Heuvelman.'Tubular Linear Induction Motor for Use As Servo Actuator',Proc.IEE,Vol.137,Pt.B,No.4,July,1990
[2] Freeman,E.M.'Travelling Waves in Induction Machines:Input Impedance and Equivalent Circuits',Proc.IEE,1968,115,(12), pp. 1772-1776
[3] S.A.Nasar & I.Boldea,'Linear Electric Motors ; Theory, Design, and Practical Applications', Prentice-Hall, 1987.
[4] G.N.Watson,'A Treatise on the Theory of Bessel Functions',Cambridge University Press, 1966
[5] Leonhard W. 'Control of Electric Drives', Springer-Verlag, 1985

Simulation and Design of Applied Electromagnetic Systems
T. Honma (Editor)

Development of the ropeless elevator with linear synchronous motor

Daiki Ebihara, Hong Joong Kim, Yorikazu Tasaki, Masaya Watada

Musashi Institute of Technology
1-28-1 Tamazutsumi Setagaya-ku, Tokyo 158 JAPAN

Abstract

When speaking of skyscrapers, it becomes more significant to search the high technological elevators, but the other hand it creates many problems. First, the number of the elevators has to increase in order to support the heavy usage. Second, the space of elevator shaft has to become larger. And finally, it is said that the limited length of rope cannot exceed eight hundred meters.

Considering from these point of views, we have been studying to develop a rope-less elevator with linear synchronous motor("LSM") that doesn't require ropes. This paper presents a basic model of vertical linear drive system and plans for rope-less elevator system run by LSM.

1. INTRODUCTION

Linear motors have been used for many different purposes since they were first developed in the 1840.[1] Many have been used to move direct drive systems lately, but now a great deal of attention is being turned toward their future use in vertical linear motor drives. As part of that, we have begun research on their use in elevators, which play such an important part in our daily lives.

Elevator technology and elevator shaft area have developed in close connection with the development skyscrapers. Elevators are indispensable to many-storied buildings and the number of elevators and the elevator shaft area increases as taller buildings are constructed. Japan's big construction contractors are planning a skyscrapers. Since the elevator has to travel a longer distance, and the overload caused by rope weight alone limits existing elevator drive methods to exceed eight hundred meters. Considering from these point of views, an examination is desirable for methods that will not use ropes. Also, many cages are able to run smoothly in a one shaft, instead of one cage in the shaft, as in the normal elevators. It was for these reasons that we have tested a rope-less elevator system that uses a LSM as its drive mechanism.

The reasons for choosing the LSM as the preferable type are as follows. First, it provides a large amount of propulsion power. Second, it is no slip as there is with linear induction motors, which makes it easy to synchronize and control several movers in relation to one stator. And finally, it ensures stability during power failures by using the coercive force of permanent magnets, or super conductive magnets, for braking.

2. VERTICAL LINEAR DRIVE SYSTEM

Fig. 1 shows the two drive types, roped and rope-less, that can be used with the vertical linear motors. As we explain two types about driving theory simply:

· Roped type : The well-bucket setup is used for connecting the cage and counterbalance by the rope. The propulsion force of the linear motor attached to the counterbalance, moves the counterbalance up and down, and that moves the cage connected to the ropes up or down, to take the cage to the desired floor. It has ability to drive only one cage in one vertical elevator shaft, and the weight of lift ropes becomes overload because the long track of elevator shaft in the high building.

· Rope-less type : The propulsion force of a linear motor installed on the side of, or behind the cage moves directly. Many cages can be moved independently in one vertical shaft, allowing the most economical shaft width.

From this comparison, we have designated the roped type to be most suitable for low to mid-level general buildings and rope-less type for skyscrapers.

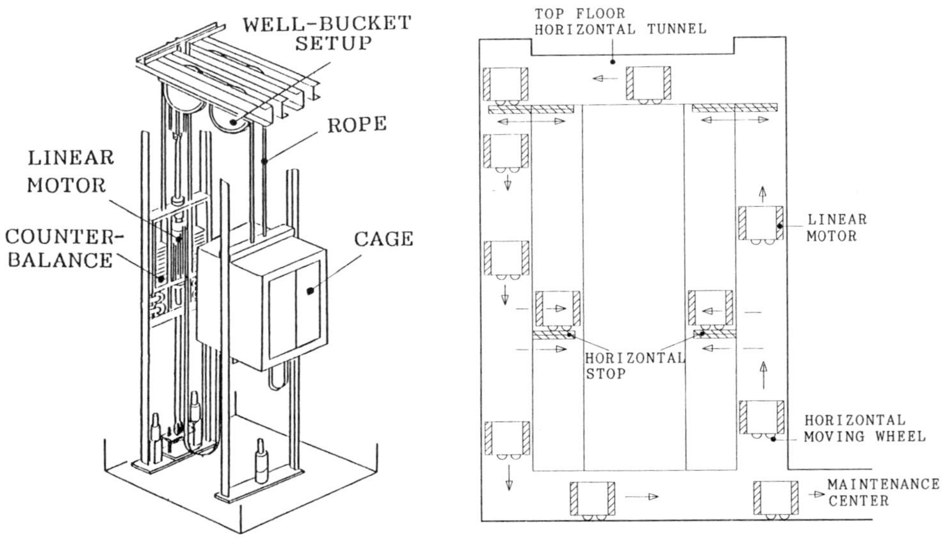

(a) Roped type (b) Rope-less type

Fig. 1. Basic structure of linear motor elevators [2]

3. BASIC STRUCTURE OF LINEAR MOTOR ROPE-LESS ELEVATOR

The advantage of using the rope-less vertical linear motor method with elevator, is the ability to directly and independently move more than one cage in the same shaft. We cogitated on ways in which many cages can be used more efficiently. The results are shown in the three models in Fig. 2. This is a concept for skyscraper.

Fig. 2 (A) and (B) shows models with multiple elevator entrances on the departure floor but fewer vertical drive paths. Several elevator cages can operate in the same shaft at the same time, but a separate vertical shaft is needed for descent. Increased efficiency is achieved by building floors on which cages can pass each other so that when one cage stops, it will not obstruct the transit of another coming behind. And local elevators will use the space between passing floors. Before and after departure and passing floors, the cages will merge into or leave, the main shafts via diagonal shafts. The method in Fig. 2 (C) has a horizontal drive mechanism at each floor station, which allows one cage to pass another. This model requires separate position maintaining mechanisms and horizontal drive mechanisms for stops. As we explain cage movement of three models simply, the running speed is fastest of model A, as it rises and enters the main shaft. In model B, from the start floor, the cage moves horizontally to the main shaft before it gets raised up. In model C, the cage branches from the main shaft horizontally in each floors for a passengers who are getting on and off.

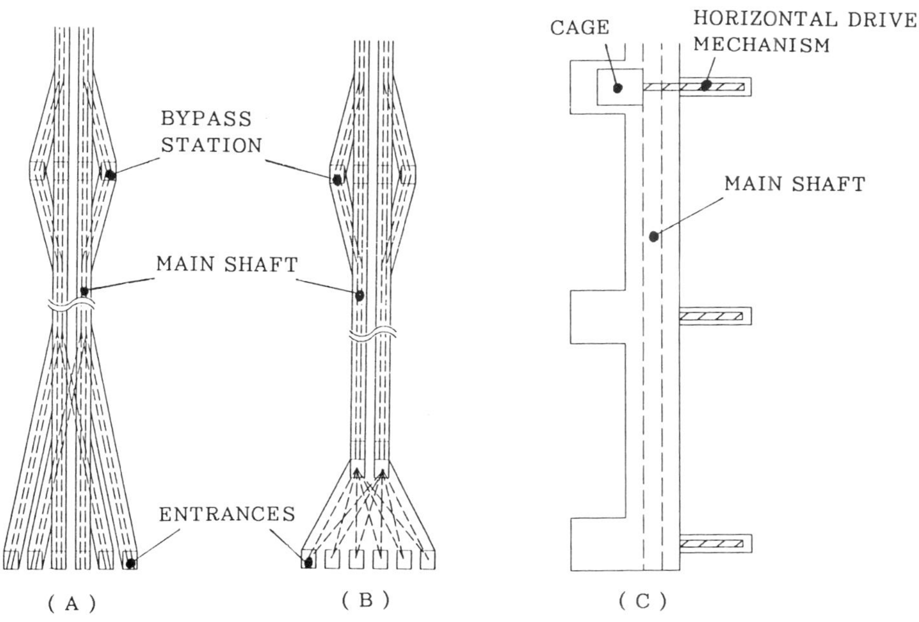

Fig. 2. Schematic diagram of the track for rope-less elevator

4. THE SIMULATION DEVICE

At present, the simulation device based on this idea is under production. We are going to measure a variety of quantity by using the simulation device and than examine the data from this device collected, in order to create the rope-less elevator with vertical linear motor. Basic structure of vertical LSM at our laboratory is shown in Fig. 3. Specifications of vertical LSM is shown in Table 1. Structurally, in order to provide a large amount of propulsion power, a stator have two of double side types, and mover consists of four magnetic blocks and a guide bearing.

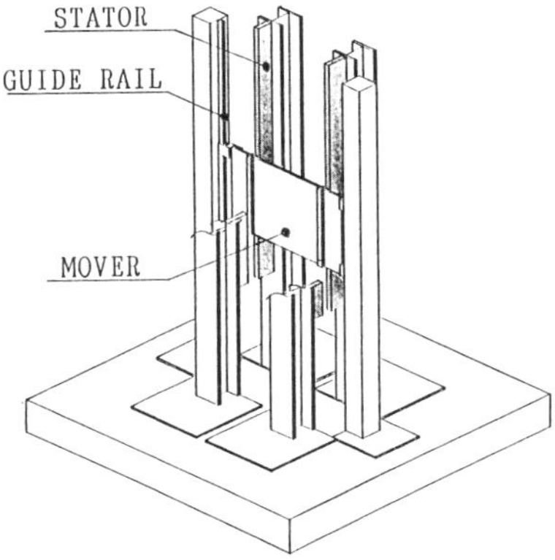

Fig. 3. Basic structure of vertical LSM

Table 1.
Specifications of vertical LSM

Stator		Mover	
Total Length	1200 [mm]	Total Length	360 [mm]
Width	30 [mm]	Width	500 [mm]
Pole Pitch	75 [mm]	Thickness	40 [mm]
Pole Number	16	Magnet Material	Pr-Fe-B-Cu
Winding Number	100 [turn]	Thickness	3 [mm]
Rated Current	2.5 [A]	Width	30 [mm]
		Length	75 [mm]

5. CONCLUSION

The new linear motor rope-less elevator system, can independently move multiple elevator cages in two limited areas, in the space-saving vertical shafts. That gives great promise as a high speed, large-capacity transport system in skyscrapers. Although the future research and development will be necessary for means of dealing with power failures, technology for building larger and more practical units, building of different types of equipment needed, and devising of proper maintenance techniques.

Ongoing development of peripheral technology based on flexible thinking and fresh concepts will lead to the practical application of a LSM rope-less elevator system.

6. REFERENCES

1. E.MASADA : Technical of Linear Drive and this Application. Ohm CO. 1991
2. The Journal of the Institute of Electrical Installation Engineers of Japan. Vol.12. No.10. 1992

Simulation and Design of Applied Electromagnetic Systems
T. Honma (Editor)

A basic study of a transportation system for magnetically levitated thin iron plates

D.Ebihara[a], H.Kawaguchi[a], Y.Muraguchi[a], M.Watada[a], T.Nakagawa[b], S.Komura[c] and T.Ohhara[c]

[a]Department of electrical and electronic engineering, Musashi Institute of Technology, 1-28-1, Tamazutsumi, Setagaya-ku, Tokyo 158, JAPAN

[b]Department of electrical engineering, Tokyo Denki University, kandanisikicho, Chiyoda-ku, Tokyo 101, JAPAN

[c]Chugai Ro Co.,LTD., 2-4-7, Kyomachibori, Nishi-ku, Osaka 550, JAPAN

Abstract

The conventional contact suspension has many problems in the surface processing of thin iron plates. To solve these problems the authors have turned our attention to magnetically levitated suspension, proposed the simplified model which an iron plate is levitated by using two electromagnets and gotten many experimental data.

1. INTRODUCTION

There are many kinds of iron plate transportation system, including the rora type or arm type. However such contact suspension types have various problems namely scratches, noise, vibration and friction. Therefore, the authors take an interest in a magnetically levitated system, and have studied non-contact transportation system to solve these problems. We have already succeeded in the magnetic levitation experiments with a cavity iron ball which we assumed to be iron plate.[1] Taking many data from these experiments into consideration, we would like to propose an actual magnetic levitation system which transports a thin iron plate.

As for a magnetically levitated system for iron plates, it needs to have many electromagnets to control their vibration modes.[2] Therefore, as the first step, the authors propose the simplified model which an iron plate is levitated by using two electromagnets. That required giving close attention to two points. One is the peculiar characteristics of the inductance and the attractive force between the thin iron plate and the electromagnet. The other is mode-analysis for thin plate vibration.

In this paper, we report the results of measurements for the inductance and the attractive force and how to control for such a magnetically levitated system.

2. THE SIMPLIFIED MODEL OF IRON PLATE LEVITATION

We succeeded in controlling and maintaining a cavity iron ball like figure 1 in constant position last year. As the first step, the authors assumed the plate to be a beam to study one-axis vibration mode and designed the simplified model like figure 2 which an iron plate is levitated by using two electromagnets.

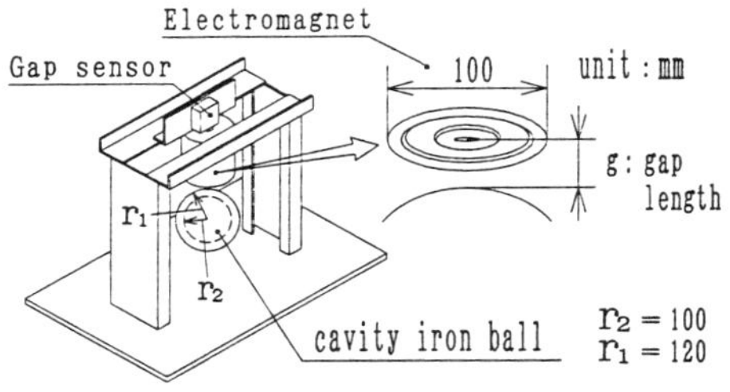

Figure 1. Simplified apparatus (first version)

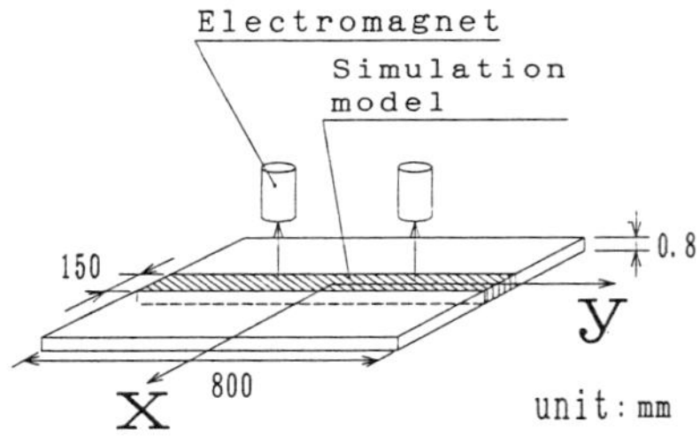

Figure 2. Simplified apparatus (second version)

3. MEASURING INDUCTANCE AND ATTRACTIVE FORCE

We measured the characteristics of inductance and attractive force between the thin iron plate and the electromagnet.

First, we measured current-inductance characteristics having plate thickness δ and gap length g as parameter. The results are shown in figure 3. Figure 3(c) shows the inductance does not depend on the magnet-current within the bound which g is from 5mm to 8mm. As results, we can assume that the inductance is a function of only the gap length if we take advantage of this tendency.

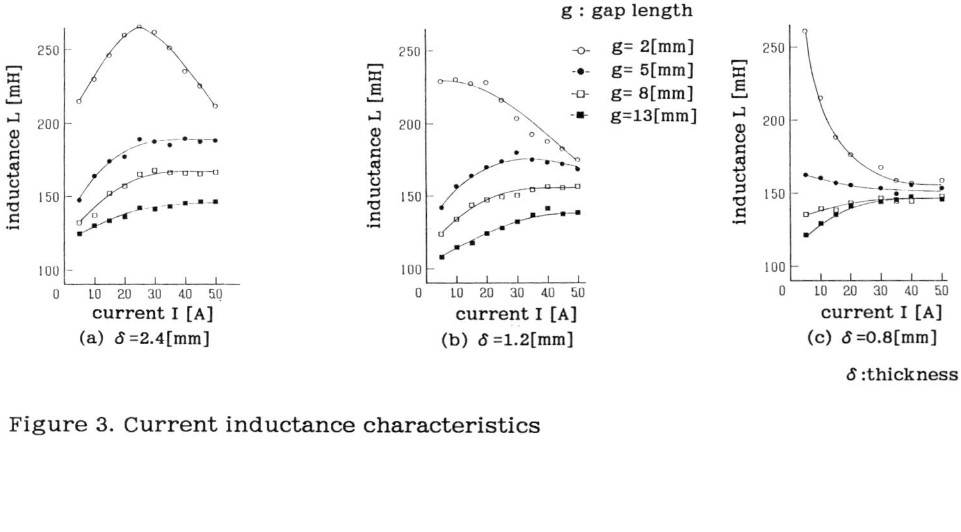

Figure 3. Current inductance characteristics

Second, we measured attractive force with plate thickness δ as parameter. The result is shown in figure 4. From these figures, we can get that the thinner the iron plate becomes, the lower the attractive force will be. As the rate of increase of attractive force is nearly constant in the case of 2A over, magnetic saturation occurs in the plate.

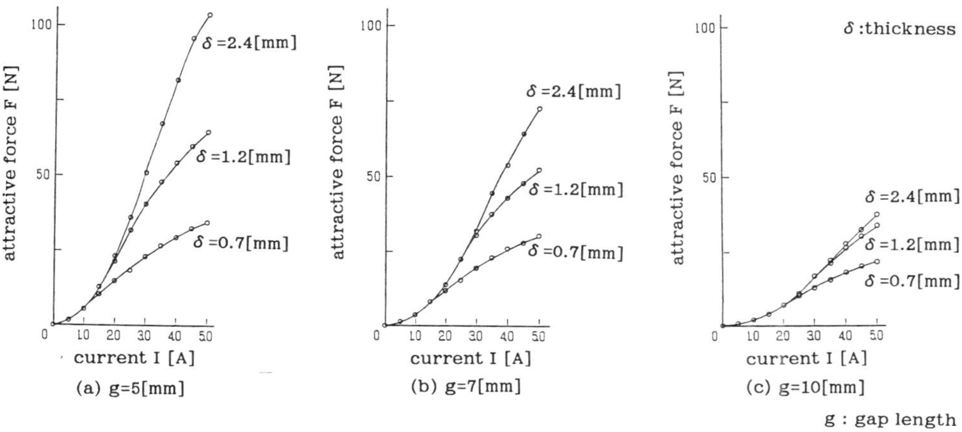

Figure 4. Current attractive force characteristics

4. DETERMINATION OF EQUATIONS FOR L AND F

A simplified model was designed taking into consideration the characteristics between the electromagnets and the thin iron plate.

From the characteristics mentioned above, we found that the inductance dose not depend on current and is a function of only the gap length. Therefore, steady state gap length is determined 8mm.

Attractive force for each of the electromagnets in this case needs about 3.5N because of weight of the thin plate. To satisfy this requirement, state current is determined 1.0A.

From the gap-inductance characteristic at steady state current, inductance and attractive force between thin iron plate and electromagnet are as follows.

$$L = a x^{b} \tag{1}$$

$$F = -\frac{d}{dx}\left(\frac{1}{2} L I^{2}\right) = -\frac{1}{2}\, a b I^{2} x^{b-1} \tag{2}$$

where a and b are respectively constants.

These equations are different from the usual ones in the electro-magnetic theory at least and comes from special characteristics of the magnetically levitated system for the thin plate.

5. DESIGNING OF CONTROL SYSTEM

We designed control system of the simplified model which is approximated a beam. The dynamic equation and electrical equation of this model are

$$M_g \ddot{x} = M_g - F \tag{3}$$

$$Ri + L\frac{di}{dt} + i\frac{dL}{dt} = e \tag{4}$$

where M_g is mass of iron plate, x is gap length, F is attractive force, L is inductance and R is resistance. The state equation and the output equation are obtained by using the steady point.

$$\begin{pmatrix} \dot{x} \\ \ddot{x} \\ \dot{i} \end{pmatrix} = \begin{pmatrix} 0 & 1 & 0 \\ \frac{ab(b-1)}{2M_n} I_0 x_0^{b-2} & 0 & \frac{ab}{M_n} I_0 x_0^{b-1} \\ 0 & -\frac{b}{x_0} I_0 & -\frac{R}{a x_0^b} \end{pmatrix} \begin{pmatrix} x \\ \dot{x} \\ i \end{pmatrix} + \begin{pmatrix} 0 \\ 0 \\ \frac{1}{a x_0^b} \end{pmatrix} e \tag{5}$$

$$y = \begin{pmatrix} 1 & 0 & 0 \end{pmatrix} \begin{pmatrix} x \\ \dot{x} \\ i \end{pmatrix} \tag{6}$$

From equation of natural frequency, we found first and second mode-frequencies of iron plate are 0.19Hz and 0.52Hz, respectively.

It is necessary that response frequency of the control is larger enough than the natural frequencies of the thin iron plate. Therefore, feedback gain K_1,K_2 and K_3 based on optimal control theory are determined as K_1=2,420,000, K_2=18,300, K_3=-26,100.

6. CONCLUSION

Because of many problems in conventional contact suspension systems, we have studied how to apply magnetic levitation to the transportation system for the thin iron plate.

By measuring the characteristics of the inductance and the attractive force between the thin iron plate and the electromagnet, we were able to get the simplified model composed of two electromagnets and design a compensator based on optimal control theory.

We intend to carry on the levitation experiments continuously and show better control method to suppress the vibration.

7. REFERENCES

1 D.Ebihara: "Control of magnetically levitated iron plate", IEEJ No.775 (1992)

2 T.Nakagawa: "Modeling of a lightweight vehicle for H∞ control to get robust stability and high ride quality", IEE of Japan, IAS, No.28 (1992)

Simulation and Design of Applied Electromagnetic Systems
T. Honma (Editor)
1994 Elsevier Science B.V.

LQI Optimal Control of Electro-dynamic Suspension

S. Nakadai and M. Nagai

Tokyo University of Agriculture and Technology
Koganei-shi, Tokyo 184, Japan

Abstract

Magnetically levitated (MAGLEV) vehicle systems have been researched for future high-speed transportation. It is especially important to investigate the dynamic characteristics from the viewpoint of running stability, safety and ride quality at high speed. This paper deals with repulsive-type MAGLEV systems using electro-dynamic suspension (EDS). Although the EDS is inherently stable, the damping force is not large enough to suppress the vibration excited by guideway irregularity and other disturbances.

Therefore, this paper investigates theoretically and experimentally the magnetic force control of EDS systems with a hybrid magnet configuration of a permanent magnet and an additional electromagnet, using LQI optimal control theory. The dynamical characteristics are analyzed by an experimental set-up in which a one degree of freedom mass is suspended above an aluminum rotating plate. The computer simulation and experiment results show that the vibration characteristics are greatly improved by the active control, and the optimal control with integral compensation eliminates the offset caused by the stationary disturbance.

1. INTRODUCTION

Repulsive magnetic levitation systems using super conductive or permanent magnets have low-damping characteristics. Particularly, it is known that electro-dynamic suspensions have slightly negative damping on certain operating conditions. Therefore, it is necessary in such case to control the vibration to prevent the continuous unstable oscillations.

An experimental system has been constructed. The system is a one degree-of-freedom vehicle model of the EDS system with a hybrid configuration of permanent magnet and additional electromagnet which is suspended above an aluminum rotating plate. The levitation characteristics under various running conditions are show by the experimental system, and the damping characteristics are greatly improved by controlling the electromagnetic current [1]. In this paper, an optimal controller with integral compensation is designed using the optimal regulator theory to eliminate the offset caused by the stationary disturbance. Computer simulation together with experiment is carried out to examine the improvement of vibration characteristics.

2. THEORETICAL MODEL

2.1 Fundamental Equations

The equation describing the suspension dynamics around the nominal operating point of the one degree-of-freedom model shown in Fig.1 is expressed by the following linearized equation.

$$m\ddot{z} + c\dot{z} + kz = hi \tag{1}$$

where m is the equivalent mass of the levitated body, z is the levitation gap variation from the nominal operating point and i is the electromagnetic current. The parameters c, k and h in Eq.(1), which are determined by the running speed and magnetic characteristics, are the magnetic damping coefficient, the magnetic stiffness and the current coefficient, respectively. From Eq.(1) assuming that the state variables are the levitation gap and the vertical velocity of the levitated body, the following state-space equation can be formulated.

$$\frac{d}{dt}\begin{bmatrix} z \\ \dot{z} \end{bmatrix} = \begin{bmatrix} 0 & 1 \\ -\frac{k}{m} & -\frac{c}{m} \end{bmatrix}\begin{bmatrix} z \\ \dot{z} \end{bmatrix} + \begin{bmatrix} 0 \\ \frac{h}{m} \end{bmatrix} i \tag{2}$$

The total electro-magnetic current i is decomposed into controlled current i_c and disturbance current i_d, that is,

$$i = i_c + i_d \tag{3}$$

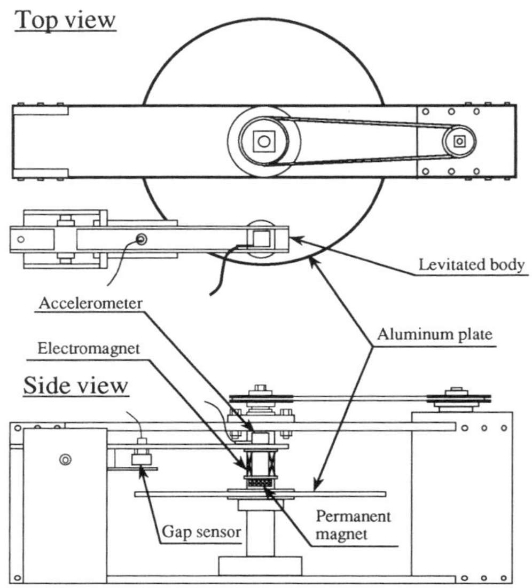

Fig.1 Experimental set-up of EDS

2.2 Optimal control system

The feedback controller with integral compensation is designed to eliminate the offset caused by stationary disturbances. This paper applies a linear feedback control law defined as

$$i_c = -f_1 z - f_2\dot{z} - f_i\int_0^t z\ dt \tag{4}$$

where the feedback gain of state variables f_1 and f_2, and the gain of the integral compensation f_i are calculated by minimizing a prespecified performance index according to the LQI optimal control theory. In this paper the following performance index is defined as

$$J = \mathrm{E}\left[q_1\dot{z}^2 + q_2\ddot{z}^2 + q_i z^2 + \dot{i}_c^2\right] \tag{5}$$

where q_1, q_2 and q_i are weighting coefficients. The closed-loop system of the state-space equation Eq.(2) together with the control law Eq.(4) becomes the third-order system.

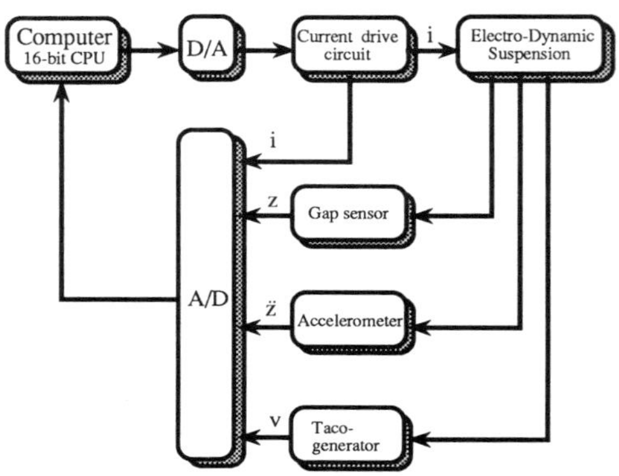

Fig.2 Configuration of the control circit

3. EXPERIMENT

3.1 EDS structure

Fig.1 shows the top and side views of the experimental set-up of Electro-Dynamic Suspension. The levitated body can rotate around the horizontal axis and behave as a one-degree-of-freedom model. The hybrid magnet configuration of permanent and electro-magnets is adopted and attached to the edge of the levitated body, which is suspended above the 20mm thick aluminum rotating plate. The plate is driven by an electric motor to change the horizontal speed from 20km/h to 100km/h.

The permanent magnet used is a rare-earth magnet which has a cylindrical form with the cross-sectional area 1000mm^2 and the flux density 1.22T. The electromagnet is attached on the permanent magnet. The number of copper coils is 1100 turns around the iron core which is 27mm in diameter and 60mm in height. The magnetic current is allowed up to 2 ampere.

3.2 Control circuit

The control circuit is shown in Fig.2. The electromagnetic current is compensated by a feedback control loop, measuring the vertical acceleration and the gap variations of the levitated body. The measured signals are input to a 16-bit microcomputer through an A/D converter as shown in Fig.2. According to the optimal regulator principle described in section 2.2, the microcomputer produces an output signal which actuates the electromagnet through a D/A converter. Considering the resonant frequency of the levitated body, the sampling interval is determined to 20msec. The current is driven in a way of pulse width modulation (PWM).

4. RESULTS

4.1 Step response (LQ control)

Figs.3 show the passive and the active transient responses of the levitation gap and the electromagnetic current to disturbance current 1A, when the running velocity is 50km/h. In this figure, (a) shows the calculated result, (b) shows the experimental result. Although the response of the passive system continues to vibrate with low damping, the response of the active system presents significantly reduced overshoot and improved settling time. Furthermore, it is found that there is a good agreement between the calculated and experimental results.

4.2 Step response with integral compensation (LQI control)

Figs.4(a) and (b) respectively show the calculated and experimental step responses with integral compensation of the levitation gap and the electromagnetic current for different weighting coefficient q_i when the running velocity is 50km/h. In these figures, when q_i equals zero, the controller means optimal control without integral compensation. From the result in Fig.4, it is clear that the settling time of the levitation gap in the case of optimal control(LQ) is the same as that in the case of optimal control with integral compensation, and the offset is eliminated by the integral operation. The magnetic current becomes zero when the levitation gap is controlled to the nominal operating point. From the results shown in Figs.4(a) and (b), the results of computer simulation are almost the same as those of the experiment when the feedback gain is chosen as small values, but the results of computer simulation and experiment differs when the feedback gain is chosen as big values. This difference can be explained by the time delay in the electromagnetic circuit.

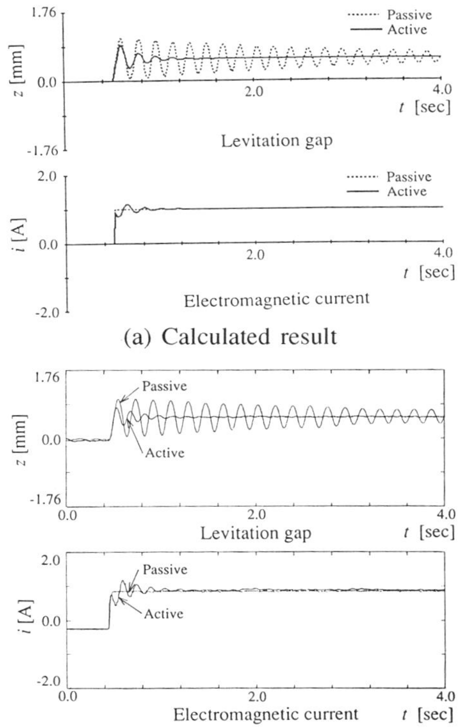

(b) Experimental result

Fig.3 Step responses (LQ)

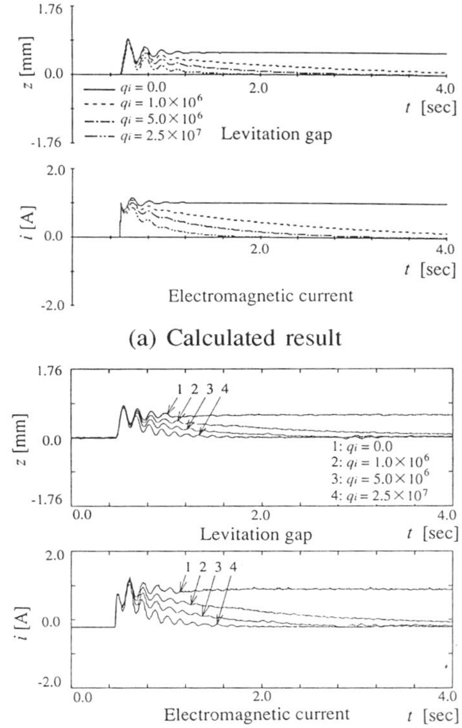

(b) Experimental result

Fig.4 Step responses (LQI)

5. CONCLUSIONS

The effectiveness of actively controlled electromagnetic current applied to electro-dynamic suspension system has been analyzed theoretically and experimentally. The optimal control system with integral compensation designed by optimal regulator theory, eliminates the offset caused by the stationary disturbance.

According to the computer simulation and experiment using an electro-dynamic suspension set-up in which a one degree of freedom mass is suspended above an aluminum rotating plate, the following findings are obtained;

· The vibration characteristics are greatly improved by the proposed active control.

· The optimal control with integral compensation using the accelerometer and the gap sensor eliminates the offset caused by the stationary disturbance.

REFERENCE

[1] H.Mori, S.Nakadai, H.Yamazaki and M.Nagai: "Active damping control of electro-dynamic suspension using electromagnets", Proceedings of First International Conference on Motion and Vibration Control, Yokohama, (1992), pp.669-674

Control and Magnetic Bearing

Simulation and Design of Applied Electromagnetic Systems
T. Honma (Editor)

Design of a robust magnetic levitation system

T. Eisaka[a], N. Hanajima[b], Y. Yanagita[b], Y. Kamiya[a] and R. Tagawa[c]

[a] Dept. of Information Engg., Kitami Institute of Technology, Kitami, 090, Japan

[b] Dept. of Electrical Engg., Hokkaido University, Sapporo, 060, Japan

[c] Asahikawa National College of Technology, Asahikawa, 071, Japan

Abstract: This paper presents a novel design method for the control of a magnetic levitation system based on *robust model matching* using inter-sample output signal of a plant.

Notation: $r \in R^m$: reference inputs, $u \in R^m$: control variables, $y \in R^m$: controlled variables on sampling time, $z \in R^m$: controlled variables at inter-sample time. I: unit matrix, P_{ab}, W_{ab}, denotes transfer functions of the plant, the closed loop system from signal a to b, respectively. In particular, ${}_MW$ and ${}_RW$ means the transfer functions of Model-matching and Robust-model matching, respectively.

1. INTRODUCTION

Magnetic levitation system possess a lot of advantages as actuators, however, model uncertainties lose their excellent performance, hence, it is important to obtain robust control systems against the uncertainties.

Recently, a novel design method of a robust control system for minimum-phase plants has been developed, and named it *Robust Model Matching* (RMM) [1]. As for digital control, however, the conventional RMM can not be applied to magnetic levitation systems because of unstable limiting zeros [2]. Lately, it has shown that the problem occurred by unstable zeros can be avoided by *Two Delay Output Control* (TDOC) which use inter-sample outputs [3]. In this paper, we show a novel design method of a digital controller within the RMM framework using the TDOC technique, and apply it to a magnetic levitation system.

2. CONTROLLER DESIGN

First, we will summarize the conventional RMM and point out the problem caused by unstable zeros. Next, we will show the way how to solve the problem.

2.1. Robust model matching

The influence of model uncertainties can be replaced by conceptual signals called *equivalent disturbances*: d added to the nominal model. Insensitive control systems against the model uncertainties can be obtained by *zeroing* the overall-transfer functions from the equivalent disturbances to the controlled variables: W_{dy} .

The RMM control system is composed of a model matching part and a robust compensator, separately. The robust compensator is composed of the following three factors.

$R_1(z)$: The observer of the equivalent disturbances.

$R_2(z)$: The zeroing factor which makes ${}_RW_{dy}$ strictly zero.

$F(z)$: The low pass filter which makes the controller proper, and adjusts the degree of robustness.

The W_{dy} of the RMM can be written with that of model matching in the form:

$${}_RW_{dy}(z) = (I - F(z)) \cdot {}_M W_{dy}(z) \tag{1}$$

Therefore, arbitrarily low sensitive control system over any bounded frequency range can be achieved with the filter $F(z) \rightarrow I$.

However, the element $R_2(z)$ becomes unstable for non-minimum-phase plants, and it leads closed-loop system unstable, thus the conventional RMM can not be applied to non-minimum-phase plants.

2.2. Stabilization of the R_2

The robust compensator includes the inverse system of the plant zeros: $N_{uz}(z)$ whose input is z_i. (Note: Not y_i but z_i is used as the controlled variables, and y_i is used to stabilize the $R_2(z)$)

The part $l_i = N_{uz}^{-1}(z) \cdot z_i$ can be realized in the controller canonical form as Eq. (2).

$$\bar{x}_{i+1} = \bar{A}_{co} \cdot \bar{x}_i + \bar{B}_{co} \cdot z_i \tag{2}$$

$$l_i = \bar{x}_{1i}$$

Here, $\bar{x}_i = [\bar{x}_{1i}, \bar{x}_{2i}, \cdots, \bar{x}_{ni}]^T = [1, z, \cdots, z^{n-1}]^T \bar{x}_{1i}$

On the other hand, the state equation of the plant is described as follows:

$$x_{i+1} = A \cdot x_i + B \cdot u_i \tag{3}$$

$$y_i = C \cdot x_i$$

The x_i and z_i are given by the next equations.

$$x_i = \{(adj(zI - A)B) / \det(zI - A)\} u_i \tag{4}$$

$$z_i = \{N_{uz} / \det(zI - A)\} u_i \tag{5}$$

Then, the state x_i can be written by z_i as

$$x_i = \{adj(zI - A)B / N_{uz}\} z_i \tag{6}$$

From Eq.(2) and (6), we get the following relationship between the state x_i and the state $\bar{x}_i$.

$$x_i = H \cdot \bar{x}_i \tag{7}$$

Here, H is regular constant matrix given by

$$adj(zI - A)B = H[1, z, \cdots, z^{n-1}]^T \tag{8}$$

When we have the following observer expressed as Eq.(9), the error e_i holds Eq.(10).

$$\hat{x}_{i+1} = (\overline{A}_{co} + L \cdot C \cdot H)\hat{x}_i - L \cdot y_i + \overline{B}_{co} \cdot z_i \tag{9}$$

$$e_{i+1} = \hat{x}_{i+1} - \overline{x}_{i+1} = (\overline{A}_{co} + L \cdot C \cdot H)e_i \tag{10}$$

Thus we see the factor $R_2(z)$ can be replaced by the stable observer (9) with an adequate coefficient vector L.

3. MAGNETIC LEVITATION SYSTEM

The overview of a magnetic levitation system is shown in Figure 1, and the parameters of it are given in Table 1.

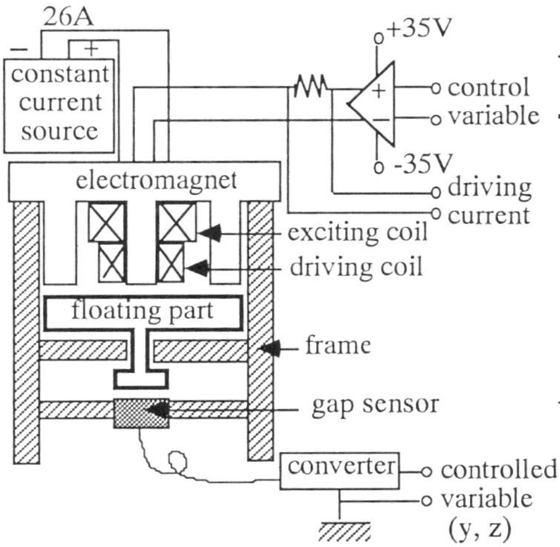

Figure 1. Magnetic levitation system

Table 1 Parameters of the plant

Parameter	Value(unit)
Statinary gap of the electromagnet	1(mm)
Mass of the floating part	0.45(Kg)
Resistance of the exciting coil	0.32(Ω)
Inductance of the exiting coil	1.4(mH)
Resistance of the driving coil	3.24(Ω)
Inductance of the driving coil	18.8(mH)

The control objective is to control the position of the metal floating part by electromagnetically sucking force. The exciting circuit generates electromagnetic force to balance the floating part with gravitational pull on it, and the driving circuit controls the position of it. The driving current and position of the metal floating part are available for feedback signals.

The pulse transfer function of the linealized model of the plant has unstable zero as follows:

$$P_{uz}(z) = \frac{.0026(z+20.2)(z+.92)(z+.042)}{(z-1.2)(z-.75 \pm j.148)} \tag{11}$$

Here, sampling time is settled as 2 (msec) and inter sample point is decided as middle of the sampling period.

4. SIMULATIONS

A controller is designed by the proposed method. Here, the control objective was settled by the following W_{rz}.

$$W_{rz}(z) = \frac{4.9e^{-4}(z+20.2)(z+.92)(z+.042)}{(z-.67)(z-.65)(z-.6)(z-.55)} \tag{12}$$

The filter $F(z)$ was settled as $.6/(z-.4)$, and the poles of the observer (9) were settled as $.05$, $.06$, and $.07$.

We will illustrate the validity of proposed system (RMM servo) comparing with model-matching (MM servo), by computer simulations.

Figure 2 shows step reference responses of each servo with the floating part whose mass is 0.7 times the nominal one. Figure 3 shows step reference responses of each servo with the driving coil whose resistance is 1.2 times the nominal one. In the results, response of the MM servo oscillates or it is unstable, though the responses of the RMM servo converge to the nominal one.

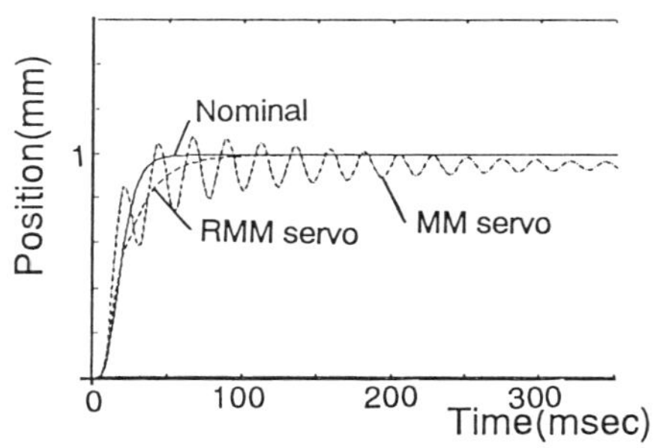

Figure 2. Step reference responses. (Mass changes)

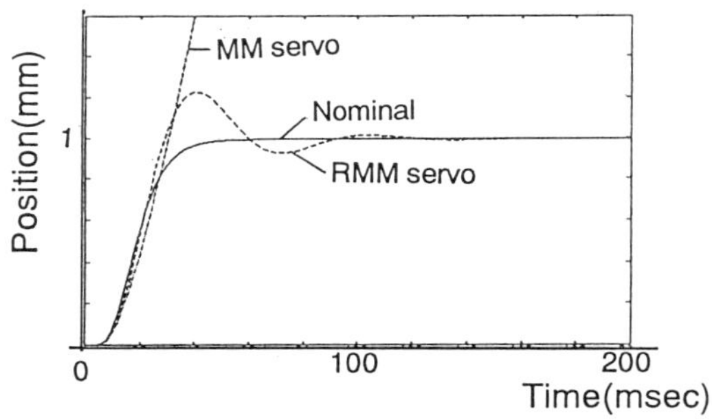

Figure 3. Step reference responses. (Resistance changes)

5. CONCLUSION

A novel design method based on *Robust-model matching* framework using *Two-delay output control* technique has been proposed. By this method, we can obtain arbitrarily low sensitive control system over any bounded frequency range even for non-minimum-phase plants.

6. REFERENCES

1 R. Tagawa, Computrol, Corona publishing Co., Ltd. (1986) 53 (in Japanese)

2 K. J. Astrom, P. Hagander, and J. Sternby, Automatica, 20-1, (1984) 31.

3 T. Mita, Y. Chida, Y. Kaku and H. Numasato, IEEE Trans. AC-35-8, (1990) 962.

Simulation and Design of Applied Electromagnetic Systems
T. Honma (Editor)

Self-sensing electrostatic suspension

Ju Jin[a] and Toshiro Higuchi[a,b]

[a] The 7th Laboratory, Kanagawa Academy of Science and Technology, 3-2-1 Sakato, Takatsu-ku, Kawasaki 213, Japan.

[b] Department of Precision Machinery Engineering, University of Tokyo, 7-3-1 Hongo, Bunkyo-ku, Tokyo 113, Japan.

Abstract

This paper proposes a new type of active electrostatic suspension technique without using position sensors. Voltage which varies with the gap between the electrode and the suspended object is used as feedback signal to control supply voltage. Simulation results are given to confirm the proposed method.

1. Introduction

In recent years, electrostatic suspension techniques have revived again. Not only is their usefulness in the elimination of friction and physical contact in microactuators and micromotors being widely recognized, but also many advantages are becoming more and more apparent. Traditionally, in order to achieve a stable suspension, the position signal of the suspended object is used as the feedback signal to control the applied voltage. Independent position sensors are necessary in such systems. The employment of position sensors yields two problems. First, position sensors are relatively expensive to the overall cost. Secondly, the presence of position sensors brings an increase in total size. This is not suitable to micro-devices.

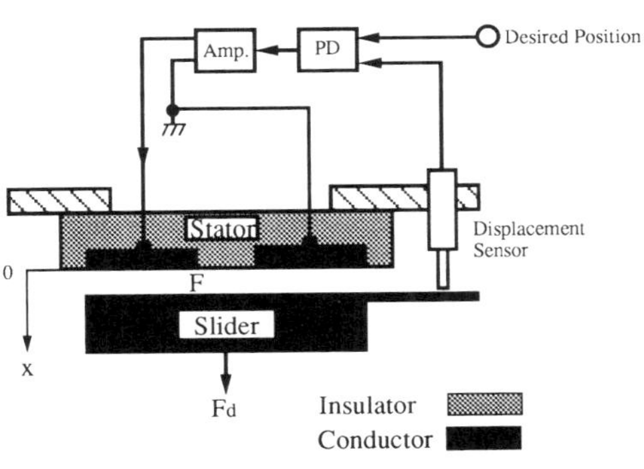

Fig. 1 Electrostatic Suspension System

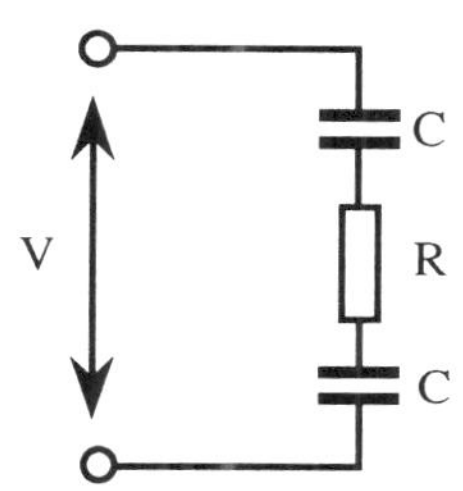

Fig.2 Equivalent Circuit

2. Principle of Suspension

Fig. 1 shows the one degree of freedom model of self-sensing electrostatic suspension systems. It is known that with unvarying supply voltage, when the gap changes, current is generated and the voltage across the electrode and the suspended object also changes. This very property provides the possibility of stablizing a suspension system using voltage feedback instead of direct position feedback.

3. System Descriptions

The equivalent circuit of the CR circuit in Fig. 1 is shown in Fig. 2. It can be described by

$$(2+R\frac{d\tilde{C}}{dt})\tilde{V}_c + R\tilde{C}\frac{d\tilde{V}_c}{dt} = \tilde{V} \tag{1}$$

where R denotes the resistance; $\tilde{V}_c$, the voltage across the electrode and the suspended object; $\tilde{C}$, the capacitance of each capacitor; and $\tilde{V}$, the supply voltage.

When the overlapping area of the electrode and the suspended object is large enough compared to the gap, to a close degree of approximation, the attractive force can be represented as

$$\tilde{F} = -\frac{1}{2}\frac{d\tilde{C}}{d\tilde{x}}\tilde{V}_c^2 = \frac{\varepsilon_0 A\tilde{V}_c^2}{2\tilde{x}^2} \tag{2}$$

where ε_0 denotes the permittivity of air and is equal to 8.855×10^{-12}; A, the overlapping area of each electrode; $\tilde{x}$, the gap between the electrode and the suspended object; and $\tilde{F}$, the attractive force.

The equation of motion of the suspended object is

$$m\ddot{\tilde{x}} = mg - 2\tilde{F} + F_d \tag{3}$$

where m denotes the mass of the suspended object and F_d is the disturbance force applied upon the suspended object.

4. Voltage Feedback Control

It is supposed that in order to achieve a stable suspension, the supply voltage should be controlled as

$$\tilde{V} = V_e - V \tag{4}$$

where V_e denotes the value at the equilibrium state, and V is the control voltage.

The control law we have chosen is

$$V(s)=\frac{Y_2 s^2+Y_1 s+Y_0}{T_2 s^2+T_1 s+T_0}V_c(s)=G(s)V_c(s) \tag{5}$$

where s denotes the Laplace transformation.

It is obvious from Eqs. (1)-(3) that the suspension system is nonlinear. We can linearize them near the equilibrium state using the small signal theory. Each of the variables can be expressed as the sum of the value at the equilibrium state, denoted by subscript e, plus a small change.

$$\tilde{x} = x_e + x \qquad \tilde{C} = C_e + C \qquad \tilde{F} = F_e + F \qquad \tilde{V}_c = V_{ce} + V_c \tag{6}$$

With Eqs. (5) and (6) inserted, neglecting the nonlinear components of small change Eq. (1) can be simplified to

$$V_c(s) = \frac{RC_e \dfrac{V_{ce}}{x_e} s}{RC_e s + 2 - G(s)} X(s) \tag{7}$$

In the same way, the force equation (2) can be rewritten as

$$\tilde{F} = F_e + \left.\frac{\partial \tilde{F}}{\partial \tilde{V}_c}\right|_{x_e} V_c + \left.\frac{\partial \tilde{F}}{\partial \tilde{C}}\right|_{x_e} x = F_e + K_v V_c - K_x x \tag{8}$$

where

$$K_V = \left.\frac{\partial \tilde{F}}{\partial \tilde{V}_c}\right|_{x_e} = \frac{\varepsilon_0 A V_{ce}}{x_e^2} \qquad K_x = -\left.\frac{\partial \tilde{F}}{\partial \tilde{C}}\right|_{x_e} = \frac{\varepsilon_0 A V_{ce}^2}{x_e^3}$$

From Eq. (7), we can easily find the relationship between force and the displacement,

$$F(s) = \frac{\sum_{i=0}^{2} \gamma_i s^i}{\sum_{i=0}^{3} \beta_i s^i} X(s) \tag{9}$$

where

$$\begin{cases} \beta_3 = aT_2 \qquad \beta_2 = aT_1 + 2T_2 - Y_2 \qquad \beta_1 = aT_0 + 2T_1 - Y_1 \qquad \beta_0 = 2T_0 - Y_0 \\ \gamma_1 = 2K_x(Y_1 - 2T_1) \qquad \gamma_2 = 2K_x(Y_2 - 2T_2) \qquad \gamma_0 = 2K_x(Y_0 - 2T_1) \end{cases}$$

Inserting Eq. (9) into Eq. (3), give its Laplace transformation as

$$X(s) = \frac{\sum_{i=0}^{3} \beta_i s^i}{\sum_{i=0}^{5} \alpha_i s^i} F_d(s) \tag{10}$$

where

$$\alpha_5 = m\beta_3 \qquad \alpha_4 = m\beta_2 \qquad \alpha_3 = m\beta_1 \qquad \alpha_2 = m\beta_0 + \gamma_2 \qquad \alpha_1 = \gamma_1 \qquad \alpha_0 = \gamma_0$$

Here it is assumed that the weight of the suspended object is only the steady load.

It is obvious from the definitions of α_i, β_i and γ_i that for an appropriate choice of the control parameters Y_2, Y_1, Y_0, T_1, T_0, it is possible to assure the coefficients α_i, (i = 0, 1,

2, 3, 4, 5), all positive, satisfying the Routh-Hurwitz stability criterion, and hence stabilize the suspension system.

5. Numerical Analysis

Now let us examine the analysis made above through a numerical example. In order to verify the validity of the proposed method, here we use system parameters obtained from a real physical system rather than pure academic values. These parameters are listed in Table 1. System response to a step disturbance F_d = 1.0 (gf) is shown in Fig. 3.

6. Conclusion

In this paper, we have proposed a new active electrostatic suspension technique, self-sensing electrostatic suspension. It can realize a stable suspension without using any position sensors. This always leads to lowering the system cost and downsizing the total system. We believe that this technique can find wide industrial application in the future.

Table 1. System Parameters

x_e	100(μm)	C_e	211.8(pF)	V_e	341.0(V)
V_{ce}	170.5(V)	R	1000(Ω)	m	3.33(g)
T_2	2.0	T_1	5.0×10^6	T_0	1.0×10^{11}
Y_2	5.0	Y_1	1.0002×10^7	Y_0	2.0×10^{11}

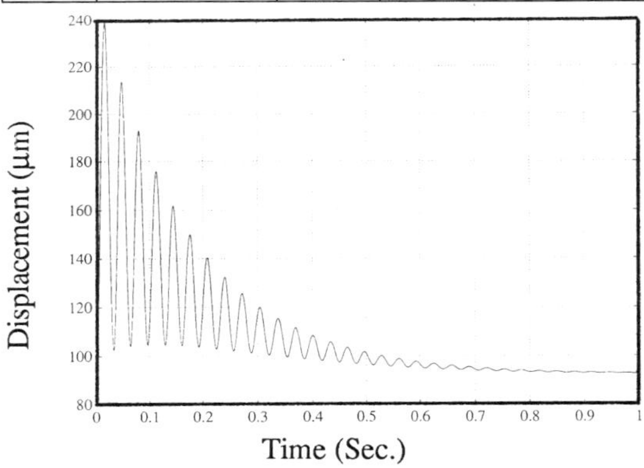

Fig. 3 System Response to Step Disturbance

7. Reference

[1] Knoble, H. W., "Electric Vacuum Gyro", *Control Engineering,* Vol. 11, pp. 70-73, Feb. 1964.

Simulation and Design of Applied Electromagnetic Systems
T. Honma (Editor)

Magnetic levitation system by reluctance control — Levitation by permanent magnet and linear actuator —

T. Higuchi and K. Oka

Department of Precision Machinery Engineering, Faculty of Engineering, University of Tokyo, 7-3-1, Hongo, Bunkyo-ku, Tokyo, 113, Japan

Abstract

A new type of active magnetic levitation system is proposed. The features of this system are that the forces of permanent magnets are used for levitation and controlled by adjusting the reluctance of the magnetic circuit. We describe the principle of the levitation system and typical reluctance control methods, and we introduce an experimental device whose reluctance is controlled by adjusting the air gap between the permanent magnet and the levitated object. Linear control theory and experimental results verify the feasibility of the reluctance control levitation system.

1. INTRODUCTION

Noncontact magnetic levitation systems without super conductors need position feedback loops and control of magnetic forces. Position sensors and electromagnets with current feedback control are normally installed. However electromagnet coils sometimes cause the problem of heat generation.

We propose a new type of active levitation system with permanent magnets and a reluctance control mechanism in place of electromagnets and a current control mechanism. As no electromagnets are used, it is effective for saving energy and avoiding heat generation. In this paper, we will explain the principle of such a system and introduce two types of reluctance control. A prototype levitation system with a reluctance control mechanism is introduced. Linear control theory is applied to stabilize the levitation. Experimental examinations verify the feasibility of our approach and the validity of the theoretical analysis.

2. RELUCTANCE CONTROL MAGNETIC LEVITATION

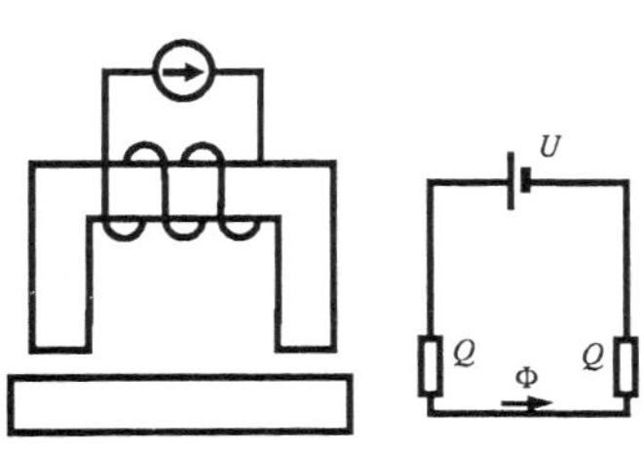

Figure 1. System with electromagnet

We investigate active magnetic levitation systems which control attractive forces of magnets. We are making the following assumptions: (i)magnetic flux does not leak in magnetic circuits, (ii)reluctance of iron core can be neglected and (iii)attractive force is proportional to the square of the magnetic flux which flows through the air gap.

A typical *electromagnetic* suspension system is shown in Figure 1. U stands for magnetomotive force of a coil, and Q stands for the reluctance of the air gap. Magnetic flux Φ through the air gap of this system is

$$\Phi = \frac{U}{2Q} \qquad (1).$$

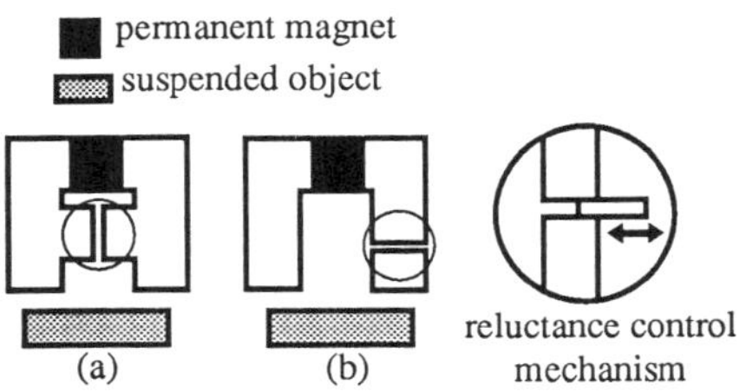

Figure 2. Reluctance control system (type I)

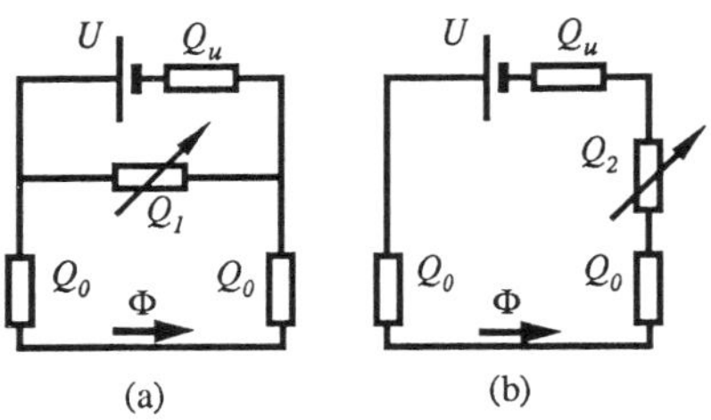

Figure 3. Equivalent magnetic circuits

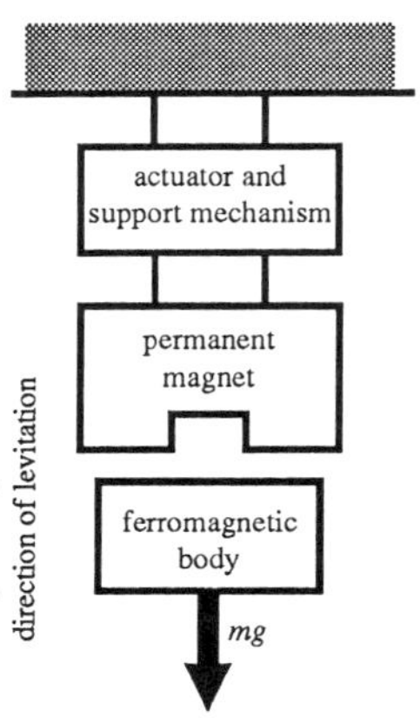

Figure 4. Levitation system II

The magnetomotive force of an electromagnet can be controlled by adjusting the coil current. From assumption (iii), the attractive force can be controlled.

2.1 Reluctance control method I (new reluctance control mechanism in the magnetic circuit)

We propose that the *reluctance* is adjusted for attractive force control. Two typical suspension systems with reluctance control mechanism are shown in Figure 2. The circles in Figure 2 (a)(b) are reluctance control mechanism such as shown in the right of the figure. This reluctance control mechanism uses magnetic saturation of a ferromagnetic body. Equivalent magnetic circuits of the systems in Figure 2 are shown in Figure 3. Reluctance control mechanisms are represented as variable resistors Q_1 and Q_2. The magnetic flux through the air gap of (a) and (b) are respectively

$$\Phi_a = \frac{Q_1 U}{2Q_u Q_0 + Q_u Q_1 + 2Q_0 Q_1}$$
$$\Phi_b = \frac{U}{Q_u + 2Q_0 + Q_2} \qquad (2)$$

where, Q_u is internal reluctance of a permanent magnet. From equation (2) and assumption iii), attractive forces can be controlled by varying the reluctance in the magnetic circuit.

2.2 Reluctance control method II (air gap control by permanent magnet and linear actuator)

Variation of the length of the air gap can also control reluctance. Levitation systems can be implemented by the application of varying gap mechanism to the circle parts of Figure 2.

We have proposed another type of levitation system with adjusting gap mechanism. Construction of the proposed system is shown in Figure 4. In this system, the direction of levitation is vertical. The reluctance of the magnetic circuit is controlled by the adjustment of the air gap between a magnet and a suspended object (ferromagnetic body). To adjust the air gap, the permanent magnet is driven by a actuator in the vertical direction as follows. If the object moves towards the magnet from equilibrium position, the actuator drives the magnet upwards. Oppositely, if the object moves away, the actuator drives the magnet downwards.

3. EXPERIMENTAL SYSTEM AND ITS THEORETICAL FEASIBILITY ON LEVITATION

Following the concept of section 2.2, an experimental device was build. The configuration of the experimental system is shown in Figure 5, and its photograph is shown in Figure 6. The shaded area in Figure 5 moves in concert with a permanent magnet and is supported by parallel plate springs and driven by a voice coil motor. The suspended object is an iron ball. Its position is sensed by an eddy current gap sensor located under the ball. The resolution of the sensor is 0.5μm. The control system has a PD feedback loop and drives the voice coil motor by a voltage amplifier.

A model of this system is shown in Figure 7. Variables in the figure are: k_e = spring constant of parallel plate springs (4.57kN/m), ξ = damping factor of springs (0.27Ns/m), m_0 = mass of the iron ball (96.4g), m_1 = mass of the hatched area (28.2g), x_0 = position of the iron ball, x_1 = position of the permanent magnet, f_a = propulsive force of the voice coil motor. Other variables are: i = current of the voice coil, V = applied voltage, R = resistance (6.35Ω), L = inductance (0.01H), k_t = propulsive constant(propulsive force about coil current) (33.5N/A), f_m = attractive force of magnet. The equations of motion of this system are

$$\begin{aligned} m_0\ddot{x}_0 &= m_0 g - f_m \\ m_1\ddot{x}_1 + \xi\dot{x}_1 + k_e x_1 &= f_a + f_m + m_1 g \\ L\dot{i} + Ri + k_t\dot{x}_1 &= V \\ f_a &= k_t i \end{aligned} \qquad (3).$$

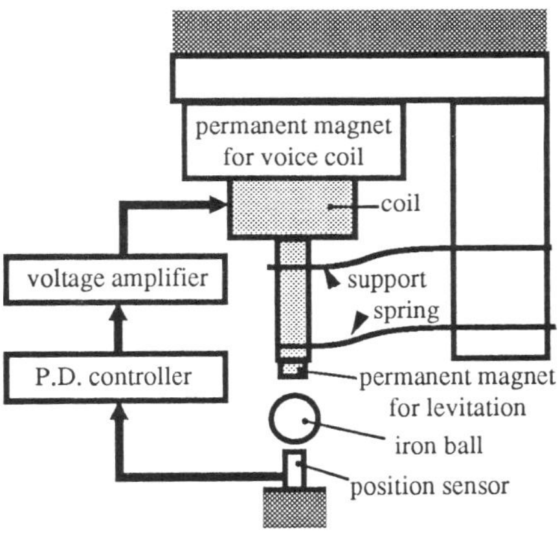

Figure 5. Configuration of experimental system

Considering that attractive force is represented as a function of the length of the air gap and becomes larger as the gap decreases, we obtain

$$-\left.\frac{\partial f_m}{\partial x_0}\right|_{\substack{x_0=x_{00}\\x_1=x_{10}}} = \left.\frac{\partial f_m}{\partial x_1}\right|_{\substack{x_0=x_{00}\\x_1=x_{10}}} \equiv k_m\,, \quad (k_m > 0) \qquad (4)$$

where x_{00} and x_{10} are equilibrium position. From equation (3) and (4), the state equation is

$$\dot{x} = Ax + bu\,, \quad y = Cx \qquad (5)$$

where,

$$x' = (\, x_0, x_1, \dot{x}_0, \dot{x}_1, i \,)\,, \quad u = V\,,$$

$$B' = (0\ 0\ 0\ 0\ 1/L\,)\,, \quad C = (\,1\ 0\ 0\ 0\ 0\,)\,,$$

Figure 6. Photograph of experimental setup

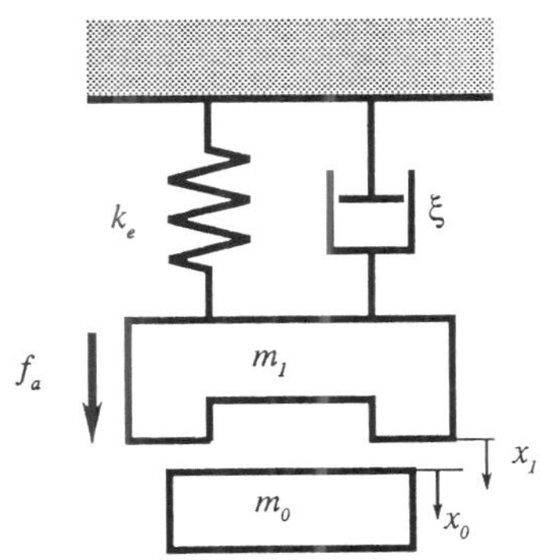

Figure 7. Model of the experimental system

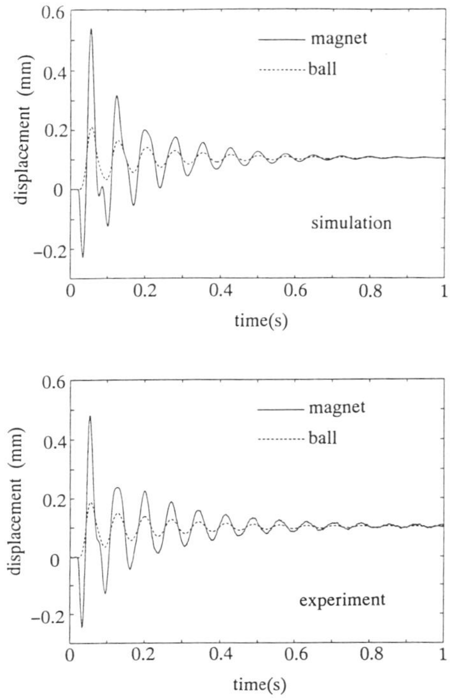

Figure 8. Step responses

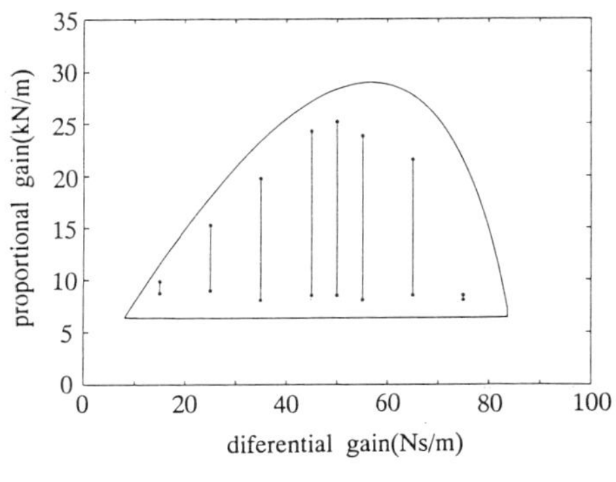

Figure 9. Stability range of feedback gains

$$A = \begin{pmatrix} 0 & 0 & 1 & 0 & 0 \\ 0 & 0 & 0 & 1 & 0 \\ k_m/m_0 & -k_m/m_0 & 0 & 0 & 0 \\ -k_m/m_1 & (k_m-k_e)/m_1 & 0 & -\xi/m_1 & k_t/m_1 \\ 0 & 0 & 0 & -k_t/L & -R/L \end{pmatrix}.$$

The system of equation (5) is controllable and observable.

4. EXPERIMENTAL AND THEORETICAL EXAMINATION

We examined a levitation experiment for the system shown in Figure 5. It was verified that the iron ball can be levitated without any mechanical contact. The precision of levitation was better than 1μm.

We examined experiment and numerical simulation of step response. The results are shown in Figure 8. The displacements of the ball and the magnet from initial position were recorded. The reference movement was 0.1mm downward (see the direction of x in Figure 7). In the figure, we can see that first the magnet moves upward to make the attractive force smaller, after that it moves as the ball converges on the reference points. Both numerical and experimental results are similar, so the model in Figure 7 is assumed to be correct.

We investigated the range of feedback gains which make the levitation system stable. The result is shown in Figure 9. The part surrounded with a horizontal line and a slanted parabola is the theoretical stability range. Vertical lines show the experimental stability range. Both ranges agree well. The theoretical analysis is considered to be very useful.

5. CONCLUSION

Reluctance control magnetic levitation was proposed and its principle was explained. An experimental levitation system with a permanent magnet and a voice coil motor was introduced. The feasibility of the system was verified from theoretical and experimental examinations. As further study, non-contact conveyor devices with reluctance control will be investigated.

Simulation and Design of Applied Electromagnetic Systems
T. Honma (Editor)

Frequency-shaped optimal control with a feedforward link on magnetically levitated vibration isolation system

S. K. Kang, K. Yoshida and S. Hara

Dept. of Mechanical Engineering, Keio University, 3-14-1 Hiyoshi, Kohoku-ku, Yokohama 223, JAPAN

Abstract

The object of this paper is the application of the optimal control with a feedforward link and the frequency-shaped optimal control for the vibration isolation system using magnetic levitaion. By carrying out experiment, it was experimentally verified that the control efficiency can be improved by applying the optimal control with a feedforward link and can be further improved by adopting the frequency-shaped optimal control.

1. INTRODUCTION

Recently, as demands for the high accuracy of the measuring technology and the production manufacturing technology grows, the development of the active vibration isolation system make rapid process. The actuator using magnetic levitation[(1)] which is composed of permanent magnet and electromagnet is paid attention to as a new actuator since it dosen't make friction and is a type of energy saving.

For the case of appling the optimal control theory to the active vibration isolation, as the objective function in the criterion function is not always described by only the state variables, it is necessary to perform the optimal control with a feedforward link[(2)]. Furthermore, to improve the control efficiency of vibration isolation, it is better to change the physical weighting of the objective function in the frequency range.

2. MODEL OF VIBRATION ISOLATION SYSTEM

In this study, we deal with a single-degree-of-freedom model, which is composed of permanent magnets and electromagnets as shown in Fig.1. Its disk table is levitated by the repulsion force of permanent magnets and the vibration of the table is controlled by the attraction force of electromagnets.

We have the following state equation by using the relative coordinate system, which consists of the equations of motion and voltage.

$$\dot{\boldsymbol{x}} = \boldsymbol{A}\boldsymbol{x} + \boldsymbol{b}u + \boldsymbol{D}_x z \tag{1}$$

$$\boldsymbol{x} = [x \quad \dot{x} \quad i]^{\mathrm{T}} \quad z = [z \quad \dot{z} \quad \ddot{z}]^{\mathrm{T}}$$

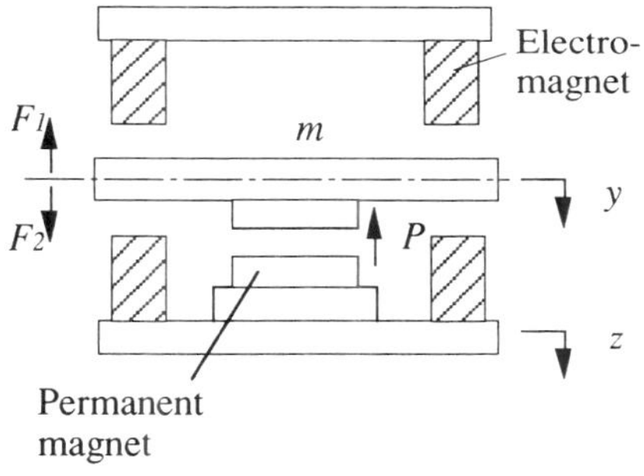

Figure 1. Single D.O.F model of vibration isolation system

$$A = \begin{bmatrix} 0 & 1 & 0 \\ -\omega_s^2 - \dfrac{2F_{20}}{mX_{20}}(B+1) & -2\zeta_s\omega_s & \dfrac{4F_{20}}{mI_{20}} \\ 0 & \dfrac{I_{20}}{X_{20}}(1-\dfrac{L_a}{L_0}) & -\dfrac{R}{L_0} \end{bmatrix} \quad D_x = \begin{bmatrix} 0 & 0 & 0 \\ 0 & 0 & -1 \\ 0 & 0 & 0 \end{bmatrix}$$

$$b = \begin{bmatrix} 0 & 0 & 1/L_0 \end{bmatrix}^{\mathrm{T}}$$

The disturbance might be assumed as a white noise in the case that its information is not known. For such case, since the feedforward control law cannot be derived, we assume that, by restricting the frequency range of disturbance, the disturbance is such a colored noise that has a tansfer function of a low pass filter. In this case, the state equation of the disturbance is written by

$$\dot{z} = A_z z + dw \tag{2}$$

$$A_z = \begin{bmatrix} 0 & 1 & 0 \\ 0 & 0 & 1 \\ 0 & -\omega_z^2 & -2\zeta_z\omega_z \end{bmatrix} \quad d = \begin{bmatrix} 0 & 0 & \omega_z^2 \end{bmatrix}^{\mathrm{T}}$$

Utilizing Eqs.(1) and (2), we have the following augmented state equation.

$$\dot{x}_p = A_p x_p + B_{up} u + B_{wp} w \tag{3}$$

$$x_p = \begin{bmatrix} x & z \end{bmatrix}^{\mathrm{T}} \quad A_p = \begin{bmatrix} A & D_x \\ 0 & A_z \end{bmatrix} \quad B_{up} = \begin{bmatrix} b & 0 \end{bmatrix}^{\mathrm{T}} \quad B_{wp} = \begin{bmatrix} 0 & d \end{bmatrix}^{\mathrm{T}}$$

3. DISTURBANCE-ACCOMMODATION OPTIMAL CONTROL

3.1 Optimal control law

By adopting the following criterion function,

$$J(u) = \lim_{tf \to \infty} \frac{1}{tf} \mathrm{E}\left\{ \int_0^{tf} \left[x_p^{\mathrm{T}} Q x_p + r u^2 \right] dt \right\} \qquad Q = \begin{bmatrix} Q_1 & G \\ G^{\mathrm{T}} & Q_2 \end{bmatrix} \tag{4}$$

we derive the following optimal control law by applying the opimal control theory to Eqs.(3) and (4)

$$u = -F_b x - F_f z \quad : F_b = r^{-1} b^{\mathrm{T}} P_{11},\ F_f = r^{-1} b^{\mathrm{T}} P_{12} \tag{5}$$

where P_{11} and P_{12} are the solutions of the Riccati equation.

3.2 Frequency-shaped optimal control law

In order to perform most effective vibration isolation, it is necessary to adopt such a frequency-shaped criterion that the displacement is chosen below the natural frequency of the system and the acceleration is done above it. Accordingly, we try to apply the frequency-shaped optimal control with a feedforward link. In the criterion function, the absolute displacement and the absolute acceleraton of the disk table of the vibration isolation system and control force are chosen as follows.

$$J = \int_{-\infty}^{\infty} [\boldsymbol{x}_p^*(s)\boldsymbol{Q}_d(s)\boldsymbol{x}_p(s) + \boldsymbol{x}_p^*(s)\boldsymbol{Q}_a(s)\boldsymbol{x}_p(s) + u^*(s)r(s)u(s)]_{s=j\infty} d\omega \quad (6)$$

where * implies complex conjugate. Fig.2 shows the physical weighting of objective function changed in the frequency range. Utilizing the state space equation by Eqs.(6) and Eqs.(2), we have the augmented state equation. By applying the optimal control theory, we can have the optimal control law.

$$u = -r_n^{-1} B_n^{\mathrm{T}} \boldsymbol{P}_n \boldsymbol{x}_n \quad (7)$$

where $\boldsymbol{P}_n$ is the solution of Riccati equation for the augmented state equation.

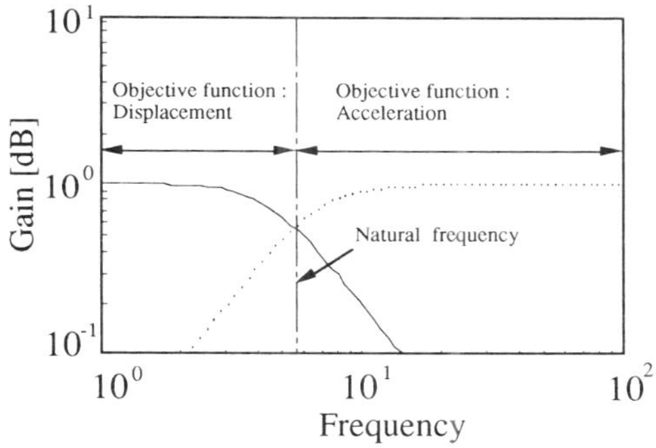

Figure 2. Filter feature of weighting on objective function

4. EXPERIMENT

4.1 Experimental setup

The experiment was performed by using experimental setup as shown in Fig.3. The frequency responses were measured for the relative coordinate system. Its input is the absolute displacement of disturbance and its output is the absolute displacement of the disk table. The white noise which has band width 0-50Hz was used as the disturbance. Before performing the control, the initial voltage of 0.7V was given to the upper electromagnet and the lower electromagnet to keep up static load. The sampling period of control was 3ms.

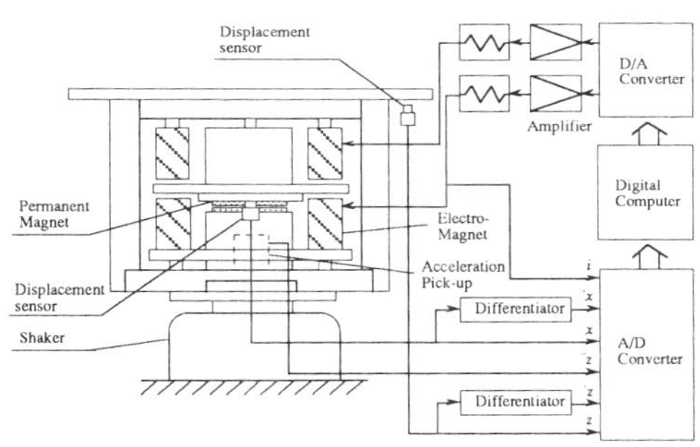

Figure 3. Experimental setup

4.2 Experimental result

The value of parameters used in the experiment are as follows

m=1.53kg, k=1890.4N/m, c=1.474kg/s, X_{20}=1.20e-3m, B=1.0, F_{20}=1.27e-1N, L_o=9.65e-2H, L_a=3.35e-2H, I_{20}=1.4e-1, R=5.04Ω, $\omega_z=5\omega_s$, ζ_z=0.7

Figure 4 shows the results of frequency responses of the numerical calculation and the experiment. From these results, it is seen that the control property corresponds to the objective function irrespective of adopting the coordinate of the state variables.

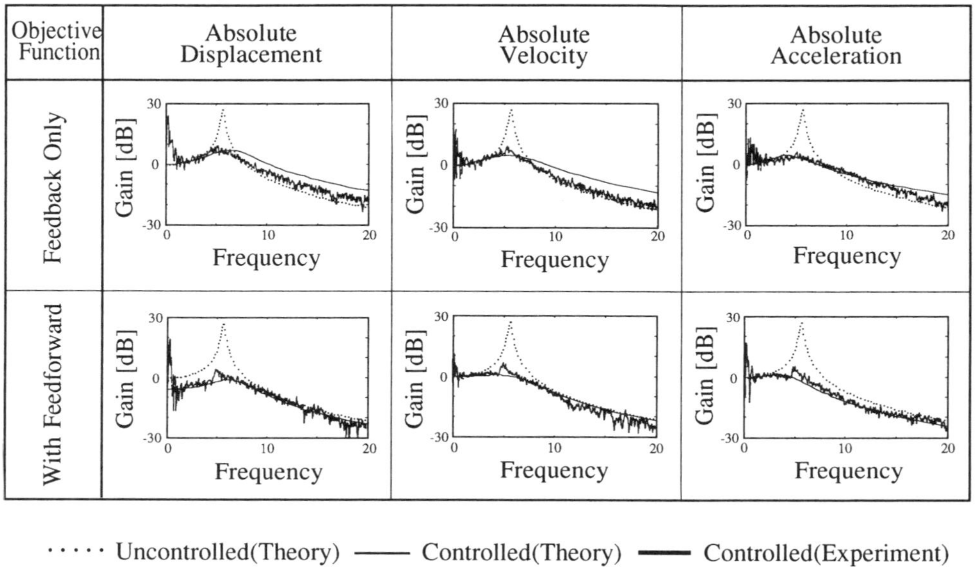

Figure 4. Frequency response by disturbance-accommodation optimal control

Figure 5 shows the result of the frequency-shaped optimal control. The value of weighting coefficient on the control is r_n=1.0e-6. From this result, the control characteristics of the displacement adopted into the objective function appear below the natural frequency, those of the acceleration are realized above the natural frequency.

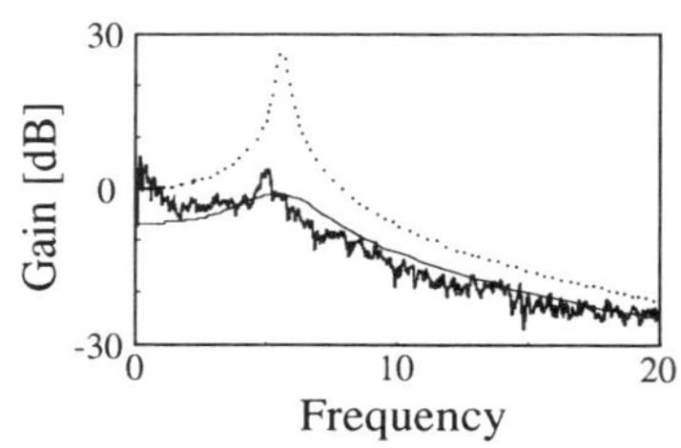

Figure 5. Frequency response by frequency-shaped optimal control

5. CONCLUSIONS

The optimal control with a feedforward link and the frequency-shaped optimal control were attempted to apply to the vibration isolation system using magnetic levitaion. In this study, it was experimentally verified that the control efficiency can be improved by applying the optimal control with a feedforward link and can be further improved by adopting the frequency-shaped optimal control.

6. REFERENCES

1. Nagaya, K., Arai, N., Trans. Japan Soc. Mech. Eng. 56-525C(1990), pp.1230
2. Kang, S.-K., Yoshida, K., The 1st International Conf. on Motion and Vibration Control, Yokohama, Sept. 1992,Vol. 1 pp.291.

Simulation and Design of Applied Electromagnetic Systems
T. Honma (Editor)

Disturbance observer based controller for magnetic bearings

Y. Okada[a], Y. Ueda[a], B. Nagai[a] and K. Ohnishi[b]

[a]Department of Mechanical Engineering, Ibaraki University, 4-12-1 Nakanarusawa, Hitachi, 316 JAPAN

[b]Department of Electrical Engineering, Keio University, 3-14-1 Hiyoshi, Kouhoku-ku, Yokohama, 223 JAPAN

Abstract

A disturbance observer based controller is introduced which is simple, but very adequate to control magnetic bearings. The controlled magnetic bearings have high stiffness and robustness. A simple test apparatus is made to clarify its capability.

1. INTRODUCTION

Magnetic bearing is gradually used because of its noncontact supporting capability. It has the merits of no friction, no wear and clean bearing [1]. However, there are some problems; for example, poor supporting force, highspeed instability problem and system robustness. A stiff and robust controller is highly requested.

This paper introduces a simple disturbance observer based controller to magnetic bearings [2] [3]. A second order observer can identify the disturbing force and the velocity from the measured displacement and the demand signal. The disturbing force can be canceled with the estimated one, while the estimated velocity is used to improve the dynamic property of the system. The proposed system has a well damped high stiffness and robustness compared to its simple control algorithm. The capability of this method is tested with a simple experiments.

2. DISTURBANCE OBSERVER BASED CONTROL SYSTEM

One dimensional controller of magnetic bearing is shown schematically in Fig. 1. The dynamical equation of motion is given by

$$m\frac{d^2x(t)}{dt^2} = f(t) - f_d(t) \tag{1}$$

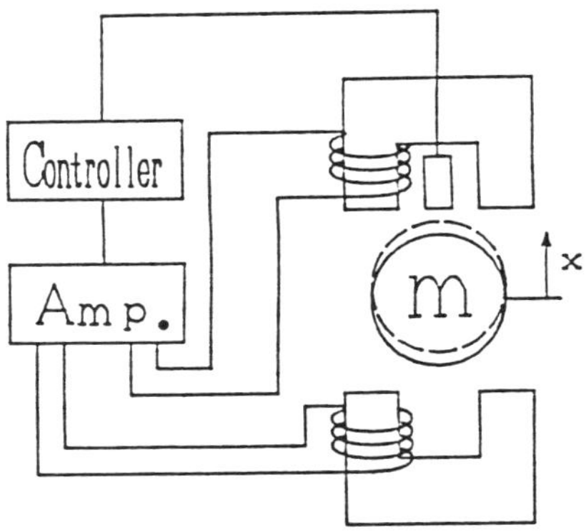

Fig. 1. Rotor and the magnetic bearing system

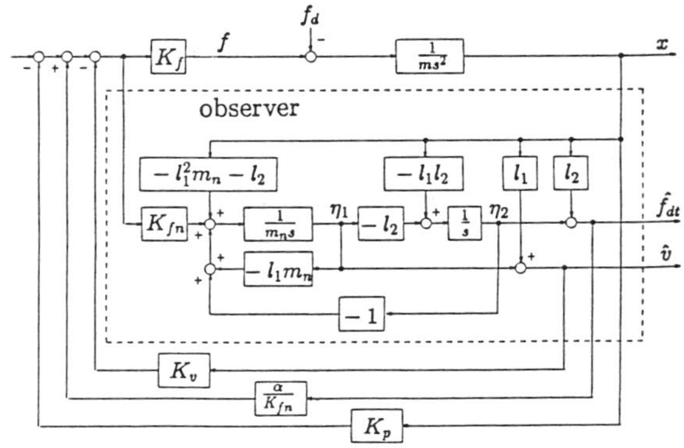

Fig. 2. Observer Based Controller for Magnetic Bearings

$$f(t) = K_f i(t) + k_m x(t) \tag{2}$$

Where, m is the mass [kg], $x(t)$ is the bearing displacement [m], $f(t)$ is the control force [N], $f_d(t)$ is the disturbance force [N], k_m is the negative stiffness of magnet [N/m], $i(t)$ is the demand current [A] and K_f is the force factor [N/A].

Let us introduce the estimation error of coefficients, because the exact parameters are impossible to estimate.

$$m = m_n + \triangle m, \qquad K_f = K_{fn} + \triangle K_f \tag{3}$$

Where, m_n and K_{fn} are the nominal mass and force factor, while $\triangle m$ and $\triangle K_f$ are the variation of them, respectively. Substituting eqs. (2) and (3) into eq. (1), we have

$$m_n \frac{d^2x(t)}{dt^2} = K_{fn} i(t) - \{f_d(t) + \triangle m \frac{d^2x(t)}{dt^2} - \triangle K_f i(t) - k_m x(t)\} \tag{4}$$

The second term on the right hand side of eq.(4) represents the disturbance effect except of the desired acceleration force. This is defined as a total disturbing force.

$$f_{dt}(t) = f_d(t) + \triangle m \frac{d^2x(t)}{dt^2} - \triangle K_f i(t) - k_m x(t) \tag{5}$$

Then eq.(4) can be modified by using the nominal coefficients and the total disturbing force.

$$m_n \frac{d^2x(t)}{dt^2} = K_{fn} i(t) - f_{dt}(t) \tag{6}$$

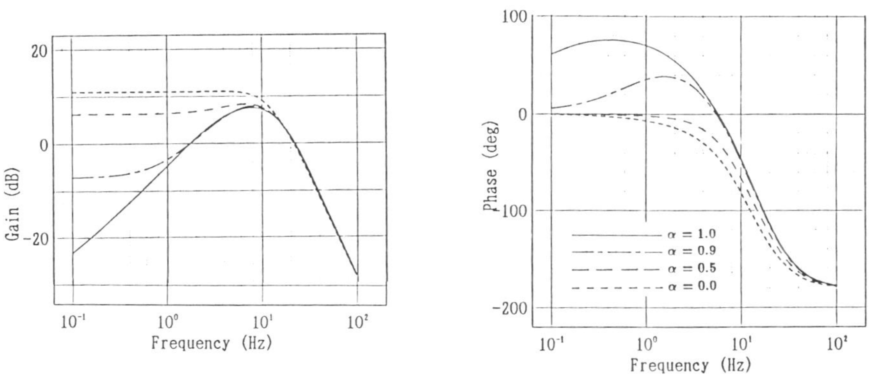

Fig. 3. Frequency response of the rotor from the disturbing force

When the displacement $x(t)$ is measured, the state equation of this system is written by,

$$\frac{d}{dt}\begin{bmatrix} x \\ v \\ f_{dt} \end{bmatrix} = \begin{bmatrix} 0 & 1 & 0 \\ 0 & 0 & -\frac{1}{m_n} \\ 0 & 0 & 0 \end{bmatrix}\begin{bmatrix} x \\ v \\ f_{dt} \end{bmatrix} + \begin{bmatrix} 0 \\ \frac{K_{fn}}{m_n} \\ 0 \end{bmatrix} i(t) \tag{7}$$

$$y(t) = \begin{bmatrix} 1 & 0 & 0 \end{bmatrix}\begin{bmatrix} x \\ v \\ f_{dt} \end{bmatrix} \tag{8}$$

The disturbing force f_{dt} is uncontrollable, but observable. Hence, f_{dt} and v can be estimated by means of the minimum order observer, which is shown schematically in the dotted block of Fig. 2.

The levitated object has the transfer function of 2nd order integrator. To stabilize this plant, the position and velocity feedbacks K_p and K_v are necessary. Addition to them, disturbance cancellation control is used to get high stiffness and the robustness of the control performance. The effect is clarified by changing the gain α $(0 < \alpha < 1)$: $\alpha = 0$ corresponds to no cancellation, while $\alpha = 1$ means to the perfect cancellation.

3. THE EXPERIMENTAL RESULTS AND CONSIDERATIONS

To confirm the proposed technique, the system shown in Fig. 2 is simulated on an analog computer. The frequency responses are calculated and shown in Fig. 3. The dotted line indicates the the case of $\alpha = 0$, where only the position and velocity signals are fedback. The response is well controlled with the estimated velocity feedback. Increasing α, low frequency response is further reduced. This means that the disturbance cancellation control increases the static stiffness and improves robustness of the performance.

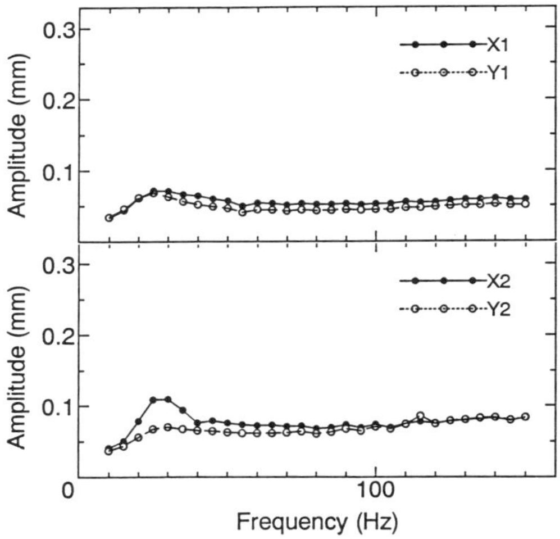

Fig. 4. Unbalance response controlled by observer based controller

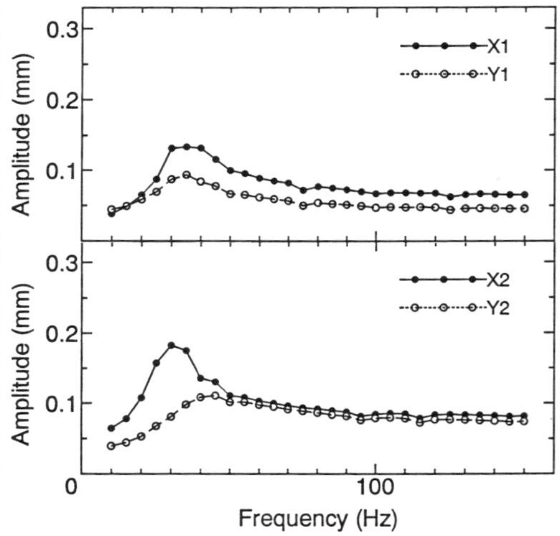

Fig. 5. Unbalance response controlled by PD controller

A simple experimental setup is made and the four radial magnetic bearings are individually controlled with the proposed controller. The unbalance response is shown in Fig. 4. The similar PD controller is installed and tested, and the resulting unbalance response is shown in Fig. 5. The response shown in Fig. 4 indicates the better performance, which clarifies the superiority of the proposed algorithm.

4. CONCLUSIONS

A simple disturbance observer based controller is proposed which is applied to the active magnetic bearings. The following merits can be obtained.

- All the parameters in modeled rotor appears to be fixed to its nominal values.
- The effect of the road to the system output is almost zero.
- The bearing stiffness looks infinity.
- The feedback of the disturbance realizes an acceleration controller.

References

[1] Allaire, P., ed., Proc. of the 3rd Intrn. Symp. on Magnetic Bearings, July 29-31, 1992, Alexandria, VA, USA

[2] Nakao, M., et al, IEEE Proc. Robotics and Automation, Vol. 1 (1987), 326

[3] Ueda, Y., et al, Preprint of the JSME, No. 920-17, Vol. C (1992), 64

Simulation and Design of Applied Electromagnetic Systems
T. Honma (Editor)

Motor bearing control of permanent magnet(PM) type rotor

T.Ohishi[a] ,Y.Okada[b], and K.Dejima[b]

[a]R&D Laboratory,Shinko Electric Co.Ltd.,Ise Works,
100 Takegahana,Ise,516 JAPAN

[b]Department of Mechanical Engineering,Ibaraki University,
4-12-1 Nakanarusawa,Hitach,316 JAPAN

Abstract

A method of levitation control of PM type motor is proposed. It is assumed that the rotor has a sinusoidally distributed magnetic poles, while the stator has a current sheet to produce arbitrary distributed flux. The levitational force can be independently controlled with the proposed PM^2 algorithm[1]. A simple experimental apparatus clarifies the validity of the proposed method.

1. INTRODUCTION

In the applications of active magnetic bearings, a high-speed motor produces undesirable drag force. Hence, the size of magnetic bearing becomes relatively big which causes slow dynamic response. The structure of magnetic bearing is similar to that of the AC motor, hence the combined control theory of rotation and levitation for the motor is highly requested.

This paper introduces a new theory which gives a solution of levitated PM type motor. The rotor have the P magnetic poles which is produced by permanent magnets(PM). Usually the same number of poles (P poles) of the stator current distribution gives the torque to the rotor. In addition, we produce P±2 poles to the current sheet which produces the pure drag force to the rotor. Controlling the magnitude and phase of this P±2 pole current distribution, the drag force can be controlled independently from the torque. To confirm the capability of this PM^2 pole algorithm, a simple experimental apparatus is made. The results of the experiment are showing the high possibility of the complete levitated rotating motor.

2. LEVITATION CONTROL OF PERMANENT MAGNET ROTOR

First, a 4 pole PM motor is considered which is shown in Fig. 1. A rotor is assumed to have 4 permanent magnetic poles which produces the following flux density.

$$B_r(\theta, t) = B_R \cos(\omega t - 2\theta) \quad (1)$$

where B_R is the peak density of magnetic flux, ω is the rotating speed of the motor, θ is the angular coordinate and is assumed $\theta = 0$ at the center of N pole. The stator current is assumed to have the following current distribution to produce the torque

$$I_m(\theta, t) = I_M \cos(\omega t - 2\theta + \phi) \quad (2)$$

where I_M is the peak current, ϕ is the phase difference. The motor is
a synchronous motor when $\phi = 90$,or
a servomotor when $\phi = 0$ or 180 .

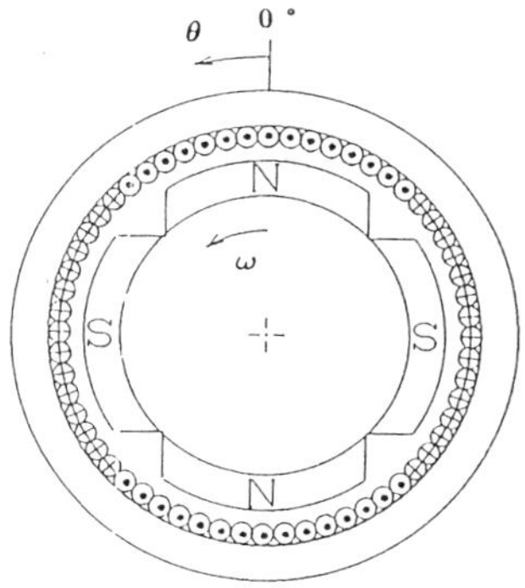

Fig.1. Scheme of 4 pole PM motor.

In addition to the torque current of eq.(2), levitation control current is requested. The 6 pole current gives the following flux density,

$$B_f(\theta, t) = B_F \cos(\omega t - 3\theta) \quad (3)$$

where B_F is the peak flux. Then the total magnetic flux density in the airgap is given by

$$B = B_R \cos(\omega t - 2\theta) + B_F \cos(\omega t - 3\theta) \quad (4)$$

This flux produces the attractive force to each radial direction. Hence the total levitational force to $\theta = 0$ direction is

$$F = \int_0^{2\pi} \frac{B^2 rL}{2\mu_0} \cos\theta \, d\theta$$

$$= \frac{B_R B_F rL}{2\mu_0} \pi = constant \quad (5)$$

This means that the levitational force is constant and independent from phase angle ωt. This is shown schematically in Fig.2.

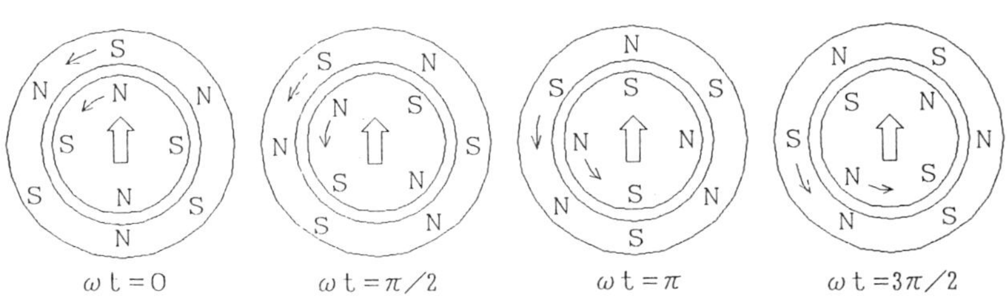

Fig.2. Levitation control of 4 pole PM motor

The direction of the levitational force can be controlled by changing the phase angle of the distributed current.

$$I_f(\theta, t) = I_F \sin(\omega t - 3\theta + \psi) \tag{6}$$

Which produces the following distributed flux density

$$B_f(\theta, t) = B_F \cos(\omega t - 3\theta + \psi) \tag{7}$$

where ψ is the phase angle. By changing ψ, the direction of the levitational force can be controlled, as shown schematically in Fig.3.

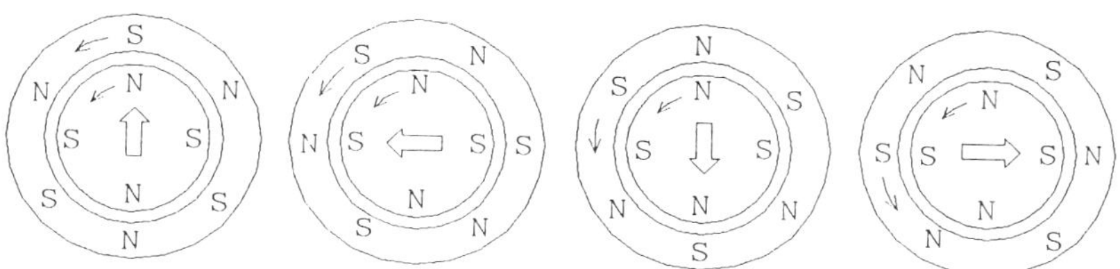

Fig.3. Axial direction control of levitational force.

3. GENERALIZATION OF LEVITATION CONTROL ALGORITHM

Let us introduce the general solution of arbitrary pole number. Suppose that the pole pair number (half of the pole number) of the stator and rotor are M and N respectively. Then the magnetic flux distribution in the airgap can be given by

$$B(\theta, t) = B_R \cos(\omega t - M\theta) + B_F \cos(\omega t - N\theta) \tag{8}$$

This flux produces the levitational force to $\theta = 0$ direction

$$F = \frac{rL}{2\mu_0} \int_0^{2\pi} \{ B_R \cos(\omega t - M\theta) + B_F \cos(\omega t - N\theta) \}^2 \cos\theta \, d\theta$$

$$= \frac{B_R B_F rL}{2\mu_0} \int_0^{2\pi} [\cos\{(M-N-1)\theta\} + \cos\{(M-N+1)\theta\}] \, d\theta \tag{9}$$

When $M-N = \pm 1$, Equation (9) becomes a constant force. That is, the pure levitation control can be obtained if

the stator pole number = the rotor pole number $\pm$ 2 (10)

We call eq.(10) as PM^2 theory. This results are showing generally the pole arrangements which produce the levitational force.

4. EXPERIMENTAL RESULTS AND CONSIDERATIONS

To confirm the previous levitation control algorithm, a simple experimental apparatus is made as shown schematically in Fig.4. One end of the rotor shaft is supported by a ball bearing and the other end is the levitated motor. The rotating torque is given by a 4 pole distributed current control. Two gap sensors are installed near the rotor to produce the levitation control signal. The levitation control algorithm used is the classical PD controller.

With the levitation control, as shown in Fig.5, the rotating speed is gradually increased by the 4 pole synchronous motoring control and reached to 8,850 rev./min. The validity of the proposed levitation control algorithm is experimentally proved.

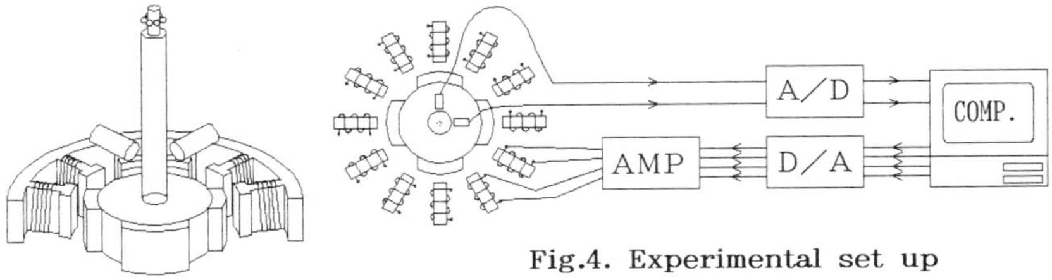

Fig.4. Experimental set up

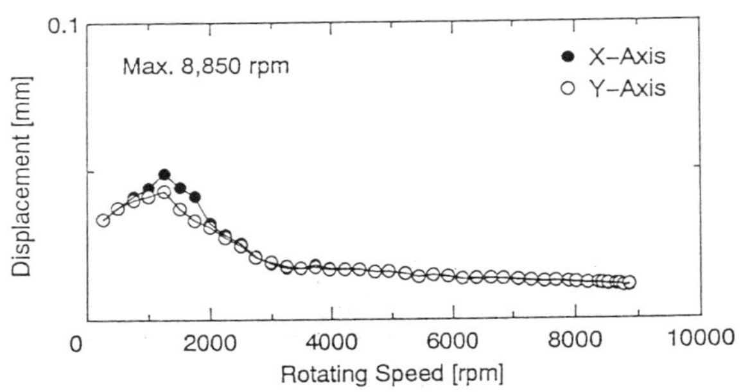

Fig.5. Displacement under levitated state

5. CONCLUSIONS

A new method of levitation control algorithm of PM motor is proposed. Theoretically the proposed levitation control is independent from the motoring control. This method is applicable either to synchronous motor or servomotor. A simple experimental apparatus is made and successfully controlled to achieve the levitated synchronous motor. Further work is continuing to apply this idea to PM servomotor.

6. REFERENCES

[1] Y.Okada, T.ohishi et.al., Prc. of Magnetic Bearings, Magnetic Drives, and Dry Gas Seales Conf.& Exhib., July 29-31, 1992, pp. 157

Simulation and Design of Applied Electromagnetic Systems
T. Honma (Editor)

Slip frequency control for high efficiency drive of single-sided linear induction motor

D. H. Im[a] , G. T. Kim[b] , O. M. Kwon[a]

[a]Dept. of Electrical Engineering, Hanyang University, Seoul, 133-791, KOREA

[b]Dept. of Electrical Engineering, Changwon Nat'l University, 641-773, KOREA

Abstract

In this paper, variable slip-frequency control method is presented to the LIM(which is used for vehicle) control method for the sake of high efficiency drive. This method is compared with existing constant slip-frequency control method. When the presented method and conventional method are experimented respectively, the result shows that the presented method in this paper is more useful than conventional method for high efficiency drive.

1. INTRODUCTION

When LIM is used for power machines, acceleration and deceleration chracteristics is very important[1]. As a result, many studies on the acceleration and the deceleration chracteristics have been taken up to date[2,3]. One of the driving method for LIM at accelerating and decelerating time is constant slip-frequency control. In this case, slip-frequency is mainly estimated by the bases which are put emphasis on thrust and normal force. If a reference slip-frequency is estimated by above criteria, efficiency and power factor aren't so good. Variable slip-frequency control is presented under constant V/f ratio in order to improve low efficiency of constant slip-frequency control.

Variable slip-frequency pattern is obtained from the equivalent circuit based on Maxwell's equation[4]. Dynamic characteristics of the proposed method and constant slip-frequency method are analyzed and compared. To show that variable slip-frequency control is an effective efficiency promotion method, experiment is made by a controller with DSP(digital signal processor).

2. THE STEADY-STATE CHARACTERISTICS OF LIM

The LIM test facility is an endless track type. Fig. 1~4 represent the results of steady-state characteristic analysis using the equivalent circuit[4].

Fig. 1 shows thrust-speed characteristics. Since the motor is operated at a constant air gap flux(In this case, the ratio of V/f is 5), the figure shows that maximum thrust has proportional shifting according to frequency. Fig. 2 is normal force-speed characteristics. As frequency increases, attration force becomes repulsion force in high slip. While eddy current decreases in low slip , normal force is attration force. Fig. 3 is the curves of efficiency-speed. As frequency increases, the figure shows that maximum efficiency values are increased. Fig. 4 shows the curves of power factor-speed. This figure is plotted to consider the capacity of Inverter.

Table 1. The specification of test facility

	Electrical specification	Mechanical specification
Primary	Type : short primary single-sided Number of phase : 3 [phases] Number of pole : 4 [poles]	Size of core(width*height*length) 175*110*908 [mm^3] Size of slot (width*depth*pitch) 175*50*12 [mm^3] Pole pitch : 182 [mm] Turns/phase : 96 [turns] Slots/pole/phase : 3
Secondary		Thickness : (Al) 5 [mm] (Fe) 50 [mm] Size of rotor(width*diameter) 280*2020 [mm^2] Moment of inertia : 958.4[kgm^2]

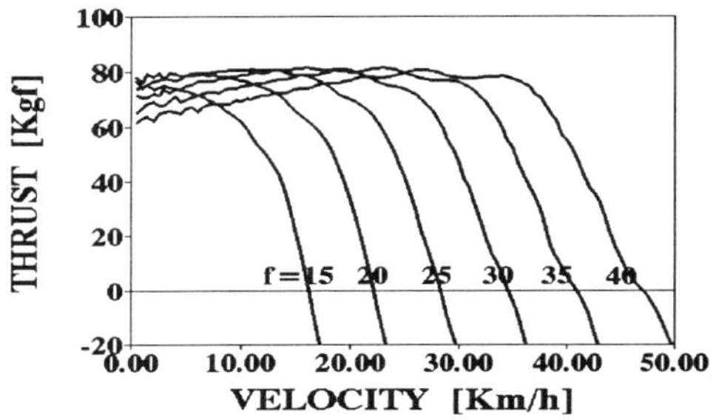

Fig. 1 Thrust vs. velocity

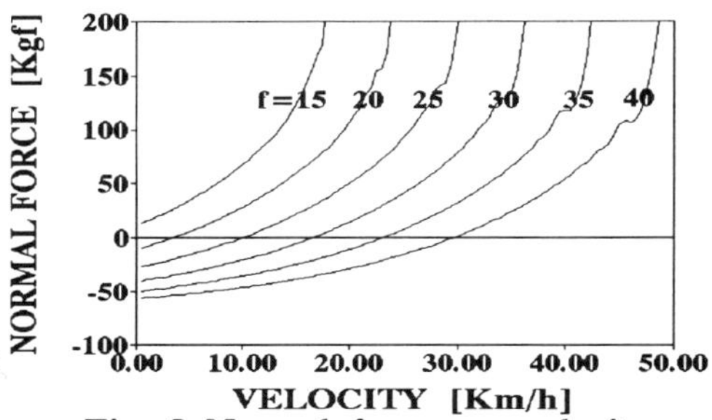

Fig. 2 Normal force vs. velocity

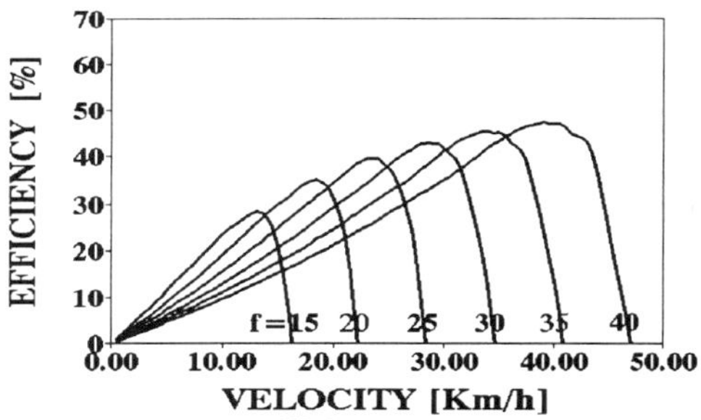

Fig. 3 Efficiency vs. velocity

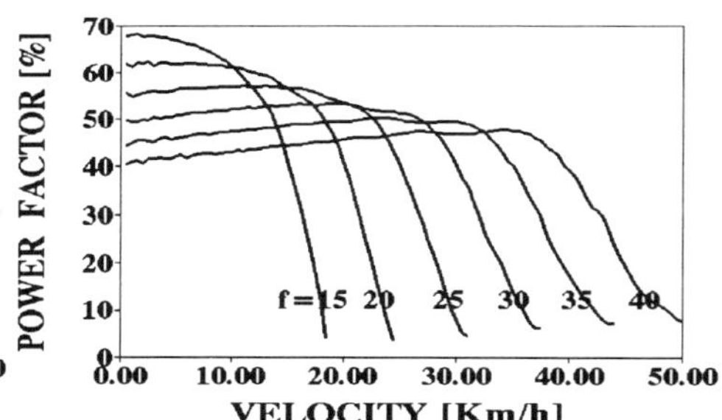

Fig. 4 Power factor vs. velocity

3. THE SELECTION OF SLIP-FREQUENCY

To select the slip-frequency improving efficiency, Constraints are given as follows.

(1)Thrust is larger than 60[Kgf]

(2)Normal force is smaller than 1.5x thrust limit(60 [Kgf])

(3)Efficiency and power factor must be neighboring maximum point(slip - frequency with respect to efficiency is extended to the point of 90% of maximum efficiency value). As a result, slip-frequency pattern for high efficiency drive is written

sf = 8.0 (V2 ≤ 20[Km/h])
sf = 0.113 x V2 + 5.74(V2>20[Km/h])

And, to prove improvement of efficiency, Constant slip-frequency (sf='11[Hz]') is compared with the presented method.

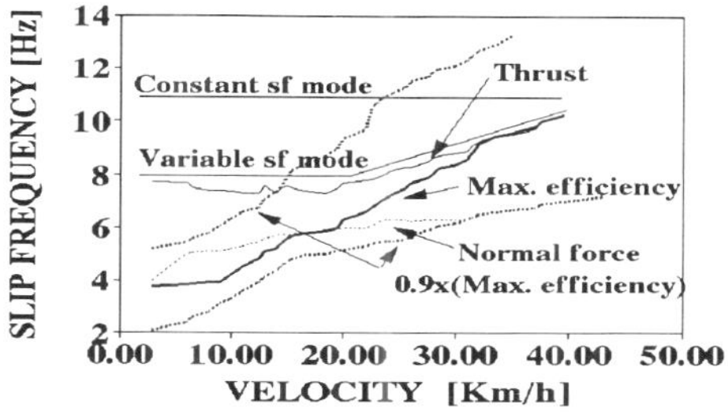

Fig.5 Slip-frequency slection

4. DYNAMIC CHARACTERISTICS ANALYSIS[5]

The motion of LIM is simulated using the equations of motion. When the velocity of vehicle is v_i[km/h], and the time of driving is t_i[sec], acceleration $A(v_i)$[km/h/s] and driving resistance $R(v_i)$[Kgf] are written by

$$A(v_i) = (F_x(v_i) - R(v_i))/28.35W$$
$$R(v_i) = (3 + 0.03v_i + 0.00168v_i^2) \times (W + F_y(v_i)) \quad (1)$$

where, W:the weight of vehicle[ton], F_x:thrust[kgf], and F_y:attraction force[Kgf] Because attraction force leads to increase equivalent weight of vehicle, it is included in a vehicle weight.

If we assume for a short time Δt(s), velocity(v_{i+1}) and time step t_{i+1} after Δt can be written

$$v_{i+1} = v_i + A(v_i)\Delta t, \qquad t_{i+1} = t_i + \Delta t \quad (2)$$

In this paper, the weight of vehicle is 1[ton], and time step is 0.03 [sec].

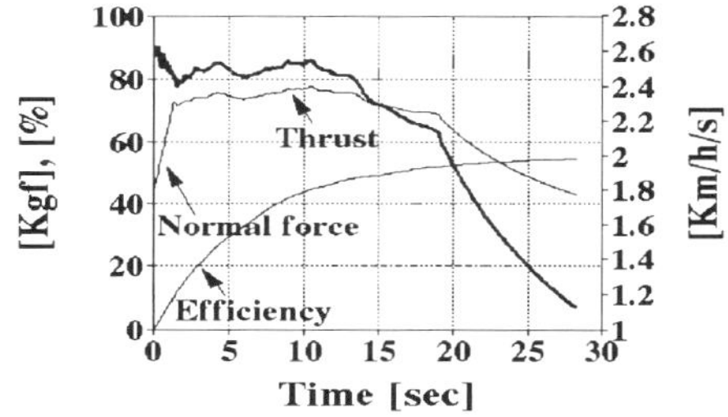

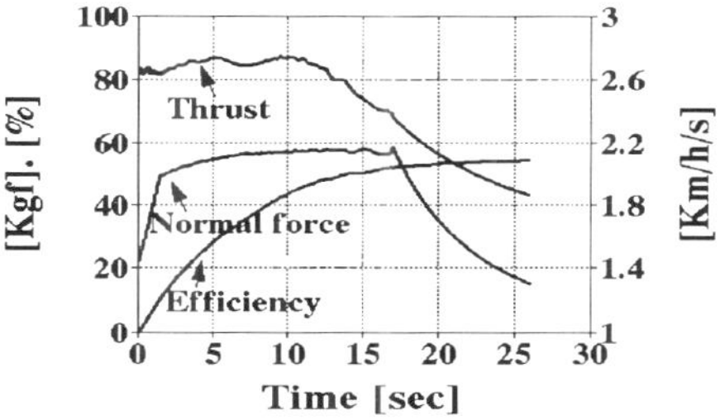

Fig.6 Dynamic characteristics(sf=vari.) Fig.7 Dynamic characteristics(sf=cont.)

5. EXPERIMENT

The controller using TMS320C25 DSP is installed in IBM PC . IBM PC is the host, it inputs the data(which is control algorithm and slip-frequency, and etc)to the DSP through common RAM. After that operation, DSP drives I/O port(A/D, D/A) for itself, does feedback velocity to the controller, and after calculation, outputs control reference to Inverter.

In experimental results on constant slip-frequency control, the normal force is confined within 90 [Kgf], and the thrust comes out beyond thrust limit. Because the thrust generates near constant force, acceleration becomes constant within slip-frequency control region. Also, efficiency and power factor by experiment are similar to those by simulation. As for the variable slip-frequency control, though the normal force generates a bit more than limiting value, it exists in a feasible region. But, the thrust generates beyonds constraint value, stable acceleration is achived. When the variable slip-frequency control is compared with the constant one, efficiency increases about 5 %, and power factor is similar.

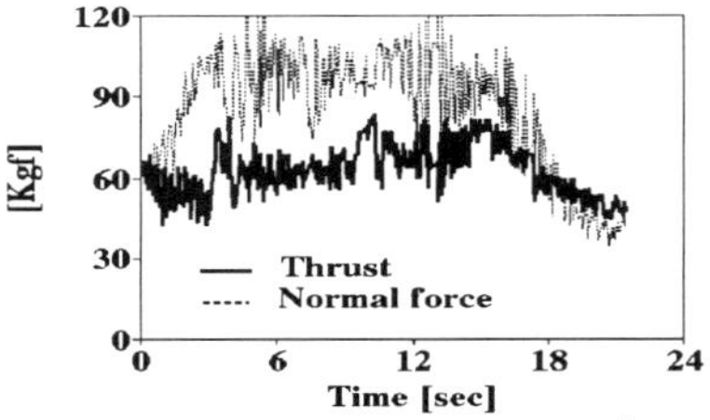

Fig.8 Thrust & Normal force(sf=vari.)

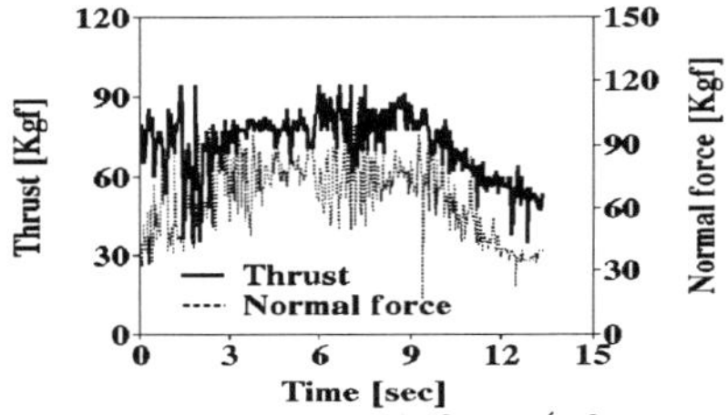

Fig.9 Thrust & Normal force(sf=const.)

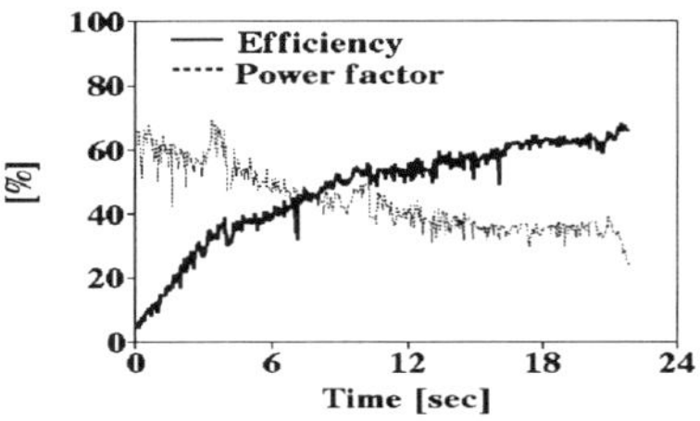

Fig.10 Efficiency & Power factor(sf=vari.)

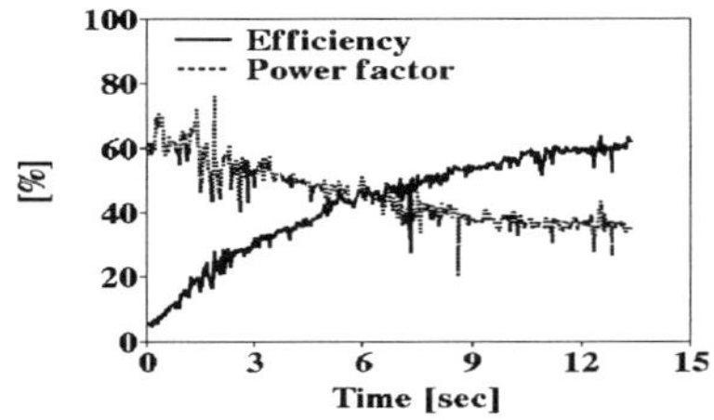

Fig.11 Efficiency & Power factor (sf=const.)

6. CONCLUSION

In this paper, the authors try to find a method improving efficiency at the time of accelerating. Simulation results correspond with experimental results. Efficency is increased 5[%] compared with the constant slip-frequency method. If the method is utilized by Urban transit vehicle, the method is to be very useful driving method.

REFERENCES

[1] S. A. Nasar, I. Bolder, "Linear Motion Electric Machines", John Wiley & Sons, 1976

[2] S. Nonaka, N. Fujii, "Dynamic characteristics of Linear Induction Motor for Urban Transit under Constsant V/f Drive", proc. IEEJ vol. 109, no. 7, pp. 493-499, 1989

[3] S. Nonaka, N. Fujii, "Dynamic characteristics of Urban Transit LIM with Feed-Back Speed Control",Int. Conference Maglev '89, pp. 327 - 332, 1989

[4] S. Nonaka, T. Higuchi, "Approximate Equations for Calculation of Characteristics of The single-sided Linear Induction Motors", Proc. IEEJ, vol. 57-B69, pp.565 - 572, 1982

[5] S. Nonaka, T. Higuchi,"On the Reduction of Energy Consumption of Linear Induction Motor for Urban Transit", Int. Conference Maglev '89, pp. 333-338, 1989

Acknowledgement : This work was supported in part by the institute of the Korean Science and Engineering Foundation.(No. 92-47-00-02)

Electromagnetic Mass Accelerators

Simulation and Design of Applied Electromagnetic Systems
T. Honma (Editor)

Improvement in launching velocity of linear accelerator

Daiki Ebihara[a] Kenji Nagashima[a] Masaya Watada[a]
Shogo Tanaka[b] Yoshihiko Kuriyama[b]

[a]Depertment of Electrical and Electronics Engineering, Musashi Institute of Technology, 1-28-1, Tamazutsumi, Setagaya-ku, TOKYO 158, JAPAN

[b]Hitachi Metals, Ltd.
2-2-1, Marunouchi, Chiyoda-ku, TOKYO 100, JAPAN

Abstract

This paper describes the improvements in launching velocity and launching stability of a Linear Accelerator ("LAC"). The authors have designed and produced a railgun using an LAC for weaving loom shuttles. An LAC is an electromagnetic launcher which enable a projectile to be launched at a speed of more than 50 meters per second.

This report outlines the basic structure of an LAC and the results of many experiments.

1. INTRODUCTION

Accelerators using linear motors have long been a subject of research. In its early stages, this research was concerned with the development of weaving loom shuttles using a linear induction motor in USA.[1] On the other side, research on the electronic guns for military use had a try at Japanese Army Scientific Institute in JAPAN.[1] Current research is concentrating on the use of linear motors for non-military purposes such as launchers for space rockets and for the disposal of nuclear waste in space[1][2]. Collision nuclear fusion is also being researched[1][3] using a railgun with a Linear DC Motor("LDM"). Particles are caused to accelerate at high speeds of several kilometers per second and collide with other particles. The extremely high pressure and temperatures instantly generated inside the particles create new matter.

In view of this situation, the authors have been conducting basic research into a railgun as instantaneous acceleration devices with an LAC. An LAC basically consists of an uni-pole LDM, the most difficult problem in this system being the stabilization of the collection of current on the projectile. In order to solve this problem, we have conducted launch experiments using two projectiles that were different in materials.[4]

In the next place, aiming at further improvements in launching velocity, we optimized the circuit constants and thought out a multi-stage type LAC.

This paper describes the basic structure and operating principles of an LAC, and the results of many launch experiments.

2. BASIC STRUCTURE OF LAC

An LAC operates on the same principle as DC rotary motors, using the force between the magnetic field and electric current. The experimental circuit is an R-L-C serial circuit. Firstly, a capacitor (capacitance C[F]) is charged sufficiently to a voltage V[V] by a DC power supply, and change-over switch is switched to discharge electric charge Q[C] stored in the capacitor via a coil (inductance L[H]). At this moment, the projectile is launched by force F[N], which has been generated in the direction that satisfies Fleming's left-hand rule, in relation to the magnetic flux density B[T] acting on the projectile and the current I[A] passing through it.

A cutaway view of an LAC is shown in Figure 1, and the specifications of an LAC are given in Table 1. Four carbon rails, two above and two below the projectile, are installed so that the contact between a projectile and rails is always maintained. The advantages of using carbon for the rails are that it prevents the generation of sparks and minimizes friction by virtue of its smooth surface, and has a high melting point.

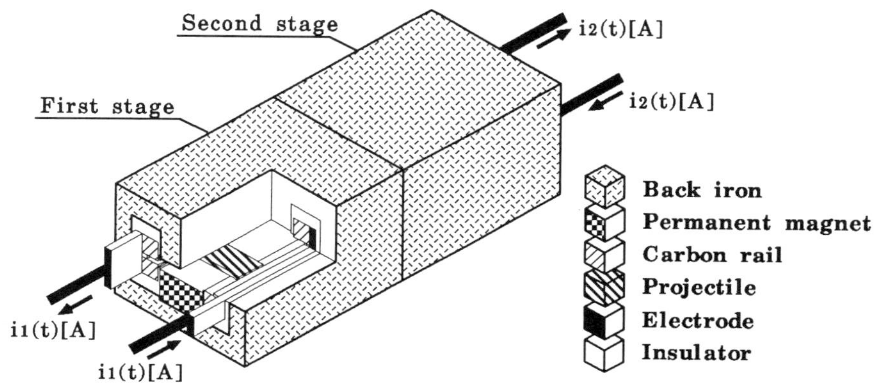

Figure 1. Cutaway View of LAC

Table 1.
Specifications of LAC

Mechanical				
Items			Values and Unit	
Launcher	Size		140×96×320	[mm]
	Rail length	ℓ_r	300	[mm]
	Rail material		Carbon	
	Flux density	B_g	0.678	[T]
Projectile	Size		50×3.7×21.3	[mm]
	Material		Carbon	
	Mass	m	6.97	[g]

Electrical			
Items		Values and Unit	
		First Stage	Second Stage
Circuit resistance	R_1/R_2	57.7 [mΩ]	41.6 [mΩ]
Inductance	L_1/L_2	240 [μH]	70 [μH]
Capacitance C/Withstand voltage		38.0/350	[mF]/[V]
Supply voltage	V	0～350	[V]

3. STABILIZATION OF LAUNCHING PERFORMANCE

The greatest problem with this system is related to the collection of current on the projectile. Therefore, we conducted launch experiments using the different the projectiles of material. These are the aluminium projectile and the carbon projectile. The following is each results of launch experiments.

3-1. Aluminium projectile

Firstly, we experimented using the aluminium projectile. The maximum velocity is 43.408[m/s]. There were some sparks and the explosion was loud when the aluminium projectile was launched. In addition, a contact surface of the aluminium projectile between the carbon rails and the projectile has melted with heat. Therefore the aluminium projectile couldn't use for the launch experiments many times.

3-2. Carbon projectile

Secondly, we experimented using the carbon projectile. The maximum velocity is 36.092[m/s]. There were also fewer sparks and the explosion was nothing when the carbon projectile was launched. A contact surface of the carbon projectile has never melted with heat, so that the carbon projectile could use for the experiments repeatedly.

3-3. Comparison of both the projectiles

When the measured velocities are compared, it can be observed that the aluminium projectile launched faster than the carbon projectile. This is probably because the aluminium projectile is lighter than the carbon projectile. However, a contact surface of the aluminium projectile has melted with heat, so that it has never used for experiments many times. On the other hand, a contact surface of the carbon projectile didn't melt. Therefore, a carbon is the more suitable for the projectile material.

4. OPTIMIZATION OF THE CIRCUIT CONSTANTS

The experimental circuit is an R-L-C serial circuit. Changing the circuit constants can increase the current flowing through the projectile. This method can therefore advance the launching velocities of the projectile.

To increase the charging voltage V[V] of capacitor raises the peak value of the discharge current, so that the launching velocities increase.

To increase in quantity of the capacitance C[F] raises the peak value of the current and lengthens the discharge time, and in consequence the velocities increase.

To change the inductance L[H] of a coil changes the peak value and the discharge time. As a result, the discharge current wave form is changed, so that it can spend the input energy without the waste.

Then we change the circuit constants according to the above means. As a result, when V is equal to 300[V], C is 38[mF] and L is 240[μH] , we succeeded that the carbon projectile was launched at a speed of 36[m/s] on the present system.

5. IMPROVEMENTS IN LAUNCHING VELOCITY USING DOUBLE-STAGE TYPE LAC

We were almost able to solve the most difficult problem which is the stabilization of the collected current on the projectile using the carbon projectile. However, further improvement of the launching velocity is still required. At the present time, the maximum velocity is only 36[m/s] using the carbon projectile. In order to further improve the launching velocity, we thought out a multi-stage type LAC. Firstly, we designed and produced a Double-stage type LAC as the basic study. This Double-stage type LAC is joined two launcher in series and causes the projectile to accelerate in two stages. The first stage and the second stage are the same structure, however these electric circuit are separate respectively.

Then we conducted launch experiments using a Double-stage type LAC with the carbon projectile. Table 2 shows the measurement of launching velocities. When Table 2 is compared with the results of the previous launcher, it can be observed that the velocity has increased. The maximum velocity using a Double-stage type LAC is about 75[m/s].

Table 2.
Measurement of velocities

	Velocity [m/s]	
Shot No.	First Stage v_1	Second Stage v_2
1	46.352	73.801
2	48.965	71.572
3	47.752	75.082
4	50.132	72.926
5	50.863	72.976
Average	48.813	73.271

6. CONCLUSION

This research has been conducted in order to further improve the launching stability and velocity. To improve the stability of the collection of current on the projectile, then, we have performed many experiments using the two projectiles. As a result of many experiments, we were almost able to solve this problem using the carbon projectile. However, the maximum velocity is only 36[m/s]. We optimized the circuit constants and produced a Double-stage type LAC, therefore, to further improve the velocity. We have achieved that the carbon projectile was launched at a speed of 75[m/s] using Double-stage type LAC.

7. REFERENCES

1 H.YAMADA, et al. :"Linear Motor Application Handbook",Kogyo Chosakai, JAPAN, (1986)

2 E.MASADA, et al. :"Linear Drive Technique and The Applications",Ohm Co., Tokyo, Chap.9, pp.244-245, (1991)

3 S.USUBA, et al. :"Development of Railgun Accelerator and Impact Fusion", PHYSICAL SOCIETY OF JAPAN, Vol.38, No.8, pp.660-670,(1983)

4 D.EBIHARA, et al. :"Basic Study On The Linear Accelerator" IEE of JAPAN, IAS, No.E.2-2, pp.26~29,(1992)

Simulation and Design of Applied Electromagnetic Systems
T. Honma (Editor)

Calculation of Rail Gun Parameters and Performance Prediction

J. Lee, J.S. Kim, C.H. Ahn, G.T. Moon, S.G. Moon

Agency for Defence Development, P.O. Box 35 Yusong, Taejeon,305-600, Korea

Abstract

This paper presents the methods to calculate the required parameters for the analysis of rail gun system. It also includes dynamic analysis method to predict its performance. Nonlinear programming method is used for optimization. For the performance prediction, mathematical model for rail gun is derived from the equivalent circuit and Newton's law. Numerical and analytical method is used to calculate the rail inductance and resistance, respectively. It is confirmed that the required basic data can be obtained prior to rail gun experiment by this proposed method.

1. Introduction

ADD(Agency for Defence Development) has been done significant efforts to develop Electromagnetic Launcher Technology which is considered as the most attractive alternative to chemical propellants for various reasons. To confirm the possibility of the technology, experimental research is performed from 1989 cooperated with KERI(Korea Electrotechnology Research Institute) and Seoul Nationl University. At 1991 the experiment was succeeded to propell the 1g projectile up to the velocity of 4.3km/sec. The work for development of HPG(Homopolar Generator) is done 1992. Rail gun analysis method for which 2D, 3D FEM and other algorithm are used, are also actively done[1]. The objective of this paper is to obtain the basic data prior to real experiment. For this purpose, optimal design model is proposed and the circuit parameters for this model are calculated. The mathematical model is derived from the electric circuit and Newton's law to predict the performance of the rail gun.

2. Optimal design model

To obtain the optimal design parameters the numerical optimal design method is used. The objective function and constraint are formulated as follows[2]

Objective function : Minimize weight (1)

Subject to : Maximum stress < Allowable stress
Maximum Ceramic Width < 0.12
Preload : $Ps < (1-b^2 / C^2)Tmax$

$$\frac{CERT}{STET} < 3.0 \tag{2}$$

The design variables are shown in Fig. 1

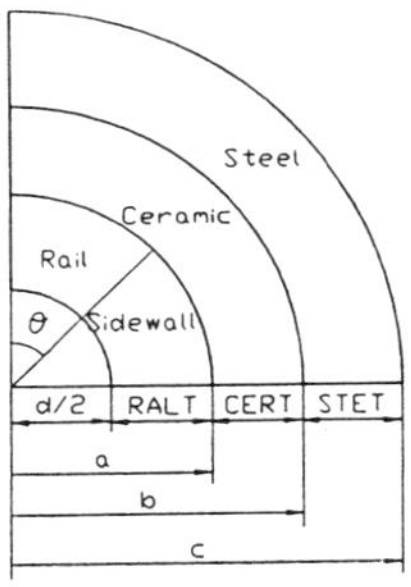

Fig. 1 Design Variables

From this optimal design technique, the following optimal design parameters are obtained.

Table 1
Optimal Design values of Rail Gun

DESIGN PARAMETERS	θ	RALT	CERT	STET
	45	0.0721	0.1218	0.0479

3. Calculations of Circuit Parameters

Rail inductance and resistance are varied with the projectile position and the variation of parameters significantly affect to the system performance. FEM is used to calculate the rail inductance. From the Maxwell equation and constitutive relationship, the governing equation with vector potential in transient magnetic field is obtained as

$$-\nabla \frac{1}{\mu} \nabla A = -\sigma \left(\frac{\partial A}{\partial t} + \nabla V \right) + Js \tag{3}$$

With the Galerkin's Method, magnetic vector potential can be obtained. Magnetic flux density, magnetic energy and inductance can be calculated from the magnetic vector potential.

The depth that the current has to diffuse into the rails is not uniform, but a function of the time at each point along the rails. So resistance are varied with time according to the position of projectile.

Assume that the total current I behind the projectile is constant, then the total resistance gradient of each rail is derived as

$$R' = \frac{1}{p} \left| \frac{\rho \pi \mu}{2} \right|^{1/2} \left| \frac{a}{2} \right|^{1/4} \left| \frac{8}{3} X^{-1/4} \right| \tag{4}$$

4. Mathematical Modeling

The electric circuit of a capacitor bank-driven rail gun is shown in Fig. 2 as the lumped parameters.

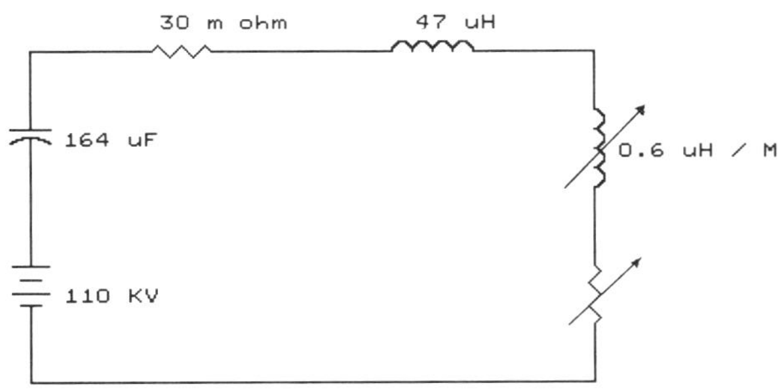

Fig 2. Equivalent Electric Circuit

By applying Kirchoff's law and Newton's law, the differential equations governing the circuit behavior are obtained as

$$i' = \frac{V_c - (Ro + R'x + L'v)i}{L_1 + L'x} \tag{5-a}$$

$$V_c' = -\frac{i}{C} \tag{5-b}$$

$$v' = \frac{F}{m} \tag{5-c}$$

$$x' = v \tag{5-d}$$

where v is velocity, F is force and m is mass of projectile respectively.
The total force is determined by the sum of Lorentz force, friction force and ablation etc. The related equation is obtained as

$$F = F_L + F_T - F_P - F_A - F_D = (m_a + m_P)\frac{dv}{dt} \tag{6}$$

where the various force(F_L, F_T, F_P, F_A, F_D) can are explained in reference[1].

5. Results and Discussion

Fig. 3 shows flux density distribution and inductance. Fig. 4 shows stress distribution and deformation of optimal design model. Fig. 5 shows dynamic behavior of Rail Gun with the 167μF capacitor bank of 1MJ energy.

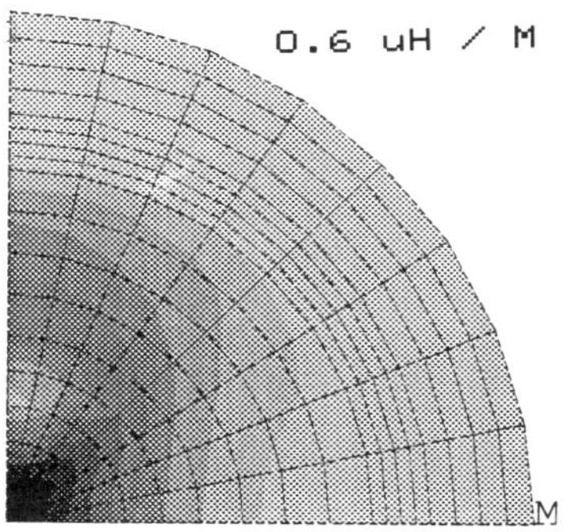

Fig.3 Flux Density Distribution of Optimal Design Model

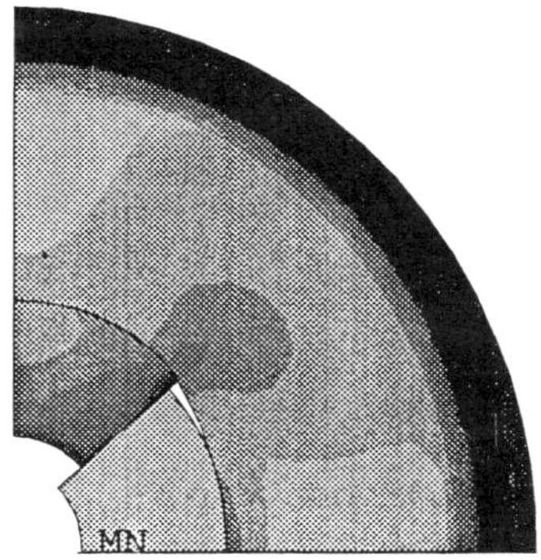

Fig. 4 Stress Distribution and Deformation of Optimal Design Model

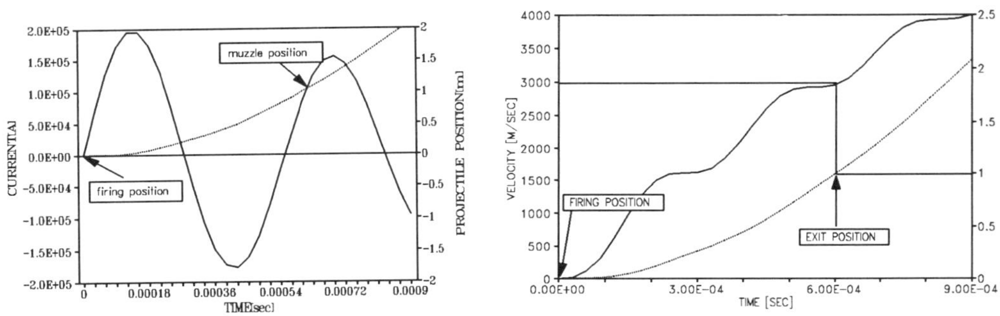

Fig. 5 Dynamic Behavior of Rail Gun (external inductance : 47μH)

From this results it is known that the optimally designed rail gun can achieve 3Km/sec. If the external inductance are reduced from 47μH to 10μH, the system can achieve up to the velocity of 5.7Km/sec.

6. Some Concluding Remarks

In this study the methods to calculate the optimal design parameters and circuit parameters are presented. By using equivalent circuit and Newton's law, dynamic mathematical model for rail gun is derived. It concludes that the basic data can be obtained by the proposed method prior to real experiment.

7. Reference

[1] J. Lee, J.S. Kim, C.H. Ahn G.T.Moon, S.K. Moon, " A Study on the performance analysis of Electromagnetic Launcher" , Summer Conference of KIEE, 1992.

[2] C.H. Ahn, "Optimal structural Design for Electromagnetic Launcher", Master Thesis, 1992.

Simulation and Design of Applied Electromagnetic Systems
T. Honma (Editor)

Development of Railgun Pellet Injector using a Laser-induced Plasma Armature

M. Onozuka[a], Y. Oda[a], K. Azuma[a], M. Ogino[a], S. Kasai[b], and K. Hasegawa[b]

[a]Mitsubishi Heavy Industries, Ltd., Advanced Nuclear Plant Engineering Department, Wadasaki-cho, Hyogo-ku, Kobe, 652, JAPAN

[b]Japan Atomic Energy Research Institute, Naka-gun, 319-11, JAPAN

Abstract

The application of the electromagnetic railgun system for high-speed pellet injection for fusion plasmas has been investigated. A pellet pre-acceleration technique is used in order to reduce rail erosion. The plasma armature is induced by a laser beam. This unique feature reduces the supplied voltage to the rails, in order to avoid unnecessary breakdown between the rails and to reduce rail erosion. The power supply employs Pulse-Forming-Networks, which provides a rapid current-rise to the rails and control of the current-hold time, thereby enhancing acceleration efficiency. The ignited plasma armature is accelerated by an electromagnetic force that accelerates the pellet to high velocity. A series of experiments using dummy pellets have been conducted to examine the system.

1. INTRODUCTION

Recent magnetically confined fusion experiments have demonstrated better plasma confinement using a pellet injection technique for fueling particles into plasma. This fueling technique provides high-speed injections of a solid-isotope pellet into the plasma, which allows high plasma density inside the core plasma and active control of the density profile, as well as plasma confinement.

For larger volume and higher temperature of the plasma, higher speed pellets are required in order to reach the core of the plasma. The application of the electromagnetic railgun system for pellet injection is one of the most feasible technologies for accelerating a brittle, solid hydrogen-isotope pellet to a high speed (up to 5km/sec or over) without a sabot (protector of the pellet) [1-3]. Using this method, an armature, a moving conductor formed behind a pellet, is gradually accelerated by the electromagnetic (Lorentz) force, that then accelerates the pellet to high velocity.

Mitsubishi Heavy Industries, Ltd. has been investigating the application of the railgun system for the pellet injector in collaboration with Japan Atomic Energy Research Institute. The objective of our research is to establish a railgun system that can achieve a 5km/sec speed-class pellet injection.

2. MAIN FEATURES OF RAILGUN SYSTEM

2.1. Pellet Pre-acceleration Technique

In conventional railgun systems, a high electrical current is supplied to the armature to accelerate the pellets, and the initial armature is produced generally by a high supplied voltage and current. However, in these conventional systems, the damage to the rails, caused by the initial formation of the armature and ablation from the armature, is significant. Therefore, the rail materials can be used only a few times. In order to solve this problem, our railgun system employs a pellet pre-acceleration technique, which reduces the current passing time through the plasma armature between the rails. Using this method, a pellet is accelerated initially up to 1km/sec by a pneumatic injector.

2.2. Laser-induced Plasma Armature Technique

In the railgun system, the armature must be induced behind the pellet. A metal armature is usually used in railgun systems. However, it cannot be used for the pellet injector because of the impurity restriction. Since the pellet is initially accelerated by the pneumatic method, the propellant gas can be ionized to form a plasma armature. A tungsten needle is usually installed to induce the plasma armature. However, it is necessary for the propellant gas pressure to be adjusted to the preferable condition for the propellant gas to breakdown. Since the ionization by breakdown forms a limited number of ionized particles under reduced pressure, a high supplied current must be applied to the rails in order to maintain the plasma armature. In addition, unnecessary breakdown is sometimes caused between the rails under reduced pressure.

Our system employs a pulse laser beam. Using this method, ionization of the propellant gas is optically initiated by an intense laser beam. This method provides the following merits:
(1) Reduction of the supplied voltage to the rails, in order to avoid unnecessary breakdown between the rails and to reduce rail erosion.
(2) Production of plasma armature under high pressure gas (unnecessary to reduce the propellant gas pressure), to provide high pre-acceleration pellet speed and to avoid unnecessary breakdown between the rails.

2.3. Pulse-Forming-Networks Power Supply

In our railgun system, Pulse-Forming-Networks (PFN) are used for the power supply. The PFN power supply provides a rapid current-rise to the rails and control of the current-hold time, thereby enhancing acceleration efficiency.

3. EXPERIMENTAL DEVICE

A schematic diagram of the current railgun system is shown in Figure 1. The specifications of the system are summarized in Table I. The system consists of a pneumatic pre-accelerator and railgun for its primary acceleration. A pellet is pre-accelerated by helium gas, then enters the bore of the rails. A Q-switched YAG laser is triggered to induce the plasma armature just behind the pellet, forming a conducting path between the rails. Electric power is supplied from PFN to the rails to accelerate the plasma armature, that then accelerates the pellet.

Table I Specifications of Railgun

Type:	Laser-Prearc Railgun
Railgun:	Size: 3mm(Diameter)x1m(Length)
	Materials: Cu or W-Alloy
Pellet:	Size: 3mm(Diameter)x3mm(Length)
	Materials: Poly-Carbonate, Lauan, Balsa
Propellant Gas:	Material: Helium
	Pressure: up to 4MPa
Laser Beam:	Type: Pulsed YAG Laser
	Pulse Energy: 900mJ for 6nsec
	Wavelength: 1064nm
Power Supply:	Type: Pulse-Forming-Networks (23.5mFx4)
	Charging Voltage: up to 350V
	Supplied Current: up to 25kA

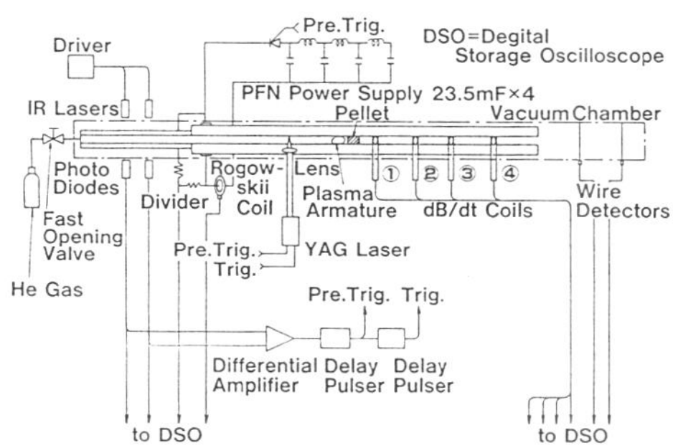

Figure 1 Schematic Diagram of Railgun System

4. EXPERIMENTAL RESULTS

4.1. PFN Power Supply

The newly built PFN power supply has been tested. Four capacitors are connected in parallel to provide a rapid current-rise to the rails and control of current-hold time, aiming at higher acceleration efficiency. It was found that the number of capacitors in use controls the current-hold time.

4.2. Dummy Pellet Acceleration

More than one hundred experiments using various kinds of dummy pellets have been conducted to examine the characteristics of the railgun system. In addition to conventional copper rails, newly made tungsten-alloy is used as a rail material, aiming at higher resistance against plasma ablation and longer durability. Figure 2 shows the relation between the charging voltages to the capacitors and the final speeds of the accelerated pellets, as well as the supplied currents to the rails. Wooden (lauan) pellets were accelerated using three capacitors. The relation between pellet mass and pellet velocity is presented in Figure 3.

The experimental results are summarized as follows:

(1) The final pellet velocities and the supplied currents are linearly proportional to the charging voltages to the capacitors. This means that the pellet velocity can be easily controlled by the charging voltage.

(2) The final pellet velocities can also be controlled by the number of capacitors in use, i.e., by the current hold-time. The

PFN power supply provides easy control of the pellet velocity.
(3) Tungsten-alloy rails yield higher pellet velocities and currents than the copper rails under the same conditions. This is because the tungsten-alloy has a lower working function and produces more thermal electrons than copper does. Therefore, the resistivity of the plasma between the tungsten-alloy rails is considered to be lower than that between the copper rails.
(4) Under these experimental conditions, copper rails were used for 25 operations before they were damaged due to ablation from the plasma, while tungsten-alloy rails lasted 40 operations. This is because the tungsten-alloy has a higher melting point (3683K) than copper has (1356K). Thus, tungsten-alloy is a superior rail material. The durability of the rails can be also extended significantly by optimizing the supplied current, rail length, and pre-accelerated pellet speed.
(5) The initial and final pellet velocities are inversely proportional to the square roots of the pellet mass. Using this relation, the hydrogen pellet (around 2mg) is estimated to be accelerated to more than 4km/sec. The acceleration energy was confirmed to be constant, as shown in Figure 3.

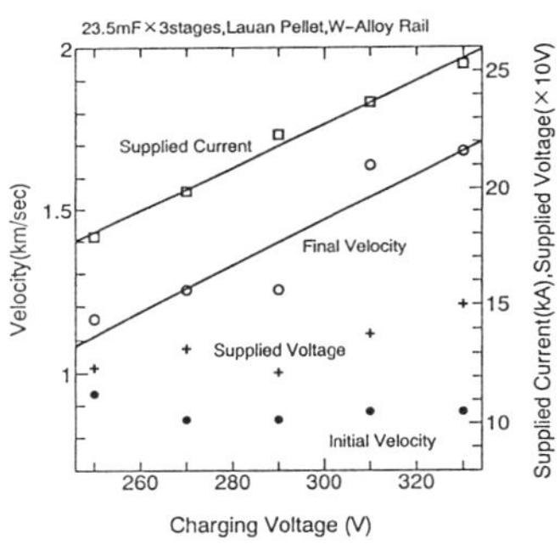

Figure 3 Pellet Acceleration Test Results

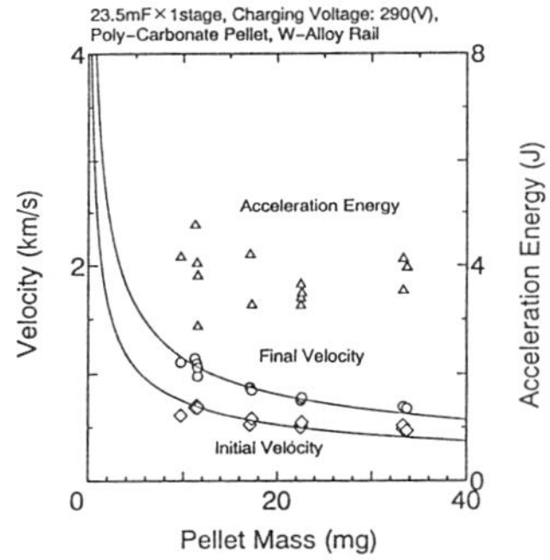

Figure 4 Effect of Pellet Mass

5. CONCLUSION

The application of electromagnetic railgun system for high-speed pellet injection has been investigated. Although development is still in progress, current research results indicate that the railgun system can be one of the optimal methods to provide higher speed pellets for fusion plasmas.

6. REFERENCES

1 H.Tamura, A.Sawaoka, Y.Oda, M.Onozuka, S.Kuribayashi, and S.Shimizu, Rev. Sci. Instrum., 63(5), pp. 3102-3107, 1992.

2 M.Onozuka, Y.Oda, S.Kuribayashi, K.Azuma, K.Satake, S.Kasai, and K.Hasagawa, Proc. 17th Symp. Fusion Tech., 1992.

3 K.Kim, J.Zhang, T.L.King, W.C.Manns, and R.G.Haywood, Proc. 6th Symp. Electromagnetic Launch Tech., 1992.

Simulation and Design of Applied Electromagnetic Systems
T. Honma (Editor)

Experimental study of rail-type electromagnetic mass accelerator

T. Maruo, K. Fujioka, and K. Ikuta

The Japan Steel Works, Ltd., 1-3, Takanodai, Yotsukaido, Chiba, 284, Japan

Abstract

Acceleration study of fairly big projectiles and small(gram order) projectiles was carried out using 3 m long rail-type electromagnetic accelerator with 4MJ capacitor bank. Velocities of 4.5km/s with a 50g projectile and 8.3km/s with a 3.2g projectile were obtained, whose energy efficiencies were 13.7 and 8.6%, respectively. The experimental results are reported and compared with calculation briefly.

1. INTRODUCTION

Researchers of hypervelocity impact phenomena request a mass accelerator who can attain about 14km/s, although a velocity limit of the most commonly used mass accelerators such as the two-stage light gas gun is 8-10km/s. The electromagnetic mass accelerator has the unique potentiality to overcome the velocity limit of the light gas gun.

A number of electromagnetic mass accelerator have been developed. Especially many efforts have been concentrated to develop a rail-type electromagnetic mass accelerator (railgun), because it is suitable for accelerating a fairly light projectile with the structural simplicity. The railgun, however, have failed to achieve higher performance reproduciblly than that of two-stage light gas gun due to complicated interaction between plasma armature and railgun structure such as plasma drag and secondary arc, etc. [1]

The experimental study of railgun by JSW started in 1987 using 200kJ capacitor bank. Since then, the most important finding is that both rail augmentation and preacceleration of projectile were effective for reducing the rail erosion and secondary arc(restrike) and that JSW railgun achieved velocity of 5.6 km/s with a 1.3g polycarbonate projectile in the atmospheric pressure using 10 mm square bore 1 m long railgun. We, also, found that the erosion of the rail electrodes is fairly mitigated when the combination of tungsten-copper alloy rail electrode and zinc armature fuse is employed.

Recently we initiated the experimental work using 3 m long accelerator with 4MJ capacitor bank for accelerating bigger projectiles and for higher velocity with small (gram order) projectiles. The square bore railgun of 35 mm was used for accelerating bigger projectiles and 10 mm square bore for small projectiles. We, however, could not attain a velocitiy over 8km/s with 10 mm square bore accelerator because of severe restrike. Instead a study on Highly Elongeted Rectangular Bore (HERB) accelerator (railgun) was started. We here report this case, where the velocity over 8km/s was attained.

2. FACILITY

A schematic of the facility is shown in Figure 1. The power source is 4MJ capacitor bank which is consist of eight 500kJ(0.01F,10kV) modules.

Bore size of the accelerator is varied by the selection of rail elctrodes and insulators. Experiments were carried out using both the classical and the augmented rail geometry for 35 mm square bore in order to compare acceleration performance. In the cases of 10 mm square bore and HERB accelerator only augmented configuration is employed because it is desirable to operate with reduced current density. The cross sections for various case are illustrated in Figure 2. Bore sizes of 20x10 and 29x10 mm, where 20 and 29 mm are rail space and 10 mm is insulator space respectively, are used for HERB railgun. A metal sheet is attached to the back of projectile as a seed of an plasma armature. The projectile is loaded at some distance from the breech and accelerated in the atmospheric pressure.

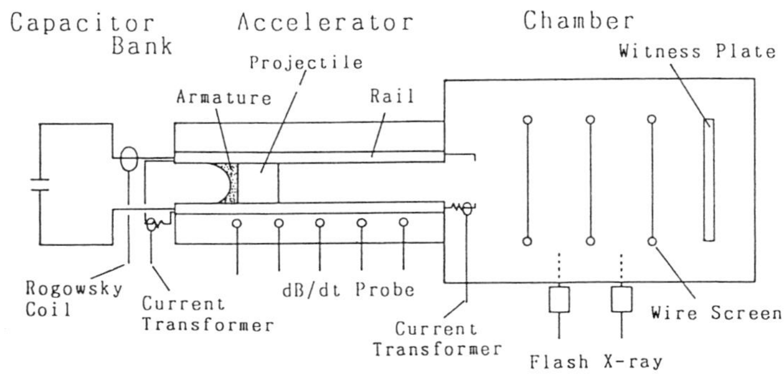

Figure 1. Schematic of the facility

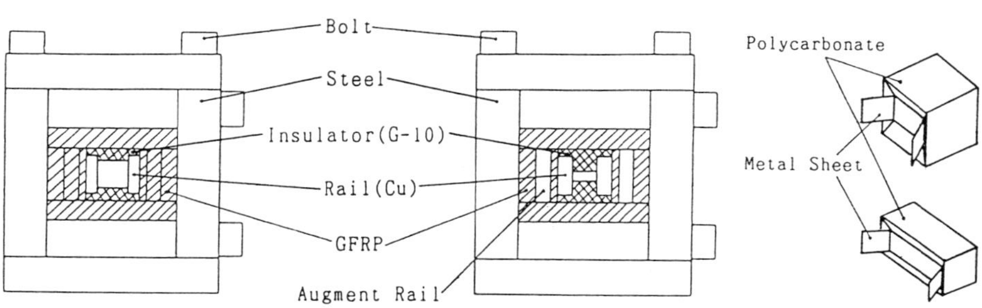

Square bore, Classical HERB, Augmented

Figure 2. Cross sections of the accelerator and projectiles

3. RESULTS

3.1. 35 mm Square Bore

The experimental results for 35 mm square bore accelerator are summarized in Table 1 with numerical calculation(velocity) which was conducted using the following equation of motion[3],

$$m(dV/dt) = I^2L'/2 - fI^2L'/2 - F_{air}$$

where m is projectile mass, V is velocity, t is time, I is rail current, L' is inductance per unit length, f is coefficient related to loss such as plasma armature as well as friction between a projectile and launcher, and F_{air} is retarding force of the air in front of the projectile.

We see that the observed velocities coincide with the numerical calculation and rail augmentation is effective for the enhancement of velocity and efficiency with lower current compared to classical rail geometry. Maximum velocity and efficiency achieved were 4.5km/s and 13.7% with a 50g projectile, respectively.

Table 1 Results of 35 mm square bore accelerator

Projectile Mass g	Rail Type	Acceler. Length m	Charging Energy MJ	Velocity measured km/s	Efficiency %	Velocity calculated km/s
50	Classical	2.75	1.28	1.9	7.1	1.92
	Classical	2.85	1.84	2.6	9.2	2.56
	Classical	2.65	3.69	4.5	13.7	4.72
	Augmented	2.85	1.84	2.7	9.9	3.00
100	Classical	2.60	0.98	1.2	7.3	0.90
	Classical	2.70	1.84	1.7	7.6	1.67
	Augmented	2.58	0.98	1.3	8.6	1.23
	Augmented	2.80	1.84	2.0	10.9	2.10

3.2. 10 mm Square Bore

Table 2 summarizes the experimental results of the 10 mm square bore accelerator. Observed velocities are all extremely slower than those of numerical calculation in this series. Efficiency decreases with raising charging energy of capacitor bank. This may be attributed, from magnetic probe signals, from severe restrike of arc plasma near the breech of the the electrodes.

Table 2 Results of 10 mm square bore accelerator

Projectile Mass g	Acceler. Length m	Charging Energy MJ	Velocity measured km/s	Efficiency %	Velocity calculated km/s
1.2	2.75	0.41	3.7	2.0	8.3
	2.20	0.85	5.2	1.9	11.1
	2.30	1.81	6.2	1.3	14.7

3.3. Highly elongated rectangular bore (HERB)

Results of HERB accelerator are listed in Table 3. Differences between obtained and calculated velocities are still considerablly large. We, however, see that high aspect ratio of the bore(or expansion of rail space) is effective for increasing efficiency and makes the differences with small ones. This seems, due to high aspect ratio of the bore, that the occurrence of restrike is delayed. The magneic probe signal supports this suggestions.

Maximum velocity attained is 8.3km/s with a 3.2g projectile and the efficiency is close to that of 35 mm square bore augmented accelerator in which no restrikes have occurred.

Table 3 Results of HERB accelerator

Bore Size mm	Projectile Mass g	Acceler. Length m	Charging Energy MJ	Velocity measured km/s	Efficiency %	Velocity calculated km/s
20x10	2.2	2.75	1.13	6.0	3.5	10.9
20x10	2.0	2.80	1.28	7.0	3.8	11.5
29x10	3.2	2.70	1.28	8.3	8.6	9.7

Figure 2 plots all observed velocities against numerically calculated ones. It should be noted here that our numerical results described very well the previous experimental observations of 10 mm square bore accelerator with 200kJ capacitor bank, in which no restrike occurred, up to about 5km/s.

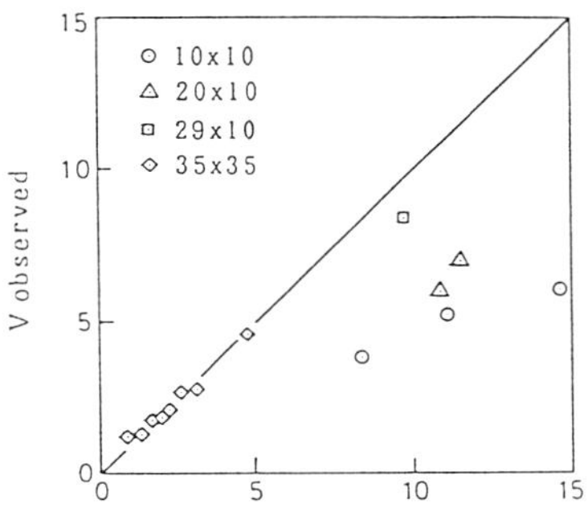

Figure 3. Observed velocities vs. calculated ones.

4. CONCLUSIONS

Experiments have been carried out using 3 m long rail-type mass accelerator with 4MJ capacitor bank. We recognized that experimental results of fairly big projectiles with 35 mm square bore accelerator coincide with the numerical calculation. On the other hand, the results of small projectiles with 10 mm square bore were extremely different from the calculation. We also claim that HERB railgun attained the velocity over 8km/s.

5. REFERENCES

1 J.V. Parker, IEEE Trans. on Mag., No.1 (1989) 418.

2 T.Maruo, K.Fujioka, N.Nagaoka, A.Okamoto, K.Ikuta, and K.Nemoto, IEEE Trans. on Mag., No.1 (1991) 65.

3 T.Maruo, K.Fujioka, S.Nakamura, and K.Ikuta, National Convention Record I.E.E. Japan, 1 (1992) S.1-13.

Application of Electromagnetic Forces

Simulation and Design of Applied Electromagnetic Systems
T. Honma (Editor)

The Superconducting MHD-propelled Ship YAMATO-1

K. Imaichi and H. Tamama

Ship & Ocean Foundation, 15-16, Toranomon 1-chome, Minato-ku, Tokyo 105, Japan

Abstract

This paper describes the outline of the superconducting magneto-hydrodynamic(MHD) propelled ship YAMATO-1 and the operation and performance of the testcraft.

1. INTRODUCTION

MHD ship propulsion is an application of Fleming's left-hand rule, one of the basic theorems of electromagnetism. When electric currents flow square to magnetic fields created in the seawater by use of inboard electromagnets, the resulting electromagnetic force operates upon seawater and the ship is moved by its reaction.

Since an MHD-propelled ship is driven by electromagnetic force, it is free from noise and vibration stemming from screw propellers; it is not subjected to oil leakage from shaft bearing, because the ship has none of such gears as a drive shaft and a stern tube. With little possibility of cavitation, MHD ship propulsion is suitable for high-speed watercraft.

MHD thruster efficiency improves in proportion to the square of magnetic field strength. Superconducting electromagnets capable of generating a strong magnetic field are, therefore, advantageous to MHD ship propulsion.

Stimulated by substantial advancements in superconducting technology like Maglev, the Ship & Ocean Foundation (SOF) organized in 1985 a Comittee under the chairmanship of Mr. Yohei Sasakawa and began researches into MHD ship propulsion. Based upon results of theoretical studies and experiments, construction began in 1989 of a 30 m-long testcraft having a displacement of 185 tons. The testcraft christened YAMATO-1 underwent sea trials in June 1992.

2. THE TESTCRAFT

The testcraft YAMATO-1 was so designed as to clear the government's inspections. Taking into account performance and the arrangement of equipment, a unique, streamlined hullform was worked out for the futuristic ship. Each of the twin thrusters has an independent system having an individual electrode power supply unit. Table 1 is the principal particulars of the testcraft, and Fig. 1 her general arrangement.

Table 1
Principal particulars of YAMATO-1

Length	26.4 m	Breadth	10.4 m	Depth(max)	3.7 m
Draft(max)	2.7 m	Displacement	185 t	Speed	abt.8 kt
Complement	10 (including three crews)				

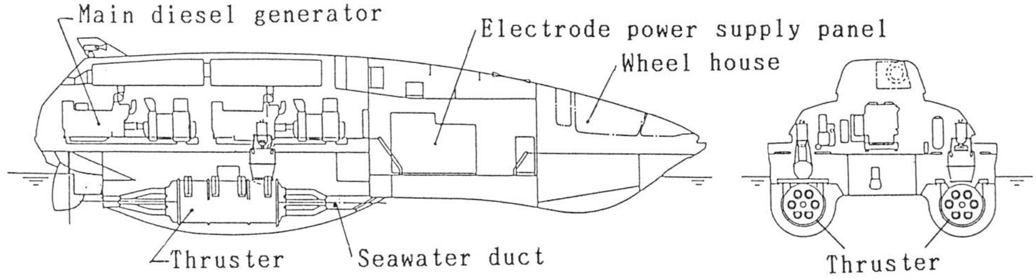

Figure 1. General arrangement of YAMATO-1

2.1 The testcraft structure

The testcraft consists of: ① Hull ② Superconducting MHD thrusters comprising superconducting magnets, electrodes, ducts, etc. ③ Helium refrigiration system ④ Electrode power supply units ⑤ Auxiliary equipment.

To reduce the weight of the testcraft, back-up facilities were installed on land which accomodate ⑥ large-sized helium refrigeration system and ⑦ energizing power supply units.

Procedures of the testcraft operation are as follows. Pure helium gas cooled by the land-based helium refrigeration system is transferred aboard to cool the superconducting magnets to 20 K (-254 ℃), and then liquid helium is pumped into the magnets for storage. After liquid helium injection, the land-based helium refrigeration system is disconnected with the ship, and on-board refrigerators reliquefy helium evaporated due to heat intrusion, keeping the magnets at ultra-low temperature.

Just before test runs, the superconducting magnets are energized by means of energizing power supply units on land and put into persistent current mode. And the ship is disconnected from the power supply units, while on-board electrode power supply units apply currents to magnetic fields in the duct seawater; thus electromagnetic force is generated. The magnitude of electromagnetic force or thrust is controlled by changing electrode current density.

After completing test runs, the superconducting magnets are demagnetized by use of the back-up facilities.

2.2 MHD thrusters

The superconducting magnets were designed not only to project strong magnetic fields into the duct seawater, but to minimize magnetic leakage. Two double-layered saddle-shaped coils surrounding a normal temperature bore forms a dipole coil unit; six dipole coil units are accomodated in a single cryostat. The principal particulars of the superconducting magnet are shown in Table 2, and its schematic diagram in Fig. 2.

Table 2
Particulars of the right thruster

(Superconductor)			
Type : Keystone shape compacted strand cable		Material : NbTi/Cu	
Strand diameter	0.763 mm	Copper ratio	1.0
Dimension	1.35/1.43 ×9.84 mm	Number of strands	26
(Unit coil)			
Coil shape : Double shell type dipole		Number of turns	253× 2
Inner diameter	360 mm	Outer diameter	401 mm
Length (str. part)	2950 mm	Length (overall)	3884 mm
(Fully assembled six coils)			
Operation current	3288 A	Magnetic flux	4 T
Inductance	3.7 H	Stored energy	20 MJ
(Thruster)			
Weight	17 t		
Lorentz force	8000 N (at electrode current 2000A)		

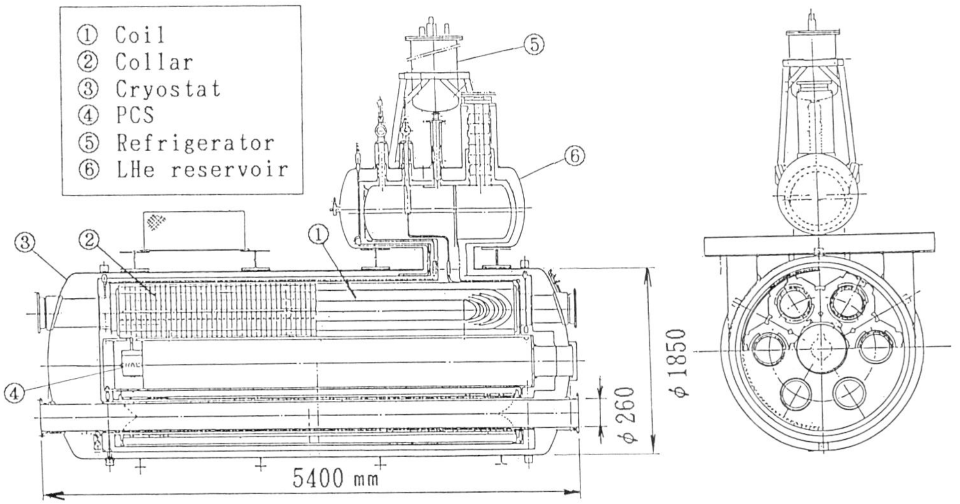

Figure 2. Cross section of superconducting magnet for the right thruster

Fig. 3 illustrates contour lines of magnetic field, and indicates that a uniform magnetic flux is generated in the duct and that a magnetic field is formed into a loop, which leads to magnetic leakage reduction.

A seawater duct made of FRP is inserted into the normal temperature bore, so that the duct is inscribed to the internal surface of the bore; a pair of electrodes are incorporated into the duct. In the YAMATO-1, two sets of three serially-linked electrodes are connected in parallel, thus realizing low voltage and less leakage current.

2.3 The results of sea trials

Passing speed tests and maneuverability tests, the YAMATO-1 was given a Certificate of Ship's nationality. Evaluation tests were pursued after delivery, under different test conditions.

Fig. 4 is the result of speed tests where magnetic flux density is kept constant, while electrode current changed; the speed is achieved as expected.

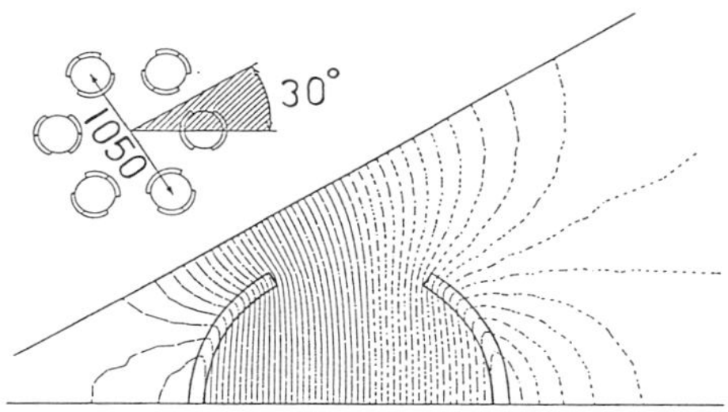

Figure 3. Contour lines of magnetic flux

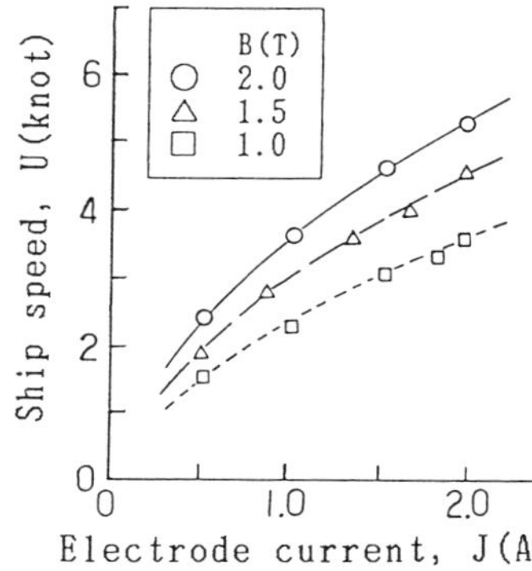

Figure 4. Result of speed trials

3. AFTERWORD

The YAMATO project has purposed to demonstrate that ships can be powered by superconducting MHD thrusters. The trials that took place in June 1992 ended in a success. MHD ship propulsion is a mode of propulsion utilizing seawater conductivity. A larger part of electricity passing between electrodes is lost as Joule's heat because of great resistance. Therefore, MHD-propelled ships at this stage of development cannot vie in efficiency with watercraft powered by screw propellers.

To boost the efficiency of MHD ship propulsion, large-sized superconducting magnets capable of projecting strong magnetic fields are a must, because thruster efficiency is in proportion to both the square of magnetic field strength and working space volume.

4. REFERENCE

The Proceedings for the International Symposium on Superconducting MHD Ship Propulsion (MHDS 91) Kobe, Oct. 1991

Simulation and Design of Applied Electromagnetic Systems
T. Honma (Editor)

Nonlinear oscillations of a string embedded in the electromagnetic field.

Jan Awrejcewicz [a] and Tsuneo Someya [b]

[a] Technical University of Lodz, Division of Dynamics and Control (K-13), Stefanowskiego 1/15, 90-924 Lodz, Poland.

[b] Musashi Institute of Technology, Department of Mechanical Engineering, Tamazatsumi 1-28-1, Setagaya-ku, Tokyo 158, Japan

Abstract.
A general symbolic-numerical computational approach is applied to a study of the oscillations of the string type generator with a time-delay amplifier. First the analytical perturbation method (supported by symbolic computation) which yields the averaged differential equations is used and then the obtained averaged equations are analyzed numerically showing some surprising phenomena.

1. INTRODUCTION

Our report has two aims. On the one hand we analyze nonlinear dynamics of the electromechanical system, whereas on the other hand based on this example a systematic strategy of solving many other relative problems is given. The investigation includes a few steps. 1. Equations of dynamics are derived. 2. The averaging method is proposed and a written program for symbolic computation yields the averaged differential equations. 3. A further systematical study of the obtained equations is developed.

2. ANALYZED SYSTEM AND GOVERNING EQUATIONS

The electromechanical model under consideration is rather simple. It consists of a distributed mass system (string), whose oscillations are governed by a partial differential equation. The dynamics of an amplifier is modelled as a simple linear oscillator with the damping c and frequency $\sqrt{k}$. The amplifier supplies a current to the string, which is embedded into a magnetic field. The amplitude of the current undergoes control changes from the amplifier with time delay.

The electromagnetic induction B(x) acting along the string generates stresses at the ends of the string according to the following equation

$$E(t) = \int_0^l B(x)\frac{\partial u(t,x)}{\partial t}dx, \tag{1}$$

where x is a spatial coordinate, t denotes time, u(t,x) is the amplitude of oscillations of the string in the (t,x) point and l is the length of the string. The generated stresses are responsible for the force excitation

$$Y(t) = h_1 E(t) - h_2 E^3(t), \tag{2}$$

where h_1, h_2 are constant coefficients. Dynamics of the amplifier is governed by the equation

$$\ddot{I}(t) + 2\lambda\dot{I}(t) + kI(t) = \dot{Y}(t-\mu), \tag{3}$$

where a "dot" denotes differentiation with respect to t, λ is damping coefficient, and μ is a time delay. The changes in time of I(t) and the changes in x of B(x) causes the oscillations of the string due to the equation

$$\frac{\partial^2 u(t,x)}{\partial t^2} - c^2\frac{\partial^2 u(t,x)}{\partial x^2} = -\frac{\epsilon}{\rho}\left(2h_0\frac{\partial u(t,x)}{\partial t} - B(x)I(t)\right), \tag{4}$$

where h_0, ρ are constants, and ϵ is a small positive parameter. The frequencies of free oscillations of the string are given by $\omega_s = \Pi cs/l$, and the homogeneous boundary conditions are as follows

$$u(t,0) = u(t,l) = 0. \tag{5}$$

3. AVERAGING METHOD

For $\epsilon = 0$ the solution to the equation (4) is given by

$$u_0 = a_1\cos(\omega_1 t + \theta_1)\sin\left(\frac{\pi x}{l}\right) + a_3\cos(3\omega_1 t + \theta_3)\sin\left(\frac{3\pi x}{l}\right), \tag{6}$$

where a_1, a_3 are the amplitudes, and θ_1, θ_3 the phases.

For small enough $\epsilon \neq 0$ the solution to the equation (4) is expected to be of the form

$$u = u_0 + \epsilon u_1(x, a_1, a_3, \theta_1, \theta_3) + higher\ \ order\ \ terms, \tag{7}$$

We take

$$B = B_1\sin(\Pi x/l) + B_3\sin(3\Pi x/l). \tag{8}$$

For small μ the right hand side of equation (3) can be approximated by $\dot{Y} - \mu\ddot{Y}$. The solution to the linear equation (3) has a form

$$I_0(t) = M_1 \cos\omega_1 t + N_1 \sin\omega_1 t + M_3 \cos 3\omega_1 t + N_3 \sin 3\omega_1 t, \qquad (9)$$

where M_1, N_1, M_3, N_3 can be obtained and are rather complicated terms.

Further analysis is typical of the perturbation technique and details can be found elsewhere [1-3]. Because B(x) and I(t) are defined, therefore equation (4) can be solved using a classical perturbation approach. Substituting equation (7) to equation (4) and taking into account that $a_i = a_i(t)$ and $\theta_i = \theta_i(t)$ (i=1,3) the following averaged equations are obtained

$$\frac{da_i}{dt} = \epsilon P_i(a_1, a_3, \eta),$$

$$\frac{d\theta_i}{dt} = \epsilon Q_i(a_1, a_3, \eta). \qquad (10)$$

where

$$P_i = -\frac{1}{2}\frac{B_i(N_i \cos\theta_i + M_i \sin\theta_i)}{i\rho\omega_1},$$

$$Q_i = -\frac{1}{2}\frac{B_i(M_i \cos\theta_i - N_i \sin\theta_i)}{i\rho a_i \omega_1}. \qquad (11)$$

As can be seen from equation (10), we have one variable η instead of θ_1 and θ_3. This is the key point of the averaging procedure presented here. Variable η results from

$$\eta = \theta_3 - 3\theta_1, \qquad (12)$$

and again the averaging is applied to the equation (10) in a somewhat special way, e.g., for i=1 we take $\theta_3 = \eta + 3\theta_1$, whereas for i=3 we take $\theta_1 = (1/3)(\theta_3 - \eta)$. (It must be acknowledged that the first attempt to derive an averaged set of equations has been made by Rubanik [4]).

As it is assumed by the averaging procedure, amplitudes a_i and θ_i change with a time very slowly, and a long numerical integration to trace a behaviour of the system is required.

4. ANALYSIS OF THE AVERAGED EQUATIONS

First the stationary solutions of the equation (10) are considered. This leads to the question of solving the following nonlinear algebraic equations. To solve the problem the Powell hybrid method and an variation of Newton's method have been used. It takes a finite-difference approximation to the Jacobian with high precision arithmetics.

It is rather very difficult task to proof existance of time dependent solutions in the analyzed system of the averaged equations. The most expected situation is to find stable fixed points, which correspond to the oscillations with constant amplitudes in the original system. One of the important questions is: in which way the phase flow reaches stable fixed points, i.e. exponential or oscillatory manner. Theoretically the system can also exhibit unbounded solutions as well as stationary long time aperiodic behaviour can be found. We have found the various, and sometimes very surprising phenomena discovered during the numerical analysis.

5. CONCLUDING REMARKS

The obtained results are briefly summarized below.
1. With the increase of the B_1 constant amplitude a_1 decreases, whereas a_3 and η remain constant.
2. For each value of B_1 there are two solutions of a_1 and one of a_3 and a reach set of η solutions which possess characteristic structure.
3. In the case of time dependent solutions the occurrence of time delay causes loss of stability of $a_1 = a_3 = 0$ and sudden jump to the constant positive values, where they remain constant in spite of $\theta_4(t)$ time evolution (the new frequency has $\dot{\theta}_3$ occurred).
4. Amplitude a_1 is constant, whereas amplitude $a_3(t)$ monotonously decreases in time, and the phases remain constant (time independent).
5. A transitional phase modulated state can be realized.
6. After a passage with ω through resonance a sudden extinguishing of oscillation occurred.
7. Unbounded solutions exhibited by a_3 has been found.
8. Steady state with the oscillation of a_1, a_3, θ_3 and constant $\dot{\theta}_1$ is also detected.

6. REFERENCES

1 I. G. Malkin, Some Problems in the Theory of Nonlinear Oscillation, Nauka, Moscow, 1956, in Russian.
2 V. I. Zubov, Theory of Oscillations, High School, Moscow, 1979, in Russian.
3 J. Guckenheimer and P. Holmes, Nonlinear Oscillations, Dynamical Systems and Bifurcations of Vector Fields, Springer, New York, 1983.
4 W. P. Rubanik, Oscillations in Complex Quasilinear Systems with Delay, University Press, Minsk, 1985, in Russian.

Simulation and Design of Applied Electromagnetic Systems
T. Honma (Editor)

Analysis of permanent magnet type automotive solenoid valve for fast response using finite element method

Y. Ohdachi [a], Y. Kawase [b] and Y. Murakami [c]

[a] R & D Center, Toyoda Automatic Loom Works, Ltd.
2-1 Toyoda, Kariya, Aichi, 448, Japan

[b] Department of Electronics and Computer Engineering, Gifu University,
1-1, Yanagido, Gifu, 501-11, Japan

[c] Electronics Department, Aichi Steel Works, Ltd.
1, Arao, Tokai, Aichi, 476, Japan

Abstract

Recently, many automotive electromagnetic components require small-size, fast response solenoid valves. The design factor are dimension, components, characteristics of the material of an actuator and the applied voltage waveform. And the effects of design factor have been clarified in detail. In this paper, the response time of the two typical types, which are the plunger type and the permanent magnet type, are examined by a dynamic analysis using finite element method. The effect of the return spring for fast response is especially analyzed, and the advantage of the permanent magnet type clarified.

1. INTRODUCTION

Recent automobiles are equipped with many electromagnetic components which require small-size, fast response solenoid valves. Conventionally, many plunger types are used in automotive solenoid valve, because of the cost and the simple structure.

Now, the optimum design of the fast response is required. Usually, magnetostatic field analysis is applied to the design of such solenoid valves. However, for accurate calculation of the response time, the finite element analysis is necessary, which takes into account the motion of an armature[1], [2], [3]. We already developed the method for the optimum design of the plunger type solenoid valve[4], [5]. However, the fast response is limited by a return spring, which is one of the disadvantageous factors in the plunger type. In this paper, the response time of the permanent magnet type, which dose not have a return spring, is examined by the dynamic analysis. And we analyzed about the response time, comparing a plunger type solenoid valve with a permanent magnet type solenoid valve.

2. CALCULATION MODEL AND METHOD

The calculation models are axisymmetric dc solenoid valves as shown in Figs.1 and 2. The types are the plunger type and the permanent magnet type. They are almost the same dimension. The calculation method is coupled with a magnetic circuit, an electric circuit and a motion analysis[1].

The calculation is performed by a step-by-step method and couples problems using the fundamental equations. In the next step, the subdivision, which is a transferred element of airgap without changing the connection with nodes, must be re-generated for the motion of an armature.

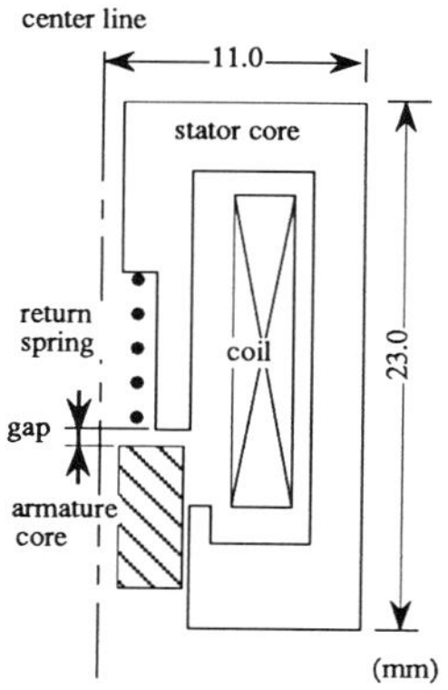

Fig. 1 Plunger type solenoid valve.

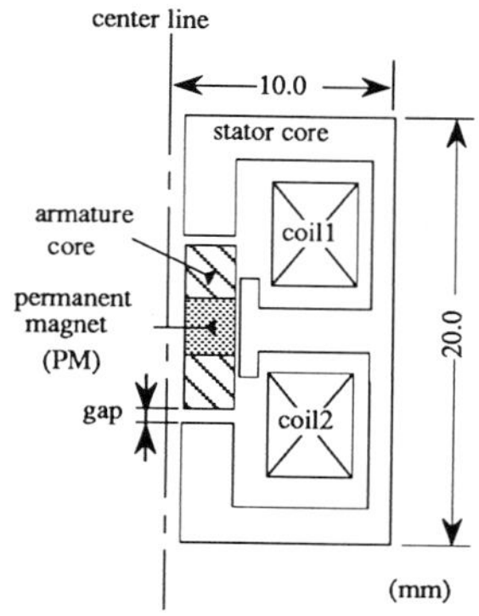

Fig. 2 Permanent magnet type solenoid valve.

Figures 3 and 4 show the typical time-stroke characteristics of the plunger type and the permanent magnet type, respectively. The plunger type has a single exciting winding and is applied a single voltage. The permanent magnet type has two exciting windings, which are reverse-serious connection. Two exciting windings of the plunger type are applied the forward voltage or the reverse voltage, so the return spring is not necessary.

The response time of a plunger type can be evaluated by T_{on} and T_{off} as shown Fig. 3. T_{on} is the time required for transition from a free state to a closed state and T_{off} is the opposite. The response time of a permanent magnet type can be evaluated by T_{on} as shown in Fig. 4. T_{on} is the time required for transition from an upper state to a lower state and the opposite time T_{off} also is T_{on}.

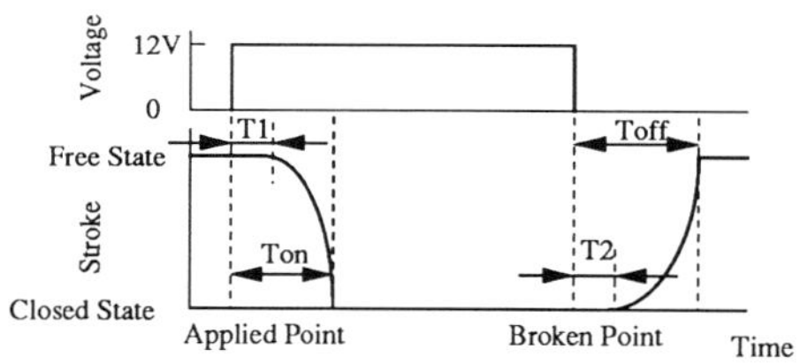

Fig. 3 Typical time-stroke characteristics. (Plunger type)

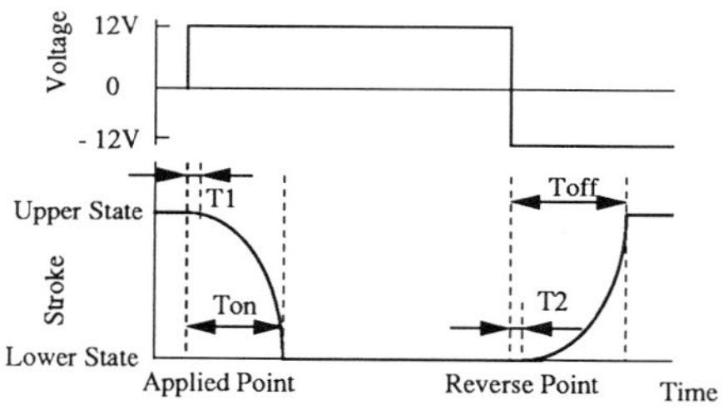

Fig. 4 Typical time-stroke characteristics. (PM type)

3. CALCULATION RESULTS

Figure 5 shows examples of the flux distributions in the plunger type solenoid valve, when the voltage is applied and the current is broken, respectively. Figure 6 shows examples of the flux distributions in the permanent magnet type solenoid valve, when the voltage is applied and the voltage is reversed, respectively. The skin effect is shown on each flux distribution.

T_1 and T_2 of the plunger type shown in Fig. 3 are a dead time, which is not response time, by the influence of a reverse force. The reverse forces are caused by a return spring and the eddy currents. On the other hand, the dead time of the permanent magnet type shown in Fig.4 is short because of no return spring.

Figures 7 and 8 show the time-stroke characteristics of the plunger type and the permanent magnet type when the voltage is applied and the current is broken, respectively. Table 1 shows T_{on} , T_{off} and T_{sum}, which is the sum of T_{on} and T_{off}.

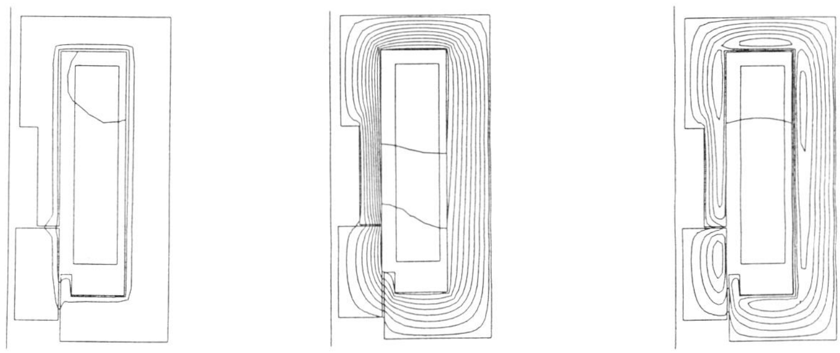

(a) (0.2msec. after voltage is applied) (b) (3.0msec. after voltage is applied) (c) (0.2msec. after current is broken)

Fig. 5 Flux distributions. (Plunger type)

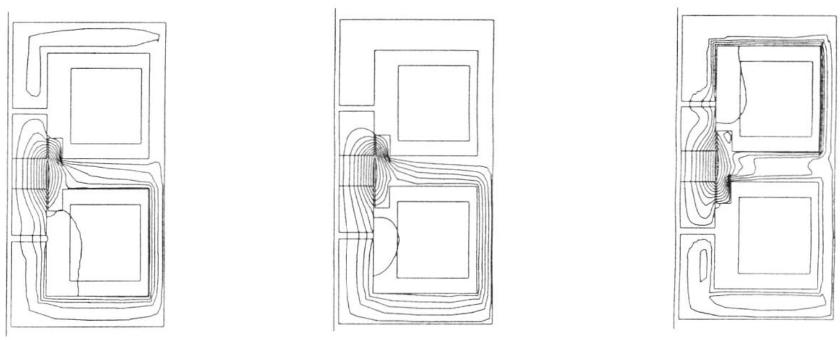

(a) (0.2msec. after voltage is applied) (b) (3.0msec. after voltage is applied) (c) (0.2msec. after current is broken)

Fig. 6 Flux distributions. (PM type)

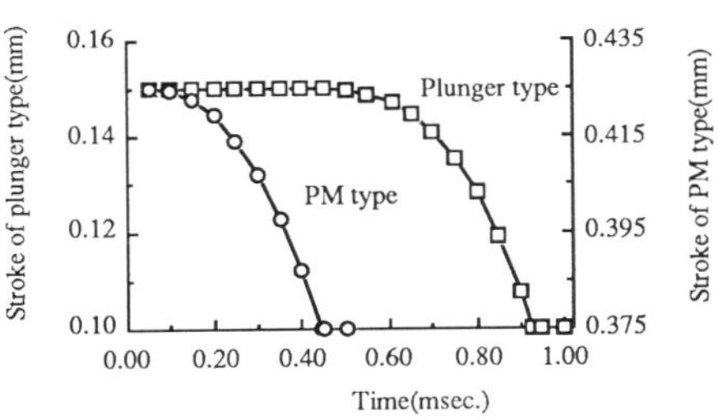

Fig. 7 Time-stroke characteristics(T_{on}).

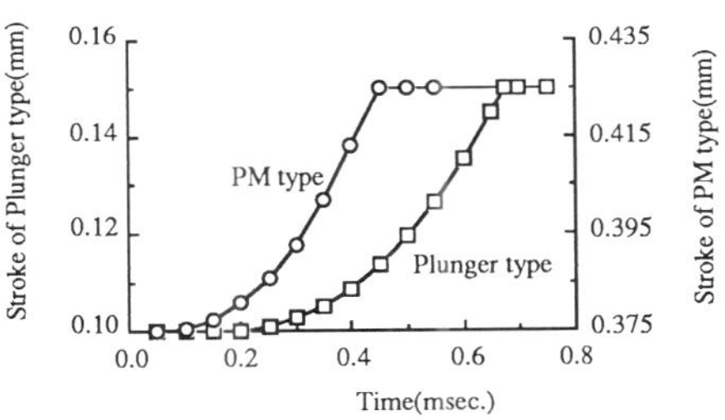

Fig. 8 Time-stroke characteristics(T_{off}).

Table 1 Response time(msec.).

	Ton	Toff	Tsum
Plunger type	0.93	0.67	1.60
PM type	0.45	0.45	0.90

Figure 9 shows the relationships between the spring constant k and T_{sum} in the plunger type. When k is small, T_{on} is short but T_{off} is long. On the other hand, when k is large, the contrary results are produced. In the plunger type, k is optimum at 500 N/m for the minimum T_{sum}, which is 1.6 msec. The response time of the PM type are aloso shown in Fig. 9. In the permanent magnet type, T_{on} and T_{off} are same values, which are 0.45msec.

Therefore, the permanent magnet type solenoid valve is much better than the plunger type solenoid valve for fast response.

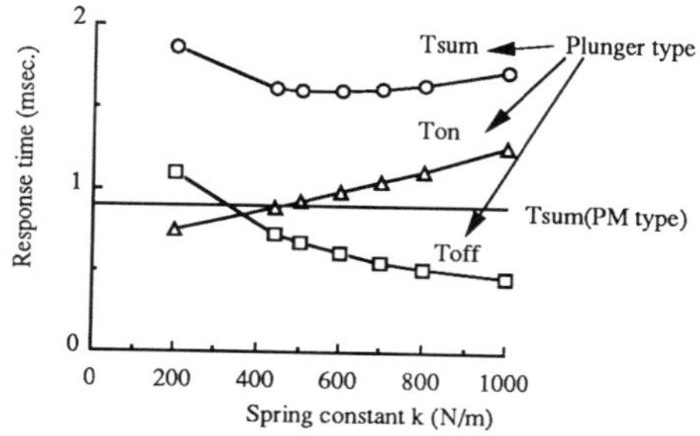

Fig. 9 Relationships between spring constant k and T_{sum}.

4. CONCLUSIONS

The calculation of the response time on the two type solenoid valves produced the following results.

(1) The dynamic response of the solenoid valves can be accurately calculated.

(2) The permanent magnet type is much better than the plunger type for fast response because of the reverse forces of return spring. The minimized response time of the plunger type was 1.6 msec, and the minimized response time of the permanent magnet type was 0.9msec.

5. REFERENCES

[1] Y. Kawase and Y. Murai: "Numerical Analysis of Transient Dynamic Characteristics of Plunger Type Electromagnet" Trans. of IEICE Japan, J71-C,1,25 (1988).

[2] A. Kuromiya, K. Takeuchi, T. Nakata, N. Takahashi and K. Fujiwara: "Analysis of Electronic Fuel Injector taking into Account External Electric and Valve Motion" Proc. of 11th Numerical Electrical and Electronic Symposium by the Institute of Simulation of Japan, I-12, 89, (1990).

[3] M. Pawlak: "Transient Design Optimization of Solenoid Actuators using Finite Elements" Proc. of COMPUMAG, Tokyo, AP-33, 163, (1989).

[4] Y. Kawase and Y. Ohdachi: "Dynamic Analysis of Automotive Solenoid Valve using Finite Element Method" IEEE Trans. on Magn., MAG-27,5, 3939(1991).

[5] Y. Ohdachi, Y. Kawase, Y. Murakami and Y. Inaguma: "Optimum Design of Dynamic Response in Automotive Solenoid Valve" IEEE Trans. on Magn., MAG-27,6, 5226(1991).

Simulation and Design of Applied Electromagnetic Systems
T. Honma (Editor)

Analysis of speed characteristics of electromagnetic suspension system with lateral displacement by hybrid FE and BE method

D.H. Im, J.P. Hong, J.W. Im and G.H. Lee[a]

[a]Dept. of Electrical Engineering, Hanyang University, 133-791 Seoul, Korea

Abstract

In this paper, hybrid method for 3D eddy current calculation in magnetic system with velocity is proposed. The analysis space is partitioned into three regions, which contain saturable magnetic region with magnetizing currents, which contain eddy currents, and free space. In the hybrid method, we adopt the current vector potential **T** and the magnetic scalar potential Ω as physical quantities. The satuable magnetic and currents region are analyzed by FEM using **T** and Ω, on the other hand, the free space by BEM formulation using Ω. We apply the hybrid method to electromagnet for levitation system and analyze the charateristics of attractive, lateral, and magnetic drag force according to velocity.

1. INTRODUCTION

In recent years, the computation 3D eddy current problems has become very increasingly important in order to achieve the rational design of electromagnetic apparatus. There have been several method of eddy current compution [1,3,5]. In the **T**-Ω method, the current vector potential **T** is only defined current carring region, and so the **T**-Ω method has advantage in computational memory and time compared with those of other methods, when the proportion of the current carring region to the whole region is small.

Especially, hybrid method can reduce the many unknown variables. The BEM does not require mesh divisions in analysis space except on the boundary of the surface, but it is difficult to apply to the non-uniform region, since BE formulation is applied to uniform region like as free space. In this paper, the magnetic field produced by magnetizing current is calculated by the magnetic vector potential $\mathbf{T}_0$, and so the unknown variables are reduced[2].

In magnetc levitation system, eddy current decreases its performance with increasing speed. Previous some articles has treated this characteristics. However, in 2D analysis, it can not calculate drag force taking into account the eddy current and can not consider leakage flux into the traveling direction.

In this paper, we adopt 3D hybrid method using **T**-Ω and analyze attractive, lateral, and magnetic drag forces as a function of speed.

2. BASIC EQUATION

In order to analyze, we divide the space into three regions. The first is saturable magnetic region containing magnetizing current. The second is conductive region containing eddy current. And the third is free space with constant permeability μ_o.

2.1 Determination of T_o

In order to determine $\mathbf{T}_o$ from magnetizing current $\mathbf{J}_o$, if we assume the y-component of magnetizing current does not exist, $\mathbf{J}_{ox}$ and $\mathbf{J}_{oz}$ can be represented as forrows[2,4].

$$J_{ox} = -\frac{\partial T_{oy}}{\partial Y} , \qquad J_{oz} = -\frac{\partial T_{oy}}{\partial X} . \tag{1}$$

If the magnetizing current is distributed uniformly in the winding, $\mathbf{T}_{oy}$ has constant value $|\mathbf{J}_o| * \mathbf{L}$.

2.2 FEM analysis region

The basic equation in saturable magnetic region is given by[5]

$$\mathrm{div}[\ \mu(\mathbf{T}_o - \ \mathrm{grad}\ \Omega)\ ] = 0 \tag{2}$$

where μ is the permeability and the current vector potential $\mathbf{T}_o$ is defined by

$$\mathbf{J}_o = \mathrm{rot}(\mathbf{T}_o) . \tag{3}$$

The magnetic scalar potential Ω is defined by

$$\mathbf{H} - \mathbf{T}_o = -\ \mathrm{grad}\ \Omega . \tag{4}$$

In the conductive region in which eddy current is existed, the field can be described as

$$\mathbf{H} = \mathbf{H}_o + \mathbf{H}_e = \mathbf{H}_o + \mathbf{rot(J_e)} \tag{5}$$

$$\mathrm{rot}(\mathbf{E}) = \mathrm{rot}(\mathbf{V} \times \mathbf{B}) = \mathrm{rot}[\ \mathbf{V} \times (\mathbf{B}_o + \mathbf{B}_e\ )] \tag{6}$$

$$\mathbf{J}_e = \mathrm{rot}(\mathbf{T}) \tag{7}$$

$$\mathbf{J}_e = \sigma\ \mathbf{E} \tag{8}$$

$$\mathbf{B}_e = \mu \mathbf{H}_e \tag{9}$$

$$\mathrm{div}\mathbf{B} = 0 \tag{10}$$

where $\mathbf{J}_e$ is eddy curent density, and $\mathbf{V}$ is velocity.

$\mathbf{H}_o$ is calculated from eq.(2), in magnetostatic case.

From equation (5)-(10), the basic equation in this region are represented by eq.(11,12).

$$\mathrm{div}[\mu_o\mathbf{H}_o + \mu(\mathbf{T} - \mathrm{grad}\ \Omega)] = 0 \tag{11}$$

$$\mathrm{rot}\ [\ \frac{1}{\sigma}\ \mathrm{rot}\ (\mathbf{T})\ ] = \mathrm{rot}\ [\ \mathbf{V} \times \{\ \mathbf{B}_o + \mu(\mathbf{T} - \mathbf{grad}\ \Omega)\ \}\ ] \tag{12}$$

2.3 BEM analysis region

In this region, free space with permeablity μ_o, the field can be described as

$$\mathbf{H} = -\text{gad}\ \Omega \tag{13}$$

$$\text{divB} = 0\ . \tag{14}$$

From eqation (13)-(14), the basic equation in this region are given by

$$\text{div}[\mu(-\text{grad}\ \Omega)] = 0\ . \tag{15}$$

3. BOUNDARY CONDITION

In the interface area between each regions, following boundary condition should be satisfied.

$$\Omega_1 = \Omega_2 \tag{16}$$

$$\mathbf{B}_1 \circ \text{n} = \mathbf{B}_2 \circ \text{n} \tag{17}$$

$$\mathbf{H}_1 \text{ x } \mathbf{t} = \mathbf{H}_2 \text{ x } \mathbf{t} \tag{18}$$

where the subscript 1 and 1 denote each regions, respectively. **n** and **t** denote outward normal and tangential direction vectors of the surface, respectively.

On the other hand, inteface condition between eddy current region and free space region must follow[5,6].

$$\mu[\text{H}_o + (\text{T} - \text{grad}\ \Omega\ )] = \mu[\text{H}_o - \text{grad}\ \Omega\ ]\ . \tag{19}$$

Eq.(19) is used for the calculation of the surface integral of eq.(11).

4. HYBRID ANALYSIS

Figure 1. shows the analysis model. Taking into account the symmetry, we have analyzed only 1/2 region of a magnet. In this model, conductivity(σ) of railis is $5.5\text{x}10^5$[s/m], relative permeability of rail and magnet is 500 and 800 resfectively. Magnetizing current density is $1\text{x}10^6[\text{A/m}^2]$.

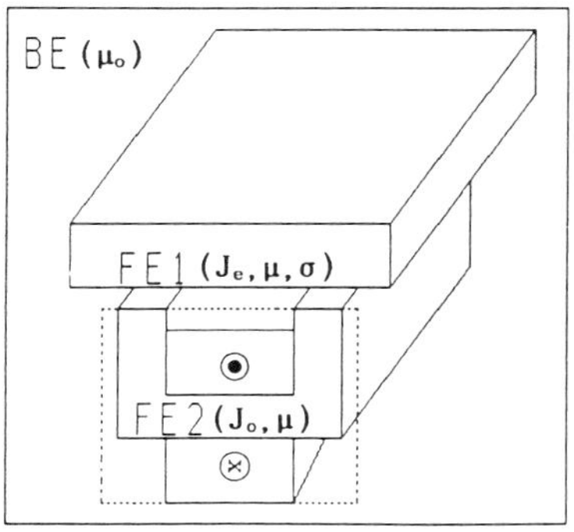

Figure 1. Analysis model.

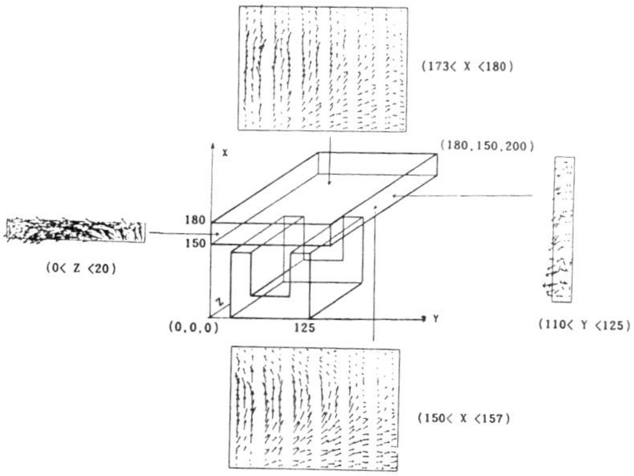

Figure 2. Flux distribution.
(v=100[km/h], d=0)

The attractve, lateral, and magnetic drag force depending on speed and lateral displacement are shown in Figures 3. and 4.

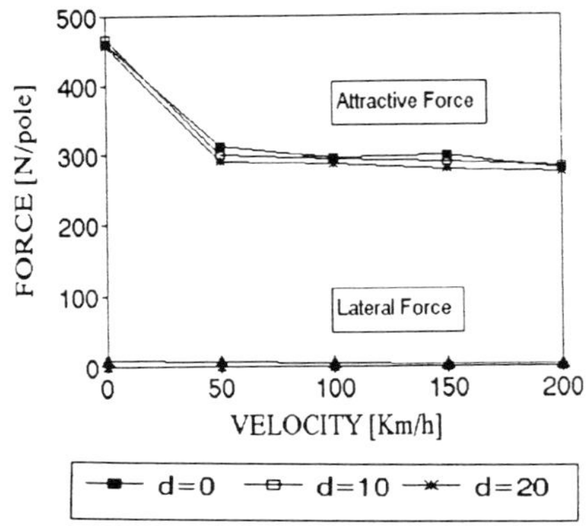

Figure 3. Attrative and lateral force. (d:lateral displacement[mm])

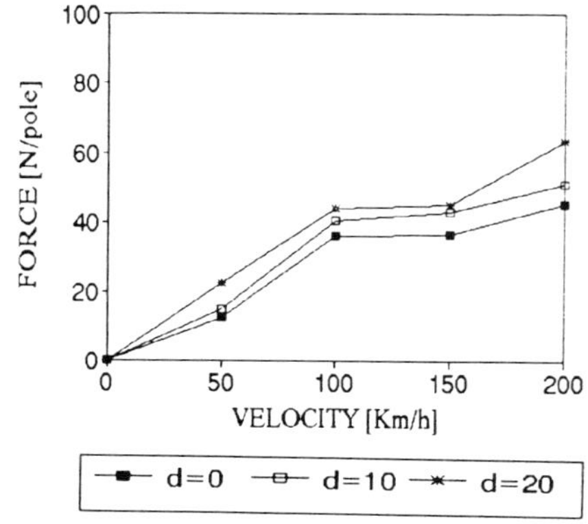

Figure 4. Magnetic drag force. (d:lateral displacement[mm])

5. CONCLUSION

We applied the 3D hybrid T−Ω method to electromagnetic levitation system considering speed. In the results of analysis, the performance of electromagnet, with increasing speed, are decreased as expected. It has been shown that speed characteristics should be considered in system design.

In the open boundary problems, hybrid method using T−Ω was very effective , and the simplification of magnetizing current distribution by T_o was too.

6. REFERENCES

1. T. Onuki and S. wakao, Hybrid finite element and boundary element method utilizing scalar potential for 3D electromagnetic field analysis, T.IEE ,Japan,112−A, 5, 1992.
2. T.Nakata, N.takahashi, K.Fujiwara and Y.Okada, Improvements of the T−Ω method for 3−D eddy current analysis, IEEE Trans. on Mag. 24, 1, 1988.
3. Y. Tanaka, Tree−dimensional magnetic field analysis by T− Ω method using the finite element method coupled with integral equation, IEE, Japan, 1991.
4. S.Hasebe, Y.kano and H.Miyazi, Relation between T−Ω method and a new current model, IEE Japan, 1988.
5. F.Bouillault, Z.Ren and A.Razek, Calculation of 3D eddy curreent problems by an hybrid T−Ω method, IEEE Trans. on Mag. 26, 2, 1990.
6. T.Nakata, N.Takahashi, K.Fujiwara, M.Miura and Y.Okada, Boundary conditions for finite element analysis of 3−D fields, THE INTERNATIONAL WORKSHOP FOR EDDY CURRENT CODE COMPARASION, TOKYO, 1986.

Acknowledgement : This work was supported in part by the institute of the Korean Science and Engineering Foundation. (No. 92−47−00−02)

Simulation and Design of Applied Electromagnetic Systems
T. Honma (Editor)

Study on Flux Distribution to Inductor Specification in Claw Pole Linear Pulse Motor

D.Ebihara[a], Y.Nakata[a], M.Watada[a], T.Watanabe[b] and H.Nakano[b]

[a]Musashi Institute of Technology
1-28-1, Tamazutumi, Setagaya-Ku, Tokyo 158, Japan

[b]Fuji Electrochemical Co.,Ltd.
2281, Washizu, Kosai-Shi, Shizuoka 431-04, Japan

Abstract

The use of linear pulse motors(LPM) have increased in the industry applications because of the desirability of direct drive for control. The authors have developed the Claw-Pole Linear Pulse Motor (CP-LPM) which is a linear version of a rotating PM type stepping motor. In this paper, we investigate inductor specifications in the CP-LPM.

1.INTRODUCTION

The adoption of direct drive for control have increased in the industry applications because of the desirability of high speed and prevention of ageing to positioning accuracy. For the linear pulse motor(LPM) is possible to drive for open loop control, it is suitable for this situation. In general, these motors have been desired high power, high resolution, lightweight and flat type, therefore specially modified motors have also been developed. However, these motors raise production costs and require that machines have higher processing precision.

The authors have developed the Claw-Pole Linear Pulse Motor (CP-LPM) which lay out flat a rotating PM type stepping motor. The CP-LPM features a simple and inexpensive-to-build structure, magnetic poles that can be press-formed. However, it also has such structural disadvantages as a large flux leakage and stator poles that magnetically saturates quite easily[1].

This study is purposed at improving thrust and remedying structural deficiencies of magnetic saturation and flux leakage. Difficulties of measuring flux characteristics in a prototype motor require designing and building a simulation model with which we can find out how inductor specifications affect flux distribution. We also designed several inductors to test flux characteristics to those specifications.

We report here on flux density characteristics in the simulation model, of inductor specification parameters of sawtooth shape, mechanical clearance and pole pitch, and thrust characteristics in the prototype motor. And we investigate relationships between parameters.

2.BASIC STRUCTURE OF CP-LPM

Figure 1 shows the CP-LPM's basic structure and Table 1 lists its specifications. The stator consists of exciter coil, stator yoke and claw inductor. The stator yoke and inductor are made of soft magnetic material and the coil wraps around them by concentrated winding. The mover consists of two magnetic blocks and a cart. [1]

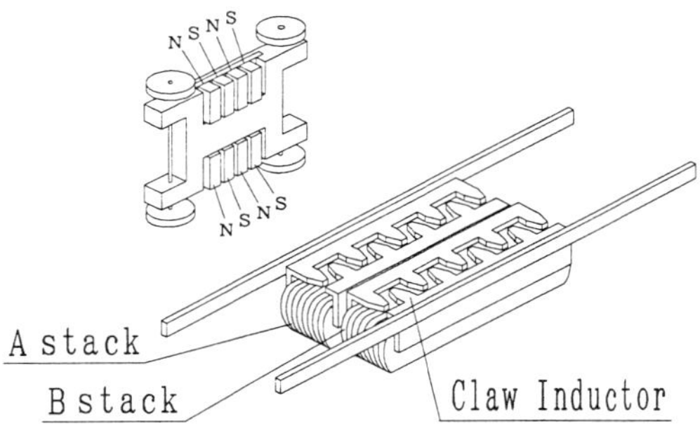

Figure 1. Basic structure of CP-LPM

Table 1
Specifications of CP-LPM

Stator	
Total Length	35.75 [mm]
Pole Pitch	2.5 [mm]
Teeth Thickness	1.0 [mm]
Teeth Number	14
Phase Number	2
Winding Number	120 [turn]
Mover	
Pole Pitch	2.5 [mm]
Magnet Number	8
Material	Ferrite
Thickness	1.0 [mm]
Width	2.5 [mm]
Depth	3.0 [mm]

3.STRUCTURE OF SIMULATION MODEL

To overcome CP-LPM's deficiencies, we should examine the flux density characteristics between the related inductor specification parameters. To measure flux density, we built a one-phase simulation model, about 10 times actual size. Figure 2 shows the structure of simulation model and Table 2 lists its specifications. And we built inductors based on the assumption that differences in the shape of inductor saw-teeth and pole pitch affect flux distribution. Figure 3 shows the inductor shapes and Table 3 lists their specifications.

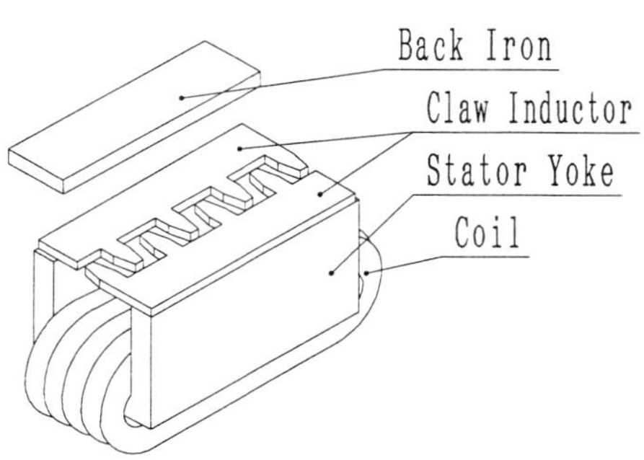

Figure 2. Structure of simulation model

Table 2
Specifications of simulation model

Stator Yoke		Back Iron	
Terms	Specifications	Terms	Specifications
Length	280 [mm]	Length	300 [mm]
Height	125 [mm]	Width	85 [mm]
Width	150 [mm]	Thickness	15 [mm]
Thickness	25 [mm]		
Winding Number	400 [turn]		

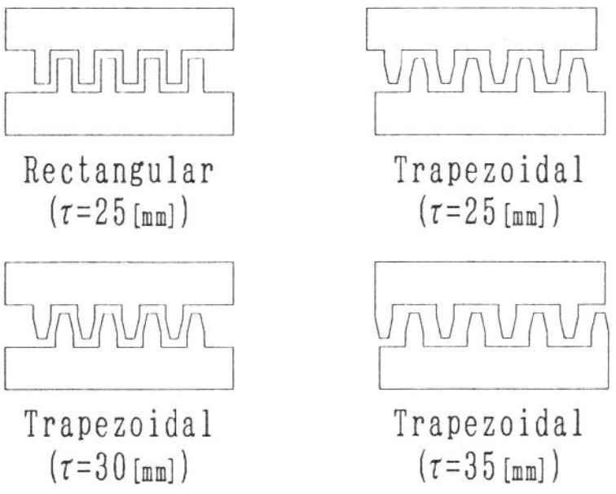

Figure 3. Inductor shapes

Table 3
Specifications of inductors

Terms	Specifications			
Inductor	Rectangular	Trapezoidal		
Total Length	280[mm]	280[mm]	280[mm]	280[mm]
Pole Pitch	25[mm]	25[mm]	30[mm]	35[mm]
Salient Pole Width	17[mm]	20[mm]	20[mm]	20[mm]
Slot Width	33[mm]	30[mm]	40[mm]	50[mm]
Teeth Thickness	10[mm]	10[mm]	10[mm]	10[mm]
Teeth Number	8	8	8	8

4.FLUX DENSITY CHARACTERISTICS

Because flux density distribution affects the LPM's generated thrust, effective flux density was measured using a Hall sensor with the inductor central surface as the point of measurement.

Figures 4 and 5 show the results of flux density measurement. Figure 4 shows the characteristics with a mechanical clearance g of 5.0 [mm]. Figure 5 shows the characteristics with a magnetomotive force NI of 1000 [A].

In Figures 4 and 5, this confirmed that effects of magnetic saturation and flux leakage are dependent on inductor shape, and that enlarging the cross-sectional root area of the trapezoidal inductor gives a shape that makes magnetically saturation difficult. It also shows that increasing pole pitch τ to decrease flux leakage on the sides improves flux density in the trapezoidal inductor.

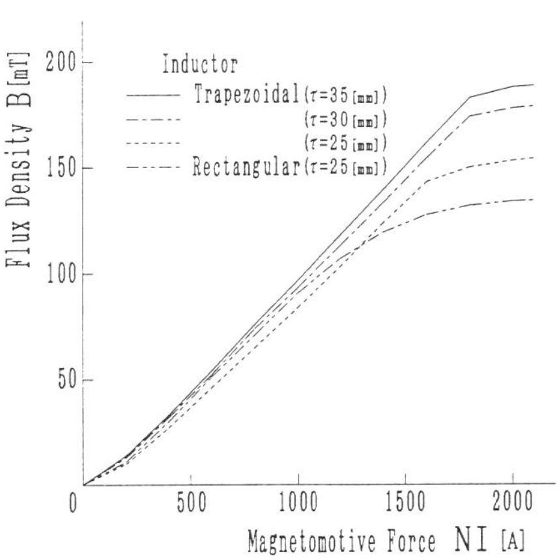

Figure 4. Magnetomotive force vs. flux density

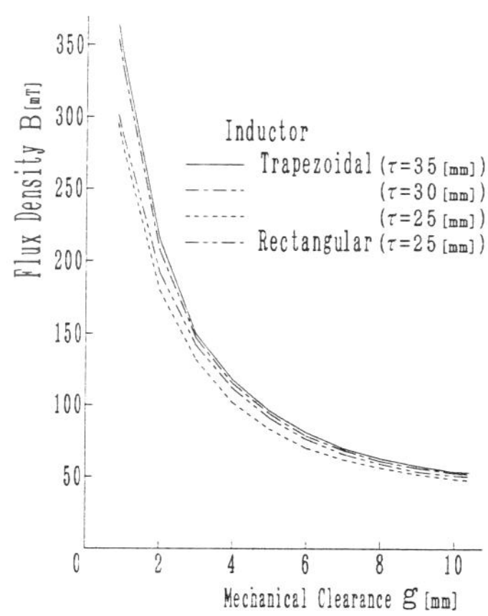

Figure 5. Mechanical clearance vs. flux density

5. STATIC THRUST CHARACTERISTICS

Static thrust characteristics is influenced by magnetic saturation of inductor's salient poles. We measured the thrust characteristics in a prototype motor to study the effect of sawtooth shape. Figures 6 and 7 show the thrust characteristics in the rectangular and trapezoidal inductors. Generally, rectangular inductor gives a shape that makes magnetic saturation easy because magnetic flux concentrates on salient poles. On the other hand, trapezoidal inductor give a shape that makes magnetic saturation difficult, and that makes detent thrust little. However, the results of thrust measurement indicate that there are hardly any differences on characteristics beyond the limits of magnetic pole's saturation. As a result, considering the no saturation areas, we can select the rectangular inductor to be expected of positioning precision in a state of detent.

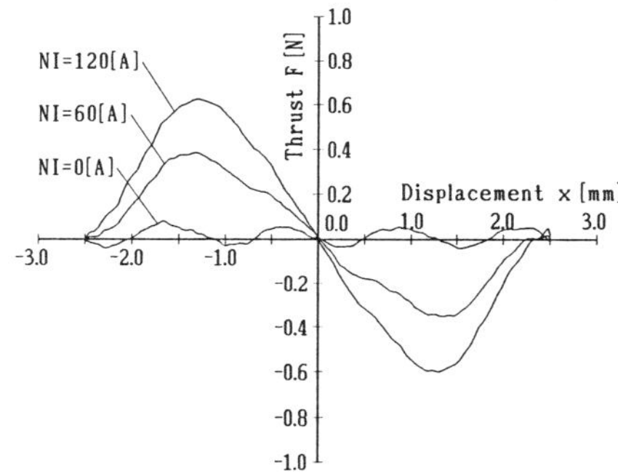

Figure 6. Thrust characteristics (rectangular inductor)

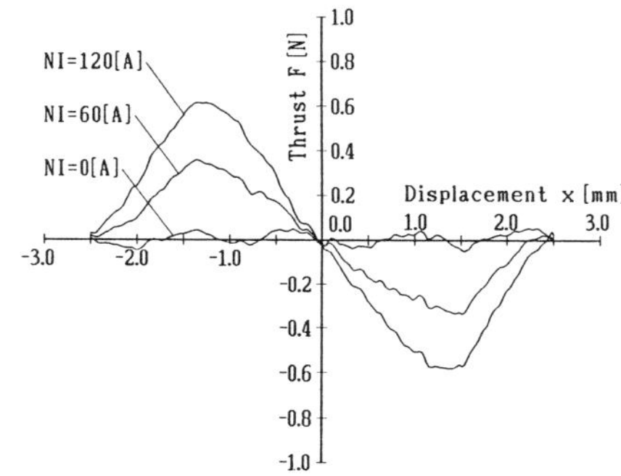

Figure 7. Thrust characteristics (trapezoidal inductor)

6.CONCLUSION

To improve the defects of CP-LPM structure, we measured flux density in the simulation model and static thrust in the prototype motor.

Measurements of flux density confirmed the assumption that modifying inductor shape from rectangular to trapezoidal makes magnetic saturation difficult. Increasing pole pitch τ also improved magnetic characteristics. The results also show that changing inductor shape from rectangular to trapezoidal and increasing pole pitch were appropriate improvements. However, the results of thrust measurement indicate that there are hardly any differences on characteristics beyond the limits of magnetic pole's saturation.

7.REFERENCE

1 D.Ebihara and Y.Kuno, "The Development of Claw Pole linear Pulse Motor with Permanent Magnet", IEEE TRANSACTIONS ON MAGNETICS Vol.25 No.5 (1989)

Simulation and Design of Applied Electromagnetic Systems
T. Honma (Editor)

Analytical Method of Surface Motor

Daiki Ebihara[a] , Takashi Watanobe[a] , Masaya Watada[a] and Kyotaro Shimbo[b]

[a] Musashi Institute of Technology
1-28-1, tamazutsumi, Setagaya-ku, Tokyo 158, Japan

[b] Shin-etsu Chemical Co.,Ltd.
2-6-1, Ohtemachi, Chiyoda-ku, Tokyo 100, Japan

Abstract

Recently, in X-Y drive systems, the direct drive method incorporating linear motors has been introduced and applied to practical use. However, the direct-drive Surface Motor ("SFM"), which has only one motor and moves in both the X and Y axes, has never yet to be practically applied. In view of this situation, we proposed an SFM that enables bidirectional motion using only one motor. In this paper, we examine a method for analyzing the SFM.

1. INTRODUCTION

Recent advances in office and factory automation equipment have led to the development and commercialization of direct linear motor drive systems. In fact, it is reported that an X-Y plotter, using two linear motors, is now in practical use.[1] However, the direct-drive SFM, which has only one motor and moves in both the X and Y axes, has never yet to be practically applied.

To make such an application a reality, we proposed an SFM that enables bidirectional motion using only one motor.[2] However, this SFM has some problems, which require its characteristic analysis. For this analysis, we must set up a method.

Therefore, considering a computer with the necessary data storage capacity, we examined the three-dimensional approximation to the SFM analysis.[3] This report describes the method of SFM analysis and the verification on simulation model.

2. BASIC STRUCTURE OF SFM

Figure 1 and Table 1 show the SFM's basic structure. The SFM's stator is a checkerboard arrangement of N- and S-pole magnets, laid atop a back iron plate. The mover is a bakelite truck platform and has eight core coils, each of which is magnetically independent. The basic SFM operation principle is the same as that for the PM-type Linear Pulse Motor. [2]

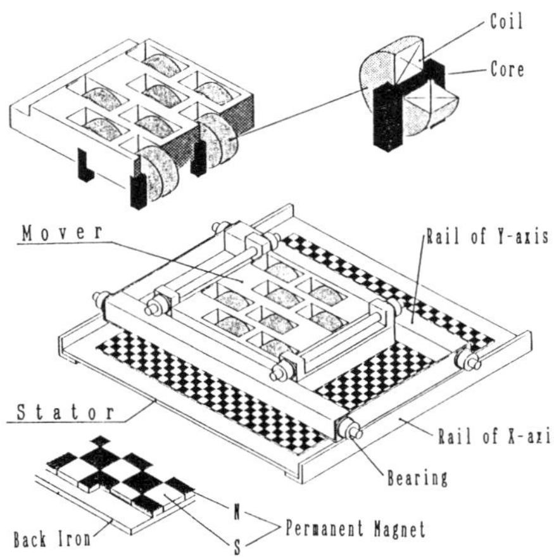

Figure 1. Structure of SFM

Table 1
SFM Specifications

Mechanical			
	Item		Value and unit
Stator	Size		160 x 160 [mm]
	Pitch τ		4 [mm]
	Magnet size		4 x 4 [mm]
		thickness	2 [mm]
	Back iron thickness		2 [mm]
		Material	Fe
Mover	Size		83 x 59.6 [mm]
	Phase number		8 [phase]
	Pole number		16 [pole]
	Material	Core	Fe
		Truck	Bakelite

Electrical	
Item	Value and unit
Winding number	900 [turn/magnet]
Rated exciting current	350 [mA]
Magnetomotive force	315 [A/magnet]
Winding resistance	17 [Ω/magnet]
PM material	Nd-Fe-B
stroke (X x Y)	100 x 116 [mm]
Excitation method	Bipolar
Gap length g	0.98 [mm]

3. APPROXIMATE THREE-DIMENSIONAL ANALYSIS

Figure 2 shows the differences between the real motor and the analytical model for two-dimensional Finite Element Method. Comparing the diagrams of the models, we think that, electromagnetically, the differences are in the behavior of the electromagnet and the permanent magnets. Particularly, the disagreement in the Z-direction flux density is caused by the difference in the electromagnet shapes. Therefore, we correct the way in which magnetomotive force is applied.

In making corrections, we first use the following Biot-Savart's law to derive both the flux densities.

$$dB = \frac{\mu_0 I}{4\pi r^2} \sin\theta \; dl \tag{1}$$

If I_r[A] and I_a[A] are the magnetomotive force of the real motor and the analytical model, then the respective generated flux density B_r[T], B_a[T] are calculated as follows.

$$B_r = \frac{\mu_0 I_r}{2r} \tag{2}$$

$$B_a = \frac{\mu_0 I_a}{\pi r} \tag{3}$$

Correcting the analysis requires that equation (2) and (3) are equal. Thus, we obtain the following relation.

$$I_a = \frac{\pi}{2} I_r \tag{4}$$

Figure 3 compares the measurements with the values obtained from corrected analysis. This shows that the corrective method is appropriate.[3]

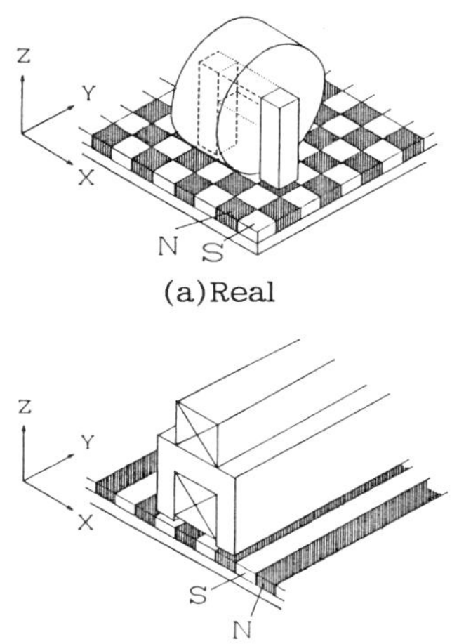

Figure 2. Real and Analytical Model

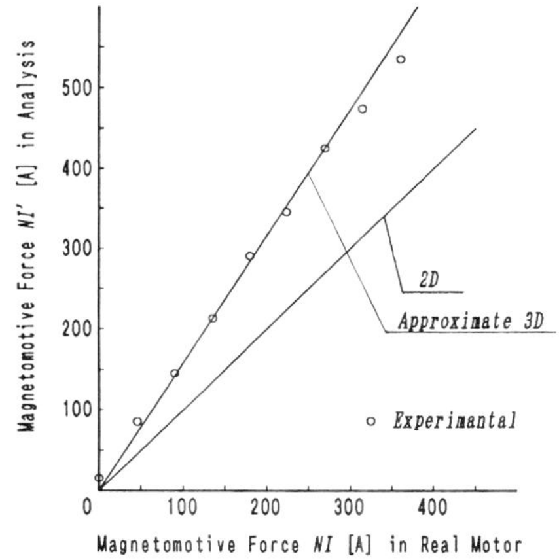

Figure 3. Correction of Magnetomotive Force

4. VERIFICATION ON SIMULATION MODEL

The pattern of the SFM's flux distribution is complex in three-dimension, so we must examine the behavior of the flux. And we must check for coincidence in three-dimensional approximation. However, it is impossible to directly measure the flux density in the air gaps. Therefore, we build a one-core simulation model, 5 times actual size. Figure 4 and Table 2 show the simulation model.

We used the simulation model to measure Z-direction flux density and attempt to compare with the values obtained from two-dimensional and approximate

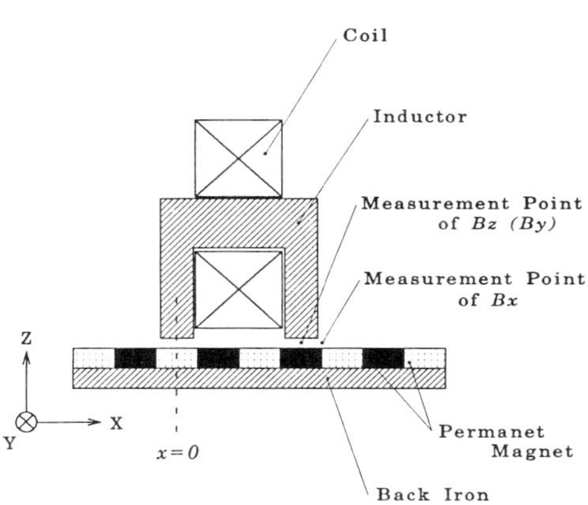

Figure 4. Simulation Model

Table 2
Simulation Model Specifications

Mechanical		
	Item	Value and unit
Stator	Size	120 x 180 [mm]
	Pitch τ	20 [mm]
	Magnet size	20 x 20 [mm]
	thickness	10 [mm]
	Back iron thickness	10 [mm]
	Material	Fe
Inductor	Height	70 [mm]
	Width	76 [mm]
	Thickness	16 [mm]
	Depth	45 [mm]
	Teeth pitch	60 [mm]
	width	16 [mm]
	Material	Fe

Electrical	
Item	Value and unit
Winding number	2400 [turn]
Exciting current	600 [mA]
Magnetomotive force	1440 [A]
PM meterial	Nd-Fe-B
Gap length g	5.00 [mm]

three-dimensional analysis. Figure 5 shows the results of that comparison. Here, inductor displacement x is zero. Figure 5(a) shows coincidence in the behavior of the electromagnet on approximate three-dimensional analysis. However, Figure 5(b) doesn't obtain coincidence. This disagreement is caused by the differences in the permanent magnets between the real motor and the analytical model.

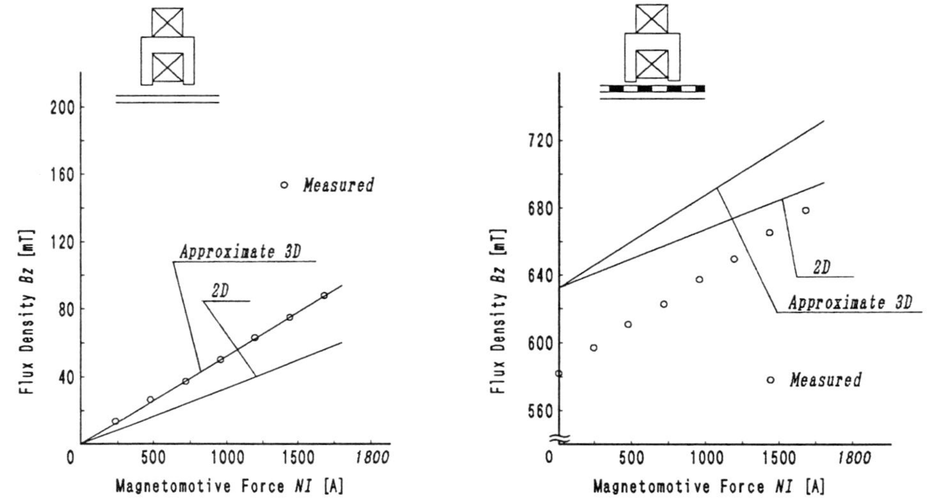

(a)Inductor (b)Inductor + Stator

Figure 5. Flux Density Characteristics

5. CONCLUSION

Considering the difficulty of obtaining a three-dimensional analysis usable for SFM, we examined a method of approximate three-dimensional analysis. For the approximation, we corrected the distributions of magnetic field, and then checked for coincidence on simulation model.

As a result, coincidence in the behavior of the electromagnet on approximate three-dimensional analysis has been obtained. In the future, we are going to examine the behavior of the permanent magnet and establish the method of SFM analysis.

6. REFERENCES

[1] B.A.Sawyer, U.S. Patent 3,376,578 (1968)
[2] D.Ebihara and M.Watada, "Study of a Basic Structure of Surface Actuator", *IEEE Trans. Magn.*, Vol.25, No.5, pp.3916-3918 (1989)
[3] D.Ebihara, T.Watanobe and M.watada, "Characteristic Analysis of Surface Motor", *IEEE Trans. Magn.*, Vol.28, No.5, pp.3033-3035 (1992)

Simulation and Design of Applied Electromagnetic Systems
T. Honma (Editor)

A study on high power of linear oscillatory actuator for artificial heart

Masaya Watada[a], Kenji Yanashima[a], Daiki Ebihara[a] and Koh Imachi[b]

[a]Department of Electrical and Electronics Engineering, Musashi Institute of Technology, 1-28-1, Tamazutsumi, Setagaya-ku, Tokyo 158, JAPAN

[b]Department of Medicine, University of Tokyo, 7-3-1, Hongo, Bunkyo-ku, Tokyo 113, JAPAN

Abstract

The development of a compact and high output actuator in total implantable artificial hearts is especially desirable. The authors developed a simple structured, permanent magnets single-winding Linear Oscillatory Actuator (LOA) for use as artificial heart actuator. In order to use the LOA as artificial heart actuator, one of requirements is to increase in starting thrust. This paper reports the exciting current control for the purpose.

1. INTRODUCTION

As a countermeasure of a heart disease, developments of a total implantable artificial heart actuator are being advanced.[1] The authors developed for use as a compact and high output actuator in artificial hearts, a simple structured, permanent magnets single-winding Linear Oscillatory Actuator (LOA) in which a single electromagnet performs reciprocal movement. The LOA is designed to satisfy all required structural specifications in dimension and form. Measurements of static thrust characteristics, frequency response and other characteristics showed static satisfaction of those specifications.[2][3] Further, we tested with a mock circulatory system, which simulates a human circulatory system, and measured the LOA's dynamic characteristics for use in artificial heart implants. Measurements showed that the LOA's characteristics, in standard drive method, were unable to satisfy requirements of aortic pressure and cardiac output necessary for human hearts. The main reason was the inadequacies to be starting thrust, for an incomplete understanding of blood circulation's load characteristics.[3][4]

For a LOA's practically use for artificial heart actuator, improvement on starting thrust is needed. As for a way to improve on this thrust, we designed a method of exciting current control, that was, to excite the LOA with required current. The current derived from this control is used to actually drive the LOA, and the measurements showed that any value desired thrust was get, therefore, increasing starting thrust is achieved.

This paper reports the exciting current control method and the results.

2. STRUCTURE OF LOA

Figure 1 shows a cutaway view of LOA. Table 1 shows specifications of LOA. Structurally, a stator has an exciting coil inside itself and a mover is a structure to put two axially-magnetized permanent magnets into three inductors. The mover goes to forward and reverse, when the stator coil is excited alternately.[2][3]

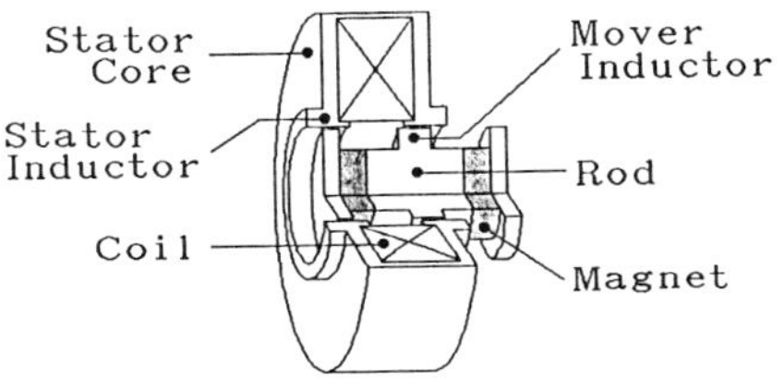

Figure 1. Cutaway View of LOA

Table 1
Specifications of LOA

Mechanical			Electrical		
Items	Values	[Unit]	Items	Values	[Unit]
Stator			Turn of Coil	1300	[Turn]
Size	φ70×20	[mm]	M. M. F.	1300	[A]
Material	SPC, SGP		Resistance of Coil	36	[Ω]
Mover					
Size	φ26×34	[mm]	Stroke	14	[mm]
Material	SUM-22		Gap Length	0.5	[mm]
Thick of Magnets	5	[mm]	Material of Magnets	Nb-Fe-B	

3. EXCITING CURRENT CONTROL

3.1. Characteristics of LOA

Table 2 shows comparison between requirements for artificial heart actuator and some measurements. From table 2, LOA satisfied the requirements at steady state. Accordingly, we tested with a mock circulatory system, which simulated a human circulatory system, and measured the LOA's characteristics. As results, the LOA couldn't satisfy performances of human heart. One of these causes was short of starting thrust, because of an incomplete understanding of blood circulation's load characteristics, and we tried to increase in starting thrust.

Table 2
Comparison between requirements and measurements

Items	Requirements		Measurements	
Size	φ85×30	[mm]	φ70×20	[mm]
Capacity	350	[mL]	57.1	[mL]
Mass	350	[g]	510	[g]
Stroke	10~20	[mm]	14	[mm]
Frequency	0.5~3.4	[Hz]	0~3.4	[Hz]
Force	50~100	[N]	63	[N]
Input power	2~10	[W]	36	[W]
Rising temperature	5~10	[°C]	47	[°C]
Life time	1	[year]	—	[year]
Aortic pressure	120~140	[mmHg]	40	[mmHg]
Cardiac output	2~5	[L/min]	1.6~1.9	[L/min]

3.2. Method of exciting current control

Figure 2 shows static thrust characteristics with a dotted line, when exciting current I=0A and I=1A were used. For improvement on starting thrust, we attempted a way to modify exciting current, namely, exciting current control. A way of the control is explained as follows.

If LOA's static thrust characteristics can be broken down into one term that depends on current and another that does not, this thrust can be written as

$$F_x(x) = a \times I(x) + F_d \tag{1}$$

where $F_x(x)$ is static thrust, F_d is thrust of independent on current, and a is coefficient. From equation (1),

$$I(x) = \frac{F_x(x) - F_d}{a} \tag{2}$$

When $F_x(x)$ is replaced by any value thrust on equation (2), current to get the thrust is shown.

We designed a drive system in figure 3, which is chopping controller with Pulse Width Modulation circuit.

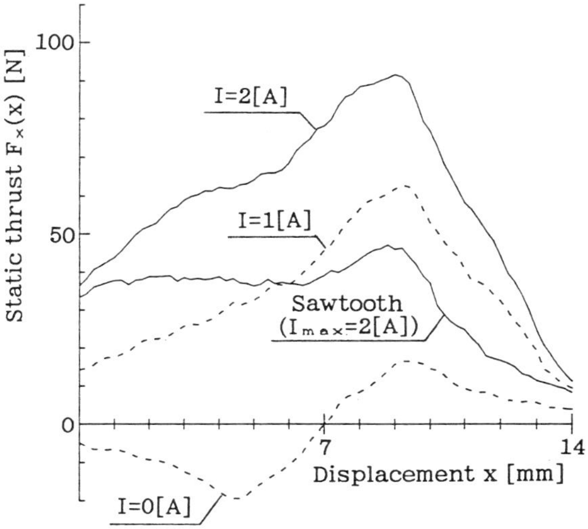

Figure 2. Static thrust characteristics

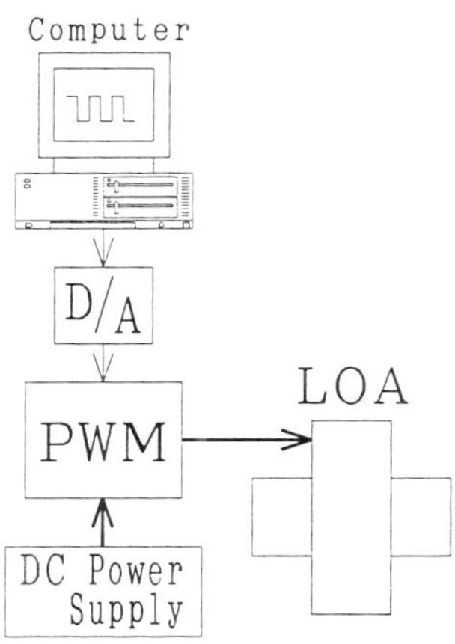

Figure 3. Drive system

3.3. Adaptation to the method

We try to apply this idea to the actual state. First, so as to transform static thrust $F_x(x)$ in figure 2 as style of equation (1), harmonic components in $F_x(x)$ must be taken into consideration, for that reason, $F_x(x)$ is needed to open to Fourier series. Second, $F_x(x)$ was solved about exciting current I(x) as equation (2), and I(x) was shown equation (3).

$$I(x) = \frac{F_x(x) - \left\{ \sum_{n=0} F_{\alpha n} \cos(n\omega x) - \sum_{n=0} F_{\beta n} \sin(n\omega x) \right\}}{\left\{ k_0 + \sum_{n=0} k_{\alpha n} \cos(n\omega x) + \sum_{n=0} k_{\beta n} \sin(n\omega x) \right\}} \tag{3}$$

where n is positive integer, $k_{\alpha n}$ and $k_{\beta n}$ are harmonic thrust constants, $F_{\alpha n}$ and $F_{\beta n}$ are amplitude of harmonics in detent force, ω is $2 \cdot \pi / \tau$, and τ is stroke. When $F_x(x)$ was replaced by required thrust on equation (3), the current given the requirement was got.

Measurements of static thrust with exciting currents, which were rectangular waveform's current (I=2A) and sawtooth waveform's current (I_{max}=2A), were added in figure 2 with solid lines. Resultants of starting thrust 35N and aortic pressure 85mmHg by mock test, which are twice over than before improvement were got.

From the above-mentioned, exciting current control make get any value static thrust, and make increase starting thrust.

In the future, if thrust characteristics which is suited for moving artificial heart pomps was found, the current given the characteristics can be led by the method.

4. CONCLUSIONS

The authors tested the LOA with mock circulatory system. As results, the LOA didn't move for low starting thrust, as a countermeasure to improve on starting thrust, we designed the exciting current control which LOA could be get any value static thrust. As measurements of static thrust to excite current to be get with the control, both improvement on starting thrust and increasing in aortic pressure with mock test were shown.

5. REFERENCES

1 H.YAMADA:"Research on Development of Implantable Artificial Heart",Task NO.01306022, Research Report Supported by Grant-in-Aid of The Ministry of Education in 1989,JAPAN (1990)

2 D.EBIHARA,M.WATADA :"Development of Single-winding Linear Oscillatory Actuator" IEEE TRANSACTIONS ON MAGNETICS, Vol.28, No.5,3030-3023(1992)

3 D.EBIHARA,M.WATADA,T.KASUGAI :"Development and High-power Design of Single-winding Linear Oscillatory Actuator"TRANSACTIONS OF IEEJ,Vol.113, No.1(1993)

4 D.EBIHARA,K.YANASHIMA,M.WATADA:"The Study of High Power of Linear Oscillatory Actuator" NATIONAL CONVENTION RECORD IEEJ INDUSTRY APPLICATIONS SOCIETY, No.205(1992)

Electromagnetic Propulsion

Simulation and Design of Applied Electromagnetic Systems
T. Honma (Editor)

Magnetoplasmadynamic thrusters - design criteria and numerical simulation

M. Auweter-Kurtz, C. Boie, H.J. Kaeppeler, H.L. Kurtz, H.O. Schrade
P.C. Sleziona, H.P. Wagner, Th. Wegmann

Institut für Raumfahrtsysteme, Universität Stuttgart, Fed. Rep. Germany

ABSTRACT

In this report the status of research on steady state self-field MPD thrusters, both experimental and theoretical, with particular emphasis on design criteria is discussed. In the section on experimental investigations, the two MPD thruster types operated at the IRS, nozzle type DT and cylindrical thruster ZT, are presented and their characteristic features are analyzed. The problems associated with cathode phenomena are treated. The second part of this report concerns analytical studies of thruster phenomena and general instability. Specific instability phenomena, particularly space charge instabilities limiting thruster performance, are presented in Section 3 with the aid of normal mode analysis and nonlinear development. In the last section, numerical codes developed and used in calculating MPD thruster parameters and a new development of a high order explicit TVD scheme permitting accurate calculation of local gradients are discussed.

INTRODUCTION

The research on MPD thrusters in Stuttgart (Germany) started two an a half decades ago at the DFVLR [1] and is carried on at the IRS (Institut für Raumfahrtsysteme) since the beginning of the 1980s [2]. Particular attention was placed on the development of design criteria, description of fundamental processes and numerical simulation for comparison with experiments [3].

EXPERIMENTAL INVESTIGATION OF STEADY STATE SELF-FIELD MPD THRUSTERS

The main goal of the MPD program at IRS is the investigation of stationary operated MPD thrusters for power levels from 50 kW to 1 MW. Therefore different MPD thrusters are examined at two test stands, both equipped with pendulum thrust balances. A separate test stand is available which is in terms of power supply, thrust balance and measurement equipment specifically designed for the investigation of quasi-steady MPD thrusters and for the measurement of the current distribution in these devices.

Two types of self-field MPD thrusters were developed, the nozzle type thruster series DT with a diverging discharge area downstream (see Fig. 1) and the cylindrical thruster of type ZT (see Fig. 2). Various modifications of these MPD thrusters were built, tested and analyzed.

The nozzle type MPD thrusters of the DT-series shown in Fig. 1 are hybrids between pure MPD and thermal arcjet thrusters [4] and differ mainly in the throat diameter: DT2 24 mm, DT5 30 mm and DT6 36 mm, that is without any constriction. The anode is made of water-cooled copper whereas the cathode is made of thoriated tungsten and during continuous operation is glowing. The nozzle contour is formed by water-cooled copper segments, each on floating potential. The thrusters are operated with argon (Ar), hydrogen (H_2) or nitrogen (N_2)

as propellants which are fed in the arc chamber tangentially along the cathode. Additional propellant can be injected tangentially to the anode.

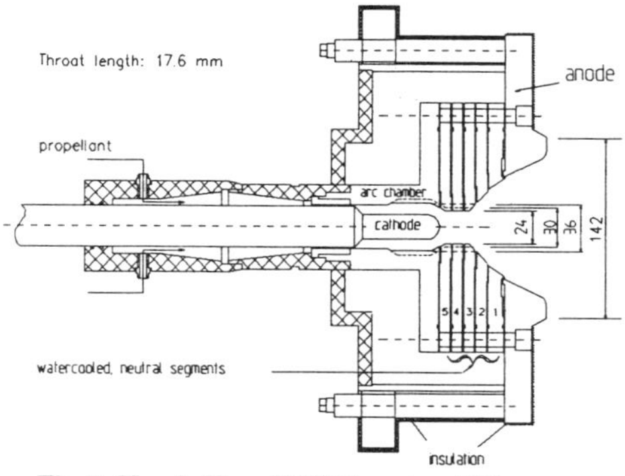

Fig. 1: Nozzle Type MPD thrusters DT

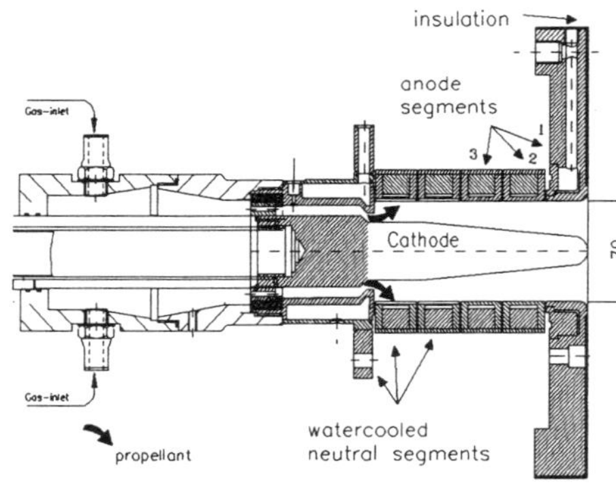

Fig. 2: Scheme of the ZT3-thruster

Figure 3 shows typical voltage-current characteristics of the DT2-thruster for different propellants. High frequency oscillations, whose onset with rising current are marked on each curve by an arrow, limit the application of these nozzle type thrusters.

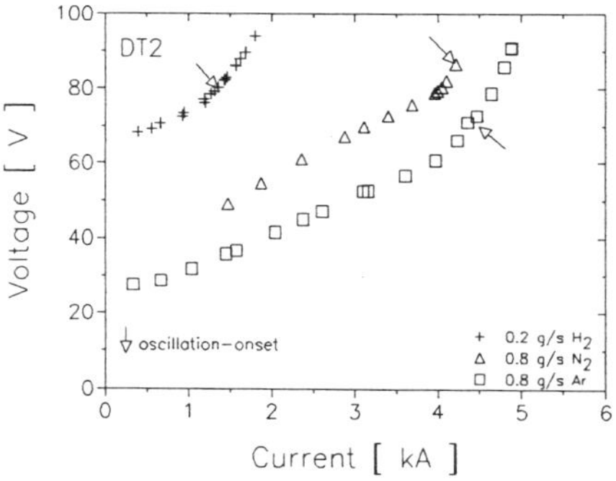

Fig. 3: Voltage-current characteristics of the DT2

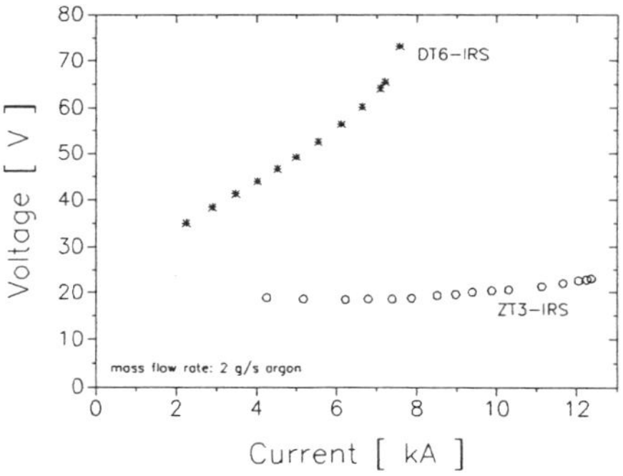

Fig. 4: Voltage-current characteristics of the ZT3- and DT6-thrusters

Beside the nozzle type MPD thrusters a series of cylindrical MPD thrusters ZT were designed. With these MPD devices, which should have a mainly radial current distribution and a low discharge voltage, an attempt was made to achieve a maximum relation magnetoplasmadynamic thrust to thermal thrust and a low pinch effect at a certain power level.

The cylindrical MPD-thruster ZT3, shown in Fig. 2, consists of three anode segments, a backplate segment and the thoriated tungsten cathode. The discharge chamber has a diameter of 70 mm and a total length of 150 mm. The current through each anode segment can be measured separately. The cathode has been specially designed to counterbalance the thermal damage effect experienced with former cathodes with a constant diameter of 18 mm [5]. In the rear part of this new thoriated tungsten cathode the diameter was enlarged to 40 mm with a subsequent conical reduction in diameter in the anode region to a tip radius of 10 mm. The voltage dependence of the ZT3-thruster is compared to the results of the DT6-thruster in Fig. 4, both running with the same mass flow rates. The voltage of the cylindrical thruster is low and increases only slowly with the current, while for the nozzle type thruster it is a factor 2 to 4 higher and rises steeply. At about 7450 A the plasma instabilities occur for the DT6-thruster, what is also recognizable by the additional ascent of the voltage. For the ZT3-thruster no indication for an occurrence of instabilities could be detected up to 12.7 kA where an $I^2/\dot{m}$ - value of more than $8 \cdot 10^{10} A^2 s/kg$ was reached, whereas for the nozzle type MPD-Thruster a critical $I^2/\dot{m}$ - value of about $2.7 \cdot 10^{10}$ $A^2 s/kg$ was found, with argon as propellant, from experiments with all nozzle throats investigated. The ZT3-thruster is designed for higher current levels, but then the anode cooling problems become severe.

THEORETICAL AND NUMERICAL INVESTIGATIONS

Theoretical investigations of MPD thruster phenomena are carried out at the IRS in three research groups: 1) Analytical studies of MPD thruster performance and instability onset, 2) normal mode analysis and nonlinear development of instabilities in MPD thruster flows, and 3) numerical simulation of MPD thruster flows.

The first topic has been essentially treated by Schrade and collaborators [6, 7]. The second topic was initiated by Kaeppeler in 1987 and has received attention as a joint venture of the IRS with the Plasma Focus Group of IPF (Institut für Plasmaforschung, University of Stuttgart) and a research team of MAN Technologie AG in Munich [8, 9, 10]. To handle this problem, investigations using normal mode analysis and a nonlinear development called QAM [8] yielded instabilities, the most important of which is a space charge instability leading to current chopping [11].

The third topical section of work at the IRS concerns numerical simulation of MPD thruster flows. In the numerical code, an extended Ohm's law is used to calculate the discharge current distribution and a two-dimensional flow code is used to obtain velocity, pressure and temperature distribution of heavy particles. The electron temperature distribution is calculated by means of the electron energy equation. These calculated temperature distributions show a characteristic thermal nonequilibrium plasma flow [12]. Argon is considered as a propellant in these calculations. Experimental investigations with the MPD thrusters at the IRS show argon III lines in addition to the common argon lines. Thus, in order to achieve conformity with actual physical phenomena of the plasma flow in MPD thrusters, the ionization of argon to higher levels is taken into account in the computations [13]. The calculated current contour distribution and the velocity vector distribution for the ZT3 thruster at a current of 6 kA with a mass flow rate of 2 g/s are shown in Figs. 5 and 6. This calculation is performed under the assumption that the cathode is hot glowing and is emitting electrons thermionically along its complete length, which is not the case in the experiment at such low current level.

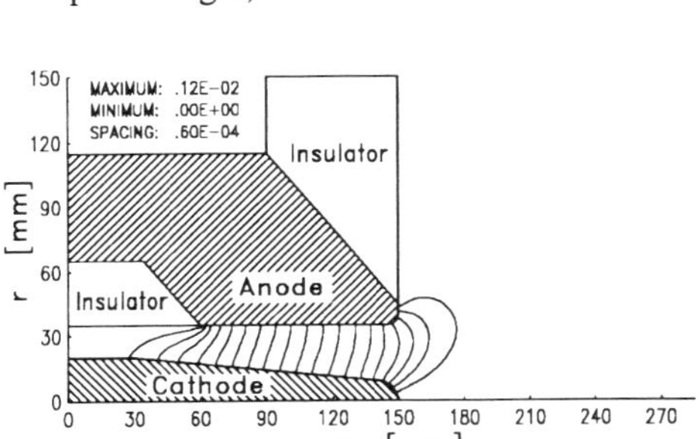

Fig. 5: Calculated current contour lines for ZT3

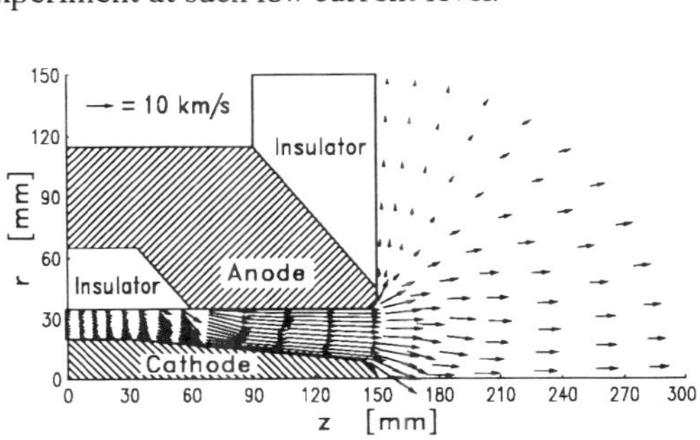

Fig. 6: Velocity vector distribution for ZT3

For the flow data required for the stability criteria obtained from the solutions of the linear dispersion relation, a numerical simulation by means of a high resolution non-oscillatory scheme is required to obtain strong gradients and geometrical influences with the desired accuracy. An MPD flow code, based on a higher order explicit TVD scheme permits high order accuracy while eliminating numerical oscillations [14].

CONCLUSIONS

The aim of the self-field MPD thruster development program at IRS is to identify and avoid critical regimes of operation and predict the performance of MPD thrusters up to the power level of multi-MW. Therefore, basic technology studies are being performed experimentally as well as theoretically using different thruster geometries. For the optimization of these devices computer codes have been developed.

A series of nozzle type thrusters with different nozzle throat diameters have been investigated. The operation region of these hybrid thrusters, where the thrust is produced by magnetic forces as well as thermodynamically, is limited due to plasma instabilities. An investigation using normal mode analysis and a non-linear quasi analytical method (QAM) showed that space charge instabilities leading to current chopping are likely to occur at these conditions, possibly due to the steep gradients in the anode region caused by the pinch effect.

With the cylindrical MPD thrusters an attempt has been made to come up with a device distinguished by a radial current distribution and low voltage level in order to achieve a high portion of magnetic thrust and a low pinch effect. But this thruster has to be operated at very high current levels. Both electrodes have to be designed with respect to the high heat load at this current level. First experimental results show low voltages even at a very high current level compared to the operation region of the nozzle type thrusters and no instabilities could be detected so far even at an operation point with a $I^2/\dot{m}$ more than twice the value where the instabilities set in with the nozzle type thrusters. So it is hoped to achieve higher specific impulses with this thruster geometry.

ACKNOWLEDGEMENT

The authors would like to thank the AFOSR, the DARA and the DFG who enabled the work on MPD thrusters at IRS with several grants during the last years.

REFERENCES

[1] Hügel, H., DFVLR-Report, DFVLR-FB-80-20, 1980
[2] Bühler, R.D., AFRPL-TR-86-013, Edwards AFB, Ca., 1986
[3] Auweter-Kurtz, M., "Lichtbogenantriebe für Weltraumaufgaben", Teubner Verlag, Stuttgart, 1992
[4] Wegmann, T., et al, AIAA92-3464, 1992
[5] Auweter-Kurtz, M., et al, AIAA-90-2662, 1990
[6] Schrade, H.O, et al, Final Scientific Report, AFOSR Grant 91-0118, April 1992
[7] Auweter-Kurtz, M., et al, Final Report AFOSR Grant 89-0535, August 1992
[8] Maurer, R., Hayd, A., Kaeppeler, H.J., Journal of Computational Physics **66**, 151, 1986
[9] Kaeppeler, H.J., Report IPF-83-3, Inst. für Plasmaforschung, Stuttgart 1983
[10] Wagner, H.P., Auweter-Kurtz, M., Kaeppeler, H.J., Report IRS-89-S89, IRS, Stuttgart, 1989
[11] Maurer, M., Kaeppeler, H.J., Richert, W., Proc. (ISEM), Sapporo, 1993.
[12] Sleziona, P.C., et al, Int. Numerical Methods in Engineering 34, 759, 1992
[13] Sleziona, P.C., Auweter-Kurtz, M. Schrade, H.O., AIAA-92-3296, 1992
[14] Boie, C., et al, Proc. (ISEM), Sapporo, 1993.

Simulation and Design of Applied Electromagnetic Systems
T. Honma (Editor)

Review of Design Considerations in Very High Power Electromagnetic Propulsion and Power Systems

P.J. Turchi

Department of Aeronautical and Astronautical Engineering, The Ohio State University, 328 Bolz Hall, 2036 Neil Avenue, Columbus, Ohio, 43210-1276, United States of America

Abstract

For applications such as controlled thermonuclear fusion, advanced space propulsion, and various systems of defense interest, it is often useful to deliver large amounts of electromagnetic energy at very high power levels. Typical energy levels will exceed one megajoule, supplied in single pulses of a microsecond or less. Usually, manipulation of magnetic flux and energy occurs in a sequence of multi-megajoule events that can result in power multiplication to levels above 100 terawatts. Principles and techniques used in successfully designing very high power, high energy electromagnetic systems are briefly discussed. Applications of these techniques to somewhat less adventurous electromagnetic propulsion devices (e.g., space thrusters) are also described.

1. INTRODUCTION

There have been continuing connections among quite distinct application areas that all involve the utilization of magnetic flux and energy in dynamic conductor arrangements. Thus, for example, early work on pinch discharges for controlled nuclear fusion offered a basis for large-radius, plasma implosions for space propulsion. These implosions in turn led to the development of high power (>terawatt), soft X-ray sources for nuclear weapon effects simulators [1]. For power multiplication to achieve very high energy sources, electromagnetic acceleration along coaxial electrodes is used in both the plasma flow switch [2], and in compact toroid generation, compression, and acceleration systems [3]. These devices closely resemble pulsed and steady-state plasma thrusters, which operate at much lower powers (0.1-10 MW _vs_ 1-10 TW).

The present discussion is intended to provide a brief indication of some of the activities in which modeling and simulation are used to design high power, electromagnetic propulsion and power systems. Two situations are considered as examples both of the techniques developed, and of the synergisms between experiment, modeling, simulations, and applications: multi-megampere, multi-megajoule pulsed plasmas for inductive storage and switching, and kiloampere, steady-state, magnetoplasma-dynamic thrusters for space propulsion.

2. PULSED PLASMAS FOR INDUCTIVE STORAGE AND SWITCHING

A principal difficulty in achieving very high power soft X-ray sources at multi-megajoule energy outputs has been the transfer of energy into the vacuum region surrounding the implosion load [1]. An approach [2,4] that has been successfully followed is the plasma flow switch, a form of coaxial plasma gun in which a thin, annular plasma is electromagnetically-accelerated between a pair of coaxial electrodes. As the discharge current increases, energy is stored in the near-vacuum, coaxial inductive store upstream of the plasma. Peak current is attained as the discharge, acting as a plasma "armature", sweeps off the abrupt end of the center conductor, releasing the stored magnetic energy radially-inward toward an implosion load.

The dynamics of the plasma armature were initially studied numerically using slug-models, coupled to circuit-solving calculations. Experimental evidence indicated that this approach was inadequate to describe simple arrangements (i.e., initial plasma armature density uniform with radial position), so the experiment was modified to make the modeling easier; the initial density was distributed to match the r^{-2} magnetic pressure variation. An unsteady, one-dimensional, method-of-characteristics model [2] was developed analytically that successfully indicated the behavior of the plasma discharge motion in the evacuated coaxial electrode channel. This analysis was later extended to an unsteady, quasi-two-dimensional, computer-graphical solution of the magnetized-plasma flow during switch operation [2].

Success with experiments designed using these techniques encouraged the development of the MACH2 computer code [5]. MACH2 is a 2-1/2 dimensional, unsteady, magnetohydrodynamics code that includes all three components of vector fields (velocity, magnetic field) with variations in two dimensions (e.g., axisymmetric, rotational flow). The code follows fluid motion with an arbitrary Lagrangian-Eulerian formulation, and handles complex geometries by transformations to a set of rectangular blocks that exchange information through "ghost" cells at the block boundaries. It is a single-fluid, multi-temperature code, in which a generalized Ohm's law, including Hall effect, is used to calculate electron motion. Plasma radiation is handled by options for optically-thin, volumetric radiation loss, or grey-body, radiation diffusion, based on opacity coefficients. Options also are included for analytical equations of thermodynamic and transport properties, or access to the SESAME tables of material properties. Phenomenological models for anomalous transport coefficients, based on plasma instabilities, can be invoked whenever local discharge conditions satisfy criteria for such instabilities.

MACH2 has successfully modeled the plasma flow switch [5], including accurate prediction of the current pulse and voltage spike (amplitude, timing, and duration) obtained experimentally [2]. Furthermore, MACH2 indicated leakage of magnetized-plasma past the plasma armature during switching, and correctly suggested the proper placement of a small, metal extension to block the leak, enabling efficient switch operation [4,5].

More recently, MACH2 has been used extensively in the design and analysis of compact toroid systems [3]. A pulse of plasma and azimuthal magnetic field in a coaxial plasma gun can be used to distend initially radial magnetic field lines (tied to the coaxial electrodes) until these field lines reconnect around the plasma. MACH2 has successfully modeled the generation of compact toroids by this technique, and has also correctly predicted the subsequent equilibrium magnetic field distribution, and motion of the toroid. The use of compact toroids for switching of magnetic energy accumulated upstream of the toroid (in a manner similar to plasma flow switch operation) has also been examined with MACH2 [6].

3. STEADY-STATE PLASMA PROPULSION FOR ORBIT-TRANSFER

In the 1960's, calculational tools were not available to study electromagnetically-accelerated flows in plasma thrusters. Over the last few years, however, the increase in computer capability, and progress in the development of large-scale MHD codes has permitted their use for plasma propulsion problems. For example, megawatt, quasi-steady, self-field, magnetoplasmadynamic (MPD) thrusters have been modeled using the MACH2 code [7]. The approach is simply to run the calculation with the appropriate boundary condition of steady mass input, and a current waveform that quickly achieves a constant value, until all magnetic and flow field quantities attain steady-state conditions. With a strictly azimuthal magnetic field, the magnetic field boundary conditions are directly related to the discharge current.

For the situation of an applied-field MPD thruster, which may need to operate at more modest powers (< 100 kW), it has been necessary to modify the MACH2 code for two distinct reasons: a basic difficulty in establishing proper boundary conditions for the solenoidal field, and the increased importance of viscous effects at the relatively low densities associated with lower power operation. The eventual steady-state solution includes the azimuthal currents generated by the plasma flow interacting with the solenoidal magnetic field distribution. This distribution depends on the boundary conditions for the field, which depend, in part, on the azimuthal plasma currents. The approach taken [8] has been to compute the magnetic field components that would exist at the calculational boundaries in steady-state (i.e., fully-diffused, magnetostatic solution) due to all the azimuthal currents (plasma and external coils) at any instant. The computation of azimuthal plasma currents, therefore, has the opportunity to relax to a solution that is consistent with the steady-state boundary conditions. In calculations to date, steady-state solutions are indeed achieved in physical times that are not unreasonably large compared to characteristic times for convection and magnetic diffusion in the plasma (10^{-4} vs 10^{-5} sec).

The interaction of the discharge current with the solenoidal magnetic field causes the plasma to swirl about the thruster axis. This swirl then reduces the particle density near the center conductor, further increasing the importance of viscous effects in lower power MPD thrusters. The MACH2 code, which uses an implicit computation of inviscid flow, has therefore recently been altered to provide the option for an explicit calculational step to include (real) viscous forces. (This alteration was based on an earlier option for elastic-plastic dynamics). Future work will be needed to account for the substantial velocity slip that is possible at low densities, in an attempt to stretch the range of MACH2 as a useful calculational tool. It is anticipated that such development for plasma thrusters will then find application in very high power systems, particularly for boundary layers flows that can interfere with the delivery of electromagnetic power.

4. REFERENCES

1 P.J. Turchi and W.L. Baker, J. Appl. Phys., 44 (1973) 4936.

2 P.J. Turchi, et al, IEEE Trans. Plasma Sci.,PS-15 (1987) 747.

3 R.E. Peterkin, Jr., et al, in Proc. of 6th Intl. Conf. on Megagauss Magnetic Fields, Albuquerque, NM. M. Cowan, ed., Nova Science, NY, 1993.

4 J.H. Degnan, et al, IEEE Trans. Plasma Sci.,PS-15 (1987) 760.

5 J. Buff, et al, ibid., 766.

6 R.E. Peterkin, Jr., et al, in Proc. of XVth Intl. Symp. on Discharges and Electrical Insulation in Vacuum, Darmstadt, 1992.

7 P.J. Turchi, J.F. Davis, N.F. Roderick, AIAA Preprint 90-2664 (1990).

8 P.G. Mikellides, P.J. Turchi, N.F. Roderick, AIAA Preprint 92-3740 (1992).

Acknowledgments

In preparing the present article and associated talk, the author has had the kind assistance of P.G. Mikellides at The Ohio State University, and R.E. Peterkin, Jr., at the USAF Phillips Laboratory, Albuquerque, NM, and many years of useful discussions with M.H. Frese and N.F. Roderick.

Simulation and Design of Applied Electromagnetic Systems
T. Honma (Editor)
1994 Elsevier Science B.V.

MPD Thruster Simulation Using High-Resolution Schemes

C. Boie, P. C. Sleziona, H. J. Kaeppeler and M. Auweter-Kurtz

Institut für Raumfahrtsysteme, Universität Stuttgart,
Pfaffenwaldring 31, D-7000 Stuttgart 80, Germany

Abstract

In this paper, several elements for the simulation of flow in high power magnetoplasmadynamic thrusters are discussed including grid generation (resolution in space) and flow solver (resolution of the solution in time). The main goal of this project is to obtain exact values of the gradients of plasma parameters in the flow. This is important to characterize the plasma parameters necessary for instability investigation.

1. Introduction

MPD thruster simulations based on Euler and Navier-Stokes equations have been carried out at our institute for many years under the aspect of obtaining reasonable results for flow variables for comparison with experimental data [1].

Parallel to this, plasma instabilities were investigated with linear and nonlinear methods. This demands highly resolved numerical solutions of flow equations near steep gradients [2]. Since available computer capacities and speeds have remarkably increased during the last years and improved numerical schemes are now available, the development of a new simulation software written in `C` with an `XWindows` interface has been started.

2. The Grid generator

As the geometrical influence on the behavior of MPD thrusters is to be studied for several different thruster geometries in use at IRS, a flexible grid is advantageous. In order to employ automatic grid refinement techniques, an unstructured grid is used for the simulation.

The grid generator is an advancing front generator which works in two steps. Step one is the triangle generation step, step two is an optimizing step to minimize grid stress. An example is shown in Fig. 1.

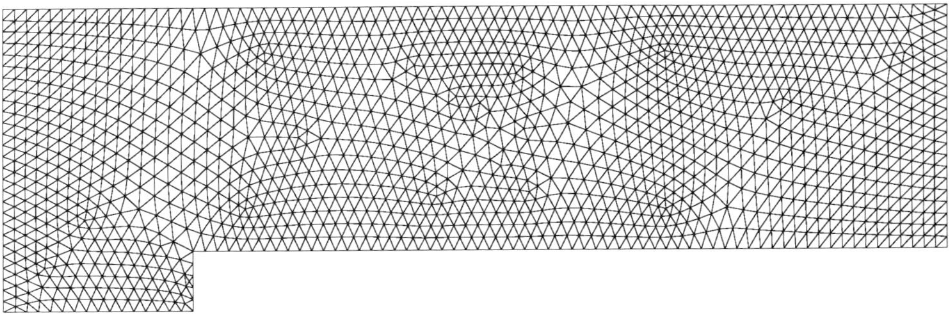

Figure 1: Unstructured grid for M=3 test problem(M=Mach number)

3. The governing equations

The computation described in this paper uses a simple one fluid model for the plasma. The plasma (m_e=electron mass, T_e=electron temperature) is treated as an ideal gas with specific resistance

$$\eta \approx \frac{\pi e^2 m_e^{1/2}}{(4\pi\varepsilon_0)^2 (kT_e)^{3/2}} \log\Lambda \approx \frac{1212.0}{T_e^{3/2}} \left[\Omega m K^{3/2}\right] . \tag{1}$$

The particle mass for Argon is $m_{Ar} = 6.633 \cdot 10^{-26}\ kg$, the ratio of specific heats $\gamma = c_p/c_v = 1.65$. Excluding viscosity and heat conduction, the computation of flow is based on the Euler equations written in integral form,

$$\frac{\partial}{\partial t}\iiint_V \rho\, dxdydz = -\iint_{\partial V} \rho\,(\vec{v}^T\vec{n})\ da\ , \tag{2}$$

$$\frac{\partial}{\partial t}\iiint_V \rho\vec{v}\, dxdydz = -\iint_{\partial V} \rho\vec{v}(\vec{v}^T\vec{n}) + p\vec{n}\ da\ + \iiint_V \left[\vec{J}\times\vec{B}\right]\ dxdydz, \tag{3}$$

$$\frac{\partial}{\partial t}\iiint_V e\, dxdydz = -\iint_{\partial V} (e+p)(\vec{v}^T\vec{n})\ da + \iiint_V \eta(\vec{J}^T\vec{J}) + \vec{v}^T\left[\vec{J}\times\vec{B}\right]\ dxdydz, \tag{4}$$

$$e = \frac{p}{\gamma-1} + \frac{\rho(\vec{v}^T\vec{v})}{2} \qquad \text{(ideal gas)}\ . \tag{5}$$

(ρ= mass density, $\vec{v}$=mean mass velocity, $e = e_{th} + e_{kin}$=total energy, p=pressure, $\vec{n}$=surface normal)

Since the time scale for changes in the electromagnetic field is considerably smaller than for the dynamics of the flow, a stationary $\vec{J}$- and $\vec{B}$-field is assumed at every time step with

$$\oint_{\partial S} \vec{E}^T d\vec{s} = 0\ , \tag{6}$$

$$\oint_{\partial S} \vec{B}^T d\vec{s} = \mu_0 \iint_S \vec{J}^T\vec{n}\ da\ , \tag{7}$$

$$\eta\vec{J} = \vec{E} + \left[\vec{v}\times\vec{B}\right]\ . \tag{8}$$

($\vec{B}$=magnetic induction, $\vec{E}$=electric field, $\vec{J}$=current density)

It is assumed that $\vec{B} = (0, B_\varphi(r,z), 0)^T$, so $\nabla\vec{B} = 0$ is always fulfilled. The computation of the current distribution and magnetic field is performed by a relaxation method (SOR) at every timestep.

4. The flow solver

The flow field is now solved with a finite volume scheme using an exact Riemann solver. This approach has the advantage that no parameters are introduced. Furthermore, total mass, momentum and energy are fully conserved during the computation and physically impossible states (i.e. negative mass or energy) cannot occur by definition of the scheme. In Fig.2, a computation of the M=3 test problem of [3] is shown.

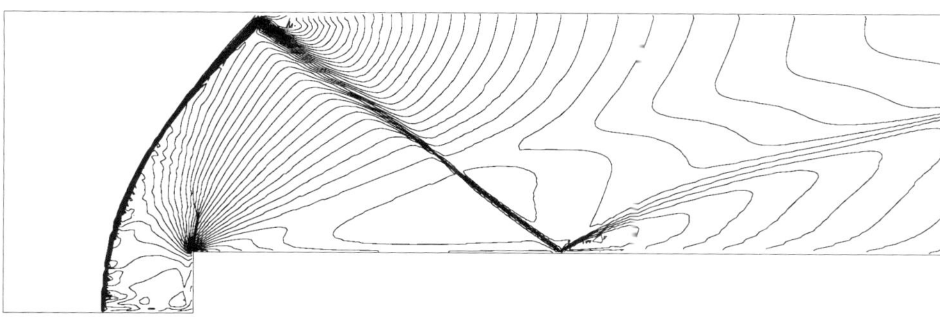

Figure 2: Exact Riemann-solver computing the M=3 test problem of [3] ($\rho = 1.4, p = 1, v = 3, \gamma = 1.4$)

5. Results and Conclusions

The implementation of the high order concepts and the extension of the physical model is currently under work. The grid generator and the flow solver constitute the starting-point for the development of a high-resolution simulation system. In Fig. 3, first results are shown for the mach number distribution of the ZT3-thruster after $99\mu s$. A detailed description of the full simulation system with comparison of the different thruster geometries will shortly be presented elsewhere. The role of the investigation presented in this paper with regard to the theoretical and experimental work on the onset of performance limiting instability in MPD thrusters at the IRS is explained in [2].

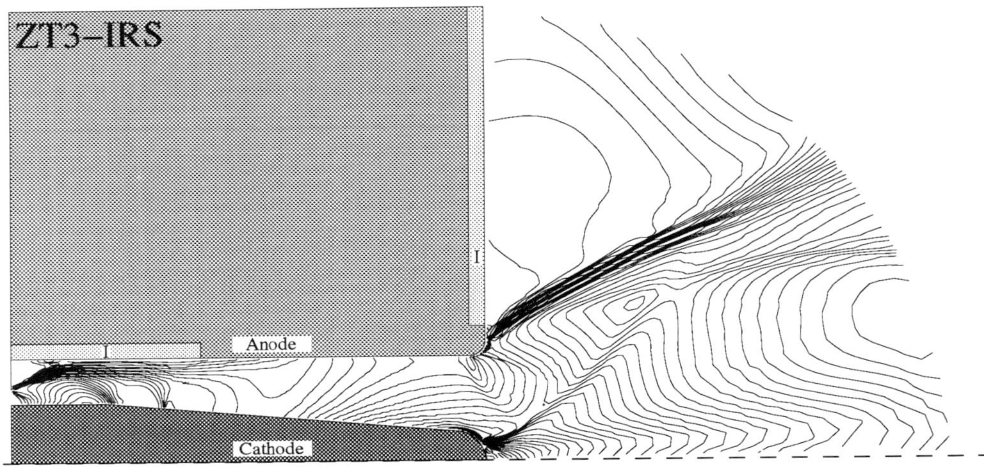

Figure 3: Simulation example of ZT3 thruster run

References

[1] P. C. Sleziona, M. Auweter-Kurtz, and H. O. Schrade. Computation of MPD-flows and comparison with experimental results. *International Journal for Numerical Methods in Engineering*, 34, 1992.

[2] M. Auweter-Kurtz, C. Boie, H. J. Kaeppeler, H. L. Kurtz, H. O. Schrade, P. C. Sleziona, H. P. Wagner, and T. Wegmann. Magnetoplasmadynamic thrusters–design criteria and numerical simulation. *Proceedings of the International Symposium on Simulation and Design of Applied Electromagnetic Systems (ISEM-93), Sapporo, Japan*, January 1993.

[3] Paul Woodward and Phillip Colella. The numerical simulation of two-dimensional fluid flow with strong shocks. *Journal of Computational Physics*, 54:115–173, 1984.

[4] Chi-Wang Shu and Stanley Osher. Efficient implementation of essentially non-oscillatory shock-capturing schemes. *Journal of Computational Physics*, 77:439–471, 1988.

Simulation and Design of Applied Electromagnetic Systems
T. Honma (Editor)

Electromagnetic particle-in-cell simulation of plasma thruster exhaust plumes

H. Matsumura, H. Tahara, Y. Tsubakishita and T. Yoshikawa

Department of Mechanical Engineering, Faculty of Engineering Science, Osaka University, Machikaneyama 1-1, Toyonaka, Osaka 560, Japan

Abstract

An electromagnetic particle-in-cell (EM-PIC) code was developed to simulate numerically a collisionless plasma thruster plume released in a magnetized vacuum. The simulation results showed that the axial magnetic field was weakened on the plasma boundary near the thruster exit with a diamagnetic current just after discharge start. However, the field was strengthened inside the plasma because of an $E_r \times B_z$ drift current of electron owing to charge separation.

1. INTRODUCTION

When a high-velocity plasma exhausted from an electric thruster such as a magnetoplasmadynamic thruster, ion thruster and arcjet thruster [1-3] expands radially and axially into space, electric and magnetic fields are expected to be induced in the plasma plume, and also in Earth's magnetic field or a magnetic field applied for improvement of thruster performance the complicated interaction between the plasma plume and the magnetic field is expected to occur. From the viewpoint of electromagnetic wave environment in space and study of the interaction between the plasma plume and space vehicle bodies, it is important to understand the process of plasma expansion and the property of electromagnetic waves generated by it. Computational simulation is a powerful tool for them.

In the present study, an electromagnetic particle-in-cell (EM-PIC) code [4] is developed. A collisionless plasma exhausted from an electric thruster, which is released in a magnetized vacuum, is simulated numerically using the EM-PIC code.

2. SIMULATION MODEL AND METHOD

A PIC simulation method is a popular particle simulation method, in which the equations of motion of each particle are time-integrated directly and governed by quantities calculated on spatial grids around the particle. Some weighting is conducted for connection between quantities at particle positions and on grids. In EM-PIC code [4], Maxwell's equations of Faraday's law and the Ampere-Maxwell law are time-integrated on grids to yield the electric and magnetic fields, and the position and velocity of a charged particle are determined by the Newton-Lorentz equations of motion.

The EM-PIC code developed in this study run in two spatial dimensions with three momentum components per particle. The leap-frog method is used as time-integration schemes. First-order weighting is adoptted. Cylindrical axial coordinates with a magnetic field along the axis of symmetry are used to study the particle motion along the magnetic field and variation of the field as a plasma expands against it. The computational volume is circular cylinder. Plasma particles with some temperature and a radial distribution of axial velocity, which corresponds to a plasma thruster plume, begin being introduced from a slit parpendicular to the center axis, and then the simulation starts. The slit corresponds to the thruster exit.

In the simulation, there exist two characteristic parameters of beta (the particle-to-field energy density ratio) and a ratio of Larmor radius to the slit radius. These parameters are expected to govern the behavior of a plasma expanding into a vacuum.

3. RESULTS AND DISCUSSION

The computations are carried out with parameters of ion and electron temperatures of 10-15 eV, a plasma density of 10^{14}-10^{16} m^{-3}, a Gaussarian radial distribution of axial velocity with an averaged axial velocity of 10^5 m/s, i.e. a specific impulse of 10^4 sec, an ion-to-electron mass ratio of 20 and an uniform axial magnetic field strength of 10^{-3}-10^{-6} T, i.e. a beta of 10^{-3}-10^{3} and a radii ratio of 0.4-40 ,and a slit radius of 0.05-0.1 m [1]. The magnetic field strength is mainly varied. The number of particles used is 10^4 to 10^5. The time step and distance between grids are 0.02 and 0.1, respectively, normalized with a plasma frequency and the speed of light in vacuum.

Figures 1 and 2 show the typical ion and elecron distributions after 80 and 120 periods of plasma oscillation, i.e. just after discharge start. Both ions and electrons emitted from the slit expand axially and radially. Particularly, with low plasma densities electrons expand extremely compared with ions because of its lighter mass though the electric field which is induced by charge separation prevents them from expanding separately.

Figure 3 shows the radial profiles of axial magnetic field. Near the slit of the thruster exit, the magnetic field slightly decreases from the original one B_o near the radial boundary of the plasma because of a diamagnetic current of a $\mathbf{B}\times\mathbf{grad}p$ drift current, and however the magnetic field is higher inside the plasma on account of an $E_r\times B_z$ drift current of electron, in which the radial electric field E_r is due to charge separation. Also, the magnetic field does not decreases at the much more downstream region of the slit; that is, the $E_r\times B_z$ drift current is dominant compared with the diamagnetic current.

In this analysis, the magnetic field is found to be slightly pushed out and a cylindrical zone in which the field is weakened to be formed near the thruster exit when the plasma plume has a large radial gradient of pressure enough or the magnetic field is weak enough, though the field is strengthened inside the zone. The axial expansion of the plasma is essentially free though the radial expansion is slightly restricted

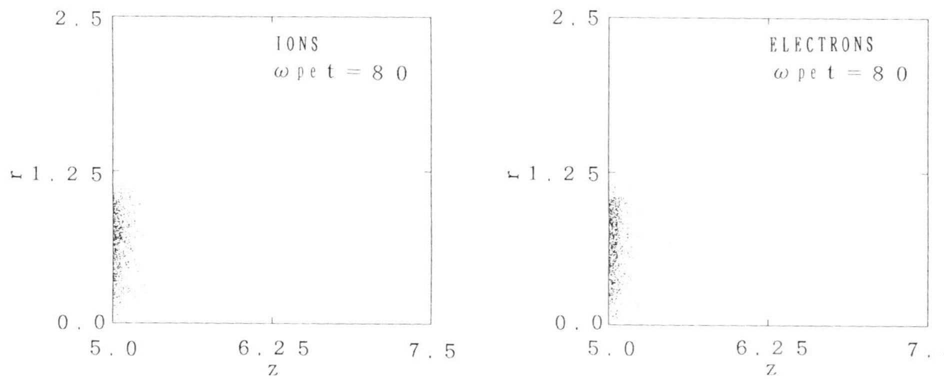

(a) Ion (b) Electron

Figure 1. Typical ion and electron distributions after t=80/ ωpe.

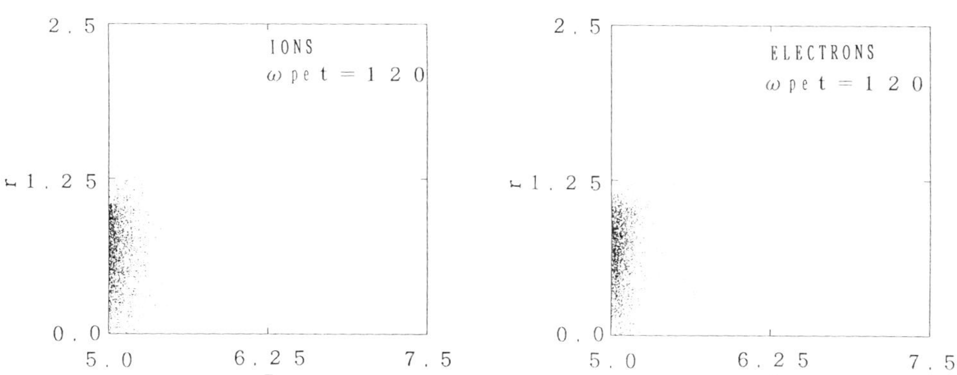

(a) Ion (b) Electron

Figure 2. Typical ion and electron distributions after t=120/ ωpe.

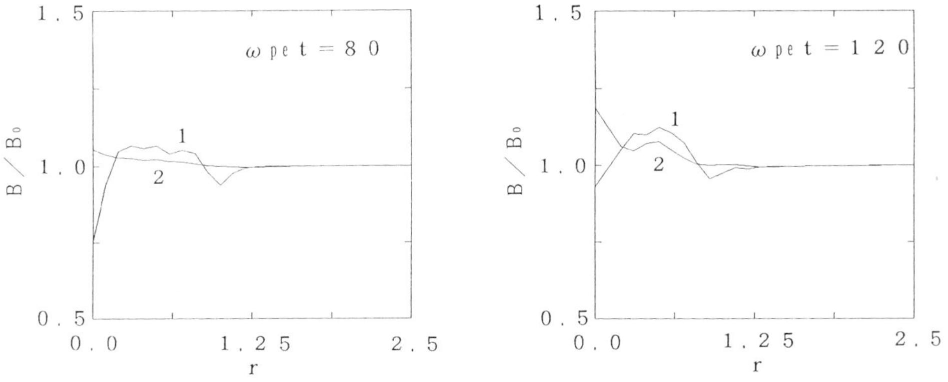

(a) t=80/ ωpe (b) t=120/ ωpe

Figure 3. Radial profiles of axial magnetic field on thruster exit symbol 1 and on 20-mm downstream of exit symbol 2.

by the pressure of the ambient magnetic field.

When the beta value is high enough like in nuclear thruster plumes controlled by strong nozzle-shaped magnetic fields, the plasma is able to expand radially resitting to the field. Then, the magnetic field is reduced near the plasma edge, and the diamagnetic cavity forms. Expansion front has an edge of field to the ambient vacuum region, and electromagnetic waves launched from plasma are also generated.

4. CONCLUSIONS

An axisymmetric electromagnetic particle-in-cell (EM-PIC) code was developed to simulate numerically a collisionless plasma thruster plume released in a magnetized vacuum. The simulation resuits showed that the axial magnetic field was weakened on the radial plasma boundary near the thruster exit with a diamagnetic current just after discharge start. However, the field was strengthened inside the plasma or in the more downstream region because of an $E_r \times B_z$ drift current of electron owing to charge separation.

5. REFERENCES

1 H. Tahara et al., Proc. of AIDAA/AIAA/DGLR/JSASS 22nd Int. Electric Propulsion Conf. (IEPC), Italy, 91-073 (1991).

2 H. Tahara et al., Proc. of 22nd IEPC, Italy, 91-015 (1991).

3 T. Yasui et al., Proc. of 22nd IEPC, Italy, 91-085 (1991).

4 C. K. Birdsall and A. B. Langdon, Plasma Physics via Computer Simulation, McGraw-Hill, New York (1985).

Simulation and Design of Applied Electromagnetic Systems
T. Honma (Editor)

Current distributions of a self-field magnetoplasmadynamic arcjet

T. Tsubaki, H. Tahara, Y. Kagaya, Y. Tsubakishita
and T. Yoshikawa

Department of Mechanical Engineering, Faculty of Engineering Science, Osaka University, Machikaneyama 1-1, Toyonaka, Osaka 560, Japan

Abstract

A quasisteady magnetoplasmadynamic (MPD) arcjet with 30-cm-class long electrodes was developed to examine current distributions in the discharge chamber. The current distributed on the segmented anode concentrated near the downstream end regardless of gas species though the fraction for molecular gases was much larger than that for monoatomic gases.

1. INTRODUCTION

A quasisteady magnetoplasmadynamic (MPD) arcjet is an efficient propulsion device characterized by high specific impulse and large thrust density, which is suitable for future missions such as planetary exploration and construction of large stations. One or two dimensional analyses of MPD arcjet flowfields have been made by many researchers [1], in which the electric field is determined by the point of the sonic or the magnetosonic singularity. These calculated results are significant for optimum design of MPD discharge chambers and for understanding of discharge feature and acceleration mechanism. Particularly, the streamwise discharge current distribution is important because the processes of heating and acceleration are determined positively by the current pattern.

In the present study, current distributions in MPD arcjets are examined using an MPD arcjet chamber with long electrodes, i.e., with a long acceleration zone almost realizing an one-dimensional duct flow. The experimental results are compared with analytical ones.

2. EXPERIMENTAL APPARATUS

The coaxial MPD arcjet newly designed, as shown in Fig.1, is provided with a straight anode 50-mm in diameter × 300-mm deep. The anode made of copper is divided axially into 15 anode parts, each of which is electrically insulated to one another. The current entering each anode segment is measured by a Rogowski coil to infer the current pattern in the inter-electrode region. A cylindrical cathode 9.5-mm in diameter ×

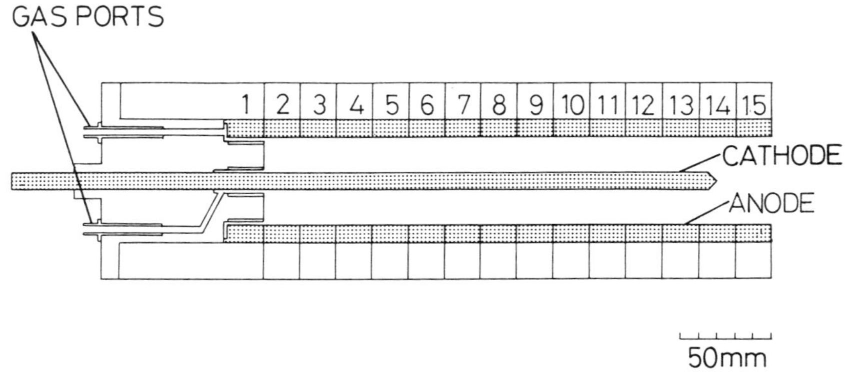

Figure 1. Cross section of MPD arcjet with long electrodes.

270-mm long is made of thoriated tungsten.

Gases are injected into the discharge chamber through a fast acting valve (FAV) fed from a high pressure reservoir. The rise time and width of the gas pulse, measured with a fast ionization gauge, are 0.5~ 1.0-msec and 6-msec, respectively. The mass flow rates are controlled by adjusting the reservoir pressure and the orifice diameter of the FAV.

A pulse forming network delivers a single nonreversing quasisteady current with a pulse width of 0.6-msec. A vacuum tank 0.6-m in diameter × 5.75-m in length, where the arcjet is fired, is evacuated to some 10^{-3} Pa prior to each discharge by diffusion and rotary pumps.

Discharge currents are measured by a Rogowski coil. Voltage measurements are performed with a current probe, which detected the small current bled through a known resistor between the electrodes.

3. RESULTS AND DISCUSSION

The present experiments are conducted with five gases (He, Ar, H_2, N_2 and mixture of N_2+$2H_2$ simulating hydrazine N_2H_4). The mass flow rate for each gas is set up at a corresponding critical current of about 10 kA.

Figures 2 and 3 show the current fractions distributed on the segmented anode. The current fractions on the anode segment No.15 are highest for all gases; i.e., the current concentrates near the downstream end. Furthermore, the current fractions of No.15 for molecular gases are much larger than those for monoatomic gases regardless of molecular weight, and in other words for monoatomic gases the current distributes in the intermediate region of the anode though for molecular gases the current hardly exist in the region. Also, the current fractions of No.15 decrease with increasing discharge current, and the current distributes in more upstream segments.

Hence, the current fractions on the anode are found to depend strongly on gas species and discharge current levels.

From the experimental and analytical results [1], it is expected that the current patterns in the interelectrode region are divided into two classes regardless of molecular weight. For monoatomic gases the discharge occurs mainly inside the discharge chamber, and the current flows almost uniformly over the side surface of the cathode. For molecular gases the current conduction region moves downstream, and most of the discharge current enters the cathode tip and the end surface of the anode. This is because the ionization process of molecular gases is slower than that of monoatomic ones owing to a time lag due to dissociation process. Also, current concentration near the upstream end is not extreme compared with the analytical results.

4. CONCLUSIONS

The self-field MPD arcjet with the segmented anode 30-cm long was developed to examine current distributions in the discharge chamber. The current distributed on the anode concentrated

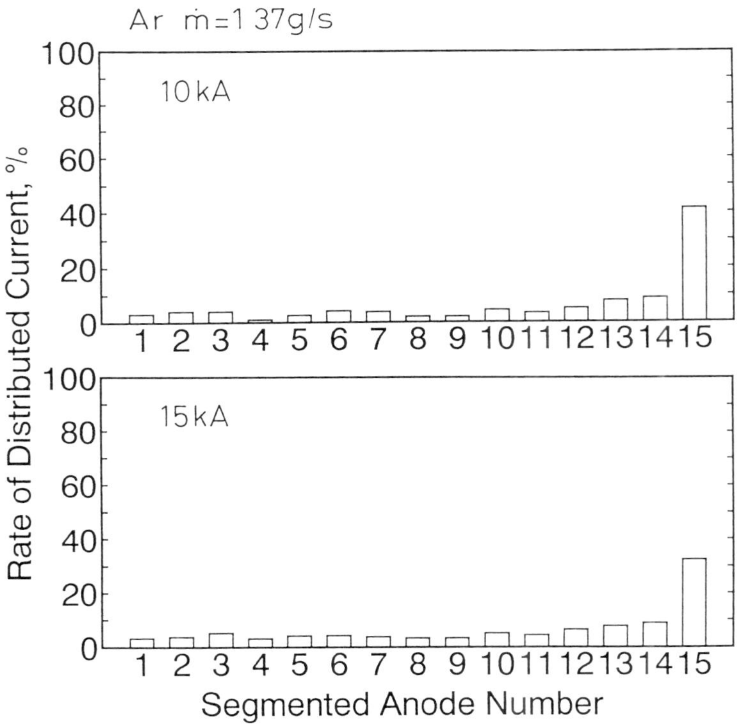

Figure 2. Current fractions on segmented anode for Ar.

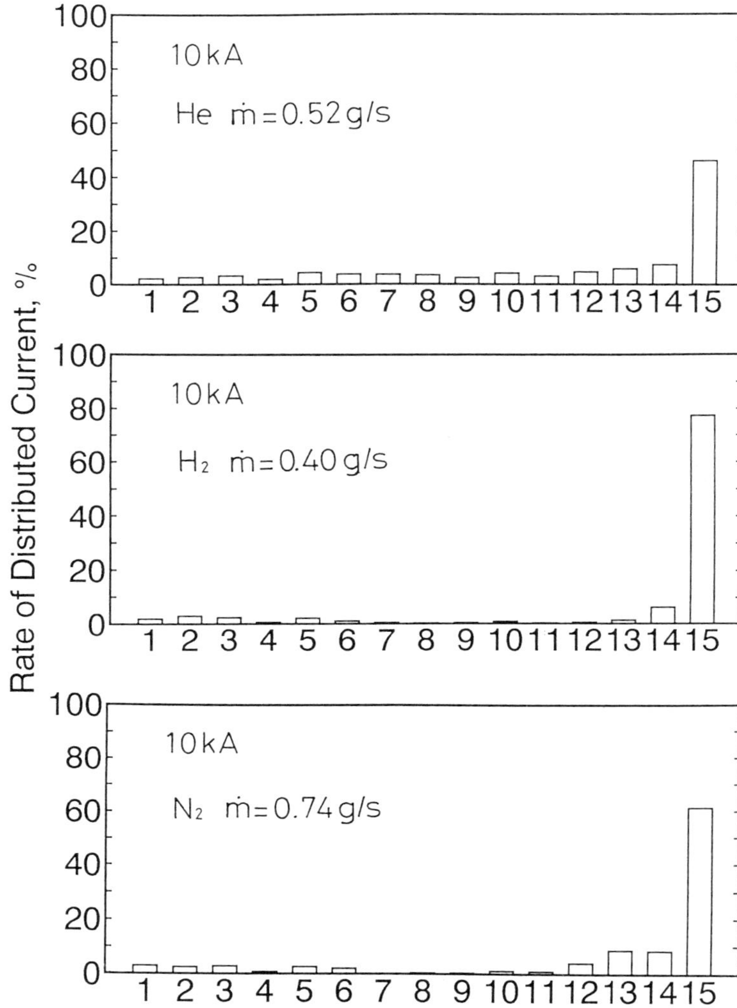

Figure 3. Current fractions on segmented anode at 10 kA.

near the downstream end regardless of gas species though the fraction for molecular gases was much larger than that for monoatomic gases. Hence, the current distributions were expected to depend on gas species and discharge current levels.

5. REFERENCES

1 H. Tahara et al., AIAA Paper 87-1093 (1987).

Simulation and Design of Applied Electromagnetic Systems
T. Honma (Editor)

Continuous operational test and thermal design of magnetoplasmadynamic arcjet thruster

Y. Kagaya, H. Tahara, Y. Tsubakishita and T. Yoshikawa

Department of Mechanical Engineering, Faculty of Engineering Science, Osaka University, Machikaneyama 1-1, Toyonaka, Osaka 560, Japan

Abstract

Quasisteady magnetoplasmadynamic (MPD) thruster periodical operations were demonstrated in 1-kW-class averaged input powers, i.e., 1-MW-class input powers during arc discharge. Powers deposited into the electrodes and plasma flow were examined. The energy balance characteristics depended on propellant species and arc current levels. Also, the thermal analytical results showed that only cathode tip was very severe in 10-kW-class averaged-power operations.

1. INTRODUCTION

A magnetoplasmadynamic (MPD) arcjet is an applicable device to a future propulsion system in space. The MPD thruster must be operated in high discharge current levels of kiloamperes to induce a strong magnetic field for effective electromagnetic acceleration. Therefore, so as to establish such high current discharge, the MPD thruster system is usually equipped with a pulse forming network (PFN) of a LC ladder circuit. Thus, the MPD thruster is operated in periodically and repetitively quasi-steady firing mode. From the standpoint of flight model design of MPD thruster systems, it is important to discuss technological problems in high power operations.

In the present study, a quasi-steady MPD arcjet system is developed to obtain a preliminary assessment of thruster performance and thermal characteristics in repetitive firing mode at an averaged input power of about 1 kW, realizing in 0.3-1.2 MW of input power during arc discharge. Powers deposited into the electrodes are measured calorically, and energy balance is examined. Furthermore, thermal analysis of a 10-kW-class MPD thruster head is carried out for a future propulsion system, and the possibility is discussed from the standpoint of the thermal design.

2. EXPERIMENTAL APPARATUS

The quasi-steady MPD thruster system [1,2], as shown in Fig.1, consists of a thruster head, propellant supply unit, power supply unit and operation control unit. The water-cooled MPD thruster head, as shown in Fig.2, is provided with a ring coil connected in series with a main discharge circuit for application of an axial magnetic field [3]. The power supply unit is composed of a PFN, discharge trigger driver and FAV

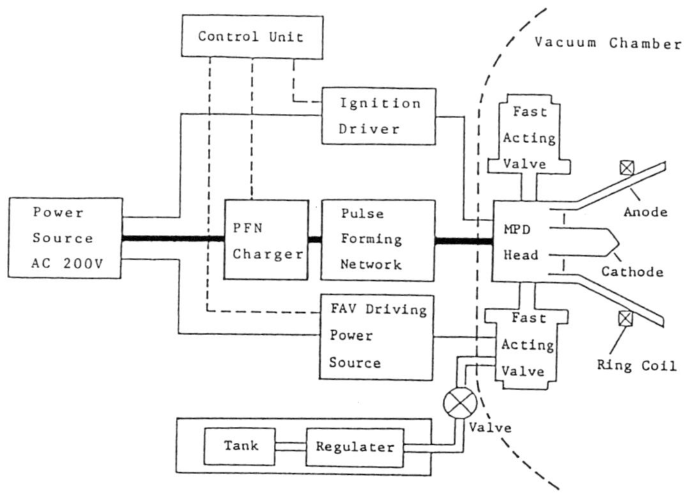

Figure 1. MPD thruster system diagram.

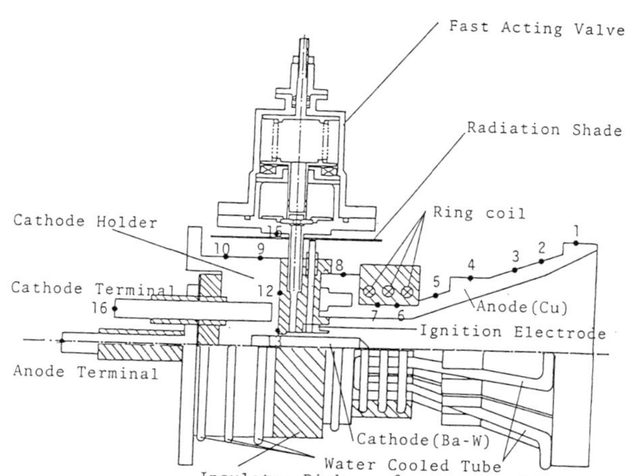

Figure 2. MPD thruster head with fast acting valves.

(fast acting valve) driver. The PFN, which is capable of storing 2.7 kJ at 300 V, is charged up within 0.3 sec with a constant charging current. Propellant gases are injected into the discharge chamber from a couple of FAVs. The width of the gas pulse is matched with that of the arc current pulse.

The thruster head is installed on a ceramic thruster stand in a vacuum tank 0.5-m in diameter × 1-m long. The tank pressure is kept to be 2-5 Pa in periadical operations. The firing test starts with charging of the PFN at a presettable voltage. After that, the FAVs are triggered; an ignition pulse of 2 kV is just applied between the cathode and a ignition electrode at the half rise time of the gas pulse, and the main discharge is generated by the ignition firing. Thus, this procedure repeats in a frequency of 0.5-1.0 Hz up to about 2×10^3 shots, i.e., until the outer surface of the thruster head reaches a thermally steady-state condition.

Arc currents and voltages are measured with a Rogowski coil and isolation amplifier, respectively. Electrode power depositions, i.e., heats to the electrodes are evaluated from temperatures and flow rates of cooling water. A heat exchanger or target is used for plume diagnostic measurements [1,2].

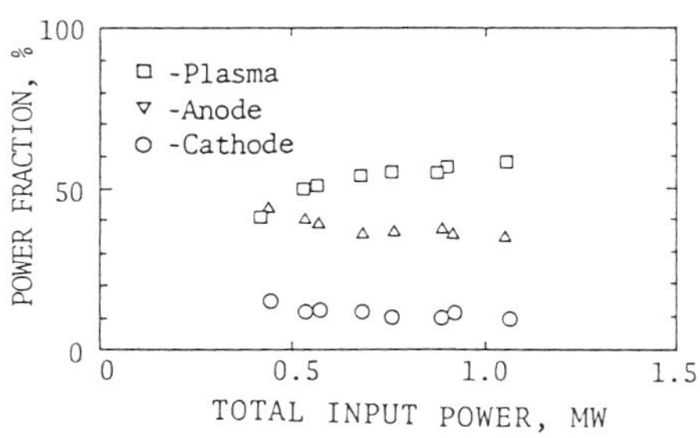

a) Mixture of N_2+2H_2, 1.6 g/s. Simulated hydrazine N_2H_4.

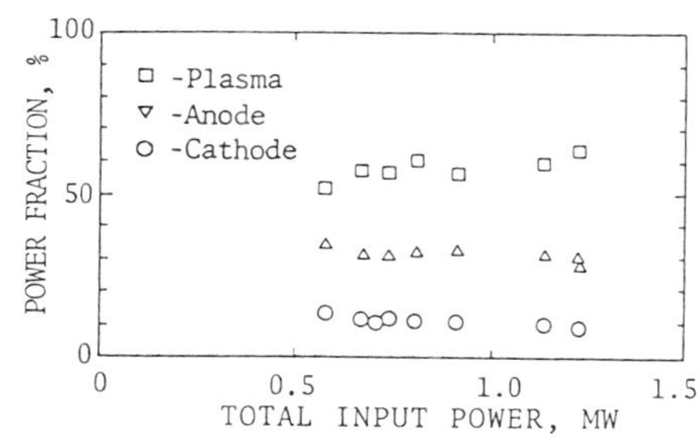

b) Mixture of N_2+3H_2, 2.0 g/s. Simulated ammonia NH_3.

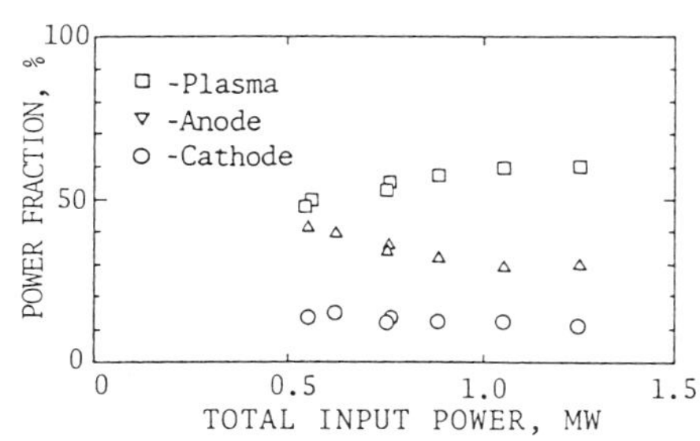

c) Hydrogen, 0.7 g/s.

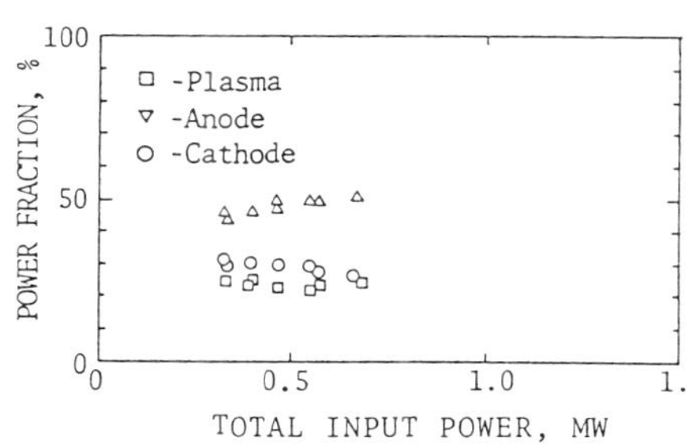

d) Argon, 3.1 g/s.

Figure 3. Fractions of deposited powers vs input power characteristics.

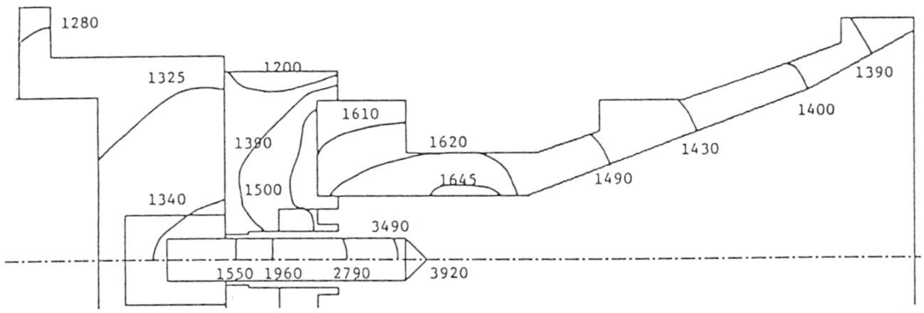

Figure 4. Calculated temperature distribution in MPD thruster head at 15-kW averaged input power.

3. RESULTS AND DISCUSSION

Figure 3 shows the fractions of deposited powers vs total input power characteristics [2]. The results show that the fractions of input power lost to the anode decrease with increasing total input power except for argon and that they are 30 to 40 % for nitrogen/hydrogen mixtures and hydrogen and 40 to 50 % for argon. On the other hand, the fractions of input power deposited into the cathode are 10 to 15 % for nitrogen/hydrogen mixtures and hydrogen and 35 to 40 % for argon. Hence, the power depositions in the electrodes and plasma flow depend on gas species and arc current levels.

Thermal analysis of the MPD thruster head is conducted under assumptions as described in Refs.1 and 2. Typical temperature distribution calculated in 15-kW averaged input power is shown in Fig.4. The local temperatures are below the melting temperature of the tungsten electrodes except for the cathode tip, though the temperatures at the cathode tip and the cylindrical part of the anode where current concentrated increased drastically with the input power. Thus, the key technology for higher power operations is found to be how to reduce the cathode tip temperature.

4. CONCLUSIONS

Quasisteady MPD thruster periodical firing tests were conducted. The energy balance characteristics depended on gas species and arc current levels. The thermal analytical results showed that only cathode tip was very severe in 10-kW-class averaged-power operations.

5. REFERENCES

1 K. Kagaya et al., Proc. of 17th ISTS, Tokyo, 1 (1990) 431.
2 K. Kagaya et al., Proc. of 18th ISTS, Kagoshima, 1 (1992) 309.
3 H. Tahara et al., J. Propulsion and Power, 5 (1989) 713.

Simulation and Design of Applied Electromagnetic Systems
T. Honma (Editor)

Magnetosonic Point in MPD Flow

I.Funaki, K.Toki, and K.Kuriki

Institute of Space and Astronautical Science

Yoshinodai 3-1-1, Sagamihara, Kanagawa 229, Japan

Abstract

A quasi one-dimensional model was constructed in order to compare the flowfield of MPD arcjets for different configurations. This model involves acceleration process and thrust generation mechanism near the cathode tip. Calculated results have shown that the position of magnetosonic point features MPD flowfield depending on the channel geometries.

1.Introduction

MPD(Magnetoplasmadynamic) arcjet is a plasma rocket propulsion expecting a high power operation and high specific impulse (1000 to 5000sec). Inside the discharge chamber, plasma is accelerated by means of Lorentz force that results from arc current of several thousands of amperes coupled with its induced magnetic field (Fig.1).

Two different forces contribute to thrust generation mechanism: one is blowing force which generates axial thrust in the interelectrode region; the other is pumping force which compresses plasma near the cathode tip. In order to treat these two mechanisms as simple as possible, we analyzed two types of quasi one-dimensional models. One is pure quasi one-dimensional model (Ref.1), in which the acceleration process by the blowing force can be discussed, the other is divided region model, which we provide in the following.

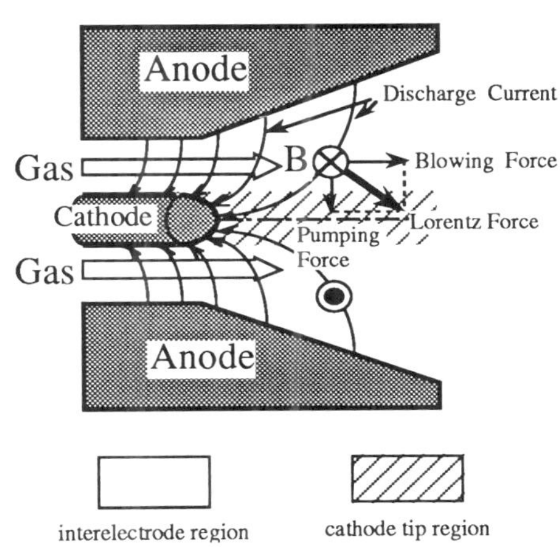

Fig.1 Flowfield of MPD Arcjet

2.Divided Region Model

Flow model

In pure quasi one-dimensional model (Fig.2), we introduce velocity:u, magnetic flux density:B and current density:j, having only one component in x, z and y direction, respectively. Divided region model involves axial current near the cathode tip, and incorporates two effects to the former model.

- pumping force which causes mass transfer from interelectrode region to cathode tip region (and vice versa)
- Joule heating by axial current

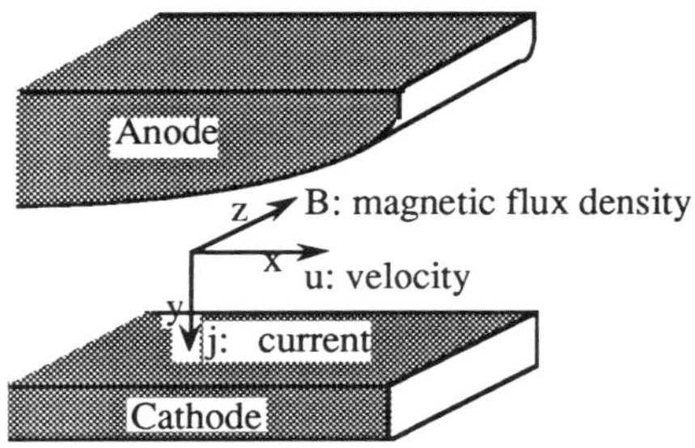

Fig.2 Pure Quasi One-Dimensional Flow

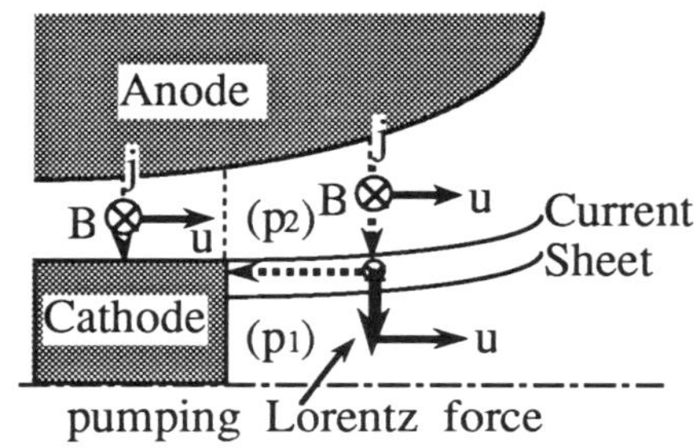

Fig.3 Divided Region Model

Let regard a part of cathode in Fig.2 as plasma region with zero magnetic flux density (see Fig.3). In each region quasi one-dimensional assumption is still valid with permission of mass transport between region 1 and region 2. Because mass flow rates of each quasi one-dimensional region: $\dot{m}_1$ and $\dot{m}_2$, are new unknown variables, total mass conservation $\dot{m} = \dot{m}_1 + \dot{m}_2$ and balance equation of both pumping force and pressure difference: $p_2 = p_1 + B^2/2\mu_0$ were added. Here p_2 ,p_1 means static pressure in region 2 and region 1, and μ_0 denotes magnetic permeability.

Mass transport between region 1 and region 2 is accompanied by energy and momentum transfer. Here we assume P_{12}: momentum transfer per unit mass transport from region 1 to region 2 and Q_{12}: energy transfer as

$P_{12} = u_i + p_i$, $Q_{12} = u_i^2/2 + e_i$

where $P_{12} = -P_{21}$ and $Q_{12} = -Q_{21}$. u_i and e_i represents velocity and internal energy of incoming gas from one region to another. Only ideal gas is treated here. The energy equation subsequently contains Q_{add}: kinetic energy addition by pumping force

$Q_{add} = DS/\rho_i$

where DS is mass transport from region 1 to region 2 and ρ_i is the density of incoming gas.

Electric field model

In pure quasi one-dimensional model, voltage between two electrodes is constant along the channel. An axial current in the divided region model was assumed so as to flow on the boundary between the two regions. We assume axial electric field E_X is determined by only Joule heating. With axial current J, conductivity σ, and the width of axial current d_{cat}, E_X is expressed as $E_x = J/\sigma d_{cat}$ and

$$E_x + \frac{\partial E_y A}{\partial x} = \text{const. (equals to discharge voltage)}$$

3.Calculated Results

Calculated results are shown in Fig.4. Velocity and temperature distribution of flared with short cathode (Fig.4-1) indicate that there is a steep rise near the cathode tip. From mass flow rate distribution (Fig.4-2) it is understood that such phenomena near the cathode tip is attributed to large energy addition to region 2 by pumping force.

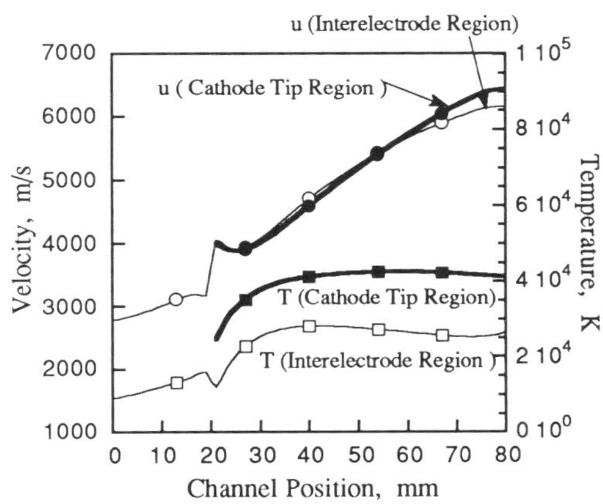

Fig.4-1

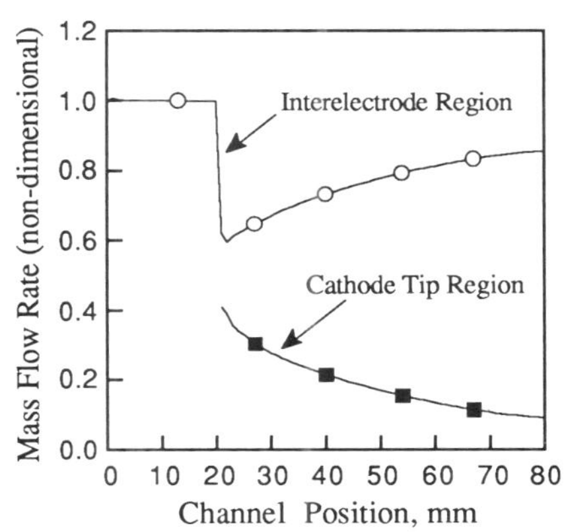

Fig.4-2

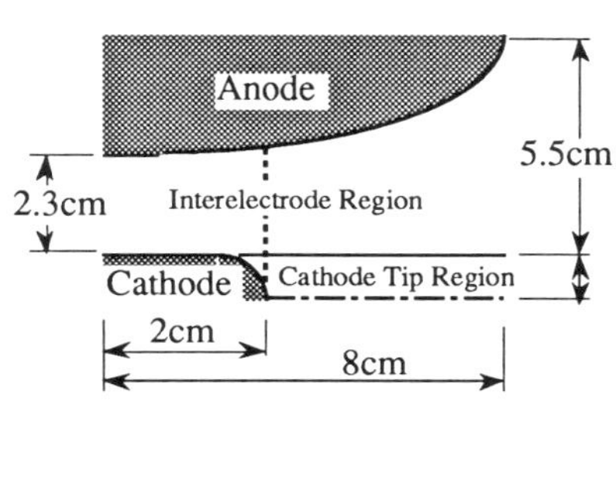

$\dot{m} = 12.5$g/sec/m
$T_{in} = 5000$K
$B_0 = 0.6$T
σ : Spitzer-Härm

Fig.4 Example of Simulation Result

Thrusts of long and short cathode cases are compared at Table 1. There found a large pressure thrust contribution in case of short cathode.

Table1 Thrust comparison of long and short cathode cases (flared,Bo=0.05T, $\dot{m}$=12.5g/s/m)

	thrust (/m)	pressure thrust at cathode tip
long cathode (pure quasi one-dimensional)	81.4N	–
short cathode (divided region model)	84.1-90.6N	7N

Figure 5 shows the position of magnetosonic point, calculated by pure quasi one-dimensional model, which is a characteristic speed of MPD flowfield. Magnetosonic point occurs near the throat in case of converging-diverging channels, near the inlet for flared channels. One can find a gas dynamical sonic point only immediately adjacent to the inlet.

Location of magnetosonic point calculated by divided region model is almost the same as the one by pure one-dimensional model. It is concluded that the position of magnetosonic point features the flowfield of MPD arcjets.

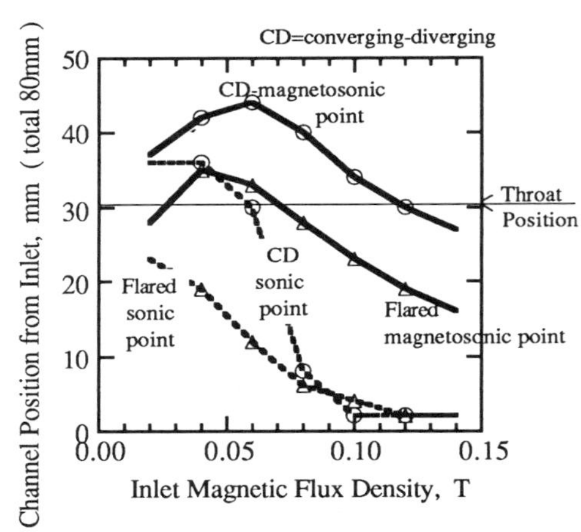

Fig.5 Location of Magnetosonic Point

4.Conclusion

From the analysis of quasi one-dimensional models of MPD arcjet, a large contribution of pressure-thrust was found at the tip of short cathode, that is, near magnetosonic point. Flowfield dependence on discharge chamber geometries was characterized by the position of magnetosonic points, i.e., near the throat for converging-diverging channel, and near the inlet for flared channel.

Reference

1. Kuriki, K., and Nakayama, T., "Magnetosonic Condition in Magnetoplasmadynamic Flow", Journal of Propulsion and Power, Vol.8, No.6, 1992, pp.1208-1211

Simulation and Design of Applied Electromagnetic Systems
T. Honma (Editor)

A finite difference analysis of the two dimensional MPD thruster

H. Itoh[a], I. Funaki[b], H. Igarashi[a], K. Toki[b], T. Honma[a] and K. Kuriki[b]

[a]Department of Electrical Eng., Faculty of Eng., Hokkaido University
Kita 13, Nishi 8, Kita-ku, Sapporo, 060 Japan

[b]Institute of Space and Astronautical Science
Yoshinodai 3-1-1, Sagamohara, Kanagawa, 229 Japan

Abstract

This paper describes a quasi-steady two-dimensional analysis of an MPD (magnetoplasmadynamic) thruster. The fluid equations and magnetic field equation are numerically solved by an iterative method to obtain self-consistent solutions. The finite difference method is employed to solve the fluid and magnetic field equations. The resultant current-thrust characteristics are shown to agree well with experimental results.

1. INTRODUCTION

An MPD (magnetoplasmadynamic) thruster, which is one of the electric propulsion systems, gets the power from the exhaust of the plasma accelerated by the Lorentz force [1]. It is of importance for improvement of the thruster to numerically analyze the MPD flows involving complicated interactions between the electromagnetic field and the plasma flow. This paper presents a numerical analysis of the multichannel two-dimensional MPD thruster [2] which was developed in order to observe the discharge by an optical method, avoiding the chaotic concentration of arc spots on the electrodes. Figure 1 shows the numerical model of this MPD thruster.

In this paper, the finite difference method (FDM) utilized to solve the fluid equations is based on the explicit TVD (Total Variation Diminishing) scheme [3]. On the other hand, the magnetic field equation derived from Maxwell's equations and Ohm's law is solved by the SOR (successive over-relaxation) method. To find the consistent relation between the electromagnetic field and the plasma flow, we iteratively solve the above equations.

2. FORMULATION

The simplifying assumptions employed in the analysis are summarized as follows : (a) The quasi-steady state, two-dimensional and self-induced magnetic fields are considered, where the quasi-steady state holds under the condition that the characteristic time of the plasma (~μsec) is shorter than the duration of the

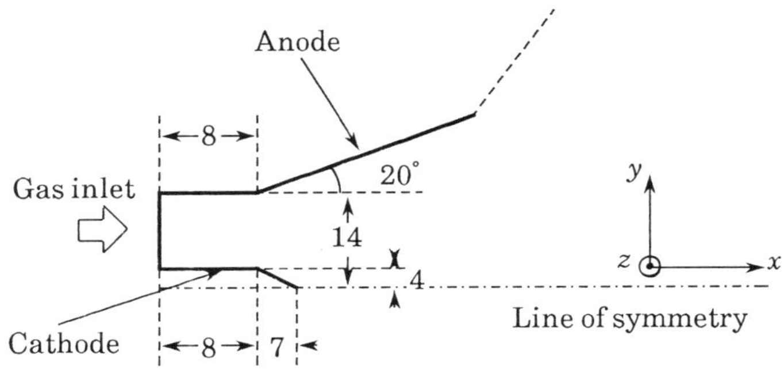

Figure 1. Numerical model of the two dimensional MPD thruster

plasma gas and power supply (~*msec*). (b) The plasma is composed of fully ionized argon. (c) The electrical conductivity is constant. (d) The hall and viscous effects, ionization processes, chemical reactions and the voltage drop across the plasma sheath and the electrode erosion are neglected.

The equation governing the plasma flow is written in the conservative form

$$\frac{\partial \mathbf{Q}}{\partial t} + \frac{\partial \mathbf{F}}{\partial x} + \frac{\partial \mathbf{G}}{\partial y} + \mathbf{S} = 0 \ , \tag{1}$$

where

$$\mathbf{Q} = [\rho, \rho u, \rho v, e]^t , \tag{2}$$

$$\mathbf{F} = [\rho u, \rho u^2 + p, \rho uv, (e + p) u]^t , \tag{3}$$

$$\mathbf{G} = [\rho v, \rho uv, \rho v^2 + p, (e + p) v]^t , \tag{4}$$

$$\mathbf{S} = \left[0, 2B\frac{\partial B}{\partial x}, 2B\frac{\partial B}{\partial y}, 2E_y\frac{\partial B}{\partial x} - 2E_x\frac{\partial B}{\partial y} \right]^t , \tag{5}$$

and ρ is mass density, p the pressure, e the energy density, B the z component of the magnetic field , and $(E_x\ E_y)$ and (u, v) denote the x and y components of the electric field and the velocity, respectively. All the above field quantities were normalized by the width of the thruster, the cross-sectional area at the inlet, the magnetic acoustic speed, the magnetic pressure at the inlet, the mass flow rate $\dot{m}$, magnetic flux density at the inlet, the electrical conductivity, the total current and the magnetic Reynolds number R_m .

After the fluid equation (1) is rewritten in the general curvilinear coordinates to use the boundary fit technique, it is solved by the explicit symmetric TVD scheme.

On the other hand, the electromagnetic field is governed by the equation of induction,

$$\frac{\partial B}{\partial t} + \frac{\partial (uB)}{\partial x} + \frac{\partial (vB)}{\partial y} - \frac{1}{R_m}\left(\frac{\partial^2 B}{\partial x^2} + \frac{\partial^2 B}{\partial y^2} \right) = 0 \ . \tag{6}$$

The plasma flow is assumed here to be frozen in the analysis of the electromagnetic field because the characteristic velocity of the electromagnetic field is much larger than that of the plasma flow. Equation (6) is then reduced to a time-independent equation which can be solved by SOR.

3. NUMERICAL RESULTS AND CONCLUSION

The iterative solutions of (1) and (6) give the field quantities in the region between the anode and cathode shown in Figure 1. The resultant current paths, velocity distributions and mass density contours are shown in Figures 2, 3 and 4, respectively. The applied currents I_{app} are here chosen as 8 and 16 [kA]. The increase in the discharge current presses the current paths downstream and results in the concentration of the current at the cathode tip, which are consistent with experiments. While the Lorentz force accelerates the plasma flows mainly above the cathode, the Joule heating in the vicinity of the cathode tip also gives rise to the acceleration near the line of symmetry, as shown in Figure 3. Moreover, the shock waves appear in both current conditions as shown in Figure 4. The current-thrust characteristics are shown in Figure 5, in which we see that the

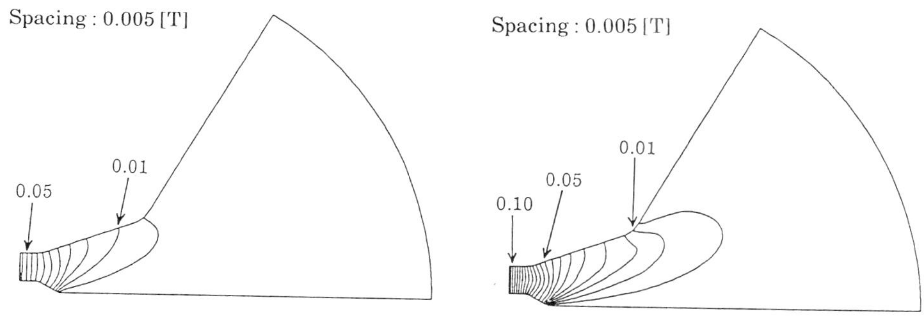

(a) $I_{app}=8$ [kA] (b) $I_{app}=16$ [kA]
Figure 2. Current paths

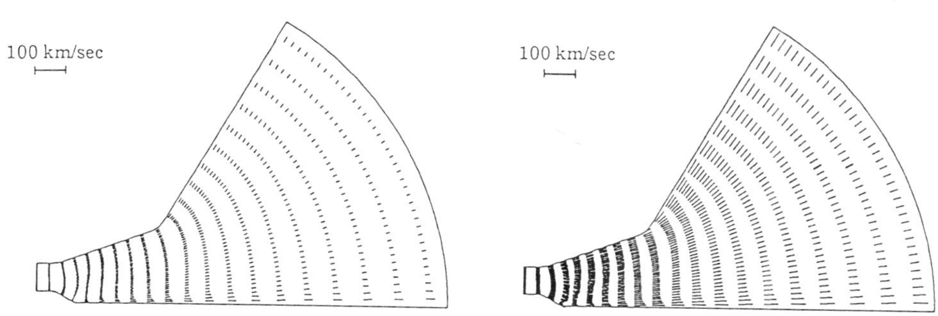

(a) $I_{app}=8$ [kA] (b) $I_{app}=16$ [kA]
Figure 3. Velocity distributions

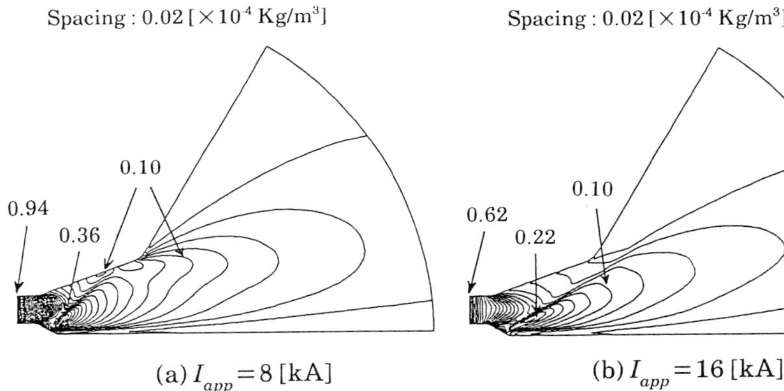

Figure 4. Mass density contours

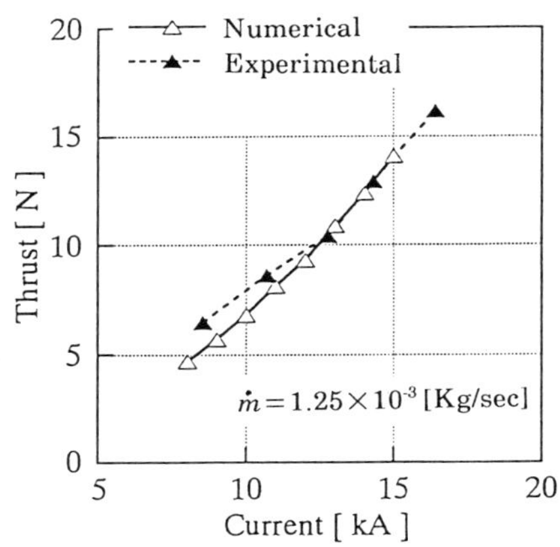

Figure 5. Current-thrust characteristics

numerical results are in good agreement with experimental results. In order to assess the reliability of the present computational method in detail, we plan to analyze the plasma flow in a wider parameter space.

4. ACKNOWLEDGMENTS

This work was partially supported by the Ministry of Education, Science and Culture of Japan under Grant-in-Aid for Scientific Research (04302025).

5. REFERENCES

1 R. G. Jahn, Physics of Electric Propulsion, McGrow-Hill, New York, 1964.
2 K. Toki, M. Sumida and K. Kuriki, J. Propulsion, 8 (1992) 93.
3 H. C. Yee, J. Comp. Phys. 68 (1987) 151.

Simulation and Design of Applied Electromagnetic Systems
T. Honma (Editor)

Numerical analyses of MHD flows in a self-field MPD thruster

Takeshi Miyasaka and Toshi Fujiwara

Department of Aeronautical Engineering,
Nagoya University, Nagoya 464-01, Japan

Abstract

The calculation of MHD flows in an MPD thruster was performed using a explicit upwind TVD scheme, which is effective in capturing sudden changes in physical quantities of the flow in the MPD thruster. The plasma gas is treated by a one-fluid model with the corresponding MHD equations. In this paper, we introduced a zonal method in which a grid fitted the cathode shape is employed. As a result of this study, the phenomena around the cathode was caught.

1. INTRODUCTION

For many years an MPD thruster has been studied from grobal experimental points of view. However, plasma flows in an MPD thruster are essentially difficult to be experimentally observed. In our laboratory numerical analyses have been performed on MHD flows in an MPD thruster [1,2]. The flow had been calculated using TVD method with a general coordinate system adapted to the flow. However, it was difficult to calculate the current density distribution where the current density increases extraordinary due to its geometical singularity. In this paper, in order to catch physical phenomena around the cahtode tip , the round cathode and the zonal method [3] are employed.

2. ASSUMPTIONS

1. Propellant gas is argon, which is fully ionized at the inlet of a thruster. (Ionization process is not considered.)
2. Plasma is perfect gas and its specific heats are constant.
3. Magnetic field is entirely self-induced and consists only of azimuthal component.
4. The electric conductivity is given by the Spitzer-Haärm formula.

3. FUNDAMENTAL EQUATIONS

In general coordinate system,

$$\frac{\partial \hat{U}}{\partial t} + \frac{\partial \hat{F}}{\partial \xi} + \frac{\partial \hat{G}}{\partial \eta} + \hat{H} = 0 \ .$$

$$U = \begin{bmatrix} \hat{\rho}\hat{u} \\ \hat{\rho}\hat{v} \\ \hat{q} + \hat{B}^2 \\ \hat{B} \end{bmatrix} , \quad F = \begin{bmatrix} \hat{\rho}\hat{u} \\ \hat{\rho}\hat{u}^2 + \hat{\rho}\hat{T} + \hat{B}^2 \\ \hat{\rho}\hat{u}\hat{v} \\ (\hat{q} + \hat{\rho}\hat{T})\hat{u} + 2\hat{u}\hat{B}^2 \\ \hat{u}\hat{B} \end{bmatrix} , \quad G = \begin{bmatrix} \hat{\rho}\hat{v} \\ \hat{\rho}\hat{u}\hat{v} \\ \hat{\rho}\hat{v}^2 + \hat{\rho}\hat{T} + \hat{B}^2 \\ (\hat{q} + \hat{\rho}\hat{T})\hat{v} + 2\hat{v}\hat{B}^2 \\ \hat{v}\hat{B} \end{bmatrix} ,$$

$$H = \begin{bmatrix} \hat{\rho}\hat{v}/\hat{r} \\ \hat{\rho}\hat{u}\hat{v}/\hat{r} \\ \hat{\rho}\hat{v}^2 + 2\hat{B}^2/\hat{r} \\ (\hat{q} + \hat{\rho}\hat{T})\hat{v}/\hat{r} + h_1 \\ h_2 \end{bmatrix} ,$$

$$\hat{q} = \frac{1}{2}\hat{\rho}(\hat{u}^2 + \hat{v}^2) + \frac{1}{\gamma - 1}\hat{\rho}\hat{T} .$$

$$h_1 = -\frac{\partial}{\partial \hat{z}}\left(\frac{2\hat{B}}{(Re)_m}\frac{\partial \hat{B}}{\partial \hat{z}}\right) - \frac{\partial}{\partial \hat{r}}\left(\frac{2\hat{B}}{(Re)_m}\frac{1}{\hat{r}}\frac{\partial}{\partial \hat{r}}(\hat{r}\hat{B})\right) .$$

$$h_2 = -\frac{\partial}{\partial \hat{z}}\left(\frac{1}{(Re)_m}\frac{\partial \hat{B}}{\partial \hat{z}}\right) - \frac{\partial}{\partial \hat{r}}\left(\frac{1}{(Re)_m}\frac{1}{\hat{r}}\frac{\partial}{\partial \hat{r}}(\hat{r}\hat{B})\right) .$$

$$(Re)_m = \mu_0 \sigma V_m r_c .$$

, where

$$\xi = \xi(x, y) , \quad \eta = \eta(x, y) ,$$

$$\hat{U} = J^{-1}U , \quad \hat{F} = y_\eta F - x_\eta G , \quad \hat{G} = -y_\xi F + x_\xi G , \quad \hat{H} = J^{-1}H .$$

$$(Re)_m = \mu_0 \sigma V_m r_c .$$

4. NUMERICAL METHOD

Figure 1 shows the schematic picture of zonal grid. At each time step, conservative variables at the points $\eta = 34, 35$ in zone 1 are given by the values in zone 2, then conservative variables in the region occupied with zone 2 in zone1 are given by the values in zone 2. This operation is done at every step.

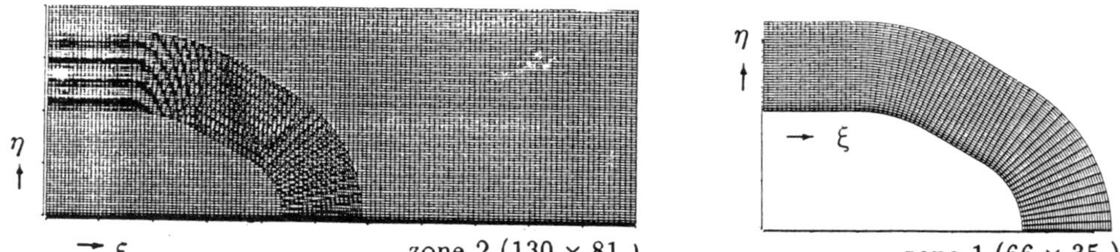

(a) zone 1 + zone 2 (b) zone 1

Figure 1. Grid system

A gasdynamic flow without any electromagnetic effect is assumed as the initial condition. And the initial magnetic flux density distribution is given by the calculation of the equation of induction derived from Maxwell equations with a constant electric conductivity.

In the calculation of MHD flows, there are two different time steps. First, the flowfield is calculated with frozen magnetic flux density and temperature by one step using the time step Δt_1 which is determined from the equations of mass and momentum continuity equations). The full MHD equations are calculated by several steps using the time step Δt_2 which is determined from the energy and induction equations). These calculations are repeated by turns.

Table 1. Inlet condition parameters

Density	ρ_{in}	1.0×10^{-3}	g/cm^3
Velocity	u_{in}	1300	m/sec
Pressure	P_{in}	575	Pa
Temperature	T_{in}	2760	K
Mass flow rate	$\dot{m}$	0.553	g/sec
Mach number	M_{in}	1.3	
Total discharge current	J	5	kA
$J^2/\dot{m}$		45.2	$kA^2/(g/sec)$

5. RESULTS

Initial density and temperature contours and initial magnetic flux density contour are shown in Figs. 2-4. Initial current density is calculated as shown in Fig. 5.

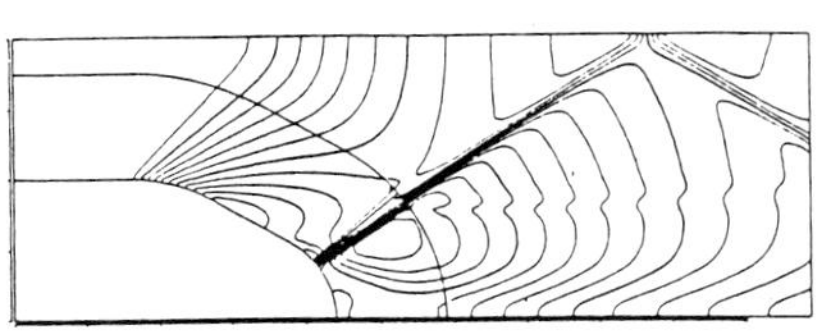

Figure 2. Density contour

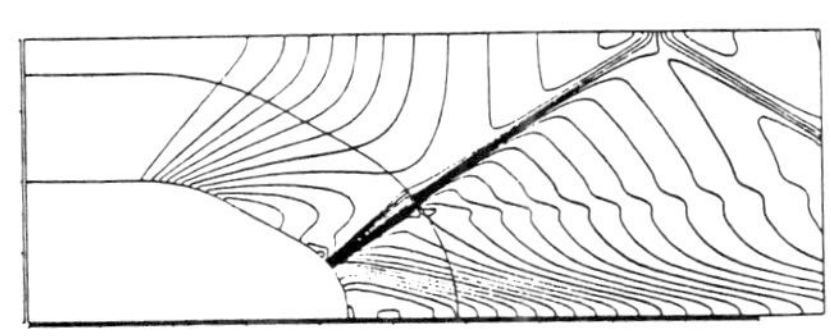

Figure 3. Temperature countour

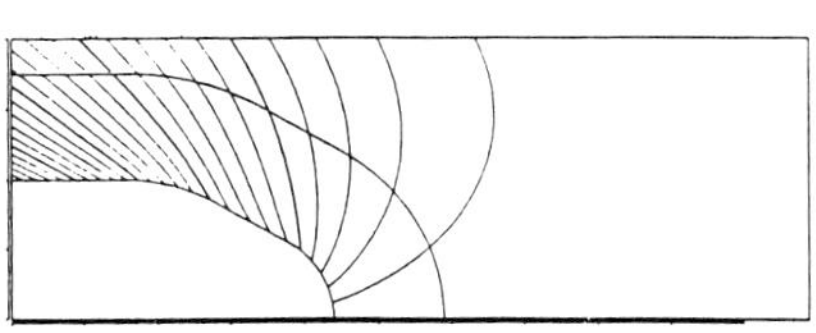

Figure 4. Magnetic flux density contour

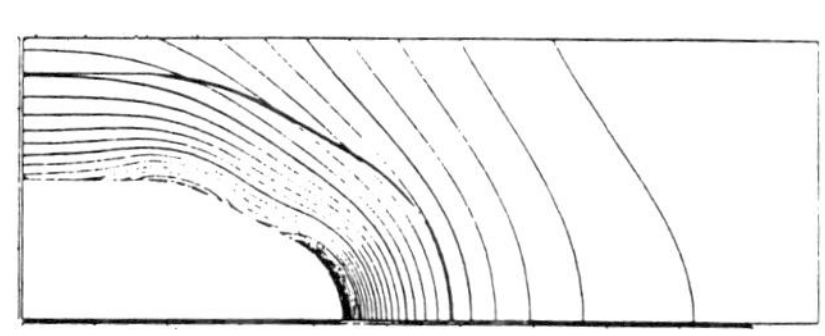

Figure 5. Current density contour

The results of calculation are shown in Figs. 6-9. This calculation is not perfectly converged. However, some characteristic phenomena are perceived. As shown in Fig. 6, the argon gas is mainly accelerated near oblique region of the cathode. The consentration of the temperature around the cathode tip is observed in Fig. 7. This is due to the concentration of the current density as shown in Fig. 9.

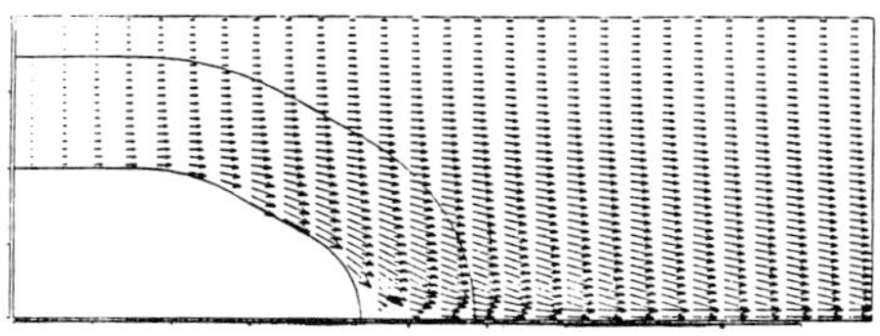

Figure 6. Velocity distribution

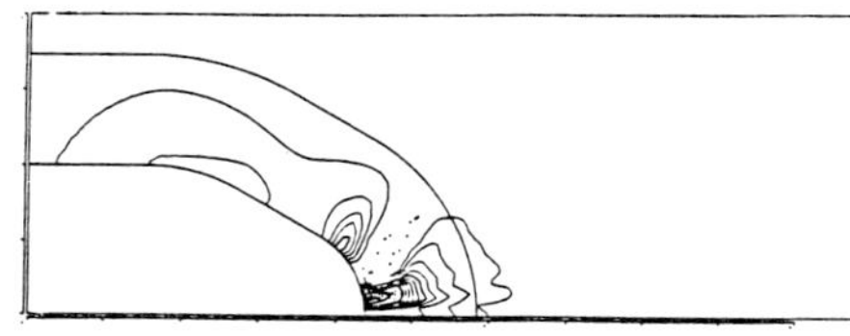

Figure 7. Temperature countour

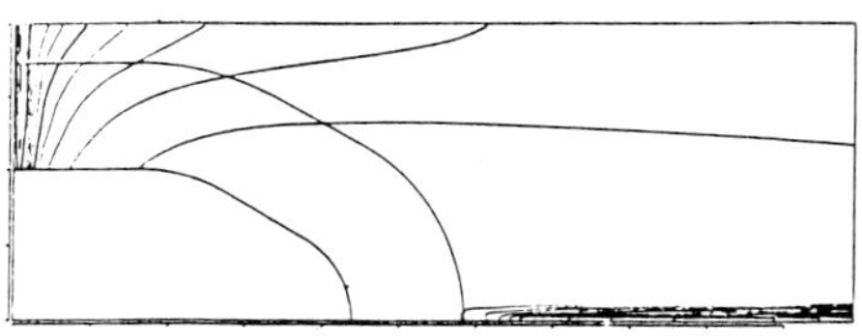

Figure 8. Density contour

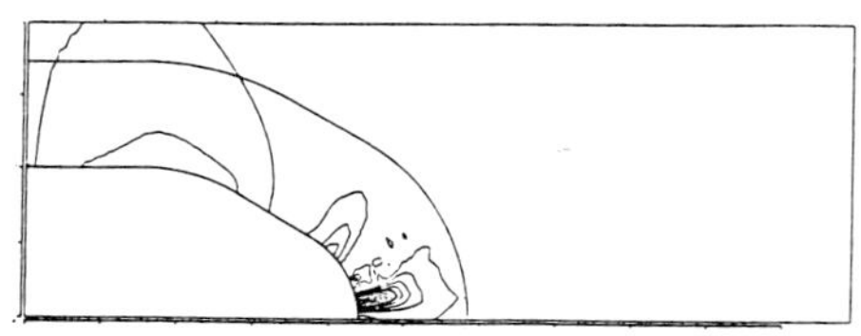

Figure 9. Current density contour

6. CONCLUSION

Physical phenomena around the cathode was caught well using the zonal method. However, to analize the consentration of the temperature more accurately, the effect of the heat conduction must be taken into account.

7. REFERENCES

1 H.Yamada, T.Ao, and T.Fujiwara, "Two-Dimensional Analysis of a Flow in an MPD Thruster," Proceedings of the Symposium on Mechanics for Space Flight, No. 3 (1985)67.

2 O.Kimura, and T.Fujiwara, "An Efficient Numerical Technique for Flows in an MPD Thruster," Proceedings of 17th Internatinal Symposium on Space Technology and Science, (1990).

3 S.Ogawa, Y.Wada, and T.Ishiguro, "Numerical Computations Using Multi-Domain Technique," NAL SP-16 (1991) 21.

Simulation and Design of Applied Electromagnetic Systems
T. Honma (Editor)

Dynamics of electromagnetic/static/thermal hybrid space propulsion devices

A. Sasoh

Institute of Fluid Science, Tohoku University 2-1-1 Katahira, Aoba, Sendai 980, Japan

Abstract

A general formula on the performance of electric space propulsion devices which utilizes arc discharge energy input has theoretically been derived. In this study, the effects of electromagnetic, electrostatic and electrothermal accelerations are taken into account. The total specific impulse is obtained from energy conservation relation. The specific impulse is expressed by two characteristic parameters; $X = (\mu_0 J_d^2/\dot{m})/u_A$ and $Y = B_z/B_\theta|_{r_a}$ (μ_0; permeability in vacuum, J_d; discharge current, $\dot{m}$; propellant mass flow rate, u_A; Alfvén speed, B; magnetic flux density, r_a; anode radius).

1. INTRODUCTION

When gas flows through an interelectrode region which surrounded by coaxial electrodes, thrust can be obtained by electrical energy input – typically by arc discharge. Such a propulsion device, which hereafter will be referred to as an electric propulsion 'thruster,' has its promising application field for space propulsion. The major advantage of this type of thruster is that the exhaust velocity of its propellant, which is in proportion with specific impulse I_{sp}, can be made much higher than that by a chemical rocket because the kinetic energy is in principle not limited as is in chemical rockets by the intrinsic energy of the propellant.

A typical electric propulsion thruster has its coaxial electrodes with axisymmetric structure. Electric power is supplied through arc discharge in its interelectrode region. In general, there are three types of acceleration mechanisms:[1]

[1] Electromagnetic acceleration, wherein energy is input to propellant through work done by electromagnetic force (1-1 by external magnetic field, 1-2 by self-induced magnetic field).
[2] Electrostatic acceleration, wherein the energy is input through electric field (Hall acceleration is included in this category)
[3] Electrothermal acceleration, wherein the energy is input through Joule heating.

In order to satisfy conditions required to a thruster, and hence, to optimize the thruster performance, it is important to obtain a general formula on the performance of thrusters in which all the above-mentioned acceleration mechanisms are combined with one another.

The difficulty in calculating their thrust is based on the fact that the total thrust can not obtained just by summing up thrusts which are individually calculated based on the respective thrust production mechanisms.

Sasoh and Arakawa[2] derived thrust formula for electromagnetic/electrostatic (Hall) acceleration from the energy conservation relation. The purpose of this paper is to extend their results, and hence to obtain a general formula on the performance of electromagnetic/static/thermal hybrid thrusters.

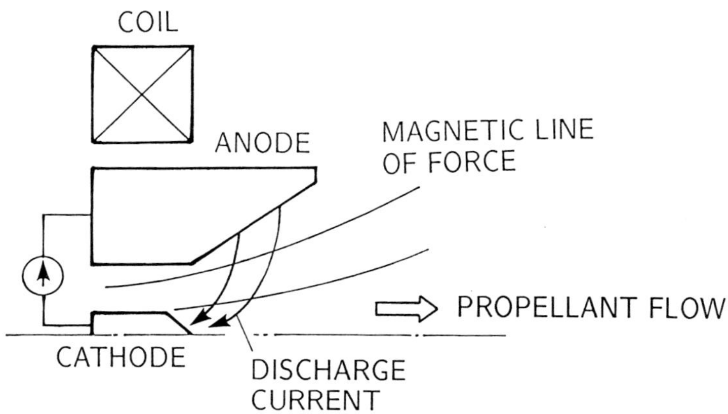

Fig. 1 Thruster model

2. MODEL AND BASIC EQUATIONS

Here, a simple model of a coaxial thruster is assumed (see Fig. 1). Current density $\mathbf{j}$, magnetic flux density $\mathbf{B}$, fluid velocity $\mathbf{u}$ and electric field $\mathbf{E}$ are assumed to have the following components, respectively.

$$\begin{array}{lcl} \mathbf{j} & = & [\ j_r \quad j_\theta \quad 0\ ] \\ \mathbf{B} & = & [\ B_r \quad B_\theta \quad B_z\ ] \\ \mathbf{u} & = & [\ 0 \quad u_\theta \quad u_z\ ] \\ \mathbf{E} & = & [\ E_r \quad 0 \quad 0\ ] \end{array}$$

Basic equations are as follows:

[1] Total energy conservation

$$\int \mathbf{j}\cdot\mathbf{E}dV = (\frac{Mu_{ex}^2}{2e} + \alpha V_{ion})J_b + \frac{\bar{h}_e}{e}J_d \tag{1}$$

[2] Generalized Ohm's law

$$\mathbf{j} = \sigma(\mathbf{E}' - \beta\mathbf{j}\times\mathbf{B}), \qquad \mathbf{E}' = \mathbf{E} + \beta\nabla p_e + \mathbf{u}\times\mathbf{B} \tag{2}$$

[3] Assumption on energy conversion

$$\int \frac{j^2}{\sigma} dV = \frac{\bar{h_e}}{e}(J_b + J_d) + \alpha V_{ion} J_b \qquad (3)$$

(u_{ex};exhaust velocity, e;electricity, α;degree of ionization, V_{ion};ionization potential, J_b;propellan flow rate in the unit of electric current, J_d;discharge current, $\bar{h_e}$;representative value of electron specific enthalpy in the acceleration region).

3. SPECIFIC IMPULSE AND THRUST EFFICIENCY

Transforming Eqs. (1) to (3) in the same way as in Ref. 2 yields a quadratic equation with respect to u_{ex}/J_d, thereby obtaining a specific impulse formula:

$$M_A = \frac{M_{A,hall} + M_{A,self}}{2} + [(\frac{M_{A,hall} + M_{A,self}}{2})^2 + M^2_{A,swirl} + M^2_{A,thermal} - q]^{\frac{1}{2}} \qquad (4)$$

$$M_A \equiv \frac{u_{ex}}{u_A}, \qquad u_A \equiv (\frac{2eV_{ion}}{M})^{\frac{1}{2}} \qquad (5)$$

$$X \equiv \frac{\mu_0 J_d^2/\dot{m}}{u_A}, \qquad Y \equiv \frac{B_z}{\mu_0 J_d/(2\pi r_a)} = \frac{B_z}{B_\theta|_{r_a}} \qquad (6)$$

$$M_{A,hall} = c_1(\omega_e \tau_e)XY \qquad (7)$$

$$M_{A,self} = c_2 X \qquad (8)$$

$$M_{A,swirl} = c_3 XY \qquad (9)$$

$$M_{A,thermal} = c_4 X^{\frac{1}{2}} \qquad (10)$$

$$q = \alpha c_5 \qquad (11)$$

$$c_1 = \frac{1-\psi'}{1+\psi} \frac{\pi\bar{\phi} c_{hall}}{\bar{r}/r_a}, \quad c_2 = \frac{1}{1+\psi}\frac{b}{4\pi}, \quad c_3 = 2\pi \frac{1}{1+\psi} c_s''^{\frac{1}{2}},$$

$$c_4 = (\frac{1}{1+\psi})^{\frac{1}{2}} (\frac{2b_t\xi}{R_m})^{\frac{1}{2}}, \quad c_5 = \frac{1}{1+\psi}\xi$$

$$R_m = \sigma \mu_0 u_A l \qquad (12)$$

$$\psi = c_\psi XY^2 \quad \psi' = c_{\psi'} XY^2 \qquad (13)$$

$$c_\psi = \frac{\phi^2 c_{hall} R_m}{8\pi}, \quad c_{\psi'} = \frac{2c_s'(\bar{r}/r_a)^2}{c_{hall}} c_\psi \qquad (14)$$

$$c_s' = \frac{1-(r_c/r_a)^2}{1+(r_c/r_a)^2}, \quad c_s'' = \frac{[1-(r_c/r_a)^2]^2}{2[1+(r_c/r_a)^2]} \qquad (15)$$

$$\xi \equiv \frac{J_b + c_p J_d}{J_b + J_d} \qquad (16)$$

(M_A; Alfvén Mach number, $c_p \simeq 1$ is a coefficient which contains the effect of electron pressure gradient on equivalent electric field.)

Here, M_A is equivalent to a normalized specific impulse. As seen in Eqs. (4) and from (7) to (10), M_A is expressed using two characteristic parameters; X and Y. X scales with the square root of the ratio of energy input through arc discharge to ionization energy. Y is the ratio of applied to self-induced magnetic fields. Hall parameter $\omega_e\tau_e$ is also a function of X and Y.

4. SUMMARY

A general theoretical formula for the specific impulse of electromagnetic/static/thermal hybrid thrusters has been established. The specific impulse is expressed using two characteristic parameters; $X = (\mu_0 J_d^2/\dot{m})/u_A$ and $Y = B_z/B_\theta|_{r_a}$. The formula derived here enables one to determine thruster operation parameters which satisfy required thruster performance, see Fig. 2.

REFERENCES

1 A. Sasoh and Y. Arakawa, 22nd Intl. Elect. Propul. Conf., IEPC 91-062, Viareggio, Italy, 1991.

2 R. G. Jahn, Physics of Electric Propulsion, McGraw-Hill, New York, 1968.

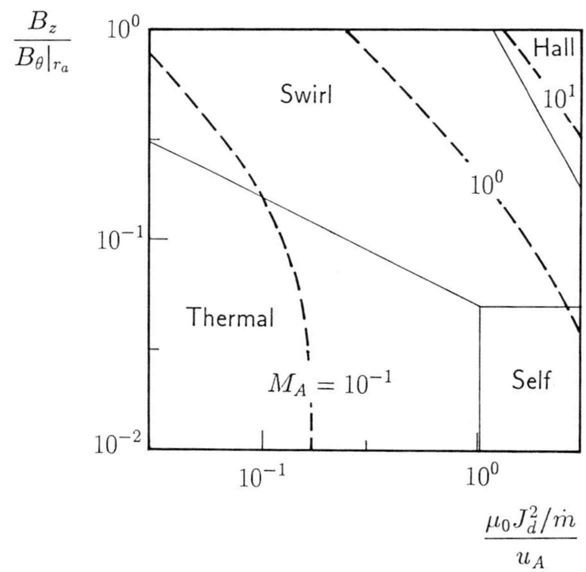

Fig. 2 Contour map of normalized specific impulse M_A (Alfvén Mach number), $\phi = B_r/B_z = 0.1$, $r_c/r_a = 0.3$, $c_{hall} = 0.5$, $(\bar{r}/r_a) = 0.5$, $b = 2$, $\xi = 1$, $R_m = 10$, $\omega_e\tau_e = (\omega_e\tau_e)_{X=1,Y=1}X^2Y$, $(\omega_e\tau_e)_{X=1,Y=1} = 10$).

MHD Generators

Simulation and Design of Applied Electromagnetic Systems
T. Honma (Editor)

MHD GENERATOR OPTIMIZATION

J. T. Lineberry

The University of Tennessee Space Institute
Tullahoma, TN 37388

Abstract

MHD power technology is being researched and developed for applications to science and industry. The design of the MHD generator for optimum performance is one major technical challenge of this work. The basic concepts for optimization of the generator to maximize power or efficiency are evaluated in this paper. It is stressed that generator optimization is dependent upon its mission. Trade-offs between performance, system requirements and economics must be made.

1. INTRODUCTION

Magnetohydrodynamic (MHD) power is being researched and developed on an international scale with MHD power concepts being conceived for use in diverse applications. Premier among the potential applications for MHD are central power production for utilities and pulse power where compact systems with high power density are demanded. In both disciplines, it is the MHD generator that lies at the heart of the MHD power system. The design and optimization of the generator is the principal technical challenge in MHD power research and development.

1.1 The MHD Generator

The MHD generator is a versatile power device that is capable of producing electric power with high energy conversion efficiency. Figure 1 illustrates a simple MHD generator along with the physics of the MHD power process.

The MHD generator has three features, a working fluid, a generator channel, and a magnet. The working fluid of the MHD generator can be any electrically conducting fluid but in large scale power systems the conducting fluid is a plasma produced by ionization of atomic species. A high temperature and pressure plasma is necessary to drive the MHD generator - to provide the thermal energy required to produce and sustain thermal ionization and provide the flow total pressure required to force the plasma through the magnetic field.

The generator channel is constructed as a duct to contain a high velocity plasma flow. Opposing walls of the channel form electrodes for electrical power extraction. The channel is positioned in a magnet with the field oriented orthogonal to the plasma flow direction - the magnetic field allows for electromagnetic induction.

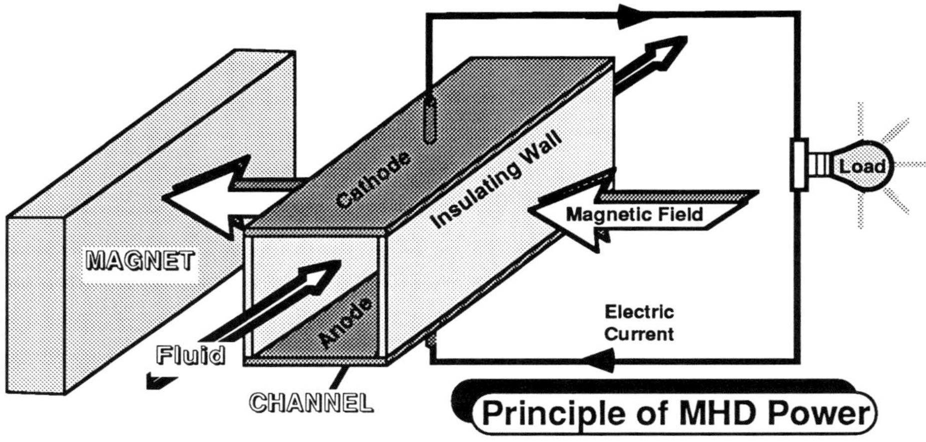

Figure 1. Schematic of a Simple MHD Generator

The plasma MHD generator produces useful power by the converting flow kinetic energy to work following the same premise as other expansion devices, e.g., a turbine. However, the MHD generator has no moving parts and contains no physical structure that is subjected to flow stagnation stress. The MHD generator can operate in a higher temperature regime than competing power technologies and consequently can extract energy at higher efficiency in accordance with standard thermodynamic measures, e.g., Carnot Cycle efficiency.

1.2 MHD Generator Performance

The MHD power process is described according to the following equation:

$$P_d = \mathbf{J} \cdot \mathbf{E} = \mathbf{U} \cdot (\mathbf{J} \times \mathbf{B}) + \frac{J^2}{\sigma}$$

where **J** is current density, **E** is electric field intensity, **U** is flow velocity and σ is fluid conductivity. The power density, P_d, is output electric power per generator unit volume. This equation illustrates how energy in the MHD generator is liberated; the total flow energy, $U \cdot J \times B$ (MHD Push Power), is dissipated as heat, J^2/σ (Joule Heating), in the working fluid and as external electrical power production, P_d.

Two key measures of the MHD process and generator design can be as:

$$P_d \propto \sigma U^2 B^2 \quad ,\text{or,} \quad P_d \propto \sigma U^2$$

The second proportionality contains only properties of the "dynamic" fluid state. It identifies the power producing capacity of the working fluid. For a plasma, conductivity is inversely proportional to velocity, consequently, maximizing power density requires design operation at a plasmadynamic condition where the product σU^2 maximizes. A second performance measure is the efficiency in converting flow energy to electrical power, enthalpy extraction, η_{ext}. Enthalpy extraction is the ratio of power output to the total energy state, H_o, of the plasma entering the generator,

$$\eta_{ext} = \frac{\int (\mathbf{J} \cdot \mathbf{E})\, dV}{H_o} = \frac{<J \cdot E> V}{H_o}$$

with integration over the generator volume, *V*. Furthermore, enthalpy extraction per unit length is proportional to plasma dynamic properties as follows:

$$\eta_{ext} \propto \frac{\sigma U}{\rho} \quad ,\text{or,} \quad \eta_{ext} \propto \frac{P_d}{\rho U}$$

where ρ is fluid density. Therefore, to maximize enthalpy extraction requires design operation at a plasmadynamic condition where the quantity $\sigma U/\rho$ maximizes.

The two parameters defined above provide first order criteria to scale and optimize the MHD generator. Both parameters identify traits of any MHD system that are desirable to maximize performance, e.g., high conductivity, high velocity flow, and high magnetic field strength. At the same time, the dependency of performance on the dynamic of the working fluid (plasma) is evident.

2.0 MHD GENERATOR OPTIMIZATION

In all applications for MHD power, minimizing the size of the MHD generator is fundamental for its optimization. The small generator is economical and eases operation and maintenance requirements. Maximizing P_d and η_{ext} will minimize generator size, however, P_d and η_{ext} cannot normally be maximized at a common operating condition. This fact is shown in Figure 2 which gives plots of P_d and η_{ext} for a fossil-fuel combustion plasma as functions of combustion pressure, Pc, and

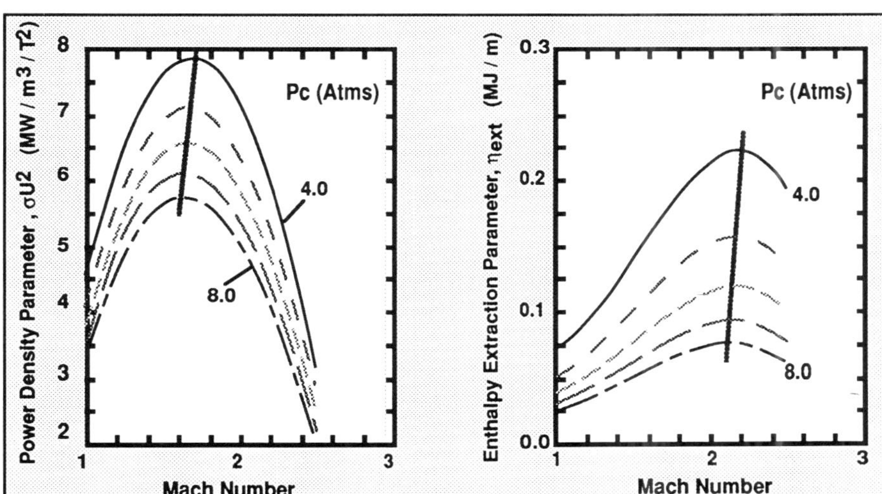

Figure 2. Typical Character of Pd and η_{ext} for a Combustion Plasma

Mach number. For combustion plasmas, both P_d and η_{ext} maximize in the high supersonic flow regime. In Figure 2, a difference in the level of Mach number required to maximize these is seen, i.e., greater area expansion is needed to optimize η_{ext}. For the same power rating, a generator optimized for η_{ext} will have a greater internal volume than one optimized for P_d. For the example of Figure 2, an η_{ext} optimized generator will be 80% as long with a 1.65 times larger flow area than a P_d optimized generator. Optimization for η_{ext} yields a generator with a larger length to diameter ratio (L/D).

What criteria govern optimization of the MHD generator is mostly dependent on the power system's mission. An MHD pulse power system is an attractive power supply from the viewpoint of maximum power density. In MHD pulse power, a high conductivity plasma can easily be produced from high energy fuels and system compactness and portability take precedence over efficiency. Alternatively, the optimum central power MHD generator is best judged by maximum efficiency. High efficiency is the promise that MHD holds for the commercial power industry.

The optimum design of an MHD generator for the commercial power plant is a much more complex problem than just maximizing performance through efficiency. The commercial generator design is conditional to economic and operational criteria imposed upon it by the needs of the overall power plant. Some crucial subjects in optimizing the fossil-fuel fired MHD generator are stated in Table 1. Meeting these issues while simultaneously maintaining maximum MHD and overall plant efficiency is the principal developmental issue facing MHD commercialization.

Table 1
Fossil Fuel Central Power MHD Generator Optimization Issues

DESIGN ISSUES	PERFORMANCE / OPERATIONAL ISSUES
• Simple Design	• MHD Performance
Mechanical and Electrical	Maximum MHD Efficiency
Manufacturing / Packaging	Integrated Cycle Efficiency
• Ecomonical Design	Base & Part Load Operation
Inexpensive Materials	Controllability
Efficient Bore Utilization	
	• Plant Operations
• System Compatibility	MHD Generator Reliability
Magnet	Maximum Service Lifetime
High Voltage Isolation	Minimum Maintenance
Power Conditioning	Plant Emissions

3.0 REFERENCES

1. R.J. Rosa, Magnetohydrodynamic Energy Conversion, McGraw-Hill, 1968.
2. J.T. Lineberry, "Design Criteria for High Performance MHD Generators", AIAA-87-0382, 1987.
3. J.T. Lineberry, MHD Plant Turndown Considerations, IECEC (1991), Vol.2, 47.

Simulation and Design of Applied Electromagnetic Systems
T. Honma (Editor)

Estimation of leakage current of coal-fired MHD generators

M. Ishikawa and J. Umoto

Dept. of Electrical Engineering Kyoto University
Yoshida Honmachi Sakyo-ku, Kyoto 606, Japan

Abstract

Effects of leakage current on the coal-fired MHD generators are studied through numerical analyses. Three channels are examined, of which scales are from an experimental to a pilot-plant size. A 2-D equivalent circuit method is applied for electrodynamics, while the gasdynamical behavior is treated with time-dependent 1-D approximation. The following results are obtained: (1) The predicted overall channel performance yields excellent agreement with experimental results. (2) The leakage current becomes about 70 A, 150 A and 200 A for channels with about 500 kWe, 1.5 MWe and 10 MWe output, respectively. (3) The loss due to slag leakage becomes about 30 % of generated power for the CDIF channel, and 20 % for the generator of 10 MWe.

1. INTRODUCTION

In coal-fired MHD channels, the electrical nonuniformities were observed experimentally [1], where the coal slag deposits on walls and is polarized by the current transportation through the slag layers, leading to strong current leakage and electrical nonuniformities on cathode walls. The presumed strong leakage current should have large effects both on overall channel performance and on local phenomena [2-4]. The present paper will reveal, through numerical analyses, effects of leakage current through polarized slag layers on the coal-fired channels as well as effects of the electrical nonuniformities. Three channels are analyzed: (a) AVCO Mark VI Faraday channel, (b) CDIF 1-B channel with Faraday and diagonal modes, (c) 10 MWe diagonal channel designed for a pilot-plant and controlled with external circuits [5].

2. MATHEMATICAL MODEL AND NUMERICAL SCHEMES

The basic equations for electrodynamics are the Maxwell equations and the generalized Ohm's law. The two-dimensional electric equivalent circuit method is applied, where MHD channels are divided into small regions [6,5]. The equivalent circuit method can treat simultaneously the electric field within the MHD channel and the behavior of external circuits. One electrode pitch is divided by two nodes and channel height in the y direction is done by 9 nodes including slag layers.

The one-dimensional flow fields are calculated and the two-dimensional gasdynamical properties are estimated from the one-dimensional flow field by using assumption of turbulent boundary layer.

The flow field in AVCO Mark VI channel and CDIF channel is taken from the other papers [7,4] and direct coupling between gasdynamics and electrodynamics is not taken into account, since the MHD interaction seems still small in these two channels. When the 10 MWe channel is treated, the direct coupling is taken into account. The basic equations used for the gas dynamical part are time-dependent one-dimensional equations [8], which are solved with the 1969 MacCormack two step explicit method [9].

The slag layers are divided into two regions where each layer has high or low electrical conductivity. The electrical nonuniformities are modeled by making the slag layer be lost on the inter-cathode wall. In the present study the position of "open gaps" is decided by the critical Hall voltage δV_c between some cathodes; i.e., when the accumulated Hall voltage exceeds δV_c, the slag is set to be lost there [10].

3. EFFECTS OF CURRENT LEAKAGE AND ELECTRICAL NONUNIFORMITIES

3.1 AVCO Mark VI Faraday Channel

This section treats the AVCO Mark VI Faraday channel operated with B = 4 T and subsonic flow condition. A number of open-gaps calculated with the present model is roughly proportional to the Hall voltage, corresponding to the experimental observation. With δV_c = 70 V, the power output is 517 kW, the Hall potential is 2.75 kV and the number of open-gaps is 35 in the calculation, whereas the experiment showed 515 kW, 2.7 kV and 34 respectively, showing good agreement between computation and experiment. The analyses show that rather large leakage current is induced along the cathode wall: about 70 A which is about three to four times as large as each load current.

3.2 CDIF Faraday and Diagonal Channels

The CDIF 1-B channel with Faraday and diagonal modes is analyzed in this section [2-4]. The peak value of magnetic field is 2.95 T, the electrode pitch is 1.55 cm, and the channel length is 4.35 m with 281 electrode pairs. The critical value of δV_c is modified into 100 V, as the channel size is taken into account.

The calculations predict that the power output is 1.6 MW, the Hall potential is 4.04 kV and the total power generated by the channel is 2.16 MW with the anode slag loss of 40.3 kW and the cathode slag loss of 524 kW for the Faraday mode, whereas the experiment showed the output of 1.6 MW and the Hall potential of 4.1 kV, showing a good agreement between the present computation and the experiment (88-FARA-5).

Figure 1 illustrates comparison of distributions of power output along the channel. It is found that slight discrepancy between the analyses and the experiments exists in the Hall potential, although the distribution of power output shows excellent agreement. The calculated Faraday current and Faraday voltage are about 18 A and 430 V at mid-channel, whereas about 20 A and 450 V were measured in the experiment, showing a good agreement, though the experimental results showed larger fluctuation. It is shown that the leakage current induced along the cathode wall is about 140 A in the mid-channel.

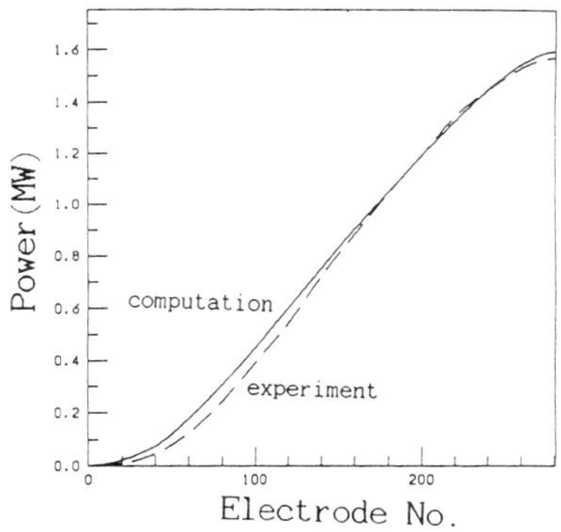

Figure 1. Comparison of computational and experimental power for CDIF Faraday channel.

The power output of the diagonal mode is 1.54 MW where 0.91 MW in the main load and 0.63 MW in the diagonal links, and the total power generated by the channel is 2.36 MW with the anode slag loss of 55.9 kW and the cathode slag loss of 730 kW. The experiment (88-DIAG-11) [11] showed the power output of 1.6 MW, showing also good agreement between the computation and experiment. Figure 2 depicts the leakage current induced along the cathode wall. It is about 150 A in the mid-channel, which is slightly larger than in the Faraday channel.

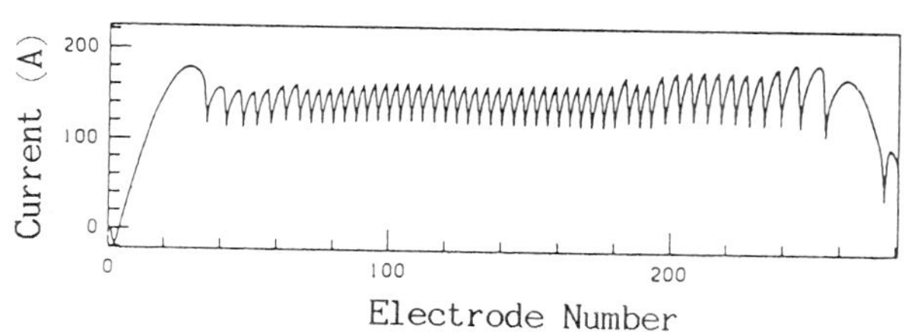

Figure 2. Leakage current distribution on cathode wall of CDIF diagonal channel.

3.3 10 MWe Diagonal Channel

The MHD channel studied in this section is the supersonic diagonal channel designed for a pilot plant with thermal input of about 100 MW. The designed channel is 5m long, while the peak value of magnetic flux density is 4.8 T. The whole channel is controlled with the master/slave current controller and current consolidation are implemented at end regions [5]. The calculated total power generated in the channel is 10.88 MW, and the power delivered to the load is 8.59 MW with the slag loss of 2.09 MW and the circuit loss of 163 kW. Large leakage current over 200 A is induced along the cathode wall, as shown in Fig.3. The power loss due to the strong leak current becomes about 21 % of the power output.

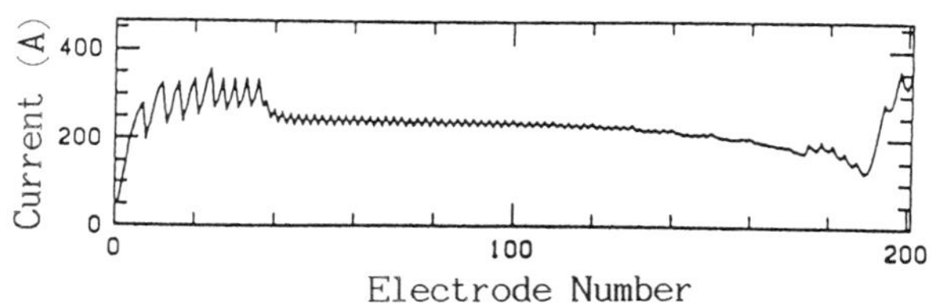

Figure 3. Leakage current distribution on cathode wall of diagonal channel with 10 MWe.

4. CONCLUDING REMARKS

(1) The overall channel performance predicted with the present model shows excellent agreement with the experimental data for the medium and large experimental channels.
(2) The leakage current becomes about 70 A for a channel with about 500 kW output, about 150 A for a channel with about 1.5 MW output, and about 200 A for a large channel with about 10 MW output.
(3) The loss due to slag leakage becomes about 30 % of generated power for the CDIF channel, and about 20 % for the generator of 10 MWe.

REFERENCES

1 S.W. Petty, Proc. 21st Engng. Aspects MHD, 4.5 (1983).
2 G.L. Nelson et al., Proc. 27th Engng. Aspects MHD, 4.3 (1989).
3 V.W. Daniel et al., ibid, 6.3 (1989).
4 J.T. Lineberry al., Proc. 28th Symp. Engng. Aspects MHD, VI.6 (1990).
5 M. Ishikawa et al., Proc. 29th Symp. Engng. Aspects MHD, IX.1 (1991).
6 M. Yoshida et al., Proc. 7th Int. Conf. MHD, II (1980) 524-529.
7 A.W. McClaine et al., Proc. 19th Symp. Engng. Aspects MHD, 1.1 (1981).
8 M. Ishikawa, Reports Inst. Atomic Energy, Kyoto Univ., 192 (1982).
9 R.W. MacCormack, AIAA Hypervelocity Impact Conference, No.69-354 (1969).
10 C.C.P. Pian et al., Proc. 26th Symp. Engng. Aspects MHD, 3.1 (1988).
11 MSE, Test Reports No.88-DIAG-9 (1988).

Simulation and Design of Applied Electromagnetic Systems
T. Honma (Editor)

Numerical studies on the shape optimization of a coupled 1300MWth coal combustion MHD channel and magnet coil system

R. Nishimura, Y. Aoki, S. Oikawa, S. Yatsu and N. Kayukawa

Advanced MHD Research Institute, Faculty of Engineering,
Hokkaido University
N13 W8 Kita-ku, Sapporo 060, Japan

Abstract

This paper discusses the effect of the cross-sectional shape of the magnet coil upon the reduction of the MHD channel length. Employing a coal-combustion plasma with semi-perfect gaseous characteristics, the generator output performance of various magnet coil shapes is evaluated by the quasi 1-D gasdynamic model. The shape optimization of a magnet coil cross section is carried out for a 1300 MWth supersonic Faraday-type MHD generator. It is shown that the channel length can be shortened more than 30% by the optimization of the coil shape in comparison with the case of the crescent shaped coil producing a uniform magnetic field, where the enthalpy extraction is effectively kept unchanged. Also, it is estimated that the capital cost for a stand alone commercial MHD plant can be decreased more than 6% by this coil shape optimization.

1. INTRODUCTION

The cost of electricity of an MHD generator, as well as the generator output performance, is crucially dependent upon the distribution of the magnetic field over the cross section of the channel. Therefore, an optimal design of the applied magnetic field distribution across the plasma flow is needed in order to improve the generator characteristics. In this paper, aiming to minimize the capital cost of a commercial open-cycle MHD plant, as well as to increase in the enthalpy extraction per unit length of the MHD channel, we examined the effect of the magnet coil shape, as well as nonuniform design of the applied magnetic field, upon the length of the channel, with specified gaseous conditions at the inlet and a given stagnation pressure at the outlet.

2. METHODOLOGY

The length of an MHD channel was calculated by solving the one-dimensional equations (the set of three conservative equations) for a turbulent flow, including the MHD effects, e.g. the Lorentz force [1]. These equations are solved by the Runge-Kutta-Fehlberg method with the initial condition specified at the inlet. The channel length is determined by the axial point, where the stagnation pressure has reached the specified level. The electric field and the current density inside the MHD channel cross section are calculated by using Maxwell's equations and the generalized Ohm's law, neglecting the effects of the ion slip [1]. The effects of the plasma nonuniformity over the channel cross section, i.e. the effects of the boundary layer, on electrical values were considered at each axial point.

First, the calculation was carried out assuming a uniform magnetic field produced by a crescent coil (Fig. 1) [1]. This crescent coil with the specified cross-sectional area and the coil current, was designed to maximize its magnetic field.

Next, the coil current and the coil cross-sectional area being fixed, a coil-shape optimization was carried out in order to improve the generator performance. The crescent coil was segmented by rectangular elements. Then, the elements were relocated to find the channel with the shortest length and the highest enthalpy extraction by using the steepest gradient method. This optimization of the coil shapes was carried out for various inlet channel aspect ratio defined by the distance (D) between the anode and the cathode to that between the insulator walls faced each other (W).

3. RESULTS

The above optimization of an MHD generator channel including a magnet coil was carried out for a 200 MWe coal combustion supersonic MHD generator. We assumed a coal-combustion plasma with the semi-perfect gaseous characteristics and a supersonic constant-velocity Faraday-type generator. The parameters for numerical evaluation are shown in Table 1. The thermodynamic and transport properties of the working fluid are estimated by the equations described in Ref. 2.

Table 1
Parameters for numerical evaluation.

Channel cross sectional area at the inlet	4800 cm^2
Coil cross sectional area (Total)	4600 cm^2
Static pressure at the inlet	2.3 atm
Stagnation pressure at the outlet	1 atm
Core flow velocity at the inlet	1500 m/s
Core flow temperature at the inlet	2800 K
Boundary layer thickness at the inlet	5 cm
Minimum distance between the coil and the plasma	$\Delta_y = \Delta_z$ = 10 cm (See Fig. 1.)
Coil current density	10^8 A/m^2
Slag-layer thickness	1.5 mm
Slag surface temperature	1870 K
Wall temperature	1000 K
Load factor	0.8

The electrical power output and MHD channel length for the crescent coil (uniform magnetic field) are shown in Table 2. Those for the optimal coil are shown in Table 3. It was shown that, under an approximately same enthalpy extraction, the channel length can be shortened more than 30% by the optimization of both the coil shape and the channel aspect ratio at the inlet. It was also shown that the optimal inlet channel aspect ratio, which minimize the channel length without decreasing the enthalpy extraction, exists between 0.5~1. Figure 2 shows the arrangement of the optimal coil and the MHD channel for the case of D/W = 1 at the inlet.

Table 2
The electrical power output and the length of the MHD channel for the crescent coil.
(Uniform magnetic field)

Inlet Channel Aspect Ratio	Power Output [MW]	Channel Length [m]	Magnetic field [T]
0.33	197	3.8	5.6
0.5	200	3.6	5.4
1	201	4.1	4.7
2	195	5.2	4.0

Table 3
The electrical power output and the length of the MHD channel for the optimal coil.

Inlet Channel Aspect Ratio	Power Output [MW]	Channel Length [m]
0.33	199	3.0
0.5	202	3.0
1	204	2.7
2	198	4.3

CONCLUSION

In order to shorten the length of the Faraday-type MHD channel, optimization of the coil-cross sectional shape was carried out assuming a supersonic constant-velocity channel. The conditions obtained are as follows:

With this coil-shape optimization, the channel length can be shortened more than 30% without a decrease of the enthalpy extraction in comparison with the uniform-magnetic-field case. This means that the enthalpy extraction per unit length of the channel was improved by this coil-shape optimization.

It was also shown that the optimal inlet channel aspect ratio, which minimize the channel length without decreasing the enthalpy extraction, exists between 0.5~1.

The pair of the magnet and the channel represent about 18% of the total plant cost [3]. This estimate is based on a stand alone MHD plant. For a retrofit configuration, the percents of plant cost attributable to the magnet and the channel is larger due to the decrease in total plant cost of a retrofit configuration. Because of these, It was also estimated that the plant cost for a commercial MHD plant using the present type of the channel model can be decreased more than 6% by the design of the coil-shape optimization.

REFERENCES

1 R. J. Rosa, Magnetohydrodynamic Energy Conversion, (Hemisphere, New York, 1987).
2 M. Ishikawa, et al., INTERACTION BETWEEN FLUID AND ELECTRO-MAGNETIC FIELD IN MHD GENERATORS, Sapporo, (1983) pp.226-235.
3 W.R. Owens, Performance and Economics of MHD Power Systems, Proc. of the International workshop on MHD SUPERCONDUCTING MAGNET, Bologna, (1991) pp.103-111.

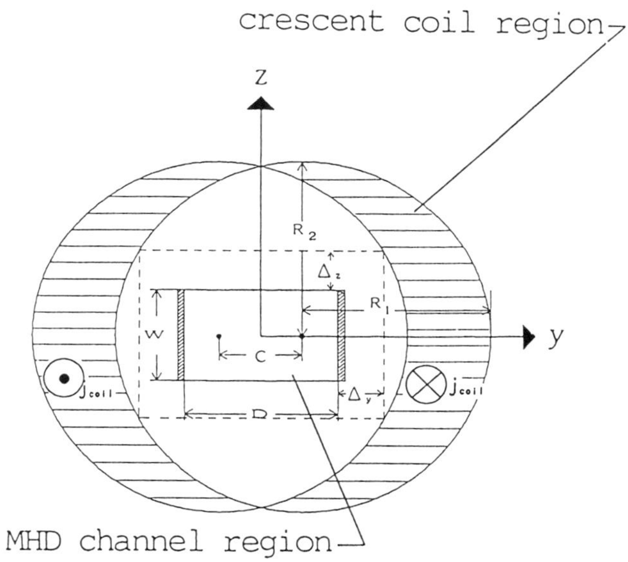

Fig. 1. Cross section of the crescent coil and MHD channel, and the coordinate system.

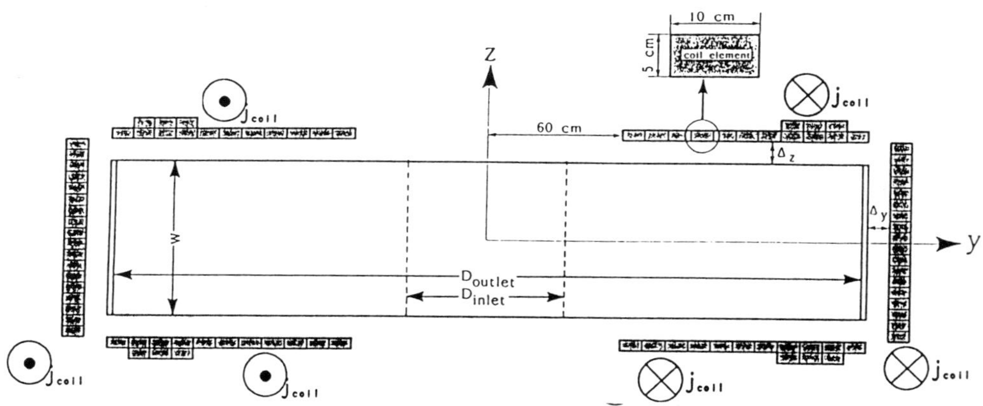

Fig. 2. Cross section of the optimal coil and MHD channel for D/W=1 at the inlet.

Simulation and Design of Applied Electromagnetic Systems
T. Honma (Editor)

Instability and generator performance in nonequilibrium MHD disk generators

S. Kabashima, K.Sasao and V.M.Zubtsov

Tokyo Institute of Technology, 4259 Nagatsuta, Midori-ku, Yokohama 227, Japan

1. Introduction

In the nonequilibrium MHD power generation a noble gas containing a small amount of alkali metal is used as the working gas. For the power generation circumstance the electrons are mainly supplied from seeded alkali metal and have a different temperature from the noble gas. The situation makes plasma in a nonequilibrium state, which is described by the two temperature model of MHD equations. In the nonequilibrium plasma the instability due to the weak ionization of alkali metal (seed material) and noble gas and it brings the decrease in the generator performance.

In this report the instability has been investigated by the numerical simulations based on the two temperature model equations and examined the generator performance.

2. The two temperature model equations

The two temperature model equations which are conventionally used for the analyses of nonequilibrium MHD plasma in the MHD power generation are given as follows.

< Equations for electron >

ionization equation

$$\frac{\partial n_i^+}{\partial t} + \nabla \cdot (n_i^+ \vec{u}) = \dot{n}_i^+ = k_{fi} n_e n_i - k_r n_e^2 n_i^+ \tag{1}$$

$$n_e = \sum_i n_i^+ , \quad (i = Seed, \, Noble\, Gas)$$

generalized Ohmic equation

$$\vec{j} + \frac{\beta}{|\vec{B}|} \vec{j} \times \vec{B} = \sigma(\vec{E} + \vec{u} \times \vec{B}), \tag{2}$$

where the electric conductivity σ and Hall parameter β are expressed,

$$\sigma = \frac{e^2 n_e}{m_e \nu_e}, \qquad and \qquad \beta = \frac{eB}{m_e \nu_e}.$$

energy equation It is assumed the steady state equation as

$$\frac{|\vec{j}|^2}{\sigma} = 3n_e m_e k(T_e - T_g)\sum_i \frac{\nu_{ei}}{m_i} + \sum_i \dot{n}_i^+ \left(\frac{3}{2}kT_e + \epsilon_i\right). \tag{3}$$

< Gas dynamic equations >

$$\frac{\partial \rho}{\partial t} + \nabla \cdot (\rho \vec{u}) = 0 \tag{4}$$

$$\frac{\partial}{\partial t}(\rho \vec{u}) + \nabla \cdot (\rho \vec{u}\vec{u}) = \vec{j} \times \vec{B} - \nabla p - P_L \tag{5}$$

$$\frac{\partial E_s}{\partial t} + \nabla \cdot \{(E_s + p)\vec{u}\} = \frac{|\vec{j}|^2}{\sigma} - \vec{u} \cdot (\vec{j} \times \vec{B}) + Q_L \tag{6}$$

Here the notations commonly used are adopted. The meaning of the notation is shown in reference1.[1] In addition to the foregoing equations the field equations and the state equation of gas is used:

$$\nabla \cdot \vec{j} = 0, \qquad \nabla \times \vec{E} = 0 \tag{7}$$

$$p = \rho R T \tag{8}$$

3. Procedure and Results

Here we concentrate ourselves to the plasma flow in the disk type generator. The principle of the generator is schematically shown in Fig.1. The working gas flows in radial direction and the magnetic field is applied perpendicular to the disk plane. In the configuration circular Faraday current is induced and Hall voltage appears between the electrodes located upstream and and downstream.

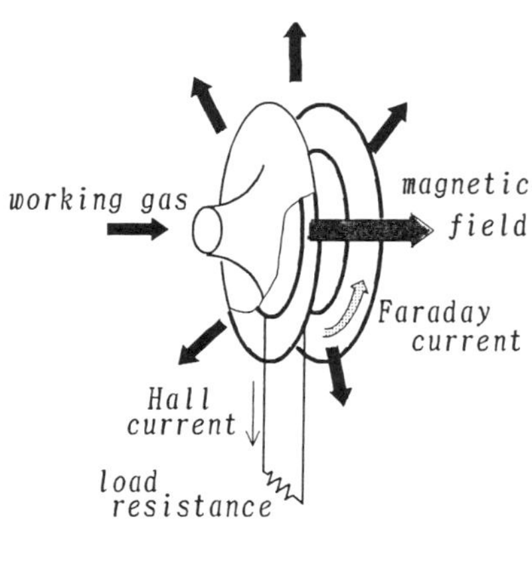

Fig.1 disk generator

Table1. Generator parameters and conditions.

Geometry	
Inlet Radius	0.160 m
Inlet Height	1.47 $\times 10^{-2}$ m
Exit radius	0.380 m
Exit height	2.09 $\times 10^{-2}$ m
Operation Conditions	
Working gas	Helium
Seed material	Cesium
Stagnation temperature	2000 K
Stagnation pressure	3.0 atm
Thermal input	10 MW
Seed fraction	5.0 $\times 10^{-5}$
Magnetic flux density	4.0 T (Constant)
Inlet Mach number	2.12
Design values	
Electron temperature	5000 K
Enthalpy extraction	30.4 %
Load resistance	3.27 Ω

The equations have been solved two dimensionally for a disk generator in the r-θ plane. The explicit MacCormac method is used in the numerical treatments of gas dynamic equations, and the Galerkin finite-element method is employed in solving field equations.

Dimensions and operation conditions of the generator dealt in the calculations are shown in Table 1. The generator works with the fully ionized seed plasma for the load resistance of 3.27 Ω.

The plasma structures obtained from calculations for several load resistances are shown with electron temperatures in Fig.2, where the calculation is performed under the inlet condition of electron temperature as 5000 K.

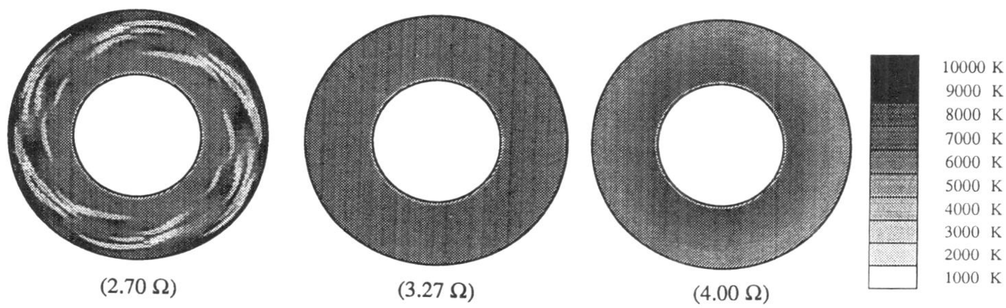

Fig.2 Plasma structure for several load resistance

It is found that the uniform plasma corresponding to the fully ionized seed appears for designed value of load resistance, 3.27 Ω. For the lower load resistance the spiral structure of the plasma is observed. The plasma showsthe instability due to weak ionization of seed material. It is observed for the higher load resistance that the plasma with high electron temperatures arises near the inlet. The plasma is related to the weak ionization instability of noble gas and has strong interactions with the gas flow.

The instability is excited by the low inlet electron temperature. Figure 3 shows the growth of the instability when the inlet electron temperature is changed from 5000 K to 3000 K at time 0. The plasma observed in the case of low load resistance grows in whole region of the generator. The spiral structure is a characteristics of the instability with low electron temperature.

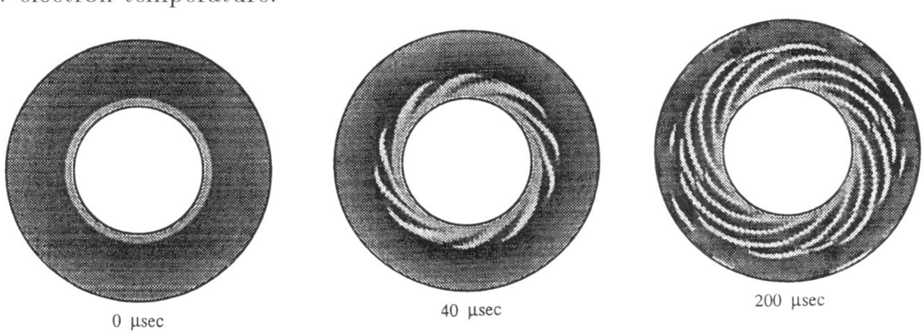

Fig.3 Growth of the instability for low inlet electron temperature.

It is well known from linear analyses of the two temperature model equations [2] that several instabilities are included and the fully ionized seed plasma is stable and homogeneous for small disturbances. However, nonlinear analyses of the energy equation of electron [3] indicate that even the fully ionized seed plasma can include the instabilities for certain amounts of big disturbances. It is understood from Fig.3 that the instabilities excited by the low electron temperature at the inlet coexist in the generator with the fully ionized seed state. In this case the electron temperature 3000 K at the inlet is the big disturbance for the fully ionized seed plasma. When the low electron temperature condition is given for a part of the inlet, the instability excited transports in the generator channel as shown in Fig.4(a), where the low temperature plasma grows in azimuthal direction. The instability profile shown in Fig.4(a) is called "streamer". On the other hand if the part of inlet plasma has high electron temperature, the instabilities flow in the generator as shown in Fig.4(b), where we can see the quite different behavior of the plasma as compared to the Fig.4(a). The plasma with high temperature arises and transports in the radial direction, and the plasma with low temperature is also excited and behaves similarly to the streamer. The shape of the high temperature plasma is called "domain". In both cases the gas pressure is disturbed by the instability and the output power decreases [4].

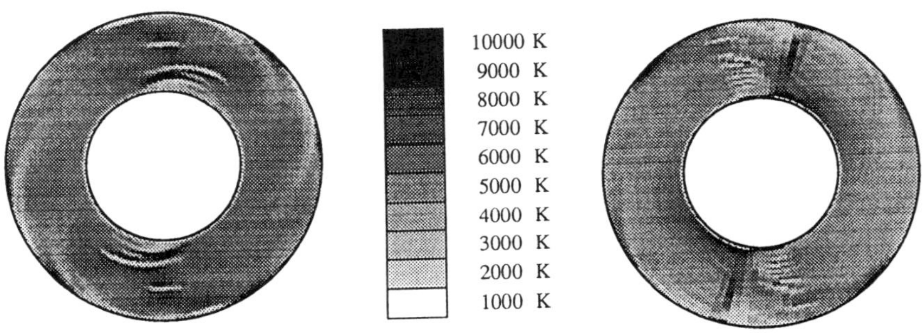

Fig.4 (a) streamer structure (b) domain structure

4. Concluding Remarks

The instabilities are found in the case where the conditions of generator inlet and/or loading parameter do not suit to produce the fully ionized seed plasma. The instabilities coexist with the fully ionized seed plasma and give the characteristic structure for low and high load conditions and both instabilities influences on the generator performance.

Reference

[1]M.Mitchner and C.H.Kruger, *Partially Ionized Gases*, Wiley New York (1973)

[2] J.L.Kerrebrock:AIAA J. **2**(1964) 1072-1080

[3] V.M.Zubtsov:Sov. Phys. Dokl **26**(1981) 56-58.
and Sov.Phys. Tech.Phys.**26**(1981)1261-1262

[4] S.Kabashima, K.Sasao and K.Nakata: 3rd Japan-CIS MHD Symposium, Hokkaido Univ.1993 March, p8.5.1-8.5.8

Simulation and Design of Applied Electromagnetic Systems
T. Honma (Editor)

Electrogasdynamical correlations of seeded combustion plasma in a Faraday-type magnetohydrodynamic channel

N. Kayukawa, S. Oikawa, Y. Aoki and S.Yatsu

Advanced MHD Research Institute, Hokkaido University, Sapporo, 060 Japan

Abstracts

Effects of the temporal fluctuation in an anisotropic combustion plasma on both the instantaneous and mean electrical behavior of the Faraday-type MHD channel were studied. The 2D MHD electrical equations were solved over the cross section of the channel with simulated turbulent data of each the velocity and the temperature. It was clarified that the correlation of the conductivity and the velocity fluctuation and that of the conductivity and the transverse electric field are positive and amount to several percents of the corresponding mean core values. We showed that the mean output performance is enhanced by the turbulence, although the current and the electric field correlation results in the local electrode potential drops.

1.INTRODUCTION

In a magnetohydrodynamic channel, the velocity, the temperature, and the electrical quantities fluctuate statistically in time and space due to the gasdynamical turbulence. Owing to the coupled nonlinear electrogasdynamical behaviour, the mean electrical processes may be affected by the correlations among the conductivity, the flow velocity, the electric field and the current. In solving the MHD phenomena, the turbulent correlations have been considered only in the Reynolds stress in the flow equations [1-2], but those in the mean Ohm's law as well as in the MHD source term have been neglected so far.

We evaluated the effects of the fluctuation on the mean MHD process dealing with simulated turbulent fluctuations superimposed on the mean 1/7th velocity and temperature profiles in a Faraday-type MHD channel with segmented electrodes. The Saha equilibrium was assumed to calculate the instantaneous conductivity. This holds for the combustion plasma with the mean temperature higher than about 1800 K and the frequency being less than about few kHz [3].

Assuming all gaseous and electrical variables can be divided into mean and turbulent parts as $A = \langle A \rangle + A'$, we can write the mean generalized Ohm's law and the MHD term in the energy conservation equation for a plasma with low magnetic Reynolds number as follows:

$$\langle J \rangle = \langle \sigma \rangle (\langle E \rangle + \langle V \rangle \times B) - \mu_e \langle J \rangle \times B + \langle \sigma' E' \rangle + \langle \sigma' V' \rangle \times B \qquad (1)$$

$$\langle J \cdot E \rangle = \langle J \rangle \cdot \langle E \rangle + \langle J' \cdot E' \rangle, \qquad (2)$$

where **B**, **J** and **E** are the applied magnetic field, the current density and the electric field. The fluctuation in the mobility μ_e was neglected. Equations (1) and (2) mean that the correlations $\langle\sigma'\mathbf{E}'\rangle$, $\langle\sigma'\mathbf{V}'\rangle$ and $\langle\mathbf{J}'\cdot\mathbf{E}'\rangle$ should be evaluated to deal with the mean electrical behavior. As the conductivity of an MHD plasma is a strong nonlinear function of the temperature, the mean conductivity $\langle\sigma\rangle$ is usually higher than the value $\sigma(\langle T\rangle)$ at the mean temperature $\langle T\rangle$ by approximately,

$$\langle\sigma\rangle - \sigma(\langle T\rangle) = \frac{1}{2}\frac{\partial^2\sigma(\langle T\rangle)}{\partial\langle T\rangle^2}\langle T'^2\rangle, \tag{3}$$

where $\langle T'^2\rangle$ is the temperature variance. Because $\partial^2\sigma(\langle T\rangle)/\partial\langle T\rangle^2$ is large at low temperature, we note that the effect of the turbulence is significant in the region, where $\langle T\rangle$ is low, e.g. in the boundary layers.

2.MODELLING OF THE GASEOUS DATA AND NUMERICAL METHOD

We consider the axial and the transverse velocity components u', v', and w' in x, y, and z directions and the temperature T as given turbulent data. Employing the Markov form of the spectrum ; $S_g(f) = \langle g'^2\rangle\alpha/(\alpha^2 + 4\pi^2 f^2)$, we calculate the fluctuation $\mathbf{g}'=(u', v', w', T')$ in time as,

$$g'(t) = C_g \sum_{j=1}^{n} [4\pi S_g(f_j)\Delta f_j]^{1/2}\cos(2\pi f_j t + \theta_j). \tag{4}$$

Here, α is an arbitrary constant and θ_j is the random phase angle. The constant C_gs were specified as C_u=6.5, C_v=4 C_w=4 and C_T=5 each for u', v', w', and T' in order for the mean amplitudes to coincide approximately with the empirical level. The sign of **g'** and the spatial distribution of the variance were also statistically considered referring the sign of $\nabla\langle u\rangle$ and $\nabla\langle T\rangle$ in the cross section of the channel. The instantaneous electrical quantities were calculated by the following potential equation derived from the instantaneous Ohm's law, $\mathbf{J} = \sigma(\mathbf{E}+\mathbf{V}\mathbf{x}\mathbf{B}) - \mu_e\mathbf{J}\mathbf{x}\mathbf{B}$, $\nabla\cdot\mathbf{J} = 0$ and $\nabla \times \mathbf{E}=0$ for the gaseous data generated by the above way.

$$\Phi_{yy} + (1+\beta^2)\Phi_{zz} + \frac{\sigma_y}{\sigma}\Phi_y + (1+\beta^2)\frac{\sigma_z}{\sigma}\Phi_z + Bu_y - \beta Bv_y - \frac{\sigma_y}{\sigma}(\beta E_x - uB + \beta vB) = 0 \tag{5}$$

The subscripts y and z mean the partial differentiation. The axial variation has been neglected for simplicity because the axial convection is much larger than in the transverse convection in the MHD channel. The Hall field E_x is constant with respect to y and z but fluctuates in time. It is calculated by the condition, where the net axial current is zero. The Hall parameter $\beta = \mu_e B$ is independent of the fluctuation. The boundary conditions for Eq.(5) are $\Phi_z = 0$ on the electrodes and $J_z = 0$ on the insulators. The finite difference method was adopted with variable meshes to allow small scale inhomogeneities near the walls and large scale ones in the core region. The pressure was 1 atm and uniform. The mean core plasma temperature, the wall temperature, the magnetic field and the boundary layer thickness for the mean 1/7th profile were assumed to be 2800 K, 1800 K, 3 Tesla and $\delta/H = 0.206$, respectively, where H is the channel height.

4.RESULTS

The distribution of $<\sigma'u'>$ over the cross section is given in Fig.1. It has a positive correlation and amounts to several percents of $\sigma(<T_c>)<u_c>$, which results in the increase in the mean current density. The correlation $<\sigma'v'>$ is negative in the cathode side and positive in the anode side. The magnitude is about 1/10 of $<\sigma'u'>$, so that the effect on the mean phenomena can be neglected.

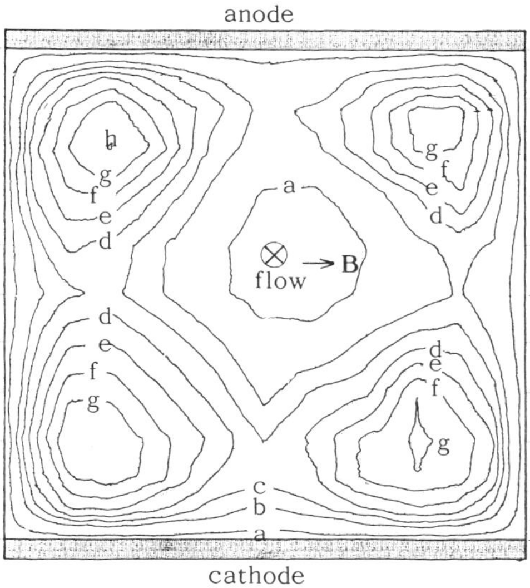

Figure 1. The distribution of the correlation $<\sigma'u'>/\sigma(T_C)<u_C>$, a:0.005, b:0.01, c:0.0125, d:0.015 e:0.0175, f:0.02, g:0.0225, h:0.0275.

Table 1
Comparison of the mean open voltage, short current and effective resistance with those of the 1/7th model.

	Turbulent (mean)	1/7th
$\frac{<V_o>}{<u_c>BH}$	0.9240	0.9163
$\frac{<Is>}{\sigma(T_c)<u_c>BS}$	0.1732	0.1670
$\frac{<R_p>}{R_p(T_c)}$	5.3550	5.4870

The instantaneous potential is shown in Fig.(2). We note that the plasma turbulence does not affect the core potential, while it does significantly in the electrode boundary layers. The mean output values and the effective internal resistance are compared with those calculated by the 1/7th velocity and the temperature profiles in Table 1, which indicates that the mean open voltage and the short current, e.g. the average power, can be increased and the mean plasma resistance is decreased due to the turbulence by an order of several percents of those obtained without fluctuation. This is mainly attributable to the mean conductivity increase as given by Eq.(3), although the mean correlation $<\sigma'E_y'>$ affects to deteriorate the generator performance because of its positive characteristics as shown in Fig.(3).

In contrast with $<\sigma'E_y'>$, the correlations $<\sigma'E_x'>$ and $<\sigma'E_z'>$ are effectively zero over the cross section of the channel. With regard to the correlation $<\mathbf{J}'\cdot\mathbf{E}'>$, $<J_x'E_x'>$ and $<J_z'E_z'>$ are negligible. However, $<J_y'E_y'>$ has a large positive value at localized points in the vicinity of the electrodes. This results in the local potential drop. The correlation does not affect significantly the gloss generator power. However, its effect on the local damage of the electrodes could be considerable.

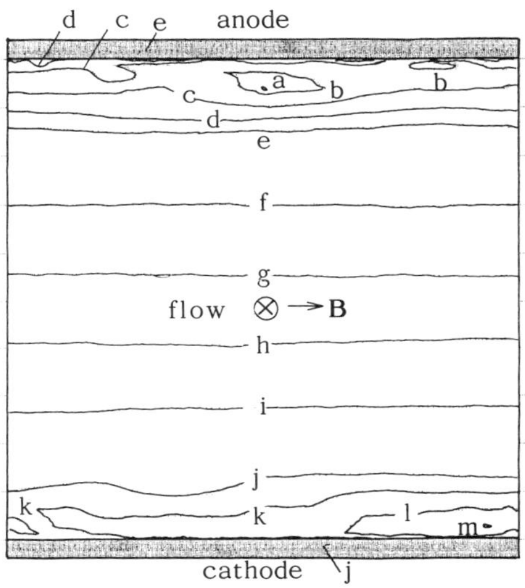

Figure 2. The instantaneous plasma potential $\Phi/\langle u_c \rangle BH$. a:-0.066, b:-0.06, c:-0.04, d:-0.02, e:0.0, f:0.133, g:0.266, h:0.399, i:0.530, j:0.663, k:0.683, l:0.703, m:0.710.

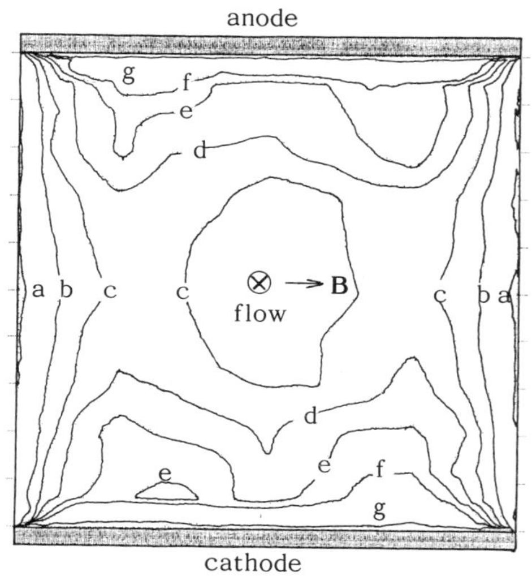

Figure 3. Distribution of the correlation $\langle \sigma' E_y' \rangle / \sigma(T_c) \langle u_c \rangle B$.
a:0.0, b:0.01, c:0.02, d:0.03, e:0.04, f:0.05, g:0.10.

5. CONCLUSIONS

The effects of the turbulent fluctuation of the combustion plasma on the mean conductivity and the electrical parameters were discussed. We showed that the correlations $\langle \sigma' u' \rangle$, $\langle \sigma' E_y' \rangle$ and $\langle J_y' E_y' \rangle$ have substantial effects on the mean electrical behavior of the MHD generator, while the other correlations can be neglected in dealing with the mean electrical phenomena. Because the electric field behavior, especially that in the vicinity of the electrodes depends on the generator type, the results presented above may be valid for the MHD device of the Faraday-type. For the Hall or the diagonal type, the effects of the turbulent correlation should be evaluated separately.

6. REFERENCES

[1] C.D. Maxwell, D.M. Mrkham, S.T. Demetriades and D.A. Oliver, Proc. 16th SEAM, Pittsburgh, May (1977), p.VII-3.

[2] S. Oikawa, N. Kayukawa, S. Yatsu and Y. Aoki, Int. J. Appl. Electromagn. Mater. 3(1992), 17-27.

[3] n. Kayukawa, S. Oikawa, Y. Aoki and S. Yatsu, Proc. 11th Int. Confer. MHD Elect. Pow. Gen., 2, Beijing(1992), IX.6.754-761.

Simulation and Design of Applied Electromagnetic Systems
T. Honma (Editor)

Turbulence damping in a weakly ionized plasma with a temperature gradient

S. Oikawa[a], N. Kayukawa[b], Y. Aoki[b] and T. Enoto[a]

[a]Dept. of Nuclear Engineering, Hokkaido Univ., N-13 W-8, Sapporo 060, Japan.

[b]Advanced MHD Research Institute, Hokkaido Univ., N-13 W-8, Sapporo 060, Japan.

Abstract

The Reynolds' stress tensor and its damping coefficients tensor based on the Prandtl's mixing length hypothesis in the presence of a magnetic field for a weakly ionized plasma were derived. The approximate expressions for the steady flow due to anisotropic turbulence damping are also presented.

1. INTRODUCTION

The authors have derived[1-2] the generalized Ohm's law for turbulent plasmas partially ionized with the gradient of electrical conductivity, $\nabla\sigma$. This equation suggests that the application of a magnetic field perpendicular to $\nabla\sigma$ induces a steady mass flow due to anisotropic turbulence damping. In this paper, we derive approximate expressions for this steady flow and the Reynolds stress damping coefficient tensor.

We consider a turbulent fluid lump with velocity $\mathbf{v}$ at a point P and time $t = t$, and suppose this fluid lump is at a point Q when time $t = t - \Delta t$. If the Prandtl's mixing length hypothesis holds, the fluid lump retains an individual identity during the lifetime so that the instantaneous electrical conductivity $\sigma_P(t)$ is the same as $\sigma_Q(t - \Delta t)$.

$$\sigma_P(t) = \overline{\sigma}_P\{1 + \frac{\delta\sigma_P(t)}{\overline{\sigma}_P}\} \tag{1}$$

$$\delta\sigma_P(t) = \delta\sigma_Q(t) - \Delta t\mathbf{v}\cdot\nabla\overline{\sigma}(T_P). \tag{2}$$

where T_P denotes the long time average of temperature at P, σ_P the conductivity evaluated by average temperature at point P, and $\delta\sigma_P(t)$ is the instantaneous conductivity deviation from σ_P. It should be noted that the long time average of $\delta\sigma_P$ is non-zero and positive since the electrical conductivity in the partially ionized plasma is a direct nonlinear function of temperature as follows[3].

$$\sigma(T) \approx \sigma_0\left(\frac{T(t)}{T_0}\right)^N, \tag{3}$$

with $N \approx 12$ for a temperature range of 2000 K $< T <$ 3000 K, and σ_0 stands for a reference conductivity at a reference temperature $T = T_0$.

Assumed that the force exerted on the fluid lump is the Lorentz force $\mathbf{j} \times \mathbf{B}$ only, the equation of motion in a Lagrangian coordinate system for the fluid lump and the instantaneous fluctuating current density $\mathbf{j}$ are as follows.

$$\frac{d\mathbf{v}}{dt} = \frac{\mathbf{j} \times \mathbf{B}}{\overline{\rho}} = -\frac{1}{\tau}\left(1 + \frac{\delta\sigma_P(t)}{\overline{\sigma}_P}\right)\left(\mathbf{v}(t) - \frac{\mathbf{BB}\cdot\mathbf{v}(t)}{\mathbf{B}^2}\right), \tag{4}$$

$$\mathbf{j}(t) = (1-\mathrm{K})\sigma_P(t)\mathbf{v}(t) \times \mathbf{B}, \tag{5}$$

$$\tau \equiv \frac{\overline{\rho}}{(1-\mathrm{K})\sigma_P(t)}$$

where K denotes the effective loading factor of the fluid lump with $0 < K < 1, B$ the externally applied magnetic field (we are dealing with plasmas of the magnetic Reynolds number $R_m \ll 1), \rho$ the mass density of the working fluid, τ the time constant of turbulence damping by magnetic field.

We choose a coordinate system (x^1, x^2, x^3) in which (x^1, x^2) is the plane perpendicular to the magnetic field with x^1 being the direction of temperature gradient on the plane, and x^3 is along the field. For example, in electrode boundary layers in a linear MHD generator, (x^1, x^2, x^3) corresponds to $(y, x, -z)$ in which x is the direction of the main flow, y is normal to the electrodes and z is normal to the side walls.

The instantaneous conductivity fluctuation, $\delta\sigma_P(t)$, is caused mainly by the temperature fluctuation, $\theta(t)$, which has a correlation with the velocity fluctuation, $\mathbf{v}(t)$. This correlation term is expressed in the ordinary fluid dynamics as

$$\overline{\theta(t)\mathbf{v}(t)} = -\frac{\nu_t}{P_t}\nabla T, \tag{6}$$

where ν_t and P_t are the eddy diffusivity and the turbulent Prandtl number, respectively, and $X(t)$ denotes the long time average of a quantity X. Similarly, the correlation of fluctuations between the electrical conductivity and the velocity can be expressed as

$$\overline{\delta\sigma(T)\mathbf{v}(t)} = -\frac{\nu_t}{P_t}\nabla\sigma(\overline{T}), \tag{7}$$

Equation(6) tells us that when considering the correlations between fluctuations of temperature and velocity, the most important component of the fluctuating velocity vector which lies along the gradient of the time averaged temperature, i.e. x^1. So that we can assume that $\delta\sigma_P(t)$ is approximately proportional to $v_1(t)$, or

$$\frac{\delta\sigma_P(t)}{\overline{\sigma}_P} \approx -\frac{v_1}{v_b} \tag{8}$$

where v_b corresponds to the cross correlations between $\delta\sigma_P(t)$ and $v_1(t)$ such as spatial gradients of the conductivity and the eddy diffusivity in the presence of a magnetic field. Then the governing equation for the fluid lump becomes the so called Bernoulli differential equation, and it can be solved analytically with the initial condition of $\mathbf{v}(t=0) = \mathbf{v}_0$ as

$$\mathbf{v}(t) = \mathbf{V}_0\mathbf{a}(t) = \begin{bmatrix} v_1^0 & 0 & 0 \\ 0 & v_2^0 & 0 \\ 0 & 0 & v_3^0 \end{bmatrix} \begin{bmatrix} ((1-\lambda)\exp(t/\tau)+\lambda)^{-1} \\ ((1-\lambda)\exp(t/\tau)+\lambda)^{-1} \\ 1 \end{bmatrix} \tag{9}$$

where λ is defined as

$$\lambda \equiv v_1^0 / v_b, \tag{10}$$

V^0 is a 3×3 diagonal tensor with $\mathsf{V}^0_{ii} = v^0_i$, and $\mathbf{a}(t)$ is the instantaneous part of the velocity $\mathbf{v}(t)$.

2. Mean fluctuating velocity

According to the Prandtl's mixing length hypothesis, a fluctuating fluid lump retains its individual identity during a time interval Δt, or lifetime, during which the fluid lump moves a distance ℓ. After that, it loses its identity by mixing with other fluid lumps and merges into its final surroundings. In the absence of a magnetic field, the instantaneous velocity fluctuation is constant during the lifetime ($\mathbf{v}(t) = \mathbf{v}^0$). Contrary to this, in the presence of a magnetic field the velocity components on the plane perpendicular to the magnetic field experience damping during the lifetime. We define the mean value of instantaneous quantity $X(t)$ during the lifetime by following expression;

$$< X > \equiv \frac{1}{\Delta t} \int_0^{\Delta t} X(t) dt. \tag{11}$$

The mean velocity of the fluid lump during its lifetime, in the presence of a magnetic field, becomes

$$< \mathbf{v} > = \mathsf{V}^0 < \mathbf{a} > = \begin{bmatrix} v_1^0 & 0 & 0 \\ 0 & v_2^0 & 0 \\ 0 & 0 & v_3^0 \end{bmatrix} \begin{bmatrix} -\frac{1}{\lambda\chi} \ln(1 - \lambda\chi D(\chi)) \\ -\frac{1}{\lambda\chi} \ln(1 - \lambda\chi D(\chi)) \\ 1 \end{bmatrix}, \tag{12}$$

where χ and $D(\chi)$ are the magnetic interaction parameter of the fluid lump and the damping coefficient of the fluctuating velocity, respectively. They are defined as

$$\chi \equiv \frac{\Delta t}{\tau}, \text{ and } D(\chi) \equiv \frac{1 - \exp(\chi)}{\chi}. \tag{13}$$

3. Ensemble Average of Mean Fluctuating Velocity

The magnitudes of the fluctuation velocity $\mathbf{v}_i^0$ of fluid lumps are not the same for each fluid lump, and have a probability density function (pdf). Thus we can replace the long time average X by the ensemble average using a probability density function F, as

$$\hat{X} \equiv \int_\Omega X(\mathbf{v}^0) F(\mathbf{v}^0) d^3\mathbf{v}^0. \tag{14}$$

The ensemble average of fluctuating velocities in the presence of a magnetic field is as follows:

$$< \hat{\mathbf{v}} > = \mathsf{V}^0 < \hat{\mathbf{a}} > \approx \frac{\chi D^2(\chi)}{2} \begin{bmatrix} \frac{\nu_t}{P_t} \frac{\partial \ln \sigma}{\partial x^1} \\ \frac{\nu_t}{P_t} \frac{\partial \ln \sigma}{\partial x^2} \\ 0 \end{bmatrix} \approx \frac{\chi}{2} \begin{bmatrix} \frac{\nu_t}{P_t} \frac{\partial \ln \sigma}{\partial x^1} \\ 0 \\ 0 \end{bmatrix} \tag{15}$$

where $\chi^2 \ll 1$ is assumed; this is the case of combustion gas plasmas in open cycle MHD generators(e.g. $\chi \approx \frac{\sigma B^2 \ell}{\rho v} \approx \frac{10 \times 5^2 \times 0.01}{0.3 \times 1000} \approx 0.08$). In the case of no magnetic field, the ensemble average of the *fluctuating* velocity is zero by definition. But when a magnetic field is applied to a plasma with the spatial gradient of the electrical conductivity in the plane perpendicular to the magnetic field, steady mass flows are induced due to the turbulence suppression as in equation(15). In this case, the instantaneous *fluctuating* velocity should be re-defined as

$$\mathbf{v}^B(t) = \mathbf{v}(t) - < \hat{\mathbf{v}} >, \tag{16}$$

such that

$$< \hat{\mathbf{v}}^B > = 0. \tag{17}$$

4. Reynolds Stress Tensor

From equations (9),(11) and (14), the Reylnolds stress tensor, $< \widehat{\mathsf{T}^{\mathsf{B}}} >$ in the presence of a magnetic field becomes:

$$\begin{aligned} < \widehat{\mathsf{T}^B} > &= \left\langle \mathsf{V}^0 \begin{bmatrix} D(2\chi) & D(2\chi) & D(\chi) \\ D(2\chi) & D(2\chi) & D(\chi) \\ D(\chi) & D(\chi) & 1 \end{bmatrix} \mathsf{V}^0 \right\rangle \\ &+ \frac{\chi D^2(\chi)}{2} \left\langle \lambda \mathsf{V}^0 \begin{bmatrix} 2 & 2 & 1 \\ 2 & 2 & 1 \\ 1 & 2 & 0 \end{bmatrix} \mathsf{V}^0 \right\rangle + \mathcal{O}(\chi^2) \end{aligned} \tag{18}$$

The Reynolds' stress tensor damped by the magnetic field in the absence of the spatial gradient of the electrical conductivity was formulated by Olin[4]. The coefficients derived here coincide with his results when $1/v_b \to 0$, i.e. there is no electrical conductivity fluctuation.

ACKNOWLEDGEMENTS

Computation have been carried out with the FACOM M-382 in the National Institute for Fusion Science (NIFS), Japan. This work was supported in part by Scientific Research Fund from the Ministry of education, JAPAN, under contract No. 02855056.

References

1. S. Oikawa, N. Kayukawa, Y. Aoki and T. Enoto, 1991 National Convention Record, I.E.E. Japan, Vol.12, pp.143-144, 1991.

2. S. Oikawa, N. Kayukawa and T. Enoto *Trans. of Japan Society of Plasma Science and Nuclear Fusion Research.* **68**, 409-414(1992). (in Japanese).

3. T. R. Brogan, *Progress in Astronautics*, ed. H. G. Walfhard, I. Glassman and L. Green Jr., (Academic Press Inc., New York, 1964), Vol.12, p.319.

4. J. G. Olin, SU-IPR Report No. 85, Institute for Plasma Research, Stanford University, July 1966.

Simulation and Design of Applied Electromagnetic Systems
T. Honma (Editor)

Plasma instability in a cesium thermionic energy converter

D.Y. Lee[a], K.H. Song[a], and D.L. Jacobson[b]

[a]Department of Materials Engineering, Daelim Junior College, 526-7. Bisan-Dong, Anyang, Kyungki-Do, Korea

[b]Department of Chemical, Bio and Materials Engineering, Arizona State University, Tempe, AZ 85287, U.S.A.

Abstract

The phenomenon of current oscillation is investigated with a spectrum analyzer in a bounded plasma of thermionic energy converter (TEC). The variable-spacing cesium thermionic diode is in the form of a guard-ringed plane-parallel geometry in which a polycrystalline rhenium emitter of 2 cm^2 area faces a radiation-cooled polycrystalline rhenium collector of 1.9 cm^2 area. At emitter temperatures above 1900 K and emitter-collector gaps less than 1 mm, oscillations with radio-frequencies are observed in the range of 0.7 to 13 MHz.

1. INTRODUCTION

TEC is emerging as the method of choice for space nuclear power system and communication applications. Because of a renewed emphasis in developing TEC for high efficiency operation in space, frequency studies are required to evaluate the stability of plasma and nuclear reactor. The indirect conversion of relatively low voltage direct current (d.c.) output to alternating current (a.c.) requires a large additional heat rejection weight and imposes a major problem for TEC in space nuclear power applications [1]. However, the direct internal generation of a.c. outputs can render power conditionings obsolete in certain regimes.

Fixed-spacing TEC of planar electrode geometry has been designed and fabricated for precise performance specifications. However, the variable-spacing TEC has an advantage of achieving data at each stage of spacing without disturbing the internal and external structures. A variable-spacing TEC test station was designed and fabricated for the present study. A variable-spacing diode used for this research is a guard-ringed polycrystalline rhenium emitter and collector. The diode was built by Electro-Optical System, Inc. (EOS). The theory and design of TEC has been presented in detail in [2].

The frequency of the oscillations depends primarily on the diode spacing and is affected by relative ion and electron concentrations at the emitter surface. In a low pressure diode the frequency is inversely proportional to the distance between the electrodes [3]. Even though there are many studies of these parametric relationships, there is little data for the close spacing diode (less than 1 mm). The variation of frequency with cesium vapor pressure for interelectrode gaps in the range of 1 mm to 0.08 mm has been studied using a spectrum analyzer. The experimental results are then compared with the theory describing the dependence of frequency on the cesium vapor pressure. The theoretical dependency was based on the beam instability approach discussed by Stakhanov and Stepanov [3].

2. EXPERIMENTAL DEVICES

The main advantage of this vehicle is the precise in-situ adjustment of the interelectrode gap by using three linear motion flanges with precision micrometers as shown in Figure. 1. The vehicle is designed to control the interelectrode gap in the range of 0.02 mm to 1 mm with an accuracy of ± 0.015 mm. Emitter temperatures are varied from 1300 K to 2200 K, collector temperatures from 300 K to 1100 K, and cesium reservoir temperatures from 300 K to 600 K. The liquid cesium reservoir is positioned far from the electrodes so that the low cesium reservoir temperature can be achieved by controlling a shielded resistance cesium heater. Then, thermocouples are attached around the cesium reservoir wall with a copper clamp and monitored by a thermocouple digital meter. The cesium vapor pressure is then calculated by measuring temperature of the liquid cesium reservoir with an accuracy of ±5 K. The electron bombardment gun is composed of well-shielded, counter wound, pancake shape tungsten filament (1.9 cm diameter) to provide uniform heating of the diode's emitter. The emitter temperature is measured through hohlraum (10:1, length to diameter ratio) on the side of the emitter by means of an optical pyrometer within ±7 K. The collector is cooled by the mechanical attachment of a copper radiator (3.3 cm x 15.7 cm). A nichrome cable heater is brazed on the copper radiator. The collector temperature is then sensed by sheathed chromel-alumel (type K) thermocouples.

The TEC station is evacuated by a turbo-molecular pump as well as an ion pump. Vacuum better than 5×10^{-7} torr is obtained at room temperature before the experiments. After the spacing calibration, vacuum better than 5×10^{-7} torr is maintained during experimentation. Also, ceramic sleeves and washers are employed to provide electrical isolation between the diode and the test station. The a.c. output were monitored across a 30 Ω resistor using a spectrum analyzer. Then, an oscilloscope camera was employed to obtain photographs of the output spectrum. The frequency accuracy was ±5 %.

3. RESULTS AND DISCUSSION

The effect of interelectrode gap on the frequency is shown in Figure. 2. Each data point indicates the average of the results obtained from at least 3 experiments under identical test condition. The results indicated that, as the spacing decreased, the frequency increased when the spacing was greater than 0.08 mm. At low cesium pressure (below 10^{-2} torr) and high emitter temperature (above 1900 K), the frequency was inversely proportional to the interelectrode gap. Then, no correlation was found when the spacing was less than 0.08 mm due to the occurrence of d.c. power output. These experimental results were in good agreement with the theoretical frequency dependency, f α 1/d [3].

The variation of frequency with cesium vapor pressures (up to 10^{-2} torr) and emitter temperatures (1900 to 2200 K) was studied. The signal first appeared at the cesium vapor pressure of 10^{-6} torr and then disappeared gradually into the original background frequency with increasing cesium vapor pressure (10^{-2} torr). The sequence of spectrum damping indicates that Coulomb collisions (long-range) are a key collision process. The emitter temperature also had the same trend as that of the cesium vapor pressure. At the same spacing, the same spectrum was obtained at different emitter temperatures. It is experimentally clear that the cesium vapor pressure and the emitter temperature do not cause the appreciable change of the oscillating frequency.

The geometry limitation of the apparatus made it difficult to measure the electron temperature, mobility, cross section, and plasma density because the employed diode was vacuum sealed with ceramic. Therefore, physical values can not be measured directly in this work. Two types of output from heat are produced depending on the plasma stability, either power generation (stable palsma) or electromagnetic energy (unstable plasma). A maximum frequency of 13 MHz was experimentally observed at spacings less than 0.08 mm and emitter temperatures above 1900 K. A calculated theoretical plasma-ion frequency in ionized gases at a plasma density of $5x10^{11}$ cm^{-3} and an electron temperature of $2x10^4$ K using the equation in Ref. 4 was 13 MHz. The electron temperature and plasma density data were taken from Refs. 4 and 5 at a cesium reservoir temperature of 383 K and a tungsten emitter material. Therefore, the observed frequency of 13 MHz may be the maximum frequency of the positive ion oscillation at plasma densities around 10^{11} cm^{-3}. A power output of 13.47 watt/cm^2 [6] was obtained above plasma densities of $5x10^{11}$ cm^{-3} due to two-body collisions (short-range) that resulted in loss of organized plasma behavior (no frequency generation).

4. CONCLUSIONS

The frequency was inversely proportional to the interelectrode gap in the high emitter temperature regime when the interelectrode gap was greater than 0.08 mm. No effect of cesium vapor pressure and emitter temperature on the frequency was observed. No effect of the cesium vapor pressure, the interel--ectrode gap, and the emitter temperature on the frequency was found when the spacing was less than 0.08 mm.

In the high emitter temperature regime (above 1900 K), the effect of Coulomb collision played a significant role in oscillations. The broad spectra (0 MHz to 13 MHz) were experimentally observed and the spectra disappeared gradually to the original background frequency. The virtual emitter formation due to the Pierce instability (electron-ion interaction) may be the main reason for oscillations in this regmie [7].

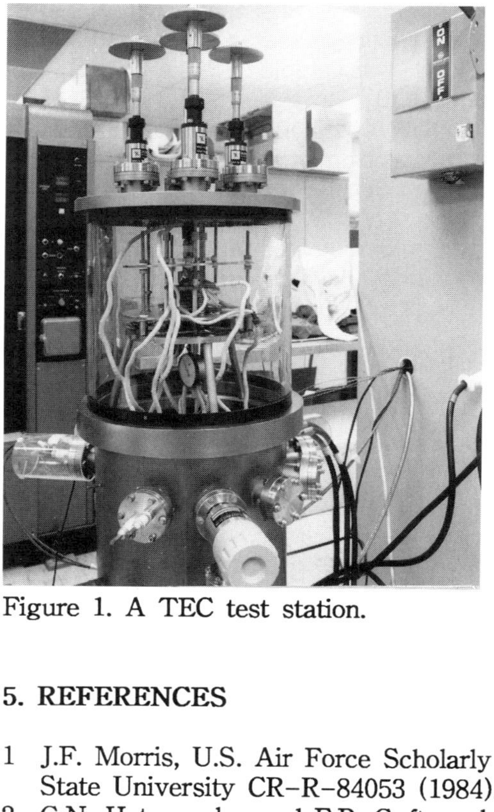

Figure 1. A TEC test station.

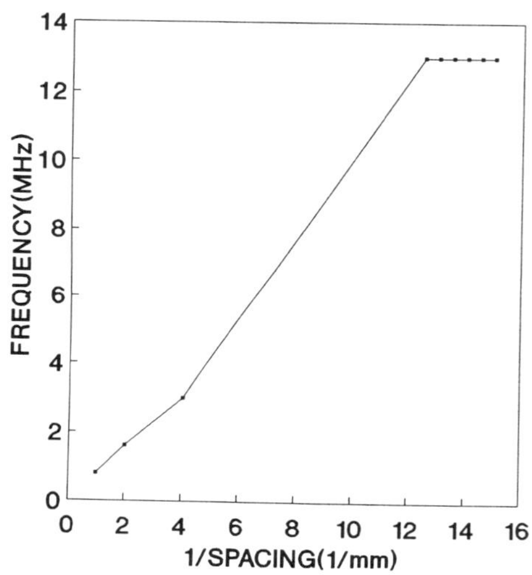

Figure 2. The dependence of frequency on the spacing.

5. REFERENCES

1 J.F. Morris, U.S. Air Force Scholarly Research Program Report, Arizona State University CR-R-84053 (1984).
2 G.N. Hatsopoulos and E.P. Gyftopoulos (eds.), Thermionic Energy Conversion, The MIT Press, Cambridge, 1973.
3 I.P. Stakhanov and A.S. Stepanov, Sov. Phys. Tech. Phys., 10 (1961) 100.
4 L. Tonks and I. Langmuir, Phys. Rev., 33 (1929) 195.
5 J.S. Chivian, J. Appl. Phys., 35 (1963) 302.
6 D.Y. Lee, Trans. KIEE., 41 (1992) 1085.
7 J.R. Pierce, J. Appl. Phys., 15 (1944) 721.

AUTHOR INDEX